BIOENGINEERING AND MOLECULAR BIOLOGY OF PLANT PATHWAYS

Advances in Plant Biochemistry and Molecular Biology

Volume 1 - Bioengineering and Molecular Biology of Plant Pathways

Hans J. Bohnert, Henry Nguyen, and Norman G. Lewis

Advances in Plant Biochemistry and Molecular Biology

VOLUME 1

BIOENGINEERING AND MOLECULAR BIOLOGY OF PLANT PATHWAYS

Edited by

HANS J. BOHNERT
Urbana, IL, USA

HENRY NGUYEN
Columbia, MO, USA

NORMAN G. LEWIS
Pullman, WA, USA

ELSEVIER

Amsterdam • Boston • Heidelberg • London • New York • Oxford
Paris • San Diego • San Francisco • Singapore • Sydney • Tokyo
Pergamon is an imprint of Elsevier

Pergamon is an imprint of Elsevier
525 B Street, Suit 1900, San Diego, CA 92101–4495, USA
Linacre House, Jordan Hill, Oxford OX2 8DP, UK
Radarweg 29, PO Box 211, 1000 AE Amsterdam, The Netherlands

First edition 2008

British Library Cataloguing in Publication Data
A catalogue record for this book is available from the British Library

Library of Congress Cataloging-in-Publication Data
A catalog record for this book is available from the Library of Congress

ISBN: 978-0-08-044972-2
ISSN: 1755-0408

For information on all Pergamon publications
visit our Web site at www.books.elsevier.com

Printed and bound in the United Kingdom

Transferred to Digital Print 2011

DEDICATION

The editors and contributing authors dedicate this first volume of the Advances in Plant Biochemistry and Molecular Biology" entitled "**Bioengineering and Molecular Biology of Plant Pathways**" to the memory of Paul Stumpf, who sadly passed away on February 10, 2007. Plant biochemistry benefited immensely from Paul's life-long passion to this subject, as well as his scientific rigor and insight. The scientific community is indebted to both he and Eric Conn for their dedication in helping advance the very basis of plant biology/plant biochemistry.

CONTENTS

CONTRIBUTORS

Rachel Amir
Plant Science Laboratory, Migal Galilee Technological Center, Rosh Pina 12100, Israel.

R. Malcolm Brown Jr.
Section of Molecular Genetics and Microbiology, School of Biological Sciences, The University of Texas at Austin, Austin, Texas 78712.

Hans J. Bohnert
Departments of Plant Biology and of Crop Sciences, University of Illinois at Urbana-Champaign, Urbana, Illinois 61801.

Ray A. Bressan
Department of Horticulture and Landscape Architecture, Purdue University, Horticulture Building 1165, West Lafayette, Indiana 47907-1165.

Edgar B. Cahoon
USDA-ARS Plant Genetics Research Unit, Donald Danforth Plant Science Center, 975 North Warson Road, St. Louis, Missouri 63132.

Laurence B. Davin
Institute of Biological Chemistry, Washington State University, Pullman, Washington 99164.

Susanne Frick
Donald Danforth Plant Science Center, St. Louis, Missouri 63132.
Leibniz Institut für Pflanzenbiochemie, Weinberg 3, 06120 Halle/Saale, Germany.

Gad Galili
Department of Plant Sciences, The Weizmann Institute of Science, Rehovot 76100, Israel.

Shmuel Galili
Institute of Field and Garden Crops, Agricultural Research Organization, Bet Dagan 50250, Israel.

P. Michael Hasegawa
Department of Horticulture and Landscape Architecture, Purdue University, Horticulture Building 1165, West Lafayette, Indiana 47907-1165.

David R. Holding
Department of Plant Sciences, University of Arizona, Tucson, Arizona 85721.

Nicholas J. Kruger
Department of Plant Sciences, University of Oxford, Oxford OX1 3RB, United Kingdom.

Toni M. Kutchan
Donald Danforth Plant Science Center, St. Louis, Missouri 63132.
Leibniz Institut für Pflanzenbiochemie, Weinberg 3, 06120 Halle/Saale, Germany.

Brian A. Larkins
Department of Plant Sciences, University of Arizona, Tucson, Arizona 85721.

Norman G. Lewis
Institute of Biological Chemistry, Washington State University, Pullman, Washington 99164.

Christiane Nawrath
Département de Biologie Moléculaire Végétale, Biophore, Université de Lausanne, CH-1015 Lausanne, Switzerland.

W. David Nes
Department of Chemistry and Biochemistry, Texas Tech University, Lubbock, Texas 79409.

Henry T. Nguyen
Division of Plant Sciences, National Center for Soybean Biotechnology, University of Missouri-Columbia, Columbia, Missouri 65211.

Yves Poirier
Département de Biologie Moléculaire Végétale, Biophore, Université de Lausanne, CH-1015 Lausanne, Switzerland.

R. George Ratcliffe
Department of Plant Sciences, University of Oxford, Oxford OX1 3RB, United Kingdom.

Fumihiko Sato
Department of Plant Gene and Totipotency, Graduate School of Biostudies, Kyoto University, Kyoto 606-8502, Japan.

Inder M. Saxena
Section of Molecular Genetics and Microbiology, School of Biological Sciences, The University of Texas at Austin, Austin, Texas 78712.

Katherine M. Schmid
Department of Biological Sciences, Butler University, 4600 Sunset Avenue, Indianapolis, Indiana 46208.

John Shanklin
Biology Department, Brookhaven National Laboratory, Upton, New York 11973.

Shigeru Shigeoka
Department of Advanced Bioscience, Faculty of Agriculture, Kinki University, 3327-204 Nakamachi, Nara 631-8505, Japan.

Daniel G. Vassão
Institute of Biological Chemistry, Washington State University, Pullman, Washington 99164.

Marion Weid
Leibniz Institut für Pflanzenbiochemie, Weinberg 3, 06120 Halle/Saale, Germany.

Yasuyuki Yamada
Graduate School of Biological Sciences, Nara Institute of Science and Technology, 8916-5 Takayama, Ikoma, Nara 630-0192, Japan.

Akiho Yokota
Graduate School of Biological Sciences, Nara Institute of Science and Technology (NAIST), 8916-5 Takayama, Ikoma, Nara 630-0101, Japan.

Wenxu Zhou
Department of Chemistry and Biochemistry, Texas Tech University, Lubbock, Texas 79409.

INTRODUCTION TO THE SERIES AND ACKNOWLEDGEMENTS

This new series was initiated conceptually and organizationally by W. David Nes with the assistance of Norman G. Lewis, with the first volume commissioned by W.D. Nes. Sadly, Dr. Nes was unable to oversee the completion of the volume as originally planned.

This particular volume has as its origin an U.S. National Science Foundation (NSF) workshop entitled "Realizing the Vision: Leading Edge Technologies in Biological Systems". In this regard, we are deeply grateful to NSF for supporting this most exciting workshop, in helping identifying critical barriers to ongoing biological endeavors, and thus in initiating this series. This volume, addresses several of the critical areas from the workshop, such as metabolic flux regulation, and the challenges and opportunities that still remain as humanity attempts to understand the blueprints of life and the opportunities that this new knowledge now gives us (see attached preface by Bohnert and Nguyen).

The reader is strongly encouraged to comprehensively review all of the 13 chapters/topics within the volume. In so doing, it becomes rapidly evident that while the rate of genomic sequencing in animal, microbial and plant systems has occurred very rapidly, this knowledge is not, however, matched by any comparable levels of discovery of gene and/or protein function, i.e. and thus of yet gaining a deep understanding of the "blueprints of life". This series is therefore designed to focus upon leading edge and emerging technologies, as well as critical barriers that face various areas in the plant sciences. Overcoming these will bring the field of metabolic plant biochemistry to new levels of scientific excellence and societal influence.

The reader should also note that we commissioned both Eric Conn and Paul K. Stumpf to write a Prologue as regards their "Comprehensive Treatise". Sadly at the time of this publication, Prof. Paul K. Stumpf passed away (February 10, 2007). We are nevertheless grateful to have this volume graced by both of these remarkable plant biochemistry pioneers. We are also indebted to both Ms. Hiroko Hayashi who worked tirelessly in coordinating and correcting the various manuscripts, as well as to the many reviewers of these contributions.

Respectfully,
Norman G. Lewis

PREFACE TO VOLUME 1

Volumes published during the 1980s that made up the series on *"The Biochemistry of Plants–A Comprehensive Treatise"*, edited by Eric Conn and Paul K. Stumpf, covered many of the then known aspects of plant biochemistry. During the last two decades, however, our knowledge on plant biochemistry, physiology, and molecular genetics has been augmented to an astonishing degree. This remarkable revolution has been brought about by new techniques, new concepts that are now summarized as "genomics", "proteomics" and "metabolomics," as well as to a large degree by new forms of instrumentation for each type of application. This volume has been designed to incorporate new concepts and insights in plant biochemistry and biology as part of a new series titled *"Advances in Plant Biochemistry and Molecular Biology"* edited by Professor Norman Lewis. To put this into suitable context, attached is a Foreword by Eric Conn and the late Paul K. Stumpf as regards the need for this new series.

The increased knowledge about the structure of genomes in a number of species, about the complexity of their transcriptomes, and the nearly exponentially growing information about mutant phenotypes have now set off the large scale use of transgenes to answer basic biological questions, and to generate new crops and novel products. This volume includes thirteen chapters, which to variable degrees describe the use of transgenic plants to explore possibilities and approaches for the modification of plant metabolism, adaptation or development. The interests of the authors of these chapters range from tool development, to basic biochemical know-how about the engineering of enzymes, to exploring avenues for the modification of complex multigenic pathways, and include several examples for the engineering of specific pathways in different organs and developmental stages.

Kruger and Ratcliffe focus on the tools for analyzing metabolic network structures and provide a conceptual framework about the challenges faced in engineering pathways. Sections on metabolic flux and control analysis as well as kinetic modeling that measure the impact of changes on network structure, with excellent discussion of the literature, are destined to set a standard. Enzyme engineering with theoretical and practical considerations is discussed by Shanklin with a focus on structure models as the guiding light. Examples of success from the author's laboratory provide lucid documentation.

The engineering potential for altering photosynthetic performance, discussed by Yokota and Shigeoka, addresses a fundamental set of pathways, whose improvements would be of great importance, although complexity and barriers to change have shown to be still considerable. The authors, nevertheless, provide an overview of the failures and discuss prospects provided by the emerging new biology. In another example on the engineering of primary metabolism, Galili and colleagues describe approaches and progress with respect to altering amino acid metabolism. The conspicuous successes in this area are discussed with respect to individual amino acid families and with respect to metabolic fluxes.

Three chapters discuss progress and potential in the engineering of metabolic end-products that are of vast economical importance: the genetic engineering of cellulose by Saxena and Brown, of seed storage proteins by Holding and Larkins, and of content and composition of edible and industrial oils by Cahoon and Schmid. Owing to the different complexities that these three "pathways" present to engineers, these chapters present views of how to go about in dissecting complexity into manageable partitions. Nawrath and Poirier focus on pathways for the synthesis of polyesters in plants, with examples for the engineering of existing plant pathways, cutin and suberin, and the engineering of a foreign pathway, leading to polyhydroxyalkanoates. As in many of the chapters in this volume, the authors point to the necessity for more fundamental research into plant metabolic pathways. Addressing a problem of yet higher complexity, Bressan and coworkers tackle genetic engineering for salinity tolerance. They point to the multitude of pathways, developmental ages, and tissues that must be integrated to achieve a goal that can stand the test of performance in the real world.

Finally, four chapters are devoted to the engineering of secondary metabolism. Kutchan and coworkers, on the progress and prospects of plant alkaloid biosynthetic pathways, discuss the substantial progress in the identification of pathways and metabolites. Similarly, Sato and Yamada provide an overview on the engineering and use of cells in culture for the biosynthesis of secondary metabolites as a source for medicinal compounds. Zhou and colleagues describe strategies for bioengineering of sterol methyltransferases. The chapter covers enzyme and pathway structure and proceeds to the ecology of sterol functions. Lewis and colleagues discuss prospects of engineering allylphenols, lignins and lignans, based on tremendous progress made in recent years. This theme, in combination with the discussion on cellulose biosynthesis and engineering by Saxena and Brown, is of particular relevance in the light of efforts to develop energy from renewable lignocellulosic materials.

The challenges that lie ahead for genetic manipulation of plant pathways to become truly productive are several. Minimizing unexpected detrimental, pleiotropic effects on plant growth and development, owing to complex regulation of biochemical pathways is one of these challenges. To achieve the desired levels of metabolites and end-products will require that the information, presently in part available for a few model species, on genome structure, transcript abundance and regulation, on pathway and protein regulation, and on metabolic flux become understood on a more fundamental mechanistic level. This volume presents concepts and strategies that are required to overcome limitations that obstruct coordinated pathway regulation.

The older volumes on the biochemistry of plants contained the sum of our knowledge at the time. They have provided basic knowledge, much of it still useful, that many plant scientists used as a start point and springboard for creative new approaches. It is hoped that the present volume with its emphasis on plant engineering will have a similarly inspiring influence such that, in the future, we can proceed from the modification of individual genes or a few proteins and enzymes to metabolic pathway engineering on a fundamental scale.

Hans Bohnert
Henry Nguyen
January 2007

PROLOGUE

A good way to introduce the new series of volumes entitled *Advances in Plant Biochemistry and Molecular Biology* is to examine the state of plant biochemistry in 1980, when an earlier series was initiated. At that time, Paul Stumpf and Eric Conn undertook the task of organizing a collection of volumes edited and written by leaders in the field of plant biochemistry. The General Preface to that collection, which we wrote in 1980, explained why we thought it was time for a series entitled *The Biochemistry of Plants*.

General Preface to *The Biochemistry of Plants*[1]

In 1950, James Bonner wrote the following prophetic comments in the Preface of the first edition of his *Plant Biochemistry*, published by Academic Press.

There is much work to be done in plant biochemistry. Our understanding of many basic metabolic pathways in the higher plant is lamentably fragmentary. While the emphasis in this book is on the higher plant, it will frequently be necessary to call attention to conclusions drawn from work with microorganisms or with higher animals. Numerous problems of plant biochemistry could undoubtedly be illuminated by the closer application of the information and the techniques that have been developed by those working with other organisms. . . .

Certain important aspects of biochemistry have been entirely omitted from the present volume because of the lack of pertinent information from the domain of higher plants.

The volume had 30 chapters and a total of 490 pages. Many of the biochemical examples cited in the text were derived from studies on bacterial, fungal, and animal systems. Despite these shortcomings, the book had a profound effect on a number of young biochemists, since it challenged them to enter the field of plant biochemistry and to correct "the lack of pertinent information from the domain of higher plants."

Since 1950, an explosive expansion of knowledge in biochemistry has occurred. Unfortunately, the study of plants has had a mixed reception in the biochemical community. With the exception of photosynthesis, biochemists have avoided tackling, for one reason or another, the incredibly interesting problems associated with plant tissues. Leading biochemical journals have frequently rejected sound manuscripts for the trivial reason that the reaction had been well described in *E. coli* and liver tissue and was of little interest to again describe its presence in germinating pea seeds! Federal granting agencies, the National Science Foundation excepted, have also been reluctant to fund applications when

[1] Stumpf, P. K., and Conn, Eric E., eds. in chief. (1980). The Biochemistry of Plants: A Comprehensive Treatise, Vol. 1, pp. xiii–xiv. Academic Press, New York.

it was indicated that the principal experimental tissue would be of plant origin despite the fact that the most prevalent illness in the world is starvation.

The second edition of *Plant Biochemistry* had a new format in 1965 when J. Bonner and J. Varner edited a multiauthored volume of 979 pages; in 1976, the third edition containing 908 pages made its appearance. A few textbooks of limited size in plant biochemistry have been published. In addition, two continuing series resulting from the annual meetings and symposia of photochemical organizations in Europe and North America provided the biological community with highly specialized articles on many topics of plant biochemistry. Plant biochemistry was obviously growing.

Although these publications serve a useful purpose, no multivolume series in plant biochemistry has been available to the biochemist trained and working in different fields who seeks an authoritative overview of major topics of plant biochemistry. It therefore seemed to us that the time was ripe to develop such a series. With the encouragement and cooperation of Academic Press, we invited six colleagues to join us in organizing an eight-volume series to be known as *The Biochemistry of Plants: A Comprehensive Treatise*. Within a few months, we obtained commitments from more than 160 authors to write authoritative chapters for these eight volumes.

Our hope is that this Treatise not only will serve as a source of current information to researchers working in plant biochemistry, but equally important will provide a mechanism for the molecular biologist who works with *E. coli*, or for the neurobiochemist to become better informed about the interesting and often unique problems that the plant cell provides. It is hoped too that the senior graduate students will be inspired by one or more comments in chapters of this Treatise and will orient their future career to some aspect of this science.

Despite the fact that many subjects have been covered in this Treatise, we make no claim to have been complete in our coverage or to have treated all subjects in equal depth. Notable is the absence of volumes on phytohormones and on mineral nutrition. These areas, which are more closely associated with the discipline of plant physiology, are treated in multivolume series in the physiology literature and/or have been the subject of specialized treatises. Other topics (e.g., alkaloids, nitrogen fixation, flavonoids, plant pigments) have been assigned single chapters even though entire volumes, sometimes appearing on an annual basis, are available.

These sixteen volumes, covering many aspects of plant biochemistry as was known at that time, were published during 1980 and 1990. Since then, a remarkable revolution has occurred as the techniques of molecular biology burst on the scene and extended our knowledge on many aspects of plant growth and development. With this new approach, a large number of transgenic plants have been designed specifically to function well under harsh environments of drought and salinity as well as withstand attacks by microbial, fungal, viral, and insect populations. Highly sophisticated techniques can now probe the secrets of the plant life cycle and identify genes involved in germination, growth, flowering seed formation, and other processes. Thus, it is appropriate that a new series will again summarize the recent advances in plant biochemistry and molecular biology.

It will be most welcome as plants continue to affect the many aspects of life in this ever more complicated world.

The overall goals and aims of Volume 1 of the present series are summarized in the following overview by Hans Bohnert and Henry Nguyen.

Paul K. Stumpf
Eric E. Conn

CHAPTER 1

Metabolic Organization in Plants: A Challenge for the Metabolic Engineer

Nicholas J. Kruger and **R. George Ratcliffe**

Contents

Abstract

Predictive models of plant metabolism with sufficient power to identify suitable targets for metabolic engineering are desirable, but elusive. The problem is particularly acute in the pathways of primary carbon metabolism, and ultimately it stems from the complexity of the plant metabolic network and the plethora of interacting components that determine the observed fluxes. This complexity is manifested most obviously in the

Department of Plant Sciences, University of Oxford, Oxford OX1 3RB, United Kingdom

Advances in Plant Biochemistry and Molecular Biology, Volume 1
ISSN 1755-0408, DOI: 10.1016/S1755-0408(07)01001-6

remarkable biosynthetic capacity of plant metabolism, and in the extensive subcellular compartmentation of steps and pathways. However it is argued that while these properties provide a considerable challenge at the level of identifying enzymes and metabolic interconversions - indeed the definition of the plant metabolic network is still incomplete - the real obstacle to predictive modelling lies in identifying the complete set of regulatory mechanisms that influence the function of the network. These mechanisms operate at two levels: one is the molecular crosstalk between effectors and enzymes; and the other is gene expression, where the relationship between fluctuations in expression and network performance is still poorly understood.

The tools that are currently available for analysing network structure and performance are described, with particular emphasis on constraints-based network analysis, metabolic flux analysis, kinetic modelling and metabolic control analysis. Based on a varying mix of theoretical analysis and empirical measurement, all four methods provide insights into the organisation of metabolic networks and the fluxes they support. Specifically they can be used to analyse the robustness of metabolic networks, to generate flux maps that reveal the relationship between genotype and metabolic phenotype, to predict metabolic fluxes in well characterised systems, and to analyse the relationship between substrates, enzymes and fluxes. No single method provides all the information necessary for predictive metabolic engineering, although in principle kinetic modelling should achieve that goal if sufficient information is available to parameterize the models completely.

The level of sophistication that is required in predictive models of primary carbon metabolism is illustrated by analysing the conclusions that have emerged from extensive metabolic studies of transgenic plants with reduced levels of Calvin cycle enzymes. These studies highlight the intricate mechanisms that underpin the responsiveness and stability of carbon fixation. It is argued that while the phenotypes of the transgenic plants can be rationalised in terms of a qualitative understanding of the components of the system, it is not yet possible to predict the behaviour of the network quantitatively because of the complexity of the interactions involved.

Key Words: Constraints-based network analysis, Elementary mode analysis, Enzyme regulation, Kinetic modeling, Metabolic compensation, Metabolic control analysis, Metabolic engineering, Metabolic flux analysis, Photosynthetic carbon metabolism, Subcellular compartmentation.

1. INTRODUCTION

Although many plants with interesting phenotypes have been generated by genetic manipulation, the central metabolic objective of being able to make predictable changes to specified fluxes generally remains elusive. The numerous reports of engineered plants with metabolic phenotypes that are not usefully different from the wild type, for example, in starch metabolism (Fernie *et al.*,

2002), show that the rational manipulation of plant metabolism is far from straightforward, and that in many instances our understanding of plant metabolic networks is insufficient to permit accurate predictions about the metabolic consequences of genetic manipulation. Unexpected metabolic phenotypes are interesting in their own right since they often provide information about the structure and regulatory properties of the network, but from an engineering perspective, they are undesirable since they consume resources and reduce the efficiency of the process.

If the production of unwanted metabolic phenotypes is to be avoided, then metabolic engineering has to be based on a detailed quantitative understanding of the capabilities of the metabolic network. Essentially this requires: (1) definition of the network of reactions, (2) definition of all the molecular interactions in the system that have an impact on the functioning of the network, and (3) specification of the intracellular and external environments in which the network is functioning. Unfortunately, each of these requirements is potentially very demanding: the plant metabolic network is of necessity complex, reflecting the demands placed on sessile organisms that live in a fluctuating environment; this complexity increases the scope for regulation of the network through changes in enzyme level (via changes in gene expression and protein turnover) and enzyme activity (via covalent modification, effector binding, and changes in substrate and product concentrations); and for most purposes, plants have to be grown under non–steady-state conditions, thus complicating any prediction of metabolic performance. The net result of these complications is that models of plant metabolism (Giersch, 2000; Morgan and Rhodes, 2002) tend to be relatively limited in scope and to fall some way short of the virtual cell that is required if accurate predictions are to be made of the impact of genetic manipulation on metabolic fluxes.

Three topics central to the development of a quantitative understanding of the metabolic capabilities of plant cells are discussed in this chapter. First, the complexity of the plant metabolic network is described and the prospects for obtaining a complete description of the network are assessed. Second, a review is provided of some of the tools that are now available for understanding the structure and performance of the network. Finally, to emphasize the level of sophistication that is required for models with real predictive value, we review some landmark studies that highlight the complexity of the system-wide mechanisms that permit the integration of plant metabolism. The emphasis is on the primary pathways of carbon metabolism since these pathways are fundamentally important for the functioning and manipulation of the network.

2. PLANT METABOLIC NETWORKS AND THEIR ORGANIZATION

The first characteristic feature of plant metabolism is its biosynthetic capacity (Croteau *et al.*, 2000; Wink, 1999). While bacterial and yeast metabolisms encompass only a few hundred metabolites, the number of known plant secondary products is estimated to be 100,000 (Schwab, 2003), and the actual number may be as high as

200,000 (Sumner *et al.*, 2003). Obviously individual species synthesize only a particular subset of these compounds, but any attempt to define the metabolic network in a plant cell has to include substantially more biosynthetic pathways than in a typical microorganism. Moreover, since the manipulation of the fluxes through these pathways can be of agronomic and commercial interest (Dixon and Sumner, 2003), the definition of the secondary pathways in the metabolic network may be as important as the definition of the pathways of central metabolism in generating predictive models appropriate for metabolic engineering.

Another characteristic and well-known feature of plant metabolism is the extensive subcellular compartmentation that occurs within a typical plant cell (ap Rees, 1987). The cytosolic, plastidic, peroxisomal, and mitochondrial compartments are all metabolically important, with the plastids in both heterotrophic and photosynthetic cells having a notable role in biosynthesis. In some cases, particular metabolic steps occur uniquely in one compartment, for example, the synthesis of starch from ADPglucose is exclusively plastidic, but there are many instances where a particular step occurs in more than one compartment, and in extreme cases this leads to the duplication of whole pathways in two or more compartments. For example, there is considerable duplication of the pathways of carbohydrate oxidation between the cytosol and the plastids of heterotrophic tissues (Neuhaus and Emes, 2000) and many of the reactions of folate-mediated one carbon metabolism can occur in three compartments—the cytosol, mitochondria, and plastids (Hanson *et al.*, 2000). Subcellular compartmentation has two major consequences for defining the metabolic network and constructing a predictive model of plant metabolism, and these are discussed in the following paragraphs.

First, it is necessary to identify all the transport steps that link the subcellular metabolite pools as well as the subcellular location(s) of each metabolic step. New plastidic transporters are still being identified (Weber *et al.*, 2005), and when added to the multiple metabolite transporters in the inner mitochondrial membrane (Picault *et al.*, 2004), the result is to add considerably to the complexity of the plant metabolic network. Moreover, identifying the subcellular location(s) of particular steps can be difficult because of the uncertainties associated with the preparation of sufficiently pure subcellular fractions from tissue extracts, and the result in any case is often both species and tissue specific. For example, the extent to which all the enzymes of the pentose phosphate pathway are present in the cytosol is variable (Debnam and Emes, 1999; Kruger and von Schaewen, 2003), and our understanding of the pathway of starch synthesis in cereal endosperm has had to be revised following the characterization of a cytosolic isoform of the normally plastidic ADPglucose pyrophosphorylase (Burton *et al.*, 2002; Denyer *et al.*, 1996).

Second, identical steps in different compartments are generally catalyzed by isozymes with distinct properties. Thus, duplication of pathways complicates the construction of predictive models by increasing the amount of kinetic and regulatory information that is required for the network. Moreover, the subcellular concentrations of substrates, coenzymes, and effectors will usually be different in different compartments (Farré *et al.*, 2001), increasing the information that is required for the construction of a realistic model. A further complication is that even when an activity has been localized to a compartment, it may be distributed nonuniformly and in this situation there is the possibility that the effective concentrations of the substrates, coenzymes, and effectors will differ from their overall values. Thus, in the case of several cytosolic enzymes, there is good evidence for a membrane-associated subfraction that can be expected to have distinct kinetic properties and presumably a specific functional role within the network. Examples include nitrate reductase (Lo Piero *et al.*, 2003; Wienkoop *et al.*, 1999) and sucrose synthase (Amor *et al.*, 1995; Komina *et al.*, 2002), both of which have forms associated with the plasma membrane, and the recent demonstration of an extensive association of the enzymes of glycolysis with the outer mitochondrial membrane in *Arabidopsis* (Giegé *et al.*, 2003).

Another important feature of the plant metabolic network is that much remains to be discovered before a definitive map can be drawn. This assertion is supported by the discovery of several major pathways in recent years, for example, the pathway for the synthesis of ascorbate (Smirnoff *et al.*, 2001) and the methylerythritol pathway for the synthesis of terpenes (Eisenreich *et al.*, 2001), and even apparently well-characterized areas of the network, such as the pathway to ADPglucose in leaves, can become candidates for reevaluation in the light of new data (Baroja-Fernandez *et al.*, 2004, 2005; Munoz *et al.*, 2005; Neuhaus *et al.*, 2005). Moreover, the introduction of new techniques for probing plant metabolism invariably provides new information about the architecture and regulation of the plant metabolic network. For example, the development of insertional mutagenesis for gene silencing has generated a powerful method for probing the redundancy of the network, and this technique has been used to investigate the interaction between peroxisomes and mitochondria in plant lipid metabolism (Thorneycroft *et al.*, 2001). There is also a very strong indication from the *Arabidopsis* and rice genomes that much remains to be identified before a complete metabolic network can be constructed. It is already apparent from the incompletely annotated genomes that many of the identified enzymes exist in multiple isoforms, and a notable example of this phenomenon is provided by pyruvate kinase, which appears to be represented by up to 14 genes in *Arabidopsis* (Fig. 1.1). Presumably different isoforms play significant roles in particular compartments of particular cell types at appropriate stages in the plant life cycle, and incorporating this level of detail into a predictive metabolic model is likely to be a major challenge.

While the complexity of the plant metabolic network is an obstacle to predictive modeling, it is also a fundamental characteristic of plant metabolism and it would be unrealistic to imagine that it can be ignored. An analysis of the metabolic network in *Escherichia coli* suggests that increased complexity is a desirable property for cells exposed to uncontrollable external conditions, conferring robustness and the ability to function at near optimal rates over a range of physiological conditions (Stelling *et al.*, 2002). This fundamental property of complex systems undermines the central objective of attempting to manipulate the performance of the network through genetic engineering, and it emphasizes the importance of establishing as complete a description of the network as possible. Fortunately, annotation of the *Arabidopsis* and other plant genomes should provide a complete

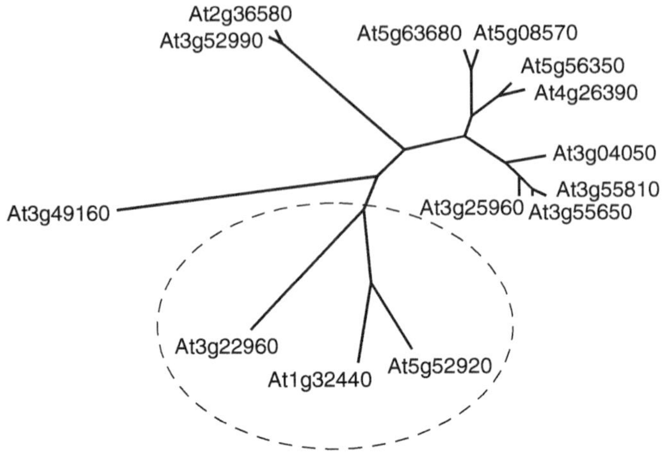

FIGURE 1.1 Unrooted phylogenetic analysis of putative pyruvate kinase genes from *Arabidopsis thaliana*. Each gene is identified by its AGI gene code. The deduced amino acid sequences of predicted pyruvate kinase isoforms were compared using CLUSTAL W. Genes proposed to encode plastid isoforms of the enzyme were identified using ChloroP and are enclosed within the broken ellipse. Predicted transit peptides were removed prior to sequence comparison.

inventory of the catalytic components of various plant metabolic networks in due course, and while this will not lead to the immediate clarification of the complex relationships that determine the way in which the enzymes function in such networks, it will at least define the scale of the problem.

Assuming that the enzymes and their locations can be identified, there is still much that needs to be determined to define the metabolic network at a level that is suitable for predictive modeling of fluxes. In particular, as well as defining the levels of the enzymes and their substrates, it is also necessary to identify all the regulatory mechanisms that operate in the network. At one level, this requires the characterization of all the molecular crosstalk that allows the components of the system to influence enzyme activity through effector-binding interactions; and at a higher level, and particularly in a system that will generally not be maintained in a steady state, it is also necessary to define the relationship between gene expression and the performance of the network, for example, to include the effects of circadian rhythms, light–dark transitions, and developmental triggers on enzyme levels. Clearly, the information required to define a metabolic network at this level of precision is not available for the cells of an organism as complicated as a higher plant, and indeed it is arguable that the emerging discipline of systems biology is unlikely to provide it, since the methodological focus is analytical, concentrating on genome-scale datasets for transcripts, proteins, and metabolites rather than mechanistic (Sweetlove *et al.*, 2003). It is also interesting to note that transcriptomic and proteomic analysis of simpler systems has not revealed direct quantitative correlations with metabolic fluxes (Oh and Liao, 2000; Oh *et al.*, 2002; ter Kuile and Westerhoff, 2001), demonstrating that high-throughput methods are not yet able to provide an effective alternative to the detailed kinetic and regulatory characterization of a metabolic network if the aim is predictive metabolic engineering.

While this section has emphasized the importance and difficulty of defining a complete plant metabolic network, the analysis of even an incompletely specified metabolic network can be informative. For example, genome-scale models of metabolism have been developed that allow reliable predictions of the growth potential of mutant phenotypes in *E. coli*, even though the analysis is based on genome annotation that is only 60–70% complete (Edwards and Palsson, 2000a; Edwards *et al.*, 2001; Price *et al.*, 2003). Similarly, a metabolic flux analysis of the principal pathways of carbon metabolism in *Corynebacterium glutamicum* was sufficiently detailed to identify a substantial diversion of resources into a cyclic flux involving the anaplerotic pathways (Petersen *et al.*, 2000). This observation provided the basis for a rational manipulation of the system and indeed the production of a strain lacking phosphoenolpyruvate (PEP) carboxykinase had the desired effect of decreasing metabolic cycling and increasing lysine production (Petersen *et al.*, 2001). Thus, while it is always possible that an incomplete metabolic model lacks the key feature that determines a relevant property of the system, worthwhile predictions of metabolic performance can often be made with such models. Moreover, even incorrect predictions are useful because they may suggest ways in which the model can be improved.

3. TOOLS FOR ANALYZING NETWORK STRUCTURE AND PERFORMANCE

In general, individual metabolic fluxes are the net result of the coordinated activity of the whole network and so rational manipulation of these fluxes requires tools that can analyze the network as a system rather than focusing on individual steps. The available modeling approaches can be classified on the basis of their underlying assumptions (Wiechert, 2002), and the resulting hierarchy matches the usefulness of the models for metabolic engineering.

The simplest models are the structural network models that are based on the metabolites and reaction steps that make up the network (Wiechert, 2002). Models of this kind are useful for exploring the architecture of the network, but they are of rather limited use in a physiological context because they lack quantitative information about the metabolites and reaction steps. This deficiency is remedied in stoichiometric models by assuming constant fluxes and intracellular pool sizes. Stoichiometric models provide the basis for determining intracellular fluxes (Bonarius *et al.*, 1997), as well as permitting the identification of fundamental network properties such as elementary flux modes and extreme pathways (Klamt and Stelling, 2003). Stoichiometric modeling can also be applied at the level of the individual carbon atoms in metabolites, and this leads to a more general method of determining intracellular fluxes based on the steady-state analysis of the redistribution of ^{13}C labels (Kruger *et al.*, 2003; Wiechert, 2001; Wiechert *et al.*, 2001).

Models that provide an explanation of the empirically derived flux distribution can be obtained by incorporating a kinetic description of each reaction step into a stoichiometric model (Wiechert, 2002). These mechanistic (kinetic) models require detailed information about the *in vivo* kinetic properties of the enzymes in the network, and this is a major obstacle in developing useful models. However, kinetic modeling is now well developed in yeasts (Teusink *et al.*, 2000) and red blood cells (Mulquiney and Kuchel, 2003). Accurate mechanistic models are expected to have predictive value in the context of metabolic engineering, and they can also be used to investigate the distribution of control within the conceptual framework of metabolic control analysis (Fell, 1997). Mechanistic models can be used to analyze both steady-state and transient fluxes and in the longer term it may also be possible to allow for fluctuations in enzyme level by incorporating the regulatory networks for gene expression (Wiechert, 2002).

It is clear from this survey that the analysis of the properties of metabolic networks can be approached using a variety of model-based strategies. Some of these approaches aim to make deductions about the performance of the network from an analysis of the constraints imposed by its structure and stoichiometry alone, whereas others are heavily dependent on direct measurements of metabolic fluxes and the kinetic properties of the enzymes that define the network. The aim here is to describe four of these methods in more detail and to comment on their utility as predictive tools for plant metabolic engineering.

3.1. Constraints-based network analysis

Constraints-based network analysis aims to reveal the function and capacity of metabolic networks without recourse to kinetic parameters (Bailey, 2001). The development and scope of the method has been reviewed (Covert *et al.*, 2001; Papin *et al.*, 2003; Price *et al.*, 2003, 2004), and its current importance as a modeling strategy owes much to the successful completion of numerous microbial genome sequencing projects. The analysis follows a three-step procedure: construction of a network, application of the constraints to limit the solution space of the network, and extraction of physiologically relevant information about network performance. The first step draws heavily on genome annotation, but biochemical and physiological data can provide complementary information that helps to improve the accuracy of the deduced network (Covert *et al.*, 2001). Ideally, the reconstructed network should also include regulatory elements at the level of gene expression to allow the model to be applicable under non–steady-state conditions (Covert and Palsson, 2002). The next step is to use reaction stoichiometry, directionality, and enzyme level to constrain the network and to work out the full set of allowed flux distributions (Price *et al.*, 2004). Finally, these solutions are analyzed to identify the flux distribution that optimizes a particular outcome, for example, growth rate (Price *et al.*, 2003).

Constraints-based genome-scale models have been constructed for several microorganisms and their utility for probing the relationship between genotype and phenotype is now well established (Price *et al.*, 2003). Assessing the impact of gene additions and deletions on predicted growth rate turns out to be a powerful test of the validity of the model as well as an effective way of identifying useful targets for genetic manipulation (Edwards and Palsson, 2000a; Price *et al.*, 2003).

Moreover, network robustness can be modeled by constraining the maximum flux through particular reactions, and this has demonstrated how effectively the network can sustain growth despite quite severe restrictions on central carbon metabolism (Edwards and Palsson, 2000b). The response to genetic modification and pathway robustness can also be assessed in terms of elementary flux modes—the set of nondecomposable fluxes that make up the steady-state flux distributions in the network (Klamt and Stelling, 2003; Schuster *et al.*, 1999). Thus, changes in network topology brought about by the addition or deletion of genes have an immediate effect on the set of elementary flux modes, and the impact on the synthesis of a particular metabolite and the efficiency with which it can be produced can be predicted (Schuster *et al.*, 1999). For example, an analysis of a metabolic network linking 89 metabolites via 110 reactions in *E. coli* revealed over 43,000 elementary flux modes, and from an *in silico* exploration of the consequences of gene deletion, it was concluded that the relative number of elementary flux modes was a reliable indicator of network function in mutant phenotypes (Stelling *et al.*, 2002), suggesting that elementary mode analysis could be a major asset in identifying targets for metabolic engineering (Cornish-Bowden and Cardenas, 2002).

The extent to which constraints-based network analysis succeeds in generating realistic and useful models of metabolism can be assessed directly from work on red blood cells. Much effort has been put into developing a comprehensive kinetic model of red blood cell metabolism (Jamshidi *et al.*, 2001; Mulquiney and Kuchel, 2003), and the question arises as to whether network analysis can make accurate predictions about the performance of the network. In fact, the complete set of the so-called extreme pathways (essentially a subset of the elementary modes for the network) has been worked out for the red blood cell network and after suitable classification it was shown that these pathways could be used to make physiologically sensible predictions about ATP:NADPH yield ratios (Wiback and Palsson, 2002). Thus, it has been concluded that network analysis can indeed generate metabolically important insights without the need for the labor-intensive measurement of a multitude of kinetic parameters (Papin *et al.*, 2003). Interestingly, network analysis has recently been combined with *in vivo* measurements of concentrations and a simplified representation of enzyme kinetics to calculate the allowable values of these kinetic parameters, and this novel approach may well facilitate the construction of kinetic models in the absence of the full characterization of the enzymes in the network (Famili *et al.*, 2005).

In the light of this conclusion, and particularly given the utility of network analysis in guiding metabolic engineering (Papin *et al.*, 2003; Price *et al.*, 2003; Schuster *et al.*, 1999), there would appear to be a strong case for extending the constraints-based approach to the analysis of plant metabolic networks. However, there appear to have been few attempts to do so, and the only substantial contribution is a paper describing an elementary modes analysis of metabolism in the chloroplast (Poolman *et al.*, 2003). This analysis highlighted the interaction between the Calvin cycle and the plastidic oxidative pentose phosphate pathway, and the potential involvement of the latter in sustaining a flux from starch to triose phosphate in the dark.

3.2. Metabolic flux analysis

Metabolic flux analysis takes a stoichiometric model of a metabolic network and aims to quantify all the component fluxes (Wiechert, 2001). In simple systems, these fluxes can be deduced from steady-state rates of substrate consumption and product formation, but in practice this approach of metabolite flux balancing is unable to generate sufficient constraints to provide a full flux analysis in most cases (Bonarius *et al.*, 1997). In particular, metabolite flux balancing is largely defeated by the substrate cycles, parallel pathways, and reversible steps that are commonly encountered in metabolic networks (Wiechert, 2001), and for these and other reasons discussed elsewhere metabolite flux balancing is unlikely to be useful in the quantitative analysis of plant metabolism (Morgan and Rhodes, 2002; Roscher *et al.*, 2000).

A more powerful approach for measuring intracellular fluxes, again developed using microorganisms, is to analyze the metabolic redistribution of the label from one or more ^{13}C-labeled substrates (Wiechert, 2001). While flux information can be deduced from the time course of such a labeling experiment, constructing and analyzing time courses can be demanding, and so it is usually preferable to analyze the system after it has reached an isotopic steady state. Typically, a metabolic flux analysis using this approach would therefore involve incubating the tissue or cell suspension with a ^{13}C-labeled substrate for a period that is sufficient to allow the system to reach a metabolic and isotopic steady state; a mass spectrometric and/or nuclear magnetic resonance analysis of the isotopomeric composition of selected metabolites in tissue extracts; and finally construction of the flux map based on the stoichiometry of the network and the measured redistribution of the label (Wiechert, 2001). The number of fluxes in the final map depends on the labeling strategy, the structure of the network, and the extent to which the redistribution of the label is characterized, but the usual objective in microorganisms is to generate a flux map that covers all the central pathways of metabolism (Szyperski, 1998; Wiechert, 2001; Wiechert *et al.*, 2001).

Metabolic flux analysis generates large-scale flux maps in which forward and reverse fluxes are defined at multiple steps in the metabolic network. This manifestation of the metabolic phenotype provides a quantitative tool for comparing the metabolic performance of different genotypes of an organism, as well as for assessing the metabolic consequences of physiological and environmental perturbations (Emmerling *et al.*, 2002; Marx *et al.*, 1999; Sauer *et al.*, 1999). Most of these studies lead to the conclusion that metabolic networks are flexible and robust, in agreement with much larger-scale theoretical studies (Stelling *et al.*, 2002), and thus emphasize the point that targets for metabolic engineering have to be selected rather carefully if they are to have the intended effect on the flux distribution. The investigation of lysine production in *C. glutamicum* mentioned earlier provides a good illustration of the way in which an analysis of the flux distribution can be used to identify a rational target for metabolic engineering (Petersen *et al.*, 2000, 2001).

Although the extension of steady-state metabolic flux analysis to plants is complicated by subcellular compartmentation, by duplication of pathways, and by the difficulty of establishing an isotopic and metabolic steady state (Roscher *et al.*, 2000), there is increasing evidence that such analyses are both feasible and physiologically useful (Kruger *et al.*, 2003; Schwender *et al.*, 2004; Ratcliffe and Shachar-Hill, 2006). Some of these investigations measure only a small number of fluxes through specific steps or pathways, while others emulate the large-scale analyses of central metabolism that were pioneered on microorganisms. Examples in the small-scale category include an analysis of the relative contribution of malic enzyme and pyruvate kinase to the synthesis of pyruvate in maize root tips (Edwards *et al.*, 1998); an assessment of the impact of elevated fructose 2,6-bisphosphate levels on pyrophosphate: fructose-6-phosphate 1-phosphotransferase in transgenic tobacco callus (Fernie *et al.*, 2001); and the many applications of retrobiosynthetic flux analysis for assessing the relative importance of the mevalonate and methylerythritol phosphate pathways in terpenoid biosynthesis (Eisenreich *et al.*, 2001).

While these small-scale analyses provide useful information about specific aspects of the metabolic phenotype that may well be directly relevant, as in the case of the transgenic tobacco study (Fernie *et al.*, 2001), to the characterization of engineered genotypes, large-scale analyses of multiple fluxes in extensive networks have the potential to provide a much broader assessment of the impact of genetic manipulation on the metabolic network. It is therefore encouraging to note that steady-state stable isotope labeling is now being used to generate flux maps for central carbon metabolism in several plant systems. The first extensive flux map of this kind, based on the measurement of 20 cytosolic, mitochondrial, and plastidic fluxes, was obtained in a study of excised maize root tips (Dieuaide-Noubhani *et al.*, 1995). This map proved to be useful in physiological experiments, for example, in assessing the impact of sucrose starvation on carbon metabolism (Dieuaide-Noubhani *et al.*, 1997). It also led to the development of a more detailed flux map for a tomato cell suspension culture (Rontein *et al.*, 2002), from which it was concluded that the relative fluxes through glycolysis, the tricarboxylic acid cycle, and the pentose phosphate pathway were unaffected by the progression through the culture cycle, whereas the generally smaller anabolic fluxes were more variable. Steady-state flux maps have also been published for the pathways of primary metabolism in developing embryos of oilseed rape (Schwender *et al.*, 2003) and soybean (Sriram *et al.*, 2004). An interesting feature of the oilseed rape model is that the labeling patterns showed rapid exchange of key intermediates between the cytosolic and plastidic compartments, thus simplifying the analysis and the resulting flux map. This result is in contrast to the situation in maize root tips and tomato cells, where the labeling of the unique products of cytosolic and plastidic metabolism showed that the cytosolic and plastidic hexose and triose phosphate pools were kinetically distinct.

The conclusion to be drawn from these studies is that large-scale flux maps can be generated for plant metabolic networks using steady-state stable isotope labeling and that the problems inherent in the complexity of these networks are not necessarily insuperable. These maps have been mainly used to gain further understanding of the operation of wild-type pathways, but, as already seen in microorganisms, it can only be a matter of time before they are also used to assess the impact of genetic manipulation and to propose potentially useful engineering strategies.

3.3. Kinetic modeling

Kinetic models provide the most powerful method for understanding flux distributions under both steady-state and non–steady-state conditions, but they are totally dependent on the availability of accurate kinetic data for each enzyme-catalyzed step in the network (Wiechert, 2002). The difficulty of assembling such information means that kinetic models are generally restricted to fragments of the metabolic network, for example, glycolysis in yeast (Pritchard and Kell, 2002; Teusink *et al.*, 2000), and to date the only kinetic models that attempt to cover the complete network of a cell have been set up for the metabolically specialized red blood cell, with its greatly reduced metabolic network (Jamshidi *et al.*, 2001; Mulquiney and Kuchel, 2003). Small-scale kinetic models are a more realistic target for the analysis of plant metabolism and, as documented elsewhere (Morgan and Rhodes, 2002), there has been sustained interest in the development of such models since the publication of an influential model of C_3 photosynthesis (Farquhar *et al.*, 1980).

One application of such models in a metabolic engineering context is in rationalizing and understanding the behavior of transgenic plants with altered levels of particular enzymes. Kinetic models can be used to predict the flux control coefficients of individual enzymes, and these can be compared with the values obtained empirically. This approach can be illustrated by an analysis of the Calvin cycle that included starch synthesis, starch degradation, and triose phosphate export from the chloroplast to the cytosol (Poolman *et al.*, 2000). The calculated flux control coefficients showed that the control distribution varied between fluxes—for example, the CO_2 assimilation flux was predicted to be largely determined by the activities of ribulose 1,5-bisphosphate carboxylase/oxygenase (Rubisco) and sedoheptulose-1,7-bisphosphatase (SBPase), and to be largely independent of the activity of the triose phosphate translocator—and it was concluded that the predictions were broadly consistent with the observations that have been made on transgenic plants. This conclusion provides some reassurance that the model is a reasonable, though still imperfect, representation of the experimental system, but the real value of the approach probably lies not so much in how close the fit can be, but in providing insights into the operation of the pathway. Thus, this modeling exercise highlighted the previously largely neglected role of SBPase in the assimilation process, and it reinforced the view that the manipulation of a single selected enzyme is unlikely to increase the assimilatory capacity of the pathway (Poolman *et al.*, 2000).

This leads to the second major application for kinetic models in metabolic engineering, which is their use as predictive tools for generating hypotheses about flux limitation in a metabolic network and thus providing the basis for a rational engineering strategy. A good example of this approach can be found in an analysis of the synthesis of glycine betaine in transgenic tobacco expressing choline monooxygenase (McNeil *et al.*, 2000a,b). In this work, the aim was to identify the constraints on the synthesis of glycine betaine as part of a program to engineer stress tolerance into tobacco through the production of an osmoprotectant. The first stage in the analysis was to establish which of three parallel, interconnected pathways were used for the synthesis of choline from ethanolamine in tobacco (McNeil *et al.*, 2000a). This objective was achieved by incubating the system with ^{14}C- and ^{33}P-labeled precursors and monitoring the time course for the redistribution of the label into the intermediates of choline synthesis. With a knowledge of the corresponding pool sizes, it was then possible to construct a flux model that described the labeling kinetics for each precursor and thus to deduce that the predominant pathway involved N-methylation of phosphoethanolamine (McNeil *et al.*, 2000a). This led to the suggestion that overexpression of phosphoethanolamine *N*-methyltransferase would be a rational target for improving the endogenous choline supply for glycine betaine synthesis. Subsequently, further modeling of [^{14}C]choline-labeling experiments revealed two more constraints—inadequate capacity for choline uptake into the chloroplast and excessive choline kinase activity—both of which work against the provision of substrate for choline monooxygenase. It was concluded that the failure of the engineered plants to accumulate significant levels of glycine betaine was due to multiple causes and that it would be necessary to address all of them to obtain a glycine betaine concentration comparable to that found in natural accumulators (McNeil *et al.*, 2000b).

These examples demonstrate the utility of kinetic modeling as a procedure for probing relatively small metabolic networks. They also highlight the way in which the properties of the network conspire against simple engineering solutions, a conclusion that is consistent with the wealth of empirical data on flux control coefficients that has been accumulated in recent years and the theoretical predictions of metabolic control analysis (see next section).

3.4. Metabolic control analysis

Metabolic control analysis provides a theoretical framework for analyzing the control and regulation of metabolism (Fell, 1997). At a practical level, the introduction of metabolic control analysis has had two important consequences for the empirical analysis of plant metabolism. First, by providing a new set of fundamental parameters for characterizing metabolic pathways, particularly flux control coefficients, elasticities, and response coefficients, metabolic control analysis has stimulated a substantial effort to measure these quantities in an attempt to put the description of the control and regulation of plant metabolism on a firm foundation (Stitt and Sonnewald, 1995). Inevitably, this has involved the characterization of many transgenic lines since genetic manipulation provides the most versatile way of altering the endogenous level of specific enzymes for the measurement of flux control coefficients; and as discussed in the following section, this rigorous approach has provided ample evidence for the delocalized control of flux and for the complexity of the regulatory interactions in plant metabolic networks. Second, as illustrated by the modeling of the Calvin cycle described in the previous section (Poolman *et al.*, 2000), metabolic control analysis provides a tool for analyzing steady-state kinetic models and for deducing flux control coefficients. This indirect approach to the determination of flux control coefficients further emphasizes the way in which control is distributed throughout the network and the dependence of this distribution on the prevailing physiological state of the organism.

These practical applications of metabolic control analysis are complemented by the important theoretical conclusions that have emerged concerning the feasibility of flux manipulation or metabolic engineering. First, overexpression of a single enzyme in a pathway is likely to have only a limited impact on flux because even if the chosen enzyme has a significant flux control coefficient in the wild-type plant, control will be redistributed to other steps in the pathway as the level of the enzyme is increased. The validity of this conclusion, and its challenging message for the plant metabolic engineer, has been borne out by a large body of experimental evidence from genetically engineered plants, including the notable and early failure to increase glycolytic flux in potato tubers via the overexpression of phosphofructokinase (Burrell *et al.*, 1994). Second, overexpressing multiple pathway enzymes may lead to an increased flux, as demonstrated for tryptophan synthesis in yeast (Niederberger *et al.*, 1992). In effect, this strategy can be seen to mimic the coordinated upregulation of gene expression that occurs in many physiological responses, for example, in the mobilization of storage lipid during the germination of *Arabidopsis thaliana* (Rylott *et al.*, 2001), but it poses the problem of how to produce a coordinated change in the expression of several genes in a transformed plant. Third, the success of any attempt to increase the flux through a pathway also depends on maintaining the supply of the necessary substrates and ensuring that there is an increased demand for the product. In support of this conclusion, recent investigations have shown that the starch content and yield of potato tubers can be increased by downregulating the plastidic isoform of adenylate kinase, apparently as a direct result of increasing the availability of plastidic ATP for ADPglucose synthesis (Regierer *et al.*, 2002); and the glycolytic flux in *E. coli* has been enhanced by introducing a soluble F_1-ATPase to provide a sink for ATP (Koebmann *et al.*, 2002; Oliver, 2002). Both these investigations are notable for their manipulation of a coenzyme that is necessarily involved in multiple reactions, and establishing the extent to which the observed phenotypes can be attributed exclusively to the direct effect of changes in ATP level and turnover may be problematic. However, the success of these manipulations emphasizes just how widely control is distributed in metabolic networks and hence the difficulty in selecting targets for manipulation.

The relationship between the substrates, enzymes, and fluxes in complex metabolic networks revealed by metabolic control analysis emphasizes the intrinsic difficulty of rational metabolic engineering. Moreover, while it is possible to predict that some strategies are likely to be successful—for example, diverting a small proportion of a flux into a novel product or eliminating the formation of a toxic product (Morandini and Salamini, 2003)—there is no certainty in the outcome. Moreover, engineering objectives that require extensive redirection of the fluxes through the central pathways of metabolism are likely to be particularly challenging and may be too ambitious or even intrinsically impossible without wholesale restructuring of the network (Morandini and Salamini, 2003). Despite this assessment, the recent progress in engineering increased starch production in potato tubers (Regierer *et al.*, 2002) highlights the importance of sustained empirical investigations that are guided by a rigorous understanding of metabolic control.

4. INTEGRATION OF PLANT METABOLISM

The complexity of the plant metabolic network and its regulatory mechanisms has been amply confirmed by the compelling body of experimental evidence that has accumulated over the past decade from studies of the primary pathways of carbohydrate metabolism. In particular, there have been numerous studies of photosynthetic carbon assimilation and it is the aim of this section to present the principal conclusions about network performance that can be drawn from investigations of transgenic plants with reduced levels of Calvin cycle enzymes. The analysis highlights the robustness of the metabolic network and the complexity that needs to be incorporated into realistic models of plant metabolism.

4.1. Relationship between enzyme properties and network fluxes

At the most fundamental level, ***the kinetic properties of an enzyme and the displacement of its reaction from thermodynamic equilibrium in vivo do not provide a reliable indicator of the effect on pathway flux of a reduction in the amount of the enzyme***. Thus, although Rubisco, plastidic fructose-1,6-bisphosphatase, and phosphoribulokinase have traditionally been considered to be important in the control of photosynthesis on the basis that they catalyze irreversible reactions and are subject to regulation by effectors and reversible posttranslational modification (Macdonald and Buchanan, 1997), a moderate decrease in the amount of any of these enzymes usually has little effect on the rate of CO_2 fixation under normal growth conditions (Stitt and Sonnewald, 1995). This tendency for metabolic pathways to compensate for a decrease in the amount of an enzyme arises from the inevitable complementary changes that occur in the concentrations of metabolites throughout the reaction network. These changes may be sufficient to compensate for decreased expression of an enzyme by increasing the proportion of its catalytic capacity that is realized *in vivo*, as observed in tobacco lines with an 85–95% decrease in expression of phosphoribulokinase (Paul *et al.*, 1995), or by altering the activation state of the targeted enzyme, thus increasing the catalytic capacity of the residual protein, as observed for Rubisco (Stitt and Schulze, 1994).

4.2. Limitations on metabolic compensation within a network

The capacity of the metabolic network to compensate for alterations in the amount of an enzyme depends on the impact of the associated changes in metabolite concentrations on all the steps in the network. Enzymes that are sensitive to modulation by effectors, particularly metabolites from within the pathway, can compensate for decreased expression because small changes in the concentrations of substrates, products, inhibitors, and activators are likely to be sufficient to stimulate the activity of the residual enzyme. However, for enzymes that lack such regulatory properties, compensation can occur only through alterations in the concentrations of the immediate substrates and products of the enzyme. The extent to which this can occur is constrained *in vivo* by the effect that such changes can have on the operation of the other enzymes in the network. Thus, flux can be reduced because the changes in metabolite concentration that would be required to prevent the decrease have adverse effects on other sections of the pathway, rather than because the manipulated enzyme has insufficient catalytic capacity to support the flux. This explains why a moderate decrease in either plastidic aldolase (Haake *et al.*, 1998, 1999) or transketolase (Henkes *et al.*, 2001) inhibited the rate of CO_2 fixation even though the maximum catalytic capacity of the residual enzyme was seemingly still in excess of that required to accommodate the normal rate of photosynthesis. The mechanisms that restrict flux through the pathway in these examples are considered in more detail below.

4.3. Impact of physiological conditions on network performance

The metabolic impact of altering the amount of an enzyme depends on the physiological state of the system. Extensive analysis of transgenic tobacco lines possessing decreased amounts of Rubisco has established that the flux control coefficient of the enzyme on photosynthesis varies in response to both the immediate conditions and the conditions under which the plant developed (Stitt and Schulze, 1994). For plants grown and analyzed under moderate irradiance, photosynthesis was only slightly inhibited when Rubisco was decreased to about 60% of the wild-type amount. However, stimulation of photosynthesis by an immediate increase in light intensity resulted in a near-proportional relationship between the amount of Rubisco and the rate of photosynthesis. In contrast, when photosynthesis was measured at saturating CO_2 levels, Rubisco content could be decreased by as much as 80% without any appreciable effect on the rate of assimilation. Thus, the metabolic impact of modifying the amount of Rubisco depended on the conditions under which the flux was measured. Moreover, the response to reduced Rubisco also depended on the conditions under which the plants were grown: a moderate decrease in Rubisco had a relatively minor effect on photosynthesis in plants grown at high irradiance, in contrast to the near-proportional decrease in photosynthesis for plants grown at low irradiance prior to transfer to a higher light intensity. Similarly, growth of plants on low nitrogen fertilizer increased the extent to which photosynthesis was impaired by a decrease in the amount of Rubisco. This extensively investigated example emphasizes that any assessment of the potential of a specific enzyme as a target for metabolic manipulation must take into consideration both the conditions in which flux is being measured and the conditions in which the plant is grown (Stitt and Schulze, 1994).

4.4. Network adjustments through alternative pathways

Manipulating the amount of a particular enzyme can influence a metabolic process through more than one route. Currently, the clearest demonstration of this point is provided by studies of transgenic potato plants in which the amount of aldolase was selectively decreased (Haake *et al.*, 1998, 1999). When grown under low irradiance, a 30–50% decline in aldolase expression led to an accumulation of triose phosphates and a decrease in ribulose 1,5-bisphosphate (RuBP) and 3-phosphoglycerate (3PGA). These changes are consistent with restrictions in the capacity of the two reactions of the Calvin cycle catalyzed by aldolase (Fig. 1.2A). Under these conditions, photosynthesis is inhibited because of a limitation in the regeneration of RuBP,

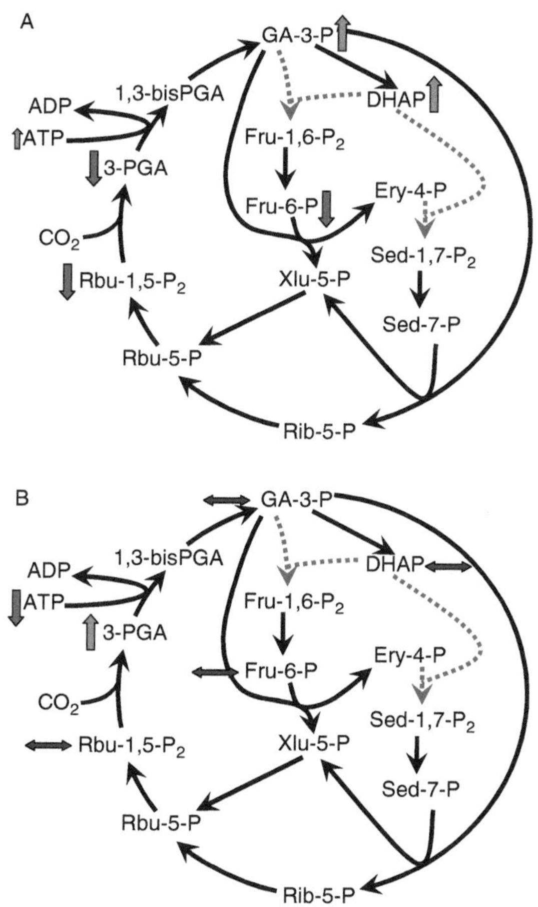

FIGURE 1.2 Effect of a decrease in aldolase content on photosynthetic intermediates in potato plants (Haake *et al.*, 1999). Changes in the steady-state levels of Calvin cycle intermediates in aldolase-antisense lines grown under low irradiance (A) or high irradiance in the presence of elevated CO_2 (B) are compared with those in wild-type plants grown under the same conditions. The reactions catalyzed by aldolase are indicated by dotted lines. Symbols refer to the following changes in metabolite content: ↑, increase; ↓, decrease; ↔, roughly similar. (See Page 1 in Color Section.)

presumably resulting from a decrease in the steady-state concentration of pentose phosphates downstream of the reactions catalyzed by aldolase. However, when grown under high irradiance, and especially in the presence of elevated CO_2, triose phosphates remained very low, RuBP remained high, and 3PGA levels were higher in the transformants than in wild-type plants. Under these circumstances, the inhibition of photosynthesis cannot be attributed to a lack of CO_2 acceptor since the steady-state concentration of RuBP remained high, but instead appears to result from Pi-limitation arising from a restricted capacity for starch synthesis. This limits ATP production and restricts the conversion of 3PGA to triose phosphates. Thus, under these conditions, the immediate cause for the decrease in photosynthesis is product inhibition of Rubisco by the increase in 3PGA (Fig. 1.2B).

An important corollary of this point is that the relative importance of the mechanisms by which a metabolic process is affected may vary. In the aldolase investigation, it is likely that the apparent switch between the two mechanisms for inhibiting photosynthesis reflects the extent to which regeneration of RuBP or end-product (starch) formation dominated control of photosynthesis under the chosen experimental conditions. However, there is nothing to suggest that these mechanisms are mutually exclusive, and it is likely that the relative significance of the two processes will shift gradually as their relative importance in determining the rate of photosynthesis varies. These considerations imply that in order to predict the consequences of manipulating an enzyme, it is necessary to identify all possible mechanisms by which a change in the amount of the enzyme can influence flux through the network, and to quantify the relative contribution of each of these mechanisms to the control of metabolic flux under the relevant physiological conditions.

4.5. Propagation of metabolic perturbations through networks

The metabolic consequences of altering the amount of an enzyme are unlikely to be confined to a single pathway. A clear illustration of the extent of the interactions that occur between pathways is provided by a study of transgenic tobacco lines in which the amount of transketolase was selectively decreased (Henkes *et al.*, 2001). These lines displayed a near-proportional decrease in the maximum rate of photosynthesis in saturating CO_2 and a smaller inhibition of photosynthesis under normal growth conditions. This inhibition was accompanied by large decreases in the steady-state levels of RuBP and 3PGA, smaller decreases in the amounts of triose phosphates and fructose 1,6-bisphosphate, and a large increase in the amount of fructose 6-phosphate. These changes are entirely consistent with restrictions in the two reactions of the Calvin cycle catalyzed by transketolase and suggest that the immediate cause for the decrease in photosynthesis is a restriction in the ability to regenerate RuBP (Fig. 1.3). Thus, the effect of reduced transketolase appears to be similar to that obtained when the aldolase content was decreased under low light (Fig. 1.2A). However, in contrast to the consequences of manipulating aldolase content, a decrease in transketolase also caused a disproportionately large decrease in the levels of aromatic amino acids, intermediates of the phenylpropanoid pathway, and secondary products such as chlorogenic acid and lignin. These observations suggest that the level of transketolase has a major impact

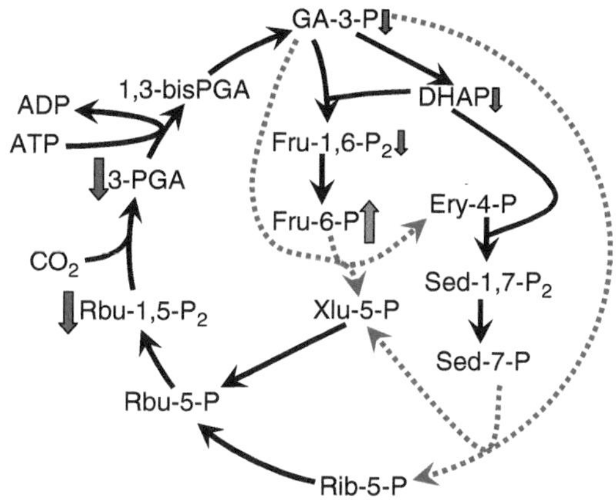

FIGURE 1.3 Effect of decreased transketolase content on photosynthetic intermediates in tobacco plants (Henkes *et al.*, 2001). Changes in the steady-state levels of Calvin cycle intermediates in transketolase-antisense lines are compared with those in wild-type plants grown under the same conditions. The reactions catalyzed by transketolase are indicated by dotted lines. Symbols refer to the following changes in metabolite content: ↑, increase; ↓, decrease. (See Page 2 in Color Section.)

on the channeling of intermediates into the shikimic acid pathway and the likely explanation for this effect is that the metabolic network responds to a decrease in the amount of transketolase by decreasing the amount of erythrose 4-phosphate (Fig. 1.3). Consequently, flux into the shikimic acid pathway is restricted by the supply of erythrose 4-phosphate and phenylpropanoid metabolism is constrained by the corresponding decreased provision of aromatic amino acids.

The multiple responses to reducing transketolase highlight the extent of integration within the central metabolic pathways and the potential difficulties in attempting to modify flux through a specific section of the metabolic network. In particular, the results suggest that major changes in secondary metabolism may require appropriate reprograming of primary pathways to ensure an adequate supply of the necessary precursors. Corroborative evidence that the formation of secondary products may be limited by the availability of primary precursors is provided by a report that a decrease in the levels of aromatic amino acids due to ectopic expression of tryptophan decarboxylase led to decreases in the amounts of chlorogenic acid and lignin in transgenic potato plants (Yao *et al.*, 1995).

In fact both the structure and chemical organization of metabolic networks suggest that transketolase is unlikely to be unique in the manner in which changes in its activity influence other metabolic processes. This view is supported by a theoretical analysis of the potential metabolic interactions for each of the intermediates of glycolysis and the oxidative pentose phosphate pathway (Table 1.1). Although there is considerable variation between compounds, on average each metabolite is a reactant for about 20 enzymes, and either activates or inhibits a further 22 enzymes. These values provide only a crude estimate of the complexity that arises through the multiplicity of ligand-binding interactions and the estimate

TABLE 1.1 Metabolic reactivity of intermediates of primary pathways of carbohydrate oxidation

	Number of enzymes for which specified metabolite is:		
Metabolite	Reactant	Activator	Inhibitor
UDP-glucose	74	3	19
Glucose 1-phosphate	25	7	10
Glucose 6-phosphate	17	16	32
Fructose 6-phosphate	19	9	22
Fructose 1,6-bisphosphate	7	13	37
Dihydroxyacetone phosphate	18	5	10
Glyceraldehyde 3-phosphate	18	3	15
1,3-Bisphosphoglycerate	10	0	2
3-Phosphoglycerate	13	9	25
2-Phosphoglycerate	4	0	9
Phosphoenolpyruvate	19	12	43
Pyruvate	106	9	61
6-Phosphoglucono-1,5-lactone	2	0	0
6-Phosphogluconate	5	4	19
Ribulose 5-phosphate	8	1	2
Ribose 5-phosphate	17	2	12
Xylulose 5-phosphate	6	1	1
Erythrose 4-phosphate	6	4	9
Sedoheptulose 7-phosphate	6	0	3

The number of enzymes for which each metabolite is a substrate or product was taken from the Kyoto Encyclopedia of Genes and Genomes (KEGG) database at GenomeNet (Kanehisa *et al.*, 2002), and the number of enzymes activated or inhibited by the compound was obtained from the Braunschweig Enzyme Database (BRENDA) (Schomburg *et al.*, 2002).

is in any case very dependent on the extent to which all potential inhibitory and stimulatory responses have been identified for the selected enzymes. Even so, the analysis suggests that perturbation of the level of any metabolite within the central pathways of carbohydrate oxidation has a very strong likelihood of affecting several other reactions, thus allowing the consequences of the initial change to propagate widely throughout the network. Such considerations further emphasize the integrated nature of the central metabolic pathways and the difficulties that are likely to be encountered in attempting to modify individual processes selectively.

4.6. Enzyme-specific responses within networks

Individual reactions in a pathway may affect the same process in different ways. Although antisense inhibition of each of several Calvin cycle enzymes ultimately restricts the rate of CO_2 assimilation, the mechanisms by which photosynthesis is affected differ for the different enzymes. This is revealed by considering the impact of the decrease in the rate of CO_2 assimilation on the two major photosynthetic end-products, sucrose and starch. In Rubisco antisense lines, the decrease in photosynthesis led to proportional decreases in the rate of sucrose and starch synthesis (Stitt and Schulze, 1994), whereas inhibition of CO_2 fixation due to decreased expression of aldolase (Haake *et al.*, 1998), plastid fructose-1,6-bisphosphatase (Kossmann *et al.*, 1994), or SBPase (Harrison *et al.*, 1998) was accompanied by a far greater inhibition of starch synthesis and preferential retention of sucrose synthesis. In contrast, decreased expression of transketolase led to preferential retention of starch accumulation and a decrease in sucrose content, suggesting a shift in allocation in favor of starch relative to sucrose (Henkes *et al.*, 2001).

The difference in assimilate partitioning may be explained in part by the position of the selected enzyme within the Calvin cycle relative to fructose 6-phosphate, the immediate precursor for starch synthesis. Transketolase operates downstream of fructose 6-phosphate, which is therefore likely to increase when expression of the enzyme is decreased, hence stimulating starch synthesis (Fig. 1.3). In contrast, aldolase and plastid fructose 1,6-bisphosphatase are both upstream of fructose 6-phosphate and decreased expression of either of these enzymes is likely to result in lower levels of this intermediate, reducing the availability of precursors for starch synthesis.

However, the availability of fructose 6-phosphate cannot provide the complete explanation because SBPase is also downstream of fructose 6-phosphate and yet a decrease in expression of this enzyme led to a preferential restriction of starch production rather than enhancement (Harrison *et al.*, 1998). This apparent anomaly arises because erythrose 4-phosphate is a potent inhibitor of phosphoglucoisomerase, the enzyme catalyzing the conversion of fructose 6-phosphate to glucose-6-phosphate in the pathway of photosynthetic starch synthesis. Both aldolase and SBPase are involved directly in the catabolism of erythrose 4-phosphate, and decreased expression of either of these enzymes is likely to result in an increase in the level of this intermediate, leading to increased inhibition of starch synthesis. In contrast, decreased expression of transketolase presumably leads to lower erythrose 4-phosphate, which relieves inhibition of phosphoglucoisomerase and thus favors starch synthesis despite a decline in the concentration of 3PGA, an important activator of ADPglucose pyrophosphorylase, which might otherwise be predicted to restrict starch production. This implies that the metabolic consequences of adjusting the amount of a specific enzyme must be assessed on their own merits and that any similarity to the changes produced by different target enzymes should not be taken to imply that the manipulations are affecting the process through a common route.

4.7. Impact of metabolic change on network structure

Finally, although most efforts to manipulate metabolism focus on the immediate metabolic effects of adjusting the amount of a specific enzyme, there is appreciable evidence that ***altering the amount of an enzyme can also influence metabolism through its impact on network structure brought about via metabolite signaling or perturbation of nutritional status.*** The complexity of the physiological and metabolic responses brought about by regulation at this level is highlighted in a study that examined the relationship between photosynthesis, nitrogen assimilation, and secondary metabolism (Matt *et al.*, 2002). This investigation showed that inhibition of photosynthesis by decreasing Rubisco led to a preferential decrease in the amounts of amino acids relative to sugars, a disproportionate decline in the absolute levels of secondary metabolites, and a shift in the proportions of carbon- and nitrogen-rich secondary metabolites. Many of these effects were most apparent in plants grown in high nitrate. Under these conditions, the fall in amino acid levels despite the availability of nitrate can be explained, at least in part, by a reduction in nitrate reductase activity occurring as a consequence of a decrease in the levels of sugars that are required to maintain expression of the genes encoding nitrate reductase and to promote posttranslational activation of the enzyme (Klein *et al.*, 2000). In turn, the reported decrease in chlorogenic acid was probably a direct consequence of low levels of phenylalanine restricting flux into phenylpropanoid metabolism, while the decrease in nicotine was presumably related to the general inhibition of primary nitrogen metabolism and associated decreases in amino acids. The disproportionately large decrease in amino acid levels in the lines in which Rubisco expression was suppressed may also provide the explanation for the seemingly counterintuitive observation that accumulation of nitrogen-rich nicotine was preferentially inhibited relative to carbon-rich chlorogenic acid when photosynthetic carbon assimilation was inhibited under nitrogen-replete conditions (Matt *et al.*, 2002).

Analysis of the response of nitrogen metabolism and the consequential changes in secondary metabolism to decreased photosynthesis in plants grown under conditions of low nitrogen availability revealed a further layer of complexity. Many of the effects seen in high nitrate were obscured under limiting nitrogen conditions. The likely explanation for this is that because of lower rates of photosynthesis, and hence a decreased requirement for organic nitrogen, the Rubisco antisense lines were less nitrogen-limited than wild-type plants when grown in low nitrogen. This indirect amelioration of nitrogen deficiency masked the direct inhibitory effects of low Rubisco activity on nitrogen assimilation. Thus, wild-type tobacco grown on low nitrogen had low levels of nitrate and glutamine, and a low glutamine:glutamate ratio typical for nitrogen-limited plants, whereas the plants with decreased Rubisco had increased nitrate and glutamine and a higher glutamine:glutamate ratio. As a result of these differences, the decrease in nicotine accumulation in the transgenic lines relative to wild type observed under nitrogen-replete conditions was diminished or even reversed in low nitrogen fertilizer (Matt *et al.*, 2002). Such considerations provide a compelling reminder of the difficulties in interpretation of metabolic comparisons between plant lines even under seemingly carefully defined growth conditions and of the danger in ascribing a metabolic change to a single direct effect.

5. SUMMARY

The metabolic organization of plant cells poses a severe challenge for the development of the predictive models that are required for the rational manipulation of plant metabolism. While constraints-based network analysis, metabolic flux analysis, kinetic modeling, and metabolic control analysis provide a powerful complementary set of theoretical and empirical approaches for analyzing the structure and performance of plant metabolic networks, these tools have not yet led to easy solutions in the quest for useful targets for plant metabolic engineering. The task is particularly daunting in relation to the central pathways of carbon metabolism, where the metabolic characterization of transgenic plants reveals a remarkably robust metabolic network. These investigations indicate that the network can often compensate for alterations in the amounts of enzymes through changes in the steady-state levels of pathway intermediates and the activation state of the enzymes. Moreover, investigations of transgenic plants have revealed numerous instances of effects that arise as a secondary consequence of the original enzymic modification or that arise in pathways that seem at first sight to be quite separate from the pathway that is being manipulated. While it is clear that our qualitative understanding of primary plant metabolism is sufficient to rationalize the response of the metabolic network to changes in expression of a specific enzyme, it is difficult to believe that most of the responses that have been observed could be predicted with any degree of certainty with the currently available models. To do so would require a complete, quantitative understanding of all the relevant interactions between the components of the metabolic network and much further work will be required to achieve this goal.

ACKNOWLEDGEMENTS

The authors thank Dr. Y. Shachar-Hill for a critical reading of the chapter and they acknowledge the financial support of the Biotechnology and Biological Sciences Research Council. R.G.R. also thanks the Université de Picardie Jules Verne for financial support and hospitality.

REFERENCES

Amor, Y., Haigler, C. H., Johnson, S., Wainscott, M., and Delmer, D. P. (1995). A membrane-associated form of sucrose synthase and its potential role in synthesis of cellulose and callose in plants. *Proc. Natl. Acad. Sci. USA* **92,** 9353–9357.

ap Rees, T. (1987). Compartmentation of plant metabolism. *In* "The Biochemistry of Plants" (D. D. Davies, ed.), vol. 12, , pp. 87–115. Academic Press, New York.

Bailey, J. E. (2001). Complex biology with no parameters. *Nat. Biotechnol.* **19,** 503–504.

Baroja-Fernandez, E., Munoz, F. J., Zandueta-Criado, A., Moran-Zorzano, M. T., Viale, A. M., Alonso-Casajus, N., and Pozueta-Romero, J. (2004). Most of ADP-glucose linked to starch biosynthesis occurs outside the chloroplast in source leaves. *Proc. Natl. Acad. Sci. USA* **101,** 13080–13085.

Baroja-Fernandez, E., Munoz, F. J., and Pozueta-Romero, J. (2005). Response to Neuhaus *et al.*: No need to shift the paradigm on the pathway to transitory starch in leaves. *Trends Plant Sci.* **10,** 156–158.

Bonarius, H. P. J., Schmid, G., and Tramper, J. (1997). Flux analysis of underdetermined metabolic networks: The quest for the missing constraints. *Trends Biotechnol.* **15,** 308–314.

Burrell, M. M., Mooney, P. J., Blundy, M., Carter, D., Wilson, F., Green, J., Blundy, K. S., and ap Rees, T. (1994). Genetic manipulation of 6-phosphofructokinase in potato tubers. *Planta* **194,** 95–101.

Burton, R. A., Johnson, P. E., Beckles, D. M., Fincher, G. B., Jenner, H. L., Naldrett, M. J., and Denyer, K. (2002). Characterization of the genes encoding the cytosolic and plastidial forms of ADP-glucose pyrophosphorylase in wheat endosperm. *Plant Physiol.* **130,** 1464–1475.

Cornish-Bowden, A., and Cardenas, M. L. (2002). Metabolic balance sheets. *Nature* **420,** 129–130.

Covert, M. W., and Palsson, B. O. (2002). Transcriptional regulation in constraints-based metabolic models of *Escherichia coli*. *J. Biol. Chem.* **277,** 28058–28064.
Covert, M. W., Schilling, C. H., Famili, I., Edwards, J. S., Goryanin, I. I., Selkov, E., and Palsson, B. O. (2001). Metabolic modelling of microbial strains *in silico*. *Trends Biochem. Sci.* **26,** 179–186.
Croteau, R., Kutchan, T. M., and Lewis, N. G. (2000). Natural products (Secondary metabolites). *In* "Biochemistry & Molecular Biology of Plants" (B. B. Buchanan, W. Gruissem, and R. L. Jones, eds.), pp. 1250–1318. American Society of Plant Physiologists, Rockville.
Debnam, P. M., and Emes, M. J. (1999). Subcellular distribution of enzymes of the oxidative pentose phosphate pathway in root and leaf tissues. *J. Exp. Bot.* **50,** 1653–1661.
Denyer, K., Dunlap, F., Thorbjornsen, T., Keeling, P., and Smith, A. M. (1996). The major form of ADP-glucose pyrophosphorylase in maize endosperm is extra-plastidial. *Plant Physiol.* **112,** 779–785.
Dieuaide-Noubhani, M., Raffard, G., Canioni, P., Pradet, A., and Raymond, P. (1995). Quantification of compartmented metabolic fluxes in maize root tips using isotope distribution from ^{13}C- or ^{14}C-labeled glucose. *J. Biol. Chem.* **270,** 13147–13159.
Dieuaide-Noubhani, M., Canioni, P., and Raymond, P. (1997). Sugar-starvation-induced changes of carbon metabolism in excised maize root tips. *Plant Physiol.* **115,** 1505–1513.
Dixon, R. A., and Sumner, L. W. (2003). Legume natural products: Understanding and manipulating complex pathways for human and animal health. *Plant Physiol.* **131,** 878–885.
Edwards, J. S., and Palsson, B. O. (2000a). Robustness analysis of the *Escherichia coli* metabolic network. *Biotechnol. Prog.* **16,** 927–939.
Edwards, J. S., and Palsson, B. O. (2000b). The *Escherichia coli* MG1655 *in silico* metabolic genotype: Its definition, characteristics, and capabilities. *Proc. Natl. Acad. Sci. USA* **97,** 5528–5533.
Edwards, J. S., Ibarra, R. U., and Palsson, B. O. (2001). *In silico* predictions of *Escherichia coli* metabolic capabilities are consistent with experimental data. *Nat. Biotechnol.* **19,** 125–130.
Edwards, S., Nguyen, B.-T., Do, B., and Roberts, J. K. M. (1998). Contribution of malic enzyme, pyruvate kinase, phospho*enol*pyruvate carboxylase and the Krebs cycle to respiration and biosynthesis and to intracellular pH regulation during hypoxia in maize root tips observed by nuclear magnetic resonance and gas chromatography-mass spectrometry. *Plant Physiol.* **116,** 1073–1081.
Eisenreich, W., Rohdich, F., and Bacher, A. (2001). Deoxyxylulosephosphate pathway to terpenoids. *Trends Plant Sci.* **6,** 78–84.
Emmerling, M., Dauner, M., Ponti, A., Fiaux, J., Hochuli, M., Szyperski, T., Wüthrich, K., Bailey, J. E., and Sauer, U. (2002). Metabolic flux responses to pyruvate kinase knockout in *Escherichia coli*. *J. Bacteriol.* **184,** 152–164.
Famili, I., Mahadevan, R., and Palsson, B. O. (2005). *k*-cone analysis: Determining all candidate values for kinetic parameters on a network scale. *Biophys. J.* **88,** 1616–1625.
Farquhar, G. D., von Caemmerer, S., and Berry, J. A. (1980). A biochemical model of photosynthetic CO_2 assimilation in leaves of C_3 species. *Planta* **149,** 78–90.
Farré, E. M., Tiessen, A., Roessner, U., Geigenberger, P., Trethewey, R. N., and Willmitzer, L. (2001). Analysis of the compartmentation of glycolytic intermediates, nucleotides, sugars, organic acids, amino acids and sugar alcohols in potato tubers using a nonaqueous fractionation method. *Plant Physiol.* **127,** 685–700.
Fell, D. (1997). "Understanding the Control of Metabolism," 301 p. Portland Press, London.
Fernie, A. R., Roscher, A., Ratcliffe, R. G., and Kruger, N. J. (2001). Fructose 2,6-bisphosphate activates pyrophosphate: fructose-6-phosphate 1-phosphotransferase and increases triose phosphate to hexose phosphate cycling in heterotrophic cells. *Planta* **212,** 250–263.
Fernie, A. R., Swiedrych, A., Tauberger, E., Lytovchenko, A., Trethewey, R. N., and Willmitzer, L. (2002). Potato plants exhibiting combined antisense repression of cytosolic and plastidial isoforms of phosphoglucomutase surprisingly approximate wild type with respect to the rate of starch synthesis. *Plant Physiol. Biochem.* **40,** 921–927.
Giegé, P., Heazlewood, J. L., Roessner-Tunali, U., Millar, A. H., Fernie, A. R., Leaver, C. J., and Sweetlove, L. J. (2003). Enzymes of glycolysis are functionally associated with the mitochondrion in *Arabidopsis* cells. *Plant Cell* **15,** 2140–2151.
Giersch, C. (2000). Mathematical modeling of metabolism. *Curr. Opin. Plant. Biol.* **3,** 249–253.

Haake, V., Zrenner, R., Sonnewald, U., and Stitt, M. (1998). A moderate decrease of plastid aldolase activity inhibits photosynthesis, alters the levels of sugars and starch, and inhibits growth of potato plants. *Plant J.* **14,** 147–157.

Haake, V., Geiger, M., Walch-Liu, P., Engels, C., Zrenner, R., and Stitt, M. (1999). Changes in aldolase activity in wild-type potato plants are important for acclimation to growth irradiance and carbon dioxide concentration, because plastid aldolase exerts control over the ambient rate of photosynthesis across a range of growth conditions. *Plant J.* **17,** 479–489.

Hanson, A. D., Gage, D. A., and Shachar-Hill, Y. (2000). Plant one-carbon metabolism and its engineering. *Trends Plant Sci.* **5,** 206–213.

Harrison, E. P., Willingham, N. M., Lloyd, J. C., and Raines, C. A. (1998). Reduced sedoheptulose-1,7-bisphosphatase levels in transgenic tobacco lead to decreased photosynthetic capacity and altered carbohydrate accumulation. *Planta* **204,** 27–36.

Henkes, S., Flachmann, R., Sonnewald, U., and Stitt, M. A. (2001). A small decrease of plastid transketolase expression in antisense tobacco transformants has dramatic effects on photosynthesis and phenylpropanoid metabolism. *Plant Cell* **13,** 535–551.

Jamshidi, N., Edwards, J. S., Fahland, T., Church, G. M., and Palsson, B. O. (2001). Dynamic simulation of the human red blood cell metabolic network. *Bioinformatics* **17,** 286–287.

Kanehisa, M., Goto, S., Kawashima, S., and Nakaya, A. (2002). The KEGG databases at GenomeNet. *Nucleic Acids Res.* **30,** 42–46.

Klamt, S., and Stelling, J. (2003). Two approaches for metabolic pathway analysis? *Trends Biotechnol.* **21,** 64–69.

Klein, D., Morcuende, R., Stitt, M., and Krapp, A. (2000). Regulation of nitrate reductase expression in leaves by nitrate and nitrogen metabolism is completely overridden when sugars fall below a critical level. *Plant Cell Environ.* **23,** 863–871.

Koebmann, B. J., Westerhoff, H. V., Snoep, J. L., Nilsson, D., and Jensen, P. R. (2002). The glycolytic flux in *Escherichia coli* is controlled by the demand for ATP. *J. Bacteriol.* **184,** 3909–3916.

Komina, O., Zhou, Y., Sarath, G., and Chollet, R. (2002). *In vivo* and *in vitro* phosphorylation of membrane and soluble forms of soybean nodule sucrose synthase. *Plant Physiol.* **129,** 1664–1673.

Kossmann, J., Sonnewald, U., and Willmitzer, L. (1994). Reduction of the chloroplastic fructose-1,6-bisphosphatase in transgenic potato plants impairs photosynthesis and plant growth. *Plant J.* **6,** 637–650.

Kruger, N. J., Ratcliffe, R. G., and Roscher, A. (2003). Quantitative approaches for analysing fluxes through plant metabolic networks using NMR and stable isotope labelling. *Phytochem. Rev.* **2,** 17–30.

Kruger, N. J., and Von Schaewen, A. (2003). The oxidative pentose phosphate pathway: Structure and organisation. *Curr. Opin. Plant. Biol.* **6,** 236–246.

Lo Piero, A. R., Cultrone, A., Monachello, D., and Petrone, G. (2003). Different kinetic and regulatory properties of soluble and membrane-bound nitrate reductases in tomato leaves. *Plant Sci.* **165,** 139–145.

Macdonald, F. D., and Buchanan, B. B. (1997). The reductive pentose phosphate pathway and its regulation. *In* "Plant Metabolism" (D. T. Dennis, D. H. Turpin, D. D. Lefebvre, and D. B. Layzell, eds.), pp. 299–313. Longman, Harlow.

Marx, A., Eikmanns, B. J., Sahm, H., De Graaf, A. A., and Eggeling, L. (1999). Response of the central metabolism in *Corynebacterium glutamicum* to the use of an NADH-dependent glutamate dehydrogenase. *Metab. Eng.* **1,** 35–48.

Matt, P., Krapp, A., Haake, V., Mock, H. P., and Stitt, M. (2002). Decreased Rubisco activity leads to dramatic changes of nitrate metabolism, amino acid metabolism and the levels of phenylpropanoids and nicotine in tobacco antisense RBCS transformants. *Plant J.* **30,** 663–677.

McNeil, S. D., Nuccio, M. L., Rhodes, D., Shachar-Hill, Y., and Hanson, A. D. (2000a). Radiotracer and computer modelling evidence that phospho-base methylation is the main route of choline synthesis in tobacco. *Plant Physiol.* **123,** 371–380.

McNeil, S. D., Rhodes, D., Russell, B. L., Nuccio, M. L., Shachar-Hill, Y., and Hanson, A. D. (2000b). Metabolic modelling identifies key constraints on an engineered glycine betaine synthesis pathway in tobacco. *Plant Physiol.* **124,** 153–162.

Morandini, P., and Salamini, F. (2003). Plant biotechnology and breeding: Allied for years to come. *Trends Plant Sci.* **8,** 70–75.

Morgan, J. A., and Rhodes, D. (2002). Mathematical modelling of plant metabolic pathways. *Metab. Eng.* **4,** 80–89.

Mulquiney, P. J., and Kuchel, P. W. (2003). "Modelling Metabolism with Mathematica," 309 p. CRC Press, Boca Raton.

Munoz, F. J., Baroja-Fernandez, E., Moran-Zorzano, M. T., Viale, A. M., Etxeberria, E., Alonso-Casajus, N., and Pozueta-Romero, J. (2005). Sucrose synthase controls both intracellular ADP glucose levels and transitory starch biosynthesis in source leaves. *Plant Cell Physiol.* **46,** 1366–1376.

Neuhaus, H. E., and Emes, M. J. (2000). Nonphotosynthetic metabolism in plastids. *Annu. Rev. Plant Physiol. Plant Mol. Biol.* **51,** 111–140.

Neuhaus, H. E., Häusler, R. E., and Sonnewald, U. (2005). No need to shift the paradigm on the pathway to transitory starch in leaves. *Trends Plant Sci.* **10,** 154–156.

Niederberger, P., Prasad, R., Miozzari, G., and Kacser, H. (1992). A strategy for increasing an *in vivo* flux by genetic manipulations—the tryptophan system of yeast. *Biochem. J.* **287,** 473–479.

Oh, M.-K., and Liao, J. C. (2000). Gene expression profiling by DNA microarrays and metabolic fluxes in *Escherichia coli. Biotechnol. Prog.* **16,** 278–286.

Oh, M.-K., Rohlin, L., Kao, K. C., and Liao, J. C. (2002). Global expression profiling of acetate grown *Escherichia coli. J. Biol. Chem.* **277,** 13175–13183.

Oliver, S. (2002). Demand management in cells. *Nature* **418,** 33–34.

Papin, J. A., Price, N. D., Wiback, S. J., Fell, D. A., and Palsson, B. O. (2003). Metabolic pathways in the post-genome era. *Trends Biochem. Sci.* **28,** 250–258.

Paul, M. J., Knight, J. S., Habash, D., Parry, M. A. J., Lawlor, D. W., Barnes, S. A., Loynes, A., and Gray, J. C. (1995). Reduction in phosphoribulokinase activity by antisense RNA in transgenic tobacco: Effect on CO_2 assimilation and growth in low irradiance. *Plant J.* **7,** 535–542.

Petersen, S., De Graaf, A. A., Eggeling, L., Möllney, M., Wiechert, W., and Sahm, H. (2000). *In vivo* quantification of parallel and bidirectional fluxes in the anaplerosis of *Corynebacterium glutamicum. J. Biol. Chem.* **275,** 35932–35941.

Petersen, S., Mack, C., De Graaf, A. A., Riedel, C., Eikmanns, B. J., and Sahm, H. (2001). Metabolic consequences of altered phosphoenolpyruvate carboxykinase activity in *Corynebacterium glutamicum* reveal anaplerotic regulation mechanisms *in vivo. Metab. Eng.* **3,** 344–361.

Picault, N., Hodges, M., Palmieri, L., and Palmieri, F. (2004). The growing family of mitochondrial carriers in *Arabidopsis. Trends Plant Sci.* **9,** 138–146.

Poolman, M. G., Fell, D. A., and Thomas, S. (2000). Modelling photosynthesis and its control. *J. Exp. Bot.* **51,** 319–328.

Poolman, M. G., Fell, D. A., and Raines, C. A. (2003). Elementary modes analysis of photosynthate metabolism in the chloroplast stroma. *Eur. J. Biochem.* **270,** 430–439.

Price, N. D., Papin, J. A., Schilling, C. H., and Palsson, B. O. (2003). Genome-scale microbial *in silico* models: The constraints-based approach. *Trends Biotechnol.* **21,** 162–169.

Price, N. D., Reed, J. L., and Palsson, B. O. (2004). Genome-scale models of microbial cells: Evaluating the consequences of constraints. *Nat. Rev. Microbiol.* **2,** 886–897.

Pritchard, L., and Kell, D. B. (2002). Schemes of flux control in a model of *Saccharomyces cerevisiae* glycolysis. *Eur. J. Biochem.* **269,** 3894–3904.

Ratcliffe, R. G., and Shachar-Hill, Y. (2006). Measuring multiple fluxes through plant metabolic networks. *Plant J.* **45,** 490–511.

Regierer, B., Fernie, A. R., Springer, F., Perez-Melis, A., Leisse, A., Koehl, K., Willmitzer, L., Geigenberger, P., and Kossmann, J. (2002). Starch content and yield increase as a result of altering adenylate pools in transgenic plants. *Nat. Biotechnol.* **20,** 1256–1260.

Rontein, D., Dieuaide-Noubhani, M., Dufourc, E. J., Raymond, P., and Rolin, D. (2002). The metabolic architecture of plant cells. Stability of central metabolism and flexibility of anabolic pathways during the growth cycle of tomato cells. *J. Biol. Chem.* **277,** 43948–43960.

Roscher, A., Kruger, N. J., and Ratcliffe, R. G. (2000). Strategies for metabolic flux analysis in plants using stable isotope labelling. *J. Biotechnol.* **77,** 81–102.

Rylott, E. L., Hooks, M. A., and Graham, I. A. (2001). Co-ordinate regulation of genes involved in storage lipid mobilization in *Arabidopsis thaliana. Biochem. Soc. Trans.* **29,** 283–287.

Sauer, U., Lasko, D. R., Fiaux, J., Hochuli, M., Glaser, R., Szyperski, T., Wüthrich, K., and Bailey, J. E. (1999). Metabolic flux ratio analysis of genetic and environmental modulations of *Escherichia coli* central metabolism. *J. Bacteriol.* **181,** 6679–6688.

Schomburg, I., Chang, A., and Schomburg, D. (2002). BRENDA, enzyme data and metabolic information. *Nucleic Acids Res.* **30,** 47–49.

Schuster, S., Dandekar, T., and Fell, D. A. (1999). Detection of elementary flux modes in biochemical networks: A promising tool for pathway analysis and metabolic engineering. *Trends Biotechnol.* **17,** 53–60.

Schwab, W. (2003). Metabolome diversity: Too few genes, too many metabolites. *Phytochemistry* **62,** 837–849.

Schwender, J., Ohlrogge, J. B., and Shachar-Hill, Y. (2003). A flux model of glycolysis and the oxidative pentose phosphate pathway in developing Brassica napus embryos. *J. Biol. Chem.* **278,** 29442–29453.

Schwender, J., Ohlrogge, J., and Shachar-Hill, Y. (2004). Understanding flux in plant metabolic networks. *Curr. Opin. Plant. Biol.* **7,** 309–317.

Smirnoff, N., Conklin, P. L., and Loewus, F. A. (2001). Biosynthesis of ascorbic acid in plants: A renaissance. *Annu. Rev. Plant Physiol. Plant Mol. Biol.* **52,** 437–467.

Sriram, G., Fulton, D. B., Iyer, V. V., Peterson, J. M., Zhou, R., Westgate, M. E., Spalding, M. H., and Shanks, J. V. (2004). Quantification of compartmented metabolic fluxes in developing soybean embryos by employing biosynthetically directed fractional ^{13}C labelling, two-dimensional [^{13}C, ^{1}H] nuclear magnetic resonance, and comprehensive isotopomer balancing. *Plant Physiol.* **136,** 3043–3057.

Stelling, J., Klamt, S., Bettenbrock, K., Schuster, S., and Gilles, E. D. (2002). Metabolic network structure determines key aspects of functionality and regulation. *Nature* **420,** 190–193.

Stitt, M., and Schulze, E.-D. (1994). Does Rubisco control the rate of photosynthesis and plant growth? An exercise in molecular ecophysiology. *Plant Cell Environ.* **17,** 465–487.

Stitt, M., and Sonnewald, U. (1995). Regulation of metabolism in transgenic plants. *Annu. Rev. Plant Physiol. Plant Mol. Biol.* **46,** 341–368.

Sumner, L. W., Mendes, P., and Dixon, R. A. (2003). Plant metabolomics: Large-scale phytochemistry in the functional genomics era. *Phytochemistry* **62,** 817–836.

Sweetlove, L. J., Last, R. L., and Fernie, A. R. (2003). Predictive metabolic engineering: A goal for systems biology. *Plant Physiol.* **132,** 420–425.

Szyperski, T. (1998). ^{13}C-NMR, MS and metabolic flux balancing in biotechnology research. *Q. Rev. Biophys.* **31,** 41–106.

ter Kuile, B. H., and Westerhoff, H. V. (2001). Transcriptome meets metabolome: Hierarchical and metabolic regulation of the glycolytic pathway. *FEBS Lett.* **500,** 169–171.

Teusink, B., Passarge, J., Reijenga, C. A., Esgalhado, E., Van Der Weijden, C. C., Schepper, M., Walsh, M. C., Bakker, B. M., Van Dam, K., Westerhoff, H. V., and Snoep, J. L. (2000). Can yeast glycolysis be understood in terms of *in vitro* kinetics of the constituent enzymes? Testing biochemistry. *Eur. J. Biochem.* **267,** 5313–5329.

Thorneycroft, D., Sherson, S. M., and Smith, S. M. (2001). Using gene knockouts to investigate plant metabolism. *J. Exp. Bot.* **52,** 1593–1601.

Weber, A. P. M., Schwacke, R., and Flugge, U. I. (2005). Solute transporters of the plastid envelope membrane. *Annu. Rev. Plant Biol.* **56,** 133–164.

Wiback, S. J., and Palsson, B. O. (2002). Extreme pathway analysis of human red blood cell metabolism. *Biophys. J.* **83,** 808–818.

Wiechert, W. (2001). 13C metabolic flux analysis. *Metab. Eng.* **3,** 195–206.

Wiechert, W. (2002). Modeling and simulation: Tools for metabolic engineering. *J. Biotechnol.* **94,** 37–63.

Wiechert, W., Möllney, M., Petersen, S., and De Graaf, A. A. (2001). A universal framework for ^{13}C metabolic flux analysis. *Metab. Eng.* **3,** 265–283.

Wienkoop, S., Ullrich, W. R., and Stohr, C. (1999). Kinetic characterization of succinate-dependent plasma membrane-bound nitrate reductase in tobacco roots. *Physiol. Plantarum* **105,** 609–614.

Wink, M. (ed.). (1999). "Biochemistry of Plant Secondary Metabolism," 374 p. CRC Press, Boca Raton.

Yao, K., De Luca, V., and Brisson, N. (1995). Creation of a metabolic sink for tryptophan alters the phenylpropanoid pathway and the susceptibility of potato to *Phytophthora infestans*. *Plant Cell* **7,** 1787–1799.

CHAPTER **2**

Enzyme Engineering

John Shanklin

Contents

Abstract

Enzymes perform the biochemical transformations that direct metabolite flow through metabolic pathways of living cells. Metabolic engineering is made possible *via* genetic transformation of plants with genes encoding enzymes that selectively divert fixed carbon into desired forms. Genes encoding these enzymes may be identified from natural sources or may be

Biology Department, Brookhaven National Laboratory, Upton, New York 11973

Advances in Plant Biochemistry and Molecular Biology, Volume 1
ISSN 1755-0408, DOI: 10.1016/S1755-0408(07)01002-8

variants of naturally occurring enzymes that have been tailored for specific functionality. The evolution of novel enzyme activities in natural systems provides a context for discussing laboratory-directed enzyme engineering. This process, also called directed evolution, facilitates the expansion of enzyme function beyond the range identified in nature, by altering factors such as substrate specificity, regioselectivity and enantioselectivity. Changes in kinetic parameters such as k_{cat}, K_m and k_{cat}/K_m can also be achieved. Key steps in this process are described, including the selection of starting genes, methods for introducing variability, the choice of a heterologous expression system, ways to identify improved variants, and methods for combining improved variants to achieve the desired activity. Introduction of appropriately engineered proteins into plants has great potential not only for metabolic engineering of desired storage compounds but also for enhancement of productivity by improving resistance to pathogens or abiotic stresses.

Key Words: Enzyme engineering, Directed evolution, Enzyme evolution, Rational design, Sequence space, Variant enzyme, Fitness landscape, Gene shuffling.

1. INTRODUCTION

For ~10,000 years, humans have been tailoring plants to meet their needs. The vast majority of this crop development occurred as a result of conventional breeding, that is, by recombining germplasm within the natural breeding barrier. The results were spectacular improvements in terms of output (harvestable) traits like yield, and to a lesser extent input (protective) traits such as disease resistance and stress tolerance. Recently, conventional breeding has been greatly enhanced by the development of molecular tools. A second wave of improvement occurred over the past 20 years or so with the development of methods of plant transformation of genes irrespective of source, with the use of techniques such as *Agrobacterium tumefaciens*-mediated transformation and DNA particle bombardment. In contrast to conventional breeding, the major impacts thus far have been with input traits such as insect and disease resistance. The introduction of engineered enzymes can be considered as a third wave of plant improvement in which enzymes with specific tailored properties are introduced into plants with the goal of conveying specific desired traits. The first example of this was the introduction of an engineered thioesterase from *Garcinia mangostana* into canola that resulted in increased accumulation of stearic acid (Facciotti *et al.*, 1999).

With the emerging wealth of genome information, and the availability of genes from increasing numbers of organisms, one might ask why engineer genes instead of simply looking for naturally occurring genes that encode enzymes that already perform the desired transformation? The simplest answer is that a desired enzyme might not occur in any natural system. An example might be a biotransformation for which the substrate is a compound not normally found in nature. Second, one might identify an enzyme that performs the desired transformation, but does so very poorly. To make the enzyme useful, its activity would need to be optimized for the desired substrate. Third, the enzyme might have good *in vitro* activity, but may behave poorly in the metabolic context of the new host. Thus, the performance of the enzyme has the potential to be dramatically improved for use under a specific set of conditions. This could be the case if protein–protein interactions are necessary for function or if a particular concentration of cofactor is required. Enzyme engineering can modulate the K_m for substrates and cosubstrates. Finally, the fold of the enzyme may present an inherent limitation to achieving the optimal catalytic rate for a desired biotransformation, and it might be better to start with a different protein fold that will allow a higher turnover to be achieved.

The goal of this chapter is to present the rationale for plant enzyme engineering in the context of improving plants to meet the increasing and changing demands of society. To achieve this, I will first lay a conceptual framework for understanding enzyme evolution as it occurs in nature and then show how the results of this process may not be ideal for transgenic applications. Next, I will describe approaches employed for laboratory evolution of enzymes. Finally, I will summarize where I see future benefits and applications of these technologies.

2. THEORETICAL CONSIDERATIONS

2.1. Enzyme architecture is conserved

Gene sequences are commonly compared as two-dimensional alignments. It is useful to remember that significant homology between two sequences (DNA or deduced amino acid) implies general homology between their three-dimensional structures. Regions of homology within genes typically represent conserved structural features with similar relative orientations in three-dimensional space. In cases where structural information is available, the common way of displaying such information is to compare the fold, or Cα-carbon chains, from different proteins superimposed in such a way as to maximize superposition. There are thought to be ~1000 protein folds, at least an order of magnitude fewer folds than the number of enzymes (Zhang and Delisi, 1998). Typically, when the derived amino acid sequence homology is 25% or greater, the protein folds of two enzymes are likely to be very similar (Hobohm and Sander, 1995). However, there are cases in which the amino acid homology is too low to be detected by computer algorithms but the fold is highly conserved.

2.2. Genomic analysis suggests most enzymes evolve from preexisting enzymes

The determination of whole genome sequences allowed the identification of all of the gene families related by primary sequence homology within a specific organism. Figure 2.1 shows a cluster analysis of the proteins encoded by the *Arabidopsis* genome (Thomas Girke, University of California Riverside, personal

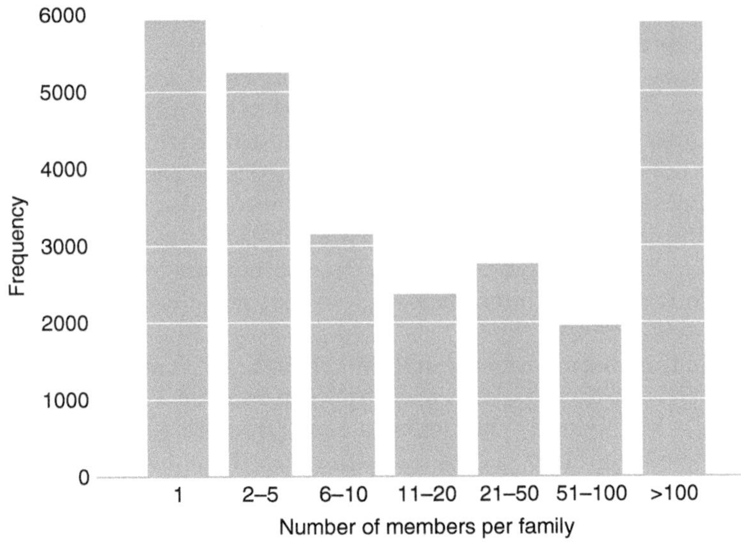

FIGURE 2.1 Frequency distribution of protein families in *Arabidopsis*.

communication). For example, of the ~27,000 individual proteins in *Arabidopsis*, ~80% of proteins are members of homology-related families, whereas only ~20% represent unique sequences. The distribution shows that approximately half of the genes are members of groups consisting of >11 members and that nearly one quarter of proteins belong to groups of >100 members. The larger families include large numbers of protein kinases and cytochrome *P450*s. This clearly illustrates that new proteins evolved one from another and that divergent evolution is a primary mechanism for achieving novel functionality.

2.3. Evolution of a new enzymatic activity in nature

Enzyme evolution in natural systems typically involves several steps: (1) gene duplication, (2) change in functionality, and (3) selection for activity/specificity (see Fig. 2.2). Duplications that occur at the individual gene level provide the starting point for enzyme evolution.

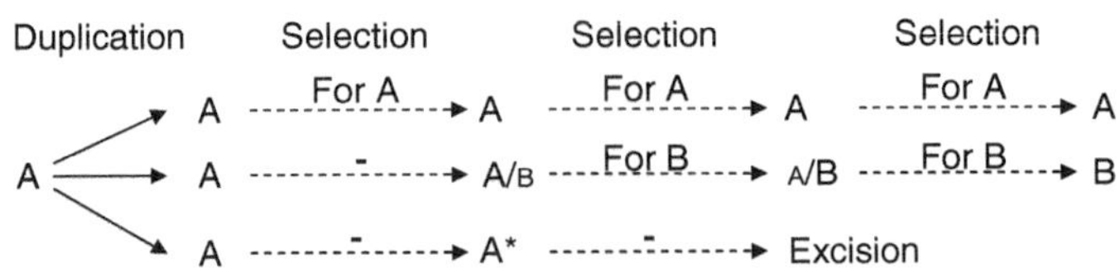

FIGURE 2.2 General scheme for natural evolution of enzyme activity. A, Parental gene; A/B gene encoding protein with dual activity that can perform activity B poorly; A/B, gene that encodes protein with dual activity where B is the major activity; B gene encoding activity B that is unable to perform activity A; A* represents a gene pseudogene that becomes excised.

Mutations constantly arise in genes, but their accumulation depends on stringency of the selection pressure for the function of the gene product. There are three common fates that befall duplicated genes (Fig. 2.2): (1) retention of function, (2) change of function (either change in activity or change in expression pattern), or (3) loss of function followed eventually by excision.

Changes in enzyme function typically follow one of the three mechanisms (Gerlt and Babbitt, 2001). The first mechanism is one in which a partial reaction or a strategy for stabilization of energetically unfavorable transition state is maintained, while the substrate specificity changes. In a second mechanism, substrate specificity is maintained, but the chemistry changes during evolution. A third mechanism involves retaining only the active site architecture, without maintaining either substrate specificity or chemical mechanism.

Whichever of the mechanisms predominate, several features are likely to be common. An initial gene duplication event is followed by the accumulation of multiple mutations in one of the copies. A prerequisite for alteration of specificity is that the original tight active site substrate specificity should relax allowing a number of potential substrates to bind, or the same substrate to bind in alternate conformations. Once an alternate substrate is capable of binding (or the same substrate in a different binding conformation), an altered enzymatic transformation may occur, resulting in the accumulation of a novel product. If the new product conveys a selective advantage, over successive generations the accumulation of further mutation/selection can lead to an increase in the new activity. This "tuning" to the new substrate often occurs at the cost of catalytic efficiency with respect to the original transformation. Thus, a characteristic of newly evolved enzymes, or enzymes caught in transition, would be the observation of relaxed specificity. Examples of this can be found in the fatty acid desaturases (Broun *et al.*, 1998; Dyer *et al.*, 2002), where enzymes that exhibit "unusual" specificity with respect to the parental enzymes are often bifunctional in that they are capable of performing the archetypal reaction, often with lower catalytic rates than the parental enzyme (Shanklin and Cahoon, 1998). Amino acid substitutions that change the geometry of the binding pocket can be either direct, that is, when the amino acid side chains directly line the binding pocket, or alternatively can be at sites remote from the binding pocket and mediate their effects via subtle changes in the relative organization of secondary structural elements. In this context, amino acid side chains have been referred to as "molecular shims"(Whittle *et al.*, 2001) that orient the substrate with respect to the active site in a very precise manner similar to the way carpentry shims are used to level furniture. The stronger the selection pressure for the improvement in activity, the faster it will progress.

Similarly, in the case of changes to the chemistry occurring on the same substrate, it is envisaged that the enzyme became bifunctional with respect to reaction outcome either by acquiring two or more alternate binding modes or by alterations in the amino acid side chains that participate in catalysis. This has been observed for the Fad2 family of fatty acid modification enzymes (Broadwater *et al.*, 2002; Broun *et al.*, 1998). Once the new reaction occurs even at low levels, selection can favor mutations that increase the new activity and lead to improved fitness at the organismal level and thus provide selective advantage. In either case where substrate specificity changes, or chemistry on the same substrate alters, the ability of an enzyme to perform alternate reactions shows it has the potential to acquire a new dominant activity.

Duplicated genes that do not provide a selective advantage are rapidly excised by unequal crossover at meiosis. Evidence for this includes studies in which subfunctionalization is shown to occur rapidly upon polyploidization in cotton (Adams *et al.*, 2003) and the observation of lower than expected occurrence of pseudogenes (Force *et al.*, 1999).

2.4. The natural evolution process initially produces poor enzymes

Changes in substrate selectivity or reaction chemistry often require amino acid substitutions at two or more specific locations along the amino acid chain. During evolution, point mutations leading to amino acid substitutions occur at random amino acid positions, so the probability of accumulating specific amino acid changes at two predefined locations with two random mutations is very low indeed. Consequently, many mutations accumulate in the gene before changes that can affect the specificity of the enzyme occur. This helps explain why related enzymes with different specificities often differ in sequence identity by >50%. If we consider any particular amino acid location, the chances of a substitution increasing stability and/or activity of the enzyme are less likely than decreasing its stability and/or activity (Taverna and Goldstein, 2002a). Thus, by the time a gene accumulates sufficient numbers of mutations to achieve a new functionality, its catalytic properties (K_m and k_{cat}), in addition to its stability, are impaired. This decline in functionality is inevitable because selection for the new functionality can only occur after the new catalysis arises. Only at this time can selection pressure for the product of the new reaction lead to subsequent selection of mutants with improved catalytic properties (Taverna and Goldstein, 2002b).

2.5. Sequence space and fitness landscapes

The concept of sequence space is used to illustrate the range of possible combinations of amino acids that compose the polypeptide chain of a protein. Sequence space is very large because there are 20 possible amino acids that could occupy every position of the polypeptide chain. Thus, for an average-sized protein composed of 300 amino acids, there are 20^{300} possible combinations of sequences. This number is so large that one can only ever sample a minute fraction of total sequence space. A corollary to this is that most of sequence space is devoid of function (see Fig. 2.3A) because many combinations of amino acids will not fold into stable structures. Active stable structures thus appear as islands among a sea of inactivity. Another way of looking at sequence space is to consider the fitness landscape (Fig. 2.3B). This shows three enzymatic activities α, β, and γ that correspond to the three sequences shown in Panel A. As we move across sequence space, we track across peaks of activity for α, then β, and then γ. Note that between activities α and β, there is an area of overlap in which the enzyme is bifunctional,

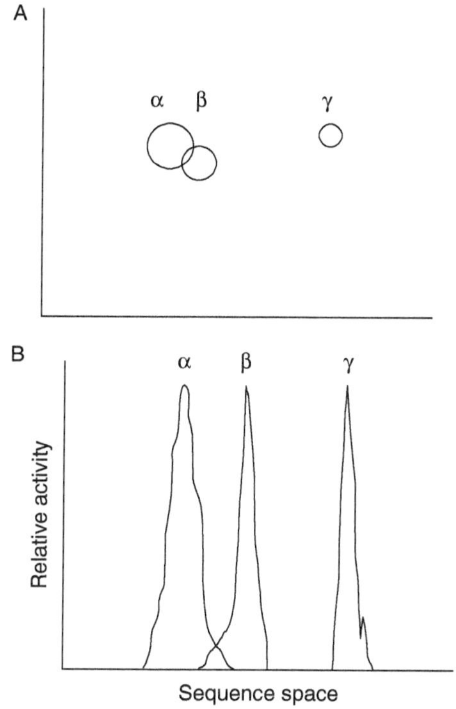

FIGURE 2.3 (A) Sequence space; (B) Fitness landscape. α, β, and γ represent enzymes with different activities.

but that between activities β and γ there is no overlapping region. As noted above, the fact that most enzymes evolve from existing enzymes, it is common for newly evolved enzymes to be bifunctional with somewhat poorer activity for one or other of the catalyzed reactions. Also, because of the tendency for duplicated genes to become excised if there is no selection pressure on them, it is far more likely for a gene to convert from function α to β because there is always function that can be selected for, rather than from α or β to γ in which a functionless intermediate must be maintained.

3. PRACTICAL CONSIDERATIONS FOR ENGINEERING ENZYMES

Over the last decade or so, enzyme engineers have developed strategies for creating variant tailored enzymes that are collectively referred to as directed evolution (Arnold, 1998). These combinatorial methods used to alter specific properties of enzymes have resulted in remarkable improvements in enzyme activity for specific substrates (Stemmer, 1994b; Whittle *et al.*, 2001), reversal of enantioselectivity (Reetz *et al.*, 1997), as well as changes in global properties such as solvent (You and Arnold, 1996) and heat (Zhao and Arnold, 1999) tolerance (see also several excellent reviews Farinas *et al.*, 2001; Powell *et al.*, 2001).

There are four key steps to engineering a desired enzyme activity successfully: (1) identification of parental enzymes to be modified, (2) introducing variation into the gene(s), (3) choice of host system to express the enzyme, and (4) method for identifying improvements in property of interest. See Fig. 2.4 for a generic scheme for altering the properties of an enzyme.

3.1. Identifying appropriate starting enzyme(s)

The first step in any enzyme engineering project is to choose a source or parental enzyme(s). Because sequence space is vast and mostly devoid of function, selecting the most appropriate starting point for a desired activity is critical. The parental enzyme should be the closest activity available to the desired enzyme because this minimizes the sequence space that needs to be traversed in order to achieve the desired property (Fig. 2.3). For any particular biotransformation, an ideal starting point would be an enzyme that performs the desired activity as a side reaction. For example, a galactosidase can also perform a fructosidase reaction albeit very inefficiently (Zhang *et al.*, 1997). Enzymes to be used for reengineering projects can be identified from the biochemical literature and genes can be isolated from the many publicly funded seed and culture collections. An alternate, and particularly appealing, strategy for identifying starting enzymes is to screen samples from multiple environments for the desired enzymatic activity (Gray *et al.*, 2003). This can be achieved by isolating total DNA from a particular environment and creating an expression library that is then screened for the desired activity. This circumvents the classical microbiological route of identification of an organism capable of performing a specific biotransformation, followed by protein purification/gene isolation of the corresponding activity. The approach has advantages in that many organisms from a particular environment are screened simultaneously, even ones for which culture conditions have not been developed. A disadvantage of this approach is that it may fail because genetic control elements from one organism may not be functional in the expression host organism. Also, multicomponent activities may be difficult to isolate in this manner if one or more of the components is unsuited to heterologous expression.

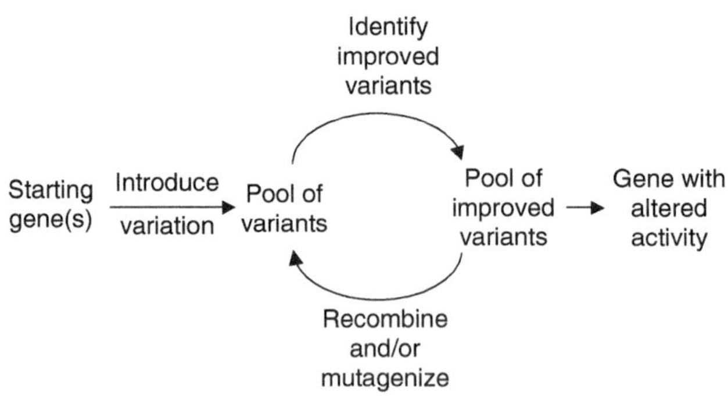

FIGURE 2.4 Generic scheme for directed evolution of an enzyme.

3.2. Ways of introducing variability into genes

There are many ways of introducing mutations into genes of interest. The most commonly used is error prone polymerase chain reaction (EP-PCR) that exploits the low proofreading fidelity of *Taq* polymerase (Cadwell and Joyce, 1992). Thus, by varying the concentration of dNTPs and the divalent cation Mn^{2+}, it is possible to obtain a range of introduced mutations typically from 0.1% to ~1% of the bases of the target DNA. Random point mutagenesis, that is, a base change at one of the three locations in the triplet that encodes a single amino acid, has an inherent limitation related to the structure of the genetic code itself. That is, depending on the degeneracy of the amino acid encoded by a particular triplet, one can only reach between three and seven amino acid substitutions per site. Compounding this problem, EP-PCR has been shown to exhibit considerable base change bias in that >70% of changes are seen from A and T, and <20% from C and G (Shafikhani *et al.*, 1997). This further reduces the number of possible amino acid changes to less than the 4–7 that would occur for random substitutions. A DNA polymerase (Mutazyme, Stratagene, La Jolla, CA) was developed in which the base change proclivity is inverted from that of *Taq*, such that the two enzymes can be used in concert to minimize bias among variants.

Another powerful method of introducing changes into genes is to perform a partial digest with DNase followed by reassembly of the fragments in an autopriming PCR reaction and amplification of reassembled product with the addition of terminal primers (Stemmer, 1994a). This method exploits lack of fidelity in the reassembly reaction in which mutations are introduced at the borders of overlap extension reactions. Because DNase cuts randomly, the positions of introduced mutations occur randomly along the length of the target DNA. This method has been successfully used to generate a population of variants starting from a single parental gene. A limited analysis of the base changes introduced by this method suggests that it is less biased than EP-PCR. All of these methods suffer from the limited range of amino acids that can be reached by point mutagenesis as described above. To circumvent this limitation, a method called gene site saturation mutagenesis was devised in which oligonucleotides encoding all possible 19 amino acid substitutions at a particular site are used to make a library of variants that can be assayed for desired related activities (Desantis *et al.*, 2003). Given sufficient resources, all possible substitutions can be made at every position along the amino acid chain to identify improved variants.

3.3. Choice of expression system

Once a library of genetic variants has been generated, the genes must be introduced into an organism so that they can be expressed and the activity of the variants estimated. Ideally, variants should be evaluated in the final desired host because heterologous expression can result in changes in measured activity. For instance, if proteinaceous cofactors are necessary for function, their interactions can be optimized leading to improved activity. Other factors that can contribute to improved activity include improvements in mRNA and protein stability, reduction in protease sensitivity, optimization for host temperature pH or osmotic conditions, interaction with available chaperone proteins, etc. However, while direct expression and evaluation of variants in plants is desirable, it should be recognized that such experiments are inherently problematic. First, plant generation times are upwards of several months, making experimental cycles long if stable expression is to be employed. However, it may be possible to reduce this time for seed phenotypes using a fluorescence-based screen (Stuitje *et al.*, 2003). Second, and perhaps more problematic, insertion of a gene encoding particular activity into the plant genome via *Agrobacterium*-mediated transformation, yields a wide spectrum of expression levels, and consequently, enzyme activity depending on the integration site of the T-DNA (Nowak *et al.*, 2001). This is particularly problematic for identifying variants with improved activities because it is difficult to determine whether changes in activity are the result of changes in the enzyme or alterations in expression between independent transformed plants. If the screen is for qualitative differences, such as the occurrence of a novel product, this problem may not be prohibitive. Transient expression in systems such as tobacco suspension cultures or soybean embryos may offer a partial solution to this problem (Cahoon *et al.*, 1999). Whether whole plant or transient expression system is employed, a major problem is attaining sufficiently high numbers of transformants to provide a reasonable probability of identifying a substantially improved activity. Typically, directed evolution experiments require the generation of 10^4–10^5 per cycle of improvement.

On the other hand, microbial systems offer generation times in hours to days (rather than months for whole plants), and it is relatively straightforward to produce sufficiently large numbers of transformants for analysis. However, in heterologous expression, often improvements in performance can be attributed to improvements in codon usage specific for the heterologous host. Such changes, while they improve the property being measured in the heterologous host, do not translate into improvements when expressed in the desired host; indeed mutations to improve expression of a plant gene in *Escherichia coli* would likely result in decreased expression when the "improved" gene is reintroduced into plants. This example underscores the genetic maxim that "you always get what you select for" and reinforces the notion that creating a screen that achieves the goals of any particular project without producing unwanted results is one of the biggest challenges facing protein engineers.

In summary, the best screens are conducted in the desired host; however, one must weigh the constraints of time and transferability when designing a strategy for improving a particular enzyme. A useful compromise for assessing plant enzymes and variants is heterologous expression in yeast (Broadwater *et al.*, 2002; Covello and Reed, 1996). Being a single-celled eukaryotic system, it has the short generation times of microbes along with the subcellular organization of eukaryotes.

3.4. Identifying improved variants

Another major challenge in directed evolution is to measure changes in the property of interest. The problem is that typically thousands of assays have to be performed to identify variants with improved properties. However, for traditional biochemical assays this can be prohibitively time consuming and reagent intensive. One appealing solution to this problem is to identify a selection system for the improved enzyme. In this scenario, the host organism is unable to survive unless a variant of the expressed enzyme attains a particular property that allows the host to survive under defined growth conditions. Such a system was reported for plant fatty acid desaturase genes. An *E. coli* strain MH13 is an unsaturated fatty acid auxotroph that has to be supplemented with unsaturated fatty acids in the growth medium for survival (Cahoon and Shanklin, 2000; Clark *et al.*, 1983). The enzyme encoded by the plant desaturase gene was specific for 18-carbon substrate, but *E. coli* contains insufficient 18-carbon substrate for the desaturase to convert to the unsaturated fatty acid necessary for survival (Cahoon and Shanklin, 2000). However, *E. coli* does contain sufficient levels of 16-carbon substrate for the enzyme to desaturate, but the enzyme was far more active on 18- versus 16-carbon substrates. So, a library of variants was constructed from the 18-carbon preferring desaturase and *E. coli* containing these variants was challenged to grow on media lacking unsaturated fatty acids. This method allowed the identification of many variants specific for 16-carbon substrates (Whittle *et al.*, 2001). When reintroduced into plants, these enzymes efficiently desaturated 16-carbon fatty acid resulting in the accumulation of unusual fatty acids in seed oils.

The benefits of such selection systems are immediately apparent, that is, that all growing colonies are "winners," and that millions of variants can be assessed in a short period of time. However, it should be noted that there are also problems using this approach. It can be very difficult or impossible to design such selection systems because the product of a desired reaction may not be essential for survival. It can also be difficult to manipulate the threshold necessary for survival. This means that one might have too tight or too loose a criterion for survival, in which cases one might get no colonies, or get too many to perform follow-up analysis. Even with the extremely powerful fatty acid auxotrophy selection described above, it proved difficult to alter the survival constraints, and so it was relatively easy to identify the first round of improved variants, but the system was of little use in identifying further improved variants after subsequent recombination experiments of the type described below.

The alternative to selection systems is to employ screening techniques. Because precise assays are relatively time consuming, the use of tiered screens has become routine for high-throughput applications. The idea of a tiered screen is that improved variants are first assessed for improvement with a fast but low precision methodology and that variants that meet some minimal criterion are subjected to a secondary screen that is more precise but more time- and reagent consuming. A final very precise biochemical assay is added in some cases to distinguish between the improved variants. Examples of such screens include fluorescent-activated cell sorting (FACS) or phage display, technologies that can quickly screen $>10^7$ variants (Crameri *et al.*, 1998; Gao *et al.*, 1997; Naki *et al.*, 1998); solid state colorimetric or fluorescence assays for between 10^4 and 10^7 variants (Moore and Arnold, 1996; Zhang *et al.*, 1997); microtiter format assays for 10^2–10^4 (Joo *et al.*, 1999; Zhang *et al.*, 1997); and individual high-precision assays involving gas chromatography, high performance liquid chromatography, or mass spectrometry for 10^1–10^2 samples (Altamirano *et al.*, 2000; Reetz *et al.*, 1997).

3.5. Recombination and/or introduction of subsequent mutations

Directed evolution experiments differ from traditional mutation-selection experiments in that they typically involve cycles of improvement. This can be done in a sequential fashion by identifying the most improved single variant and subjecting it to further cycles of mutagenesis and screening/selection until a variant that meets desired criteria is reached, or until further cycles fail to produce increases in the desired property. However, this method tends to be slow and laborious. A better method for recombining many improved variants is known as gene shuffling (Stemmer, 1994b). This method is a variation on the mutagenesis method described above in which genes are partially digested with the use of DNase and subsequently reassembled by primerless PCR, except that a pool of improved variants are used for the starting material rather than a single gene. The result of this procedure is to make a new library of variants in which mutations from different improved variants are recombined in many permutations and combinations. In some variants, different positive amino acid changes that independently improved the property of interest provide either additive or multiplicative improvements in performance. In other cases, positive mutations could be partially obscured by negative mutations, so the process of improvement involves both summation of positive mutations and, at the same time, elimination of negative mutations. The removal of negative mutations can also be achieved by backcross PCR. This technique is analogous to a traditional genetic backcross experiment, but in this case, the improved variant is recombined with a molar excess of parental gene, and the resulting variants screened for activity. By performing several cycles of this procedure, typically 4–7 rounds of screening/recombination of improved variants, improvements in performance of 10^1–10^4 have been documented. Examples of successes using this technique include conversion of a galactosidase into a fucosidase (Zhang *et al.*, 1997), increasing the activity of a thermophylic enzyme at low temperatures (Merz *et al.*, 2000), and the evolution of antibody-phage libraries (Crameri *et al.*, 1996).

While single gene shuffling is capable of generating huge changes in activity with regard to specific substrates, it is still a fairly inefficient process. The reason for this is that the variability introduced into the initial gene is somewhat limited by the particular method of mutagenesis employed. A major advance in efficiency of gene shuffling was made by incorporating several genes rather than a single gene as a starting point, a process referred to as family shuffling (Crameri *et al.*, 1998). In this method, several homologues (e.g., α, β, and γ of Fig. 2.3) with $>50\%$ sequence identity are shuffled together and the resulting variants screened for activity as described for single gene shuffling. An extension of this method is called synthetic shuffling in which information from sequence comparisons is used to generate oligonucleotides so that every amino acid from a set of parents is allowed to recombine independently (Ness *et al.*, 2002). Synthetic shuffling has the advantage that shuffling can be used to exploit sequence information and does not require all of the physical genes to be in hand to perform the experiment. Improvements employing these methods are shown to be more rapid and larger in magnitude. The rational for this is that each homologue represents a variant on the same protein fold and that during natural evolution each of the genes has accumulated different positive sequence attributes that contribute to the overall enzyme performance. During family shuffling, there is the potential to sum these positive attributes to produce rapid increases in performance. Another way to think of this is that a single gene shuffling experiment is essentially starting off at a single point and radiating from there in sequence space. For multigene shuffling, one starts with several independent points in sequence, space, and combinations of each of the genes cover a larger portion of sequence space than could be achieved from a single point (Fig. 2.3A). Subsequent rounds of recombination and screening occur as before for single gene shuffling. An important and intriguing finding from family gene shuffling experiments is that when genes are shuffled, instead of getting activities that are intermediate between the members shuffled, new activities beyond the range of the individuals are identified. This is true for not only activities but also for qualitative parameters such as range of regiospecificities. This has far-reaching implications in that new diversity of biocatalysts can actually arise for parameters previously not found in nature. The necessary criteria for exploiting this phenomenon are to identify and recombine the optimal parental genes and to have in place a robust high-throughput screen that has an excellent signal-to-noise ratio for the property of interest. In addition to DNaseI-based recombination techniques, there are other effective methods such as staggered extension process StEP PCR (Aguinaldo and Arnold, 2003).

An interesting variation on single and multiple gene shuffling is that of pathway and whole organism shuffling (Crameri *et al.*, 1997; Zhang *et al.*, 2002). These broader-scale methods allow changes in regulatory elements, in addition to changes in the coding regions to contribute to improved activity.

3.6. Structure-based rational design

With the determination of high-resolution crystal structures of enzymes came the expectation that one could make rational changes to the shape of the enzyme to make desired changes in the enzyme's activity. This expectation went largely unfulfilled because the resolution of the crystal structures was too low to allow sufficient precision in changes in the enzyme to allow the changes to achieve the desired results (Arnold, 2001). This is because small changes in relative orientation of substrate with respect to the active site cause large changes in catalytic efficiency. While the techniques of enzyme engineering via various shuffling technologies are becoming mature, other technologies such as computational rational design with powerful computer algorithms are emerging and reinvigorating the early excitement for rational design (Dahiyat and Mayo, 1997; Fox *et al.*, 2003). A particularly efficient approach to combinatorial analysis using chimeric enzymes involves identifying shemas, or fragments of proteins that can be recombined with minimal three-dimensional perturbation to structure (Meyer *et al.*,

2003; Voigt *et al.*, 2002). This approach is currently being successfully applied to versatile enzymes such as cytochrome *P450*s (Otey *et al.*, 2004). It seems likely that there will be a lot of interesting opportunities created by combining computational with combinatorial genetic methods.

4. OPPORTUNITIES FOR PLANT IMPROVEMENT THROUGH ENGINEERED ENZYMES AND PROTEINS

Using the technologies of laboratory-directed evolution and applying the methods of chemical engineering to devise efficient and robust high-throughput screens for enzyme evolution offer the promise to revolutionize biological transformations.

Input traits could be significantly improved via enzyme engineering. For instance to improve insect resistance, it may be possible to recombine protective proteins such as *Bacillus thuringiensis* toxin (BT) from multiple independent sources to create novel variant BT proteins with either increased potency, or decreased ability to induce resistance in the targeted pest. Alternatively, it may be possible to improve the efficiency of various pathway enzymes to synthesize more of a particular protective compound, or changing the chirality of an individual protective compound.

Output traits present the most easily defined targets for plant improvement. Plants synthesize a bewildering array of secondary products that have uses ranging from chemical feedstocks to foodstuffs to pharmaceuticals. By enzyme engineering, it may be possible to improve the accumulation of desired metabolites. Plants can efficiently convert CO_2, one of the only natural resources that continues to become more abundant, into reduced carbon storage compounds using sunlight as the energy source. It is easy to imagine replacing the enzymes and pathways used to synthesize storage proteins, carbohydrates, and lipids to novel pathways to make and store just about any organic molecule we can conceive. For example, three enzyme pathways for the accumulation of novel polyhydroxyalcanoates have been successfully engineered into plants (Poirier, 2001). Because plant oils are relatively inexpensive to produce, pathways designed to produce modified oils with desirable properties as industrial feedstocks are particularly attractive (Thelen and Ohlrogge, 2002).

Many of the natural enzymes with novel function in pathways such as fatty acid biosynthesis have been identified. However, alteration of biochemical regulation of enzyme activity via enzyme engineering of protein stability, sites of posttranslational modifications, and of allostery represents underexploited opportunities in plant biotechnology.

Allosteric regulation involves the positive or negative modulation of enzyme activity after binding of one or more metabolite(s). It represents a particularly interesting enzyme-engineering target in that the introduction of an enzyme with altered sensitivity to the interacting metabolite can overcome a potent metabolic block that cannot be overcome by simply controlling the abundance of the enzyme. In addition, for many cases, the introduction of an allosterically insensitive variant enzyme should overcome the metabolic block even in the presence of the endogenous allosterically sensitive enzyme. Several strategies can be used to identify enzymes with altered regulation. The first is to identify a naturally occurring enzyme from a source that does not exhibit allosteric regulation and to introduce the corresponding gene into the desired host organism. The second is to perform enzyme engineering and activity screening to identify variants in which the catalytic activity of the enzyme is maintained, but in which the binding of the allosteric regulator is disrupted. An excellent example of overcoming allostery involves starch metabolism. A nonregulated mutant of the *E. coli* ADPG pyrophosphorylase enzyme was identified and introduced into potato tubers (Ballicora *et al.*, 2003), resulting in a 25–60% increase in accumulation of starch compared to tubers containing the wild-type enzyme (Preiss, 1996). It is possible that under certain conditions, the metabolic flux into the desired end-product may not substantially increase if the allosterically regulated step was either colimiting or not limiting to the rate of product accumulation. In these cases, metabolic profiling (Graham *et al.*, 2002) can be employed to identify the new rate-limiting step, and efforts to increase the activity of this step can be undertaken. Similar approaches can conceivably be applied to other major forms of stored carbon such as lipids.

Many aspects of plant architecture, developmental programs, and signal transduction are regulated by members of families of transcription factors such as MYBs and MYCs and MAD box proteins. The cauliflower mutant of *Arabidopsis* is one of many examples of alteration in expression of a transcription factor leading to a profound alteration in morphology and development (Kempin *et al.*, 1995). One can envisage creating libraries of recombinant chimeras of transcription factors from these gene families and screening for desired changes in morphology or development. Such changes might include alterations in the amount and/or composition of cellulose for improved biomass accumulation.

4.1. Challenges for engineering plant enzymes and pathways

While much headway is being made in gene discovery and enzyme engineering efforts, the use of this basic science knowledge to develop novel crops is somewhat lagging. This is because plant metabolism is more complicated than previously assumed, with pathways containing unexpected genetic redundancy in addition to being under the control of multiple biochemical and genetic regulatory circuits (Sweetlove and Fernie, 2005). Superimposed on this complexity are cell biology issues such as the heterogeneity of tissues and developmental programs. While studies at the whole plant level pose significant challenges in terms of heterogeneity, stable-isotope metabolic flux analyses have provided new insight into the role of RuBisCO in carbon fixation in seeds (Schwender *et al.*, 2004a). Because metabolic flux analysis provides a direct way of measuring the effects of genetic perturbations on metabolism, it is envisaged that this technique will become increasingly valuable for interpreting future genetic engineering efforts (Schwender *et al.*, 2004b).

The application of engineering approaches in the emerging discipline of plant systems biology, that is, of high-throughput data collection along with direct flux measurements, computer modeling, and simulation, will undoubtedly provide the basis for integrating our knowledge and creating engineered crops designed to meet the increasing needs of mankind.

5. SUMMARY

Enzymes are biocatalysts that mediate many reactions necessary for life. They are remarkable because they perform their functions at ambient temperature and pressure in a highly substrate-selective fashion in the presence of scores of structurally related compounds. Gene sequence information, along with an increasing number of protein structures, reveals that many enzymes arose from a subset of common ancestors. This underscores the high degree of functional plasticity exhibited by individual enzyme folds and suggests that existing enzymes can be further adapted to perform desired biotransformations. The poor performance of some naturally occurring genes in transgenic settings, along with theoretical considerations suggesting newly evolved enzymes are likely to have poor kinetic properties and stability, provides a rationale for engineering enzymes to perform specific reactions *in planta*. The techniques of enzyme engineering represent a powerful new addition to the arsenal of the metabolic engineer. Over the last decade, enzymes have been tailored to perform specific transformations or to become adapted to perform efficiently under specific conditions. There are as yet few examples of the effects of such technologies being applied to plants. However, because plants represent the primary route of terrestrial fixed carbon, the potential impacts of enzyme engineering, and ultimately metabolic engineering, are far reaching. Using these techniques, plant scientists will be able to create rationally engineered crops that will suffer decreased losses from insects and disease which will accumulate desired forms of reduced carbon to meet the increasing and changing needs of society.

ACKNOWLEDGEMENTS

I am grateful to Dr. J. Setlow, Dr. K. Mayer, and Dr. M. Pidkowich for editorial suggestions. Funding was provided by the Office of Basic Energy Sciences of the U.S. Department of Energy.

REFERENCES

Adams, K. L., Cronn, R., Percifield, R., and Wendel, J. F. (2003). Genes duplicated by polyploidy show unequal contributions to the transcriptome and organ-specific reciprocal silencing. *Proc. Natl. Acad. Sci. USA* **100,** 4649–4654.

Aguinaldo, A. M., and Arnold, F. H. (2003). Staggered extension process (StEP) *in vitro* recombination. *Methods Mol. Biol.* **231,** 105–110.

Altamirano, M. M., Blackburn, J. M., Aguayo, C., and Fersht, A. R. (2000). Directed evolution of new catalytic activity using the alpha/beta-barrel scaffold. *Nature* **403,** 617–622.

Arnold, F. H. (1998). When blind is better: Protein design by evolution. *Nat. Biotechnol.* **16,** 617–618.
Arnold, F. H. (2001). Combinatorial and computational challenges for biocatalyst design. *Nature* **409,** 253–257.
Ballicora, M. A., Iglesias, A. A., and Preiss, J. (2003). ADP-glucose pyrophosphorylase, a regulatory enzyme for bacterial glycogen synthesis. *Microbiol. Mol. Biol. Rev.* **67,** 213–225, table of contents.
Broadwater, J. A., Whittle, E., and Shanklin, J. (2002). Desaturation and hydroxylation. Residues 148 and 324 of Arabidopsis FAD2, in addition to substrate chain length, exert a major influence in partitioning of catalytic specificity. *J. Biol. Chem.* **277,** 15613–15620.
Broun, P., Boddupalli, S., and Somerville, C. (1998). A bifunctional oleate 12-hydroxylase: Desaturase from *Lesquerella fendleri*. *Plant J.* **13,** 201–210.
Cadwell, R. C., and Joyce, G. F. (1992). Randomization of genes by PCR mutagenesis. *PCR Methods Appl.* **2,** 28–33.
Cahoon, E. B., and Shanklin, J. (2000). Substrate-dependent mutant complementation to select fatty acid desaturase variants for metabolic engineering of plant seed oils. *Proc. Natl. Acad. Sci. USA* **97,** 12350–12355.
Cahoon, E. B., Carlson, T. J., Ripp, K. G., Schweiger, B. J., Cook, G. A., Hall, S. E., and Kinney, A. J. (1999). Biosynthetic origin of conjugated double bonds: Production of fatty acid components of high-value drying oils in transgenic soybean embryos. *Proc. Natl. Acad. Sci. USA* **96,** 12935–12940.
Covello, P. S., and Reed, D. W. (1996). Functional expression of the extraplastidial *Arabidopsis thaliana* oleate desaturase gene (FAD2) in *Saccharomyces cerevisiae*. *Plant Physiol.* **111,** 223–236.
Clark, D. P., Demendoza, D., Polacco, M. L., and Cronan, J. E., Jr. (1983). Beta-hydroxydecanoyl thio ester dehydrase does not catalyze a rate-limiting step in *Escherichia coli* unsaturated fatty acid synthesis. *Biochemistry* **22,** 5897–5902.
Crameri, A., Cwirla, S., and Stemmer, W. P. (1996). Construction and evolution of antibody-phage libraries by DNA shuffling. *Nat. Med.* **2,** 100–102.
Crameri, A., Dawes, G., Rodriguez, E., Jr., Silver, S., and Stemmer, W. P. (1997). Molecular evolution of an arsenate detoxification pathway by DNA shuffling [see comments]. *Nat. Biotechnol.* **15,** 436–438.
Crameri, A., Raillard, S. A., Bermudez, E., and Stemmer, W. P. (1998). DNA shuffling of a family of genes from diverse species accelerates directed evolution. *Nature* **391,** 288–291.
Dahiyat, B. I., and Mayo, S. L. (1997). *De novo* protein design: Fully automated sequence selection. *Science* **278,** 82–87.
Desantis, G., Wong, K., Farwell, B., Chatman, K., Zhu, Z., Tomlinson, G., Huang, H., Tan, X., Bibbs, L., Chen, P., Kretz, K., Burk, M. J., *et al.* (2003). Creation of a productive, highly enantioselective nitrilase through gene site saturation mutagenesis (GSSM). *J. Am. Chem. Soc.* **125,** 11476–11477.
Dyer, J. M., Chapital, D. C., Kuan, J. C., Mullen, R. T., Turner, C., Mckeon, T. A., and Pepperman, A. B. (2002). Molecular analysis of a bifunctional fatty acid conjugase/desaturase from tung. Implications for the evolution of plant fatty acid diversity. *Plant Physiol.* **130,** 2027–2038.
Facciotti, M. T., Bertain, P. B., and Yuan, L. (1999). Improved stearate phenotype in transgenic canola expressing a modified acyl-acyl carrier protein thioesterase. *Nat. Biotechnol.* **17,** 593–597.
Farinas, E. T., Bulter, T., and Arnold, F. H. (2001). Directed enzyme evolution. *Curr. Opin. Biotechnol.* **12,** 545–551.
Force, A., Lynch, M., Pickett, F. B., Amores, A., Yan, Y. L., and Postlethwait, J. (1999). Preservation of duplicate genes by complementary, degenerative mutations. *Genetics* **151,** 1531–1545.
Fox, R., Roy, A., Govindarajan, S., Minshull, J., Gustafsson, C., Jones, J. T., and Emig, R. (2003). Optimizing the search algorithm for protein engineering by directed evolution. *Protein Eng.* **16,** 589–597.
Gao, C., Lin, C. H., Lo, C. H. L., Mao, S., Wirsching, P., Lerner, R. A., and Janda, K. D. (1997). Making chemistry selectable by linking it to infectivity. *Proc. Natl. Acad. Sci. USA* **94,** 11777–11782.
Gerlt, J. A., and Babbitt, P. C. (2001). Divergent evolution of enzymatic function: Mechanistically diverse superfamilies and functionally distinct suprafamilies. *Annu. Rev. Biochem.* **70,** 209–246.
Graham, I. A., Li, Y., and Larson, T. R. (2002). Acyl-CoA measurements in plants suggest a role in regulating various cellular processes. *Biochem. Soc. Trans.* **30,** 1095–1099.
Gray, K. A., Richardson, T. H., Robertson, D. E., Swanson, P. E., and Subramanian, M. V. (2003). Soil-based gene discovery: A new technology to accelerate and broaden biocatalytic applications. *Adv. Appl. Microbiol.* **52,** 1–27.

Hobohm, U., and Sander, C. (1995). A sequence property approach to searching protein databases. *J. Mol. Biol.* **251,** 390–399.

Joo, H., Arisawa, A., Lin, Z., and Arnold, F. H. (1999). A high-throughput digital imaging screen for the discovery and directed evolution of oxygenases. *Chem. Biol.* **6,** 699–706.

Kempin, S. A., Savidge, B., and Yanofsky, M. F. (1995). Molecular basis of the cauliflower phenotype in Arabidopsis. *Science* **267,** 522–525.

Merz, A., Yee, M. C., Szadkowski, H., Pappenberger, G., Crameri, A., Stemmer, W. P., Yanofsky, C., and Kirschner, K. (2000). Improving the catalytic activity of a thermophilic enzyme at low temperatures. *Biochemistry* **39,** 880–889.

Meyer, M. M., Silberg, J. J., Voigt, C. A., Endelman, J. B., Mayo, S. L., Wang, Z. G., and Arnold, F. H. (2003). Library analysis of SCHEMA-guided protein recombination. *Protein Sci.* **12,** 1686–1693.

Moore, J. C., and Arnold, F. H. (1996). Directed evolution of a para-nitrobenzyl esterase for aqueous-organic solvents. *Nat. Biotechnol.* **14,** 458–467.

Naki, D., Paech, C., Granshaw, G., and Schellenberger, V. (1998). Selection of a subtillisin-hyperproducing Bacillus in a highly structured environment. *Appl. Microbiol. Biotechnol.* **49,** 290–294.

Ness, J. E., Kim, S., Gottman, A., Pak, R., Krebber, A., Borchert, T. V., Govindarajan, S., Mundorff, E. C., and Minshull, J. (2002). Synthetic shuffling expands functional protein diversity by allowing amino acids to recombine independently. *Nat. Biotechnol.* **20,** 1251–1255.

Nowak, W., Gawlowska, M., Jarmolowski, A., and Augustyniak, J. (2001). Effect of nuclear matrix attachment regions on transgene expression in tobacco plants. *Acta Biochim. Pol.* **48,** 637–646.

Otey, C. R., Silberg, J. J., Voigt, C. A., Endelman, J. B., Bandara, G., and Arnold, F. H. (2004). Functional evolution and structural conservation in chimeric cytochromes p450: Calibrating a structure-guided approach. *Chem. Biol.* **11,** 309–318.

Poirier, Y. (2001). Production of polyesters in transgenic plants. *Adv. Biochem. Eng. Biotechnol.* **71,** 209–240.

Powell, K. A., Ramer, S. W., Del Cardayre, S. B., Stemmer, W. P., Tobin, M. B., Longchamp, P. F., and Huisman, G. W. (2001). Directed evolution and biocatalysis. *Angew. Chem. Int. Ed. Engl.* **40,** 3948–3959.

Preiss, J. (1996). ADPglucose pyrophosphorylase: Basic science and applications in biotechnology. *Biotechnol. Annu. Rev.* **2,** 259–279.

Reetz, M. T., Zonta, A., Schimossek, K., Liebeton, K., and Jaeger, K. E. (1997). Creation of enantioselective biocataalysts for organic chemistry by *in vitro* evolution. *Angew. Chem. Int. Ed. Engl.* **36,** 2830–2833.

Schwender, J., Goffman, F., Ohlrogge, J. B., and Shachar-Hill, Y. (2004a). RuBisCO without the Calvin cycle improves the carbon efficiency of developing green seeds. *Nature* **432,** 779–782.

Schwender, J., Ohlrogge, J., and Shachar-Hill, Y. (2004b). Understanding flux in plant metabolic networks. *Curr. Opin. Plant Biol.* **7,** 309–317.

Shafikhani, S., Siegel, R. A., Ferrari, E., and Schellenberger, V. (1997). Generation of large libraries of random mutants in *Bacillus subtilis* by PCR-based plasmid multimerization. *Biotechniques* **23,** 304–310.

Shanklin, J., and Cahoon, E. B. (1998). Desaturation and related modifications of fatty acids. *Annu. Rev. Plant Physiol. Plant Mol. Biol.* **49,** 611–641.

Stemmer, W. P. (1994a). DNA shuffling by random fragmentation and reassembly: *In vitro* recombination for molecular evolution. *Proc. Natl. Acad. Sci. USA* **91,** 10747–10751.

Stemmer, W. P. (1994b). Rapid evolution of a protein *in vitro* by DNA shuffling. *Nature* **370,** 389–391.

Stuitje, A. R., Verbree, E. C., Van Der Linden, K. H., Mietiewska, E. M., Nap, J. P., and Kneppers, T. J. A. (2003). Seed-expressed fluorescent proteins as versatile tools for easy (co)transformation and high-throughput functional genomics in Arabidopsis. *Plant Biotechnol. J.* **1,** 301–309.

Sweetlove, L. J., and Fernie, A. R. (2005). Regulation of metabolic networks: Understanding metabolic complexity in the systems biology era. *New Phytol.* **168,** 9–24.

Taverna, D. M., and Goldstein, R. A. (2002a). Why are proteins so robust to site mutations? *J. Mol. Biol.* **315,** 479–484.

Taverna, D. M., and Goldstein, R. A. (2002b). Why are proteins marginally stable? *Proteins* **46,** 105–109.

Thelen, J. J., and Ohlrogge, J. B. (2002). Metabolic engineering of fatty acid biosynthesis in plants. *Metab. Eng.* **4,** 12–21.

Voigt, C. A., Martinez, C., Wang, Z. G., Mayo, S. L., and Arnold, F. H. (2002). Protein building blocks preserved by recombination. *Nat. Struct. Biol.* **9,** 553–558.

Whittle, E., Cahoon, E. B., and Shanklin, J. (2001). Engineering delta 9–16:0-acyl carrier protein (ACP) desaturase specificity based on combinatorial saturation mutagenesis and logical redesign of the castor delta 9–18:0-ACP desaturase. *J. Biol. Chem.* **276,** 21500–21505.

You, L., and Arnold, F. H. (1996). Directed evolution of subtilisin E in *Bacillus subtilis* to enhance total activity in aqueous dimethylformamide. *Protein Eng.* **9,** 77–83.

Zhang, C., and Delisi, C. (1998). Estimating the number of protein folds. *J. Mol. Biol.* **284,** 1301–1305.

Zhang, J. H., Dawes, G., and Stemmer, W. P. (1997). Directed evolution of a fucosidase from a galactosidase by DNA shuffling and screening. *Proc. Natl. Acad. Sci. USA* **94,** 4504–4509.

Zhang, Y. X., Perry, K., Vinci, V. A., Powell, K., Stemmer, W. P., and Del Cardayre, S. B. (2002). Genome shuffling leads to rapid phenotypic improvement in bacteria. *Nature* **415,** 644–646.

Zhao, H., and Arnold, F. H. (1999). Directed evolution converts subtilisin E into a functional equivalent of thermitase. *Protein Eng.* **12,** 47–53.

CHAPTER 3

Genetic Engineering of Amino Acid Metabolism in Plants

Shmuel Galili,* Rachel Amir,† and Gad Galili‡

Contents

* Institute of Field and Garden Crops, Agricultural Research Organization, Bet Dagan 50250, Israel
† Plant Science Laboratory, Migal Galilee Technological Center, Rosh Pina 12100, Israel
‡ Department of Plant Sciences, The Weizmann Institute of Science, Rehovot 76100, Israel

Advances in Plant Biochemistry and Molecular Biology, Volume 1
ISSN 1755-0408, DOI: 10.1016/S1755-0408(07)01003-X

Abstract

Amino acids are not only building blocks of proteins but also participate in many metabolic networks that control growth and adaptation to the environment. In young plants, amino acid biosynthesis is regulated by a compound metabolic network that links nitrogen assimilation with carbon metabolism. This network is strongly regulated by the metabolism of four central amino acids, namely glutamine, glutamate, aspartate, and asparagine (Gln, Glu, Asp, and Asn), which are then converted into all other amino acids by various biochemical processes. Amino acids also serve as major transport molecules of nitrogen between source and sink tissues, including transport of nitrogen from vegetative to reproductive tissues. Amino acid metabolism is subject to a concerted regulation by physiological, developmental, and hormonal signals. This regulation also appears to be different between source and sink tissues. The importance of amino acids in plants does not only stem from being central regulators of plant growth and responses to environmental signals, but amino acids are also effectors of the nutritional quality of human foods and animal feeds. Since mammals cannot synthesize about half of the 20-amino acid building blocks of proteins, they rely on obtaining them from foods and feeds. Yet, the major crop plants contain limited amounts of some of these so-called "essential amino acids," which decreases nutritional value. Recent genetic engineering and more recently genomic approaches have significantly boosted our understanding of the regulation of amino acid metabolism in plants and their participation in growth, stress response, and reproduction. In addition, genetic engineering approaches have improved the content of essential amino acids in plants, particularly the contents of lysine and methionine, which are often most limiting.

Key Words: Transgenic plants, Genetic engineering, Amino acids, Essential amino acids, Biosynthesis, Catabolism, Metabolism, Seeds, Amide amino acids, Metabolic networks, Carbon/nitrogen partition, Nitrogen assimilation, Transport, Glutamate synthase, Glutamine synthase, Glutamate dehydrogenase, Glutamate, Glutamine, Aspartate, Asparagine, Aspartate family pathway, Lysine, Threonine, Methionine, Aspartate kinase, Dihydrodipicolinate synthase, Lysine-ketoglutarate reductase, Cystathionine γ-synthase, Threonine synthase, Lysine overproduction, Methionine overproduction,

Lysine-rich proteins, Sulfur-rich storage proteins, Vegetative storage proteins, Nutritional quality, Ruminant animals, Nonruminant animals, Light, Signal, Sucrose, Stress, Development, Food, Feed.

1. INTRODUCTION

Amino acids are essential constituents of all cells. In addition to their role in protein synthesis, they participate in both primary and secondary metabolic processes associated with plant development and in responses to stress. For example, glutamine, glutamate, aspartate, and asparagine serve as pools and transport forms of nitrogen, as well as in balancing the carbon/nitrogen ratio. Other amino acids such as tryptophan, methionine, proline, and arginine contribute to the tolerance of plants against biotic and abiotic stresses either directly or indirectly by serving as precursors to secondary products and hormones. Apart from their biological roles in plant growth, some amino acids, termed "essential amino acids," are also important for the nutritional quality of plants as foods and feeds. This is because humans, as well as most livestock, cannot synthesize all amino acids and therefore depend on their diets for obtaining them. Among the essential amino acids, lysine, methionine, threonine, and tryptophan are considered especially important because they are generally present in low or extremely low amounts in the major plant foods.

Studies on amino acid metabolism in plants have always benefited from the more advanced understanding of amino acid metabolism in microorganisms. Combined genetic, biochemical, molecular, and more recently genomics approaches, coupled with administration and metabolism of various precursors as major donors of carbon, nitrogen, and sulfur, have provided detailed identification of flux controls of amino acid metabolism in microorganisms (Stephanopoulos, 1999). These studies also clearly illustrated that amino acid metabolism in microorganisms is regulated by complex networks of metabolic fluxes, which are affected by multiple factors. Although the regulation of amino acid metabolism in higher plants may be analogous to that in microorganisms, the multicellular and multiorgan nature of higher plants presents additional levels of complexity that render metabolic fluxes and regulatory metabolic networks in plants much more sophisticated than in microorganisms. Plant seeds and fruits, most important organs as food sources, or as a source for the production of specific compounds like oils and carbohydrates, represent an exciting example to illustrate the higher complexity of metabolic regulation in plants compared to microorganisms. Seed metabolism is regulated not only by internal metabolic fluxes but also by the availability of precursor metabolites that depend in turn on metabolic process operating in vegetative tissues and on the efficiency of transport of these metabolites from the source to developing seeds. Thus, the regulation of seed metabolism in plants may be significantly different, responding to different signals than vegetative metabolism.

Due to space limitation, it is impossible to discuss in detail all aspects of amino acid metabolism in this chapter. We will therefore focus on relatively recent studies employing molecular/biochemical approaches, as well as tailor-made genetic engineering, metabolic engineering, and gene knockout approaches to study the regulation of amino acid metabolism in plants. Most recent studies employing these approaches have focused on the metabolism of glutamine, glutamate, aspartate, and asparagine, as well as on the essential amino acids lysine, threonine, and methionine. Hence, this chapter will focus mainly on these amino acids. We make the case that the regulatory principles that emerged from studies of these amino acids will also be valid for explaining the metabolism of other amino acids. For discussion of the metabolism of other amino acids, readers are directed to the recent book edited by B. J. Singh (1999) and several reviews (Coruzzi and Last, 2000; Morot-Gaudry *et al.*, 2001). Since improved understanding of plant amino acid metabolism enjoys significant biotechnological importance, we will also address this aspect focusing on metabolic engineering of the essential amino acids, lysine and methionine, for feeding ruminant and nonruminant animals. We then discuss future goals in studying plant amino acid metabolism.

2. GLUTAMINE, GLUTAMATE, ASPARTATE, AND ASPARAGINE ARE CENTRAL REGULATORS OF NITROGEN ASSIMILATION, METABOLISM, AND TRANSPORT

Glutamine, glutamate, aspartate, and asparagine constitute a metabolic network [hereafter termed for simplicity "amide amino acid metabolism" (AAAM) because it contains the two amide amino acids, glutamine and asparagine] that participates in numerous processes (Fig. 3.1). These include nitrogen assimilation, nitrogen metabolism into the various amino acids and other nitrogenous compounds, nitrogen transport between sources and sinks, carbon/nitrogen partitioning, and stress-associated metabolism. The AAAM network is regulated in a concerted manner by numerous metabolites and environmental signals, such as by light and phytochrome, in a manner that varies significantly between different plant tissues and organs, as well as in response to developmental, physiological, and environmental signals. Ammonium ion, derived either from nitrogen assimilation or from photorespiration, is incorporated into glutamine by a reaction catalyzed by glutamine synthase (GS), and glutamine is further converted into glutamate catalyzed by glutamate synthase (GOGAT) (Fig. 3.1). Glutamate is *trans*-aminated to aspartate by a large family of aspartate amino transferases and aspartate can be converted into asparagine and back from asparagine into aspartate by the activities of asparagine synthetase and asparaginase, respectively (Fig. 3.1). Glutamine, glutamate, and aspartate are used for the synthesis of other protein and nonprotein amino acids, as well as amides and other nitrogenous compounds. Asparagine, which is synthesized from aspartate, serves not only as a protein amino acid but is also as a major nitrogen transport agent. The regulation of nitrogen assimilation and metabolism in plants has been discussed in detail in

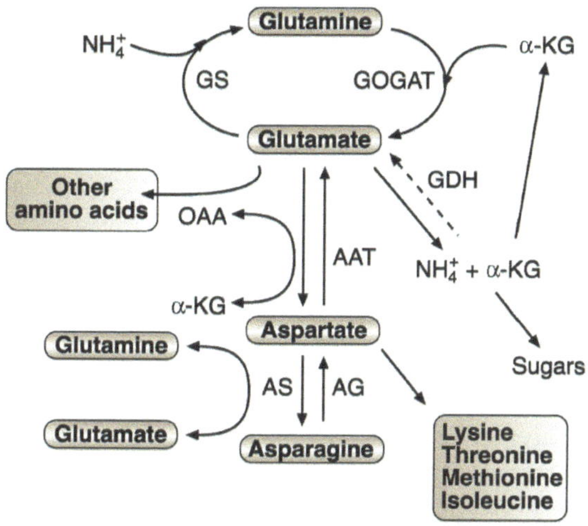

FIGURE 3.1 Schematic diagram of the network of AAAM and its connection to nitrogen assimilation, carbon metabolism, and synthesis of other amino acids. Abbreviations: GS, glutamine synthetase; GOGAT, glutamate synthase; AAT, aspartate amino transferase; GDH, glutamate dehydrogenase; AS, asparagine synthetase; AG, asparaginase; OAA, oxaloacetate; α-KG, α-ketoglutarate. The dashed arrow represents the aminating activity of GDH, which was experimentally demonstrated in plants, but its function is still a matter of debate.

a number of reviews (Hirel and Lea, 2001; Ireland and Lea, 1999; Lam *et al.*, 1995; Lea and Ireland, 1999; Miflin and Habash, 2002; Oliveira *et al.*, 2001; Stitt *et al.*, 2002).

In this chapter, we focus mainly on studies dealing with genetic engineering of enzymes associated with AAAM and analysis of plant mutants. However, several principles of AAAM are important for understanding its functional significance and the enzymes that control this metabolic network (Stephanopoulos, 1999). In this context, the synthesis of amino acids requires both carbon and nitrogen and is therefore regulated in a concerted manner by nitrogen and sugars (Singh, 1999). When nitrogen and sugar levels are not limiting, the assimilated nitrogen triggers sugar metabolism to efficiently synthesize glutamine and glutamate and the synthesis of other amino acids. However, when carbon levels are limiting (termed carbon starvation), glutamine and glutamate are efficiently converted into sugars, while the released nitrogen is stored in nitrogen-rich metabolites, such as asparagine and arginine (Coruzzi and Last, 2000). In nonsenescing tissues, amino acid metabolism is subject to a tight diurnal regulation. During daytime, when photosynthesis is active, glutamine, glutamate, and aspartate are used efficiently for synthesis of other amino acids needed for protein synthesis, while during the night these amino acids are strongly converted into asparagine serving as a nitrogen storage and transport compounds (Morot-Gaudry *et al.*, 2001). In senescing tissues, the AAAM network is used to convert the various amino acids and ammonium ion, which are derived from protein breakdown (particularly RuBisCO and other major plastid-localized photosynthetic genes), into transport competent nitrogenous compounds, such as asparagine, glutamine, and ureides (Hirel and Lea, 2001; Ireland and Lea, 1999; Lam *et al.*, 1995; Lea and Ireland, 1999; Miflin and Habash, 2002). These processes take place by the activation of many amino acid catabolism pathways as well as enzymes of AAAM. Under stress conditions, the AAAM network is used for rapid production of stress-associated metabolites, such as proline, arginine, polyamines, and γ-amino butyric acid. Hence, AAAM is a most highly controlled metabolic networks in plants.

2.1. GS: A highly regulated, multifunctional gene family

GS activity is found in many plant tissues and organs and is derived from two enzymes, GS1 and GS2. GS1 is an abundant cytosolic enzyme in vascular tissues of roots, aging leaves, and developing seeds. Equally abundant, GS2 is a plastidic enzyme in photosynthesizing leaves, in roots as well as in other tissues in a manner that varies between different plant species. Both GS1 and GS2 are encoded by small gene families (Ireland and Lea, 1999; Lam *et al.*, 1995; Oliveira *et al.*, 2001). The functions of the GS1 and GS2 gene families have been studied in a number of plant species by analysis of the spatial and temporal expression patterns of their genes as well by genetic approaches. These have been described and discussed in other reviews (Hirel and Lea, 2001; Ireland and Lea, 1999; Lam *et al.*, 1995; Lea and Ireland, 1999) and therefore will not be discussed in detail. The major function of GS2 emerging from these studies is to reassimilate ammonium ions generated by photorespiration, although GS2 also participates in the assimilation of ammonium-derived moieties from soil nitrogen (Lam *et al.*, 1995; Miflin and Habash, 2002). The major functions of GS1 are to assimilate ammonium ions into glutamine in roots, and in senescing leaves for nitrogen transport between source and sink tissues (Lam *et al.*, 1995; Miflin and Habash, 2002).

Does the GS-catalyzed assimilation of ammonium ion into glutamine represent a limiting factor for nitrogen use efficiency and plant growth? If the answer to this question is yes, three additional questions arise: (1) Does the rate-limiting effect of GS result either from insufficient nitrogen assimilation and transport between sources and sinks, or from insufficient reassimilation of ammonium ion derived from photorespiration (a fact that can cause ammonium ion toxicity), or both? (2) Can GS1 compensate for the function of GS2 and vice versa? (3) Is GS activity rate limiting in all or only in specific plant organs and tissues? These questions have been addressed by the use of recombinant gene constructs expressing GS1 and GS2 enzymes from different plants in different transgenic species and by utilizing different promoters.

Most studies on GS overexpression utilized the strong constitutive 35S promoter from the *Cauliflower mosaic virus* (CaMV), which leads to ectopic expression of the gene in most plant tissues. Genes encoding cytosolic GS1 from different plant species have been expressed in various plant species, including legumes, tobacco, and even poplar trees (Eckes *et al.*, 1989; Fei *et al.*, 2003; Fuentes *et al.*, 2001; Gallardo *et al.*, 1999; Hirel *et al.*, 1992; Lam *et al.*, 1995; Oliveira *et al.*, 2001, 2002; Ortega *et al.*, 2001; Temple *et al.*, 1993; Vincent *et al.*, 1997). These studies resulted in variable results apparently due to differential posttranscriptional and posttranslational controls of GS expression (Finnemann and Schjoerring, 2000; Miflin and Habash, 2002; Moorhead *et al.*, 1999; Ortega *et al.*, 2001). However, in many cases, GS1 overexpression caused increases in plant growth, particularly under nitrogen-limiting conditions, in total protein as well as chlorophyll content and photosynthesis. In the case of transgenic tobacco expressing a pea GS1 gene, the improved growth was dependent on light, but not on nitrogen supplementation. This suggests that the overexpressed GS1 improved photorespiratory ammonium ion assimilation in photosynthetic tissues (Oliveira *et al.*, 2002), a function generally attributed to GS2. This was supported by the fact that these transgenic tobaccos also exhibited increased levels of intermediate metabolites of the photorespiratory process, as well as an increased CO_2 photorespiratory burst (Oliveira *et al.*, 2002). Taken together, the ability of cytosolic GS1 to compensate for rate-limiting activities of the plastid-localized GS2 suggests that both ammonium ion and glutamine shuttle quite efficiently between the cytosol and the plastid. Indeed, the levels of free ammonium ion were significantly reduced in some of the transgenic plants implying that ammonium ions were more efficiently converted into glutamine.

In other studies, recombinant GS proteins were expressed in transgenic plants using nonconstitutive promoters. Expression of a soybean GS1 gene under the control of the putative root-specific *rolD* promoter in transgenic *Lotus japonicus* and transgenic pea plants resulted in reduced root ammonium ion levels as well as in reduced plant biomass (Fei *et al.*, 2003; Limami *et al.*, 1999). These interesting results suggest that the GS-catalyzed incorporation of ammonium ion into glutamine in the roots, although important for root metabolism, antagonizes plant growth. It also implies that, at least in *L. japonicus* and pea, transport of ammonium ion from roots to the shoots and its incorporation into glutamine in above ground tissues is a preferred route for efficient plant nitrogen use compared to the assimilation into glutamine in the roots.

In another study, a bean GS1 gene was expressed in wheat under control of the *rbcS* promoter (Habash *et al.*, 2001; Miflin and Habash, 2002). This promoter is highly expressed in young photosynthetic leaves, but not in roots. Although the promoter is highly expressed in young leaves, GS activity in the transgenic plants was enhanced only late in development of flag leaves, similar to the developmental pattern observed for endogenous wheat GS activity (Habash *et al.*, 2001; Miflin and Habash, 2002). This unanticipated pattern was explained by the possibility that expression of the transgenic pea GS gene was subject to post-translation control in wheat (by?) the foreign wheat host. Nevertheless, since GS activity in late wheat flag leaves is crucially involved in nitrogen transport to the developing seeds, this allowed the investigators to analyze whether GS activity also limited the incorporation of nitrogen into glutamine for source/sink nitrogen transport. Indeed, the transgenic wheat exhibited increased growth rate as well as earlier flowering and seed development than the control nontransformed plants (Habash *et al.*, 2001; Miflin and Habash, 2002), supporting a rate-limiting role for cytosolic GS activity in plant nitrogen use efficiency and transport from source to sink tissues.

These studies suggest that increasing GS activity by genetic engineering may be an important tool to improve nitrogen use efficiency and crop productivity, particularly under conditions of limiting nitrogen availability. This supposition is also supported by marker-assisted genetic studies in various crop plants in which a significant correlation was found between a number of important agronomical traits, such as nitrogen status and yield, and GS activity (Hirel and Lea, 2001; Jiang and Gresshoff, 1997; Limami and De Vienne, 2001; Masclaux *et al.*, 2000). The importance of the GS trait is not only in improving yield but also in reducing environmental damage as a result of crop overfertilization. Modern agriculture has been associated with a dramatic increase in nitrogen fertilization, much of which is not assimilated by the plants resulting in contamination of the environment (Lawlor *et al.*, 2001; Miflin and Habash, 2002; Ter Steege *et al.*, 2001).

2.2. Role of the ferredoxin- and NADH-dependent GOGAT isozymes in plant glutamate biosynthesis

Since the discovery of the GS/GOGAT-catalyzed pathway for glutamate biosynthesis, extensive studies have unequivocally shown that this pathway is the main route of soil nitrogen assimilation as well as photorespiratory ammonium ion reassimilation in plants (see for reviews Hirel and Lea, 2001; Ireland and Lea, 1999; Lam *et al.*, 1995; Lea and Ireland, 1999; Miflin and Habash, 2002; Stitt *et al.*, 2002). Plants possess two types of ferredoxin- and NADPH-dependent GOGAT isozymes (Fd-GOGAT and NADPH-GOGAT). Genes encoding Fd- and NADH-GOGAT isozymes and their regulation of expression have been extensively discussed in other reviews (Hirel and Lea, 2001; Ireland and Lea, 1999; Lam *et al.*, 1995; Lea and Ireland, 1999; Miflin and Habash, 2002; Stitt *et al.*, 2002). The Fd-GOGAT isozymes (two isoforms encoded by two different genes in *Arabidopsis*) constitute the majority of the GOGAT activity in plants, accounting for over 90% and ~70% of total GOGAT activity in *Arabidopsis* leaves and roots, respectively (Ireland and Lea, 1999; Somerville and Ogren, 1980; Suzuki *et al.*, 2001). The significant role of Fd-GOGAT in ammonium ion assimilation, particularly of photorespiratory ammonium ion, was demonstrated by a number of genetic and molecular approaches. Many plant mutants, defective in growth under photorespiratory conditions, were based on mutations in genes encoding Fd-GOGAT (Ireland and Lea, 1999; Somerville and Ogren, 1980). Notably, although *Arabidopsis* possesses two Fd-GOGAT isozymes, mutations in one are sufficient to cause sensitivity to enhanced photorespiration (Somerville and Ogren, 1980). This nonredundant function was explained by two contrasting patterns of expression of the genes encoding these isozymes (Coschigano *et al.*, 1998). The significant role of Fd-GOGAT in reassimilating photorespiratory ammonium ion was also demonstrated in transgenic tobacco plants with reduced Fd-GOGAT due to antisense expression (Ferrario-Mery *et al.*, 2000). When transferred from CO_2-rich conditions to ambient air to enhance photorespiration, the plants accumulated significantly higher levels of ammonium ion as well as the two GOGAT substrates, glutamine and α-ketoglutarate, than control plants (Ferrario-Mery *et al.*, 2000). This suggests that glutamine and α-ketoglutarate were less efficiently converted into glutamate in the transgenic plants, causing a less-efficient incorporation of photorespiratory ammonium ion into glutamine. In addition, the reduced

Fd-GOGAT expression was also associated with altered levels of leaf amino acids, implying that a number of amino acid biosynthesis pathways are affected and may be regulated in response to changes in ammonium ion and/or glutamine levels (Ferrario-Mery *et al.*, 2000).

Constituting a minor proportion of the total plant GOGAT activity, NADPH-GOGAT received less attention than the Fd-GOGAT. However, several lines of evidence indicate that, despite being a minor isozyme, the NADPH-GOGAT activity in plants is not redundant. NADPH-GOGAT is unable to compensate for Fd-GOGAT shortage, implying a distinct metabolic function (Ireland and Lea, 1999; Somerville and Ogren, 1980). Moreover, plant genes encoding NADPH-GOGAT generally exhibit contrasting expression patterns compared to Fd-GOGAT genes. While Fd-GOGAT is abundantly produced in photosynthetic leaves, NADPH-GOGAT is produced in nonphotosynthetic organs, such as roots, senescing leaves, and nodules formed in legume roots (see Lancien *et al.*, 2002 and references therein). This suggests that in contrast to the major function of Fd-GOGAT in reassimilation of photorespiratory ammonium ion, NADPH-GOGAT functions mainly in primary nitrogen assimilation and in nitrogen transport from source to sink.

To study the function of NADH-GOGAT, its activity was reduced by up to 87% in transgenic alfalfa plants, using antisense constructs controlled either by an AAT-2 promoter with enhanced expression in nodules, or by a nodule-specific leghemoglobin promoter (Cordoba *et al.*, 2003; Schoenbeck *et al.*, 2000). The transgenic plants were chlorotic and exhibited altered symbiotic phenotypes compared to controls. In addition, nodule amino acids and amides levels were lower, while sucrose levels were higher in the transgenic plants than in control plants, implying that NADPH-GOGAT represents a major rate-limiting enzyme for the incorporation of ammonium ion and sugars into amino acids in nodules.

The functional role of NADPH-GOGAT was also studied in an *Arabidopsis* T-DNA insertion within the single *Arabidopsis* gene encoding this enzyme that abolished expression of the gene (Lancien *et al.*, 2002). In contrast to plants with reduced levels of Fd-GOGAT, which exhibited metabolic and growth defects under conditions of enhanced photorespiration (see above), the *Arabidopsis* T-DNA mutant lacking NADPH-GOGAT exhibited metabolic and growth defects when photorespiration was repressed. Based on these results, NADPH-GOGAT and Fd-GOGAT appear to play nonredundant roles in the assimilation of non-photorespiratory ammonium (derived from soil nitrogen or nitrogen fixation) and photorespiratory ammonium into glutamate, respectively.

The metabolic function of NADPH-GOGAT was also studied by constitutive expression of the alfalfa enzyme in transgenic tobacco plants (Chichkova *et al.*, 2001). Shoots of the transgenic plants contained higher total carbon and nitrogen than wild-type plants when administered either nitrate or ammonium ion as sole nitrogen sources. In addition, the transgenic plants contained higher dry weight than control plants upon entering flowering. These results are consistent with the rate-limiting role of NADPH-GOGAT in nitrogen assimilation and also with the importance of nitrogen assimilation for plant growth (Chichkova *et al.*, 2001).

2.3. Glutamate dehydrogenase: An enzyme with controversial functions in plants

In microorganisms, one of the routes of glutamate synthesis is by combining ammonium ion with α-ketoglutarate in a reaction catalyzed by glutamate dehydrogenase (GDH) (Meers *et al.*, 1970). Since the major route of glutamate synthesis in plants occurs via the GS/GOGAT pathway, a parallel GDH-catalyzed route for glutamate seems highly redundant. However, plants possess GDH enzymes, whose metabolic functions have long been and still are highly controversial. The metabolic status of plants largely depends on mineral nitrogen availability from the soil (or from nitrogen fixing microorganisms) and carbon fixation from photosynthesis. Since the availability of carbon and nitrogen depends on environmental factors and may also be limiting, plants have evolved efficient ways to capture nitrogen and carbon and to regulate the partition between sugars and nitrogenous compounds to optimize plant growth and reproduction (Miflin and Habash, 2002; Stitt *et al.*, 2002). Since the GDH reaction is easily reversible leading to the release of ammonium ion from glutamate, it could function in the conversion of glutamate into organic acids under conditions of limiting carbon fixation. Indeed the catabolic function of GDH in deaminating glutamate was demonstrated directly by ^{13}[C] and ^{31}[P] nuclear magnetic resonance studies (Aubert *et al.*, 2001). This function has been indirectly implied by a number of physiological, biochemical, and molecular studies that have been discussed before (Hirel and Lea, 2001; Ireland and Lea, 1999; Lea and Ireland, 1999; Miflin and Habash, 2002).

In contrast to the well-documented catabolic functions of plant GDH, it is possible that the enzyme may also operate in parallel to GOGAT in the aminating direction of glutamate biosynthesis. Analyses of plants with reduced GOGAT activity, either due to genetic mutation or due to expression of GOGAT antisense constructs (Cordoba *et al.*, 2003; Coschigano *et al.*, 1998; Ferrario-Mery *et al.*, 2000, 2002a,b; Lancien *et al.*, 2002), suggested that GOGAT is the major enzyme responsible for glutamate biosynthesis in plants. Hence, a possible anabolic (aminating) activity of GDH, if it exists, contributes relatively little to overall glutamate biosynthesis. Nevertheless, isolated mitochondria from potato plants can combine ^{15}[N]-labeled ammonium ion and α-ketoglutarate into ^{15}[N] glutamate (Aubert *et al.*, 2001), suggesting that plant GDH can catalyze some glutamate synthesis under specific metabolic conditions. A plausible limited anabolic activity of GDH has indirectly been supported by other studies. Melo-Oliveira *et al.* (1996) found that seedlings of an *Arabidopsis gdh1* null mutant grew slower than wild-type seedlings, in particular with respect to root elongation, on media containing high levels of inorganic nitrogen. Thus, the *Arabidopsis* GDH1 appears to play a nonredundant role in assimilating ammonium ion into glutamine under conditions of excess inorganic nitrogen. Even so, the *Arabidopsis* GDH1 is likely to contribute minimally to nitrogen assimilation under regular growth conditions when nitrogen fertilization is not in excess.

Another indirect support for some compensatory aminating function of GDH was observed in transgenic tobacco plants in which Fd-GOGAT activity was significantly reduced by an antisense approach (Ferrario-Mery *et al.*, 2002a).

Under conditions of reduced photorespiration (high CO_2), reduction of the Fd-GOGAT activity affected neither the deaminating nor the aminating activity of GDH. Yet, upon transport to air, there was a significant increase in the aminating, but not the deaminating, activity of GDH in the transgenic lines, which was also correlated with increased ammonium ion levels in these plants. These results suggest that under conditions of reduced Fd-GOGAT activity and high rates of photorespiration, GDH may compensate for the reduced GOGAT activity (Ferrario-Mery *et al.*, 2002a).

Thus, the accumulating data suggest that in addition to the major catabolic activity of GDH, the enzyme may also assist GOGAT in glutamate biosynthesis under conditions of extensive photorespiration or excess nitrogen fertilization. Nevertheless, such an aminating activity of the plant GDH would be minor compared to that of GOGAT and may become important metabolically only when GOGAT activity is compromised. Additional studies, using dynamic flux, are needed to unequivocally demonstrate whether plant GDH enzymes function in the anabolic direction of glutamate biosynthesis.

In other studies, microbial GDH genes were expressed in transgenic plants, using the constitutive 35S promoter. Expression of an *Escherichia coli* GDH in transgenic tobacco plants improved plant biomass production and also rendered the plants more tolerant than wild-type plants to a glutamine synthetase inhibitor (Ameziane *et al.*, 2000). Similarly, expression of a *Neurospora intermedia* GDH in transgenic tobacco plants improved plant growth under low nitrogen (Wang and Tian, 2001). These results imply that the heterologous microbial GDH enzymes contributed to nitrogen use efficiency of the transgenic plants by operating in the aminating direction of glutamate synthesis. However, whether this function is associated with specific biochemical characteristics of the microbial GDH enzymes that are either present or not present in the plant counterparts remains to be elucidated.

2.4. The network of amide amino acids metabolism is regulated in concert by developmental, physiological, environmental, metabolic, and stress-derived signals

Amino acid metabolic pathways are connected with each other as well as to other metabolic pathways, such as nitrogen and sulfur assimilation, photosynthesis, and carbon/nitrogen balance. Essentially, AAAM provides the core of these metabolic networks and is in itself regulated by many signals, such as a number of light signals (different wavelengths), various metabolites (such as nitrogen and sugars), and photosynthesis. However, little is known about the networking of AAAM with other pathways of amino acid metabolism, and how the networks are concertedly regulated by the large number of dynamically changing signals that exert a "matrix effect" (Coruzzi and Zhou, 2001). For example, it is unknown how dynamically changing light signals of different wavelengths and intensities operate in concert with sugar and nitrogen signals to regulate amino acid metabolism in different tissues during plant development and in response to stress conditions. Do these signals either operate independently or do at least some of them operate in concert? Can some signals override others? This complex "matrix effect" has only recently been addressed, using new combinatorial tools (Thum *et al.*, 2003), on three *Arabidopsis* genes (*GLN2, ASN1,* and *ASN2*) encoding, respectively, glutamine synthetase and two asparagine synthetase enzymes. The *GLN2* and *ASN1* genes are reciprocally regulated by light as well as by sucrose that mimics the light effect (Lam *et al.*, 1995, 1996; Oliveira *et al.*, 2001), while expression of *ASN2* is reciprocally regulated with that of the *ASN1* gene being stimulated by light and sucrose like the *GLN2* gene (Lam *et al.*, 1995, 1998). To study the regulatory effects of different light signals and sucrose on the expression of the *GLN2, ASN1,* and *ASN2* genes, Thum *et al.* (2003) used *Arabidopsis* seeds germinated either in the dark or in the light (germination in the light was followed by 2 days of dark adaptation) in media containing 0% or 1% sucrose. Each of these groups was then exposed to treatments with red, blue, or far-red lights at two different intensities (2 or 100 mE/m^2s) or to white light (70 mE/m^2s) for 3 h. Sucrose attenuated the blue-light induction of the *GLN2* gene in etiolated seedlings and the white-, blue-, and red-light induction of the *GLN2* and *ASN2* genes in light grown plants. Sucrose also strengthened the far-red light induction of *GLN2* and *ASN2* in light grown plants. Depending on the intensity of the far-red light, sucrose was able to either attenuate or strengthen light repression of the *ASN1* gene in light plants. On a more general basis, sucrose exceeded light as a major regulator of *ASN1* and *GLN2* gene expression in etiolated seedlings, whereas, oppositely, light exceeded carbon as a major regulator of *GLN2* and *ASN2* gene expression in light grown plants. These results illustrate the complex interaction of light and carbon signals and apparently expose a complex interaction between signal transduction cascades that translate these signals into gene expression.

3. THE ASPARTATE FAMILY PATHWAY THAT IS RESPONSIBLE FOR SYNTHESIS OF THE ESSENTIAL AMINO ACIDS LYSINE, THREONINE, METHIONINE, AND ISOLEUCINE

3.1. The aspartate family pathway is regulated by several feedback inhibition loops

In plants, as in many bacterial species, lysine, threonine, methionine, and isoleucine are synthesized from aspartate through several different branches of the aspartate family pathway (Fig. 3.2). While one branch of this pathway leads to lysine biosynthesis, a second branch leads to threonine, isoleucine, and methionine biosynthesis. Methionine and threonine biosyntheses diverge into two subbranches and compete for *O*-phosphohomoserine as an intermediate (Fig. 3.2). The entire aspartate family pathway, except for the last step of methionine synthesis (methionine synthase), occurs in the plastid. Although methionine is often considered part of the aspartate family pathway, its biosynthesis is subject to a special regulatory pattern, apparently due to its multiple functions in plants. Therefore, we will discuss the regulation of methionine biosynthesis in a separate section.

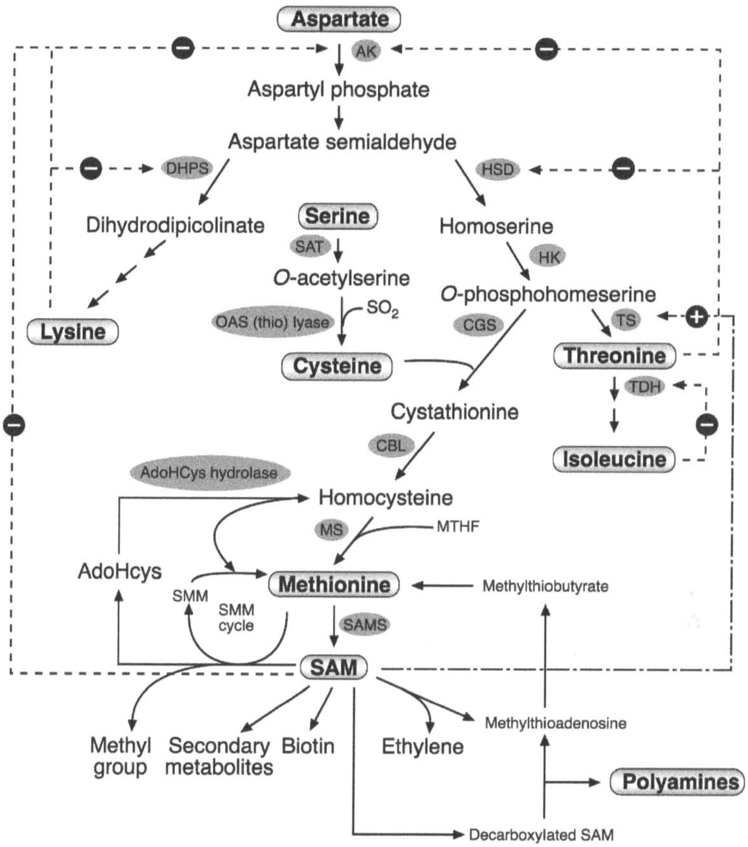

FIGURE 3.2 Schematic diagram of the metabolic network containing the aspartate family pathway, methionine metabolism, and last two steps in the cysteine biosynthesis. Only some of the enzymes and metabolites are specified. Abbreviations: AK, aspartate kinase; DHPS, dihydrodipicolinate synthase; HSD, homoserine dehydrogenase; HK, homoserine kinase; TS, threonine synthase; TDH, threonine dehydratase; SAT, serine acetyl transferase; OAS (thio) lyase; *O*-acetyl serine (thio) lyase; CGS, cystathionine γ-synthase; CBL, cystathionine β-lyase; MS, methionine synthase, SAM, *S*-adenosyl methionine; SAMS, *S*-adenosyl methionine synthase; AdoHcys, adenosylhomocysteine; SMM, *S*-methyl methionine; MTHF, methyltetrahydrofolate. Dashed arrows with a "minus" sign represent feedback inhibition loops of key enzymes in the network. The dashed and dotted arrow with the "plus" sign represents the stimulation of TS activity by SAM.

Biochemical studies showed that the aspartate family pathway is regulated by several feedback inhibition loops (see Galili, 1995 for details; Fig. 3.2). Aspartate kinase (AK) consists of several isozymes, five in *Arabidopsis*, which are feedback inhibited either by lysine or threonine. These include monofunctional polypeptides containing either the lysine-sensitive AK activity, or bifunctional AK/HSD enzymes containing both the threonine-sensitive AK and homoserine DH (HSD) isozymes linked on a single polypeptide (see Galili, 1995). Lysine also feedback inhibits the activity of dihydrodipicolinate synthase (DHPS), the first enzyme committed to its own synthesis, while threonine partially inhibits the activity of HSD, the first enzyme committed to the synthesis of threonine and methionine.

Although both the monofunctional AK and DHPS activities are feedback inhibited by lysine, DHPS is the major limiting enzyme for lysine biosynthesis, while AK is a major limiting enzyme in the second branch of the aspartate family pathway leading to threonine, isoleucine, and methionine biosynthesis. This has been concluded based on the analysis of plant mutants as well as transgenic plants expressing recombinant feedback insensitive DHPS and AK enzymes derived from either bacteria or plant sources (Galili, 1995, 2002; Galili and Hofgen, 2002; Galili *et al.*, 1995; Jacobs *et al.*, 1987; 2001). The results of these functional studies had been expected since the *in vitro* activities of plant DHPS enzymes are much more sensitive to lysine inhibition than those of the lysine-sensitive AK enzymes (see Galili, 1995 for review).

Do the lysine and threonine branches compete for the common substrate aspartate semialdehyde (Fig. 3.2)? Lysine overproduction in plants expressing a feedback-insensitive DHPS is also generally associated with reduced levels of threonine (Galili, 1995, 2002). Moreover, when the feedback-insensitive DHPS and AK were combined into the same plant, lysine levels far exceeded those of threonine levels (Ben Tzvi-Tzchori *et al.*, 1996; Frankard *et al.*, 1992; Shaul and Galili, 1993). This suggests that apart from regulation by the feedback inhibition loops of AK and DHPS, the lysine branch exerts a more powerful drain on metabolic flux than the threonine branch.

3.2. Metabolic fluxes of the aspartate family pathway are regulated by developmental, physiological, and environmental signals

Although the aspartate family pathway is subject to major regulation by feedback inhibition loops, the fluxes of this pathway also depend on the expression of genes encoding the enzymes of this pathway. Expression of the genes and activities of the encoded enzymes may be regulated by transcriptional, posttranscriptional, translational, and posttranslational mechanisms, which may respond to various developmental, physiological, and metabolic signals. One way to identify such regulatory signals is to test their effects on the steady-state levels of the aspartate family amino acids and on the expression and activity of enzymes of this pathway. However, since the aspartate family amino acids are relatively minor amino acids (Noctor *et al.*, 2002), it is difficult to draw statistically meaningful conclusions from such studies. Hence, metabolic engineering of feedback inhibition loops appears to be the appropriate strategy for functional dissection of signals that regulate the production of the aspartate family enzymes as rationalized in the following. Although feedback inhibition of DHPS and AK represents major regulators of the fluxes of the aspartate family pathway, synthesis of its end-product amino acids also depends on the expression of additional enzymes in this pathway (Fig. 3.2). Thus, if a feedback-insensitive DHPS or AK were to be constitutively expressed in transgenic plants, significant lysine or threonine overproduction would be expected only in the specific tissues or growth conditions where the genes encoding the entire set of lysine and/or threonine biosynthetic enzymes are also abundantly expressed. Indeed, lysine levels in transgenic plants constitutively expressing a feedback-insensitive bacterial DHPS fluctuated considerably under different growth conditions, being higher in young leaves and floral organs than in old leaves, and positively responding to light intensity (Shaul and Galili, 1992a; Zhu-Shimoni and Galili, 1998). In contrast, threonine levels in transgenic plants constitutively expressing a bacterial feedback-insensitive AK showed much less fluctuations than lysine levels in plants expressing the *E. coli* feedback-insensitive DHPS (O. Shaul and G. Galili, unpublished information). The results imply that metabolic fluxes of the aspartate family pathway are regulated by developmental, physiological, and environmental signals and that fluxes in the lysine and threonine branches respond differently to the signals.

The regulation of synthesis of the aspartate family amino acids was studied further by analyzing the expression patterns of two *Arabidopsis* genes encoding AK/HSD and DHPS enzymes, using Northern blot analyses and promoter fusion to the β-glucuronidase (GUS) reporter gene. The developmental expression pattern of both genes was very similar, that is, they were highly expressed in germinating seedlings, actively dividing and growing young shoot and root tissues, various organs of the developing flowers, as well as in developing embryos (Vauterin *et al.*, 1999; Zhu-Shimoni *et al.*, 1997). Exposure of etiolated seedlings to light results in an altered pattern of GUS staining in the hypocotyls and cotyledons, suggesting that expression of the AK/HSD and DHPS genes is also regulated by light (Vauterin *et al.*, 1999; Zhu-Shimoni *et al.*, 1997). This was supported by studies showing that the levels and activities of the barley AK isozymes are increased by light and phytochrome (Rao *et al.*, 1999). The similarities in the developmental and light-regulated patterns of expression of the AK and DHPS genes suggest some coordination of expression of genes encoding enzymes of the aspartate family pathway. However, this clearly does not account for the entire set of the aspartate family genes as deduced from the differential expression pattern of two of the three *Arabidopsis* genes encoding lysine-sensitive monofunctional AK isozymes. Based on an analysis of transgenic plants expressing promoter-GUS constructs, expression of one of these genes was more predominant than the other in vegetative tissues (Jacobs *et al.*, 2001). Both genes were highly expressed at the reproductive stage, but only one of these genes was expressed in fruits (Jacobs *et al.*, 2001). Whether this variation in expression pattern reflects a nonredundant function of the different AK isozymes or association with developmentally regulated variations in metabolic fluxes of the lysine and threonine branches, discussed above, remains to be elucidated.

3.3. Metabolic interactions between AAAM and the aspartate family pathway

Aspartate, the substrate of AK, serves not only as the precursor for the aspartate family pathway but is also the immediate precursor for the amide amino acid asparagine via the activity of asparagine synthetase (Fig. 3.1). As discussed before, aside from being a building block of proteins, asparagine also possesses several additional important functions in nitrogen assimilation and transport

(Lam *et al.*, 1995, 1998). How then is either the metabolic channeling of aspartate into asparagine or the aspartate family amino acids regulated? Molecular analyses suggest that this channeling may be regulated by the expression of genes encoding asparagine synthetase and AK. Plants possess two forms of asparagine synthetase genes. The expression of one is induced by light and sucrose (similar to the gene encoding AK/HSD) to enable asparagine synthesis during the day, while expression of the other is repressed by light and sucrose and is induced during the night (Lam *et al.*, 1995, 1998). Notably, expression of at least one of the *Arabidopsis* AK/HSD genes is stimulated by light and sucrose in a very similar manner to that of the asparagine synthase gene that is expressed during the daytime (Zhu-Shimoni and Galili, 1998; Zhu-Shimoni *et al.*, 1997). Thus, assuming that other genes of the aspartate family pathway respond to light and sucrose similarly to this AK/HSD gene, one can hypothesize that during the day aspartate is apparently channeled both into asparagine and into the aspartate family pathway to allow synthesis of all of its end-product amino acids. During the night, the aspartate family pathway is relatively inefficient and aspartate channels preferentially into asparagine. Indeed, asparagine levels are much higher, while lysine levels are lower at night than during daytime (Lam *et al.*, 1995).

Channeling of aspartate into the aspartate family pathway may not only be regulated by photosynthesis and "day/night" cycles. An unexpected observation supporting such a possibility was recently reported following the analysis of an *Arabidopsis* knockout mutant in one of its two DHPS genes (Craciun *et al.*, 2000; Sarrobert *et al.*, 2000). In this mutant, threonine levels increased. However, the extent of the increase (between 10- and 80-fold, depending on growth conditions) far exceeded the slight ~50% reduction in lysine levels, implying that the reduction in DHPS activity triggered an enhanced channeling of aspartate into the threonine branch of the aspartate family pathway (Fig. 3.1). This enhanced channeling may be due to increased activity of the lysine-sensitive AK isozymes as a result of their lower feedback inhibition by the reduced lysine levels. Alternatively, the DHPS knockout mutation may have triggered enhanced expression of the AK genes and perhaps other genes of the threonine branch of the aspartate family pathway.

3.4. Metabolism of the aspartate family amino acids in developing seeds: A balance between synthesis and catabolism

Genetic engineering approaches possess the advantage that gene manipulation can include coding regions as well as regulatory elements such as promoters. Hence, to study the regulation of lysine and threonine metabolism specifically in developing seeds, the *E. coli* feedback-insensitive AK and DHPS enzymes were expressed in transgenic plants under the control of a seed-specific promoter derived from a gene encoding a seed storage protein. The choice of a storage protein gene promoter was based on the assumption that lysine biosynthesis is spatially and temporally coordinated with storage protein production during seed development. Whether storage protein gene promoters are the best choice to manipulate amino acid metabolism specifically in developing seeds is still unknown and awaits detailed studies of seed development. The first studies included the seed-specific expression of the bacterial feedback-insensitive AK and DHPS in transgenic tobacco plants. Expression of the bacterial AK resulted in significant elevation in free threonine in mature seeds (Karchi *et al.*, 1993), but no increase in free lysine was evident in mature seeds of transgenic plants expressing the bacterial DHPS (Karchi *et al.*, 1994). Developing seeds of these transgenic plants also possessed over tenfold higher activity of lysine-ketoglutarate reductase (LKR), the first enzyme in the pathway of lysine catabolism (Galili *et al.*, 2001), suggesting that the low lysine level in mature seeds of the transgenic tobacco plants resulted from enhanced lysine catabolism (Karchi *et al.*, 1994).

To study the significance of lysine catabolism in regulating free lysine accumulation in seeds under conditions of regulated and deregulated lysine synthesis, Galili and associates have isolated an *Arabidopsis* T-DNA knockout mutant lacking lysine catabolism (Zhu *et al.*, 2001). This knockout mutant was crossed with transgenic *Arabidopsis* plants expressing a bacterial feedback-insensitive DHPS in a seed-specific manner (Zhu and Galil, 2003). Although both parental plants contained slightly elevated levels of free lysine compared to wild type in mature seeds, combining both traits into the same plant synergistically boosted free seed lysine levels by ~80-fold, rendering lysine as the most prominent free amino acid (Zhu and Galil, 2003). Moreover, total seed lysine in these plants was nearly doubled compared to wild-type plants (X. Zhu and G. Galili, unpublished results). Notably, the dramatic increase in free lysine in seeds expressing the bacterial DHPS but lacking lysine catabolism was associated with a significant difference in the levels of several other amino acids. The most pronounced differences were significant reductions in the levels of glutamate and aspartate and a dramatic increase in the level of methionine (Zhu and Galil, 2003), exposing novel regulatory networks associated with AAAM and the aspartate family pathway.

A feedback-insensitive DHPS derived from *Corynebacterium glutamicum* was expressed in a seed-specific manner in two additional transgenic dicotyledonous crop plants, soybean and rapeseed (Falco *et al.*, 1995; Mazur *et al.*, 1999). Seeds of these transgenic plants accumulated up to 100-fold (rapeseed) and several hundred-fold (soybean) higher free lysine than wild-type plants, values that are significantly higher than those obtained in transgenic tobacco plants expressing the *E. coli* DHPS (Karchi *et al.*, 1994). Whether this is due to the different plant species or to the different bacterial DHPS enzymes is still not clear, but seeds of the lysine-overproducing soybean and rapeseed plants also contained significantly higher levels of lysine catabolic products than wild-type nontransformed plants (Falco *et al.*, 1995; Mazur *et al.*, 1999).

In contrast to dicotyledonous plants in which storage protein synthesis typically takes place in the developing embryo, the synthesis of storage proteins in cereal seeds occurs mostly in the endosperm (Shotwell and Larkins, 1989). Also, based on *in situ* analysis, the lysine catabolism pathway was suggested to function mostly in the outer layers of the cereal endosperm (Kemper *et al.*, 1999). It is thus expected that expression of a bacterial DHPS, under control of an endosperm-specific storage protein gene promoter, will result in enhanced lysine production and perhaps also accumulation of catabolic products of lysine. This expectation was found to be incorrect because lysine overproduction in transgenic maize seeds was observed only when the bacterial DHPS was expressed under an embryo-specific, but not an endosperm-specific promoter (Mazur *et al.*, 1999). Whether the lack of increase in lysine levels upon expressing the bacterial DHPS in the endosperm tissue is due to factors associated with either lysine synthesis or catabolism or both provides an interesting topic for future research.

What then are the functions of lysine catabolism during seed development and why is this pathway stimulated by lysine? The fact that seeds of transgenic soybean, rapeseed, and *Arabidopsis* can accumulate very high levels of free lysine without a major negative effect on seed germination (only extreme lysine accumulation retards germination) (Falco *et al.*, 1995; Mazur *et al.*, 1999) suggests that lysine catabolism is not required to reduce lysine toxicity. Also, these studies show that the flux of lysine synthesis in developing seeds can become very extensive when the sensitivity of DHPS activity to lysine is eliminated. It is thus possible that during the onset of seed storage protein synthesis, lysine catabolism and likely other amino acids catabolic pathways are stimulated to convert excess-free lysine and other amino acids into sugars and lipids, and also back into glutamate in the case of the lysine catabolism pathway.

The significant research advances in the regulation of lysine metabolism in plants has made this pathway a major biotechnological target for improving the nutritional quality of crop plants. Indeed, a high-lysine corn variety (Mavera™, Monsanto Inc., St. Louis, Missouri), obtained via embryo-specific expression of a bacterial feedback-insensitive DHPS, has recently been approved for commercial growth for livestock feeding. It is highly likely that additional varieties with higher seed lysine content in which lysine catabolism is reduced and lysine-rich proteins are expressed specifically in seeds will appear in the near future.

4. REGULATION OF METHIONINE BIOSYNTHESIS

Methionine is a sulfur-containing essential amino acid, a building block of proteins that also plays a fundamental role in many cellular processes. Through its immediate catabolic product *S*-adenosyl methionine (SAM), methionine is a precursor for the plant hormones ethylene and polyamines as well as for many important secondary metabolites and vitamin B_1. SAM is also a donor of a methyl group to a number of cellular reactions, such as DNA methylation (Amir *et al.*, 2002 and references therein). In plants, methionine can be converted into *S*-methylmethionine (SMM), a metabolite that is believed to participate in sulfur transport between sink and source tissues (Bourgis *et al.*, 1999), and also to control the intracellular levels of SAM (Kocsis *et al.*, 2003; Ranocha *et al.*, 2001). Due to its vital cellular importance, the methionine level is tightly regulated both by its synthesis and catabolism. Methionine is an unstable amino acid with a very fast half-life (Giovanelli *et al.*, 1985; Miyazaki and Yang, 1987).

Methionine receives its carbon and amino groups from *O*-phosphohomoserine, an intermediate metabolite in the aspartate family pathway, and its sulfur atom from cysteine (Fig. 3.2). These two skeleta are first combined by the enzyme cystathionine γ-synthase (CGS) to form cystathionine. This is then converted by cystathionine β-lyase into homocysteine, and converted by methionine synthase into methionine, incorporating a methyl group from *N*-methyltetrahydrofolate (Fig. 3.2). Hence, the complex biosynthesis nature of methionine depends on many regulatory metabolic steps, including the aspartate family pathway, cysteine biosynthesis, and *N*-methyltetrahydrofolate metabolism. Nevertheless, molecular genetic and biochemical studies suggest that methionine biosynthesis is regulated primarily by CGS as well as by a compound metabolic interaction with threonine synthesis through a competition between CGS and threonine synthase (TS) on their common substrate *O*-phosphohomoserine (Fig. 3.2).

4.1. Regulatory role of CGS in methionine biosynthesis

Being the first enzyme specific for methionine biosynthesis, CGS is expected to play an important regulatory role in methionine metabolism. Nevertheless, there is no evidence for the regulation of CGS activity by feedback inhibition loops (Ravanel *et al.*, 1998a, 1998b). Instead, the level of CGS is regulated by either methionine, or its catabolic product(s), through posttranscriptional and posttranslational mechanisms (Amir *et al.*, 2002; Chiba *et al.*, 1999; Hacham *et al.*, 2002; Onouchi *et al.*, 2005). CGS polypeptides (without their plastid transit peptides) in mature plants contain a region of ~100 amino acids at the N-terminus, which is not present in bacterial CGS enzymes and is also not essential for CGS catalytic activity (Hacham *et al.*, 2002). A series of *Arabidopsis mto1* mutants, which accumulates up to 40-fold higher methionine in young tissues than in wild-type plants, were shown to be attributed to mutations in the region encoding this N-terminal domain of CGS (Chiba *et al.*, 1999; Inaba *et al.*, 1994). The *mto1* mutations are located in a specific subdomain (called the MTO1 region), which is conserved in the CGS genes of all plant species analyzed so far. This region apparently acts to downregulate CGS mRNA level when either the level of methionine or any of its catabolic products rise, via a mechanism that apparently involves specific nascent amino acids translated from this mRNA region (Chiba *et al.*, 1999; Inaba *et al.*, 1994).

Several lines of evidence suggest that the control of methionine synthesis cannot be solely explained by the posttranscriptional regulation through the MTO1 region. No inverse correlation between methionine and CGS mRNA levels were evident in transgenic *Arabidopsis* plants overexpressing the endogenous CGS, as well as in an *Arabidopsis* mutant with reduced methionine catabolism (Goto *et al.*, 2002; Kim *et al.*, 2002). Moreover, in contrast to *Arabidopsis*, no evidence supporting a control of CGS mRNA level by methionine was obtained in potato plants, although the MTO1 region in the potato CGS gene is highly conserved with that of the *Arabidopsis* counterpart (Kreft *et al.*, 2003). These observations suggest that the regulatory function of the MTO1 region requires interactions with additional factors that are not present in all tissues and/or are not conserved in all plant species. Notably, *Arabidopsis* and potato also differed in their response to constitutive CGS overexpression. While CGS overexpression in transgenic

Arabidopsis plants caused an approximately 4–20-fold increase in methionine (Gakiere *et al.*, 2000; Kim *et al.*, 2002), no increase in methionine was obtained in transgenic potato plants (Kreft *et al.*, 2003). Whether these differences are due to genetic or physiological factors remains to be elucidated.

The regulatory role of the N-terminal region of the mature plant CGS enzyme was also studied by either constitutive expression of a full-length *Arabidopsis* CGS or its deletion mutant lacking this region, but still containing the plastid transit peptide, in transgenic tobacco plants (Hacham *et al.*, 2002). Expression of the *Arabidopsis* CGS without its N-terminal region caused significant increases of ethylene and dimethyl sulfide, two catabolic products of methionine, over plants expressing the full-length *Arabidopsis* CGS (Hacham *et al.*, 2002). However, methionine and SMM levels, although increased over wild-type plants, did not differ significantly between transgenic plants expressing the different CGS constructs. Since the expression levels of the transgenic CGS polypeptides were comparable between the two sets of these transgenic plants, it was suggested that the N-terminal region of CGS might also regulate methionine metabolism by a post-translational mechanism (Hacham *et al.*, 2002).

4.2. Interrelationships between threonine and methionine biosynthesis

Biochemical studies suggest that methionine biosynthesis is regulated by a competition between CGS and TS for their common substrate *O*-phosphohomoserine (Amir *et al.*, 2002 and references therein). Plant TS enzymes possess approximately 250–500-fold higher affinity for *O*-phosphohomoserine than the plant CGS enzymes as measured by *in vitro* studies (Curien *et al.*, 1998; Ravanel *et al.*, 1998b). This indicates that most of the carbon and amino skeleton of aspartate should be channeled toward threonine rather than to methionine. Indeed, when the flux into the threonine/methionine branch of the heaspartate family was increased by overexpressing a bacterial feedback-insensitive AK in transgenic plants, threonine levels were greatly increased but methionine levels hardly changed (Ben Tzvi-Tzchori *et al.*, 1996; Karchi *et al.*, 1993; Shaul and Galili, 1992b). SAM, the immediate catabolic product of methionine, may buffer the competitive fluxes of threonine and methionine biosynthesis because it positively regulates TS activity (Curien *et al.*, 1998).

Studies using transgenic plants support the biochemical studies for a competition between the threonine and methionine branch of the aspartate family pathway (Fig. 3.2). However, they also show that this competition is not simple. Reduction of CGS level by gene silencing or antisense approaches resulted in a 3.3–8.3-fold increase in threonine levels in transgenic *Arabidopsis* plants, while methionine levels were only slightly reduced (Kim and Leustek, 2000; Kim *et al.*, 2002). In addition, reduction of TS activity due to a mutation in the TS gene (*mto2–1* mutant) caused an ~16-fold reduction in threonine as well as a comparable ~22-fold increase in methionine in rosette leaves compared to wild-type *Arabidopsis* plants (Bartlem *et al.*, 2000). More remarkable results were obtained when the TS levels were reduced by an antisense approach both in transgenic potato and *Arabidopsis* plants (Avraham and Amir, 2005; Zeh *et al.*, 2001). In the TS antisense transgenic potato plants, threonine levels were only moderately reduced by up to ~45%, whereas methionine levels were dramatically increased by up to ~239-fold compared to nontransformed plants (Zeh *et al.*, 2001). Similarly, in the TS antisense transgenic *Arabidopsis* plants, threonine levels were only moderately reduced by approximately 1.5–2.5-fold, while the levels of methionine increased by up to ~47-fold than in wild-type plants (Avraham and Amir, 2005). The results imply that the reduction in TS levels, rather than its activity as observed in the *Arabidopsis mto2* mutant, causes either an increased flux of the carbon and amino skeleton from aspartate to methionine or a reduced rate of methionine catabolism.

The complex competition between the methionine and threonine branches of the aspartate family pathway was supported by additional studies. In the *mto1–1* mutants, the significant increases in methionine were not associated with a significant reduction in threonine (Kim and Leustek, 2000). In addition, constitutive overexpression of CGS in transgenic *Arabidopsis*, potato, and tobacco plants caused significant increases in methionine levels, but no significant compensatory decreases in threonine levels (Gakiere *et al.*, 2000; Hacham *et al.*, 2002; Kim *et al.*, 2002; Kreft *et al.*, 2003). These results may be explained by a differential rate-limiting effect of *O*-phosphohomoserine, the common substrate for CGS and TS (Fig. 3.2), for threonine and methionine biosynthesis. The steady-state level of *O*-phosphohomoserine may be more rate limiting for methionine than for threonine biosynthesis. In addition, increased *O*-phosphohomoserine utilization by CGS may trigger an increase in the synthesis of this intermediate metabolite, rendering it nonlimiting for threonine biosynthesis. This assumption is supported by the analysis of *Arabidopsis* and potato plants expressing the antisense form of CGS. The level of *O*-phosphohomoserine in these plants was increased by ~22-fold in *Arabidopsis*, and from an undetectable level to 6.5 nmol/g fresh weight in potatoes, while the level of threonine increased only by ~8-fold in *Arabidopsis*, or was not increased in potato plants (Gakiere *et al.*, 2000; Kreft *et al.*, 2003).

5. ENGINEERING AMINO ACID METABOLISM TO IMPROVE THE NUTRITIONAL QUALITY OF PLANTS FOR NONRUMINANTS AND RUMINANTS

The aspartate family amino acids, lysine, methionine, and threonine, and the aromatic amino acid tryptophan are the most important essential amino acids required in human foods and livestock feeds. They are the most limiting essential amino acids in the major crop plants that serve as human foods and animal feeds, particularly cereals and legumes that are supplied as grain and/or as forage (Galili *et al.*, 2002). Cereals are deficient mainly in lysine and tryptophan, while legumes are mainly deficient in methionine (Syed Rasheeduddin and Mcdonald, 1974). Thus, many of the commonly used diet formulations based on these crops contain limiting amounts of these essential amino acids.

Livestock that are consumed as human foods are nonruminant animals, such as poultry or pigs, and ruminants, such as cattle or sheep, which differ in feed requirements for optimal incorporation of essential amino acids. The nonruminants or monogastric animals, like humans, cannot synthesize essential amino acids and thus depend entirely on the external supply of essential amino acids. Ruminant animals also cannot synthesize these essential amino acids; however, the microbial flora inhabiting their rumen can metabolize nonessential into essential amino acids and incorporate them into microbial proteins that later become nutritionally available. Nevertheless, these microbial proteins, although of better nutritional quality than plant proteins, do not provide sufficient essential amino acids for optimal growth and milk production (Leng, 1990). Moreover, although the rumen microflora can produce essential amino acids, it can also oppositely metabolize essential amino acids into nonessential ones. Hence, in contrast to nonruminant animals that can utilize either free or protein-incorporated essential amino acids, ruminant feeds should contain the essential amino acids in proteins that are highly stable in the rumen to minimize their degradation by the rumen microflora.

5.1. Improving lysine levels in crops: A comprehensive approach

Although free lysine content could be significantly improved in legume and cereal grain crops by expression of a bacterial feedback–insensitive DHPS (Avraham and Amir, 2005), such transgenic plants may not be optimal foods and feeds. These plants accumulate relatively high levels of intermediate products of lysine catabolism, such as α-amino adipic acid, which may act as neurotransmitters in animals and can be toxic at high levels (Bonaventure *et al.*, 1985; Karlsen *et al.*, 1982; Reichenbach and Wohlrab, 1985; Welinder *et al.*, 1982). In addition, these plants overaccumulate free lysine rather than lysine-rich proteins and are therefore not suitable for feeding of ruminant animals (National Research Council, 2001). To address this issue, Jung and Falco (2000) used a composite approach to generate lysine-overproducing transgenic maize grains. This included combined expression of two transgenes. One encoded a bacterial feedback-insensitive DHPS under an embryo-specific promoter since lysine overproduction is achieved only in maize embryos (see above). The second encoded a lysine-rich protein (either hordothionine HT12 or the barley high-lysine protein BHL8, containing 28% and 24% lysine, respectively) under an endosperm-specific promoter since the endosperm consists a major part of the maize grain. Two types of maize plants were transformed with these genes, wild-type maize and a maize mutant lacking lysine catabolism due to a knockout of the maize *LKR/SDH* gene. The HT12 and BHL8 proteins accumulated between 3% and 6% of total grain proteins, and when introduced together with the bacterial DHPS resulted in a marked elevation of total lysine to over 0.7% of seed dry weight (Jung and Falco, 2000), for example, as compared to around 0.2% in wild-type maize. Combination of these genes into a homozygous *LKR/SDH* knockout background increased grain lysine level further and alleviated the problem of high-level accumulation of lysine catabolic products (Jung and Falco, 2000).

The additive effect of free lysine overproduction in the maize embryo and its incorporation into lysine-rich proteins in the endosperm on total grain lysine content suggests that free lysine is effectively transported between the two tissues. Should the dramatic elevation of lysine levels, obtained by this composite approach, not interfere with yield and other grain quality factors, the commercial application of such high-lysine transgenic maize plants for feeding human and nonruminant livestock looks very promising. Maize is also a suitable crop for ruminant feeding because maize seed proteins are on average highly resistant to rumen proteolysis (National Research Council, 2001). Moreover, the endogenous maize seed proteins may protect transgenic high-lysine proteins from rumen degradation.

5.2. Improving methionine levels in plant seeds: A source–sink interaction

Most attempts to improve the methionine contents of seeds have focused on overexpression of methionine-rich seed storage proteins, such us Brazil nut 2S albumin, sunflower 2S albumin (SSA), and maize methionine-rich zeins (for review see Avraham and Amir, 2005). The SSA was also found highly resistance to rumen proteolysis (Mcnabb *et al.*, 1994), suggesting that transgenic plants overexpressing it may be beneficial not only for nonruminants but also for ruminant feeding. Indeed, feeding experiments with transgenic lupin grains, which expressed the SSA gene, enhanced both rat growth (Molvig *et al.*, 1997) and sheep live weight gain and wool production (White *et al.*, 2000).

Although transgenic methionine-rich proteins can accumulate to high levels in plant seeds, in most cases the total methionine is still less than necessary for optimal feeding (Avraham and Amir, 2005; Demidov *et al.*, 2003; Galili and Hofgen, 2002). This is largely because production of transgenic methionine-rich protein is associated with a compensatory decrease in the levels of endogenous sulfur-rich proteins. This phenomenon implies the presence of limiting levels of free methionine, whose synthesis in the seeds may be regulated by limited availability of its precursor metabolites cysteine, *O*-phosphohomoserine, or *N*-methyltetrahydrofolate. Combined seed-specific overexpression of a bacterial feedback-insensitive AK (apparently to increase the level of *O*-phosphohomoserine) as well as Brazil nut 2S albumin in transgenic narbon beans resulted in an additive increase of seed methionine, compared to the parental plants expressing each of these transgenes alone (Demidov *et al.*, 2003). This suggests that methionine accumulation in seeds depends on the pool size of *O*-phosphohomoserine. In addition, when the SSA was expressed in seeds of transgenic lupin and rice plants (Hagan *et al.*, 2003; Molvig *et al.*, 1997; Tabe and Droux, 2002), although seed methionine levels were increased, there was no increase in seed cysteine and the total seed sulfur content, implying that the vegetative cysteine pool and the extent of sulfur transport from the canopy to the seeds represent two additional rate-limiting factors. Thus, possible additional target genes for genetic engineering of plants with high seed methionine would be genes controlling the assimilation, metabolism, and transport of sulfur. Indeed, constitutive overexpression of serine acetyl transferase, an important regulatory enzyme in cysteine biosynthesis (Fig. 3.2), enhanced seed methionine content in transgenic maize (Tarczynski *et al.*, 2001).

Limited levels of sulfur-containing metabolites in seeds retard the synthesis of endogenous sulfur-rich proteins by negatively regulating the expression of their genes (Tabe and Droux, 2002; Tabe *et al.*, 2002). One way to overcome this negative regulation is by replacing regulatory elements of endogenous genes encoding sulfur-rich proteins with analogous elements derived from endogenous genes whose expression is not responsive to sulfur availability. In a recent study, the promoter and 5′ untranslated regions of a maize gene encoding a methionine-rich δ-zein were substituted with analogous sequences derived from another gene encoding a γ-zein gene and transformed back into transgenic maize plants (Lai and Messing, 2002). Expression of this chimeric transgene caused an ~30% increase in total seed methionine.

5.3. Improving the nutritional quality of hay for ruminant feeding

Improving the nutritional quality of hay for ruminant feeding requires the expression of proteins, which are both nutritionally balanced and resistant to rumen proteolysis in vegetative tissues. When genes encoding vacuolar methionine-rich seed storage proteins, which stably accumulate in seeds, were constitutively expressed in various transgenic plants, their encoded proteins failed to accumulate in the protease-rich vegetative vacuoles because of extensive degradation (see Avraham and Amir, 2005 for review). This was partially overcome by preventing the trafficking of these proteins from the endoplasmic reticulum (ER) to the vegetative vacuole, by engineering of an ER retention signal (KDEL or HDEL) into their C-terminus (see Avraham and Amir, 2005 for review).

Vegetative storage proteins (VSPs) may be preferred alternatives to seed storage proteins because they are nutritionally balanced and also stably accumulate in vacuoles of vegetative cells (Staswick, 1994). Galili and associates tested the potential of constitutive expression of genes encoding the α- and β-subunits of soybean VSPs to improve the nutritional quality of vegetative tissues of heterologous plants. The soybean VSPα-subunit accumulated to high levels (up to 3% of total leaf soluble proteins) and its levels remained stable also in mature leaves of transgenic tobacco plants (Guenoune *et al.*, 1999). However, this subunit was totally unstable to rumen proteolysis (Guenoune *et al.*, 2002b). The soybean VSPβ was however more resistance to rumen proteolysis (Guenoune *et al.*, 2002b), but accumulated only in young leaves and its levels declined with leaf age (Guenoune *et al.*, 2003). Coexpression of both subunits in the same transgenic plant resulted in stable accumulation of both proteins in older leaves and also improved their stability to rumen degradation (Guenoune *et al.*, 2002b).

Accumulation of transgenic proteins in vegetative tissues may be further improved by their targeting to more than one organelle. Directing the soybean VSPα to plastids resulted in a similar level to that of the vacuole-targeted counterpart (Guenoune *et al.*, 2002a). Targeting of the soybean VSPα to these two organelles in a single transgenic plant resulted in its significantly high accumulation to up to 7.5% of the total soluble proteins (Guenoune *et al.*, 2002a).

6. FUTURE PROSPECTS

Genetic engineering approaches have contributed significantly to understand the regulation of amino acid metabolism in plants. Such approaches can be expected to become major tools in future research on plant amino acid metabolism. So far, detailed studies on amino acid metabolism, using genetic engineering approaches, were limited to a narrow range of pathways, particularly the pathway of AAAM, the aspartate family pathway, and to some extent the pathways of proline and tryptophan metabolism (Kishor *et al.*, 1995; Li and Last, 1996; Nanjo *et al.*, 1999; Tozawa *et al.*, 2001; Zhang *et al.*, 2001). Similar approaches for dissecting metabolic pathways of other amino acids are needed.

Many of the studies discussed here have focused on biosynthetic pathways, while less effort has been devoted to amino acid catabolic pathways. As in the emerging progress of lysine catabolism (Galili *et al.*, 2001), amino acid catabolic pathways may be important metabolic components in plant development, reproduction, and responses to stress. Therefore, in future research, more efforts should be devoted to the dissection of amino acid catabolic pathways.

Amino acid metabolism is strongly regulated by various metabolites, many of which are non-amino acids, which serve not only as signaling molecules but also as intermediate metabolites in metabolic pathways of amino acids sugars and lipids. One example of such metabolites is pyruvate that serves as a precursor for a number of amino acid carbohydrate and lipid molecules. In microorganisms, the regulatory or rate-limiting roles of such intermediate metabolites can be studied by feeding experiments. The multicellular and multiorgan nature of higher plants does not enable proper feeding experiments in all tissues of intact plants and therefore provides additional levels of complexity that render the dissection of metabolic fluxes much more difficult to predict and study than in microorganisms. Understanding the regulation of metabolic fluxes and the importance of rate-limiting metabolites in different plant organs cannot be easily done by feeding experiments alone. Hence, such studies will depend strongly on tissue-specific and/or condition-specific genetic engineering as well as on isotope-labeling studies.

The identification of regulatory networks of amino acid metabolism as well as possible complexes of enzymes that may regulate these networks is also needed. Such studies can be strongly assisted by genetic engineering approaches. For example, identification of enzyme and complexes can be obtained by expressing chimeric genes encoding epitope-tagged enzymes in transgenic plants. It is expected that interdisciplinary approaches, such as that of the "matrix effect" will contribute to unraveling interacting molecular, metabolic, and environmental signals that regulate the networks of amino acid metabolism.

Understanding the compound regulation of metabolic networks (amino acid metabolism is included as a part of these metabolic networks) can be aided by detailed analysis of a large number of metabolites as well as by detailed analysis of the spatial, temporal and developmental patterns of expression of genes encoding enzymes and regulatory proteins associated with these networks. Thus, modern approaches such as metabolic profiling, gene expression profiling in microarrays, and proteomics will be progressively used in these studies. These issues have not been discussed in this chapter due to space limitation. Yet, several recent publications (Hunter *et al.*, 2002; Lee *et al.*, 2002; Ruuska *et al.*, 2002) illustrate how microarray analyses of gene expression in *Arabidopsis* and maize seeds uncovered specific spatial and temporal expression patterns of genes associated with the metabolism of sugars, lipids, amino acids, and storage proteins during seed development.

7. SUMMARY

Apart from serving as protein building blocks, amino acids play multiple regulatory roles in plant growth, including nitrogen assimilation and transport, carbon/nitrogen balance, production of hormones and secondary metabolites, stress-associated metabolism, and many other processes. Some of the amino acids are of particular importance not only for plant growth but also for the nutritional quality of plant foods and feeds because human and its ruminant and nonruminant livestock cannot synthesize them and depend on their availability in their diets. Genetic and metabolic engineering approaches have contributed tremendously to the understanding of the regulation of amino acid metabolism in plants. This chapter discusses how amino acid metabolism is regulated by complex regulatory networks that operate in concert with other regulatory networks of carbon and likely also lipid metabolism. These networks are, however, also subjected to concerted spatial, temporal, developmental, and environmental controls. The combined application of genomic, proteomic, and metabolomic approaches coupled with genetic and metabolic engineering, as well as analysis of dynamic fluxes in different intracellular organelles, offers a promising future for the dissection of these compound regulatory networks.

ACKNOWLEDGEMENTS

The work in the laboratory of G.G. was supported by grants from the Frame Work Program of the Commission of the European Communities, the Israel Academy of Sciences and Humanities, National Council for Research and Development, Israel, as well as by the MINERVA Foundation, Germany, The United States—Israel Binational Agricultural Research and Development (BARD). G.G. holds the Charles Bronfman Professional Chair of Plant Sciences.

REFERENCES

Ameziane, R., Bernhard, K., and Lightfoot, D. (2000). Expression of the bacterial *gdhA* gene encoding a NADPH glutamate dehydrogenase in tobacco affects plant growth and development. *Plant Soil* **221,** 47–57.

Amir, R., Hacham, Y., and Galili, G. (2002). Cystathionine γ-synthase and threonine synthase operate in concert to regulate carbon flow towards methionine in plants. *Trends Plant Sci.* **7,** 153–156.

Aubert, S., Bligny, R., Douce, R., Gout, E., Ratcliffe, R. G., and Roberts, J. K. M. (2001). Contribution of glutamate dehydrogenase to mitochondrial glutamate metabolism studied by ^{13}C and ^{31}P nuclear magnetic resonance. *J. Exp. Bot.* **52,** 37–45.

Avraham, T., and Amir, R. (2005). Methionine and threonine regulate the branching point of their biosynthesis pathways and thus controlling the level of each other. *Transgneic Res.* **14,** 299–311.

Bartlem, D., Lambein, I., Okamoto, T., Itaya, A., Uda, Y., Kijima, F., Tamaki, Y., Nambara, E., and Naito, S. (2000). Mutation in the threonine synthase gene results in an over-accumulation of soluble methionine in *Arabidopsis. Plant Physiol.* **123,** 101–110.

Ben Tzvi-Tzchori, I., Perl, A., and Galili, G. (1996). Lysine and threonine metabolism are subject to complex patterns of regulation in *Arabidopsis. Plant Mol. Biol.* **32,** 727–734.

Bonaventure, N., Wioland, N., and Roussel, G. (1985). Stereospecific effects of the α-aminoadipic acid on the retina: A morphological and electrophysiological study. *Doc. Ophthalmol.* **61,** 71–77.

Bourgis, F., Roje, S., Nuccio, M. L., Fisher, D. B., Tarczynski, M. C., Li, C., Herschbach, C., Rennenberg, H., Pimenta, M. J., Shen, T.-L., Gage, D. A., Hanson, A. D., *et al.* (1999). S-methylmethionine plays a major role in phloem sulfur transport and is synthesized by a novel type of methyltransferase. *Plant Cell* **11,** 1485–1497.

Chiba, Y., Ishikawa, M., Kijima, F., Tyson, R. H., Kim, J., Yamamoto, A., Nambara, E., Leustek, T., Wallsgrove, R. M., and Naito, S. (1999). Evidence for autoregulation of cystathionine γ-synthase mRNA stability in *Arabidopsis. Science* **286,** 1371–1374.

Chichkova, S., Arellano, J., Vance, C. P., and Hernandez, G. (2001). Transgenic tobacco plants that overexpress alfalfa NADH-glutamate synthase have higher carbon and nitrogen content. *J. Exp. Bot.* **52,** 2079–2087.

Cordoba, E., Shishkova, S., Vance, C. P., and Hernandez, G. (2003). Antisense inhibition of NADH glutamate synthase impairs carbon/nitrogen assimilation in nodules of alfalfa (*Medicago sativa* L.). *Plant J.* **33,** 1037–1049.

Coruzzi, G., and Last, R. (2000). Amino acids. *In* "Biochemistry and Molecular Biology of PlantsAmerican Society of Plant Physiologists" (B. B. Buchanan, W. Gruissem, and R. L. Jones, eds.), pp. 358–410. American Society of Plant Physiologists, Rockville, MD.

Coruzzi, G. M., and Zhou, L. (2001). Carbon and nitrogen sensing and signaling in plants: Emerging 'matrix effects'. *Curr. Opin. Plant Biol.* **4,** 247–253.

Coschigano, K. T., Melo-Oliveira, R., Lim, J., and Coruzzi, G. M. (1998). *Arabidopsis gls* mutants and distinct Fd-GOGAT genes: Implications for photorespiration and primary nitrogen assimilation. *Plant Cell* **10,** 741–752.

Craciun, A., Jacobs, M., and Vauterin, M. (2000). *Arabidopsis* loss-of-function mutant in the lysine pathway points out complex regulation mechanisms. *FEBS Lett.* **487,** 234–238.

Curien, G., Job, D., Douce, R., and Dumas, R. (1998). Allosteric activation of *Arabidopsis* threonine synthase by S-adenosylmethionine. *Biochemistry* **37,** 13212–13221.

Demidov, D., Horstmann, C., Meixner, M., Pickardt, T., Saalbach, I., Galili, G., and Muntz, K. (2003). Additive effects of the feed-back insensitive bacterial aspartate kinase and the Brazil nut 2S albumin on the methionine content of transgenic narbon bean (*Vicia narbonensis* L.). *Mol. Breed.* **11,** 187–201.

Eckes, P., Schmitt, P., Daub, W., and Wengenmayer, F. (1989). Overproduction of alfalfa glutamine synthetase in transgenic tobacco plants. *Mol. Gen. Genet.* **217,** 263–268.

Falco, S. C., Guida, T., Locke, M., Mauvais, J., Sandres, C., Ward, R. T., and Webber, P. (1995). Transgenic canola and soybean seeds with increased lysine. *Biotechnology* **13,** 577–582.

Fei, H., Chaillou, S., Hirel, B., Mahon, J. D., and Vessey, J. K. (2003). Overexpression of a soybean cytosolic glutamine synthetase gene linked to organ-specific promoters in pea plants grown in different concentrations of nitrate. *Planta* **216,** 467–474.

Ferrario-Mery, S., Suzuki, A., Kunz, C., Valadier, M. H., Roux, Y., Hirel, B., and Foyer, C. H. (2000). Modulation of amino acid metabolism in transformed tobacco plants deficient in Fd-GOGAT. *Plant Soil* **221,** 67–79.

Ferrario-Mery, S., Hodges, M., Hirel, B., and Foyer, C. H. (2002a). Photorespiration-dependent increases in phospho*enol*pyruvate carboxylase, isocitrate dehydrogenase and glutamate dehydrogenase in transformed tobacco plants deficient in ferredoxin-dependent glutamine-α-keto-glutarate aminotransferase. *Planta* **214,** 877–886.

Ferrario-Mery, S., Valadier, M.-H., Godefroy, N., Miallier, D., Hirel, B., Foyer, C. H., and Suzuki, A. (2002b). Diurnal changes in ammonia assimilation in transformed tobacco plants expressing ferredoxin-dependent glutamate synthase mRNA in the antisense orientation. *Plant Sci.* **163,** 59–67.

Finnemann, J., and Schjoerring, J. K. (2000). Post-translational regulation of cytosolic glutamine synthetase by reversible phosphorylation and 14–3–3 protein interaction. *Plant J.* **24,** 171–181.

Frankard, V., Ghislain, M., and Jacobs, M. (1992). Two feedback-insensitive enzymes of the aspartate pathway in *Nicotiana sylvestris. Plant Physiol.* **99,** 1285–1293.

Fuentes, S. I., Allen, D. J., Ortiz-Lopez, A., and Hernandez, G. (2001). Over-expression of cytosolic glutamine synthetase increases photosynthesis and growth at low nitrogen concentrations. *J. Exp. Bot.* **52,** 1071–1081.

Gakiere, B., Denis, L., Droux, M., Ravanel, S., and Job, D. (2000). Methionine synthesis in higher plants: Sense strategy applied to cystathionine γ-synthase and cystathionine β-lyase in *Arabidopsis thaliana. In* "Sulfur Nutrition and Sulfur Assimilation in Higher Plants" (C. Brunold, ed.), pp. 313–315. Paul Haupt, Bern, Switzerland.

Galili, G. (1995). Regulation of lysine and threonine synthesis. *Plant Cell* **7,** 899–906.

Galili, G. (2002). New insights into the regulation and functional significance of lysine metabolism in plants. *Annu. Rev. Plant Physiol. Plant Mol. Biol.* **53,** 27–43.

Galili, G., and Hofgen, R. (2002). Metabolic engineering of amino acids and storage proteins in plants. *Metab. Eng.* **4,** 3–11.

Galili, G., Shaul, O., Perl, A., and Karchi, H. (1995). Synthesis and accumulation of the essential amino acids lysine and threonine in seeds. *In* "Seed Development and Germination" (J. Kigel and G. Galili, eds.), pp. 811–831. Marcel Dekker, New York.

Galili, G., Tang, G., Zhu, X., and Gakiere, B. (2001). Lysine catabolism: A stress and development super-regulated metabolic pathway. *Curr. Opin. Plant Biol.* **4,** 261–266.

Galili, G., Galili, S., Lewinsohn, E., and Tadmor, Y. (2002). Genetic, molecular and genomic approaches to improve the value of plant foods and feeds. *Critical Rev. Plant Sci.* **21,** 167–204.

Gallardo, F., Fu, J., Canton, F. R., Garcia-Gutierrez, A., Canovas, F. M., and Kirby, E. G. (1999). Expression of a conifer glutamine synthetase gene in transgenic poplar. *Planta* **210,** 19–26.

Giovanelli, J., Mudd, S. H., and Datko, A. H. (1985). Quantitative analysis of pathways of methionine metabolism and their regulation in *Lemna. Plant Physiol.* **78,** 555–560.

Goto, D. B., Ogi, M., Kijima, F., Kumagai, T., Van Werven, F., Onouchi, H., and Naito, S. (2002). A single-nucleotide mutation in a gene encoding S-adenosylmethionine synthetase is associated with methionine over-accumulation phenotype in *Arabidopsis thaliana. Genes Genet. Syst.* **77,** 89–95.

Guenoune, D., Amir, R., Ben-Dor, B., Wolf, S., and Galili, S. (1999). A soybean vegetative storage protein accumulates to high levels in various organs of transgenic tobacco plants. *Plant Sci.* **145,** 93–98.

Guenoune, D., Amir, R., Badani, H., Wolf, S., and Galili, S. (2002a). Combined expression of S-VSPa in two different organelles enhances its accumulation and total lysine production in leaves of transgenic tobacco plants. *J. Exp. Bot.* **53,** 1867–1870.

Guenoune, D., Landau, S., Amir, R., Badani, H., Devash, L., Wolf, S., and Galili, S. (2002b). Resistance of soybean vegetative storage proteins (S-VSPs) to proteolysis by rumen microorganisms. *J. Agric. Food Chem.* **50,** 2256–2260.

Guenoune, D., Amir, R., Badani, H., Wolf, S., and Galili, S. (2003). Coexpression of the soybean vegetative storage protein b subunit (S-VSPb) either with the bacterial feedback-insensitive dihydrodipicolinate synthase or with S-VSPa stabilizes the S-VSPb transgene protein and enhances lysine production in transgenic tobacco plants. *Transgenic Res.* **12,** 123–126.

Habash, D. Z., Massiah, A. J., Rong, H. L., Wallsgrove, R. M., and Leigh, R. A. (2001). The role of cytosolic glutamine synthetase in wheat. *Ann. Appl. Biol.* **138,** 83–89.

Hacham, Y., Avraham, T., and Amir, R. (2002). The N-terminal region of *Arabidopsis* cystathionine γ-synthase plays an important role in methionine metabolism. *Plant Physiol.* **128,** 454–462.

Hagan, N. D., Upadhyaya, N., Tabe, L. M., and Higgins, T. J. V. (2003). The redistribution of protein sulfur in transgenic rice expressing a gene for a foreign sulfur-rich protein. *Plant J.* **34,** 1–11.

Hirel, B., and Lea, P. J. (2001). Ammonia assimilation. *In* "Plant Nitrogen" (P. J. Lea and J.-F. Morot-Gaudry, eds.), pp. 79–99. Springer-Verlag, Berlin.

Hirel, B., Marsolier, M. C., Hoarau, A., Hoarau, J., Brangeon, J., Schafer, R., and Verma, D. P. S. (1992). Forcing expression of a soybean root glutamine synthetase gene in tobacco leaves induces a native gene encoding cytosolic enzyme. *Plant Mol. Biol.* **20,** 207–218.

Hunter, B. G., Beatty, M. K., Singletary, G. W., Hamaker, B. R., Dilkes, B. P., Larkins, B. A., and Jung, R. (2002). Maize opaque endosperm mutations create extensive changes in patterns of gene expression. *Plant Cell* **14,** 2591–2612.

Inaba, K., Fujiwara, T., Hayashi, H., Chino, M., Komeda, Y., and Naito, S. (1994). Isolation of an *Arabidopsis thaliana* mutant, *mto1*, that overaccumulates soluble methionine. Temporal and spatial patterns of soluble methionine accumulation. *Plant Physiol.* **104,** 881–887.

Ireland, R. J., and Lea, P. J. (1999). The enzymes of glutamine, glutamate, asparagine, and aspartate metabolism. *In* "Plant Amino Acids: Biochemistry and Biotechnology" (B. K. Singh, ed.), pp. 49–109. Marcel Dekker, New York.

Jacobs, M., Negrutiu, I., Dirks, R., and Cammaerts, D. (1987). Selection programs for isolation and analysis of mutants in plant cell cultures. *In* "Plant Biology" (C. E. Green, D. A. Somers, W. P. Hackett, and D. D. Biesboer, eds.), pp. 243–264. Alan R. Liss, New York.

Jacobs, M., Vauterin, M., De Waele, E., and Craciun, A. (2001). Manipulating plant biochemical pathways for improved nutritional quality. *In* "Plant Biotechnology and Transgenic Plants" (Oksman-Caldentey and Barz, eds.), pp. 233–253. Marcel Dekker, New York.

Jiang, Q., and Gresshoff, P. M. (1997). Classical and molecular genetics of the model legume *Lotus japonicus. Mol. Plant Micr. Interac.* **10,** 59–68.

Jung, R., and Falco, S. C. (2000). Transgenic corn with an improved amino acid composition. *In* "Eighth International Symposium on Plant Seeds." Gatersleben, Germany.

Karchi, H., Shaul, O., and Galili, G. (1993). Seed-specific expression of a bacterial desensitized aspartate kinase increases the production of seed threonine and methionine in transgenic tobacco. *Plant J.* **3,** 721–727.

Karchi, H., Shaul, O., and Galili, G. (1994). Lysine synthesis and catabolism are coordinately regulated during tobacco seed development. *Proc. Natl. Acad. Sci. USA* **91,** 2577–2581.

Karlsen, R. L., Pedersen, O. O., Schousboe, A., and Langeland, A. (1982). Toxic effects of DL α-aminoadipic acid on muller cells from rats *in vivo* and cultured cerebral astrocytes. *Exp. Eye. Res.* **35,** 305–311.

Kemper, E. L., Neto, G. C., Papes, F., Moraes, K. C. M., Leite, A., and Arruda, P. (1999). The role of opaque2 in the control of lysine-degrading activities in developing maize endosperm. *Plant Cell* **11,** 1981–1993.

Kim, J., and Leustek, T. (2000). Repression of cystathionine γ-synthase in *Arabidopsis thaliana* produces partial methionine auxotrophy and developmental abnormalities. *Plant Sci.* **151,** 9–18.

Kim, J., Lee, M., Chalam, R., Martin, M. N., Leustek, T., and Boerjan, W. (2002). Constitutive over-expression of cystathionine γ-synthase in *Arabidopsis* leads to accumulation of soluble methionine and S-methylmethionine. *Plant Physiol.* **128,** 95–107.

Kishor, P. B. K., Hong, Z., Miao, G.-H., Hu, C.-A. A., and Verma, D. P. S. (1995). Overexpression of D^1-pyrroline-5-carboxylate synthetase increases proline production and confers osmotolerance in transgenic plants. *Plant Physiol.* **108,** 1387–1394.

Kocsis, M. G., Ranocha, P., Gage, D. A., Simon, E. S., Rhodes, D., Peel, G. J., Mellema, S., Saito, K., Awazuhara, M., Li, C., Meeley, R. B., Tarczynski, M. C., *et al.* (2003). Insertional inactivation of the methionine S-methyltransferase gene eliminates the S-methylmethionine cycle and increases the methylation ratio. *Plant Physiol.* **131,** 1808–1815.

Kreft, O., Hoefgen, R., and Hesse, H. (2003). Functional analysis of cystathionine γ-synthase in genetically engineered potato plants. *Plant Physiol.* **131,** 1843–1854.

Lai, J., and Messing, J. (2002). Increasing maize seed methionine by mRNA stability. *Plant J.* **30,** 395–402.

Lam, H.-M., Coschigano, K., Schultz, C., Melo-Oliveira, R., Tjagen, G., Oliveira, I., Ngai, N., Hsieh, M.-H., and Coruzzi, G. (1995). Use of *Arabidopsis* mutants and genes to study amide amino acid biosynthesis,. *Plant Cell* **7,** 887–898.

Lam, H.-M., Coschigano, K. T., Oliveira, I. C., Melo-Oliveira, R., and Coruzzi, G. M. (1996). The molecular-genetics of nitrogen assimilation into amino acids in higher plants. *Annu. Rev. Plant Physiol. Plant Mol. Biol.* **47,** 569–593.

Lam, H.-M., Hsieh, M.-H., and Coruzzi, G. (1998). Reciprocal regulation of distinct asparagine synthetase genes by light and metabolites in *Arabidopsis thaliana. Plant J.* **16,** 345–353.

Lancien, M., Martin, M., Hsieh, M.-H., Leustek, T., Goodman, H., and Coruzzi, G. M. (2002). *Arabidopsis glt1*-T mutant defines a role for NADH-GOGAT in the non-photorespiratory ammonium assimilatory pathway. *Plant J.* **29,** 347–358.
Lawlor, D. W., Lemaire, G., and Gastal, F. (2001). Nitrogen, plant growth and crop yield. *In* ''Plant Nitrogen'' (P. J. Lea and J.-F. Morot-Gaudry, eds.), pp. 343–367. Springer-Verlag, Berlin.
Lea, P. J., and Ireland, R. J. (1999). Nitrogen metabolism in higher plants. *In* ''Plant Amino Acids: Biochemistry and Biotechnology'' (B. K. Singh, ed.), pp. 1–47. Marcel Dekker, New York.
Lee, J.-M., Williams, M. E., Tingey, S. V., and Rafalski, J. A. (2002). DNA array profiling of gene expression changes during maize embryo development. *Funct. Integr. Genomics* **2,** 13–27.
Leng, R. A. (1990). Factors affecting the utilization of ''poor quality'' forages by ruminants particularly under tropical conditions. *Nutr. Res. Rev.* **3,** 277–303.
Li, J., and Last, R. L. (1996). The *Arabidopsis thaliana trp5* mutant has a feedback-resistant anthranilate synthase and elevated soluble tryptophan. *Plant Physiol.* **110,** 51–59.
Limami, A., Phillipson, B., Ameziane, R., Pernollet, N., Jiang, Q., Roy, R., Deleens, E., Chaumont-Bonnet, M., Gresshoff, P. M., and Hirel, B. (1999). Does root glutamine synthetase control plant biomass production in *Lotus japonicus* L.? *Planta* **209,** 495–502.
Limami, A. M., and De Vienne, D. (2001). Natural genetic variability of nitrogen metabolism. *In* ''Plant Nitrogen'' (P. J. Lea and J.-F. Morot-Gaudry, eds.), pp. 369–378. Springer-Verlag, Berlin.
Masclaux, C., Valadier, M.-H., Brugiere, N., Morot-Gaudry, J.-F., and Hirel, B. (2000). Characterization of the sink/source transition in tobacco (*Nicotiana tabacum* L.) shoots in relation to nitrogen management and leaf senescence. *Planta* **211,** 510–518.
Mazur, B., Krebbers, E., and Tingey, S. (1999). Gene discovery and product development for grain quality traits. *Science* **285,** 372–375.
Mcnabb, W. C., Spencer, D., Higgins, T. J., and Barry, T. N. (1994). *In-vitro* rates of rumen proteolysis of ribulose-1,5-bisphosphate carboxylase (Rubisco) from lucerne leaves, and of ovalbumin, vicilin and sunflower albumin 8 storage proteins. *J. Sci. Food Agric.* **64,** 53–61.
Meers, J. L., Tempest, D. W., and Brown, C. M. (1970). Glutamine/(amide): 2-oxoglutarate amino transferase oxido-reductase (NADP), an enzyme involved in the synthesis of glutamate by some bacteria. *J. Gen. Microbiol.* **64,** 187–194.
Melo-Oliveira, R., Oliveira, I. C., and Coruzzi, G. M. (1996). *Arabidopsis* mutant analysis and gene regulation define a nonredundant role for glutamate dehydrogenase in nitrogen assimilation. *Proc. Natl. Acad. Sci. USA* **93,** 4718–4723.
Miflin, B. J., and Habash, D. Z. (2002). The role of glutamine synthetase and glutamate dehydrogenase in nitrogen assimilation and possibilities for improvement in the nitrogen utilization of crops. *J. Exp. Bot.* **53,** 979–987.
Miyazaki, J. H., and Yang, S. F. (1987). The methionine salvage pathway in relation to ethylene and polyamine biosynthesis. *Physiol. Plantarum* **69,** 366–370.
Molvig, L., Tabe, L. M., Eggum, B. O., Moore, A. E., Craig, S., Spencer, D., and Higgins, T. J. V. (1997). Enhanced methionine levels and increased nutritive value of seeds of transgenic lupins (*Lupinus angustifolius* L.) expressing a sunflower seed albumin gene. *Proc. Natl. Acad. Sci. USA* **94,** 8393–8398.
Moorhead, G., Douglas, P., Cotelle, V., Harthill, J., Morrice, N., Meek, S., Deiting, U., Stitt, M., Scarabel, M., Aitken, A., and Mackintosh, C. (1999). Phosphorylation-dependent interactions between enzymes of plant metabolism and 14–3–3 proteins. *Plant J.* **18,** 1–12.
Morot-Gaudry, J.-F., Dominique, J., and Lea, P. J. (2001). Amino acid metabolism. *In* ''Plant Nitrogen'' (P. J. Lea and J.-F. Morot-Gaudry, eds.), pp. 167–211. Springer-Verlag, Berlin.
Nanjo, T., Kobayashi, M., Yoshiba, Y., Kakubari, Y., Yamaguchi-Shinozaki, K., and Shinozaki, K. (1999). Antisense suppression of proline degradation improves tolerance to freezing and salinity in *Arabidopsis thaliana*. *FEBS Lett.* **461,** 205–210.
National Research Council (2001). ''Nutrient Requirements of Dairy Cattle 7th,'' p. 381.
Noctor, G., Novitskaya, L., Lea, P. J., and Foyer, C. H. (2002). Co-ordination of leaf minor amino acid contents in crop species: Significance and interpretation. *J. Exp. Bot.* **53,** 939–945.
Oliveira, I. C., Brenner, E., Chiu, J., Hsieh, M.-H., Kouranov, A., Lam, H.-M., Shin, M. J., and Coruzzi, G. (2001). Metabolic and light regulation of metabolism in plants: Lessons from study of a single biochemical pathway. *Brazilian J. Med. Biol. Res.* **34,** 567–575.

Oliveira, I. C., Brears, T., Knight, T. J., Clark, A., and Coruzzi, G. M. (2002). Overexpression of cytosolic glutamine synthetase. Relation to nitrogen, light, and photorespiration. *Plant Physiol.* **129,** 1170–1180.

Onouchi, H., Nagami, Y., Haraguchi, Y., Nahkamoto, M., Nishimura, Y., Sakurai, R., Nagao, N., Kawasaki, D., Kadokura, Y., and Naito, S. (2005). Nascent peptide-mediated translation elongation arrest coupled with mRNA degradation in the CGS1 gene of *Arabidopsis*. *Genes Dev.* **19,** 1799–1810.

Ortega, J. L., Temple, S. J., and Sengupta-Gopalan, C. (2001). Constitutive overexpression of cytosolic glutamine synthetase (GS_1) gene in transgenic alfalfa demonstrates that GS_1 may be regulated at the level of RNA stability and protein turnover. *Plant Physiol.* **126,** 109–121.

Ranocha, P., Mcneil, S. D., Ziemak, M. J., Li, C., Tarczynski, M. C., and Hanson, A. D. (2001). The S-methylmethionine cycle in angiosperms: Ubiquity, antiquity and activity. *Plant J.* **25,** 575–584.

Rao, S. S., Kochhar, S., and Kochhar, V. K. (1999). Analysis of photocontrol of aspartate kinase in barley (*Hordeum vulgare* L.) seedlings. *Biochem. Mol. Biol. Int.* **47,** 347–360.

Ravanel, S., Gakiere, B., Job, D., and Douce, R. (1998a). The specific features of methionine biosynthesis and metabolism in plants. *Proc. Natl. Acad. Sci. USA* **95,** 7805–7812.

Ravanel, S., Gakiere, B., Job, D., and Douce, R. (1998b). Cystathionine γ-synthase from *Arabidopsis thaliana*: Purification and biochemical characterization of the recombinant enzyme overexpressed in *Escherichia coli*. *Biochem. J.* **331,** 639–648.

Reichenbach, A., and Wohlrab, F. (1985). Effects of α-aminoadipic acid on the glutamate-isolated P III of the rabbit electroretinogram. *Doc. Ophthalmol.* **59,** 359–364.

Ruuska, S. A., Girke, T., Benning, C., and Ohlrogge, J. B. (2002). Contrapuntal networks of gene expression during *Arabidopsis* seed filling. *Plant Cell* **14,** 1191–1206.

Sarrobert, C., Thibaud, M.-C., Contard-David, P., Gineste, S., Bechtold, N., Robaglia, C., and Nussaume, L. (2000). Identification of an *Arabidopsis thaliana* mutant accumulating threonine resulting from mutation in a new dihydrodipicolinate synthase gene. *Plant J.* **24,** 357–367.

Schoenbeck, M. A., Temple, S. J., Trepp, G. B., Blumenthal, J. M., Samac, D. A., Gantt, J. S., Hernandez, G., and Vance, C. P. (2000). Decreased NADH glutamate synthase activity in nodules and flowers of alfalfa (*Medicago sativa* L.) transformed with an antisense glutamate synthase transgene. *J. Exp. Bot.* **51,** 29–39.

Shaul, O., and Galili, G. (1992a). Increased lysine synthesis in tobacco plants express high levels of bacterial dihydrodipicolinate synthase in their chloroplasts. *Plant J.* **2,** 203–209.

Shaul, O., and Galili, G. (1992b). Threonine overproduction in transgenic tobacco plants expressing a mutant desensitized aspartate kinase of *Escherichia coli*. *Plant Physiol.* **100,** 1157–1163.

Shaul, O., and Galili, G. (1993). Concerted regulation of lysine and threonine synthesis in tobacco plants expressing bacterial feedback-insensitive aspartate kinase and dihydrodipicolinate synthase. *Plant Mol. Biol.* **23,** 759–768.

Shotwell, M. A., and Larkins, B. A. (1989). The biochemistry and molecular biology of seed storage proteins. *In* "The Biochemistry of Plants" (A. Marcus, ed.), pp. 297–345. Academic Press, San Diego.

Singh, B. K. (1999). Biosynthesis of valine, leucine, and isoleucine. *In* "Plant Amino Acids: Biochemistry and Biotechnology" (B. K. Singh, ed.), pp. 227–247. Marcel Dekker, New York.

Somerville, C. R., and Ogren, W. L. (1980). Inhibition of photosynthesis in *Arabidopsis* mutants lacking leaf glutamate synthase activity. *Nature* **286,** 257–259.

Staswick, P. E. (1994). Storage proteins of vegetative plant tissues. *Annu. Rev. Plant Physiol. Plant Mol. Biol.* **45,** 303–322.

Stephanopoulos, G. (1999). Metabolic fluxes and metabolic engineering. *Metab. Eng.* **1,** 1–11.

Stitt, M., Muller, C., Matt, P., Gibon, Y., Carillo, P., Morcuende, R., Scheible, W.-R., and Krapp, A. (2002). Steps towards an integrated view of nitrogen metabolism. *J. Exp. Bot.* **53,** 959–970.

Suzuki, A., Rioual, S., Lemarchand, S., Godfroy, N., Roux, Y., Boutin, J.-P., and Rothstein, S. (2001). Regulation by light and metabolites of ferredoxin-dependent glutamate synthase in maize. *Physiologia Plantarum* **112,** 524–530.

Syed Rasheeduddin, A., and Mcdonald, C. E. (1974). Amino acid composition, protein fractions and baking quality of triticale. *In* "Triticale: First Man-Made Cereal" (C. C. Tsen, ed.), pp. 137–149. American Association of Cereal Chemists, Minnesota.

Tabe, L. M., and Droux, M. (2002). Limits to sulfur accumulation in transgenic lupin seeds expressing a foreign sulfur-rich protein. *Plant Physiol.* **128,** 1137–1148.

Tabe, L., Hagan, N., and Higgins, T. J. V. (2002). Plasticity of seed protein composition in response to nitrogen and sulfur availability. *Curr. Opin. Plant Biol.* **5,** 212–217.

Tarczynski, M. C., Bo, S., Changjiang, L., Leustek, T., Falco, C., and Allen, W. (2001). Control and manipulation of sulfur amino acid metabolism in plants. *In* "Plant Foods for Human Health: Manipulating Plant Metabolism to Enhance Nutritional Quality, Keystone Sympos" pp. 6.4–11.4. Breckenridge, Colorado.

Temple, S. J., Knight, T. J., Unkefer, P. J., and Sengupta-Gopalan, C. (1993). Modulation of glutamine synthetase gene expression in tobacco by the introduction of an alfalfa glutamine synthetase gene in sense and antisense orientation: Molecular and biochemical analysis. *Mol. Gen. Genet.* **236,** 315–325.

Ter Steege, M. W., Stulen, I., and Mary, B. (2001). Nitrogen and the environment. *In* "Plant Nitrogen" (P. J. Lea and J.-F. Morot-Gaudry, eds.), pp. 379–397. Springer-Verlag, Berlin.

Thum, K. E., Shasha, D. E., Lejay, L. V., and Coruzzi, G. M. (2003). Light and carbon signaling pathways controlling genes in nitrogen assimilation: Modeling circuits of interaction. *Plant Physiol.* **132,** 440–452.

Tozawa, Y., Hasegawa, H., Terakawa, T., and Wakasa, K. (2001). Characterization of rice anthranilate synthase α-subunit genes *OASA1* and *OASA2*. Tryptophan accumulation in transgenic rice expressing a feedback-insensitive mutant of OASA1. *Plant Physiol.* **126,** 1493–1506.

Vauterin, M., Frankard, V., and Jacobs, M. (1999). The *Arabidopsis thaliana dhdps* gene encoding dihydrodipicolinate synthase, key enzyme of lysine biosynthesis, is expressed in a cell-specific manner. *Plant Mol. Biol.* **39,** 695–708.

Vincent, R., Fraisier, V., Chaillou, S., Limami, M. A., Deleens, E., Phillipson, B., Douat, C., Boutin, J.-P., and Hirel, B. (1997). Overexpression of a soybean gene encoding cytosolic glutamine synthetase in shoots of transgenic *Lotus corniculatus* L. plants triggers changes in ammonium assimilation and plant development. *Planta* **201,** 424–433.

Wang, F., and Tian, B. (2001). Neurospora NADP-glutamate dehydrogenases and its expression in *E. coli* and transgenic plants. *Chinese Sci. Bull.* **46,** 1029–1032.

Welinder, E., Textorius, O., and Nilsson, S. E. (1982). Effects of intravitreally injected DL-α-aminoadipic acid on the c-wave of the D.C.-recorded electroretinogram in albino rabbits. *Invest. Ophthalmol. Vis. Sci.* **23,** 240–245.

White, C. L., Tabe, L. M., Dove, H., Hamblin, J., Young, P., Phillips, N., Taylor, R., Gulati, S., Ashes, J., and Higgins, T. J. V. (2000). Increased efficiency of wool growth and live weight gain in Merino sheep fed transgenic lupin seed containing sunflower albumin. *J. Sci. Food Agric.* **81,** 147–154.

Zeh, M., Casazza, A. P., Kreft, O., Roessner, U., Bieberich, K., Willmitzer, L., Hoefgen, R., and Hesse, H. (2001). Antisense inhibition of threonine synthase leads to high methionine content in transgenic potato plants. *Plant Physiol.* **127,** 792–802.

Zhu-Shimoni, X. J., and Galili, G. (1998). Expression of an *Arabidopsis* aspartate kinase/homoserine dehydrogenase gene is metabolically regulated by photosynthesis-related signals but not by nitrogenous compounds. *Plant Physiol.* **116,** 1023–1028.

Zhu-Shimoni, X. J., Lev-Yadun, S., Matthews, B., and Galili, G. (1997). Expression of an aspartate kinase homoserine dehydrogenase gene is subject to specific spatial and temporal regulation in vegetative tissues, flowers and developing seeds. *Plant Physiol.* **113,** 695–706.

Zhu, X., and Galil, G. (2003). Increased lysine synthesis coupled with a knockout of its catabolism synergistically boosts lysine content and also transregulates the metabolism of other amino acids in *Arabidopsis* seeds. *Plant Cell* **15,** 845–853.

Zhu, X., Tang, G., Granier, F., Bouchez, D., and Galili, G. (2001). A T-DNA insertion knockout of the bifunctional lysine-ketoglutarate reductase/saccharopine dehydrogenase gene elevates lysine levels in *Arabidopsis* seeds. *Plant Physiol.* **126,** 1539–1545.

Zhang, X.-H., Brotherton, J. E., Widholm, J. M., and Portis, A. R., Jr. (2001). Targeting a nuclear anthranilate synthase α-subunit gene to the tobacco plastid genome results in enhanced tryptophan biosynthesis. Return of a gene to its pre-endosymbiotic origin. *Plant Physiol.* **127,** 131–141.

CHAPTER 4

Engineering Photosynthetic Pathways

Akiho Yokota* and **Shigeru Shigeoka**†

Contents

Abstract

Improvements of metabolic reactions in photosynthetic pathways, and prospects for successfully altering photosynthetic carbon reduction (PCR) cycle in particular, have become possible through technologies developed during the last decade. This chapter outlines recent strategies and achievements in engineering enzymes of primary CO_2 fixations. We emphasize antisense approaches, attempts at engineering the chloroplast genome, and the transfer into C_3 species of reactions and enzymes typical for C_4 species or cyanobacteria. In addition, we point to the importance of studying the evolutionary diversity of enzymes in primary metabolism. The resulting transgenic lines then provide material suitable for precise flux control analysis. Discussed are enzymes of the photosynthetic reaction (PCR) cycle, ribulose 1,5-bisphosphate carboxylase/oxygenase (RuBisCO), fructose

* Graduate School of Biological Sciences, Nara Institute of Science and Technology (NAIST), 8916-5 Takayama, Ikoma, Nara 630-0101, Japan
† Department of Advanced Bioscience, Faculty of Agriculture, Kinki University, 3327-204 Nakamachi, Nara 631-8505, Japan

Advances in Plant Biochemistry and Molecular Biology, Volume 1
ISSN 1755-0408, DOI: 10.1016/S1755-0408(07)01004-1

1,6-bisphosphatase (FBPase), sedoheptulose 1,7-bisphosphatase (SBPase), aldolase, and transketolase that exert control in a rate-limiting fashion. The PCR cycle, initiated by reactions that are catalyzed by RuBisCO, represents a major energy-consuming process in photosynthesis, justifying the large amount of research effort directed toward engineering this important enzyme. We also discuss progress in fine-tuning the two competing reactions catalyzed by RuBisCO, and in defining the roles and importance of PCR components, such as FBPase and SBPase. Lasting success is still elusive in improving crops by increasing primary productivity, but new tools have provided promising new avenues.

Key Words: RuBisCO, Photosynthetic carbon reduction cycle, Flux control analysis, Photorespiratory oxidation cycle, Relative specificity, RuBisCO-like protein, Enzyme engineering, Metabolic engineering, Chloroplast transformation, C_4-ization, Phosphoenolpyruvate carboxylase, Pyruvate Pi dikinase, $NADP^+$-malic enzyme.

1. INTRODUCTION

Grain availability is determined on a global level by a balance between grain production and use (Tsujii, 2000). The potential for grain production is a result of productivity of grain crops and agricultural area. Over the last century (Mann, 1999), conventional plant breeding has developed crop productivity to a level that closely approaches the maximum potential, while the global arable area reached its ceiling by the mid-1970s and is now decreasing slowly due to increasing urbanization. It is feared that the negative trend in grain production will be exacerbated by three tightly correlated factors, namely water shortage, deterioration of soils, and global warming (Vörösmarty *et al.*, 2000).

Such negative factors will severely affect photosynthesis, the primary step in grain production. Plant leaves are organs that are optimized for photosynthetic performance, this efficiency being maximal when sufficient water and nitrogen are available for the plants at moderate temperatures (Boyer, 1982). Thus, we have entered a time when we need to develop technology to maintain or increase the present productivity of crop plants to overcome grain shortage within the near future to satisfy increasing demands (Mann, 1999).

This chapter deals with challenges and initiatives for improving metabolic reactions in photosynthetic pathways, including the photosynthetic carbon reduction (PCR) cycle and other reactions in primary metabolism. The basic reaction mechanism of ribulose 1,5-bisphosphate carboxylase/oxygenase (RuBisCO) and regulation of the PCR cycle are not included in this chapter as they have been addressed in several scholarly reviews (Andersson and Taylor, 2003; Cleland *et al.*, 1998; Fridyand and Scheibe, 2000; Hartman and Harpel, 1994; Martin *et al.*, 2000; Roy and Andrews, 2000).

2. IDENTIFICATION OF LIMITING STEPS IN THE PCR CYCLE

2.1. Analysis of limiting steps in photosynthesis

The primary reactions of photosynthesis can be roughly divided into four parts: formation of NADPH and ATP, incorporation of CO_2 into ribulose 1,5-bisphosphate (RuBP) by RuBisCO to produce 3-phosphoglycerate (PGA), regeneration of RuBP in the PCR cycle (Fig. 4.1), and sucrose synthesis using triose phosphate exported into the cytosol and counterchanged with phosphate released by this synthesis. The accepted photosynthesis model (Farquhar *et al.*, 1981) is based on the prediction that the rate of synthesis of NADPH and ATP is calculated from the flux of electrons in the photosynthetic electron transport chain, with three protons transported for every ATP formed. *In situ* RuBisCO activity is calculated using the concentration of the activated catalytic site and kinetic parameters of RuBisCO (Farquhar, 1979). The steady-state concentration of RuBP is balanced both by the rate of regeneration and the utilization by RuBisCO for CO_2 fixation. Important information has been provided by simultaneous measurements of rates of gas exchange and steady-state concentrations of metabolites in the PCR cycle using part of a single attached leaf under a range of conditions. The photosynthetic rate of an attached leaf has been found to match the rate calculated with RuBisCO kinetics at CO_2 concentrations in the intercellular space below 40 Pa and at saturating light intensities, while the photosynthetic rate calculated by taking electron flux into consideration significantly exceeds the photosynthetic rate (Badger *et al.*, 1984). The intraplastidic concentration of RuBP reaches levels that are several fold higher than the concentration of the RuBisCO active site under these conditions (Badger *et al.*, 1984; Geiger and Servaites, 1994). This indicates that photosynthesis is limited by either RuBisCO or the CO_2-fixation pathway. As the intercellular CO_2 concentration increases, photosynthesis enters an RuBP-limited phase and transport of inorganic phosphate back into chloroplasts becomes rate limiting (Sage, 1990; Sage *et al.*, 1989). In contrast, the capacity for NADPH and ATP formation limits photosynthesis at nonsaturating light intensities (Farquhar *et al.*, 1981). Moreover, photosynthesis in source organs may occasionally become limited by the capacities of sink organs to accumulate photosynthates (Paul and Foyer, 2001).

2.2. Flux control analysis

Metabolic flux in a pathway is the consequence of the reactions of the enzymes involved in the pathway under a given condition, including changes in the concentration of metabolites. Generally, the contribution of any individual enzyme to the whole metabolic flux varies considerably, that is, while flux control is distributed over the entire pathway, enzymes in the pathway carry different weight. Often, the flux-limiting step is located at the first metabolic step of either a pathway or branch point and at those steps with a large free energy change that are virtually irreversible. However, the contribution to metabolic flux of an enzyme catalyzing a reversible reaction may also be high, when the catalytic

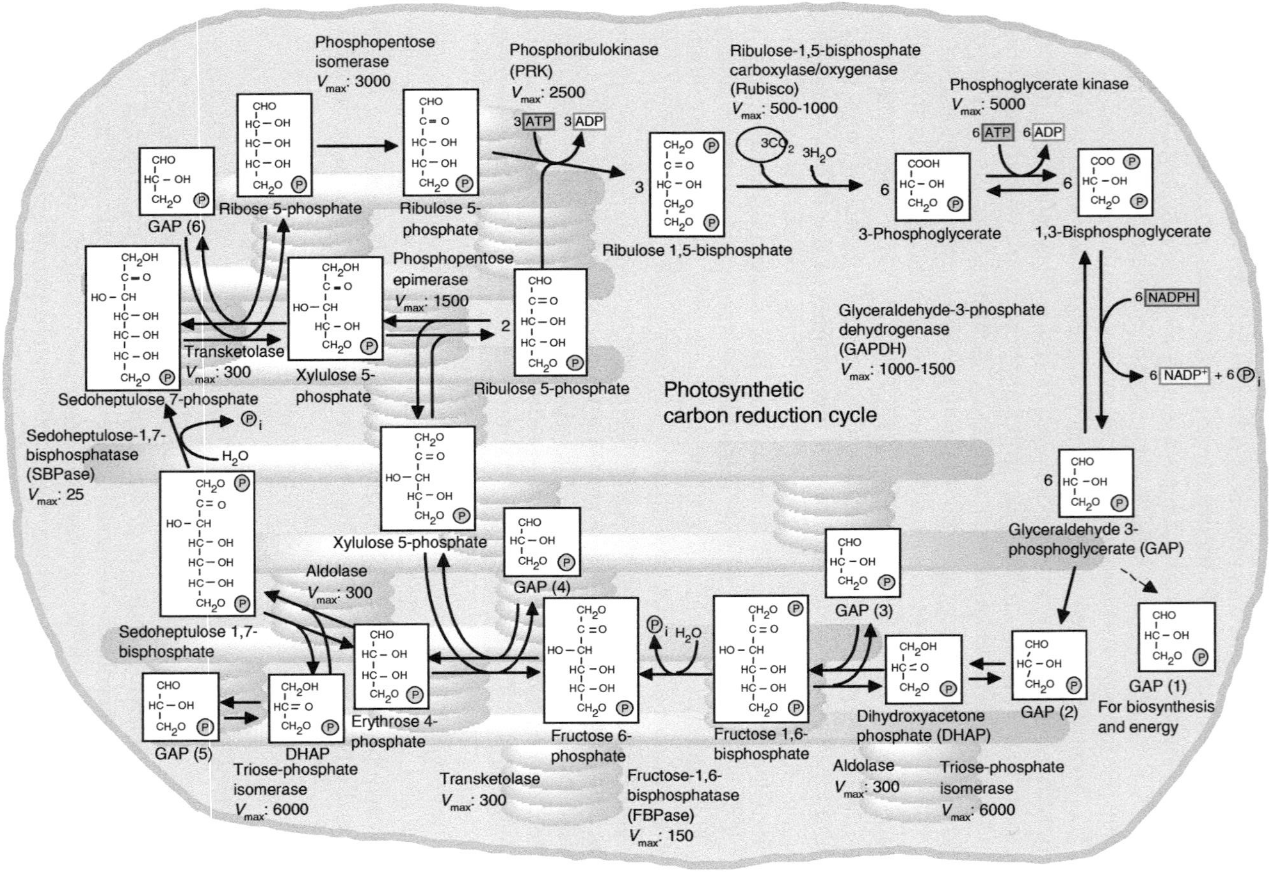

FIGURE 4.1 Photosynthetic carbon reduction cycle. V_{max} of each enzyme is given in micromoles per milligram chlorophyll per hour (Robinson and Walker, 1981). (See Page 3 in Color Section.)

efficiency of the enzyme (k_{cat}) and/or expression or steady-state amount (k_m) of an enzyme are low.

Antisense technology has provided an opportunity for precise analysis of flux control in metabolism (Stitt and Sonnewald, 1995). Metabolic flux analysis is a tool whereby metabolic flux in a system is quantified. The flux control coefficient ($C^J_E = \Delta J/\Delta E$) is the mathematical expression of the effect of a change in the relative amount of enzyme ΔE (generally corresponding to the enzyme activity) on the metabolic flux (J) (Kacser, 1987; Stephanopoulos *et al.*, 1998). An enzyme with C^J_E closer to zero contributes little to the flux and an enzyme with C^J_E closer to 1 contributes more significantly.

The PCR cycle includes 13 reactions catalyzed by 11 enzymes (Robinson and Walker, 1981). The effect of changes in the amount of these enzymes has been analyzed by downregulating the genes coding for the enzymes. Photosynthesis was not affected by decreasing the amount of RuBisCO at low light intensities over a large range of reduction but eventually its amount became limiting (Krapp *et al.*, 1994; Quick *et al.*, 1991). According to flux criteria, the C^J_E value of RuBisCO was near unity at saturating light intensities in tobacco and rice transgenic plants (Makino *et al.*, 1997; Masle *et al.*, 1993). Decreasing the enzyme level of glyceraldehyde 3-phosphate dehydrogenase in transgenic tobacco then caused the concentration of RuBP to decrease, but photosynthetic CO_2 fixation was not affected until the RuBP level had decreased to less than half the wild-type level (Price *et al.*, 1995). A reduction in fructose 1,6-bisphosphatase (FBPase) amount to below 36% of wild type lowered the rate of photosynthesis (Koßmann *et al.*, 1994). The C^J_E value of sedoheptulose 1,7-bisphosphatase (SBPase) was almost one under a wide range of conditions (Harrison *et al.*, 1998). In contrast, although phosphoribulokinase catalyzes a virtually irreversible reaction in the PCR cycle, its C^J_E was near zero until the enzyme level in transgenic tobacco plants was reduced to 20% of wild type (Paul *et al.*, 1995). Reduction in aldolase levels caused a severe decrease in photosynthesis, with the activities of FBPase and SBPase showing a proportional reduction in transgenic potato plants (Haake *et al.*, 1998, 1999). The C^J_E value of transketolase was also near unity (Henkes *et al.*, 2001). Aldolase and transketolase catalyze reversible reactions in the PCR cycle, but their activities in chloroplasts are no greater than the demand exerted by photosynthesis. Those enzymes functioning with rate-limiting activities in the PCR cycle could become targets for the genetic manipulation of crops with the aim of improving the photosynthetic performance of essential reactions in primary carbon fixation pathways.

3. ENGINEERING CO_2-FIXATION ENZYMES

3.1. RuBisCO

RuBisCO is the rate-limiting enzyme in plant photosynthesis. Under the present model for photosynthesis, it should be possible to increase CO_2 fixation in C_3 plants by about 20%, before entering RuBP- and Pi-limited phases (Sage, 1990; Sage *et al.*, 1989). Since the PCR cycle is the major consumer of energy formed at the thylakoids (Heldt, 1997), alterations of the enzyme should guarantee that the PCR cycle would siphon off and productively utilize more energy with an improved enzyme. Several directions about how to accomplish such improvement have been discussed (Andrews and Whitney, 2003; Parry *et al.*, 2003). However, another strategy would be to engineer a RuBisCO enzyme that continued to fix CO_2 under drought conditions when stomata aperture is reduced. First, we need to know which partial reaction of the enzyme constitutes the limiting step and which residues might determine the enzymatic properties (Mauser *et al.*, 2001). Second, based on the detection of naturally occurring RuBisCO enzymes that are superior to the plant enzyme, work may be directed to replace resident *rbcL* (and *rbcS*) gene in plastid and nuclear DNA with the genes coding for the superior enzyme (Andrews and Whitney, 2003; Parry *et al.*, 2003). Integration of the information from research with these superior enzymes suggests the possibility to engineer a higher plant *rbcL* gene that incorporates sequences responsible for improved RuBisCO performance. However, incorporating such engineered chimeric genes into chloroplast DNA faces challenges and obstacles that need to be addressed.

3.1.1. Enzymatic properties of RuBisCO

The turnover rate of catalysis in CO_2 fixation by plant RuBisCO is as low as 3.3 s^{-1} per site (Woodrow and Berry, 1988). The rate is less than one-thousandth of the rate of triose phosphate isomerase, the reaction of which proceeds in a diffusion-limited manner (Morell *et al.*, 1992). All RuBisCOs analyzed to date catalyze an oxygenase reaction in addition to the carboxylase reaction (Andrews and Lorimer, 1978). The K_m values of plant RuBisCO for CO_2 and O_2 are close to the concentrations of these gases in water equilibrated at normal atmospheric pressure (Woodrow and Berry, 1988). These gases compete with each other for the accepter molecule, the endiolate of RuBP (Andrews and Whitney, 2003). The relative frequency of the carboxylation and oxygenation reactions can be expressed as S_{rel}, that is, the ratio of the specificity of the carboxylase reaction to that of the oxygenase reaction (Laing *et al.*, 1974). The ratio of the velocities of both reactions can be expressed as $v_c/v_o = S_{rel} \cdot [CO_2]/[O_2]$, where v_c and v_o are the velocities of the carboxylase and oxygenase reactions, respectively, and S_{rel} is (V_{max} of carboxylase reaction/K_m for CO_2)/(V_{max} of oxygenase reaction/K_m for O_2). Since the exact concentration of CO_2 in the stroma has been estimated as 5–7 μM (Evans and Loreto, 2000), and the activation of RuBisCO in chloroplasts is not complete, only a quarter of the total RuBisCO molecules in the stroma can participate in CO_2 fixation during active photosynthesis (McCurry *et al.*, 1981). Thus, either conditions in the stroma are suboptimal with respect to the potential of RuBisCO's performance, or the intrinsic enzymatic properties of RuBisCO are inadequate with respect to stromal gas concentrations. Evolutionarily, plants have counteracted these disadvantages by investing an inordinate amount of nitrogen in RuBisCO synthesis, up to a level at which the RuBisCO concentration reaches 50% of that of total soluble proteins or 0.2 g of RuBisCO protein ml^{-1} in the stroma (equivalent to 4 mM in the concentration of its active site) (Ellis, 1979; Yokota and Canvin, 1985). However, plants must still lose water from the leaf through the open stomata in order to incorporate enough CO_2. On average, water loss through evaporation is 250- and 1000 times faster in both C_4 and C_3 plants than the rate of incorporation of CO_2 through the stomata (Larcher, 1995).

An ideal RuBisCO that could make optimal use of the global environment in C_3 plants would incorporate the following properties: a higher turnover rate, a higher affinity for CO_2, and a higher S_{rel}. In contrast, the photorespiratory carbon oxidation (PCO) cycle driven by the RuBisCO oxygenase reaction has been proposed to play an important role in several reactions that are quite possibly equally important: (1) salvaging 75% of the carbon deposited in 2-phosphoglycolate into PGA through the PCR cycle, (2) dissipating more energy than the PCR cycle during turnover and refixation of photorespired CO_2, and (3) supplying glycine and serine (Douce and Heldt, 2000; Heldt, 1997). These points apply solely to C_3 plants containing present-day RuBisCO.

To attempt to remove the oxygenase reaction from RuBisCO, even if possible, would be dangerous for plants, although a reduction in the concentration of O_2 in the atmosphere increases net photosynthesis rate (Tolbert, 1994). However, the reduction decreases J_e (RuBisCO) or the rate of utilization of electrons by the PCO cycle (Fig. 4.2). Figure 4.2B also shows that the significance of the PCO cycle increases with decreasing CO_2 concentrations and, inversely, that increasing CO_2 concentrations weaken the importance of the cycle. In addition, the fact that high CO_2 concentration in the atmosphere increases plant productivity to some degree (Sage *et al.*, 1989) supports the idea that the PCO cycle is dispensable for plants *if* the solar energy captured by chlorophyll is efficiently consumed by other metabolic events in chloroplasts. Under those conditions, serine and glycine are synthesized from PGA in metabolism through the glycolate pathway and/or phosphorylated serine pathway (Hess and Tolbert, 1966; Ho and Saito, 2001). RuBisCO of cyanobacteria does not meet two of the outlined three ideal conditions essential for desired plant photosynthesis (Badger, 1980). However, cyanobacteria grow photosynthetically, in the absence of a well-developed PCO cycle, but with the aid of an active CO_2-pumping mechanism (Kaplan and Reinhold, 1999; Shibata *et al.*, 2002).

These considerations teach us that C_3 plants are able to grow photosynthetically using RuBisCO with or without a much slower oxygenase reaction. In this case, some conditions must be met. The S_{rel} value is the ratio of specificity of the carboxylase reaction to that of the oxygenase reaction, and is varied by changing either or both of the specificities of the reactions. An increase in S_{rel} by increasing the turnover rate of the carboxylase reaction and the affinity for CO_2 twofold over that of the wild-type enzyme causes photosynthesis and J_e (RuBisCO) to increase (Fig. 4.2C and D). In contrast, RuBisCO with a higher S_{rel} value attained by lowering the specificity of the oxygenase reaction results in increased photosynthesis (Fig. 4.2C), but J_e (RuBisCO) is lowered (Fig. 4.2D). Plants containing RuBisCO manipulated to have such properties would be distressed by excess energy in high light intensities. However, this does not entail that photorespiration is completely indispensable for C_3 plants. If the excess energy caused by lowering the specificity of the oxygenase reaction could be used by the PCR cycle, that is, if the specificity of the carboxylase reaction were increased to a level equal

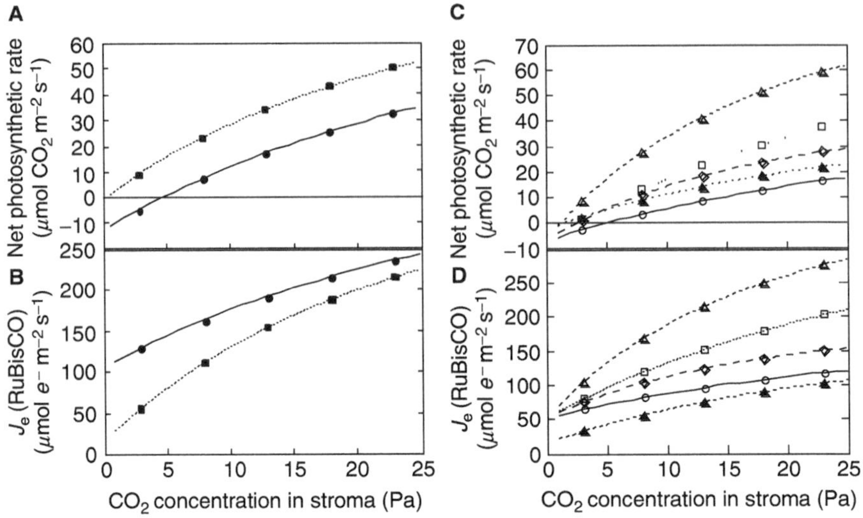

FIGURE 4.2 Simulation of the rate of net photosynthesis and flux of electrons used by PCR and PCO cycles in the electron transport chain. The rates of the carboxylase (v_c) and oxygenase (v_o) reactions of RuBisCO are expressed as $v_c = (k_c \cdot [\text{RuBisCO}]\ C_c)/\{K_c \cdot (1 + O_c/K_o) + C_c\}$ and $v_o = (k_o \cdot [\text{RuBisCO}]\ O_c)/\{K_o \cdot (1 + C_c/K_c) + O_c\}$, respectively, where k_c, k_o, K_c, and K_o are k_{cat}'s of carboxylase and oxygenase reactions and Michaelis constants for CO_2 and O_2, respectively (Miyake and Yokota, 2000). O_c and C_c are concentrations of O_2 and CO_2, respectively, around RuBisCO. [RuBisCO] is the mole number of the active sites of RuBisCO per unit leaf area. The rate of net photosynthesis (A) is expressed as follows: $A = v_c - 0.5v_o - R_d = v_c[1 - 0.5O_c/S_{rel} \cdot C_c] - \text{Rd}$, where R_d is the rate of day respiration and was assumed as 0.5 μmol CO_2 m^{-2} s^{-1}. The flux of electrons used by RuBisCO-related cycles in the electron transport chain, J_e (RuBisCO), corresponds to $4v_c + 4v_o$. Light is assumed to be saturating for photosynthesis. (A) and (B) show the effects of lowering atmospheric O_2 concentration on A and J_e (RuBisCO), respectively, in a C_3 plant undergoing photosynthesis with RuBisCO representative of the higher plant enzyme. The kinetic parameters of RuBisCO from C_3 plants were from the literature (Woodrow and Berry, 1988): S_{rel}, 89; k_c, 3.3 mol; CO_2 s^{-1} per site; k_o, 2.2 mol CO_2 s^{-1} per site; K_c, 29.5 Pa; K_o, 43.9 kPa; [RuBisCO], 18.56 μmol catalytic site m^{-2}. The concentration of O_2 in the atmosphere was assumed to be 21 (circles) and 2 kPa (squares). The effects of variations in kinetic parameters of RuBisCO on A and J_e (RuBisCO) are simulated in (C) and (D), respectively. Parameters for simulations are the same as those in (A) and (B) except that S_{rel} were varied as indicated below and [RuBisCO] was 9.28 μmol catalytic site m^{-2}. Enzymatic properties of RuBisCO are changed as follows: Circles, S_{rel}, 89, k_c, K_c, k_o, K_o; squares, S_{rel}, 180, $2k_c$, K_c, k_o, K_o; lozenges, S_{rel}, 180, k_c, $0.5K_c$, k_o, K_o; open triangles, S_{rel}, 360, $2k_c$, $0.5K_c$, k_o, K_o; closed triangles, S_{rel}, 360, k_c, K_c, $0.5k_o$, $2K_o$.

to or greater than the point where the excess energy is compensated by the PCR cycle, such a RuBisCO enzyme would improve C_3 photosynthesis without excess-light stress.

3.1.2. Naturally occurring diversity in RuBisCO kinetics

RuBisCO homologues are widely distributed among organisms and have been classified into four forms (Hanson and Tabita, 2001). Form I consists of eight large and eight small subunits of about 53 and 13 kDa, respectively, and is widely distributed among photosynthetic organisms such as higher plants, green algae, chlorophyll *b*-less eukaryotic algae, and autotrophic proteobacteria. Form II is composed only of the large subunits and is found in some eukaryotic algae, such as dinoflagellates, and photosynthetic proteobacteria. Form III is composed of only large subunits that are intermediates between Forms I and II, and is found in some Archaea (Ezaki *et al.*, 1999; Finn and Tabita, 2003). All three forms possess the amino acid residues known to be essential for catalysis of RuBisCO and, in fact, catalyze both carboxylation and oxygenation of RuBP. RuBisCO homologues found in *Bacillus subtilis*, *Chlorobium tepidum*, and *Archaeoglobus fulgidus* are classified as Form IV based on their primary sequences (Hanson and Tabita, 2001). Form IV lacks up to half of the amino acid residues essential for RuBisCO classical catalysis, and, in fact, has no RuBP-dependent CO_2-fixation activity. The exact function of Form III RuBisCO of Archaea is not known, while the RuBisCO homologue in *B. subtilis* catalyzes the 2,3-diketo-5-methylthiopentyl-1-phosphate enolase reaction in the methionine salvage pathway (Ashida *et al.*, 2003, 2005; Sekowska *et al.*, 2004). Form II RuBisCO of *Rhodospirillum rubrum* has the ability to catalyze the same reaction at a much slower rate. It has been suggested that the Form IV enzyme may be an ancestor of photosynthetic RuBisCO (Ashida *et al.*, 2003, 2005).

The S_{rel} value of Form I RuBisCO enzymes from cyanobacteria and γ-proteobacteria is around 40 (Roy and Andrews, 2000; Uemura *et al.*, 1996). The K_m for CO_2 of the cyanobacteria enzyme is 250 μM, the highest value among RuBisCO enzymes examined so far (Badger, 1980). The S_{rel} value is around 60 for RuBisCO from green microalgae, around 70 in conjugates and green macroalgae, and 85–100 in higher plants (Uemura *et al.*, 1996). β-Proteobacteria, and micro- and macroalgae in which an accessory pigment chlorophyll *b* is replaced by bile pigments, possess Form I RuBisCOs. These are developed from an ancestor separate from those that evolved into the higher plant enzyme through cyanobacterial and γ-proteobacterial ancestors in the phylogenetic tree of the primary sequence of the large subunit proteins. RuBisCOs grouped in the nongreen Form I branch have higher S_{rel} values than those grouped with the higher plant enzymes (green Form I RuBisCO) (Uemura *et al.*, 1996). One extreme is the nongreen Form I enzyme from a thermoacidophilic alga, *Galdieria partita* (Uemura *et al.*, 1997). The S_{rel} and K_m for CO_2 values are 238 and 6.6 μM at 25 °C, but the S_{rel} value decreases to 80 at 45 °C (its growth temperature). The protein structure of this enzyme has been resolved at 2.4 Å (Sugawara *et al.*, 1999). The high S_{rel} value has been proposed to be due to the stabilization of a loop partially covering the active site, loop 6, by hydrogen bonding between the main chain oxygen of Val^L-332 and amido group of Gln^L-386 (the numbering of amino acid residues follows the sequence of spinach RuBisCO, and the superscript indicates a large subunit residue) (Okano *et al.*, 2002). Generally speaking, for Form I RuBisCOs, an enzyme having a higher S_{rel} value and a lower K_m for CO_2 has a lower turnover rate and vice versa (Andrews and Lorimer, 1981). The S_{rel} value of Form II RuBisCOs is the lowest among all known enzymes, and it is possible that the assembly with small subunit proteins may be important to increase the value (Andrews and Lorimer, 1981). An exception is known in the *Pyrococcus kodakaraensis* KOD1 Form III

RuBisCO, in which five L_2 dimers make up the enzyme without any small subunits (Kitano *et al.*, 2001). The S_{rel} value in this enzyme has been reported as 300 at 90 °C but is 80 at 25 °C (Ezaki *et al.*, 1999).

The turnover rate of RuBisCO varies according to the source organism. The plant enzyme is one of the slowest catalysts, RuBisCOs from cyanobacteria and photosynthetic bacteria have a rate of 8–12 s^{-1} per site (Badger and Spalding, 2000), while the green algal enzymes occupy an intermediate position (Seemann *et al.*, 1984). The highest turnover rate has been recorded as 20–21 s^{-1} per site for a Form III RuBisCO from *A. fulgidus* (Finn and Tabita, 2003).

During the era in which photosynthetic bacteria and cyanobacteria evolved the PCR cycle and the RuBisCO enzyme, the earth's atmosphere contained high concentrations of CO_2 with a marginal level of oxygen (Badger and Spalding, 2000). Over time, CO_2 concentration decreased and the atmospheric oxygen concentration increased as a result of photosynthesis, initially by cyanobacteria and later by green algae. Cyanobacteria seem to have optimized a "CO_2-pumping mechanism" in preference over improving RuBisCO. The evolution in green algae moved partly toward improved RuBisCO properties and partly toward a mechanism that concentrated CO_2 in chloroplasts. Considering the properties of RuBisCOs of green algae, conjugates, and green macroalgae (Uemura *et al.*, 1996), and since terrestrial plants lack the CO_2-pumping system of cyanobacteria and algae, it is probable that higher plants could not be terrestrial until the S_{rel} value reached 80 and the K_m for CO_2 was lowered to 15 μM. Apparently, the turnover rate was sacrificed in favor of development of properties that improved RuBisCO properties. Evolutionarily, higher plants responded to the selection pressure imposed by a change in [CO_2] by moderately changing the structural gene sequence of *rbcL*, and compensated for the resulting disadvantages by developing a powerful promoter for the RuBisCO small subunit gene with changes in the small subunit protein that stabilized the L protein only a few hundred million years ago. Such compensation was necessarily incomplete since RuBisCO concentration in the stroma of algae was already high (Yokota and Canvin, 1985) because of the inherently slower turnover rate of this enzyme. There may still be room, however, to explore sequences of subunit proteins that exist in unexplored species, or to engineer sequence alterations that have not resulted from natural evolution. This is the research basis from which present and future protein engineering technology should succeed in improving the enzymatic properties of plant RuBisCO.

3.1.3. Engineered improvements of RuBisCO enzymatic properties

In attempts to understand the structure–function relationships of RuBisCO, many amino acid residues in both subunit proteins have been manipulated in both Forms I and II (Hartman and Harpel, 1994; Parry *et al.*, 2003; Spreitzer and Salvucci, 2002). In order to identify residues responsible for activity in one step of a sequence of partial reactions of RuBisCO, the chemical nature of the side chain of either the residue or the length of the side chain is changed. In another approach, alignments can be done of the primary sequences of more than 2000 varieties of large subunits and 300 varieties of small subunits (Spreitzer and Salvucci, 2002). This may either suggest which residue(s) or sequence(s) are responsible for a range in the S_{rel} value from 10 to 238, in K_ms for CO_2 value from 6 to 250 μM, and k_{cat}'s from 2.5 to 20 s^{-1} per site.

RuBisCO engineering depends on the synthesis of native recombinant proteins. Recombinant bacterial Forms I and II RuBisCOs can be synthesized in *Escherichia coli* (Hartman and Harpel, 1994). The genes for eukaryotic RuBisCOs can be transcribed in *E. coli*, but synthesized proteins aggregate rather than form the soluble, active enzyme (Gatenby *et al.*, 1987). This is thought to be due, at least in part, to the fact that large subunit proteins of the eukaryotic Form I RuBisCO are insoluble in the absence of the small subunit protein (Andrews and Lorimer, 1985), and partly due to *E. coli* chaperones being incompatible with large subunit proteins.

Engineering of an amino acid residue involved in a partial reaction step generally causes a loss in activity of the recombinant enzyme. Nevertheless, there are several instances in which RuBisCO properties have been successfully changed. These engineering successes could point toward rational engineering strategies for the improvement of plant photosynthesis in the near future. The recombinant Form II RuBisCO of *R. rubrum* in which Ser^L-379 is replaced by Ala shows no oxygenase activity, although the turnover rate in the carboxylase reaction decreases to less than one-hundredth of the wild-type enzyme (Harpel and Harman, 1992). The function of this residue has been confirmed using Form I RuBisCO from the cyanobacterium *Anacystis nidulans* (Lee and McFadden, 1992). The 21st and 305th residues of plant RuBisCOs are conserved lysines, which are replaced by arginine residues in many bacterial and algal enzymes (Uemura *et al.*, 1998). Simultaneously changing Arg^L-21 and Arg^L-305 of Form I RuBisCO of the photosynthetic γ-proteobacterium *Chromatium vinozum* to lysine residues resulted in an increase of the turnover rate from 8 to 15.6 s^{-1} per site with a concomitant increase in K_m for CO_2 from 30 to 250 μM (Uemura *et al.*, 2000).

The exact function of small subunit proteins in Form I RuBisCO is still unclear (Spreitzer, 2003). However, many residues in small subunits have been modified, resulting in altered catalysis of the holoenzyme, although no small subunit residue is located close to the active site on the large subunit proteins (Spreitzer, 2003). The most striking improvement was achieved by changing Pro^S-20 to alanine in the cyanobacterium *Synechocystis* sp., with the S_{rel} value increasing from 44 in wild-type to 55 in the mutated enzyme without any change in the turnover rate (Kostiv *et al.*, 1997). The engineered Ile^S-99-Val RuBisCO of the cyanobacterium had a higher affinity for CO_2 with no change in the S_{rel} value and a decrease in turnover rate (Read and Tabita, 1992a). Either Gly^S-103-Val or Phe^S-104-Leu cause small increases both in the S_{rel} value and the affinity for CO_2. RuBisCO of diatoms belongs to red-Form I with an S_{rel} value over 100. A hybrid enzyme composed of the large subunit of *Synechococcus* and the small subunit from a diatom *Cylindrotheca* exhibits a 60% increase in S_{rel} compared to the original cyanobacterial enzyme (Read and Tabita, 1992b).

3.1.4. Obstacles to be resolved for RuBisCO engineering

RuBisCO engineering has not yet succeeded in increasing S_{rel} values for cyanobacterial and *Chlamydomonas* RuBisCOs to levels observed in plant enzymes but the knowledge gained from engineering these enzymes has provided a blueprint to be applied to higher plant RuBisCO enzymes. This is expected to become possible because of our ability to manipulate the higher plant *rbcL* gene by chloroplast DNA transformation (Kanevski *et al.*, 1999; Svab and Maliga, 1993; Whitney *et al.*, 1999). Combination of this technical advance with the discovery of a RuBisCO enzyme with an extreme S_{rel} value provides an important new start point for improving plant RuBisCO and thereby alters plant productivity (Whitney *et al.*, 2001). The obstacles that still stand in the way are addressed here in a discussion of three strategies directed at changing the enzymatic properties of plant RuBisCO by genetic engineering.

The first strategy will be to introduce multiple mutations into higher plant *rbcL* genes, and then return the modified genes to their original locus in chloroplast DNA in a high-throughput fashion. This will circumvent the problem of either insolubility of large subunit proteins from higher plants in *E. coli* (Gatenby *et al.*, 1987) or the stroma of *Chlamydomonas* chloroplasts (Kato and Yokota, unpublished). While chloroplast transformation schemes are time consuming, the magnitude of the problem and the potential benefit resulting from successful engineering justify such efforts. That this is possible has been documented. Tobacco *rbcL* has already been engineered resulting in a reduction of S_{rel} and has been exchanged with the original *rbcL* in the tobacco chloroplast genome (Whitney *et al.*, 1999). The characteristics of photosynthetic CO_2 fixation of the transformant were consistent with Farquhar's photosynthetic simulation model (Whitney *et al.*, 1999).

A second strategy will be to clone genes for both large and small subunits for a RuBisCO, which is superior in S_{rel} and K_m for CO_2, and introduce them into the *rbcL* locus of chloroplast DNA of the target plant. In a pioneering study to express the Form II RuBisCO gene from *R. rubrum* in tobacco chloroplasts, the foreign gene gave rise to an active enzyme (Whitney and Andrews, 2001a). However, the genes of cyanobacterial and *Galdieria* Form I RuBisCO did not result in soluble, active enzymes (Kanevski *et al.*, 1999; Whitney *et al.*, 2001). This lack of success has been ascribed to incompatibility between the foreign large subunit peptides, the resident small subunit proteins, and the system for folding of nascent peptides in tobacco chloroplasts.

A third strategy addresses a different topic. Information on mechanisms involved in protein synthesis and folding in chloroplasts is still fragmentary (Houtz and Portis, 2003; Roy and Andrews, 2000), and our lack of knowledge of the precise mechanisms thus impedes the successful manipulation of RuBisCO genes in plants. For example, synthesis of the large subunit was formerly believed to take place on stromal free polysomes (Minami and Watanabe, 1984). However, recent work showed that a majority of the large subunits are translated by thylakoid-bound polysomes (Hatoori and Margulies, 1986). Since the large subunit itself is insoluble in an aqueous environment and translated on polysomes, one can expect the involvement of various chaperones in association with the polysomes. Otherwise, large subunit peptides in the process of translation and nascent large subunit peptides still in the process of synthesis would aggregate into an insoluble form. In this context, the observation (Amrani *et al.*, 1997) of translational pausing on polysomes is intriguing. Nascent large subunits released from polysomes assemble with lipids or membranes, the fatty acid composition of which is quite different from that of thylakoids (Smith *et al.*, 1997). Chaperonin-60 is known to bind at this stage to large subunit proteins (Gatenby and Ellis, 1990; Roy and Cannon, 1988; Smith *et al.*, 1997). The holoenzyme may then be assembled as an L_8 core to which small subunit proteins are added, as in the case of the synthesis of cyanobacterial RuBisCO (Hebbs and Roy, 1993).

The chloroplast outer and inner envelope membranes have individual translocon complexes, Toc and Tic, respectively, that recognize and transfer precursor proteins synthesized in the cytosol (Jarvis and Soll, 2002). Precursor proteins in a plastid-targeting complex with Hsp-70 and other proteins are guided to Toc and incorporated through the Toc complex in an ATP/GTP-dependent manner (Schleiff *et al.*, 2002). The precursor proteins are then passed to Tic. The transit sequence of the small subunit precursor is then cleaved and the N-terminal methionine of mature small subunits is methylated (Grimm *et al.*, 1997). One Tic component, IAP100, associates with chaperonin-60 and methylated small subunits are passed to chaperonin-60 through IAP100 (Kessler and Blobel, 1996). The L_8 core and the small subunit/chaperonin-60 complex meet to form the holoenzyme. The importance of small subunit methylation is emphasized by the fact that there is only limited incorporation into a holoenyzme of small subunits synthesized from a foreign *rbcS* gene in chloroplasts (Whitney and Andrews, 2001b; Zhang *et al.*, 2002). However, successful accumulation of the RuBisCO protein has been achieved when the promoter of the chloroplast-located *psbA* gene and the 5′-UTR-attached cDNA of a transcript encoding a small subunit protein was engineered into a transcriptionally active space of the chloroplast (Dhingra *et al.*, 2004).

When *rbcL* and *rbcS* genes are coordinately expressed in *E. coli*, even in the presence of coexpressed chloroplast chaperonin-60, no holoenzyme is formed (Cloney *et al.*, 1993). In addition to the involvement in RuBisCO assembly of known chaperonin proteins (Brutnell *et al.*, 1999; Checa and Viale, 1997; Gutteridge and Gatenby, 1995; Ivey *et al.*, 2000), there are probably several additional, still unknown, proteins in chloroplasts that participate in successful folding of the holoenzyme. Transcription and translation systems of chloroplasts are bacteria-like, and many foreign proteins can be synthesized and accumulated in an active form in chloroplasts (Daniell, 1999). One most important aspect requiring a solution is that the coordinate synthesis and assembly of RuBisCO subunit proteins is severely discriminated against by host chloroplasts of different species: chimeric RBCL/RBCS holoenzymes have not been reported.

In studying RuBisCO structure–function relationships, a tobacco *rbcL* insertion mutant has been useful (Kanevski and Maliga, 1994). In this study, the original chloroplast-localized *rbcL* gene was disrupted by insertion of a selection marker gene, *aadA*, into the gene. The *rbcL*-deficient transformant is then transformed with a different *rbcL* sequence fused at its N-terminus to a chloroplast transit peptide sequence under the control of a nuclear promoter. Another useful mutant plant is a tobacco mutant, SP25, where Gly^{L}-322 has been replaced by serine (Shikanai *et al.*, 1996), which led to dysfunctional assembly of the holoenzyme and only a small amount of RuBisCO accumulated in an aggregated form in the stroma (Foyer *et al.*, 1993). An engineered *rbcL* gene may then be introduced into chloroplast DNA of SP25.

A serious obstacle to plant RuBisCO engineering had been the difficulty in chloroplast transformation in any major crop plant. Efficient chloroplast transformation has in the past been restricted to some species in the Solanaceae, that is, tobacco (Svab and Maliga, 1993), potato (Sidorov *et al.*, 1999), and tomato (Ruf *et al.*, 2001). However, recent success appears to have been achieved with chloroplast transformation in crop species (Daniell *et al.*, 2005).

3.2. C_4-ization of C_3 plants

Water equilibrated at normal atmospheric pressure dissolves 11-μM CO_2, which forms 110-μM HCO_3^- at pH 7.2 and 25 °C (Yokota and Kitaoka, 1985). While RuBisCO fixes CO_2, phosphoenolpyruvate carboxylase (PEPC) uses HCO_3^- as the substrate. This characteristic confers a tremendous advantage to C_4 plants. Since the K_m for HCO_3^- of maize PEPC is as low as 20 μM (Uedan and Sugiyama, 1976), this enzyme can exhibit submaximal activity in the mesophyll cytosol. In the case of the C_4 plant maize, oxalacetate formed by PEPC in mesophyll cells is reduced to malate and then decarboxylated by $NADP^+$-dependent malic enzyme in the mitochondria of bundle sheath cells to give rise to CO_2 and pyruvate (Heldt, 1997; Kanai and Edwards, 1999). Pyruvate returns to mesophyll chloroplasts to be salvaged to phosphoenolpyruvate (PEP) by pyruvate Pi dikinase (PPDK). The active operation of this pathway can convert HCO_3^- in mesophyll cytosol to CO_2 concentrated in bundle sheath cells. The CO_2 concentration around RuBisCO in chloroplasts of bundle sheath cells reaches 500 μM (von Caemmerer and Furbank, 1999), causing net CO_2 fixation to be saturated at 10–15 Pa CO_2 without any detectable photorespiration (Edwards and Walker, 1983). Thus, this auxiliary metabolic CO_2-pumping system confers significantly better nitrogen investment and water-use efficiencies to C_4 plants compared with C_3 plants. If this CO_2-pumping system could be introduced into C_3 plants, the transgenic plants would be expected to show highly improved photosynthetic performance and productivity (Ku *et al.*, 1996).

The maize PEPC gene has been introduced into rice chloroplasts (Ku *et al.*, 1999). Although the severalfold higher PEPC activity in chloroplasts did not influence carbon metabolism (Häusler *et al.*, 2002), transgenic plants expressing over 50 times more PEPC activity than wild type exhibited slightly higher CO_2-fixation rates that were relatively insensitive to O_2 (Ku *et al.*, 1999). The primary CO_2-fixation product in these transgenic plants was PGA, not C_4 acid (Fukayama *et al.*, 2000). However, the introduction of single C_4 genes will not establish a metabolic CO_2-pumping system since this transgenic rice depends on glycolysis for the supply of PEP (Matsuoka *et al.*, 2001). Maize malic enzyme and PPDK have been individually introduced into rice plants, but positive effects on photosynthesis have not been observed (Fukayama *et al.*, 2001; Tsuchida *et al.*, 2001). One unexplained consequence of the ectopic expression of the maize $NADP^+$-malic enzyme in C_3 chloroplasts has been either the lack or disturbance of grana, possibly indicating altered protein–protein interactions

(Takeuchi *et al.*, 2000). The incorporation of both PEPC and PPDK into rice, generated by crossing of single-gene transformants, has been achieved and the plants appeared to behave in a more C_4-like fashion (Ku *et al.*, 2001). Introduction of more than two C_4 genes into C_3 plants has not yet been attempted.

Unlike C_4 plants, C_3 plants transgenic for all three genes may not fix CO_2 efficiently since the diffusion of CO_2 in cytosol and through membranes is rapid. An observation that seems to support this prediction is that cyanobacteria concentrate HCO_3^- within cells to a level up to 10^3 times higher than the ambient CO_2 concentration (Kaplan and Reinhold, 1999). The genes for the CO_2-pumping systems have been identified (Shibata *et al.*, 2002). Endogenous carbonic anhydrase is localized in carboxysomes where the HCO_3^- is dehydrated to CO_2 to be fixed by RuBisCO (Kaplan and Reinhold, 1999). Induction of a high level of carbonic anhydrase activity in the cytosolic space caused conversion of HCO_3^- into CO_2, which was released from the cells at a rate sufficient to nullify the pumping activity (Price and Badger, 1989). It will be important to learn more and understand how such high local concentrations of CO_2 around RuBisCO can be maintained and possibly engineered into higher plant chloroplasts. In this context, the C_4-type performance of *Borszczowia aralocaspica* (Chenopodiaceae) from the Gobi desert (Voznesenskaya *et al.*, 2001) provides another interesting example. In this plant, RuBisCO and NAD^+-malic enzyme are localized in chloroplasts and mitochondria, respectively, and are located at the proximal end of cells. Chloroplasts reside in the distal part of the cells and contain PPDK, but not RuBisCO, while PEPC is located throughout the cell. Understanding how such a spatial arrangement of enzymes is accomplished and maintained will be important for the recreation of a functional C_4 pathway in C_3 plants.

4. ENGINEERING POST-RUBISCO REACTIONS

4.1. RuBP regeneration

Flux control analysis indicated SBPase as the most likely rate-limiting step for regeneration of RuBP in the PCR cycle (Robinson and Walker, 1981; see Section 2.2). Furthermore, the two phosphatases FBPase and SBPase, as well as PRK, are light-regulated enzymes that avoid futile reactions in the dark. Regulation is exerted through the redox reaction of two SH-groups in these proteins (Buchanan, 1991). The SH-groups are also targets of hydrogen peroxide under oxidative stress that affects redox homeostasis (Shikanai *et al.*, 1998).

In contrast to the plant PCR cycle, cyanobacterial and green algal PCR pathways are insensitive to oxidation by H_2O_2 and are not subject to light/dark regulation (Tamoi *et al.*, 1998). This is because the enzyme involved in the rate-limiting step of these microorganisms lacks the functional redox-responding SH-groups (Tamoi *et al.*, 1996a,b, 2001). While the plant and algal PCR cycles include FBPase and SBPase as separate entities, both metabolic steps are catalyzed by a single enzyme, FBP/SBPase, in the PCR cycle of *Synchococcus* (Tamoi *et al.*, 1996b). The bifunctional enzyme lacks regulatory SH-groups. The gene for the cyanobacterial enzyme fused to a RuBisCO small subunit transit peptide has been introduced into tobacco (Miyagawa *et al.*, 2001; Tamoi *et al.*, 2005). The transformant created in this experiment revealed improved photosynthetic performance: transformed plants showed a 2.3-fold increase in chloroplast FBPase and SBP activities relative to wild type, accompanied by an increase in CO_2-fixation rate and dry matter to 125% and 150%, respectively, of the wild type (Fig. 4.3). The photosynthetic rates realized in these transformants may be the maximum attainable for C_3 photosynthesis because C_3 photosynthesis enters a Pi-limited state at such high CO_2-fixation rates (see section 2.1).

With the exception of FBPase and SBPase, there were no detectable changes in these transformants in either total activities or amounts of enzymes involved in the PCR cycle. The only changes observed with the transformant were increases in RuBP levels and in the activation ratio of RuBisCO by a factor of 1.8–1.2 relative to the wild type (Miyagawa *et al.*, 2001). These increases in photosynthetic rate are consistent with an increase in RuBisCO activation. Since RuBisCO activase requires a relatively high concentration of RuBP as judged from *in vitro* assays (Porits, 1990), the observed increase in activation seems to be due to the presence of the transgenic FBP/SBPase that appears to function by promoting regeneration of RuBP and, as a consequence, activating the activase. This study presents the first example of successful improvement of photosynthetic performance and productivity by the introduction of a single gene. In addition, it provides proof for the validity of the concept that single-gene transfers, based on precise knowledge of metabolic flux, its control, and enzyme activity regulation, can improve crop productivity. Similar, but smaller, effects have been reported in tobacco expressing FBPase and SBPase individually (Lefebvre *et al.*, 2005; Tamoi *et al.*, 2006).

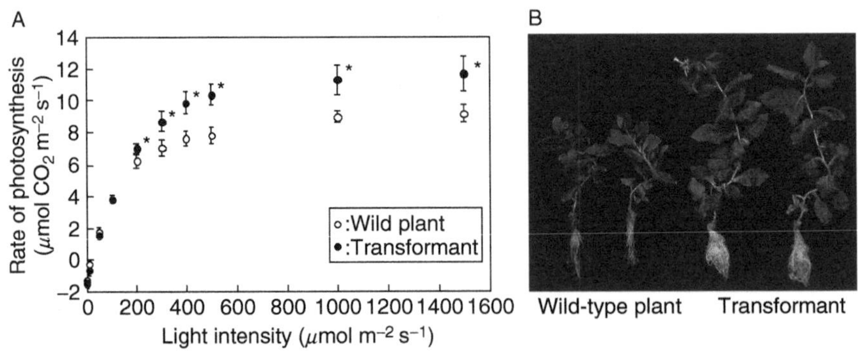

FIGURE 4.3 Phenotypes of the wild-type tobacco plant and the transformant expressing cyanobacterial FBPase/SBPase in chloroplasts. (A) Effect of increasing light irradiance on the net CO_2 assimilation at 360 ppm of CO_2, 25 °C, and 60% relative humidity. The CO_2 assimilation rate was measured using the fourth leaves down from the top of plant, after 12 weeks of culture. (B) Photographs of the wild plant and the transformant after 18 weeks of culture in 360-ppm CO_2 at 400 μmol m^{-2} s^{-1}.

4.2. Engineering carbon flow from chloroplasts to sink organs

Triose phosphate formed in the PCR cycle is transported from chloroplasts to cytosol by a phosphate transporter located in the inner membrane of the envelope. It is then used as the carbon source for sucrose synthesis (Flüge, 1998). Sucrose formed in the mesophyll cells is transferred to phloem companion cells symplastically and through the apoplastic space. The final uploading of sucrose into companion cells against the steep concentration gradient of sucrose is conducted by a sucrose transporter coupled to ATP hydrolysis (Weise *et al.*, 2000).

Transgenic tobacco plants overexpressing the phosphate transporter have been created. Sucrose synthesis is promoted in the absence of significant increases in photosynthesis (Häusler *et al.*, 2000). Sucrose phosphate synthase (SPS) is an important regulatory enzyme in sucrose synthesis in the cytosol of mesophyll cells (Huber and Huber, 1996). Overexpression of the gene for SPS has been attempted with various plants, but the effects of the transgene on productivity varied between experiments (Galtier *et al.*, 1993; Lunn *et al.*, 2003). Although more carbon was directed to sucrose in the transformants than in the wild type, photosynthesis was not enhanced in a reproducible manner. There are four family members for the sucrose transporter (SUT1–4) (Weise *et al.*, 2000). Since repression of SUT1 gave rise to severe morphological changes, it has been deduced that the transporter participated in sucrose uploading into the phloem (Riesmeier *et al.*, 1994). Potato transformants expressing SUT1 under control of the *Cauliflower mosaic virus* 35S promoter showed lower sucrose level in leaves than wild type (Leggewie *et al.*, 2003). However, no changes in either photosynthesis, starch content, or tuber yield resulted.

5. SUMMARY

The scientific challenges encountered during the last decade by attempts at improving photosynthetic productivity, even when successful, generated further questions, but even the lack of success has taught us many things. As the conclusion for this chapter, we would like to explore the approaches necessary for future achievements in improvement of crop productivity.

One most important requisite for manipulating physiology of an organism is to accumulate information about the precise mechanisms of function of the key protein(s) or enzyme(s) in question. This includes detailed knowledge on gene structure and the regulation of gene and protein expression of enzymatic properties and subcellular location. ADPglucose pyrophosphorylase, for example, had been studied extensively over a long period, from its biochemistry *in vitro* through to regulation of activity *in vivo* (Preiss *et al.*, 1991). However, only the introduction of a gene, modified to be insensitive to feedback regulation, into potato tuber amyloplasts resulted in increased starch synthesis (Preiss, 1996). FBP/SBPase from a cyanobacterium has been shown to improve productivity in tobacco (Miyagawa *et al.*, 2001). Since the functional sites of these enzymes are the chloroplast stroma, the selection of the promoter and the transit sequences for expression of these proteins could easily be accomplished based on previous knowledge. Another strategy, antisense suppression of resident genes has revealed the significance of particular enzymes in a postulated metabolic pathway.

Similar considerations are also valid for RuBisCO research. We are still ignorant, for example, about either the residues that determine the S_{rel} value, or how carbon and oxygen atoms are enabled to overcome spin prohibition on the RuBisCO protein for the oxygenation of RuBP, and about which residues limit the reaction rate in overall catalysis (Cleland *et al.*, 1998; Roy and Andrews, 2000). Translation of *rbcL* mRNA and association of RuBisCO peptides are important topics about which not enough is known (Houtz and Portis, 2003; Roy and Andrews, 2000). In general, the steps of posttranslational folding in plants and other organisms, whether *E. coli*, yeast, or human, must become known (Frydman, 2001). RuBisCO should provide an excellent model protein for study, considering that plants are able to synthesize up to 200 mg/ml of RuBisCO protein within days during the greening of leaves.

Engineering of the chloroplast genome has become the transformation strategy that promises to overcome problems encountered in the genetic manipulation of nuclear chromosomes for functions that must reside in plastids (Daniell, 1999). The technology will be indispensable for the metabolic engineering of pathways such as the PCR cycle, and starch and lipid biosyntheses. In this context, establishing methods for chloroplast genome engineering in the major crop species is an important priority.

Introduction of the cyanobacterial CO_2-pumping system into the plasma membrane of mesophyll cells or the chloroplast envelope may be one future direction. Some improvement in the photosynthetic performance of transgenic plants has already been reported with *Arabidopsis* (Lieman-Hurwitz *et al.*, 2003).

Interspecies crosses that might lead to the transfer of beneficial genes are not possible in plants or any higher organism. Attempts at improving physiological performance in diverse environments can be realized by varying the expression of genes inherited from the parents. This requires that we understand in more detail the networks of reactions that constitute the evolutionarily established reaction bandwidth and allelic plasticity of a species. Science is now beginning to elucidate the potential of natural intraspecies variation and to probe the upper limits of plants physiologically, biochemically, and at the molecular genetic levels. Furthermore, we are learning, as we have pointed out, that it is possible to raise the potential of organisms and to exceed the intrinsic limits of plant productivity by introducing genes across species barriers that of a species that cannot be crossed by traditional breeding.

ACKNOWLEDGEMENTS

The authors thank Drs. Chikahiro Miyake and Masahiro Tamoi for their help in preparing the manuscript. We also thank Miss Naoko Hamamoto for her assistance. Research in our laboratories has been supported by the "Research for the Future" programs (JSPS-RFTF97R16001 and JSPS-00L01604) of the Japan Society for the Promotion of Science.

REFERENCES

Amrani, A. E., Freire, M., Camara, B., and Couee, I. (1997). Control of the synthesis of ribulose-1,5-bisphosphate carboxylase/oxygenase large-subunit in cotyledons during dark growth of sugar beet seedlings. *Plant Mol. Biol.* **34,** 651–657.

Andersson, I., and Taylor, T. C. (2003). Structural framework for catalysis and regulation in ribulose-1, 5-bisphosphate carboxylase/oxygenase. *Arch. Biochem. Biophys.* **414,** 130–140.

Andrews, T. J., and Lorimer, G. H. (1978). Photorespiration—still unavoidable? *FEBS Lett.* **90,** 1–9.

Andrews, T. J., and Lorimer, G. H. (1981). RuBisCO: Structure, mechanisms, and prospects for improvement. *In* "The Biochemistry of Plants" (M. D. Hatch, and N. K. Boardman, eds.), vol. 10, pp. 131–218. Academic Press, San Diego.

Andrews, T. J., and Lorimer, G. H. (1985). Catalytic properties of a hybrid between cyanobacterial large subunits and higher plant small subunits of ribulose bisphosphate carboxylase-oxygenase. *J. Biol. Chem.* **260,** 4632–4636.

Andrews, T. J., and Whitney, S. M. (2003). Manipulating ribulose bisphosphate carboxylase/oxygenase in the chloroplasts of higher plants. *Arch. Biochem. Biophys.* **414,** 159–169.

Ashida, H., Saito, Y., Kojima, C., Kobayashi, K., Ogasawara, N., and Yokota, A. (2003). A functional link between RuBisCO-like protein of Bacillus and photosynthetic RuBisCO. *Science* **302,** 286–290.

Ashida, H., Danchin, A., and Yokota, A. (2005). Was photosynthetic RuBisCO recruited by acquisitive evolution from RuBisCO-like proteins involved in sulfur metabolism? *Res. Microbiol.* **156,** 611–618.

Badger, M. R. (1980). Kinetic properties of ribulose 1,5-bisphosphate carboxylase/oxygenase from *Anabaena variabilis. Arch. Biochem. Biophys.* **201,** 247–254.

Badger, M. R., and Spalding, M. H. (2000). CO_2 acquisition, concentration and fixation in cyanobacteria and algae. *In* "Photosynthesis: Physiology and Metabolism" (R. C. Leegood, T. D. Sharkey, and S. von Caemmerer, eds.), pp. 369–397. Kluwer Academic, Dordrecht.

Badger, M. R., Sharkey, T. D., and von Caemmerer, S. (1984). The relationship between steady state gas exchange of bean leaves and the level of carbon reduction cycle intermediates. *Planta* **160,** 305–313.

Boyer, J. (1982). Plant productivity and environment. *Science* **218,** 443–448.

Brutnell, T. P., Sawers, R. J., Mant, A., and Langdale, J. A. (1999). BUNDLE SHEATH DEFECTIVE2, a novel protein required for post-translational regulation of the *rbc*L gene of maize. *Plant Cell* **11,** 849–864.

Buchanan, B. (1991). Regulation of CO_2 assimilation in oxygenic photosynthesis: The ferredoxin/thioredoxin system. Perspective on its discovery, present status, and future development. *Arch. Biochem. Biophys.* **288,** 1–9.

Checa, S. K., and Viale, A. M. (1997). The 70-kDa heat-shock protein/DnaK chaperone system is required for the productive folding of ribulose-biphosphate carboxylase subunits in *Escherichia coli. Eur. J. Biochem.* **248,** 848–855.

Cleland, W. W., Andrews, T. J., Gutteridge, S., Hartman, F. C., and Lorimer, G. H. (1998). Mechanism of RuBisCO: The carbamate as general base. *Chem. Rev.* **98,** 549–561.

Cloney, L. P., Bekkaoui, D. R., and Hemmingsen, S. M. (1993). Co-expression of plastid chaperonin genes and a synthetic plant RuBisCO operon in *Escherichia coli. Plant Mol. Biol.* **23,** 1285–1290.

Daniell, H. (1999). Chloroplast genetic engineering. *Nat. Biotechnol.* **17,** 855–856.

Daniell, H., Kumar, S., and Dufourmantel, N. (2005). Breakthrough in chloroplast genetic engineering of agronomically important crops. *Trend Biotechnol.* **23,** 238–245.

Dhingra, A., Portis, A. R., and Daniell, H. (2004). Enhanced translation of a chloroplast-expressed RbsS gene reporter small subunit levels and photosynthesis in nuclear RbcS antisense plants. *Proc. Natl. Acad. Sci. USA* **101,** 6315–6320.

Douce, R., and Heldt, H. W. (2000). Photorespiration. *In* "Photosynthesis: Physiology and Metabolism" (R. C. Leegood, T. D. Sharkey, and S. von Caemmerer, eds.), pp. 115–136. Kluwer Academic, Dordrecht.

Edwards, G. E., and Walker, D. A. (1983). "C_3, C_4: Mechanisms, and Cellular and Environmental Regulation of Photosynthesis." Blackwell, London.

Ellis, R. J. (1979). The most abundant protein in the world. *Trends Biochem. Sci.* **4,** 241–244.

Evans, J. R., and Loreto, F. (2000). Acquisition and diffusion of CO_2 in higher plant leaves. *In* "Photosynthesis" (R. C. Leegood, T. D. Sharkey, and S. von Caemmerer, eds.), vol. 9, pp. 321–351. Kluwer Academic, Dordrecht.

Ezaki, S., Maeda, N., Kishimoto, T., Atomi, H., and Imanaka, T. (1999). Presence of a structurally novel type ribulose-bisphosphate carboxylase/oxygenase in the hyperthermophilic archaeon, *Pyrococcus kodakaraensis* KOD1. *J. Biol. Chem.* **274,** 5078–5082.
Farquhar, G. D. (1979). Models describing the kinetics of ribulose biphosphate carboxylase-oxygenase. *Arch. Biochem. Biophys.* **193,** 456–468.
Farquhar, G. D., von Caemmerer, S., and Berry, J. A. (1981). A biochemical model of photosynthetic carbon dioxide assimilation in leaves of 3-carbon pathway species. *Planta* **149,** 78–90.
Finn, M. W., and Tabita, F. R. (2003). Synthesis of catalytically active form III ribulose 1,5-bisphosphate carboxylase/oxygenase in archaea. *J. Bacteriol.* **185,** 3049–3059.
Flüge, U. I. (1998). Metabolite transporters in plastids. *Curr. Opin. Plant Biol.* **1,** 201–206.
Foyer, C. H., Nurmi, A., Dulieu, H., and Parry, M. A. J. (1993). Analysis of two RuBisCO-deficient tobacco mutants, H7 and Sp25: Evidence for the production of RuBisCO large subunits in the SP25 mutant that form clusters and are inactive. *J. Exp. Bot.* **44,** 1445–1452.
Fridyand, L. E., and Scheibe, R. (2000). Regulation in metabolic systems under homeostatic flux control. *Arch. Biochem. Biophys.* **374,** 198–206.
Frydman, J. (2001). Folding of newly translated proteins *in vivo*: The role of molecular chaperones. *Annu. Rev. Biochem.* **70,** 603–647.
Fukayama, H., Imanari, E., Tsuchida, H., Izui, K., Matsuoka, M., and Miyao, M. (2000). *In vivo* activity of maize phosphoenolpyruvate carboxylase in transgenic rice plants. *Plant Cell Physiol.* **41,** s112.
Fukayama, H., Tsuchida, H., Agarie, S., Nomura, M., Onodera, H., Ono, K., Lee, B., Hirose, S., Toki, S., Ku, M. S., Makino, A., Matsuoka, M., and Miyao, M. (2001). Significant accumulation of C_4-specific pyruvate, orthophosphate dikinase in a C_3 plant, rice. *Plant Physiol.* **127,** 1136–1146.
Galtier, N., Foyer, C. H., Huber, J., Voelker, T. A., and Huber, S. C. (1993). Effects of elevated sucrose-phosphate synthase activity on photosynthesis, assimilate partitioning, and growth in tomato (*Lycopersicon esculentum* var UC82B). *Plant Physiol.* **101,** 535–543.
Gatenby, A. A., and Ellis, R. J. (1990). Chaperone function: The assembly of ribulose bisphosphate carboxylase-oxygenase. *Annu. Rev. Cell Biol.* **6,** 125–149.
Gatenby, A. A., van der Vies, S. M., and Rothstein, S. J. (1987). Co-expression of both the maize large and wheat small subunit genes of ribulose-bisphosphate carboxylase in *Escherichia coli*. *Eur. J. Biochem.* **168,** 227–231.
Geiger, D. R., and Servaites, J. C. (1994). Diurnal regulation of photosynthetic carbon metabolism in C3 plants. *Annu. Rev. Plant Physiol. Plant Mol. Biol.* **45,** 235–256.
Grimm, R., Grimm, M., Eckerskorn, C., Pohlmeyer, K., Rohl, T., and Soll, J. (1997). Postimport methylation of the small subunit of ribulose-1,5-bisphosphate carboxylase in chloroplasts. *FEBS Lett.* **408,** 350–354.
Gutteridge, S., and Gatenby, A. A. (1995). Rubisco synthesis, assembly, mechanism, and regulation. *Plant Cell* **7,** 809–819.
Haake, V., Zrenner, R., Sonnewald, U., and Stitt, M. (1998). A moderate decrease of plastid aldolase activity inhibits photosynthesis, alters the levels of sugars and starch, and inhibits growth of potato plants. *Plant J.* **14,** 147–157.
Haake, V., Geiger, M., Walch, L. P., Engels, C., Zrenner, R., and Stitt, M. (1999). Changes in aldolase activity in wild-type potato plants are important for acclimation to growth irradiance and carbon dioxide concentration, because plastid aldolase exerts control over the ambient rate of photosynthesis across a range of growth conditions. *Plant J.* **17,** 479–489.
Hanson, T. E., and Tabita, F. R. (2001). A ribulose-1,5-bisphosphate carboxylase/oxygenase (RubisCO)-like protein from *Chlorobium tepidum* that is involved with sulfur metabolism and the response to oxidative stress. *Proc. Natl. Acad. Sci. USA* **98,** 4397–4402.
Harrison, E. P., Willingham, N. M., Lloyd, J. C., and Raines, C. A. (1998). Reduced sedoheptulose 1,7-bisphosphatase levels in transgenic tobacco lead to decreased photosynthetic capacity and altered carbohydrate accumulation. *Planta* **204,** 27–36.
Harpel, M. R., and Harman, F. C. (1992). Enhanced CO_2/O_2 specificity of a site-directed mutant of ribulose-bisphosphate carboxylase/oxygenase. *J. Biol. Chem.* **267,** 6475–6478.
Hartman, F. C., and Harpel, M. R. (1994). Structure, function and assembly of D-ribulose-1, 5-bisphosphate carboxylase/oxygenase. *Annu. Rev. Biochem.* **63,** 197–234.

Hatoori, T., and Margulies, M. M. (1986). Synthesis of large subunit of ribulosebisphosphate carboxylase by thylakoid-bound polyribosomes from spinach chloroplasts. *Arch. Biochem. Biophys.* **244,** 630–640.

Häusler, R. E., Schlieben, N. H., Nicolay, P., Fischer, K., Fischer, K. L., and Flügge, U. I. (2000). Control of carbon partitioning and photosynthesis by the triose phosphate/phosphate translocator in transgenic tobacco plants (*Nicotiana tabacum* L.). I. Comparative physiological analysis of tobacco plants with antisense repression and overexpression of the triose phosphate/phosphate translocator. *Planta* **210,** 371–382.

Häusler, R. E., Hirsch, H.-J., Kreuzaler, F., and Peterhänsel, C. (2002). Overexpression of C_4-cycle enzymes in transgenic C_3 plants: A biotechnological approach to improve C_3-photosynthesis. *J. Exp. Bot.* **53,** 591–607.

Hebbs, A. E., and Roy, H. (1993). Assembly of in vitro synthesized large subunits into ribulose-bisphosphate carboxylase/oxygenase. Formation and discharge of an L8-like species. *J. Biol. Chem.* **268,** 13519–13525.

Heldt, H. W. (1997). "Plant Biochemistry and Molecular Biology." p. 188. Oxford University Press, Oxford.

Henkes, S., Sonnewald, U., Badur, R., Flachmann, R., and Stitt, M. (2001). Small decrease of plastid transketolase activity in antisense tobacco transformants has dramatic effects on photosynthesis and phenylpropanoid metabolism. *Plant Cell* **13,** 535–551.

Hess, J. L., and Tolbert, N. E. (1966). Glycolate, glycine, serine, and glycerate formation during photosynthesis by tobacco leaves. *J. Biol. Chem.* **241,** 5705–5711.

Ho, C. L., and Saito, K. (2001). Molecular biology of the spastidic phosphorylated serine biosynthetic pathway in *Arabidopsis thaliana*. *Amino Acids* **20,** 243–259.

Houtz, R. L., and Portis, A. R., Jr. (2003). The life of ribulose 1,5-bisphosphate carboxylase/oxygenase-posttranslational facts and mysteries *Arch. Biochem. Biophys.* **414,** 150–158.

Huber, S. C., and Huber, J. L. (1996). Role and regulation of sucrose-phosphate synthase in higher plants. *Annu. Rev. Plant Physiol. Plant Mol. Biol.* **47,** 431–444.

Ivey, R. A., III, Subramanian, C., and Bruce, B. D. (2000). Identification of a Hsp70 recognition domain within the rubisco small subunit transit peptide. *Plant Physiol.* **122,** 1289–1299.

Jarvis, P., and Soll, J. (2002). Toc, tic and chloroplast protein import. *Biochim. Biophys. Acta* **1590,** 177–189.

Kacser, H. (1987). Control of metabolism. *In* "The Biochemistry of Plants" (D. D. Davies, ed.), vol. 11, pp. 39–67. Academic Press, San Diego.

Kanai, R., and Edwards, G. E. (1999). The biochemistry of C_4 photosynthesis. *In* "C_4 Plant Biology" (R. F. Sage and R. K. Monson, eds.), pp. 49–87. Academic Press, San Diego.

Kanevski, I., and Maliga, P. (1994). Relocation of the plastid rbcL gene to the nucleus yields functional ribulose-1,5-bisphosphate carboxylase in tobacco chloroplasts. *Proc. Natl. Acad. Sci. USA* **91,** 1969–1973.

Kanevski, I., Maliga, P., Rhoades, D. F., and Gutteridge, S. (1999). Plastome engineering of ribulose-1,5-bisphosphate carboxylase/oxygenase in tobacco to form a sunflower large subunit and tobacco small subunit hybrid. *Plant Physiol.* **119,** 133–142.

Kaplan, A., and Reinhold, L. (1999). CO_2 concentrating mechanisms in photosynthetic microorganisms. *Annu. Rev. Plant Physiol. Plant Mol. Biol.* **50,** 539–570.

Kato, K., Yokota, A. unpublished.

Kitano, K., Maeda, N., Fukui, T., Atomi, H., Imanaka, T., and Miki, K. (2001). Crystal structure of a novel-type archaeal Rubisco with pentagonal structure. *Structure* **9,** 473–481.

Koßmann, J., Sonnewald, U., and Willmitzer, L. (1994). Reduction of the chloroplastic fructose-1,6-bisphosphatase in transgenic potato plants impairs photosynthesis and plant growth. *Plant J.* **6,** 637–650.

Kostiv, R. V., Small, C. L., and McFadden, B. A. (1997). Mutations in a sequence near the N-terminus of the small subunit alter the CO_2/O_2 specificity factor for ribulose bisphosphate carboxylase/oxygenase. *Photosynth. Res.* **54,** 127–134.

Krapp, A., Chayes, M. M., David, M. M., Rodrigues, M. L., Pereira, J. S., and Stitt, M. (1994). Decreased ribulose-1,5-bisphosphate carboxylase/oxygenase in transgenic tobacco transformed with 'antisense' rbsS: VIII. Impact on photosynthesis and growth in tobacco growing under extreme high irradiance and high temperature. *Plant Cell Environ.* **17,** 945–953.

Ku, M. B. S., Kano-Murakami, Y., and Matsuoka, M. (1996). Evolution and expression of C_4 photosynthesis gene. *Plant Physiol.* **111,** 949–957.

Ku, M. B. S., Agarie, S., Nomura, M., Fukayama, H., Tsuchida, H., Ono, K., Hirose, S., Toki, S., Miyao, M., and Matsuoka, M. (1999). High-level expression of maize phospho*enol*pyruvate carboxylase in transgenic rice plants. *Nat. Biotechnol.* **17,** 76–80.

Ku, M. S., Cho, D., Li, X., Jiao, D. M., Pinto, M., Miyao, M., and Matsuoka, M. (2001). Introduction of genes encoding C_4 photosynthesis enzymes into rice plants: Physiological consequences. *Novartis Found Symp.* **236,** 100–111.

Kessler, F., and Blobel, G. (1996). Interaction of the protein import and folding machineries of the chloroplast. *Proc. Natl. Acad. Sci. USA* **93,** 7684–7689.

Laing, W. A., Ogren, W. L., and Hageman, R. H. (1974). Regulation of soybean net photosynthetic CO_2 fixation by the interaction of CO_2, O_2, and ribulose 1,5-bisphosphate carboxylase. *Plant Physiol.* **54,** 678–685.

Larcher, W. (1995). "Physiological Plant Ecology." p. 119. Springer, Berlin.

Lee, G. J., and McFadden, B. A. (1992). Serine-376 contributes to the binding of substrate by ribulose-bisphosphate carboxylase/oxygenase from *Anacystis nidulans*. *Biochemistry* **31,** 2304–2308.

Lefebvre, S., Lawson, T., Zakhleniuk, O. V., Lloyd, J. C., and Raines, C. A. (2005). Increased sedoheptulose-1,7-bisphosphatase activity in transgenic tobacco plants stimulates photosynthesis and growth from an early stage in development. *Plant Physiol.* **138,** 451–460.

Leggewie, G., Kolbe, A., Lemoine, R., Roessner, U., Lytovchenko, A., Zuther, E., Kehr, J., Frommer, W. B., Riesmeier, J. W., Willmitzer, L., and Fernie, A. R. (2003). Overexpression of the sucrose transporter *So*SUT1 in potato results in alterations in leaf carbon partitioning and in tuber metabolism but has little impact on tuber morphology. *Planta* **217,** 158–167.

Lieman-Hurwitz, J., Rachmilevitch, S., Mittler, R., Marcus, Y., and Kaplan, A. (2003). Enhanced photosynthesis and growth of transgenic plants that express *ictB*, a gene involved in $HCO^-{}_3$ accumulation in cyanobacteria. *Plant Biotechnol. J.* **1,** 43–50.

Lunn, J. E., Gillespie, V. J., and Furbank, R. T. (2003). Expression of a cyanobacterial sucrose-phosphate synthase from *Synechocystis* sp. PCC 6803 in transgenic plants. *J Exp. Bot.* **54,** 223–237.

Makino, A., Shimada, T., Takumi, S., Kaneko, K., Matsuoka, M., Shimamoto, K., Nakano, H., Miyao-Tokutomi, M., and Yamamoto, N. (1997). Does decrease in ribulose-1,5-bisphosphate carboxylase by antisense *rbs*S lead to a high N-use efficiency of photosynthesis under conditions of saturating CO_2 and light in rice plants? *Plant Physiol.* **114,** 483–491.

Mann, C. C. (1999). Crop scientists seek a new revolution. *Science* **283,** 310–314.

Martin, W., Scheibe, R., and Schnarrenberger, C. (2000). The Calvin cycle and its regulation. *In* "Photosynthesis" (R. C. Leegood, T. D. Sharkey, and S. von Caemmerer, eds.), vol. 9, pp. 9–51. Kluwer Academic, Dordrecht.

Masle, J., Hudson, G. S., and Badger, M. R. (1993). Effects of ambient CO_2 concentration on growth and nitrogen use in tobacco (*Nicotiana tabacum*) plants transformed with an antisense gene to the small subunit of ribulose-1,5-bisphosphate carboxylase/oxygenase. *Plant Physiol.* **103,** 1075–1088.

Matsuoka, M., Furbank, R. T., Fukayama, H., and Miyao, M. (2001). Molecular engineering of C_4 photosynthesis. *Annu. Rev. Plant Physiol. Plant Mol. Biol.* **52,** 297–314.

Mauser, H., King, W. A., Gready, J. E., and Andrews, T. J. (2001). CO_2 fixation by Rubisco: Computational dissection of the key steps of carboxylation, hydration, and C-C bond cleavage. *J. Am. Chem. Soc.* **123,** 10821–10829.

McCurry, S. D., Pierce, J., Tolbert, N. E., and Orme-Johnson, W. H. (1981). On the mechanism of effector-mediated activation of ribulose bisphosphate carboxylase/oxygenase. *J. Biol. Chem.* **256,** 6623–6628.

Minami, E., and Watanabe, A. (1984). Thylakoid membranes: The translational site of chloroplast DNA-regulated thylakoid polypeptides. *Arch. Biochem. Biophys.* **235,** 562–570.

Miyake, C., and Yokota, A. (2000). Determination of the rate of photoreduction of O_2 in the water-water cycle in watermelon leaves and enhancement of the rate by limitation of photosynthesis. *Plant Cell Physiol.* **41,** 335–343.

Miyagawa, Y., Tamoi, M., and Shigeoka, S. (2001). Overexpression of a cyanobacterial fructose-1,6-/sedoheptulose-1,7-bisphosphatase in tobacco enhances photosynthesis and growth. *Nat. Biotechnol.* **19,** 965–969.

Morell, M. K., Paul, K., Kane, H. J., and Andrews, T. J. (1992). Rubisco: Maladapted or misunderstood? *Aust. J. Bot.* **40,** 431–441.

Okano, Y., Mizohata, E., Xie, Y., Matsumura, H., Sugawara, H., Inoue, T., Yokota, A., and Kai, Y. (2002). X-ray structure of *Galdieria* Rubisco complexed with one sulfate ion per active site. *FEBS Lett.* **527,** 33–36.

Parry, M. A. J., Andralojc, P. J., Mitchell, R. A. C., Madgwick, P. J., and Keys, A. J. (2003). Manipulation of Rubisco: The amount, activity, function and regulation. *J. Exp. Bot.* **54,** 1321–1333.

Paul, M. J., and Foyer, C. H. (2001). Sink regulation of photosynthesis. *J. Exp. Bot.* **52,** 1383–1400.

Paul, M. J., Knight, J. S., Habash, D., Parry, M. A. J., Lawlor, D. W., Barnes, S. A., Loynes, A., and Gray, J. C. (1995). Reduction in phosphoribulokinase activity by antisense RNA in transgenic tobacco: Effect on CO_2 assimilation and growth in low irradiance. *Plant J.* **7,** 535–542.

Porits, A. R., Jr. (1990). Rubisco activase. *Biochim. Biophys. Acta* **1015,** 15–28.

Preiss, J. (1996). ADPglucose pyrophosphorylase: Basic science and applications in biotechnology. *Biotechnol. Annu. Rev.* **2,** 259–279.

Preiss, J., Ball, K., Smith-White, B., Iglesias, A., Kakefuda, G., and Li, L. (1991). Starch biosynthesis and its regulation. *Biochem. Soc. Trans.* **19,** 539–547.

Price, G. D., and Badger, M. R. (1989). Expression of human carbonic anhydrase in the cyanobacterium *Synechococcus* PCC7942 creates a high CO_2-requiring phenotype. Evidence for a central role for carboxysomes in the CO_2 concentrating mechanism. *Plant Physiol.* **91,** 505–513.

Price, G. D., Evans, J. R., von Caemmerer, S., Yu, J.-W., and Badger, M. R. (1995). Specific reduction of chloroplast glyceraldehyde-3-phosphate dehydrogenase activity by antisense RNA reduces CO_2 assimilation via a reduction in ribulose bisphosphate regeneration I transgenic tobacco plants. *Planta* **195,** 369–378.

Quick, W. P., Schurr, U., Schreibe, R., Schulze, E.-D., Rodermel, S. R., Bogorad, L., and Stitt, M. (1991). Decreased ribulose-1,5-phosphate carboxylase-oxygenase in transgenic tobacco transformed with 'antisense' rbcS. I Impact on photosynthesis in ambient growth conditions. *Planta* **183,** 542–554.

Read, B. A., and Tabita, F. R. (1992a). Amino acid substitutions in the small subunit of ribulose-1,5-bisphosphate carboxylase/oxygenase that influence catalytic activity of the holoenzyme. *Biochemistry* **31,** 519–525.

Read, B. A., and Tabita, F. R. (1992b). A hybrid ribulosebisphosphate carboxylase/oxygenase enzyme exhibiting a substantial increase in substrate specificity factor. *Biochemistry* **31,** 5553–5560.

Riesmeier, J. W., Frommer, W. B., and Willmitzer, L. (1994). Evidence for an essential role of the sucrose transporter in phloem loading and assimilate partitioning. *EMBO J.* **13,** 1–7.

Robinson, S. P., and Walker, D. A. (1981). Photosynthetic carbon reduction cycle. *In* "The Biochemistry of Plants" (M. D. Hatch and N. K. Boardman, eds.), vol. 8, pp. 193–236. Academic Press, New York.

Roy, H., and Andrews, T. J. (2000). Rubisco: Assembly and mechanism. *In* "Photosynthesis: Physiology and Metabolism" (R. C. Leegood, T. D. Sharkey, and S. von Caemmerer, eds.), pp. 53–83. Kluwer Academic, Dordrecht.

Roy, H., and Cannon, S. (1988). Ribulose bisphosphate carboxylase assembly: What is the role of the large subunit binding protein? *Trends Biochem. Sci.* **13,** 163–165.

Ruf, S., Hermann, M., Berger, I. J., Carrer, H., and Bock, R. (2001). Stable genetic transformation of tomato plastids and expression of a foreign protein in fruit. *Nat. Biotechnol.* **19,** 870–875.

Sage, R. F. (1990). A model describing the regulation of ribulose-1,5-bisphosphate carboxylase, electron transport, and triose phosphate use in response to light intensity and CO_2 in C_3 plants. *Plant Physiol.* **94,** 1728–1734.

Sage, R. F., Sharkey, T. D., and Seemann, J. R. (1989). Acclimation of photosynthesis to elevated CO_2 in five C3 species. *Plant Physiol.* **89,** 590–596.

Schleiff, E., Soll, J., Sveshnikova, N., Tien, R., Wright, S., Dabney-Smith, C., Subramanian, C., and Bruce, B. D. (2002). Structural and guanosine triphsophate/diphosphate requirements for transit peptide recognition by the cytosolic domain of the chloroplast outer envelope receptor, Toc34. *Biochemistry* **41,** 1934–1946.

Seemann, J. R., Badger, M. R., and Berry, J. A. (1984). Variations in the specific activity of ribulose-1,5-bisphosphate carboxylase (EC 4.1.1.39) between species utilizing differing photosynthetic pathways. *Plant Physiol.* **74,** 791–794.

Sekowska, A., Denervaud, V., Ashida, H., Michoud, K., Hass, D., Yokota, A., and Danchin, A. (2004). Bacterial variations on the methionine salvage pathway. *BMC Microbiol.* **4,** 9.

Shibata, M., Katoh, H., Sonoda, M., Ohkawa, H., Shimoyama, M., Fukuzawa, H., Kaplan, A., and Ogawa, T. (2002). Genes essential to sodium-dependent bicarbonate transport in cyanobacteria: Function and phylogenetic analysis. *J. Biol. Chem.* **277,** 18658–18664.

Shikanai, T., Foyer, C. H., Dulieu, H., Parry, M. A., and Yokota, A. (1996). A point mutation in the gene encoding the Rubisco large subunit interferes with holoenzyme assembly. *Plant Mol. Biol.* **31,** 399–403.

Shikanai, T., Takeda, T., Yamauchi, H., Sano, H., Tomizawa, K., Yokota, A., and Shigeoka, S. (1998). Inhibition of ascorbate peroxidase under oxidative stress in tobacco having bacterial catalase in chloroplasts. *FEBS Lett.* **428,** 47–51.

Sidorov, V. A., Kasten, D., Pamg, S. Z., Hajdukiewicz, P. T., Staub, J. M., and Nehra, N. S. (1999). Technical advance: Stable chloroplast transformation in potato: Use of green fluorescent protein as a plastid marker. *Plant J.* **19,** 209–216.

Smith, M. D., Ghosh, S., Dumbroff, E. B., and Thompson, J. E. (1997). Characterization of thylakoid-derived lipid-protein particles bearing the large subunit of ribulose-1,5-bisphosphate carboxylase/oxygenase. *Plant Physiol.* **115,** 1073–1082.

Spreitzer, R. J. (2003). Role of the small subunit in ribulose-1,5-bisphosphate carboxylase/oxygenase. *Arch. Biochem. Biophys.* **414,** 141–149.

Spreitzer, R. J., and Salvucci, M. E. (2002). RUBISCO: Structures, regulatory interactions, and a better enzyme. *Annu. Rev. Plant Biol.* **53,** 441–475.

Stephanopoulos, G. N., Aristidou, A. A., and Nielsen, J. (1998). "Metabolic Engineering: Principles and Methodologies." 461–533. Academic Press, San Diego.

Stitt, M., and Sonnewald, U. (1995). Regulation of metabolism in transgenic plants. *Annu. Rev. Plant Physiol. Plant Mol. Biol.* **46,** 341–368.

Sugawara, H., Yamamoto, H., Shibata, H., Inoue, T., Okada, S., Miyake, C., Yokota, A., and Kai, Y. (1999). Crystal structure of carboxylase-oriented ribulose-1,5-bisphosphate carboxylase/oxygenase from a thermophilic red alga, *Galdieria partita*. *J. Biol. Chem.* **274,** 15655–15661.

Svab, Z., and Maliga, P. (1993). High-frequency plastid transformation in tobacco by selection for a chimeric aadA gene. *Proc. Natl. Acad. Sci. USA* **90,** 913–917.

Takeuchi, Y., Akagi, H., Kamasawa, N., Osumi, M., and Honda, H. (2000). Aberrant chloroplasts in transgenic rice plants expressing a high level of maize NADP-dependent malic enzyme. *Planta* **211,** 265–274.

Tamoi, M., Ishikawa, T., Takeda, T., and Shigeoka, S. (1996a). Enzymic and molecular characterization of NADP-dependent glyceraldehyde-3-phosphate dehydrogenase from *Synechococcus* PCC 7942: Resistance of the enzyme to hydrogen peroxide. *Biochem. J.* **316,** 685–690.

Tamoi, M., Ishikawa, T., Takeda, T., and Shigeoka, S. (1996b). Molecular characterization and resistance to hydrogen peroxide of two fructose-1,6-bisphosphatases from *Synechococcus* PCC 7942. *Arch. Biochem. Biophys.* **334,** 27–36.

Tamoi, M., Murakami, A., Takeda, T., and Shigeoka, S. (1998). Lack of light/dark regulation of enzymes involved in the photosynthetic carbon reduction cycle in cyanobacteria, *Synechococcus* PCC 7942 and *Synechocystis* PCC 6803. *Biosci. Biotechol. Biochem.* **62,** 374–376.

Tamoi, M., Kanaboshi, H., Miyasaka, H., and Shigeoka, S. (2001). Molecular mechanisms of the resistance to hydrogen peroxide of enzymes involved in the Calvin cycle from halotolerant Chlamydomonas sp. W80. *Arch. Biochem. Biophys.* **390,** 176–185.

Tamoi, M., Nagata, M., Yabuta, Y., and Shigeoka, S. (2005). Carbon metabolism in the Calvin cycle. *Plant Biotechnol.* **22,** 355–360.

Tamoi, M., Nagaoka, M., Miyagawa, Y., and Shigeoka, S. (2006). Contribution of fructose-1,6-bisphosphatase and sedoheptulose-1,7-bisphosphatase to the photosynthetic rate and carbon flow in the Calvin cycle in transgenic plants. *Plant Cell Physiol.* **47,** 380–390.

Tolbert, N. E. (1994). Role of photosynthesis and photorespiration in regulating atmospheric CO_2 and O_2. *In* "Regulation of Atmospheric CO_2 and O_2 by Photosynthetic Carbon Metabolism" (N. E. Tolbert and J. Preiss, eds.), pp. 8–33. Oxford University Press, New York.

Tsuchida, H., Tamai, T., Fukayama, H., Agarie, S., Nomura, M., Onodera, H., Ono, K., Nishizawa, Y., Lee, B., Hirose, S., Toki, S., Ku, M. S., *et al.* (2001). High level expression of C_4-specific NADP-malic enzyme in leaves and impairment of photoautotrophic growth in a C_3 plant, rice. *Plant Cell Physiol.* **42,** 138–145.

Tsujii, H. (2000). Food shortage in the 21st century and its implications for agricultural research. *In* "Challenge of Plant and Agricultural Sciences to the Crisis of Biosphere on the Earth in the 21st Century" (K. Watanabe and A. Komamine, eds.), pp. 5–28. Landes Bioscience, Georgetown, Texas.

Uedan, K., and Sugiyama, T. (1976). Purification and characterization of phosphoenolpyruvate carboxylase from maize leaves. *Plant Physiol.* **57,** 906–910.

Uemura, K., Suzuki, Y., Shikanai, T., Wadano, A., Jensen, R. G., Chmara, W., and Yokota, A. (1996). A rapid and sensitive method for determination of relative specificity of RuBisCO from various species by anion-exchange chromatography. *Plant Cell Physiol.* **37,** 325–331.

Uemura, K., Anwaruzzaman, K., Miyachi, S., and Yokota, A. (1997). Ribulose-1,5-bisphosphate carboxylase/oxygenase from thermophilic red algae with a strong specificity for CO_2 fixation. *Biochem. Biophys. Res. Commun.* **233,** 568–571.

Uemura, K., Tokai, H., Higuchi, T., Murayama, H., Yamamoto, H., Enomoto, Y., Fujiwara, S., Hamada, J., and Yokota, A. (1998). Distribution of fallover in the carboxylase reaction and fallover-inducible sites among ribulose 1,5-bisphosphate carboxylase/oxygenases of photosynthetic organisms. *Plant Cell Physiol.* **39,** 212–219.

Uemura, K., Shibata, N., Anwaruzzaman, R., Fujiwara, M., Higuchi, T., Kobayashi, H., Kai, Y., and Yokota, A. (2000). The role of structural intersubunit microheterogeneity in the regulation of the activity in hysteresis of ribulose 1,5-bisphosphate carboxylase/oxygenase. *J. Biochem.* **128,** 591–599.

von Caemmerer, S., and Furbank, R. T. (1999). Modeling of C_4 photosynthesis. *In* "C4 Plant Biology" (R. F. Sage and R. K. Monson, eds.), pp. 173–211. Academic Press, San Diego.

Vörösmarty, C. J., Green, P., Salisbury, J., and Lammers, R. B. (2000). Global water resources: Vulnerability from climate change and population growth. *Science* **289,** 284–288.

Voznesenskaya, E. V., Franceschi, V. R., Kiirats, O., Freitag, H., and Edwards, G. E. (2001). Kranz anatomy is not essential for terrestrial C_4 plant photosynthesis. *Nature* **414,** 543–546.

Weise, A., Barker, L., Kühn, C., Lalonde, S., Buschmann, H., Frommer, W. B., and Ward, J. M. (2000). A new subfamily of sucrose transporters, SUT4, with low affinity/high capacity localized in enucleate sieve elements of plants. *Plant Cell* **12,** 1345–1356.

Whitney, S. M., and Andrews, T. J. (2001a). Plastome-encoded bacterial ribulose-1,5-bisphosphate carboxylase/oxygenase (RubisCO) supports photosynthesis and growth in tobacco. *Proc. Natl. Acad. Sci. USA* **98,** 14738–14743.

Whitney, S. M., and Andrews, T. J. (2001b). The gene for the ribulose-1,5-bisphosphate carboxylase/oxygenase (Rubisco) small subunit relocated to the plastid genome of tobacco directs the synthesis of small subunits that assemble into Rubisco. *Plant Cell* **13,** 193–205.

Whitney, S. M., Von Caemmerer, S., Hudson, G. S., and Andrews, T. J. (1999). Directed mutation of the Rubisco large subunit of tobacco influences photorespiration and growth. *Plant Physiol.* **121,** 579–588.

Whitney, S. M., Baldet, P., Hudson, G. S., and Andrews, T. J. (2001). Form I Rubiscos from non-green algae are expressed abundantly but not assembled in tobacco chloroplasts. *Plant J.* **26,** 535–547.

Woodrow, I. E., and Berry, J. A. (1988). Enzymatic regulation of photosynthetic carbon dioxide fixation. *Annu. Rev. Plant Physiol. Plant Mol. Biol.* **39,** 533–594.

Yokota, A., and Canvin, D. T. (1985). Ribulose bisphosphate carboxylase/oxygenase content determined with [^{14}C]carboxypentitol bisphosphate in plants and algae. *Plant Physiol.* **77,** 735–739.

Yokota, A., and Kitaoka, S. (1985). Correct pK values for dissociation constant of carbonic acid lower the reported K_m values of ribulose bisphosphate carboxylase to half. Presentation of a nomograph and an equation for determining the pK values. *Biochem. Biophys. Res. Commun.* **131,** 1075–1079.

Zhang, X. H., Ewy, R. G., Widholm, J. M., and Portis, A. R., Jr. (2002). Complementation of the nuclear antisense *rbcS*-induced photosynthesis deficiency by introducing an *rbcS* gene into the tobacco plastid genome. *Plant Cell Physiol.* **43,** 1302–1313.

CHAPTER 5

Genetic Engineering of Seed Storage Proteins

David R. Holding and **Brian A. Larkins**

Contents

Abstract

Seeds synthesize and accumulate variable amounts of carbohydrate, lipid, and protein to support their growth, development, and germination. The process of desiccation during seed maturation preserves these nutrients for long periods, making seeds an excellent food source and livestock feed. Over the millennia, human selection for high-yielding seed crops has resulted in dramatic increases in the accumulation of valuable nutrients and the reduction of toxic compounds and chemicals that affect the taste of foods made from seeds. However, in some cases, selection has resulted in a reduction in

Department of Plant Sciences, University of Arizona, Tucson, Arizona 85721

Advances in Plant Biochemistry and Molecular Biology, Volume 1
ISSN 1755-0408, DOI: 10.1016/S1755-0408(07)01005-3

the amount or quality of certain nutrients. Many types of seeds are adequate in one nutritional aspect but inadequate in others. Genetic engineering has created the opportunity to use the beneficial traits of certain types of seeds and ameliorate the negative aspects of others. This chapter summarizes the progress that has been made toward the improvement of seed and nonseed crops using transgenic expression of seed storage proteins. We explain the limitations of these approaches and describe promising areas of research such as reduction of allergenic seed components. We also discuss economic and ethical issues that impact this field.

Key Words: Protein quality, GM crop, essential amino acids, sulfur, methionine, lysine, glutenin, gluten, allergen, maize, soybean, wheat, prolamin, 11S globulin, 7S globulin, 2S albumin.

1. INTRODUCTION

1.1. The nature of seeds

Seeds provide a mechanism by which many types of plants propagate, and they are an important food source for many animals, including humans. The seed contains a dormant embryo and a mixture of stored metabolites (protein, starch, and lipid) that support its germination and prephotosynthetic growth. The storage proteins are a source of nitrogen and sulfur for the synthesis of new enzymes in the germinating seedling, while the starch and lipid initially provide the energy and carbon skeletons for making a variety of organic molecules. In angiosperms, which include most seed crops of agricultural importance, these storage compounds are deposited in one or more specialized tissues in the seed: the endosperm (especially in the cereals), the cotyledons of the embryo (particularly in legumes), or more rarely, the maternal perisperm tissue, as in the case of beet (Bewley and Black, 1995).

1.2. Metabolites stored in seeds and their uses

The storage proteins, carbohydrates, and lipids of particular seed crops have unique chemistries that are responsible for the physical and functional characteristics of the foods created from them. For example, the storage proteins in wheat, corn, and soybeans are responsible for the bread-making (Shewry *et al.*, 2003a), tortilla-making (Hamaker and Larkins, 2000), and tofu-making (Saio *et al.*, 1969) characteristics of their respective flours. The structure of starch, which can be altered by various mutations, allows creation of candies, sauces, or puddings with unique gelling characteristics (Orthoefer, 1987). The high contents of monounsaturated fatty acids found in olives, nuts, and rape seeds (Canola) produce the healthiest types of cooking oils (Taubes, 2001).

The nature of storage proteins, starches, and oils in seeds is subject to genetic variation and through selection, plant breeders have been able to create varieties of crop plants with unique compositions of these compounds that make them suitable for particular uses. However, there are limits to the natural qualitative and quantitative variation of these molecules, and this places restrictions on what breeders can accomplish with conventional methods of crop improvement. Furthermore, domestication and breeding of wild species for use as seed crops occurred through selective pressure for a limited number of traits, most notably improved yield. In some cases, this led to selection for one particular attribute at the expense of others. For example, the sulfur amino acid content of modern domestic corn appears to be much lower than that of its wild ancestors (Swarup *et al.*, 1995). Conventional plant breeding is sometimes analogized to working in a "black box" because it is possible to monitor only a limited number of traits during this process.

With the advent of plant genetic engineering technology, it became possible to consider novel ways of altering and enhancing seed storage metabolites. Indeed, biotechnology is currently being used to modify a number of crop traits, including the nature of the protein, starch, and lipid in seeds. In this chapter, we consider research that is being done to improve the nutritional quality and functional characteristics of seed storage proteins. Before describing this research and its potential in detail, we first provide some background information regarding the nature of seed storage proteins, how they are synthesized in seeds, and how they influence the nutritional value and the functional properties of our food and livestock feed.

1.3. Characterization of seed storage proteins

A modern classification system for seed proteins separates them into storage proteins, structural and metabolic proteins, and protective proteins, with certain proteins belonging to more than one of these classes (Shewry and Casey, 1999). Based on the knowledge of their molecular structure, the major groups of seed storage proteins are now classified as prolamins, 2S albumins, 7–8S globulins, and 11–12S globulins, where S refers to the sedimentation coefficient (Shewry and Casey, 1999). The distribution of these proteins in economically important crops is shown in Table 5.1. In general, globulins and albumins are the major components in dicotyledonous species, whereas prolamins predominate in most cultivated cereals.

Seed storage proteins are synthesized on rough endoplasmic reticulum (ER) membranes. They can be retained in the ER as localized protein accretions (protein bodies or PBs) or they can be transported, often via the Golgi complex, to specialized protein storage vacuoles (PSVs). PBs become deposited in PSVs either directly through autophagy or through the endomembrane secretory system. These pathways are illustrated in Fig. 5.1.

1.3.1. Prolamins

Prolamins were the first group of storage proteins to be widely studied. They account for about half of the grain nitrogen in most cereals, although as with other types of storage proteins, their levels vary considerably, depending on nitrogen

TABLE 5.1 Distribution of classes of seed storage protein in agronomically important seed crops[a]

	2S albumins	7–8S globulins	11–12S globulins	Prolamins
Major components	Legumes Crucifers Composites Castor bean Cottonseed Brazil nut	Legumes Cottonseed Palms Cocoa	Legumes Composites Cucurbits Oats and rice Crucifers Cannabis Brazil nut	Cereals
Minor components		Cereals	French bean	Oats Rice

[a] Adapted from Shewry and Casey (1999).

and sulfur availability (Shewry *et al.*, 1983; Tabe *et al.*, 2002). Prolamins are synthesized on rough ER membranes, and they can form accretions (PBs) directly in the ER or be transported into specialized PSVs (Fig. 5.1) (Herman and Larkins, 1999). In corn and wheat, prolamins account for about 60–70% of the endosperm protein, whereas in oats and rice they account for less than 10% of the protein (Shewry and Tatham, 1999).

Prolamins have been classified according to size and sulfur content, but no standard nomenclature exists for their classification between species. Prolamins are typically very rich in proline and glutamine, and are deficient, if not devoid, of several essential amino acids, including lysine, tryptophan, tyrosine, and threonine. As a result, monogastric animals receiving diets in which cereals are the primary protein source often develop protein deficiency disorders (Bhan *et al.*, 2003). In humans, such a deficiency is called kwashiorkor that, in addition to retarding growth and development, causes immunologic impairment and thus susceptibility to life-threatening infections (Scrimshaw, 2003). In some cereals, mutations have been found that reduce prolamin synthesis while increasing the proportion of more nutritional types of proteins (Habben and Larkins, 1995; Nelson, 2001). However, such mutants are generally associated with deleterious phenotypes, and for the most part have not been commercially developed. The fact that all classes of prolamin genes encode proteins deficient in essential amino acids means that such nutritional deficiencies are not amenable to correction by conventional plant breeding. Consequently, molecular biologists have sought to improve cereal protein quality by genetic engineering of genes encoding proteins with high levels of essential amino acids. Since prolamins also affect the functional characteristics of cereal flours, such as the bread-making quality of wheat (Shewry and Halford, 2002; Shewry *et al.*, 2003a) and the digestibility of the grain (Oria *et al.*, 2000), there is also interest in increasing or decreasing the synthesis of particular types of prolamin proteins.

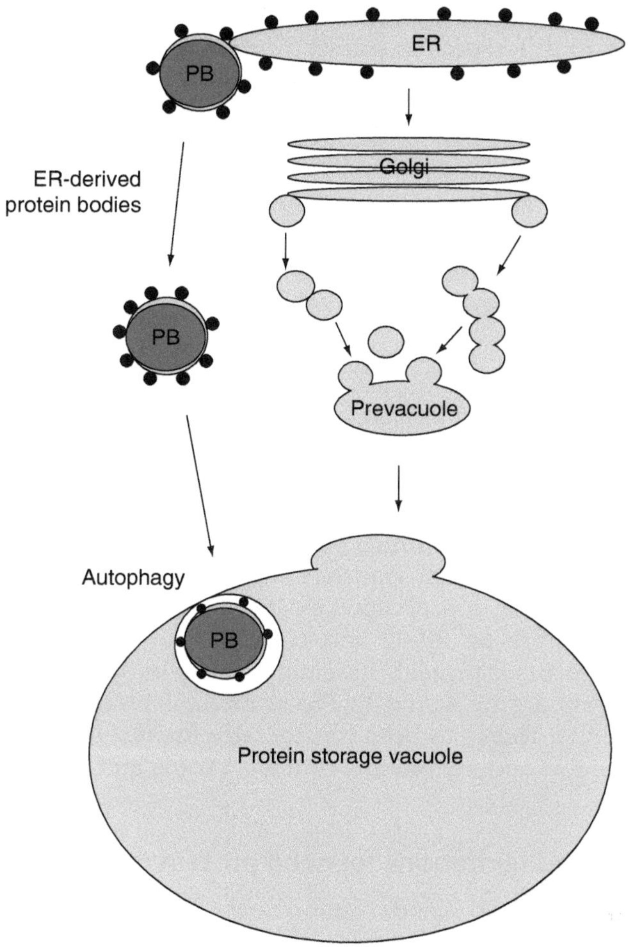

FIGURE 5.1 Diagram illustrating the ontogeny of PBs and protein storage vacuoles (PSVs). PBs form through the aggregation of storage proteins within the ER or PSVs. After formation, PBs can either remain attached to the ER or bud off and form separate organelles, that is PSVs. PBs can accumulate in the cytoplasm or become sequestered into PSVs by autophagy. PSVs are formed as the consequence of ER-synthesized storage proteins progressing through the endomembrane secretory system to specialized vacuoles (PSVs) for accumulation. Reprinted from Herman and Larkins (1999) with permission from the ASPB. (See Page 4 in Color Section.)

1.3.2. Globulins

Globulins are present to some extent in all seeds of all plants but they are the main storage proteins in most dicots and certain monocots, such as oats and rice (Table 5.1). The major storage globulins comprise the 11–12S and 7S groups and are often called legumins and vicilins, the common names given to the 11S and 7S proteins in peas. However, the 11–12S and 7S proteins typically have common names in each species (Casey, 1999). The 7S globulins exist as trimeric structures with subunit sizes of 50–70 kDa (Lawrence *et al.*, 1994), and the 11–12S globulins are hexamers with subunit sizes 60–80 kDa (Adachi *et al.*, 2003). Their size variation is due to differences in primary structure as well as posttranslational modifications. During synthesis, subunits of the proteins pass through the ER and (in some cases) the Golgi body (Fig. 5.1). They undergo partial assembly in the ER and are finally deposited in PSVs derived from the large central vacuole (Herman and Larkins, 1999; Kermode and Bewley, 1999). Dicot seeds, especially legumes, are rich sources of protein but the low levels of methionine (an essential amino acid) and cysteine in their storage globulins limit their nutritional value. Consequently, increasing the level of these sulfur-containing amino acids is a major goal for their improvement through biotechnology).

1.3.3. Albumins

Albumins were first defined as a separate group of seed proteins on the basis of their water solubility (Osborne, 1924), but it was not until the 1980s that sucrose density gradient sedimentation was used to definitively identify storage proteins of this type in seeds from a diverse range of species (Shewry and Pandya, 1999; Youle and Huang, 1981). Albumins have sedimentation coefficients of 2S, and though they exhibit substantial sequence and structural polymorphism between species, some amino acid conservation exists. Albumins typically exist in heterodimeric forms, comprising 30–40 and 60–90 amino acid subunits, which are derived from a precursor protein. Assembly occurs in the lumen of the ER, after which the proteins are delivered to PSVs for final proteolytic processing and deposition (Fig. 5.1). There has been considerable interest in 2S albumins because of their high cysteine and methionine contents (Youle and Huang, 1981).

1.4. Challenges and limitations for seed protein modification

The goal of increasing the essential amino acid content of storage proteins could be achieved by using site-directed mutagenesis to modify the coding sequences of the native storage protein genes. Alternatively, genes encoding foreign or even artificial proteins containing high levels of the deficient amino acids could be expressed. However, in order to change the essential amino acid composition of the seed, it is necessary to produce sufficient quantities of the transgenic protein to compensate for the high levels of endogenous storage proteins. Seed storage proteins are typically encoded by multigene families (Shotwell and Larkins, 1989), which partially explains the high level of storage protein synthesis in seeds. Consequently, production of sufficient quantities of a protein from a transgene that exists in one or a few copies presents a considerable technical challenge. One way of circumventing the problem of high levels of endogenous storage proteins with poor nutritional quality is to use naturally occurring mutations that reduce their level. Other possible approaches for reducing the expression of the storage protein genes are antisense gene expression, as was used to generate tomatoes with delayed ripening characteristics (Kramer and Redenbaugh, 1994) and gene silencing by cosuppression/RNA interference (RNAi) (Waterhouse *et al.*, 1998). Such techniques are also being applied toward reducing or eliminating other types of seed proteins that are antinutritional factors such as protease inhibitors, lectins, and various types of allergens.

Twenty years ago, genetic engineering of improved protein quality in seeds promised to be a straightforward process, as storage proteins were considered to have no enzymatic function and consequently appeared to be amenable to modification of primary and higher-order structures. In retrospect, this was a naive way of viewing storage proteins. It is now known that certain storage proteins have additional functions, such as protease inhibition in insect resistance. Furthermore, storage proteins possess unique structural features that direct their synthesis, secretion, and assembly into insoluble accretions in membrane vesicles. Deleterious structural modifications can create an unfolded protein response (Kaufman, 1999) that makes them unstable or creates a stress response that negatively affects the physiology of the cell.

In those early days, there was very limited knowledge of the factors affecting storage protein accumulation, including transcriptional and posttranscriptional regulation and posttranslational modifications and processing. It was thought that the relationship between amino acid biosynthesis and protein synthesis was important. For example, lysine availability in cereal endosperms was expected to influence the synthesis of lysine-containing storage proteins (Sodek and Wilson, 1970). This has yet to be demonstrated (Wang and Larkins, 2001) but the importance of sulfur availability for sulfur-containing storage protein synthesis is well documented (Tabe *et al.*, 2002). With hindsight, it appears that the processes of storage protein synthesis and deposition were not sufficiently well understood to reliably predict the effects of transgene expression.

Research during recent years has provided a great deal of fundamental information about the features of storage protein structure and synthesis, and the regulation of the genes encoding these proteins (Shewry and Casey, 1999). This knowledge has allowed progress toward improved seed protein quality. Much of this research, however, has been carried out in industrial laboratories, and consequently only a limited amount of information is publicly available. Questions about the health effects of consuming genetically modified (GM) crops have recently had an impact on this research, and this has no doubt slowed or delayed the development of these products at agricultural biotechnology companies (Dale, 1999). Hence, this overview most likely represents only a fraction of the actual research that has been done.

2. STORAGE PROTEIN MODIFICATION FOR THE IMPROVEMENT OF SEED PROTEIN QUALITY

2.1. Increasing methionine content

Perhaps the most successful approach for improving the quality of sulfur amino acids in seed crops has been the introduction of foreign genes encoding naturally sulfur-rich proteins, such as the 2S albumin from Brazil nut (BNA) (*Bertholletia excelsa*), which contains 18% methionine and 8% cysteine. BNA has been used to increase the methionine content of several crops (Tabe and Higgins, 1998). One of the first successful applications of this technology was with *Brassica napus* (rape/Canola) seeds. Rape seed is not particularly sulfur amino acid-deficient, but it was considered a good target for sulfur amino acid modification, because the processed meal is often mixed with (the more sulfur deficient) soybean in animal feeds. Altenbach *et al.* (1989, 1992) expressed BNA in transgenic Canola under control of the seed-specific *Phaseolus vulgaris* phaseolin promoter. BNA accumulated in a properly processed form up to 4% of total seed protein, resulting in up to a 33% increase in seed methionine content (Altenbach *et al.*, 1992).

Grain legumes are deficient in methionine and are consequently good candidates for protein improvement by transgenic approaches. When BNA was expressed in narbon bean (*Vicia narbonensis*) under control of the *Vicia faba* legumin B4 promoter, it was correctly processed and accumulated in the 2S albumin fraction where it accounted for up to 3% of total seed protein at maturity. This resulted in as much as a threefold increase in total seed methionine (Saalbach *et al.*, 1995), which could allow production of feedstuffs that do not require methionine supplementation (Tabe and Higgins, 1998). When expressed in soybean, BNA accumulated to more than 10% of total seed protein, resulting in up to a 50% increase in seed methionine content (R. Yung, personal communication). However, this high expression level was accompanied by downregulation of the endogenous sulfur-rich proteins, such as the Bowman-Birk proteinase inhibitor and albumins, including leginsulin. Leginsulin is a homologue of pea albumin A1 (Watanabe *et al.*, 1994), a protein that is reduced in sulfur-starved peas (Higgins *et al.*, 1986). Concomitantly, endogenous sulfur-poor storage proteins were found to be substantially increased in BNA-expressing soybean lines. The most prominent of these was the sulfur-free β-subunit of conglycinin (7S globulin), which accumulated to 30% of total seed protein, compared with 5% in control plants. These changed patterns of storage protein synthesis were similar to those observed during conditions of sulfur starvation. Furthermore, the changes could be alleviated, and even higher levels of BNA accumulated, when cotyledons of BNA-synthesizing soybean plants were cultured in the presence of exogenous methionine. Despite the observed increase of methionine in the transgenic soybean seeds, total seed sulfur remained virtually unchanged relative to control plants. Collectively, these data suggested that there is a limited pool of sulfur amino acids in soybean cotyledons, such that it is not possible to support an additional sulfur sink.

The identification of BNA as a major allergen of Brazil nut and the fact that this allergenicity was conveyed to the transgenic soybean (Nordlee *et al.*, 1996) diminished the potential transgenic use of BNA in soybean, which is used as an ingredient in many processed foods. Although BNA allergenicity may be less problematic in animal feed, this issue reduced the incentive to further develop BNA-containing seed crops for human consumption. Nevertheless, the common bean, *P. vulgaris*, a major food source in Latin America, Africa, and India, has since been targeted for methionine enrichment with BNA (Aragao *et al.*, 1999). Using the constitutive gene expression conferred by a double CaMV 35S promoter, several transgenic lines were reported that contain significantly elevated seed methionine (Aragao *et al.*, 1999).

In Australia, the grain legume, *Lupinus angustifolium*, is an important component of ruminant and nonruminant livestock feed. Lupin seed proteins are deficient in methionine and cysteine, and in order to increase animal productivity, pure methionine is routinely supplemented into the diets of pigs and poultry. Nonruminants are able to synthesize cysteine as long as adequate methionine is present. Administration of supplemental methionine has been shown to produce a 30–50% increase in wool growth in sheep (Pickering and Reis, 1993), but methionine supplementation is not practical in ruminants because it is lost due to destruction and incorporation into rumen microbial proteins. Molvig *et al.* (1997) expressed the sunflower seed albumin (SSA) protein in transgenic lupin as a means to increase methionine and cysteine intake in sheep. Lupin grain is fed to sheep in times of reduced pasture availability. SSA is reasonably stable in the rumen, and it is rich in methionine (16%) and cysteine (8%) (Kortt *et al.*, 1991; Mcnabb *et al.*, 1994). Although no overall increase in the total amount of seed sulfur was found, there was a significant increase in amino acid-bound sulfur. This consisted of a 94% increase in methionine and a 12% decrease in cysteine levels. The unexpected decrease in cysteine suggested that in the presence of a new sink for organic sulfur, the expression of other sulfur amino acid-containing proteins was altered and that, as with expression of BNA in soybean, the sulfur amino acid supply was limiting (Tabe and Droux, 2002). In preliminary feeding trials with rats, the transgenic seed was significantly better than wild type in terms of weight gain and protein digestibility (Molvig *et al.*, 1997). In subsequent trails with Merino sheep, the transgenic lupin seed diet was demonstrated to result in a 7% increase in weight gain and an 8% increase in wool growth as compared to a diet of nontransgenic lupin (White *et al.*, 2001).

The possibility of improving rice protein quality using an SSA gene as a methionine and cysteine donor was investigated (Hagan *et al.*, 2003). The SSA was modified with an ER retention signal and placed under control of the endosperm-specific wheat high-molecular weight (HMW) glutenin promoter. Although SSA accumulated to 7% of total seed protein, there was no overall change in seed sulfur amino acid content. Changes in the abundance of endogenous storage and nonstorage proteins indicated that synthesis of the transgenic protein simply caused a redistribution of limiting sulfur resources (Hagan *et al.*, 2003). It appears that rice, in common with soybean, may not have the capacity to support a transgenic sulfur sink, and that the high-level accumulation of transgenic sulfur-rich proteins creates a condition analogous to sulfur starvation in the seed. Depending on sulfur supply, the relative abundance of storage proteins that vary in sulfur content fluctuates in order to maintain nitrogen homeostasis (Tabe *et al.*, 2002). Although the intricacies of the regulatory mechanisms are only beginning to be understood (Tabe *et al.*, 2002), it is not surprising that the introduction of a new sulfur sink can cause multifaceted and unpredicted changes in protein synthesis in different plants that vary in storage protein composition.

Maize is not markedly deficient in methionine, but it is a candidate for sulfur amino acid improvement because it is often mixed in animal feeds with soybean meal. In addition, most varieties of domestic corn contain relatively low levels of the methionine-rich 10- and 18-kDa δ-zein proteins (Swarup *et al.*, 1995). The δ-zeins, which contain 23% or more methionine, are potentially useful proteins for increasing sulfur amino acid content in maize and other crop plants. The maize 10-kDa δ-zein, which is encoded by the single copy *Dzs10* gene, accumulates at low levels during endosperm development in most maize lines (Cruzalvarez *et al.*, 1991; Schickler *et al.*, 1993). This is due to a *trans*-acting posttranscriptional regulation mechanism linked to the *dzr1* locus (Benner *et al.*, 1989). Initial attempts to overexpress *Dzs10* in maize resulted in accumulation of δ-zein at up to 0.9% of total seed protein and variable increases in seed methionine (Anthony *et al.*, 1997). Similar to SSA expression in rice and BNA expression in soybean (Anthony *et al.*, 1997), potential gains from accumulation of the transgenic protein were often nullified by reduction in the levels of endogenous high-sulfur zeins. Lai and Messing (2002) created transgenic maize expressing a chimeric gene consisting of the coding region of *Dzs10* and the promoter and 5′ untranslated region of the highly expressed 27-kDa γ-zein, which is not subject to the same posttranscriptional regulation as *Dzs10*. Although the effects on endogenous high-sulfur zeins were not reported, uniformly high levels of 10-kDa δ-zein and methionine were observed and maintained over five backcross generations. Initial feeding studies with chicks suggested that the transgenic grain was as effective as nontransgenic grain supplemented with free methionine. Consequently, this product could eventually lead to corn-based rations that do not require methionine supplementation (Lai and Messing, 2002).

Coexpression of β-zein and δ-zein appears to enhance accumulation of the methionine-rich δ-zein. During PB formation in maize endosperm, the β- and γ-zeins associate in the ER, forming a continuous layer around a central core of α- and δ-zeins (Esen and Stetler, 1992; Lending and Larkins, 1989). An interaction between α- and γ-zeins was demonstrated (Coleman *et al.*, 1996), but the association of β- and δ-zeins is not well understood. Based on studies where genes encoding β- and δ-zeins were coexpressed in transgenic tobacco, there is an interaction between these proteins that helped increase δ-zein accumulation. When expressed individually, the β-zein and 10-kDa δ-zein formed unique, ER-derived, PBs in leaf cells. However, when coexpressed, 10-kDa δ-zein colocalized with β-zein and accumulated at a fivefold higher level (Bagga *et al.*, 1997). When the 18-kDa δ- and β-zeins were coexpressed, there was a 16-fold increase in δ-zein accumulation (Hinchliffe and Kemp, 2002). The increased accumulation of δ-zein was shown to result from a dramatic decrease in the rate of its degradation when β-zein was present (Hinchliffe and Kemp, 2002). There are no reports where this combination of proteins was tested in seeds. However, only modest increases in methionine and cysteine were observed when the β-zein was expressed alone in transgenic soybean (Dinkins *et al.*, 2001).

The methionine content of seeds can also be improved by reducing the abundance of endogenous sulfur-poor proteins. This strategy takes advantage of the plant's homeostasis mechanisms and results in the increased abundance of sulfur-rich proteins. An antisense approach was used to reduce the abundance of cruciferin, the main storage globulin of rape seed (*B. napus*), which is deficient in cysteine, methionine, and lysine. The transgenic plants accumulated more of the 2S albumin, napin, which has a better balance of essential amino acids. Seed lysine, methionine, and cysteine levels were increased by 10%, 8%, and 32%, respectively, over nontransformed controls (Kohnomurase *et al.*, 1995). In soybean, a cosuppression strategy was used to decrease the α- and α′-subunits of β-conglycinin, which contain low levels of sulfur amino acids (~1.4%) (Kinney *et al.*, 2001). This resulted in a concomitant increase in the accumulation of glycinin, which contains higher levels of sulfur amino acids. Notably, substantial amounts of proglycinin were shown to accumulate in novel, prevacuolar, PBs similar to those found in cereal seeds, rather than in Golgi-derived vacuolar vesicles. This may provide an alternative compartment for sequestering a variety of foreign proteins in soybeans (Kinney *et al.*, 2001).

2.2. Increasing lysine content

Perhaps the first successful research directed at improving protein quality in cereals was that of increasing the lysine content in maize (Glover and Mertz, 1987; Mertz *et al.*, 1964). The discovery that the *opaque2* (*o2*) mutation increased the lysine content of maize endosperm by decreasing the synthesis of prolamin (zein) proteins and increasing the level of other types of endosperm proteins prompted a search for similar mutants in other cereal species (Munck, 1992). Unfortunately, the low seed density and soft texture of this type of mutant were associated with a number of inferior agronomic properties, including brittleness and insect susceptibility. With only a few exceptions (Habben and Larkins, 1995), these mutants were not commercially developed. However, the subsequent identification of genetic modifiers (suppressors) that create a normal kernel phenotype while maintaining the higher lysine content caused by the *o2* mutation in maize permitted the development of a new type of *o2* mutant known as quality protein maize (QPM) (Prasanna *et al.*, 2001). QPM is currently being grown in several developing countries, where it is helping to alleviate protein deficiencies.

Other approaches to increase the lysine content of maize seed include site-directed mutagenesis of genes encoding the major prolamin proteins, α- and γ-zeins. As previously described, zeins are asymmetrically organized in ER-localized PBs, such that the most hydrophobic proteins, α-zeins, are found in the center and the more hydrophilic γ-zeins are at the periphery (Lending and Larkins, 1989). As zeins are essentially devoid of lysine (Woo *et al.*, 2001), the question arises as to whether the addition of such charged amino acids will disrupt the way in which zeins form accretions within the ER. Wallace *et al.* (1988) demonstrated the consequence of inserting lysine residues into different regions of a 19-kDa α-zein protein. When the modified proteins were synthesized in *Xenopus* oocytes, they formed accretions similar to the native proteins, suggesting that the presence of lysine was not detrimental to their aggregation and deposition. It was shown that green fluorescent protein insertions into a 22-kDa α-zein protein did not disrupt PB formation in yeast cells (Kim *et al.*, 2002). This observation suggests that α-zeins can be subjected to substantial structural modification and still aggregate into insoluble accretions.

A similar approach was taken with the sulfur-rich 27-kDa γ-zein. It was first demonstrated that 27-kDa γ-zein accumulates in ER-derived PBs in *Xenopus* oocytes and Arabidopsis (Geli *et al.*, 1994; Torrent *et al.*, 1994). When various modified versions of the protein were expressed in Arabidopsis, it was found that the N-terminal domain is necessary for ER retention and the C-terminal domain is necessary for PB formation. However, the central domain could be replaced with lysine-rich polypeptides without affecting protein stability and targeting (Geli *et al.*, 1994). These lysine-rich γ-zeins were also shown to accumulate to high levels in association with endogenous α- and γ-zeins in transiently transformed maize endosperm cells (Torrent *et al.*, 1997). Thus, the addition of lysine and other charged amino acids to α- and γ-zein proteins does not appear to alter their structural properties sufficiently to inhibit assembly into PBs. However, the consequences of these changes when the genes are expressed in stably transformed corn plants remain to be described. Another important question is whether sufficient levels of these proteins can be accumulated to make a significant increase in endosperm lysine content.

Rice contains very little prolamin; its major storage protein, a so-called glutelin, is a highly insoluble 11S globulin (Table 5.1). This protein is lysine deficient, whereas 11S globulins in legumes are deficient in sulfur-containing amino acids. Consuming both rice and legumes can provide an adequate balance of these essential amino acids, and this is especially important in vegetarian or low meat diets. Consequently, the expression of legume globulins in rice is one strategy for improving its amino acid balance. The gene encoding proglycinin, the precursor of soybean 11S globulin, was modified by replacing a variable region of amino acid sequence with a peptide encoding four contiguous methionine residues (Kim *et al.*, 1990). The genetically engineered protein was found to be stably accumulated in *Escherichia coli* cells (Kim *et al.*, 1990). In plant tissues, the modified glycinin accumulated to a similar degree as the mature protein and in the correct conformation (Utsumi *et al.*, 1993, 1994). For example, using the class 1 patatin promoter, tuber-specific expression of the modified glycinin, amounting to 0.2–1% of total protein, was achieved in transgenic potato (Utsumi *et al.*, 1994). The methionine-enriched and unmodified glycinins were transformed into rice under control of the promoter of the glutelin, *GluB-1*, which is one of the most highly expressed genes in rice endosperm (Katsube *et al.*, 1999). In transgenic rice, assembly of proglycinin into 7–8S trimeric structures, cleavage into acidic and basic subunits, and assembly into 11–12S hexameric structures in storage vacuoles all occurred in a manner similar to that in soybean. The endogenous glutelins formed 11S complexes with glycinins, indicating the transgenic protein did not adversely affect the assembly or accumulation of native storage proteins (Katsube *et al.*, 1999). Soybean glycinins have the property of lowering human serum cholesterol levels, and this fact offers an advantage for expression in rice, in addition to it being able to increase the lysine and, potentially, methionine contents (Kito *et al.*, 1993). Pea legumin, which is higher in lysine than rice glutelin, has also been expressed in rice endosperm in an effort to improve its amino acid composition (Sindhu *et al.*, 1997).

3. USE OF SEED STORAGE PROTEINS FOR PROTEIN QUALITY IMPROVEMENTS IN NONSEED CROPS

Besides seeds, a variety of other plant organs are valuable sources of protein. Potato tubers are the most important noncereal food crop, since they are consumed by humans and animals and used in the manufacture of starch and alcohol. Most transgenic research with potato has been directed toward improving yield as well as disease and pest resistance (Doreste *et al.*, 2002; Gulina *et al.*, 1994; Hausler *et al.*, 2002), rather than improving protein quality. Potato is not only protein deficient but also low in lysine, tyrosine, and sulfur amino acids (Jaynes *et al.*, 1986). Consequently, potato is a good candidate for protein improvement by genetic engineering. The possibility of using the BNA to enhance the sulfur content of potato has been investigated (Tu *et al.*, 1998). The CaMV 35S promoter was used to confer constitutive expression of the gene, and this resulted in modest levels of the protein in leaves and tubers. Significantly, it was possible to modify the variable region of the BNA gene so that the protein contains an even higher proportion of methionine. Furthermore, since the allergenicity of the protein appears to reside in this region, it may ultimately be possible to engineer nonallergenic versions of this protein (Tu *et al.*, 1998). The sulfur-rich maize δ-zein has also been expressed in potato tubers, resulting in a substantial increase in sulfur amino acid levels (Li *et al.*, 2001).

The gene encoding the storage albumin from *Amaranthus hypochondriacus* (*AmA1*) provides another potential mechanism to increase protein quality (Raina and Datta, 1992). This protein has a good balance of all the essential amino acids and apparently is nonallergenic. *AmA1* was expressed in potato under control of the CaMV 35S promoter and the tuber-specific, granule-bound starch synthase (GBSS) promoter, both of which resulted in substantial increases in all essential amino acids in the tubers (Chakraborty *et al.*, 2000). The most highly expressing transgenic lines showed a 2.5- to 4-fold increase in tuber lysine, tyrosine, methionine, and cysteine levels, whereas the GBSS lines had a 4- to 8-fold increase in these amino acids. These changes did not result in the depletion of endogenous proteins (Chakraborty *et al.*, 2000). Consequently, transgenic expression of the *AmA1* gene is a promising approach for improvement of protein quality in grain and nongrain crops.

The foliage of pasture crops is also a target for methionine enhancement and may provide a more efficient way to enrich the ruminant diet than with seeds, such as the previously described transgenic lupins. Ruminant livestock, such as cattle and sheep, require methionine in their diet. As previously noted, it is particularly important for sheep that require large amounts of sulfur amino acids for wool production. SSA is a good protein to produce in pasture crops because it is resistant to digestion in the rumen, allowing its amino acids to be absorbed in the small intestine. The subterranean clover (*Trifolium subterraneum*), which is widely cultivated in Australia, has been transformed with a gene encoding the SSA protein modified with an ER retention signal. Transgenic plants accumulated SSA up to 1.2% of total leaf protein (Khan *et al.*, 1996), but the results of sheep feeding trials have not been reported. Similar constructs were introduced into white clover (*Trifolium repens*), but much lower levels of the transgenic protein were found to accumulate in the leaves (Christiansen *et al.*, 2000).

The methionine-rich maize zein proteins have also been investigated for their ability to raise foliage methionine levels. When the δ-*zein* gene was constitutively expressed in white clover, the protein accumulated at up to 1.3% of total protein in all the tissues (Sharma *et al.*, 1998). Birdsfoot trefoil (*Lotus corniculatus*) and alfalfa (*Medicago sativa*) are two other foliage crops that have been targeted for methionine improvement by transformation with genes encoding β- and γ-zeins (Bellucci *et al.*, 2002). Earlier work showed that expression of β- and γ-*zeins* in transgenic tobacco leaves led to the colocalization of these proteins in PBs, underlining the effectiveness of exploiting natural zein interactions in accumulating the proteins in transgenic tissues (Bellucci *et al.*, 2000).

Another approach to improve amino acid deficiencies made use of artificial genes designed to correct specific amino acid deficiencies in target tissues. One strategy employed random ligation of mixtures of small oligonucleotides containing a high proportion of codons for methionine and lysine (Yang *et al.*, 1989). The product was a gene encoding a protein without any clearly defined secondary structure, and it was associated with limited protein accumulation in potato tubers (Yang *et al.*, 1989). In an attempt to produce a synthetic protein with defined secondary structure, Keeler *et al.* (1997) designed 21-base pair oligonucleotides that encode coiled-coil heptad repeats, forming polypeptides containing up to 31% lysine and 20% methionine. Several different polypeptides were produced that contained up to eight heptad repeats. Under control of the soybean β-conglycinin promoter, this gene resulted in significant increases in lysine and methionine in tobacco seeds that were stable over three generations (Keeler *et al.*, 1997). Such tailor-made proteins are potentially interesting tools for improving the protein quality of seed and nonseed crops, but it remains to be seen whether they would be acceptable to consumers.

4. MODIFICATION OF GRAIN BIOPHYSICAL PROPERTIES

In the developed world, optimization of seed protein quality is more important for livestock feed than for human diets. Indeed, the vast majority of world grain consumption is in livestock rations. With the exception of rice, human grain consumption is mainly through processed foods, and optimization of particular processing characteristics for specific end uses is of paramount importance. Wheat, in particular, is mainly used as white flour, which after removal of the germ and the bran is essentially composed of starch and gluten proteins. The amount and composition of the gluten determines end use, with high-gluten flours primarily being used for bread and pasta making (Shewry *et al.*, 2003a). The storage proteins comprising the gluten form insoluble accretions in endosperm cells of the wheat grain, but when mixed with water they create viscoelastic matrices that are essential in the bread leavening process. The HMW-glutenin subunits (HMW-GSs) are considered to be the most important components of gluten and have been subjected to structural modification for studying their function and bread-making characteristics (Shewry and Halford, 2002). For a comprehensive review of the role of glutenins in determining wheat processing properties, the reader is directed to a review by Shewry *et al.* (2003a).

Large-scale bacterial expression allowed the production of homogeneous HMW-GSs, which is necessary for detailed structure—function analyses (Dowd and Bekes, 2002; Galili, 1989). Other studies expressing modified glutenins were directed at systematically dissecting the functional domains of these proteins (Anderson *et al.*, 1996; Shimoni *et al.*, 1997).

Research aimed at upregulating HMW-GSs in wheat developed in part from the demonstration that differences in gluten properties are due to allelic variation in the composition of HMW-GS (Payne, 1987). Cultivars of hexaploid bread wheat have six genes encoding HMW-GSs, with differences in gene expression resulting in variable amounts of these proteins (Shewry and Halford, 2002). Ectopic expression of genes encoding the 1Ax and 1Dx5 subunits led to variable accumulation of the transgenic proteins and, where studied, variable effects on gluten strength (Altpeter *et al.*, 1996; Alvarez *et al.*, 2000; Barro *et al.*, 1997; Blechl and Anderson, 1996; Popineau *et al.*, 2001). Several transgenic lines exhibiting stable expression of *1Ax1* driven by its own promoter have been characterized in detail following field trials (Vasil *et al.*, 2001). There was no evidence that expression of an extra *HMW-GS* gene resulted in gene silencing or any undesirable effect on yield, protein composition, or flour functionality, and in some of the transgenic lines, mixing time, loaf volume, and water absorbance improved relative to the control cultivar (Vasil *et al.*, 2001). However, in at least one other study, gene silencing of endogenous subunits was encountered (Alvarez *et al.*, 2000). The expression of *1Ax1* and *1Dx5* transgenes caused silencing of all the endogenous HMW-GSs, and rheological analysis showed a lower dough strength (Alvarez *et al.*, 2001). In the nonsilenced lines, a direct correlation was found between the number of *HMW-GS* genes expressed and bread dough elasticity (Barro *et al.*, 1997). One line overexpressing the 1Dx5 subunit exhibited a significant improvement in dough strength. In fact, it was necessary to mix the flour with a low gluten, soft flour in order to allow adequate mixing and dough development (Alvarez *et al.*, 2001). Similarly, very strong glutens giving rise to doughs with unusual mixing characteristics were obtained with a transgenic line overexpressing *1Dx5*, in comparison to a nearly isogenic line expressing *1Ax1* that had little effect (Popineau *et al.*, 2001). While both lines accumulated the transgenic HMW-GS protein at 50–70% of total HMW glutenin and exhibited increased glutenin aggregation, only the 1Dx5 transgenic line exhibited increased dough elasticity resulting from increased glutenin cross-linking (Popineau *et al.*, 2001). The possibility of using the viscoelastic properties of glutenins to produce novel dough characteristic in maize is being investigated (Sangtong *et al.*, 2002). The 1Dx5 HMW-GS was shown to be stably expressed and genetically transmitted in maize (Sangtong *et al.*, 2002), and experiments to test the viscoelastic properties of doughs produced from such transgenic lines are under way.

There is substantial evidence to suggest that disulphide cross-linking is important in stabilizing the wheat glutenin backbone (Shewry and Tatham, 1997).

Presence of the 1Bx20 HMW-GS in pasta wheat (*Triticum durum*) is associated with poor pasta-making quality (Liu *et al.*, 1996), and when present in bread wheat, it is associated with poor bread-making quality (Payne, 1987). This subunit has been sequenced and compared to the highly similar 1Bx7 HMW-GS (Shewry *et al.*, 2003b). 1Bx7 confers increased dough strength compared with 1Bx20 and contains two N-terminal cysteines, which are substituted with tyrosine residues in 1Bx20. Therefore, the poor dough-making properties conferred by 1Bx20 are thought to be due to its reduced ability to cross-link with the gluten network (Shewry *et al.*, 2003b). This may be the reason to target this HMW-GS for transgenic down-regulation. Many studies have demonstrated the feasibility of manipulating the properties of individual glutenin subunits in order to affect gluten structure but much remains to be learned about the interactions involved.

Although the HMW-GSs form the backbone of the elastomeric gluten network, the interaction of other glutenins and gliadins is believed to be important. A new family of low-molecular weight gliadins was reported (Clarke *et al.*, 2003). Sequence analysis and genetic mapping revealed homology to a 17-kDa barley protein involved in beer foam stability and a different chromosomal location in wheat from that of the glutenins and gliadins. Purification of an *E. coli*-expressed member of this family and incorporation into a base flour produced a stronger dough with a substantial increase in bread loaf height (Clarke *et al.*, 2003). This demonstrates the importance of other types of wheat storage proteins in gluten formation and suggests that such proteins may be suitable for transgenic modification to improve bread-making characteristics.

5. TRANSGENIC MODIFICATIONS THAT ENHANCE THE UTILITY OF SEED STORAGE PROTEINS

5.1. Managing allergenic proteins

As a preliminary evaluation of the safety of transgenic plants, the verification of substantial equivalence with the genetically unmodified counterpart is now widely employed (Kuiper *et al.*, 2001). Modern, transcriptomic, proteomic, and metabolomic profiling techniques can be a vital part of such testing. Although substantial equivalence measurements are not safety assessments in themselves, they can reveal biochemical differences that can then be subjected to more rigorous toxicological and immunologic testing.

A potential consequence of the genetic modification of crop plants is introduction or creation of allergens. This could occur in several possible ways, including introduction of unknown allergens with the transgenic protein itself, modification by the host transgenic plant of the immunogenic properties of the transgenic protein, modification of the immunogenicity of endogenous proteins in the transgenic plant, and dissemination of an allergen through pollen that induces respiratory sensitization (Moneret-Vautrin, 2002). Such risks need to be evaluated prior to widespread use of a transgenic crop plant. Unfortunately, the possibility of allergen induction can be exaggerated to the general public and used to fuel the idea that genetic modification is an unpredictable and irresponsible science. It is true that the allergenicity of proteins, such as BNA, may not be widely known before their introduction into a crop plant. However, the scientific community quickly becomes aware of such potential problems (Nordlee *et al.*, 1996) and acts appropriately. For example, the transgenic soybean plants expressing BNA were never commercially developed. As we gain a better understanding of the identity and epitopic composition of common allergenic proteins, their selective modification or elimination becomes feasible, and this could lead to the development of hypoallergenic versions.

Soybean consumption is a problem for some people and animals as it contains several dominant allergenic proteins: Gly m Bd 68K, Gly m Bd 28K, and Gly m Bd 30K (P34) (Ogawa *et al.*, 2000). The widespread use of soybean in the human foods and animal feeds makes it an obvious target for genetic engineering to remove or reduce these allergens. Gly m Bd 68K and Gly m Bd 28K are seed storage proteins, and some reduction of their levels has been achieved through the development of mutant lines (Ogawa *et al.*, 2000). However, such a strategy has not been successful with P34, which is an albumin and a member of the papain family of cysteine proteases (Ogawa *et al.*, 2000). Although this protein is a minor seed constituent, it is the most dominant soybean allergen (Yaklich *et al.*, 1999). While considered an albumin, P34 partitions into oil body membranes during processing, as well as with the globulin fraction (Kalinski *et al.*, 1992). Consequently, it is almost impossible to completely remove this protein from soybean isolates. Furthermore, its ubiquitous presence in cultivated and wild soybean varieties suggests that it will not be possible to reduce its level through conventional breeding (Yaklich *et al.*, 1999). However, through the sense expression of a Gly m Bd 30K cDNA, transgenic lines have been developed in which the endogenous *Gly m Bd 30K* gene is completely silenced (Herman *et al.*, 2003). A function for this protein has not been demonstrated but no overt phenotypic change was observed in the gene-silenced plants. These transgenic soybeans are currently being further evaluated in field trials (Herman *et al.*, 2003).

Rice induces allergic reactions in some people and this is a growing problem in some countries, like Japan (Watanabe, 1993). One of the major rice allergens was identified as a 16-kDa albumin (Matsuda *et al.*, 1988; Urisu *et al.*, 1991). This protein is encoded by a multigene family composed of at least ten members (Tada *et al.*, 2003), each of which has allergenic properties (Matsuda *et al.*, 1991). An antisense strategy was used to reduce the abundance of the 16-kDa albumin as well as other gene family members (Tada *et al.*, 1996). An 80% reduction in abundance of the 16-kDa rice allergen was achieved (Tada *et al.*, 1996), and the reduction in protein levels of other family members was proportional to their degree of nucleotide sequence identity with the transgene. Highly homologous proteins were markedly lowered, and proteins with less identity were hardly reduced at all (Tada *et al.*, 2003). Highly immunogenic proteins need only be present in minute quantities in order to elicit an immune response. Thus, these results suggest that the antisense strategy may not be suitable for complete removal of allergenic proteins, especially if they are encoded by divergent multigene families. In this regard, gene silencing strategies (Waterhouse *et al.*, 1998) that require smaller regions of DNA sequence identity may prove to be more suitable.

5.2. Managing seed antinutritional characteristics

Many seeds contain components that are antinutritional and therefore restrict grain utilization for human or livestock consumption. Transgenic approaches have the potential to selectively reduce or remove these components, thereby increasing the availability of seed storage proteins for nutrition.

In order to use soybean meal in animal feed, it must be heat treated first to inactivate the endogenous trypsin inhibitor (TI) and chymotrypsin inhibitor (CI) proteins, which otherwise reduce protein digestibility. Identification of soybean lines without TI and CI activities could reduce soybean processing costs and increase amino acid availability, which can be reduced by excessive heat treatment (Herkelman *et al.*, 1993; Lee and Garlich, 1992). Screening of the USDA soybean germplasm collection led to the discovery of one line (ti) that lacked the A2 TI and manifested a 30–50% reduction in TI activity (Orf and Hymowitz, 1979). Expression of the gene encoding BNA in soybean, originally intended as a means of increasing the methionine level as described above, also resulted in a reduction in TI and CI activities (Streit *et al.*, 2001). To take advantage of both of these traits, transgenic soybean lines were created that express both BNA and the mutant ti allele of the Kunitz TI (Streit *et al.*, 2001). Compared with control plants, average reductions of 85% in TI and 61% in CI activities were observed in the absence of any significant changes in plant yield and size, maturation time, and protein and oil deposition (Streit *et al.*, 2001). While attempting to reduce seed TI and CI levels through either breeding or transgenic means, it should be considered that both proteins make substantial contributions to seed sulfur amino acid levels. Furthermore, these proteins are thought to be part of a plant defense mechanism and may need to be compensated with alternative mechanisms (Clarke and Wiseman, 2000).

In recent years, rape seed (Canola/*B. napus*) has become one of the most important oilseed crops in the world, as the healthful characteristics of the largely monounsaturated fatty acid content of its oil are widely recognized. Rape seed meal is also an important source of protein for animal feed, since its 2S albumins (napins) are rich in sulfur-containing amino acids. However, the meal is not suitable for human nutrition due to the high levels of antinutritional compounds, like sinapine esters. Sinapine is therefore a target for reduction or removal (Leckband *et al.*, 2002) and this may be achieved by careful screening and breeding of low sinapine cultivars (Velasco and Mollers, 1998) and by genetic engineering. Nair *et al.* (2000) demonstrated a 40% reduction in sinapine by expressing Cauliflower mosaic virus 35S-antisense *B. napus* ferulate-5-hydroxylase (*BNF5H*) transgene in *B. napus*. More modest reductions (17%) were achieved when the seed-specific napin promoter was used. *BNF5H* has an as yet undefined role in sinapine synthesis (Nair *et al.*, 2000).

6. SUMMARY AND FUTURE PROSPECTS

In 2001, more than 65% of soybean acres and more than 20% of corn acres in the United States were planted with GM varieties (Lusk and Sullivan, 2002), indicating that, at least in the United States, crop biotechnology has largely been accepted at the farm level. This is partly due to the fact that most US cultivation of corn and soybean is for livestock feed, so the issue of consumer acceptance has not been a problem. Furthermore, the cost and labor savings resulting from reduced pesticide or herbicide use made possible by transgenic traits is directly realized by the farmer. Improving grain nutritional quality can reduce costs for the livestock farmer and will become more important as the practice of lowering the amount of protein in livestock rations to reduce nitrogen levels in manure becomes more widely adopted (Johnson *et al.*, 2001). Corn with improved nutritional characteristics can lower the costs for the livestock producer by reducing feed supplements, assuming that the modified grain is available at a competitive price (Johnson *et al.*, 2001). Feed cost savings resulting from a variety of possible nutritional modifications to corn seed have been estimated (Johnson *et al.*, 2001). For example, lysine is the first limiting amino acid in pigs receiving corn–soybean meal diets. If the lysine level in corn were to be doubled, it was calculated that feed cost savings would range from \$4.65 to \$6.89 per ton in 2001 (Johnson *et al.*, 2001).

Considering all that has been learned about storage protein structure and gene expression, it is somewhat surprising that there are currently no GM seed storage protein products on the market. However, the development of such crops to the point where they are commercially viable is a long and expensive process. Success depends on the product providing significant value relative to its cost, and this must be carefully projected before embarking on product development. Consideration must be given to questions such as whether the cost of creating and managing a high-methionine maize feedstock that does not require amino acid supplementation would allow the grain to be grown, marketed, and distributed at a competitive price. This chapter has described preliminary research using an array of ingenious approaches for improving protein quality by genetic engineering, and in many cases, limitations to transgene expression remain to be resolved. A few types of storage proteins make up the bulk of seed proteins, and their amino acid compositions determine the protein quality of the seed. In order to improve essential amino acid balances, the transgenic proteins must be accumulated at very high levels. Even using strong, seed-specific promoters, proteins encoded by low copy number transgenes generally accumulate to less than 5% of the total seed protein, and this is usually insufficient to produce the required improvements in protein quality. In cases such as BNA expression, where high-level transgenic protein accumulation was achieved, this often resulted in changes of endogenous proteins, so that the gain in protein quality was less significant than expected.

The use of genetic engineering for the modification of grain processing characteristics in crops, such as wheat, may ultimately be useful. Presently, transgenic research is providing an increased understanding of the roles of various HMW-GSs in gluten properties. However, given the complex nature and incomplete understanding of HMW-GS interactions, identifying modifications that will have value will require more research.

One promising application of GM technology in the near term is in the reduction or removal of antinutritional components and allergens from seeds. Perhaps the most time-consuming step here is determining the identity and epitopic composition of allergenic proteins. Food hypersensitivity in children and adults is the most common type of allergy (Chandra, 2002). Furthermore, it is increasing in prevalence (Maleki and Hurlburt, 2002) and the list of foods known to elicit allergic reactions is growing. In the future it will be possible to modify allergenic domains of essential endogenous proteins or remove them completely using gene silencing. Indeed, this technique can be used to downregulate entire gene families encoding allergenic proteins. The availability of genomic, transcriptomic, and proteomic data for crops such as rice, corn, and soybean should help in identifying these proteins and the gene families that encode them.

Early research on the genetic modification of storage proteins in crop plants was initiated in the absence of knowledge of many technical constraints, such as limitations to sulfur amino acid availability. Also influencing the consummation of this research are the contentious issues of consumer perception and acceptance of GM crops. To date, the most successful GM traits in crop plants, herbicide and insect resistance, allow decreased introduction of chemicals into the environment. Some people consider these traits to have benefited the producer more than the consumer. Although the potential grain nutritional improvements described here provide the most direct benefits to the livestock producer, they would reduce food costs and improve protein nutrition for people who consume the grain directly. Unfortunately, there are limited research resources in the developing countries where the immediate benefits of grain nutritional improvements for human consumption could be realized. At present, there is little incentive for biotechnology companies to invest heavily in the development of products for primary use in developing countries, despite the humanitarian value.

Some consumers remain skeptical about GM products due to negative perceptions of the agricultural biotechnology industry and perceived environmental or personal risks. However, consumers are benefiting from the environmental effects of reduced chemical use and the more cost-effective production of commodities. The development of products with improved nutritional value, enhanced taste and appearance, and increased shelf life will surely increase consumer appreciation of the value of GM crops.

In the past, information regarding the benefits of GM technology has not been effectively communicated to the general public. In a study, it was found that consumers reading about the benefits of GM soybeans were significantly more comfortable eating them than those reading about GM soybeans with no explanation of their benefit (Brown and Ping, 2003). However, the groups did not differ in their desire for labeling foods made with these soybeans (Brown and Ping, 2003). In the United States, most consumers are not aware of the extent that GM foods have entered the marketplace. In the United Kingdom, all products containing GM ingredients must be labeled as such, but in most cases this has discouraged consumers from buying them. For example, GM tomato products were sold by several UK supermarket chains in the nineties but were withdrawn due to poor sales following anti-GMO campaigns. Information regarding the nature of the transgenic modification and its potential for flavor improvement was not readily available to the consumer. Perhaps another reason for consumer skepticism, especially in Europe, is that while GM crops are frequently cited as a vital component in sustaining the growing human population, past research is perceived to have been shrouded in secrecy and the products thought to benefit only the large agricultural biotechnology companies. It is thus becoming increasingly clear that the scientific community must place a priority on educating the public about the immediate and future benefits as well as the safety of GM crops, if their potentials are to be realized.

ACKNOWLEDGEMENTS

We are grateful to Dr. Rudolf Jung at Pioneer Hi-Bred, Inc., for sharing unpublished data on BNA expression in transgenic soybean, and to Dr. Brenda Hunter and Dr. Bryan Gibbon for critical comments on the chapter. Our work is supported by grants from the National Science Foundation (DBI-0077676) and the Energy Biosciences Program of the Department of Energy (96ER20242).

REFERENCES

Adachi, M., Kanamori, I., Matsuda, T., Yagasaki, K., Kitamura, K., Mikami, B., and Utsumi, S. (2003). Crystal structure of soybean 11S globulin: Glycinin A3B4 homopolymer. *Proc. Natl. Acad. Sci. USA* **100,** 7395–7400.

Altenbach, S. B., Pearson, K. W., Meeker, G., Staraci, L. C., and Sun, S. S. M. (1989). Enhancement of the methionine content of seed proteins by the expression of a chimeric gene encoding a methionine-rich protein in transgenic plants. *Plant Mol. Biol.* **13,** 513–522.

Altenbach, S. B., Kuo, C. C., Staraci, L. C., Pearson, K. W., Wainwright, C., Georgescu, A., and Townsend, J. (1992). Accumulation of a Brazil nut albumin in seeds of transgenic canola results in enhanced levels of seed protein methionine. *Plant Mol. Biol.* **18,** 235–245.

Altpeter, F., Vasil, V., Srivastava, V., and Vasil, I. K. (1996). Integration and expression of the high-molecular-weight glutenin subunit 1Ax1 gene into wheat. *Nat. Biotechnol.* **14,** 1155–1159.

Alvarez, M. L., Guelman, S., Halford, N. G., Lustig, S., Reggiardo, M. I., Ryabushkina, N., Shewry, P., Stein, J., and Vallejos, R. H. (2000). Silencing of HMW glutenins in transgenic wheat expressing extra HMW subunits. *Theor. Appl. Genet.* **100,** 319–327.

Alvarez, M. L., Gomez, M., Carrillo, J. M., and Vallejos, R. H. (2001). Analysis of dough functionality of flours from transgenic wheat. *Mol. Breed.* **8,** 103–108.

Anderson, O. D., Kuhl, J. C., and Tam, A. (1996). Construction and expression of a synthetic wheat storage protein gene. *Gene* **174,** 51–58.

Anthony, A., Brown, W., Buhr, D., Ronhovde, G., Genovesi, D., Lane, T., Yingling, R., Aves, K., Rosato, M., and Anderson, P. (1997). Transgenic maize with elevated 10 kDa zein and methionine. *In* "Sulphur Metabolism in Higher Plants" (W. J. Cram, L. J. De Kok, I. Stulen, C. Brunold, and H. Ronnenberg, eds.), pp. 295–297. Blackhuys, Leiden.

Aragao, F. J. L., Barros, L. M. G., De Sousa, M. V., De Sa, M. F. G., Almeida, E. R. P., Gander, E. S., and Rech, E. L. (1999). Expression of a methionine-rich storage albumin from the Brazil nut (*Bertholletia excelsa* HBK, Lecythidaceae) in transgenic bean plants (*Phaseolus vulgaris* L., Fabaceae). *Genet. Mol. Biol.* **22,** 445–449.

Bagga, S., Adams, H. P., Rodriguez, F. D., Kemp, J. D., and Senguptagopalan, C. (1997). Coexpression of the maize delta-zein and beta-zein genes results in stable accumulation of delta-zein in endoplasmic reticulum-derived protein bodies formed by beta-zein. *Plant Cell* **9,** 1683–1696.

Barro, F., Rooke, L., Bekes, F., Gras, P., Tatham, A. S., Fido, R., Lazzeri, P. A., Shewry, P. R., and Barcelo, P. (1997). Transformation of wheat with high molecular weight subunit genes results in improved functional properties. *Nat. Biotechnol.* **15,** 1295–1299.

Bellucci, M., Alpini, A., and Arcioni, S. (2000). Expression of maize gamma-zein and beta-zein genes in transgenic *Nicotiana tabacum* and *Lotus corniculatus*. *Plant Cell Tissue Organ Cult.* **62,** 141–151.

Bellucci, M., Alpini, A., and Arcioni, S. (2002). Zein accumulation in forage species (*Lotus corniculatus* and *Medicago sativa*) and co-expression of the gamma-zein: KDEL and beta-zein: KDEL polypeptides in tobacco leaf. *Plant Cell Rep.* **20,** 848–856.

Benner, M. S., Phillips, R. L., Kirihara, J. A., and Messing, J. (1989). Genetic analysis of methionine-rich storage protein accumulation in maize. *Theor. Appl. Genet.* **78,** 761–767.

Bewley, J. D., and Black, M. (1995). "Seeds: Physiology of Development and Germination," 2nd Edn., p. 445. Plenum Press, New York and London.

Bhan, M. K., Bhandari, N., and Bahl, R. (2003). Management of the severely malnourished child: Perspective from developing countries. *Br. Med. J.* **326,** 146–151.

Blechl, A. E., and Anderson, O. D. (1996). Expression of a novel high-molecular-weight glutenin subunit gene in transgenic wheat. *Nat. Biotechnol.* **14,** 875–879.

Brown, J. L., and Ping, Y. C. (2003). Consumer perception of risk associated with eating genetically engineered soybeans is less in the presence of a perceived consumer benefit. *J. Am. Diet. Assoc.* **103,** 208–214.

Casey, R. (1999). Distribution and some properties of seed globulins. *In* "Seed Proteins" (P. Shewry and R. Casey, eds.), pp. 159–169. Kluwer Academic Publishers, Dordrecht, Boston, London.

Chakraborty, S., Chakraborty, N., and Datta, A. (2000). Increased nutritive value of transgenic potato by expressing a nonallergenic seed albumin gene from *Amaranthus hypochondriacus*. *Proc. Natl. Acad. Sci. USA* **97,** 3724–3729.

Chandra, R. K. (2002). Food hypersensitivity and allergic diseases. *Eur. J. Clin. Nutr.* **56,** S54–S56.

Christiansen, P., Gibson, J. M., Moore, A., Pedersen, C., Tabe, L., and Larkin, P. J. (2000). Transgenic *Trifolium repens* with foliage accumulating the high sulphur protein, sunflower seed albumin. *Transgenic Res.* **9,** 103–113.

Clarke, B. C., Phongkham, T., Gianibelli, M. C., Beasley, H., and Bekes, F. (2003). The characterisation and mapping of a family of LMW-gliadin genes: Effects on dough properties and bread volume. *Theor. Appl. Genet.* **106,** 629–635.

Clarke, E. J., and Wiseman, J. (2000). Developments in plant breeding for improved nutritional quality of soya beans II. Anti-nutritional factors. *J. Agric. Sci.* **134,** 125–136.

Coleman, C. E., Herman, E. M., Takasaki, K., and Larkins, B. A. (1996). The maize gamma-zein sequesters alpha-zein and stabilizes its accumulation in protein bodies of transgenic tobacco endosperm. *Plant Cell* **8,** 2335–2345.

Cruzalvarez, M., Kirihara, J. A., and Messing, J. (1991). Posttranscriptional Regulation of methionine content in maize kernels. *Mol. Gen. Genet.* **225,** 331–339.

Dale, P. J. (1999). Public concerns over transgenic crops. *Genome Res.* **9,** 1159–1162.

Dinkins, R. D., Reddy, M. S. S., Meurer, C. A., Yan, B., Trick, H., Thibaud-Nissen, F., Finer, J. J., Parrott, W. A., and Collins, G. B. (2001). Increased sulfur amino acids in soybean plants overexpressing the maize 15 kDa zein protein. *In Vitro Cell. Dev. Biol. Plant* **37,** 742–747.

Doreste, V., Ramos, P. L., Enriquez, G. A., Rodriguez, R., Peral, R., and Pujol, M. (2002). Transgenic potato plants expressing the potato virus X (PVX) coat protein gene developed resistance to the viral infection. *Phytoparasitica* **30,** 177–185.

Dowd, C., and Bekes, F. (2002). Large-scale expression and purification of high-molecular weight glutenin subunits. *Protein Expr. Purif.* **25,** 97–104.

Esen, A., and Stetler, D. A. (1992). Immunocytochemical localization of delta-zein in the protein bodies of maize endosperm cells. *Am. J. Bot.* **79,** 243–248.

Galili, G. (1989). Heterologous expression of a wheat high molecular-weight glutenin gene in *Escherichia coli*. *Proc. Natl. Acad. Sci. USA* **86,** 7756–7760.

Geli, M. I., Torrent, M., and Ludevid, D. (1994). Two structural domains mediate two sequential events in gamma-zein targeting-protein endoplasmic-reticulum retention and protein body formation. *Plant Cell* **6,** 1911–1922.

Glover, D. V., and Mertz, E. T. (1987). Corn. *In* "Nutritional Quality of Cereal Grains: Genetic and Agronomic Improvement" (R. A. Olson and K. T. Frey, eds.), pp. 183–336. ASA-CSSA-SSSA, Madison, WI.

Gulina, I. V., Shulga, O. A., Mironov, M. V., Revenkova, E. V., Kraev, A. S., Pozmogova, G. E., Yakovleva, G. Y., and Skryabin, K. G. (1994). Expression of a modified gene for the delta-endotoxin of *Bacillus thuringiensis* var. tenebrionis in transgenic potato plants. *Mol. Biol.* **28,** 748–753.

Habben, J. E., and Larkins, B. A. (1995). Improving protein quality in seeds. *In* "Seed Development and Germination" (J. Kijel and G. Galili, eds.), pp. 791–810. Marcel Dekker, Inc., New York, Basel, Hong Kong.

Hagan, N. D., Upadhyaya, N., Tabe, L. M., and Higgins, T. J. V. (2003). The redistribution of protein sulfur in transgenic rice expressing a gene for a foreign, sulfur-rich protein. *Plant J.* **34,** 1–11.

Hamaker, B. R., and Larkins, B. A. (2000). Maize food and feed: A current perspective and consideration of future possibilities. *In* "Transgenic Plants and Crops" (G. G. Khachatourians, A. McHughen, R. Scorza, W. K. Nip, and Y. H. Hui, eds.), pp. 637–654. Marcel Dekker, Inc., New York and Basel.

Hausler, R. E., Hirsch, H. J., Kreuzaler, F., and Peterhansel, C. (2002). Overexpression of C-4-cycle enzymes in transgenic C-3 plants: A biotechnological approach to improve C-3-photosynthesis. *J. Exp. Bot.* **53,** 591–607.

Herkelman, K. L., Cromwell, G. L., Cantor, A. H., Stahly, T. S., and Pfeiffer, T. W. (1993). Effects of heat-treatment on the nutritional-value of conventional and low trypsin-inhibitor soybeans for chicks. *Poult. Sci.* **72,** 1359–1369.

Herman, E. M., and Larkins, B. A. (1999). Protein storage bodies and vacuoles. *Plant Cell* **11,** 601–613.

Herman, E. M., Helm, R. M., Jung, R., and Kinney, A. J. (2003). Genetic modification removes an immunodominant allergen from soybean. *Plant Physiol.* **132,** 36–43.

Higgins, T. J. V., Chandler, P. M., Randall, P. J., Spencer, D., Beach, L. R., Blagrove, R. J., Kortt, A. A., and Inglis, A. S. (1986). Gene structure, protein-structure, and regulation of the synthesis of a sulfur-rich protein in pea-seeds. *J. Biol. Chem.* **261,** 1124–1130.

Hinchliffe, D. J., and Kemp, J. D. (2002). Beta-zein protein bodies sequester and protect the 18-kDa delta-zein protein from degradation. *Plant Sci.* **163,** 741–752.

Jaynes, J. M., Yang, M. S., Espinoza, N., and Dodds, J. H. (1986). Plant protein improvement by genetic engineering - use of synthetic genes. *Trends Biotechnol.* **4,** 314–320.

Johnson, L. A., Hardy, C. L., Baumel, C. P., Yu, T. H., and Sell, J. L. (2001). Identifying valuable corn quality traits for livestock feed. *Cereal Foods World* **46,** 472–481.

Kalinski, A. J., Melroy, D. L., Dwivedi, R. S., and Herman, E. M. (1992). A soybean vacuolar protein (P34) related to thiol proteases is synthesized as a glycoprotein precursor during seed maturation. *J. Biol. Chem.* **267,** 12068–12076.

Katsube, T., Kurisaka, N., Ogawa, M., Maruyama, N., Ohtsuka, R., Utsumi, S., and Takaiwa, F. (1999). Accumulation of soybean glycinin and its assembly with the glutelins in rice. *Plant Physiol.* **120,** 1063–1073.

Kaufman, R. J. (1999). Stress signaling from the lumen of the endoplasmic reticulum: Coordination of gene transcriptional and translational controls. *Genes Dev.* **13,** 1211–1233.

Keeler, S. J., Maloney, C. L., Webber, P. Y., Patterson, C., Hirata, L. T., Falco, S. C., and Rice, J. A. (1997). Expression of *de novo* high-lysine alpha-helical coiled-coil proteins may significantly increase the accumulated levels of lysine in mature seeds of transgenic tobacco plants. *Plant Mol. Biol.* **34,** 15–29.

Kermode, A. R., and Bewley, J. D. (1999). Synthesis, processing and deposition of seed proteins: The pathway of protein synthesis and deposition in the cell. *In* "Seed Proteins" (P. Shewry and R. Casey, eds.), pp. 807–842. Kluwer Academic Publishers, Dordrecht, Boston and London.

Khan, M. R. I., Ceriotti, A., Tabe, L., Aryan, A., Mcnabb, W., Moore, A., Craig, S., Spencer, D., and Higgins, T. J. V. (1996). Accumulation of a sulphur-rich seed albumin from sunflower in the leaves of transgenic subterranean clover (*Trifolium subterraneum* L.). *Transgenic Res.* **5,** 179–185.

Kim, C. S., Kamiya, S., Sato, T., Utsumi, S., and Kito, M. (1990). Improvement of nutritional-value and functional-properties of soybean glycinin by protein engineering. *Protein Eng.* **3,** 725–731.

Kim, C. S., Woo, Y. M., Clore, A. M., Burnett, R. J., Carneiro, N. P., and Larkins, B. A. (2002). Zein protein interactions, rather than the asymmetric distribution of zein mRNAs on endoplasmic reticulum membranes, influence protein body formation in maize endosperm. *Plant Cell* **14,** 655–672.

Kinney, A. J., Jung, R., and Herman, E. M. (2001). Cosuppression of the alpha subunits of *beta*-conglycinin in transgenic soybean seeds induces the formation of endoplasmic reticulum-derived protein bodies. *Plant Cell* **13,** 1165–1178.

Kito, M., Moriyama, T., Kimura, Y., and Kambara, H. (1993). Changes in plasma-lipid levels in young healthy-volunteers by adding an extruder-cooked soy protein to conventional meals. *Biosci. Biotechnol. Biochem.* **57,** 354–355.

Kohnomurase, J., Murase, M., Ichikawa, H., and Imamura, J. (1995). Improvement in the quality of seed storage protein by transformation of *Brassica napus* with an antisense gene for cruciferin. *Theor. Appl. Genet.* **91,** 627–631.

Kortt, A. A., Caldwell, J. B., Lilley, G. G., and Higgins, T. J. V. (1991). Amino-acid and cDNA sequences of a methionine-rich 2S protein from sunflower seed (*Helianthus-annuus* L.). *Eur. J. Biochem.* **195,** 329–334.

Kramer, M. G., and Redenbaugh, K. (1994). Commercialization of a tomato with an antisense polygalacturonase gene: The Flavr Savr (Tm) tomato story. *Euphytica* **79,** 293–297.

Kuiper, H. A., Kleter, G. A., Noteborn, H., and Kok, E. J. (2001). Assessment of the food safety issues related to genetically modified foods. *Plant J.* **27,** 503–528.

Lai, J. S., and Messing, J. (2002). Increasing maize seed methionine by mRNA stability. *Plant J.* **30,** 395–402.

Lawrence, M. C., Izard, T., Beuchat, M., Blagrove, R. J., and Colman, P. M. (1994). Structure of phaseolin at 2-center-dot-2 angstrom resolution. Implications for a common vicilin/legumin structure and the genetic-engineering of seed storage proteins. *J. Mol. Biol.* **238,** 748–776.

Leckband, G., Frauen, M., and Friedt, W. (2002). NAPUS 2000. Rapeseed (*Brassica napus*) breeding for improved human nutrition. *Food Res. Int.* **35,** 273–278.

Lee, H., and Garlich, J. D. (1992). Effect of overcooked soybean-meal on chicken performance and amino-acid availability. *Poult. Sci.* **71,** 499–508.

Lending, C. R., and Larkins, B. A. (1989). Changes in the zein composition of protein bodies during maize endosperm development. *Plant Cell* **1,** 1011–1023.

Li, L., Liu, S. M., Hu, Y. L., Zhao, W. P., and Lin, Z. P. (2001). Increase of sulphur-containing amino acids in transgenic potato with 10 kDa zein gene from maize. *Chin. Sci. Bull.* **46,** 482–484.

Liu, C. Y., Shepherd, K. W., and Rathjen, A. J. (1996). Improvement of durum wheat pasta-making and bread-making qualities. *Cereal Chem.* **73,** 155–166.

Lusk, J. L., and Sullivan, P. (2002). Consumer acceptance of genetically modified foods. *Food Technol.* **56,** 32–37.

Maleki, S. J., and Hurlburt, B. K. (2002). Food allergy: Recent advances in food allergy research. *ACS Symp. Ser.* **829,** 192–204.

Matsuda, T., Sugiyama, M., Nakamura, R., and Torii, S. (1988). Purification and properties of an allergenic protein in rice grain. *Agric. Biol. Chem.* **52,** 1465–1470.

Matsuda, T., Nomura, R., Sugiyama, M., and Nakamura, R. (1991). Immunochemical studies on rice allergenic proteins. *Agric. Biol. Chem.* **55,** 509–513.

Mcnabb, W. C., Spencer, D., Higgins, T. J., and Barry, T. N. (1994). *In vitro* rates of rumen proteolysis of ribulose-1,5- bisphosphate carboxylase (rubisco) from lucerne leaves, and of ovalbumin, vicilin and sunflower albumin-8 storage proteins. *J. Sci. Food Agric.* **64,** 53–61.

Mertz, E. T., Nelson, O. E., and Bates, L. S. (1964). Mutant gene that changes protein composition and increases lysine content of maize endosperm. *Science* **145,** 279–280.

Molvig, L., Tabe, L. M., Eggum, B. O., Moore, A. E., Craig, S., Spencer, D., and Higgins, T. J. V. (1997). Enhanced methionine levels and increased nutritive value of seeds of transgenic lupins (*Lupinus angustifolius* L.) expressing a sunflower seed albumin gene. *Proc. Natl. Acad. Sci. USA* **94,** 8393–8398.

Moneret-Vautrin, D. A. (2002). The allergic risk of transgenic foods: Strategy for prevention. *Bull. Acad. Natl. Med.* **186,** 1391–1400.

Munck, L. (1992). The case of high-lysine barley breeding. *In* "Barley: Genetics, Biochemistry, Molecular Biology and Biotechnology" (P. Shewry, ed.), pp. 573–601. CAB International, Wallingford, UK..

Nair, R. B., Joy, R. W., Kurylo, E., Shi, X. H., Schnaider, J., Datla, R. S. S., Keller, W. A., and Selvaraj, G. (2000). Identification of a CYP84 family of cytochrome P450-dependent mono-oxygenase genes in *Brassica napus* and perturbation of their expression for engineering sinapine reduction in the seeds. *Plant Physiol.* **123,** 1623–1634.

Nelson, O. E. (2001). Maize, the long trail to QPM. *In* "Encyclopedia of Genetics" (E. C. R. Reeve, ed.), pp. 657–660. Fitzroy Dearboin, London and Chicago.

Nordlee, J. A., Taylor, S. L., Townsend, J. A., Thomas, L. A., and Bush, R. K. (1996). Identification of a Brazil-nut allergen in transgenic soybeans. *N. Engl. J. Med.* **334,** 688–692.

Ogawa, T., Samoto, M., and Takahashi, K. (2000). Soybean allergens and hypoallergenic soybean products. *J. Nutr. Sci. Vitaminol.* **46,** 271–279.

Orf, J. H., and Hymowitz, T. (1979). Inheritance of the absence of the kunitz trypsin-inhibitor in seed protein of soybeans. *Crop Sci.* **19,** 107–109.

Oria, M. P., Hamaker, B. R., Axtell, J. D., and Huang, C. P. (2000). A highly digestible sorghum mutant cultivar exhibits a unique folded structure of endosperm protein bodies. *Proc. Natl. Acad. Sci. USA* **97,** 5065–5070.

Orthoefer, F. T. (1987). Corn starch modification and uses. *In* "Corn: Chemistry and Technology" (S. A. Watson and P. E. Ramstad, eds.), pp. 479–500. American Association of Cereal Chemists, Inc., St. Paul, MN.

Osborne, T. B. (1924). "The Vegetable Proteins," 2nd Edn., p. 154. Longmans, Green and Co., London.

Payne, P. I. (1987). Genetics of wheat storage proteins and the effect of allelic variation on bread-making quality. *Annu. Rev. Plant Physiol.* **38,** 141–153.

Pickering, F. S., and Reis, P. J. (1993). Effects of abomasal supplements of methionine on wool growth of grazing sheep. *Aust. J. Exp. Agric.* **33,** 7–12.

Popineau, Y., Deshayes, G., Lefebvre, J., Fido, R., Tatham, A. S., and Shewry, P. R. (2001). Prolamin aggregation, gluten viscoelasticity, and mixing properties of transgenic wheat lines expressing 1Ax and 1Dx high molecular weight glutenin subunit transgenes. *J. Agric. Food Chem.* **49,** 395–401.

Prasanna, B. M., Vasal, S. K., Kassahun, B., and Singh, N. N. (2001). Quality Protein Maize. *Curr. Sci.* **81,** 1308–1319.

Raina, A., and Datta, A. (1992). Molecular-cloning of a gene encoding a seed-specific protein with nutritionally balanced amino-acid-composition from Amaranthus. *Proc. Natl. Acad. Sci. USA* **89,** 11774–11778.

Saalbach, I., Pickardt, T., Waddell, D. R., Hillmer, S., Schieder, O., and Muntz, K. (1995). The sulfur-rich Brazil nut 2s albumin is specifically formed in transgenic seeds of the grain legume *Vicia narbonensis. Euphytica* **85,** 181–192.

Saio, K., Kamiya, M., and Watanabe, T. (1969). Food processing characteristics of soybean-11S and soybean-7S proteins 1. Effect of difference of protein components among soybean varieties on formation of tofu-gel. *Agr. Biol. Chem.* **33,** 1301–1311.

Sangtong, V., Moran, D. L., Chikwamba, R., Wang, K., Woodman-Clikeman, W., Long, M. J., Lee, M., and Scott, M. P. (2002). Expression and inheritance of the wheat Glu-1DX5 gene in transgenic maize. *Theor. Appl. Genet.* **105,** 937–945.

Schickler, H., Benner, M. S., and Messing, J. (1993). Repression of the high-methionine zein gene in the maize inbred line Mo17. *Plant J.* **3,** 221–229.

Scrimshaw, N. S. (2003). Historical concepts of interactions, synergism and antagonism between nutrition and infection. *J. Nutr.* **133,** 316–321.

Sharma, S. B., Hancock, K. R., Ealing, P. M., and White, D. W. R. (1998). Expression of a sulfur-rich maize seed storage protein, delta- zein, in white clover (*Trifolium repens*) to improve forage quality. *Mol. Breed.* **4,** 435–448.

Shewry, P. R., and Casey, R. (1999b). Seed proteins. *In* "Seed Proteins" (P. Shewry and R. Casey, eds.), pp. 1–10. Kluwer Academic Publishers, Dordrecht, London, Boston.

Shewry, P. R., and Halford, N. G. (2002). Cereal seed storage proteins: Structures, properties and role in grain utilization. *J. Exp. Bot.* **53,** 947–958.

Shewry, P., and Pandya, M. (1999). The 2S albumin storage proteins. *In* "Seed Proteins" (P. Shewry and R. Casey, eds.), pp. 563–596. Kluwer Academic Publishers, Dordrecht, Boston and London.

Shewry, P. R., and Tatham, A. S. (1997). Disulphide bonds in wheat gluten proteins. *J. Cereal Sci.* **25,** 207–227.

Shewry, P. R., and Tatham, A. S. (1999). The characteristics, structures and evolutionary relationships of prolamins. *In* "Seed Proteins" (P. R. Shewry and R. Casey, eds.), pp. 11–33. Kluwer Academic Publishers, Dordrecht, London, Boston.

Shewry, P. R., Franklin, J., Parmar, S., Smith, S. J., and Miflin, B. J. (1983). The effects of sulfur starvation on the amino acid and protein compositions of barley grain. *J. Cereal Sci.* **1,** 21–31.

Shewry, P. R., Halford, N. G., Tatham, A. S., Popineau, Y., Lafiandra, D., and Belton, P. S. (2003a). The high molecular weight glutenin subunits of wheat and their role in determining wheat processing properties. *Adv. Food Nutr. Res.* **45,** 219–302.

Shewry, P. R., Gilbert, S. M., Savage, A. W. J., Tatham, A. S., Wan, Y. F., Belton, P. S., Wellner, N., D'ovidio, R., Bekes, F., and Halford, N. G. (2003b). Sequence and properties of HMW subunit 1Bx20 from pasta wheat (*Triticum durum*) which is associated with poor end use properties. *Theor. Appl. Genet.* **106,** 744–750.

Shimoni, Y., Blechl, A. E., Anderson, O. D., and Galili, G. (1997). A recombinant protein of two high molecular weight glutenins alters gluten polymer formation in transgenic wheat. *J. Biol. Chem.* **272,** 15488–15495.

Shotwell, M. A., and Larkins, B. A. (1989). The biochemistry and molecular biology of seed storage proteins. *In* "The Biochemistry of Plants; A Comprehensive Treatise" (A. Marcus, P. K. Stumpf, and E. E. Conn, eds.), pp. 297–354. Academic Press, New York.

Sindhu, A. S., Zheng, Z. W., and Murai, N. (1997). The pea seed storage protein legumin was synthesized, processed, and accumulated stably in transgenic rice endosperm. *Plant Sci.* **130,** 189–196.

Sodek, L., and Wilson, C. M. (1970). Incorporation of leucine-C-14 and lysine-C-14 into protein in developing endosperm of normal and opaque-2 corn. *Arch. Biochem. Biophys.* **140,** 29–36.

Streit, L. G., Beach, L. R., Register, J. C., Jung, R., and Fehr, W. R. (2001). Association of the Brazil nut protein gene and Kunitz trypsin inhibitor alleles with soybean protease inhibitor activity and agronomic traits. *Crop Sci.* **41,** 1757–1760.

Swarup, S., Timmermans, M. C. P., Chaudhuri, S., and Messing, J. (1995). Determinants of the high-methionine trait in wild and exotic germplasm may have escaped selection during early cultivation of maize. *Plant J.* **8,** 359–368.

Tabe, L. M., and Droux, M. (2002). Limits to sulfur accumulation in transgenic lupin seeds expressing a foreign sulfur-rich protein. *Plant Physiol.* **128,** 1137–1148.

Tabe, L., and Higgins, T. J. V. (1998). Engineering plant protein composition for improved nutrition. *Trends Plant Sci.* **3,** 282–286.

Tabe, L., Hagan, N., and Higgins, T. J. V. (2002). Plasticity of seed protein composition in response to nitrogen and sulfur availability. *Curr. Opin. Plant Biol.* **5,** 212–217.

Tada, Y., Nakase, M., Adachi, T., Nakamura, R., Shimada, H., Takahashi, M., Fujimura, T., and Matsuda, T. (1996). Reduction of 14–16 kDa allergenic proteins in transgenic rice plants by antisense gene. *FEBS Lett.* **391,** 341–345.

Tada, Y., Akagi, H., Fujimura, T., and Matsuda, T. (2003). Effect of an antisense sequence on rice allergen genes comprising a multigene family. *Breed. Sci.* **53,** 61–67.

Taubes, G. (2001). The soft science of dietary fat. *Science* **291,** 2536–2545.

Torrent, M., Geli, M. I., Ruizavila, L., Canals, J. M., Puigdomenech, P., and Ludevid, D. (1994). Role of structural domains for maize gamma-zein retention in xenopus-oocytes. *Planta* **192,** 512–518.

Torrent, M., Alvarez, I., Geli, M. I., Dalcol, I., and Ludevid, D. (1997). Lysine-rich modified gamma-zeins accumulate in protein bodies of transiently transformed maize endosperms. *Plant Mol. Biol.* **34,** 139–149.

Tu, H. M., Godfrey, L. W., and Sun, S. S. M. (1998). Expression of the Brazil nut methionine-rich protein and mutants with increased methionine in transgenic potato. *Plant Mol. Biol.* **37,** 829–838.

Urisu, A., Yamada, K., Masuda, S., Komada, H., Wada, E., Kondo, Y., Horiba, F., Tsuruta, M., Yasaki, T., Yamada, M., Torii, S., and Nakamura, R. (1991). 16-kilodalton rice protein is one of the major allergens in rice grain extract and responsible for cross-allergenicity between cereal-grains in the *Poaceae* family. *Int. Arch. Allergy Appl. Immunol.* **96,** 244–252.

Utsumi, S., Kitagawa, S., Katsube, T., Kang, I. J., Gidamis, A. B., Takaiwa, F., and Kito, M. (1993). Synthesis, processing and accumulation of modified glycinins of soybean in the seeds, leaves and stems of transgenic tobacco. *Plant Sci.* **92,** 191–202.

Utsumi, S., Kitagawa, S., Katsube, T., Higasa, T., Kito, M., Takaiwa, F., and Ishige, T. (1994). Expression and accumulation of normal and modified soybean glycinins in potato-tubers. *Plant Sci.* **102,** 181–188.

Vasil, I. K., Bean, S., Zhao, J. M., Mccluskey, P., Lookhart, G., Zhao, H. P., Altpeter, F., and Vasil, V. (2001). Evaluation of baking properties and gluten protein composition of field grown transgenic wheat lines expressing high molecular weight glutenin gene 1Ax1. *J. Plant Physiol.* **158,** 521–528.

Velasco, L., and Mollers, C. (1998). Nondestructive assessment of sinapic acid esters in brassica species: II. Evaluation of germplasm and identification of phenotypes with reduced levels. *Crop Sci.* **38,** 1650–1654.

Wallace, J. C., Galili, G., Kawate, E. E., Cuellar, R. E., Shotwell, M. A., and Larkins, B. A. (1988). Aggregation of lysine-containing zeins in protein bodies in *Xenopus* oocytes. *Science* **240,** 662–664.

Wang, X. L., and Larkins, B. A. (2001). Genetic analysis of amino acid accumulation in opaque-2 maize endosperm. *Plant Physiol.* **125,** 1766–1777.

Watanabe, M. (1993). Hypoallergenic rice as a physiologically functional food. *Trends Food Sci. Technol.* **4,** 125–128.

Watanabe, Y., Barbashov, S. F., Komatsu, S., Hemmings, A. M., Miyagi, M., Tsunasawa, S., and Hirano, H. (1994). A peptide that stimulates phosphorylation of the plant insulin-binding protein: Isolation, primary structure and cDNA cloning. *Eur. J. Biochem.* **224,** 167–172.

Waterhouse, P. M., Graham, H. W., and Wang, M. B. (1998). Virus resistance and gene silencing in plants can be induced by simultaneous expression of sense and antisense RNA. *Proc. Natl. Acad. Sci. USA* **95,** 13959–13964.

White, C. L., Tabe, L. M., Dove, H., Hamblin, J., Young, P., Phillips, N., Taylor, R., Gulati, S., Ashes, J., and Higgins, T. J. V. (2001). Increased efficiency of wool growth and live weight gain in Merino sheep fed transgenic lupin seed containing sunflower albumin. *J. Sci. Food Agric.* **81,** 147–154.

Woo, Y. M., Hu, D. W. N., Larkins, B. A., and Jung, R. (2001). Genomics analysis of genes expressed in maize endosperm identifies novel seed proteins and clarifies patterns of zein gene expression. *Plant Cell* **13,** 2297–2317.

Yaklich, R. W., Helm, R. M., Cockrell, G., and Herman, E. M. (1999). Analysis of the distribution of the major soybean seed allergens in a core collection of *Glycine max* accessions. *Crop Sci.* **39,** 1444–1447.

Yang, M. S., Espinoza, N. O., Nagpala, P. G., Dodds, J. H., White, F. F., Schnorr, K. L., and Jaynes, J. M. (1989). Expression of a synthetic gene for improved protein-quality in transformed potato plants. *Plant Sci.* **64,** 99–111.

Youle, R. J., and Huang, A. H. C. (1981). Occurrence of low-molecular weight and high cysteine containing albumin storage proteins in oilseeds of diverse species. *Am. J. Bot.* **68,** 44–48.

CHAPTER 6

Biochemistry and Molecular Biology of Cellulose Biosynthesis in Plants: Prospects for Genetic Engineering

Inder M. Saxena and **R. Malcolm Brown, Jr.**

Contents

Section of Molecular Genetics and Microbiology, School of Biological Sciences, The University of Texas at Austin, Austin, Texas 78712

Advances in Plant Biochemistry and Molecular Biology, Volume 1
ISSN 1755-0408, DOI: 10.1016/S1755-0408(07)01006-5

Abstract

Cellulose is a major component of the plant cell wall, and understanding the mechanism of synthesis of this polysaccharide is a major challenge for plant biologists. Cellulose microfibrils are synthesized and assembled by membrane-localized protein complexes that are visualized as rosettes by freeze-fracture electron microscopy. Cellulose synthase is required for cellulose synthesis. So far only this enzyme has been localized to these cellulose-synthesizing complexes. Although it has not been possible to purify and fully characterize cellulose synthase activity from plants, it has been possible to obtain cellulose synthesis *in vitro* using membranes and detergent-solubilized membrane fractions. Cellulose synthase uses uridine 5′-diphosphate (UDP)-glucose as a substrate and polymerizes glucose residues into long β-1,4-linked glucan chains in a single-step reaction. Cellulose synthases are encoded by genes belonging to a superfamily, and each plant synthesizes a number of different cellulose synthases. Genetic analysis suggests that each cellulose-synthesizing complex contains at least three nonredundant cellulose synthases and mutation in any one of these cellulose synthases results in cellulose deficiency. More interestingly, different cellulose synthases perform cellulose synthesis in the primary cell wall and the secondary cell wall. Apart from the cellulose synthases, a number of other proteins have been suggested to play a role in cellulose synthesis, but so far their functions are not clearly understood. Genetic manipulation of cellulose synthesis in plants will therefore require not only a complete understanding of the different cellulose synthases but also other proteins that regulate the temporal and spatial synthesis and assembly of this very important polysaccharide.

Key Words: Cellulose, Cellulose biosynthesis, Cellulose synthase, Cellulose synthase-like, CesA, Csl, *Arabidopsis*, Cotton, *Acetobacter xylinum*, Genetic manipulations.

1. INTRODUCTION

Cellulose is an abundant biopolymer that is synthesized by all plants, most algae, a number of bacteria including cyanobacteria, the cellular slime mold, and the ascidians (a group of animals) (Brown, 1996). The major proportion of cellulose, produced in the biosphere by plants, adds strength to the plant cell wall and helps in determining the direction of cell and plant growth. The plant cell wall itself is a complex of polysaccharides, which include cellulose and noncellulosic polysaccharides (hemicelluloses and pectins), as well as lignins and proteins. All plant cells have a primary cell wall consisting of cellulose, hemicellulose, pectin, and proteins; however, some cells additionally have a secondary cell wall consisting mainly of cellulose and lignins, and it is in these cells that the proportion of cellulose is increased considerably. The importance of cellulose as an essential component of plants and its uses in our daily lives cannot be overemphasized. Interestingly, cellulose also is the most important industrial polysaccharide, and considering its unique physical properties, it has been studied widely by chemists since its initial discovery by Anselme Payen almost 165 years ago (Klemm *et al.*, 2005).

Studies on the structure of cellulose have been crucial in developing concepts regarding the sites of cellulose synthesis and the mechanism by which it is synthesized (Preston, 1974). Although much more is known about the structure of cellulose (and these studies are still continuing) (Nishiyama *et al.*, 2003), the last decade and a half has witnessed a surge in our understanding of the biosynthesis of cellulose in plants. Many of these advances are related to the identification of genes for cellulose biosynthesis in plants (Arioli *et al.*, 1998; Pear *et al.*, 1996), analysis of mutants affected in cellulose biosynthesis (Robert *et al.*, 2004), the capability to analyze cellulose synthesis *in vitro* using cell-free extracts (Kudlicka and Brown, 1997; Lai-Kee-Him *et al.*, 2002), and visualization of enzymes involved in cellulose synthesis in living plant cells (Paredez *et al.*, 2006; Robert *et al.*, 2005). In this chapter, we will discuss the development of present-day concepts related to cellulose biosynthesis and the prospects of modifying this property in plants.

2. THE MANY FORMS OF CELLULOSE—A BRIEF INTRODUCTION TO THE STRUCTURE AND DIFFERENT CRYSTALLINE FORMS OF CELLULOSE

Unlike most known biopolymers, cellulose is a simple molecule that consists of an assembly of β-1,4-linked glucan chains. As a result, cellulose is defined less by its primary structure (β-1,4-linked glucose residues with cellobiose being the repeating unit in all chains) and more by its secondary and higher-order structure in which the chains interact via intramolecular and intermolecular hydrogen bonds, as well as van der Waals interactions, to give rise to different forms of cellulose (Fig. 6.1) (O'Sullivan, 1997). Cellulose exhibits polymorphism, and the different forms of cellulose are usually defined by their crystalline forms, although reference is also made to other forms of cellulose such as noncrystalline cellulose, amorphous cellulose, and more recently nematic-ordered cellulose (Kondo *et al.*, 2001). Whereas, the glucan chains are arranged in a specific manner with respect to each other in crystalline cellulose, no specific arrangement of the glucan chains occur in noncrystalline or amorphous cellulose. In contrast, nematic-ordered cellulose is highly ordered but not crystalline and is obtained by uniaxial stretching of water-swollen cellulose (Kondo *et al.*, 2004).

In general, cellulose produced by living organisms occurs as cellulose I and is assembled in a structure referred to as a microfibril (Fig. 6.2). The properties of the microfibril are determined by its size, shape, and crystallinity. The glucan chains in cellulose I are arranged in a parallel manner, and depending upon the arrangement of these chains, two crystalline forms of cellulose I—I_α and I_β—have been

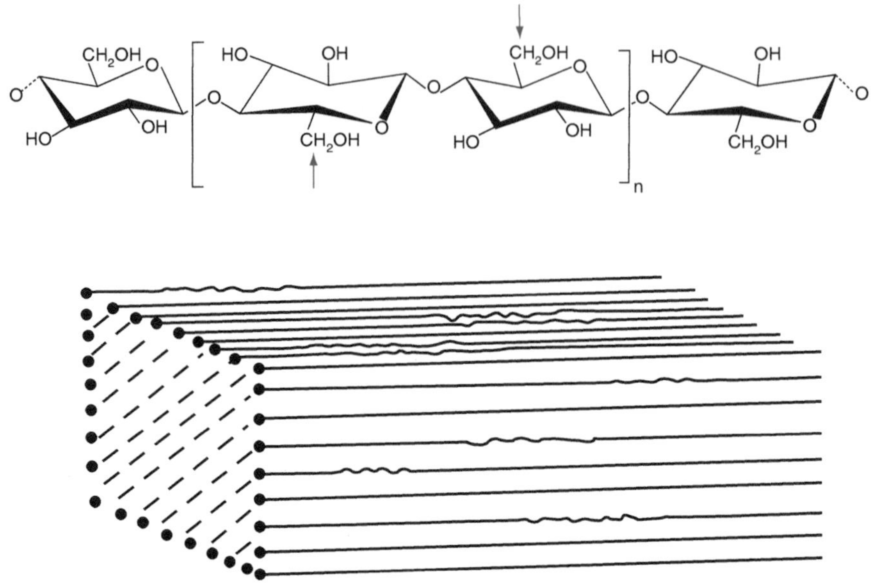

FIGURE 6.1 Top image is the structural formula for the β-1,4-linked glucan chain of cellulose. The bracketed region indicates the basic repeat unit, cellobiose, in the chain. The glucan chain has a twofold symmetry. The bottom image is a schematic representation of a crystalline cellulose I microfibril. (Reproduced from Brown, Jr. R. M., *J. Poly. Sci. Part A Poly. Chem.* **42,** 489–495.) (See Page 5 in Color Section.)

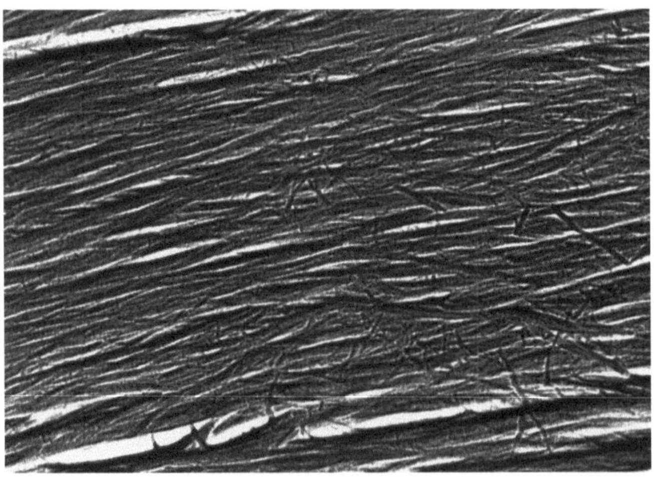

FIGURE 6.2 Freeze fracture image of cellulose microfibrils in the secondary wall of a developing cotton fiber. (*Unpublished image from R. Malcolm Brown, Jr. and Kazuo Okuda.*)

identified (Attala and Vanderhart, 1984). The more thermodynamically stable form of cellulose is cellulose II, and in this allomorph the glucan chains are arranged in an antiparallel manner. Cellulose II is produced in nature by certain organisms or under specific conditions but is generally obtained by an irreversible process upon chemical treatment (mercerization or solubilization) of native cellulose I. Furthermore, cellulose III_I and cellulose III_{II} are obtained from cellulose I and cellulose II, respectively, in a reversible process, by treatment with liquid ammonia or some amines and the subsequent evaporation of excess ammonia, and cellulose IV_I and cellulose IV_{II} are obtained irreversibly by heating cellulose III_I and cellulose III_{II} respectively to 206 °C in glycerol (O'Sullivan, 1997). Implicit in the biosynthesis of cellulose is the role of the cellulose-synthesizing machinery that allows synthesis and organization of a metastable form of cellulose (cellulose I) that is found to be desirable in living organisms in comparison to the more stable cellulose II product. Whereas the assembly of the glucan chains (crystallization) endows cellulose with its characteristic properties, it is the synthesis of these β-1,4-linked glucan chains (polymerization) that is the focus of research for most biologists.

3. BIOCHEMISTRY OF CELLULOSE BIOSYNTHESIS IN PLANTS

3.1. UDP-glucose is the immediate precursor for cellulose synthesis

Although cellulose was characterized as an aggregation of glucose units by Anselme Payen in 1839, it was in 1895 that Tollens proposed that cellulose is a chain of glucose molecules (French, 2000). While the structure of cellulose was being determined and debated, studies on its biosynthesis did not truly begin until the identification of nucleotide sugars, and specifically UDP-glucose as a glucose donor in biosynthetic reactions (Leloir and Cabib, 1953). The transfer of glucose from UDP-glucose to cellulose was first described by Glaser in 1958 using particulate fraction from cell-free extracts of the bacterium *Acetobacter xylinum* (Glaser, 1958). However, when UDP-glucose was used as the sugar donor in experiments using digitonin-solubilized fractions from various plants, the polysaccharide product obtained *in vitro* was identified as callose (β-1,3-glucan) instead of cellulose (Feingold *et al.*, 1958). Using particulate extracts from plants, the synthesis of cellulose was reported by Barber and colleagues in 1964, and from their experiments these authors concluded that the sugar donor for synthesis of cellulose was guanosine 5′-diphosphate (GDP)-glucose and not UDP-glucose (Barber *et al.*, 1964). In these experiments, the particulate extracts from plants also allowed synthesis of an alkali-insoluble polysaccharide from GDP-mannose and from a mixture of GDP-glucose and GDP-mannose. Recently, a cellulose synthase-like protein (AtCslA9), identified as a β-glucomannan synthase, has been shown to possess β-mannan synthase, β-glucan synthase, and β-glucomannan synthase activities (Liepman *et al.*, 2005). This β-glucomannan synthase can catalyze the production of β-mannan when supplied with GDP-mannose, a β-glucan when supplied with GDP-glucose or β-glucomannan when supplied with a combination of GDP-glucose and GDP-mannose. It is now clear that in the earlier experiments where GDP-glucose was used as a sugar donor with plant extracts, techniques for characterizing the *in vitro* products did not allow a clear distinction to be made between the possible β-glucomannan product and cellulose

(Barber *et al.*, 1964; Chambers and Elbein, 1970). Moreover, it was felt at the time that synthesis of the major homopolymers of glucose in plants could be regulated by using different nucleotide sugars—UDP-glucose for callose synthesis, adenosine diphosphate (ADP)-glucose for starch synthesis, and GDP-glucose for cellulose synthesis (Barber *et al.*, 1964). We now know that in plants, although ADP-glucose is the precursor for starch synthesis, the precursor for synthesis of callose and cellulose is UDP-glucose. Support for the role of UDP-glucose as a precursor of cellulose in plants came from studies tracing the flow of carbon from glucose to cellulose in developing cotton fibers (Carpita and Delmer, 1981). Evidence for the role of UDP-glucose as the precursor for cellulose synthesis in plants did not come easily, and only a brief historical account is given here to highlight one of the many difficulties encountered in dissecting the mechanism of cellulose synthesis in plants. A detailed account of the early years and the progress that has been made since then is provided by Delmer in a number of excellent review articles (Delmer, 1983, 1999). Suffice it to say that as late as 1983, in one of her reviews Delmer summarized that "convincing *in vitro* synthesis of cellulose from UDP-glucose using plant extracts has never been conclusively demonstrated" (Delmer, 1983). In plants, UDP-glucose functions as a glucose donor in a number of glucosyl transfer reactions. From genome sequencing, it is now known that plants have the largest number of carbohydrate-modifying enzymes, and consequently UDP-glucose could participate as a glucose donor in many different reactions when unpurified plant extracts are used for *in vitro* cellulose synthesis (Coutinho *et al.*, 2003). Furthermore in plants, polysaccharides, such as xyloglucan, have a backbone similar to cellulose, and it is important to distinguish the synthesis of these polysaccharides from synthesis of cellulose. Although not much has changed since the early days in the manner in which *in vitro* cellulose synthesis reactions were performed, a few modifications in the reaction conditions and better product characterization (described later) has allowed conclusive demonstration of *in vitro* cellulose synthesis from UDP-glucose using extracts from a variety of plants (Colombani *et al.*, 2004; Kudlicka and Brown, 1997; Kudlicka *et al.*, 1995, 1996; Lai-Kee-Him *et al.*, 2002; Okuda *et al.*, 1993; Peng *et al.*, 2002).

3.2. *In vitro* synthesis of cellulose from plant extracts

3.2.1. The β-1,3-glucan synthase and lessons from *in vitro* β-1,3-glucan synthesis

To understand the biochemical machinery required for cellulose synthesis in plants, it is necessary to demonstrate *in vitro* synthesis of cellulose using plant extracts. Unfortunately, much to the dismay of most researchers studying cellulose biosynthesis, the major *in vitro* polysaccharide product synthesized from plant extracts using UDP-glucose as the precursor was and is still found to be callose, the β-1,3-glucan first reported from mung bean extracts by Feingold and colleagues in 1958 (Feingold *et al.*, 1958). Observing the synthesis of this polysaccharide in place of cellulose has been both frustrating and invigorating as it brings up a number of very interesting questions, many of which have not been fully answered.

During normal development, cellulose is found in all plant cells, whereas callose generally is synthesized in response to wounding, physiological stress, or infection, and is a component of the cell plate in dividing cells apart from being present in specialized cells. As such, enzymes for synthesis of this polysaccharide are not expected to be active most of the time. The general explanation to account for the large amount of *in vitro* synthesis of callose as opposed to cellulose using plant extracts is that this occurs in response to the wounding or stress of the cells during cell breakage. Using antibodies against β-1,4-glucan synthase and β-1,3-glucan synthase, Nakashima *et al.* (2003) recently demonstrated that the activation of β-1,3-glucan synthase upon wounding may be dependent on the degradation of β-1,4-glucan synthases by specific proteases (Nakashima *et al.*, 2003). However, under appropriate conditions in the presence of UDP-glucose, plant extracts synthesize both callose and cellulose, and the optimal conditions required for synthesis of these two polysaccharides have been shown to be only slightly different. Whether the same enzyme catalyzes the synthesis of both callose and cellulose has been debated for a number of years, but so far no conclusive evidence is available in support of either the one enzyme-two polysaccharides or the one enzyme-one polysaccharide synthesis with respect to these two polysaccharides. Although it has been possible to separate the major cellulose-synthesizing and callose synthesizing activities by native gel electrophoresis, the polypeptide composition in these two fractions could not be completely analyzed (Kudlicka and Brown, 1997). Interestingly, relatively much more is known about the identity of the catalytic subunit of cellulose synthase as compared to the nature of the catalytic subunit of callose synthase. This is true, in spite of the fact that genes required for synthesis of β-1,3-glucans have been identified in yeast, and similar genes have been identified in a number of plants (Cui *et al.*, 2001; Doblin *et al.*, 2001; Hong *et al.*, 2001; Li *et al.*, 2003). Surprisingly, the proteins encoded by these genes do not show similarity to any known glycosyltransferase, much less the cellulose synthases. These proteins are classified as 1,3-β-D-glucan synthases and have been placed in family 48 of glycosyltransferases (http://afmb.cnrs-mrs.fr/CAZY/). In plants, genes encoding this protein form a gene family, and in *Arabidopsis* 10 members are identified in this gene family.

Since synthesis of β-1,3-glucans occurs much more readily when plant extracts are used *in vitro*, many more studies have reported on characterization of the conditions for β-1,3-glucan synthase activity and its purification from a variety of plants. As an example, optimal conditions for *in vitro* synthesis of β-1,3 glucan from *Arabidopsis* were defined by the presence in the reaction mixture of 50 mM 3-(*N*-morpholino) propanesulfonic acid (MOPS) buffer, pH 6.8, 1 mM UDP-glucose, 8 mM Ca^{2+}, and 20 mM cellobiose (Lai-Kee-Him *et al.*, 2001). Similar conditions, in the presence or absence of Mg^{2+} in the reaction mix, have also been shown to be optimal for the synthesis of cellulose using plant extracts (Colombani *et al.*, 2004). Since both callose synthase and cellulose synthase are membrane proteins, the choice and concentration of detergents used during extraction of the proteins have been found to be very crucial in obtaining high specific activity of both callose synthase and cellulose synthase from plant extracts. Incorporating a variety of techniques, Dhugga and Ray (1994) purified the β-1,3-glucan synthase activity 5,500-fold from pea homogenates and found two polypeptides that copurified with the enzyme activity (Dhugga and Ray, 1994). Unfortunately, the identity of these proteins could not be determined, although one of these polypeptides was shown to bind to UDP-glucose. In related sets of experiments, Kudlicka and Brown (1997) demonstrated separation of the callose synthase and cellulose synthase activities in digitonin-solubilized mung bean membranes using gel electrophoresis under nondenaturing conditions (Kudlicka and Brown, 1997). The polypeptide composition in the two fractions was analyzed by SDS-PAGE, and while three similar sized polypeptides were observed in both activities, polypeptides unique to each activity were also observed. However, the characterization of these polypeptides did not provide any further information regarding the similarities or differences between the two enzyme activities. As mentioned in this section, many of the studies for *in vitro* synthesis of callose were applicable to *in vitro* synthesis of cellulose using plant extracts. Interestingly, conclusive demonstrations of cellulose synthesis *in vitro* using plant extracts had to do more with utilizing a greater variety of techniques for product characterization than with development of novel assay methods.

3.2.2. Increasing cellulose synthase activity *in vitro* and utilizing more techniques for product characterization

Techniques to identify and characterize the cellulose product have played a crucial role in determining cellulose synthesis *in vitro*. Interestingly, many of the criteria used by Glaser in 1958 for *in vitro* cellulose production using bacterial extracts are still used for characterizing the cellulose product and determining the cellulose synthase activity, namely incorporation of ^{14}C-glucose from UDP-^{14}C-glucose into a hot alkali-insoluble fraction (Glaser, 1958). The product was further characterized by acid hydrolysis and/or enzymatic analysis using cellulases. Although less than 1% of the glucose from UDP-glucose was incorporated into the alkali-insoluble fraction in the *in vitro* reaction, the product was characterized as cellulose.

A major breakthrough in understanding cellulose biosynthesis in *A. xylinum* and increasing cellulose synthase activity in bacterial extracts came with the identification of cyclic di-guanosine monophosphate (c-di-GMP) as an allosteric activator of cellulose synthase (Ross *et al.*, 1986). This nucleotide is now recognized to be a regulator of many more bacterial functions than previously thought (D'Argenio and Miller, 2004). The addition of c-di-GMP in reaction mixtures using bacterial extracts led to a remarkable increase in incorporation of glucose from UDP-glucose into a cellulose product.

In another development, the *in vitro* product using bacterial extracts for the first time was visualized by electron microscopy, and this product was shown to bind to gold-labeled cellobiohydrolase providing evidence that this product is cellulose (Lin *et al.*, 1985). The *in vitro* product obtained using *A. xylinum* inner membrane was furthermore shown to be cellulose II (Bureau and Brown, 1987). The capability to synthesize large amounts of the *in vitro* product was crucial in performing X-ray diffraction, sugar analysis, linkage analysis and molecular weight analysis to demonstrate conclusively that the product was cellulose (Bureau and Brown, 1987).

Many of these techniques were later utilized by Okuda *et al.* (1993) using cotton fiber extracts to demonstrate the *in vitro* production of cellulose II (Okuda *et al.*, 1993). Additionally, the incorporation of glucose from UDP-glucose into an Updegraff reagent-resistant fraction was included to be a stricter criterion for the cellulose product. Although no activator comparable to c-di-GMP was identified for activation of the cellulose synthase from plant tissues, a number of nucleotides were found to increase the *in vitro* cellulose synthase activity (Li and Brown, 1993). Overall, the success in demonstrating cellulose synthesis *in vitro* is ascribed to the choice of plant tissue (cotton fibers), method of extraction, and the ability to synthesize large amounts of the *in vitro* product for characterization. Although cellulose was synthesized *in vitro* using plant extracts, the major product was still β-1,3 glucan, and this could be distinguished from cellulose using electron microscopy.

In later studies, using a variety of detergents, Kudlicka *et al.* (1995) was able to demonstrate not only an increase in the amount of cellulose synthesized *in vitro*, but also the production of cellulose I using plant extracts (Kudlicka *et al.*, 1995). Lai-Kee-Him *et al.* (2002) used detergent solubilized microsomal fractions from suspension-cultured cells of blackberry (*Rubus fruticosus*) for *in vitro* cellulose synthesis (Lai-Kee-Him *et al.*, 2002). These investigators found that the detergents Brij 58 and taurocholate were effective in solubilizing membrane proteins that allowed synthesis of both cellulose and callose given UDP-glucose as the substrate. Roughly 20% of the *in vitro* product was cellulose with taurocholate as the detergent, and no Mg^{2+} was required. The cellulose product was characterized by methylation analysis, electron microscopy, electron and X-ray synchrotron diffractions, and resistance to Updegraff reagent. Cellulose microfibrils were obtained *in vitro*, and they had the same dimensions as microfibrils isolated from primary cell walls. However, the cellulose diffracted as cellulose IV_I, a disorganized form of cellulose I that is formed when the fibrillar material contains crystalline domains that are too narrow or too disorganized to be considered real cellulose I crystals (Lai-Kee-Him *et al.*, 2002).

In related studies, but using immunoaffinity purified cellulose synthase from mung bean hypocotyls, Laosinchai (2002) also demonstrated the *in vitro* synthesis of cellulose microfibrils (Laosinchai, 2002).

3.3. Purification and characterization of cellulose synthase activity

Cellulose synthase is the enzyme that performs cellulose biosynthesis. Purification of this enzyme is a major objective for understanding its properties and in determining its structure and mode of regulation. Cellulose synthase is a membrane protein and like most membrane proteins its purification has eluded investigators interested in isolating it. However, significant progress has been made in purifying the cellulose synthase activity from *A. xylinum* using the product entrapment technique utilized for purification of the chitin synthase activity in yeast (Lin and Brown, 1989). In *A. xylinum*, using a combination of detergent solubilization and product entrapment methods, two major polypeptide bands were identified in the purified fraction. One of these polypeptides was shown to selectively bind UDP-glucose, and this polypeptide was identified as the cellulose synthase catalytic subunit (Lin *et al.*, 1990). The other polypeptide was shown to bind the activator c-di-GMP (Mayer *et al.*, 1991). Sequence information obtained from these polypeptides was useful in identifying the corresponding genes from *A. xylinum* (Saxena *et al.*, 1990, 1991). However, similar progress has not been made with purifying the cellulose synthase activity in plants. Laosinchai (2002) used immunoaffinity techniques to purify cellulose synthase activity from mung bean fractions that synthesized cellulose microfibrils *in vitro* (Laosinchai, 2002). Unfortunately, sufficient amounts of the protein could not be isolated for further characterization of this activity. The cellulose synthase activity purified from *A. xylinum* utilizes UDP-glucose as the substrate and is activated by c-di-GMP. The cellulose synthase activity in plants is also shown to use UDP-glucose as the substrate, but it is not activated by c-di-GMP. Instead, the plant activity is influenced positively in the presence of cellobiose (Li and Brown, 1993). Although no requirement for a primer has been observed for cellulose synthesis *in vitro* using bacterial or plant extracts, a proposal for the requirement of a sterol-glucoside primer has been made for cellulose synthesis in plants (Peng *et al.*, 2002). This proposal is based on the observation that cotton fiber membranes synthesized sitosterol-cellodextrins (SCDs) from sitosterol-β-glucoside (SG) and UDP-glucose under conditions that favor cellulose synthesis (Peng *et al.*, 2002). As a result, this model invokes a number of other components besides cellulose synthase and UDP-glucose, in a multistep reaction scheme, as opposed to the single-step polymerization reaction that requires only cellulose synthase and UDP-glucose. Since most of the experiments demonstrating *in vitro* cellulose synthesis do not suggest the requirement for a primer and no new evidence has been provided in support of the multistep reaction scheme, the current view is that polymerization of glucose residues from UDP-glucose occurs in a single-step reaction catalyzed by the cellulose synthase.

Interestingly, many of the features of cellulose synthases from different organisms are predicted from the derived amino sequences following identification of the genes for cellulose synthases in these organisms.

4. MOLECULAR BIOLOGY OF CELLULOSE BIOSYNTHESIS IN PLANTS

4.1. Identification of genes encoding cellulose synthases in plants

Cellulose synthase genes were first identified in *A. xylinum* and subsequently in other bacterial species (Matthysse *et al.*, 1995b; Saxena *et al.*, 1990; Wong *et al.*, 1990) before they were identified in plants (Arioli *et al.*, 1998; Pear *et al.*, 1996). *A. xylinum* produces abundant amounts of cellulose, and it has been a model organism for studies on cellulose biosynthesis, so it is not surprising that cellulose biosynthesis genes were first identified in this organism. Interestingly, the genes from this organism were not found to be useful in isolating cellulose synthase genes from other organisms by nucleic acid hybridization techniques. However, Saxena *et al.* (1995) compared the derived amino acid sequence of the bacterial cellulose synthase with other proteins and found them useful in identifying conserved amino acid residues in β-glycosyltransferases, more specifically the conserved residues and sequence motif identified as D, D, D, QXXRW in processive β-glycosyltransferases (Saxena *et al.*, 1995). Based on the deduced amino acid sequences of bacterial cellulose synthases and other β-glycosyltransferases, genes for plant cellulose synthases were first identified by random sequencing of a cotton fiber cDNA library (Pear *et al.*, 1996). Two cDNA clones (*GhCesA1* and *GhCesA2*) were identified from the cotton fiber cDNA library, and the derived amino acid sequence of GhCesA1 gave the first glimpse of the primary structure of a plant cellulose synthase (Pear *et al.*, 1996). In addition to the transmembrane regions and the conserved residues found in bacterial cellulose synthase, the cellulose synthase from plants was found to have additional features—the presence of two regions (originally referred to as CR-P and HVR) within the globular domain that contained the conserved residues and a zinc-finger domain at the N-terminus.

Around the same time that cDNA clones encoding cellulose synthases were identified in cotton by random sequencing (Pear *et al.*, 1996), a number of cDNA clones encoding amino acid sequences containing the D, D, D, QXXRW conserved residues and sequence motif were identified by sequence analysis of expressed sequence tag (EST) sequences of *Arabidopsis* and rice that were available in the public databases (Cutler and Somerville, 1997; Saxena and Brown, 1997). However, the proteins encoded by these cDNA clones did not show the additional features identified in the cotton cellulose synthases; instead these proteins resembled more the primary structure of the bacterial cellulose synthase and were referred to as cellulose synthase-like proteins with a role possibly in the synthesis of β-linked polysaccharides other than cellulose (Cutler and Somerville, 1997). Soon thereafter, a superfamily of genes encoding cellulose synthases (CesA) and cellulose synthase-like (Csl) proteins were identified in a large number of plants (Richmond and Somerville, 2000). The presence of a large number of genes belonging to the cellulose synthase superfamily in each plant was surprising at first, but the role of many of these CesA genes in cellulose biosynthesis became obvious following analyses of a number of *Arabidopsis* mutants affected in cellulose biosynthesis. Interestingly, two cellulose synthase genes were earlier identified in *A. xylinum* (Saxena and Brown, 1995). Although both genes encode a functional cellulose synthase as determined by *in vitro* cellulose synthase activities in mutants, only one gene was found to be essential for normal *in vivo* cellulose synthesis in *A. xylinum* (Saxena and Brown, 1995).

4.2. Mutant analysis allowed identification of genes for cellulose synthases and other proteins required for cellulose biosynthesis

4.2.1. Identification and functional characterization of cellulose synthases in plants by analysis of mutants and gene expression studies

Although a majority of the *CesA* and *Csl* genes have been identified from genome and EST sequences, at least six of the *CesA* genes in *Arabidopsis* were identified by mutant analysis. In a number of cellulose-deficient *Arabidopsis* mutants, the mutations were mapped to genes that encoded for cellulose synthases (Arioli *et al.*, 1998; Fagard *et al.*, 2000; Scheible *et al.*, 2001; Taylor *et al.*, 1999). Interestingly, although all the mutants exhibited different phenotypes, they all showed a deficiency in the amount of cellulose produced. The first mutant, where the mutation was identified in a gene that encoded for a cellulose synthase, was a temperature-sensitive root-swelling mutant (*rsw1*) (Arioli *et al.*, 1998). At the nonpermissive temperature, the mutant produced a larger proportion of noncrystalline cellulose in place of crystalline cellulose, and the rosette terminal complexes (TCs) normally associated with cellulose microfibrils were not observed by freeze-fracture electron microscopy. The mutation in the cellulose synthase gene (*rsw1* gene; *AtCesA1*) led to the substitution of valine for alanine at position 549 of the cellulose synthase protein and this change resulted in all the different phenotypes associated with the *rsw1* mutant (Williamson *et al.*, 2001). No biochemical changes have been characterized in the mutant protein, but it appears that at the nonpermissive temperature, the cellulose synthase is not assembled into a rosette structure. Although the mutation results in the reduction of crystalline cellulose at the nonpermissive temperature, noncrystalline cellulose still is produced suggesting that the *rsw1*-encoded cellulose synthase is able to synthesize the β-1,4-glucan chains, but does not allow for their assembly to take place, or alternatively these chains are synthesized by cellulose synthases encoded by other genes, where the assembly of these cellulose synthases is affected by the *rsw1* mutation. Changes in cell shapes and sizes suggested that the Rsw1 cellulose synthase contributed to cellulose in the primary wall. Interestingly, a number of questions still remain to be answered in terms of how the *rsw1* mutation affects cellulose biosynthesis.

A number of *irregular xylem* mutants (*irx* mutants) have been isolated by screening cross-sections of stems of *Arabidopsis* plants (Turner and Somerville, 1997). The mutations resulted in collapse of mature xylem cells in the inflorescence stems, and in many of these mutants there was a significant decrease in the amount of cellulose in the secondary cell wall of cells in the xylem. Genes mutated in some of the *irx* mutants were identified to encode for cellulose synthases. The null mutation in the *irx3* mutant results in a stop codon that truncates the cellulose synthase (Irx3; AtCesA7) by 168 amino acids (Taylor *et al.*, 1999) In two *irx1* mutants (*irx1-1* and *irx1-2*), the mutations were mapped to a different cellulose synthase gene that altered the amino acids at positions 683 ($D^{683}N$) in Irx1-1 and 679 ($S^{679}L$) in Irx1-2 (Taylor *et al.*, 2000). Both these amino acid positions reside within the conserved region of the Irx1 cellulose synthase (AtCesA8). RNA analysis indicated that *irx1* and *irx3* are highly expressed in stems but not in leaves, suggesting that both genes are involved in cellulose synthesis during secondary cell wall formation. Examination of the phenotypes of the xylem elements by electron microscopy showed that the same cell type is affected in the *irx1* and *irx3* mutants, indicating that products of both the *irx1* and *irx3* genes are required within the same cell for normal cellulose synthesis during secondary cell wall formation (Taylor *et al.*, 2000). These results allowed development of the concept regarding the nonredundant nature of cellulose synthases and the requirement of more than a single cellulose synthase in each cell for normal cellulose synthesis. Using biochemical and immunological methods, Taylor *et al.* (2000) furthermore demonstrated that the Irx1 and Irx3 cellulose synthases associate with each other, and suggested that this association is required for cellulose synthesis (Taylor *et al.*, 2000). Even as different models to explain the requirement of two different cellulose synthases for cellulose synthesis were being proposed, another gene (*irx5*) encoding for a different cellulose synthase (Irx5; AtCesA4) was identified in a further screen of *irx* mutants and it was found that the *irx1*, *irx3*, and *irx5* genes were coexpressed in the same cells (Perrin, 2001; Taylor *et al.*, 2003). Using detergent-solubilized extracts, the proteins encoded by these three genes were shown to interact with each other, and it was suggested that all three gene products probably are required for the formation of the cellulose-synthesizing complexes (rosette TCs) in plants. Interestingly, the presence of all three cellulose synthases (AtCesA8, AtCesA7, and AtCesA4), but not their activity, is required for correct assembly and targeting of the cellulose-synthesizing complex during secondary wall cellulose synthesis (Taylor *et al.*, 2004). Overall, the *irx* mutants have been crucial in not only identifying the cellulose synthase genes that are required for cellulose synthesis during secondary wall formation, but also in formulating the concept that the assembly of the cellulose-synthesizing complexes (rosette TCs) in plants requires more than a single isoform of cellulose synthase. Fig. 6.3 shows immunogold labeling of the rosette TCs from *Vigna angularis* using an antibody to a cellulose synthase.

The protein regulator of cytokinesis 1 (*PRC1*) gene in *Arabidopsis* encodes AtCesA6, and like the *rsw1* mutant of AtCesA1, mutation in this gene exhibits decreased cell elongation, especially in roots and dark-grown hypocotyls, because

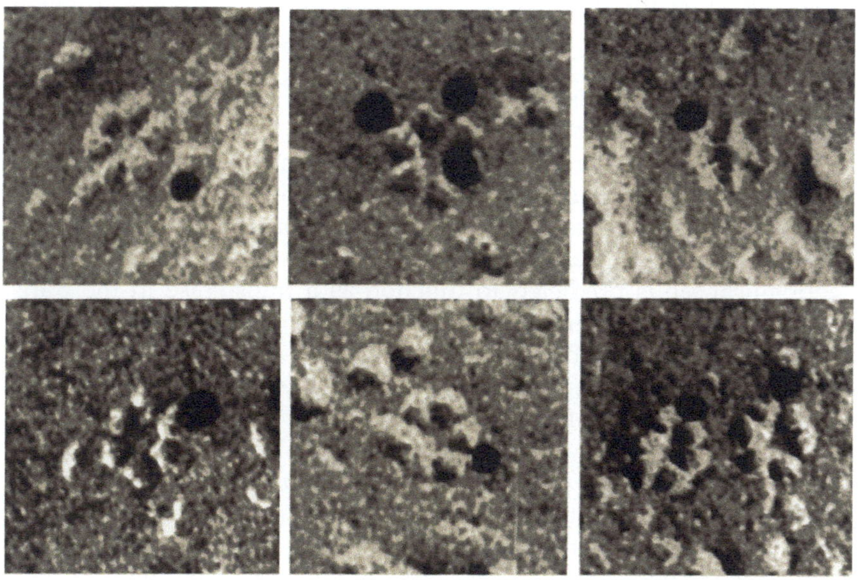

FIGURE 6.3 Rosette terminal complexes from *V. angularis* that were immunogold labeled with an antibody to cellulose synthase. (Reproduced from Kimura, S., Laosinchai, W., Itoh, T., Cui, X., Linder, R., and Brown, R. M., Jr. (1999). *Plant Cell* **11,** 2075–2085.)

of cellulose deficiency in the primary wall (Fagard *et al.*, 2000). In addition to similar mutant phenotypes, both *AtCesA1* and *AtCesA6* also show similar expression profiles in various organs and growth conditions suggesting coordinated expression of at least two distinct cellulose synthases (*AtCesA1* and *AtCesA6*) in most cells (Fagard *et al.*, 2000). However, differences were observed in the embryonic expression of these two *CesA* genes (Beeckman *et al.*, 2002). Mutations in the *ixr1* and *ixr2* genes confer resistance to the cellulose synthesis inhibitor isoxaben and these two genes encode AtCesA3 and AtCesA6, respectively (Desprez *et al.*, 2002; Scheible *et al.*, 2001). The cellulose synthases identified by analysis of the *rsw1*, *ixr1*, and *PRC1/ixr2* mutants involve members of the CesA family (AtCesA1, AtCesA3, and AtCesA6) required for primary wall cellulose synthesis. Although no physical interactions have been determined for these cellulose synthases, studies of inhibition of cellulose synthesis by isoxaben suggest that AtCesA3 and AtCesA6 together form an active protein complex in which the involvement of AtCesA1 may be required (Desprez *et al.*, 2002).

Brittle culm mutants have been identified in barley, maize, and rice. The cellulose content in the cell walls of cells in the brittle culm mutants of barley was found to be lower than the wild-type plants, but no significant differences were found in the amount of the noncellulosic components of the cell wall (Kokubo *et al.*, 1989, 1991). Brittle culm mutants in rice were useful in identifying three *CesA* genes (*OsCesA4*, *OsCesA7*, and *OsCesA9*) (Tanaka *et al.*, 2003). The three genes are expressed in seedlings, culms, premature panicles, and roots, but not in mature leaves. The expression profiles are almost identical for these three genes, and decrease in the cellulose content in the culms of null mutants of the three genes indicates that these genes are not functionally redundant (Tanaka *et al.*, 2003).

4.2.2. Identification of other genes/proteins which may be required for cellulose biosynthesis in plants

The role of β-1,4-endoglucanase during cellulose synthesis was first proposed by Matthysse *et al.* (1995a,b) during analysis of cellulose-minus mutants in *Agrobacterium tumefaciens* (Matthysse *et al.*, 1995a,b). In this bacterium, cellulose synthesis is suggested to proceed via the formation of lipid-linked intermediates, and a β-1,4-endoglucanase is predicted to function as a transferase in the transfer of β-1,4-linked glucan oligomers from a lipid carrier to the growing cellulose chain (Matthysse *et al.*, 1995a). The gene encoding β-1,4-endoglucanase is organized with the cellulose synthase gene in an operon in *A. tumefaciens*, and a similar organization of these genes is observed in a number of other bacteria (Matthysse *et al.*, 1995b; Römling, 2002). The organization of a β-1,4-endoglucanase gene with the cellulose synthase gene in the same operon in bacteria has been taken as an indication that β-1,4-endoglucanase probably has a role during cellulose synthesis. So far, there is no direct demonstration for this role in bacteria or any other organism. A gene encoding a membrane-anchored β-1,4-endoglucanase called KORRIGAN also has been identified in a dwarf mutant of *Arabidopsis* (Nicol *et al.*, 1998). In plants, the KORRIGAN protein is believed to function during primary or secondary wall cellulose synthesis (Lane *et al.*, 2001;

Mølhøj *et al.*, 2002; Nicol *et al.*, 1998; Sato *et al.*, 2001; Szyjanowicz *et al.*, 2004; Zuo *et al.*, 2000). Its exact function during cellulose synthesis remains to be determined, although various roles have been assigned to it such as terminating or editing the glucan chains emerging from the cellulose synthase complex before their crystallization into a cellulose microfibril. Alternately it could cleave sterol from the sterol-glucoside primer that is suggested to initiate glucan chain formation (Peng *et al.*, 2002). However, recent evidence does not support this role (Scheible and Pauly, 2004). A membrane-bound sucrose synthase, which converts sucrose to UDP-glucose, may be physically linked to the cellulose synthase complex for channeling UDP-glucose to the cellulose synthase in plants, and suppression of this gene has been shown to effect cotton fiber initiation and elongation (Amor *et al.*, 1995; Ruan *et al.*, 2003).

Proteins that may indirectly influence cellulose biosynthesis include those that are required for *N*-glycan synthesis and processing (Lukowitz *et al.*, 2001). One of these proteins is glucosidase I, which trims off the terminal β-1,2-linked glucosyl residue from N-linked glycans and is involved in the quality control of newly synthesized proteins that transit through the endoplasmic reticulum (ER) (Boisson *et al.*, 2001; Gillmor *et al.*, 2002). Another protein could be glucosidase II that removes the two internal β-1,3-linked glucosyl residues subsequent to the action of glucosidase I in the quality control pathway (Burn *et al.*, 2002b). Other proteins that influence cellulose production include KOBITO, a membrane-anchored protein of unknown function that is suggested to be a part of the cellulose synthase complex, and COBRA, a putative glycosylphosphatidylinositol (GPI)-anchored protein, which upon being inactivated, dramatically reduces culm strength in rice (Li *et al.*, 2003b; Pagant *et al.*, 2002; Schindelman *et al.*, 2001).

4.3. The cellulose synthase genes

As of June 2006, *CesA* and *Csl* gene sequences have been identified in 252 plant species (http://cellwall.stanford.edu/). In *Arabidopsis*, 10 *CesA* and 30 *Csl* genes have been identified. Similar numbers of *CesA* and *Csl* genes have been identified in other plants as well. In rice, at least 12 *CesA* genes have been identified by analysis of cDNA, ESTs, and genome sequencing (http://cellwall.stanford.edu/). Twelve members of the *CesA* gene family are identified in maize (Appenzeller *et al.*, 2004). In most cases, the *CesA* genes are found to be dispersed on different chromosomes and have similar numbers of exons and introns. The *CesA* genes identified in maize from cDNA analysis and mapping studies were found to be distributed to different chromosomes, similar to the *Arabidopsis CesA* genes (Holland *et al.*, 2000). In *Arabidopsis*, the genes range in size from 3.5 to 5.5 kbp and contain 9–13 introns and the CesA transcripts range in size from 3.0 to 3.5 kb, encoding proteins that are 985–1,088 amino acids in length (Richmond, 2000). Orthologs of the *Arabidopsis CesA* genes have been identified in a number of plants by phylogenetic analysis using the CesA protein sequences. Three maize CesAs, ZmCesA10–12 cluster with the *Arabidopsis* CesAs that are shown to be involved in secondary wall cellulose synthesis. *ZmCesA10*, *ZmCesA11*, and *ZmCesA12* group with *AtCesA4* (Irx5), *AtCesA8* (Irx1), and *AtCesA7* (Irx3), respectively and are probable orthologs of these genes. Based on expression patterns, these three genes appear to be coordinately expressed (Appenzeller *et al.*, 2004). Likewise, *OsCesA7*, *OsCesA4*, and *OsCesA9* are the orthologous genes in rice, as are barley *HvCesA4*, *HvCesA5/7*, and *HvCesA8* genes, respectively (Burton *et al.*, 2004; Tanaka *et al.*, 2003).

Orthologs of the *Arabidopsis CesA* genes required for secondary wall cellulose synthesis have also been identified by expression analysis of normal wood undergoing xylogenesis in hybrid aspen (Djerbi *et al.*, 2004). Four *CesA*s, *PttCesA1*, *PttCesA3–1*, *PttCesA3–2*, and *PttCesA9* were shown to exhibit xylem-specific expression, with the derived amino acid sequences and expression profiles of *PttCesA3–1* and *PttCesA3–2* being very similar, suggesting that they represent redundant copies of a *CesA* with the same function. Phylogenetic analysis indicates that the xylem-specific CesAs from hybrid poplar cluster with similar CesAs from other poplars and *Arabidopsis*. PttCesA1 is most similar to AtCesA4, PttCesA3–1, and PttCesA3–2 are closest to AtCesA8, and PttCesA9 is closest to AtCesA7 (Djerbi *et al.*, 2004). Although it has been possible to identify orthologs of CesAs required for secondary wall cellulose synthesis in various plants, the relationship between the CesAs involved in primary wall cellulose synthesis from different plants is not as clear. From phylogenetic analysis, it appears that the genes for primary wall cellulose synthesis have duplicated relatively independently in dicots and monocots (Appenzeller *et al.*, 2004).

4.4. The cellulose synthase protein

The cellulose synthase genes identified in *A. xylinum* encode either the catalytic subunit consisting of 754 amino acids and 9 potential transmembrane regions or a longer protein of approximately 1,550 amino acids consisting of the cellulose synthase catalytic domain and an activator (c-di-GMP)-binding domain with 9 potential transmembrane regions (Saxena *et al.*, 1990, 1991, 1994; Wong *et al.*, 1990). The catalytic region in these proteins was predicted to have the conserved residues and sequence motif identified as D, D, D, QXXRW (Saxena *et al.*, 1995). *CesA* genes in plants encode a large, multipass transmembrane protein with a globular region containing the D, D, D, QXXRW motif. The CesA proteins in plants have a plant-specific conserved region (CR-P) and a hypervariable region (HVR-2) within the cytosolic globular region that contains the conserved residues. A conserved, extended N-terminal region is shown to have two zinc-finger domains resembling LIM/RING domains followed by a HVR-1 region (Kawagoe and Delmer, 1997). The RING domains are predicted to mediate protein–protein interactions. Using the yeast two-hybrid system, it has been shown that the zinc-finger domain of GhCesA1 is able to interact with itself to form homodimers or heterodimers with the zinc-finger domain of GhCesA2 in a redox-dependent manner (Kurek *et al.*, 2002). This dimerization of CesAs is supposed to represent the first stage in the assembly of the rosette TC (Saxena and Brown, 2005).

5. MECHANISM OF CELLULOSE SYNTHESIS

5.1. Role of primer and/or intermediates during cellulose synthesis?

In straightforward terms, cellulose biosynthesis requires the enzyme cellulose synthase for catalyzing the polymerization of glucose residues from UDP-glucose into a β-1,4-linked glucan chain. This simple mechanism envisions direct polymerization without the need for any intermediates or a primer. Cellulose biosynthesis has been demonstrated *in vitro* using membrane and detergent-solubilized extracts from *A. xylinum* and a number of plants in the presence of only UDP-glucose (Kudlicka and Brown, 1997; Lai-Kee-Him *et al.*, 2002; Lin and Brown, 1989; Okuda *et al.*, 1993). The synthesis of cellulose *in vitro* with the minimal added components in the reaction mixture strongly supports the direct polymerization of glucose without any requirement for a primer. However, in the absence of purified cellulose synthases it is not possible to completely exclude the role of other proteins or components contributed by the membrane fraction or detergent extracts during cellulose synthesis. In 2002, Peng *et al.* proposed a model for cellulose biosynthesis in which they suggested that SG serves as a primer for synthesis of SCDs by CesA proteins (Peng *et al.*, 2002). According to their model, a membrane-associated endoglucanase Kor (encoded by the *Korrigan* gene) cleaves SCDs giving rise to SG and cellodextrins (CDs). In the next step, the CDs undergo β-1,4-glucan chain elongation catalyzed by CesA proteins. The glucose moiety of SG is found to be attached via its reducing end to sitosterol and chain elongation in the first step is predicted to proceed from the nonreducing end. Based on this model, plants deficient in sitosterol are expected to show a severe phenotype due to impairment in cellulose synthesis (Peng *et al.*, 2002). A number of mutants deficient in sitosterol content have been identified in *Arabidopsis*. However, *dwf1/dim* mutants of *Arabidopsis* that have a severe reduction in sitosterol content have been rescued to the wild type by brassinosteroid (BR) treatment suggesting that sitosterol may not have a major role in cellulose biosynthesis (Clouse, 2002). In the absence of any direct evidence for the role of sitosterol in cellulose biosynthesis, doubts have been raised regarding the proposed involvement of SG as a primer (Somerville *et al.*, 2004).

5.2. Addition of glucose residues to the growing glucan chain end

The glucose residues in the β-1,4-linked glucan chains in cellulose are arranged such that each residue is inverted with respect to its neighbor, giving rise to a twofold screw axis and a rather flat chain. If this arrangement of sugar residues is established during synthesis, it would entail either the rotation of the glucan chain or the cellulose synthase for addition of successive glucose residues to the growing end. A model suggesting that the active site of the enzyme can position two UDP-glucose molecules in an orientation such that the two glucose residues are positioned inverted to each other in the catalytic pocket was proposed by Saxena *et al.* (1995), and it was suggested that the glucose residues could be added sequentially or simultaneously to the growing end (Saxena *et al.*, 1995).

The growing end was later shown to be the nonreducing end of the β-1,4-linked glucan chain during cellulose synthesis (Koyama *et al.*, 1997). Alternatively, the twofold symmetry in the glucan chain can be obtained from a single catalytic center, based on the reasoning that there is a fairly large degree of freedom of rotation about the β-glycosidic bond. According to this proposal, the glucose residue added in one orientation relaxes into the native orientation after polymerization (Delmer, 1999). Other proposals have suggested that two catalytic centers may be present in two subunits and be organized following dimerization or two different catalytic domains within the same catalytic site participate in the dual addition (Albersheim *et al.*, 1997; Charnock *et al.*, 2001). Cellulose synthase and other processive β-glycosyltransferases have so far resisted crystal structure determination although structure of a nonprocessive β-glycosyltransferase (SpsA from *Bacillus subtilis*) has been determined (Charnock and Davies, 1999). The SpsA protein lacks the conserved QXXRW motif found in the processive enzymes, and studies with site-directed mutants of cellulose synthase have indicated a role of this motif during the synthesis of cellulose (Saxena *et al.*, 2001). The structure of the globular region of the *A. xylinum* cellulose synthase containing all the conserved aspartic acid residues and the QXXRW motif was predicted using the genetic algorithm, and it was estimated that the central elongated cavity can accommodate two UDP-glucose residues (Saxena *et al.*, 2001). The alternating orientation of the *N*-acetylglucosamine (GlcNAc) residues within the chitin chain also led to the proposal that chitin synthases possess two active sites, and this possibility was tested using UDP-derived monomeric and dimeric inhibitors of chitin synthase activity *in vitro* (Yeager and Finney, 2004). Using these inhibitors, it was found that uridine-derived dimeric inhibitors exhibited a 10-fold greater inhibition of chitin synthase activity as compared to the monomeric control, consistent with the presence of two active sites in chitin synthases (Yeager and Finney, 2004).

6. PROSPECTS FOR GENETIC ENGINEERING OF CELLULOSE BIOSYNTHESIS IN PLANTS

6.1. Manipulation of cellulose biosynthesis in plants

Genetic modifications for improvement of specific traits or the addition of new traits to economically important plants is a major objective worldwide. Not only is cellulose a constituent of all plants, a number of plants (such as cotton and forest trees) are grown specifically for their cellulose content. In general, the objective of genetic manipulation of the cellulose synthesizing capacity in these plants is to either increase the amount of cellulose or modify the physical properties of the cellulose during synthesis. For example, the secondary cell wall in cotton fibers determines the fiber properties. Considering that the secondary cell wall in cotton fibers is approximately 95% cellulose, the properties of the cotton fiber are dependent not only on the amount of cellulose deposited, but also on other features such as the structure and orientation of the cellulose microfibrils and the degree of polymerization of the glucan chains. Additionally, manipulation of cellulose synthesis in a number of crop plants may be important for improving specific agronomic traits. As an example, stalk lodging in maize results in significant yield losses, and an increase in the cellulose content in the cells in the stalk may allow improvements in stalk strength and harvest index (Appenzeller *et al.*, 2004). Apart from its importance in the growth and development of plants, cellulose is also an abundant renewal energy resource that is present in the biomass obtained from agricultural residues of major crops. Corn stover is the most abundant agriculture residue in the United States and it can be used for various applications including bioethanol production (Kadam and Mcmillan, 2003). Increasing the content of cellulose and reducing the lignin content of corn plants is therefore considered to be beneficial for ethanol production.

Cellulose biosynthesis in plants can be modified by manipulation of the cellulose synthase (*CesA*) genes or other genes that influence cellulose production. *CesA* genes have been identified in most plants, and as a result they are prime targets for directly modifying cellulose synthesis by genetic manipulation. *CesA* genes are part of a gene family, and as a result a number of features of these genes will have to be analyzed before they can be manipulated usefully. Some of these features may include understanding of the expression of the different *CesA* genes, the redundant nature of each gene in a specific cell type, and the phenotype that is generated when each gene is mutated or overexpressed (Holland *et al.*, 2000). In corn, the majority of the cellulose in the stalk is in the vascular bundles. Based on their expression patterns, 3 of the 12 *CesA* genes in corn appear to be involved in cellulose synthesis during secondary wall formation and their promoter sequences have been identified (Appenzeller *et al.*, 2004). These promoters can now be used for expression of *CesA* genes in specific cell types for increasing their cellulose content.

Direct modification of cellulose content by manipulation of the cellulose synthase genes has been performed in only a few cases so far. To improve fiber quality of cotton fibers, the *A. xylinum acsA* and *acsB* genes were transferred to cotton (Li *et al.*, 2004). The fiber strength and length of fibers were found to be greater in the transformed plants, as well as the cellulose content was found to be higher in the transformed plants as compared to the control plants. In potato, cellulose content was modified in the tuber using sense and antisense expression of the full length *StCesA3* and class-specific regions (CSR) of the four potato *CesA* cDNAs (Oomen *et al.*, 2004). The antisense and sense *StCesA3* transformants demonstrated that the cellulose content could be decreased to 43% and increased to 200% of the wild type, respectively, by modifying the RNA expression levels (Oomen *et al.*, 2004). Interestingly, the increase in cellulose content by increasing expression of a single *CesA* gene was found to be remarkable considering that multiple copies of different CesAs are believed to be required for assembly of cellulose-synthesizing complexes. The utility of antisense transgenic lines in generating a range of phenotypes is suggested to be particularly useful, especially where null mutations are potentially lethal (Oomen *et al.*, 2004). In *Arabidopsis*, the transgenic approach using antisense expression exhibited a slightly different phenotype as compared to a mutation in the corresponding gene

(Burn *et al.*, 2002a). The modulation of *CesA* RNA expression levels and concomitantly cellulose content has also been demonstrated in tobacco plants using virus-induced silencing of a cellulose synthase gene (Burton *et al.*, 2000). Apart from the *CesA* genes, genes with an indirect role in cellulose biosynthesis, such as the sucrose synthase, have been manipulated in the cotton fiber using suppression constructs. A 70% or more suppression of the sucrose synthase activity in the ovule led to a fiberless phenotype suggesting that this enzyme has a rate-limiting role in the initiation and elongation of fibers (Ruan *et al.*, 2003). In other instances, while some researchers have shown an increase in cellulose accumulation following manipulation of genes for reduced lignin synthesis in aspen trees (Hu *et al.*, 1999; Li *et al.*, 2003a), other researchers did not find any evidence in support of enhanced cellulose synthesis upon severe downregulation of lignin biosynthetic genes (Anterola and Lewis, 2002). It is believed that the synthesis of cellulose is interconnected with the synthesis of other components of the plant cell wall, and manipulation of a number of genes would therefore affect cellulose production. However, not much is known as to how the different pathways are interconnected, but a systems view of these interactions is beginning to emerge (Somerville *et al.*, 2004).

6.2. Influence of cellulose alterations in plants

Cellulose in the plant cell wall influences a number of traits, and although not much is known in terms of the effects on the plant upon increase of cellulose content in the cell wall, a number of studies have linked mutations in the genes encoding cellulose synthases and other proteins that may be required for cellulose synthesis to changes in other properties. For example, the *Arabidopsis* cellulose synthase (*AtCesA3*) mutant, *cev1*, is found to be resistant to fungal pathogens and is constitutively activated for defense pathways in a manner similar to that for the pathogen-induced pmr4 mutant (Cano-Delgado *et al.*, 2003; Ellis *et al.*, 2002; Nishimura *et al.*, 2003). Moreover, there is an accumulation of transcripts that are induced by jasmonic acid (JA) and ethylene in this mutant (Ellis and Turner, 2001; Ellis *et al.*, 2002). Increased ethylene production and/or sensitivity was observed for *cesA3*^{eli1}, *cesA6*^{prc1}, *kor1*, *elp1/pom1*, and in wild-type plants treated with 2,6-dichlorobenzonitrile (DCB) or isoxaben (Cano-Delgado *et al.*, 2003; Desnos *et al.*, 1996; Ellis and Turner, 2001; Ellis *et al.*, 2002; Zhong *et al.*, 2002). Only a brief list of changes have been mentioned here, but as is clear from these results that changes in cellulose synthesis/content in the cell wall are sensed by cells directly or indirectly through as yet unknown mechanisms.

7. SUMMARY

Cellulose is a component of all plant cells, and modification of the cellulose content or properties can have dramatic effects on the form and function(s) of specific parts or the entire plant. Cellulose synthase is the enzyme required for biosynthesis of cellulose, and a number of genes encoding this protein form part of a gene family in plants. Although plants are well endowed with genes for cellulose synthases, and expression of most of the *CesA* genes have been observed in most tissues, mutations in some of them can have very different effects. At the same time increased expression of some of the *CesA* genes may result in increased synthesis of cellulose in specific cells and tissues. More importantly, the direction in which the cellulose microfibrils are assembled in the primary cell wall helps determine the direction of cell elongation. In cells with a secondary cell wall, the orientation of the cellulose microfibrils influences the properties of the cell. Although the general view is that microtubules play a role in determining the direction of cellulose synthesis, not much is known as to how this occurs. For effective manipulation of cellulose synthesis in plant cells, it is necessary that we not only understand the machinery responsible for cellulose biosynthesis, but also as to how it is assembled, localized, and regulated.

ACKNOWLEDGEMENTS

The authors acknowledge support from the Division of Energy Biosciences, Department of Energy (Grant DE-FG03-94ER20145), and the Welch Foundation (Grant F-1217).

REFERENCES

Albersheim, P., Darvill, A., Roberts, K., Staehelin, L. A., and Varner, J. E. (1997). Do the structures of cell wall polysaccharides define their mode of synthesis? *Plant Physiol.* **113,** 1–3.

Amor, Y., Haigler, C. H., Johnson, S., Wainscott, M., and Delmer, D. P. (1995). A membrane-associated form of sucrose synthase and its potential role in synthesis of cellulose and callose in plants. *Proc. Natl. Acad. Sci. USA* **92,** 9353–9357.

Anterola, A. M., and Lewis, N. G. (2002). Trends in lignin modification: A comprehensive analysis of the effects of genetic manipulations/mutations on lignification and vascular integrity. *Phytochemistry* **61,** 221–294.

Appenzeller, L., Doblin, M., Barreiro, R., Wang, H., Niu, X., Kollipara, K., Carrigan, L., Tomes, D., Chapman, M., and Dhugga, K. S. (2004). Cellulose synthesis in maize: Isolation and expression analysis of the cellulose synthase (*CesA*) gene family. *Cellulose* **11,** 287–299.

Arioli, T., Peng, L., Betzner, A. S., Burn, J., Wittke, W., Herth, W., Camilleri, C., Höfte, H., Plazinski, J., Birch, R., Cork, A., Glover, J., *et al.* (1998). Molecular analysis of cellulose biosynthesis in *Arabidopsis*. *Science* **279,** 717–720.

Attala, R. H., and Vanderhart, D. L. (1984). Native cellulose: A composite of two distinct crystalline forms. *Science* **223,** 283–285.

Barber, G. A., Elbein, A. D., and Hassid, W. Z. (1964). The synthesis of cellulose by enzyme systems from higher plants. *J. Biol. Chem.* **239,** 4056–4061.

Beeckman, T., Przemeck, G. K. H., Stamatiou, G., Lau, R., Teryn, N., De Rycke, R., Inze, D., and Berleth, T. (2002). Genetic complexity of cellulose synthase A gene function in *Arabidopsis* embryogenesis. *Plant Physiol.* **130,** 1883–1893.

Boisson, M., Gomord, V., Audran, C., Berger, N., Dubreucq, B., Granier, F., Lerouge, P., Faye, L., Caboche, M., and Lepiniec, L. (2001). *Arabidopsis* glucosidase I mutants reveal a critical role of N-glycan trimming in seed development. *EMBO J.* **20,** 1010–1019.

Brown, R. M., Jr. (1996). The biosynthesis of cellulose. *J. Macromol. Sci. Pure Appl. Chem.* **A33,** 1345–1373.

Bureau, T. E., and Brown, R. M., Jr. (1987). *In vitro* synthesis of cellulose II from a cytoplasmic membrane fraction of *Acetobacter xylinum*. *Proc. Natl. Acad. Sci. USA* **84,** 6985–6989.

Burn, J. E., Hocart, C. H., Birch, R. J., Cork, A., and Williamson, R. E. (2002a). Functional analysis of the cellulose synthase genes *CesA1*, *CesA2*, and *CesA3* in *Arabidopsis*. *Plant Physiol.* **129,** 797–807.

Burn, J. E., Hurley, U. A., Birch, R. J., Arioli, T., Cork, A., and Williamson, R. E. (2002b). The cellulose-deficient *Arabidopsis* mutant *rsw3* is defective in a gene encoding a putative glucosidase II, an enzyme processing N-glycans during ER quality control. *Plant J.* **32,** 949–960.

Burton, R. A., Gibeaut, D. M., Bacic, A., Findlay, K., Roberts, K., Hamilton, A., Baulcombe, D. C., and Fincher, G. B. (2000). Virus-induced silencing of a plant cellulose synthase gene. *Plant Cell* **12,** 691–706.

Burton, R. A., Shirley, N. J., King, B. J., Harvey, A. J., and Fincher, G. B. (2004). The *CesA* gene family of barley. Quantitative analysis of transcripts reveals two groups of co-expressed genes. *Plant Physiol.* **134,** 224–236.

Cano-Delgado, A., Penfield, S., Smith, C., Catley, M., and Bevan, M. (2003). Reduced cellulose synthesis invokes lignification and defense responses in *Arabidopsis thaliana. Plant J.* **34,** 351–362.

Carpita, N. C., and Delmer, D. P. (1981). Concentration and metabolic turnover of UDP-glucose in developing cotton fibers. *J. Biol. Chem.* **256,** 308–315.

Chambers, J., and Elbein, A. D. (1970). Biosynthesis of glucans in mung bean seedlings. Formation of β-(1→4)-glucans from GDP-glucose and β-(1 → 3)-glucans from UDP-glucose. *Arch. Biochem. Biophys.* **138,** 620–631.

Charnock, S. J., and Davies, G. J. (1999). Structure of the nucleotide-diphospho-sugar transferase, SpsA from *Bacillus subtilis,* in native and nucleotide-complexed forms. *Biochemistry* **38,** 6380–6385.

Charnock, S. J., Henrissat, B., and Davies, G. J. (2001). Three-dimensional structures of UDP-sugar glycosyltransferases illuminate the biosynthesis of plant polysaccharides. *Plant Physiol.* **125,** 527–531.

Clouse, S. D. (2002). *Arabidopsis* mutants reveal multiple roles for sterols in plant development. *Plant Cell* **14,** 1995–2000.

Colombani, A., Djerbi, S., Bessuelle, L., Blomqvist, K., Ohlsson, A., Berglund, T., Teeri, T. T., and Bulone, V. (2004). *In vitro* synthesis of (1→3)-β-D-glucan (callose) and cellulose by detergent extracts of membranes from cell suspension cultures of hybrid aspen. *Cellulose* **11,** 313–327.

Coutinho, P. M., Stam, M., Blanc, E., and Henrissat, B. (2003). Why are there so many carbohydrate-active enzyme-related genes in plants? *Trends Plant Sci.* **8,** 563–565.

Cui, X., Shin, H., Song, C., Laosinchai, W., Amano, Y., and Brown, R. M., Jr. (2001). A putative plant homolog of the yeast β-1,3-glucan synthase subunit FKS1 from cotton (*Gossypium hirsutum* L.) fibers. *Planta* **213,** 223–230.

Cutler, S., and Somerville, C. (1997). Cloning *in silico. Curr. Biol.* **7,** R108–R111.

D'Argenio, D. A., and Miller, S. I. (2004). Cyclic di-GMP as a bacterial second messenger. *Microbiology* **150,** 2497–2502.

Delmer, D. P. (1983). Biosynthesis of cellulose. *Adv. Carbohydr. Chem. Biochem.* **41,** 105–153.

Delmer, D. P. (1999). Cellulose biosynthesis: Exciting times for a difficult field of study. *Annu. Rev. Plant Physiol. Plant Mol. Biol.* **50,** 245–276.

Desnos, T., Orbovic, V., Bellini, C., Kronenberger, J., Caboche, M., Traas, J., and Höfte, H. (1996). *Procuste1* mutants identify two distinct genetic pathways controlling hypocotyl cell elongation, respectively in dark- and light-grown *Arabidopsis* seedlings. *Development* **122,** 683–693.

Desprez, T., Vernhettes, S., Fagard, M., Refrégier, G., Desnos, T., Aletti, E., Py, N., Pelletier, S., and Höfte, H. (2002). Resistance against herbicide isoxaben and cellulose deficiency caused by distinct mutations in same cellulose synthase isoform CESA6. *Plant Physiol.* **128,** 482–490.

Doblin, M. S., De Melis, L., Newbigin, E., Bacic, A., and Read, S. M. (2001). Pollen tubes of *Nicotiana alata* express two genes from different β-glucan synthase families. *Plant Physiol.* **125,** 2040–2052.

Dhugga, K. S., and Ray, P. M. (1994). Purification of 1,3-β-D-glucan synthase activity from pea tissue. Two polypeptides of 55 kDa and 70 kDa copurify with enzyme activity. *Eur. J. Biochem.* **220,** 943–953.

Djerbi, S., Aspeborg, H., Nilsson, P., Sundberg, B., Mellerowicz, E., Blomqvist, K., and Teeri, T. T. (2004). Identification and expression analysis of genes encoding putative cellulose synthases (CesA) in the hybrid aspen, *Populus tremula* (L.) x *P. tremuloides* (Michx.). *Cellulose* **11,** 301–312.

Ellis, C., and Turner, J. G. (2001). The *Arabidopsis* mutant *cev1* has constitutively active jasmonate and ethylene signal pathways and enhanced resistance to pathogens. *Plant Cell* **13,** 1025–1033.

Ellis, C., Karafyllidis, I., Wasternack, C., and Turner, J. G. (2002). The Arabidopsis mutant *cev1* links cell wall signaling to jasmonate and ethylene responses. *Plant Cell* **14,** 1557–1566.

Fagard, M., Desnos, T., Desprez, T., Goubet, F., Refregier, G., Mouille, G., Mccann, M., Rayon, C., Vernhettes, S., and Höfte, H. (2000). *PROCUSTE1* encodes a cellulose synthase required for normal cell elongation specifically in roots and dark-grown hypocotyls of *Arabidopsis*. *Plant Cell* **12,** 2409–2423.

Feingold, D. S., Neufeld, E. F., and Hassid, W. Z. (1958). Synthesis of a β-1,3-linked glucan by extracts of *Phaseolus aureus* seedlings. *J. Biol. Chem.* **233,** 783–788.

French, A. D. (2000). Structure and biosynthesis of cellulose. Part I. Structure. *In* "Discoveries in Plant Biology" (S. D. Kung and S. F. Yang, eds.), Vol. III, pp. 163–197. World Scientific, Hong Kong.

Gillmor, C. S., Poindexter, P., Lorieau, J., Palcic, M. M., and Somerville, C. (2002). α-glucosidase I is required for cellulose biosynthesis and morphogenesis in *Arabidopsis*. *J. Cell Biol.* **156,** 1003–1013.

Glaser, L. (1958). The synthesis of cellulose in cell-free extracts of *Acetobacter xylinum*. *J. Biol. Chem.* **232,** 627–636.

Holland, N., Holland, D., Helentjaris, T., Dhugga, K. S., Xoconostle-Cazares, B., and Delmer, D. P. (2000). A comparative analysis of the plant cellulose synthase (*CesA*) gene family. *Plant Physiol.* **123,** 1313–1323.

Hong, Z., Delauney, A. J., and Verma, D. P. S. (2001). A cell plate-specific callose synthase and its interaction with phragmoplastin. *Plant Cell* **13,** 755–768.

Hu, W. J., Harding, S. A., Lung, J., Popko, J. L., Ralph, J., Stokke, D. D., Tsai, C. J., and Chiang, V. L. (1999). Repression of lignin biosynthesis promotes cellulose accumulation and growth in transgenic trees. *Nat. Biotechnol.* **17,** 808–812.

Kadam, K. L., and Mcmillan, J. D. (2003). Availability of corn stover as a sustainable feedstock for bioethanol production. *Bioresour. Technol.* **88,** 17–25.

Kawagoe, Y., and Delmer, D. P. (1997). Pathways and genes involved in cellulose biosynthesis. *In* "Genetic Engineering" (J. K. Setlow, ed.), pp. 63–87. Plenum Press, New York.

Klemm, D., Heublein, B., Fink, H.-P., and Bohn, A. (2005). Cellulose: Fascinating biopolymer and sustainable raw material. *Angew. Chem. Int. Ed.* **44,** 2–37.

Kokubo, A., Kuraishi, S., and Sakurai, N. (1989). Culm strength of barley: Correlation among maximum bending stress, cell wall dimensions, and cellulose content. *Plant Physiol.* **91,** 876–882.

Kokubo, A., Sakurai, N., Kuraishi, S., and Takeda, K. (1991). Culm brittleness of barley (*Hordeum vulgare* L.) mutants is caused by small number of cellulose molecules in cell wall. *Plant Physiol.* **97,** 509–514.

Kondo, T., Togawa, E., and Brown, R. M., Jr. (2001). Nematic ordered cellulose: A concept of glucan chain association. *Biomacromolecules* **2,** 1324–1330.

Kondo, T., Kasai, W., and Brown, R. M., Jr. (2004). Formation of nematic ordered cellulose and chitin. *Cellulose* **11,** 463–474.

Koyama, M., Helbert, W., Imai, T., Sugiyama, J., and Henrissat, B. (1997). Parallel-up structure evidences the molecular directionality during biosynthesis of bacterial cellulose. *Proc. Natl. Acad. Sci. USA* **94,** 9091–9095.

Kudlicka, K., and Brown, R. M., Jr. (1997). Cellulose and callose biosynthesis in higher plants. I. Solubilization and separation of (1 → 3)- and (1 → 4)-β-glucan synthase activities from mung bean. *Plant Physiol.* **115,** 643–656.

Kudlicka, K., Brown, R. M., Jr., Li, L., Lee, J. H., Shin, H., and Kuga, S. (1995). β-glucan synthesis in the cotton fiber. IV. *In vitro* assembly of the cellulose I allomorph. *Plant Physiol.* **107,** 111–123.

Kudlicka, K., Lee, J. H., and Brown, R. M., Jr. (1996). A comparative analysis of *in vitro* cellulose synthesis from cell-free extracts of mung bean (*Vigna radiata*, Fabaceae) and cotton (*Gossypium hirsutum*, Malvaceae). *Am. J. Bot.* **83,** 274–284.

Kurek, I., Kawagoe, Y., Jacob-Wilk, D., Doblin, M., and Delmer, D. (2002). Dimerization of cotton fiber cellulose synthase catalytic subunits occurs via oxidation of the zinc-binding domains. *Proc. Natl. Acad. Sci. USA* **99,** 11109–11114.

Lai-Kee-Him, J., Pelosi, L., Chanzy, H., Putaux, J. L., and Bulone, V. (2001). Biosynthesis of (1 → 3)-β-D-glucan (callose) by detergent extracts of a microsomal fraction from *Arabidopsis thaliana*. *Eur. J. Biochem.* **268,** 4628–4638.

Lai-Kee-Him, J., Chanzy, H., Müller, M., Putaux, J. L., Imai, T., and Bulone, V. (2002). *In vitro* versus *in vivo* cellulose microfibrils from plant primary wall synthases: Structural differences. *J. Biol. Chem.* **277,** 36931–36939.

Lane, D. R., Wiedemeier, A., Peng, L., Höfte, H., Vernhettes, S., Desprez, T., Hocart, C. H., Birch, R. J., Baskin, T. I., Burn, J. E., Arioli, T., Betzner, A. S., *et al.* (2001). Temperature-sensitive alleles of *RSW2* link the KORRIGAN endo-1,4-β-glucanase to cellulose synthesis and cytokinesis in *Arabidopsis*. *Plant Physiol.* **126,** 278–288.

Laosinchai, W. (2002). Molecular and biochemical studies of cellulose and callose synthase, Ph.D. Dissertation : The University of Texas at Austin.

Leloir, L. F., and Cabib, E. (1953). The enzymic synthesis of trehalose phosphate. *J. Am. Chem. Soc.* **75,** 5445–5446.

Li, L., and Brown, R. M., Jr. (1993). β-Glucan synthesis in the cotton fiber II. Regulation and kinetic properties of β-glucan synthases. *Plant Physiol.* **101,** 1143–1148.

Li, J., Burton, R. A., Harvey, A. J., Hrmova, M., Wardak, A. Z., Stone, B. A., and Fincher, G. B. (2003). Biochemical evidence linking a putative callose synthase gene with (1→3)-β-D-glucan biosynthesis in barley. *Plant Mol. Biol.* **53,** 213–225.

Li, L., Zhou, Y., Cheng, X., Sun, J., Marita, J. M., Ralph, J., and Chiang, V. L. (2003a). Combinatorial modification of multiple lignin traits in trees through multigene cotransformation. *Proc. Natl. Acad. Sci. USA* **100,** 4939–4944.

Li, Y., Qian, Q., Zhou, Y., Yan, M., Sun, L., Zhang, M., Fu, Z., Wang, Y., Han, B., Pang, X., Chen, M., and Li, J. (2003b). BRITTLE CULM1, which encodes a COBRA-like protein, affects the mechanical properties of rice plants. *Plant Cell* **15,** 2020–2031.

Li, X., Wang, X. D., Zhao, X., and Dutt, Y. (2004). Improvement of cotton fiber quality by transforming the *acsA* and *acsB* genes into *Gossypium hirsutum* L. by means of vacuum infiltration. *Plant Cell Rep.* **22,** 691–697.

Liepman, A. H., Wilkerson, C. G., and Keegstra, K. (2005). Expression of cellulose synthase-like (*Csl*) genes in insect cells reveals that *CslA* family members encode mannan synthases. *Proc. Natl. Acad. Sci. USA* **102,** 2221–2226.

Lin, F. C., and Brown, R. M., Jr. (1989). Purification of cellulose synthase from *Acetobacter xylinum*. *In* "Cellulose and Wood—Chemistry and Technology" (C Schuerch, ed.), pp. 473–492. John Wiley & Sons, New York.

Lin, F. C., Brown, R. M., Jr., Cooper, J. B., and Delmer, D. P. (1985). Synthesis of fibrils *in vitro* by a solubilized cellulose synthase from *Acetobacter xylinum*. *Science* **230,** 822–825.

Lin, F. C., Brown, R. M., Jr., Drake, R. R., Jr., and Haley, B. E. (1990). Identification of the uridine 5′-diphosphoglucose (UDP-Glc) binding subunit of cellulose synthase in *Acetobacter xylinum* using the photoaffinity probe 5-azido-UDP-Glc. *J. Biol. Chem.* **265,** 4782–4784.

Lukowitz, W., Nickle, T. C., Meinke, D. W., Last, R. L., Conklin, P. L., and Somerville, C. R. (2001). *Arabidopsis cyt1* mutants are deficient in a mannose-1-phosphate guanylyltransferase and point to a requirement of N-linked glycosylation for cellulose biosynthesis. *Proc. Natl. Acad. Sci. USA* **98,** 2262–2267.

Matthysse, A. G., Thomas, D. L., and White, A. R. (1995a). Mechanism of cellulose synthesis in *Agrobacterium tumefaciens*. *J. Bacteriol.* **177,** 1076–1081.

Matthysse, A. G., White, S., and Lightfoot, R. (1995b). Genes required for cellulose synthesis in *Agrobacterium tumefaciens*. *J. Bacteriol.* **177,** 1069–1075.

Mayer, R., Ross, P., Weinhouse, H., Amikam, D., Volman, G., Ohana, P., Calhoon, R. D., Wong, H. C., Emerick, A. W., and Benziman, M. (1991). Polypeptide composition of bacterial cyclic diguanylic acid-dependent cellulose synthase and the occurrence of immunologically crossreacting proteins in higher plants. *Proc. Natl. Acad. Sci. USA* **88,** 5472–5476.

Mølhøj, M., Pagant, S., and Höfte, H. (2002). Towards understanding the role of membrane-bound endo-β-1,4-glucanases in cellulose biosynthesis. *Plant Cell Physiol.* **43,** 1399–1406.

Nakashima, J., Laosinchai, W., Cui, X., and Brown, R. M., Jr. (2003). New insights into the mechanism of cellulose and callose biosynthesis: Proteases may regulate callose biosynthesis upon wounding. *Cellulose* **10,** 369–389.

Nicol, F., His, I., Jauneau, A., Vernhettes, S., Canut, H., and Höfte, H. (1998). A plasma membrane-bound putative endo-1,4-β-D-glucanase is required for normal wall assembly and cell elongation in *Arabidopsis*. *EMBO J.* **17,** 5563–5576.

Nishimura, M. T., Stein, M., Hou, B. H., Vogel, J. P., Edwards, H., and Somerville, S. C. (2003). Loss of a callose synthase results in salicylic acid-dependent disease resistance. *Science* **301,** 969–972.

Nishiyama, Y., Sugiyama, J., Chanzy, H., and Langan, P. (2003). Crystal structure and hydrogen bonding system in cellulose I_α from synchrotron X-ray and neutron fiber diffraction. *J. Am. Chem. Soc.* **125,** 14300–14306.

O'Sullivan, A. C. (1997). Cellulose: The structure slowly unravels. *Cellulose* **4,** 173–207.

Okuda, K., Li, L., Kudlicka, K., Kuga, S., and Brown, R. M., Jr. (1993). β-glucan synthesis in the cotton fiber. I. Identification of β-1,4- and β-1,3-glucans synthesized *in vitro*. *Plant Physiol.* **101,** 1131–1142.

Oomen, R. J. F. J., Tzitzikas, E. N., Bakx, E. J., Straatman-Engelen, I., Bush, M. S., Mccann, M. C., Schols, H. A., Visser, R. G. F., and Vincken, J.-P. (2004). Modulation of the cellulose content of tuber cell walls by antisense expression of different potato (*Solanum tuberosum* L.) *CesA* clones. *Phytochemistry* **65,** 535–546.

Pagant, S., Bichet, A., Sugimoto, K., Lerouxel, O., Desprez, T., Mcmann, M., Lerouge, P., Vernhettes, S., and Höfte, H. (2002). KOBITO1 encodes a novel plasma membrane protein necessary for normal synthesis of cellulose during cell expansion in *Arabidopsis*. *Plant Cell* **14,** 2001–2013.

Paredez, A. R., Somerville, C. R., and Ehrhardt, D. W. (2006). Visualization of cellulose synthase demonstrates functional association with microtubules. *Science* **312,** 1491–1495.

Pear, J. R., Kawagoe, Y., Schreckengost, W. E., Delmer, D. P., and Stalker, D. M. (1996). Higher plants contain homologs of the bacterial *celA* genes encoding the catalytic subunit of cellulose synthase. *Proc. Natl. Acad. Sci. USA* **93,** 2637–12642.

Peng, L., Kawagoe, Y., Hogan, P., and Delmer, D. (2002). Sitosterol-β-glucoside as primer for cellulose synthesis in plants. *Science* **295,** 147–150.

Perrin, R. M. (2001). Cellulose: How many cellulose synthases to make a plant? *Curr. Biol.* **11,** R213–R216.

Preston, R. D. (1974). "The Physical Biology of Plant Cell Walls." Chapman and Hall, London pp. 425–456.

Richmond, T. (2000). Higher plant cellulose synthases. *Genome Biol* **1,** 3001.1–3001.5.

Richmond, T. A., and Somerville, C. R. (2000). The cellulose synthase superfamily. *Plant Physiol.* **124,** 495–498.

Robert, S., Mouille, G., and Höfte, H. (2004). The mechanism and regulation of cellulose synthesis in primary walls: Lessons from cellulose-deficient *Arabidopsis* mutants. *Cellulose* **11,** 351–364.

Robert, S., Bichet, A., Grandjean, O., Kierzkowski, D., Satiat-Jeunemaître, B., Pelletier, S., Hauser, M.-T., Höfte, H., and Vernhettes, S. (2005). An *Arabidopsis* endo-1,4-β-D-glucanase involved in cellulose synthesis undergoes regulated intracellular cycling. *Plant Cell* **17,** 3378–3389.

Römling, U. (2002). Molecular biology of cellulose production in bacteria. *Res. Microbiol.* **153,** 205–212.

Ross, P., Aloni, Y., Weinhouse, H., Michaeli, D., Weinberger-Ohana, P., Mayer, R., and Benziman, M. (1986). Control of cellulose synthesis in *A. xylinum*. An unique guanyl oligonucleotide is the immediate activator of cellulose synthase. *Carbohydr. Res.* **149,** 101–117.

Ruan, Y. L., Llewellyn, D. J., and Furbank, R. T. (2003). Suppression of sucrose synthase gene expression represses cotton fiber cell initiation, elongation, and seed development. *Plant Cell* **15,** 952–964.

Sato, S., Kato, T., Kakegawa, K., Ishii, T., Liu, Y. G., Awano, T., Takabe, K., Nishiyama, Y., Kuga, S., Sato, S., Nakamura, Y., Tabata, S., *et al.* (2001). Role of the putative membrane-bound endo-1,4-β-glucanase KORRIGAN in cell elongation and cellulose synthesis in *Arabidopsis thaliana*. *Plant Cell Physiol.* **42,** 251–263.

Saxena, I. M., and Brown, R. M., Jr. (1995). Identification of a second cellulose synthase gene (*acsAII*) in *Acetobacter xylinum*. *J. Bacteriol.* **177,** 5276–5283.

Saxena, I. M., and Brown, R. M., Jr. (1997). Identification of cellulose synthase(s) in higher plants: Sequence analysis of processive β-glycosyltransferases with the common motif 'D, D, D35QXXRW.' *Cellulose* **4,** 33–49.

Saxena, I. M., and Brown, R. M., Jr. (2005). Cellulose biosynthesis: Current views and evolving concepts. *Ann. Bot.* **96,** 9–21.

Saxena, I. M., Lin, F. C., and Brown, R. M., Jr. (1990). Cloning and sequencing of the cellulose synthase catalytic subunit gene of *Acetobacter xylinum*. *Plant Mol. Biol.* **15,** 673–683.

Saxena, I. M., Lin, F. C., and Brown, R. M., Jr. (1991). Identification of a new gene in an operon for cellulose biosynthesis in *Acetobacter xylinum*. *Plant Mol. Biol.* **16,** 947–954.

Saxena, I. M., Kudlicka, K., Okuda, K., and Brown, R. M., Jr. (1994). Characterization of genes in the cellulose-synthesizing operon (*acs* operon) of *Acetobacter xylinum*: Implications for cellulose crystallization. *J. Bacteriol.* **176,** 5735–5752.

Saxena, I. M., Brown, R. M., Jr., Fèvre, M., Geremia, R. A., and Henrissat, B. (1995). Multidomain architecture of β-glycosyltransferases: Implications for mechanism of action. *J. Bacteriol.* **177,** 1419–1424.

Saxena, I. M., Brown, R. M., Jr., and Dandekar, T. (2001). Structure-function characterization of cellulose synthase: Relationship to other glycosyltransferases. *Phytochemistry* **57,** 1135–1148.

Scheible, W. R., and Pauly, M. (2004). Glycosyltransferases and cell wall biosynthesis: Novel players and insights. *Curr. Opin. Plant Biol.* **7,** 285–295.

Scheible, W. R., Eshed, R., Richmond, T., Delmer, D., and Somerville, C. (2001). Modifications of cellulose synthase confer resistance to isoxaben and thiazolidinone herbicides in *Arabidopsis* Ixr1 mutants. *Proc. Natl. Acad. Sci. USA* **98,** 10079–10084.

Schindelman, G., Morikami, A., Jung, J., Baskin, T. I., Carpita, N. C., Derbyshire, P., Mcmann, M. C., and Benfey, P. N. (2001). COBRA encodes a putative GPI-anchored protein, which is polarly localized and necessary for oriented cell expansion in *Arabidopsis*. *Genes Dev.* **15,** 1115–1127.

Somerville, C., Bauer, S., Brininstool, G., Facette, M., Hamann, T., Milne, J., Osborne, E., Paredez, A., Persson, S., Raab, T., Vorwerk, S., and Youngs, H. (2004). Toward a systems approach to understanding plant cell walls. *Science* **306,** 2206–2211.

Szyjanowicz, P. M., Mcminnon, I., Taylor, N. G., Gardiner, J., Jarvis, M. C., and Turner, S. R. (2004). The irregular xylem 2 mutant is an allele of korrigan that affects the secondary cell wall of *Arabidopsis thaliana*. *Plant J.* **37,** 730–740.

Tanaka, K., Murata, K., Yamazaki, M., Onosata, K., Miyao, A., and Hirochika, H. (2003). Three distinct rice cellulose synthase catalytic subunit genes required for cellulose synthesis in the secondary wall. *Plant Physiol.* **133,** 73–83.

Taylor, N. G., Scheible, W. R., Cutler, S., Somerville, C. R., and Turner, S. R. (1999). The irregular xylem3 locus of *Arabidopsis* encodes a cellulose synthase required for secondary cell wall synthesis. *Plant Cell* **11,** 769–779.

Taylor, N. G., Laurie, S., and Turner, S. R. (2000). Multiple cellulose synthase catalytic subunits are required for cellulose synthesis in *Arabidopsis*. *Plant Cell* **12,** 2529–2539.

Taylor, N. G., Howells, R. M., Huttly, A. K., Vickers, K., and Turner, S. R. (2003). Interactions among three distinct CesA proteins essential for cellulose synthesis. *Proc. Natl. Acad. Sci. USA* **100,** 1450–1455.

Taylor, N. G., Gardiner, J. C., Whiteman, R., and Turner, S. R. (2004). Cellulose synthesis in the *Arabidopsis* secondary cell wall. *Cellulose* **11,** 329–338.

Turner, S. R., and Somerville, C. R. (1997). Collapsed xylem phenotype of *Arabidopsis* identifies mutants deficient in cellulose deposition in the secondary cell wall. *Plant Cell* **9,** 689–701.

Williamson, R. E., Burn, J. E., Birch, R., Baskin, T. I., Arioli, T., Betzner, A. S., and Cork, A. (2001). Morphology of *rsw1*, a cellulose-deficient mutant of *Arabidopsis thaliana*. *Protoplasma* **215,** 116–127.

Wong, H. C., Fear, A. L., Calhoon, R. D., Eichinger, G. H., Mayer, R., Amikam, D., Benziman, M., Gelfand, D. H., Meade, J. H., and Emerick, A. W. (1990). Genetic organization of the cellulose synthase operon in *Acetobacter xylinum*. *Proc. Natl. Acad. Sci. USA* **87,** 8130–8134.

Yeager, A. R., and Finney, N. S. (2004). The first direct evaluation of the two-active site mechanism for chitin synthase. *J. Org. Chem.* **69,** 613–618.

Zhong, R., Kays, S. J., Schroeder, B. P., and Ye, Z.-H. (2002). Mutation of a chitinase-like gene causes ectopic deposition of lignin, aberrant cell shapes, and overproduction of ethylene. *Plant Cell* **14,** 165–179.

Zuo, J., Niu, Q. W., Nishizawa, N., Wu, Y., Kost, B., and Chua, N. H. (2000). KORRIGAN, an *Arabidopsis* endo-1,4-β-glucanase, localizes to the cell plate by polarized targeting and is essential for cytokinesis. *Plant Cell* **12,** 1137–1152.

CHAPTER 7

Metabolic Engineering of the Content and Fatty Acid Composition of Vegetable Oils

Edgar B. Cahoon* and **Katherine M. Schmid**†

Contents

* USDA-ARS Plant Genetics Research Unit, Donald Danforth Plant Science Center, 975 North Warson Road, St. Louis, Missouri 63132
† Department of Biological Sciences, Butler University, 4600 Sunset Avenue, Indianapolis, Indiana 46208

Advances in Plant Biochemistry and Molecular Biology, Volume 1
ISSN 1755-0408, DOI: 10.1016/S1755-0408(07)01007-7

Abstract

This chapter discusses engineering of plants for yield and composition of edible and industrial triacylglycerols (TAGs). Total oil production has been increased moderately by overexpression of genes for the first and last steps of oil synthesis, acetyl-CoA carboxylase (ACCase), and diacylglycerol acyltransferase (DGAT), respectively. However, the single enzyme approach has proved less than satisfactory, and further progress may depend on identification of regulatory genes affecting overall expression of the lipid synthesis pathways and partitioning of carbon between oil and other plant products.
The fatty acid composition of oilseeds has been more amenable to modification. Development of edible oils rich in monounsaturated fatty acids (18:1) has been achieved in several oilseeds normally dominated by polyunsaturated fatty acids such as 18:2. Approaches have included both chemical mutagenesis and transgenic alteration of the *FAD2* genes responsible for desaturation of 18:1 to 18:2. Proportions of 16:0 have been reduced substantially by reduction of *FatB*, the gene for the thioesterase that releases 16:0 from the acyl carrier protein (ACP) on which it is assembled. The last major goal in edible oil modification, production of a temperate crop sufficiently rich in saturated fatty acids for use without hydrogenation and its associated *trans*-fatty acid production, remains elusive. Mechanisms for minimizing transfer of the upregulated saturated fatty acids to plant membranes are currently lacking. Excess saturated fatty acids in plant membranes are particularly damaging in colder temperature ranges.
Finally, a wide range of genes have been identified that encode enzymes for synthesis of unusual fatty acids with potential as food additives or industrial feedstocks. Genes for production of γ-linolenic acid (GLA) and polyunsaturated ω-3 fatty acids have been introduced into plants, as have genes permitting production of 10:0 and 12:0 for the detergents industry, long-chain fatty acids for plastics and nylons, novel monounsaturated and conjugated fatty acids, and fatty acids with useful epoxy-, hydroxy-, and cyclic moieties. With the notable exception of the shorter-chain fatty acids, these efforts have been hampered by inadequate yields of the novel products. Given that plants from which many of the applicable genes were isolated do produce oils with high proportions of unusual fatty acids, increased yields in transgenic crops should be achievable. It is probable that introduction of the novel fatty acids must be coupled with appropriate modifications of the enzymes responsible for their flux into vegetable oils.

Key Words: Vegetable oil, Oilseed, Fatty acid, Triacylglycerol, Lipids, Fatty acid unsaturation, Polyunsaturated fatty acid, Saturated fatty acid, Fatty acid desaturase, Thioesterase, FAD2, Genetic engineering, Metabolic engineering.

Abbreviations: ACCase, acetyl coenzyme A carboxylase; ACP, acyl carrier protein; ARA, arachidonic acid; BCCP, biotin carboxyl carrier protein; DAG, diacylglycerol; DGAT, diacylglycerol acyltransferase; DHA, docosahexaenoic acid; EPA, eicosapentaenoic acid; ER, endoplasmic reticulum; FAD, fatty acid desaturase; FAS, fatty acid synthase; Fat, fatty acid thioesterase; GLA, γ-linolenic acid; GPAT, acyl-CoA:glycerol-3-phosphate acyltransferase; KAS,

3-ketoacyl-ACP synthase; KCS, 3-ketoacyl-CoA synthase; LPAAT, acyl-CoA: lysophosphatidic acid acyltransferase; PC, phosphatidylcholine; PDAT, phospholipid:diacylglycerol acyltransferase; RNAi, RNA interference; TAG, triacylglycerol; VLCFA, very long-chain fatty acid.

1. INTRODUCTION

Oils and fats tend to be the predominant energy reserves in mobile organisms because of their high energy value per unit weight. Plants, given a sessile lifestyle, limit oil production primarily to portable reproductive structures. Nevertheless, more than 120 million metric tons of vegetable oil reach world markets per year (United States Department of Agriculture, Foreign Agricultural Service, 2007). Oilseeds such as soybean, sunflower, and rapeseed are the major oil crops in temperate regions, although fruits of olive and especially of oil palm are significant sources on a world basis (Table 7.1).

At the molecular level, the typical oil molecule is a triacylglycerol (TAG), a glycerol molecule with a fatty acid esterified to each of the three hydroxyl groups (Fig. 7.1). The three carbon atoms of the glycerol backbone of TAG are referred to using the stereospecific numbering system as *sn*-1, *sn*-2, and *sn*-3 (Fig. 7.1). As indicated by this nomenclature, the three carbons of glycerol are stereochemically distinct. It is the fatty acid composition that determines the physical characteristics of a given oil. For example, a sufficient proportion of saturated fatty acids, which lack carbon–carbon double bonds, can raise the melting point of an oil until it is solid at room temperature, as required in some baked goods. Palmitic acid, abbreviated 16:0 because it has 16 carbons and 0 double bonds, is the most abundant of the saturated fatty acids in plants, although at least some stearic acid (18:0) occurs in most edible oils (Table 7.2). The unsaturated fatty acids of

TABLE 7.1 World production of vegetable oils in 2006

Crop plant	Tissue used for oil extraction	Vegetable oil production[1] (million metric tons)
Palm	Fruit	36.8
Soybean	Seed	36.0
Oilseed rape	Seed	17.8
Sunflower	Seed	10.8
Peanut	Seed	4.9
Cotton	Seed	4.8
Palm kernel	Seed	4.6
Coconut	Seed	3.2
Olive	Fruit	3.0

[1] United States Department of Agriculture Foreign Agricultural Service (2007). Oilseeds: World Markets and Trade, Circular Series FOP 07-07, July 2007. http://www.fas.usda.gov/psdonline/circulars/oilseeds.pdf.

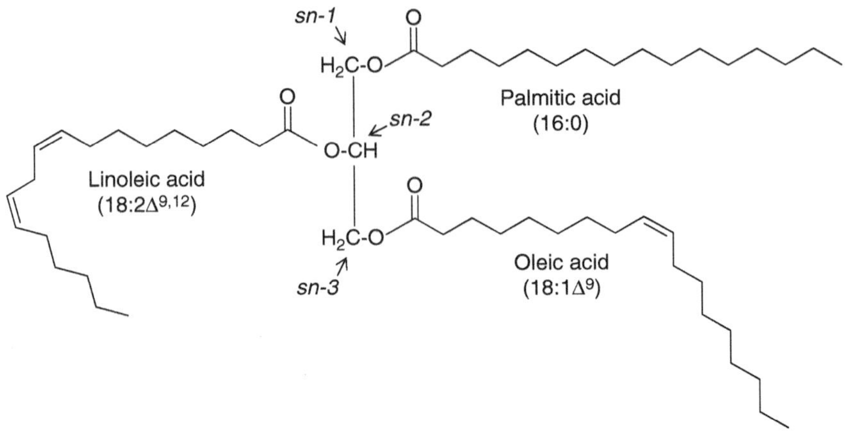

FIGURE 7.1 Structure of a typical triacylglycerol (TAG) molecule of vegetable oil. A TAG molecule consists of fatty acids attached by ester linkages to each of the three stereospecific or *sn* positions of a glycerol backbone. As shown, the *sn*-2 position of a typical plant TAG is occupied by an unsaturated fatty acid. Saturated fatty acids generally occupy only the *sn*-1 or *sn*-3 positions, but unsaturated fatty acids can be found at any of the three stereospecific positions.

TABLE 7.2 Fatty acids that commonly occur in the major vegetable oils

Fatty Acid	Abbreviation	Structure	Saturation Class	Melting Point
Palmitic Acid	16:0	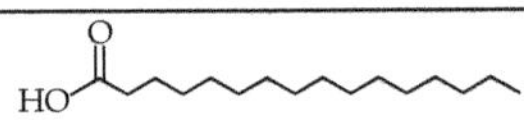	Saturated	64°C
Stearic Acid	18:0	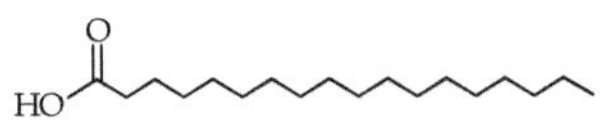	Saturated	70°C
Oleic Acid	$18{:}1\Delta^9$	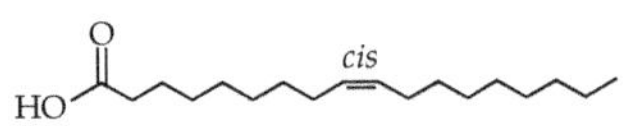	Monounsaturated	13°C
Linoleic Acid	$18{:}2\Delta^{9,12}$	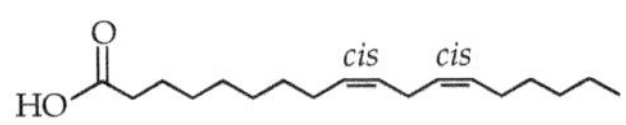	Polyunsaturated	−9°C
α-Linolenic Acid	$18{:}3\Delta^{9,12,15}$	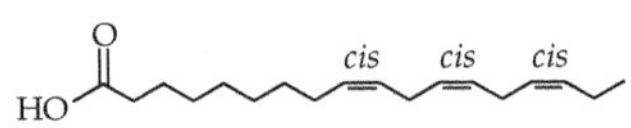	Polyunsaturated	−17°C

typical plant oils feature one or more *cis*-double bonds, which introduce kinks into the fatty acid chain and increase fluidity more effectively than would *trans*-double bonds. Oleic acid ($18{:}1\Delta^9$), the most prominent monounsaturated fatty acid, has a *cis*-double bond nine carbons from its carboxyl terminus (see Fig. 7.2 for explanation of numerical fatty acid nomenclature). It can comprise 65–85% of the olive (*Olea*) oil for which it was named, but contributes a mere 20% of traditional sunflower or soybean oils (Gunstone *et al.*, 2007). Thus, high oleic acid seed oils mimicking the qualities of olive oil as a cooking and salad oil are under development. Plant oils are also important sources of polyunsaturated fatty acids including linoleic acid ($18{:}2\Delta^{9,12}$; Fig. 7.2) and α-linolenic acid ($18{:}3\Delta^{9,12,15}$). Since

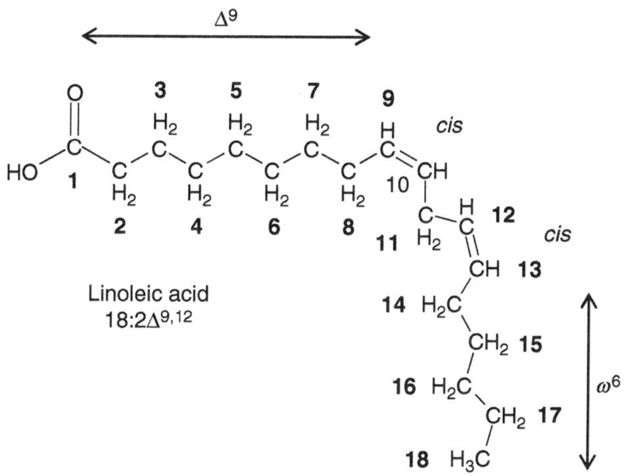

FIGURE 7.2 Structure of linoleic acid. This structure illustrates the basis for the shorthand notation that is often used for fatty acids. The 18:2$\Delta^{9,12}$ abbreviation indicates that linoleic acid contains 18 carbon atoms and 2 double bonds, which are located at the C-9 and C-12 atoms relative to the carboxyl end of the fatty acid. Linoleic acid is often referred to as an ω-6 fatty acid, which indicates that the last double bond is positioned six carbon atoms from the methyl end of the fatty acid. Vegetable oils rich in linoleic acid, such as soybean oil, are sometimes called ω-6 oils.

increasing unsaturation decreases oxidative stability, oils high in 18:3 become rancid quickly and are unsuitable for frying. However, both linoleic and α-linolenic acids are essential to the human diet. Finally, some qualities of vegetable oils reflect the arrangement of fatty acids on glycerol as well as absolute fatty acid composition. For example, the positive "mouthfeel" of cocoa butter is largely attributed to TAG having saturated fatty acids at positions 1 and 3, but 18:1Δ^9 at position 2 (Jandacek, 1992). The positional distribution of fatty acids in dietary TAG also has clinical implications (Kubow, 1996).

Although vegetable oils are primarily used in foods, they also serve as industrial feedstocks (Table 7.3). A few oils are targeted entirely to such uses. Highly unsaturated "drying oils" such as linseed oil are desirable for paints and coatings; lauric acid (12:0) in coconut and palm kernel oil is a vital component of soaps and detergents; castor oil, which contains the unusual hydroxy-fatty acid ricinoleic acid (12-hydroxy-18:1Δ^9), is used for certain plastics and lubricants; and high erucate (22:1Δ^{13}) rapeseed oil contains the raw material for Nylon 1313 and slip agents used in the manufacture of sheet plastic. Edible oils may likewise serve industrial purposes. For example, in the United States, 12% of soybean oil is currently channeled to products ranging from lubricants and biodiesel fuels to inks, polyurethane, and candles (American Soybean Association, 2007). As petroleum stocks dwindle, it is likely that vegetable oils will play a greater industrial role.

TABLE 7.3 Examples of unusual fatty acids whose biosynthetic pathways can be metabolically engineered into existing crop plants to generate vegetable oils with commercially-useful properties

Fatty Acid	Abbreviation	Structure	Potential Commercial Uses
Lauric Acid	12:0	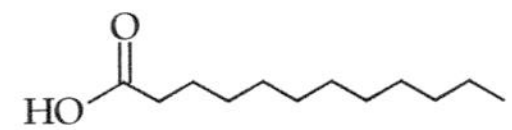	Detergents; soaps
Petroselinic Acid	$18:1\Delta^{6}$	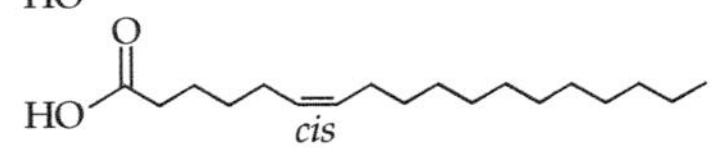	Precursor of adipic acid for nylon 6, 6 production
Ricinoleic Acid	12-hydroxy-$18:1\Delta^{9}$	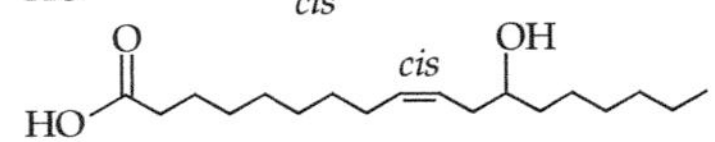	Lubricants; coatings; plastics; cosmetics
Vernolic Acid	12-epoxy-$18:1\Delta^{9}$	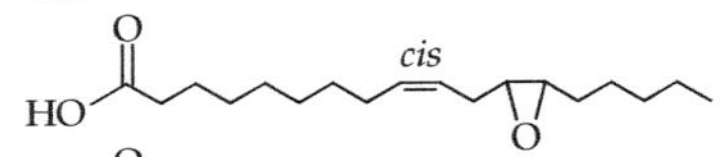	Plasitcizers; paints; adhesives; plastics
γ-Linolenic Acid (GLA)	$18:3\Delta^{6,9,12}$	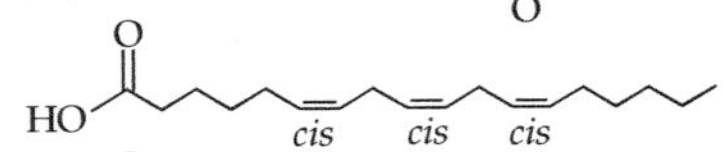	Nutraceuticals
Eleostearic Acid	$18:3\Delta^{9,11,13}$	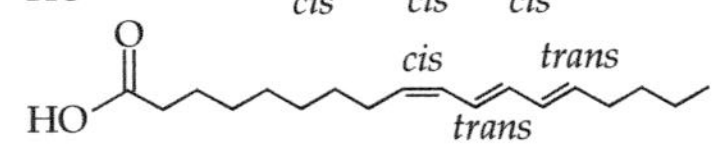	Quick-drying agent for paints, inks, and varnishes
Δ^{5}-Eicosenoic Acid	$20:1\Delta^{5}$	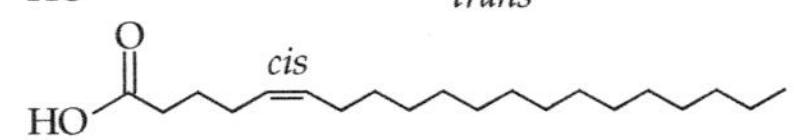	High-temperature lubricants; cosmetics
Eicosapentaenoic Acid (EPA)	$20:5\Delta^{5,8,11,14,17}$	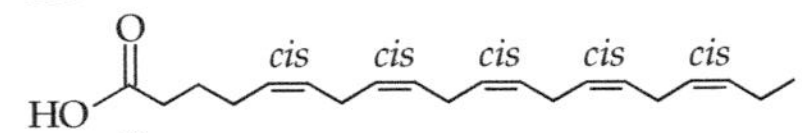	Nutraceuticals; omega-3 vegetable oils for improved cardiovascular fitness
Docosahexaenoic Acid (DHA)	$22:6\Delta^{4,7,10,13,16,19}$	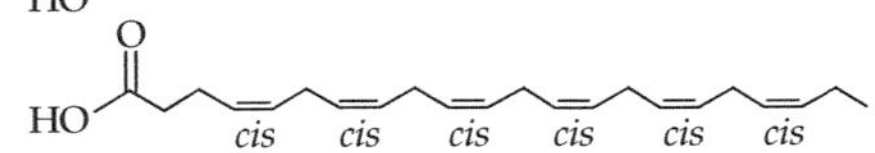	Nutraceuticals; omega-3 vegetable oils for improved cardiovascular fitness and brain development

In addition to control of oil composition, improvement of total yield of oil crops is a major goal of breeders and molecular biologists. To some extent, such improvement can involve parameters beyond the scope of this discussion. Flower number and seed set, disease resistance and fruit or seed size are only a few examples of factors indirectly affecting oil production. At a more direct level, scientists are attempting to identify control points for carbon flux into fatty acids, factors influencing partitioning of fatty acids between structural lipids and TAG, and regulatory elements determining overall expression of lipid biosynthesis genes.

2. TAG SYNTHESIS

TAG synthesis is a complex, multistep pathway involving multiple cellular compartments (Fig. 7.3). Plastids, whether the chloroplasts of photosynthetic organs or the tiny proplastids of typical oilseeds, build 2-carbon units into fatty acids with up to 18 carbons and 1 double bond. Two of these acyl units are then esterified to glycerol-3-phosphate, producing phosphatidic acid. The endoplasmic reticulum (ER) is the major site of phosphatidic acid synthesis for TAG; however, plastids likewise generate phosphatidic acid, and flow of glycerolipid backbones from the plastids into storage oils has been observed. Fatty acids ultimately incorporated into TAG can undergo further desaturation, elongation, or other modifications, often while the acyl units are esterified to phosphatidylcholine (PC) or coenzyme A. Finally, phosphatidic acid is dephosphorylated at the ER to form diacylglycerol (DAG), and a diacylglycerol acyltransferase (DGAT) adds the final fatty acid, forming TAG that is sequestered from the ER into lipid bodies for storage. Alternative mechanisms for transfer of fatty acids to TAG are also possible, as will be discussed below.

2.1. Precursors for fatty acid synthesis

The gateway to fatty acid synthesis is generally considered the plastidial acetyl coenzyme A carboxylase (ACCase), which converts acetyl-CoA to malonyl-CoA. In all plants studied other than grasses, the plastidial form of the enzyme involved in fatty acid synthesis has four dissociable subunits. A biotin carboxylase subunit first affixes a carboxyl group to the biotin of a second subunit, biotin carboxyl carrier protein (BCCP), using bicarbonate and ATP as substrates. The resulting conformational change brings the biotin arm to a carboxyltransferase domain formed by the remaining two subunits, where the biotin donates the carboxyl group to acetyl-CoA (Cronan and Waldrop, 2002; Nikolau *et al.*, 2003). Grass ACCases possess the same activities as the multisubunit form, but combine them into a multifunctional homodimer that is the primary target of herbicides targeting weedy grasses (Zagnitko *et al.*, 2001).

ACCase is considered to be the rate-limiting step in fatty acid synthesis. The multisubunit ACCase is light-activated by reduction of the carboxyltransferase

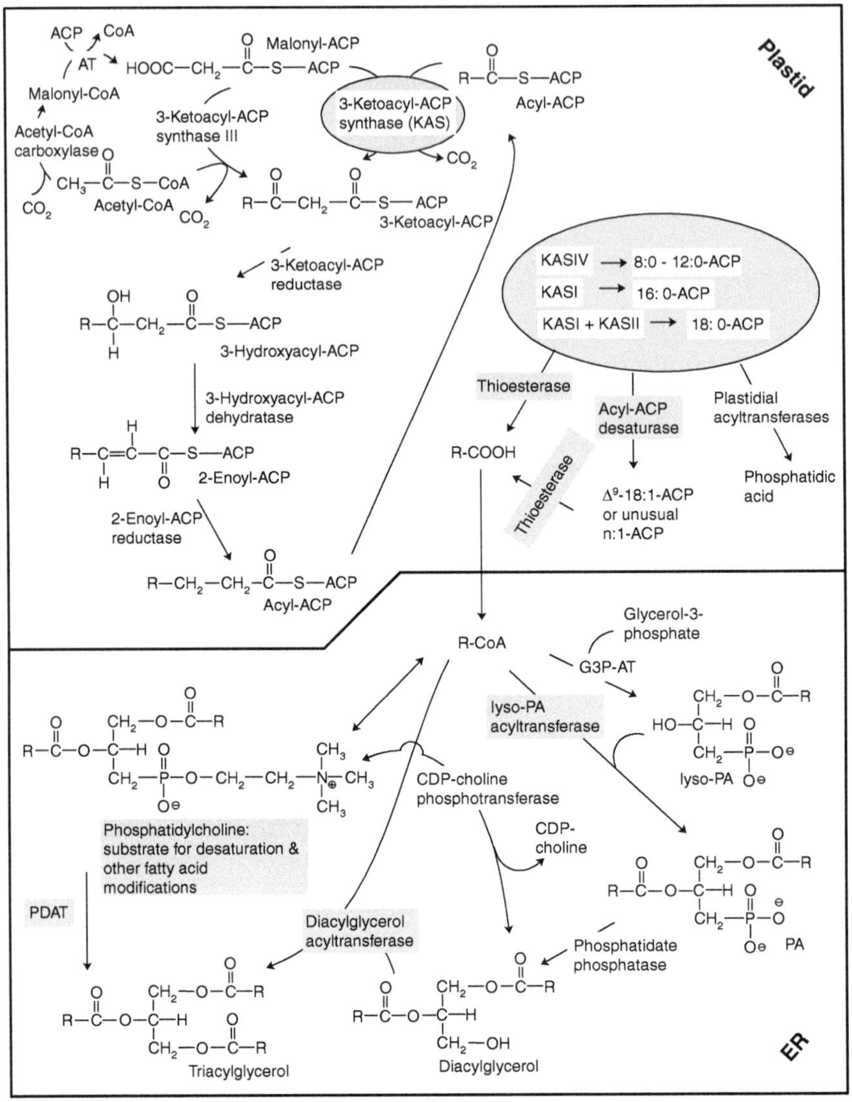

FIGURE 7.3 Triacylglycerol (TAG) synthesis, highlighting points in the pathway at which genetic engineering and/or mutagenesis have been used to modify fatty acid composition of the resulting oil ( ). The upper left portion of the diagram shows synthesis of malonyl-CoA by ACCase, and the cyclic nature of the reactions catalyzed by fatty acyl synthase (FAS). FAS is composed of malonyl-CoA:malonyl-ACP acyltransferase (AT), 3-ketoacyl-ACP synthase (KAS), 3-ketoacyl-ACP reductase, 3-hydroxyacyl-ACP dehydratase, and enoyl-ACP reductase. As shown on the right of the diagram, the products of FAS depend on the contributions of various KASes, the substrate and double bond specificities of acyl-ACP desaturases, and the substrate specificities of thioesterases that release fatty acids for export from the plastids. In the ER, phosphatidic acid (PA) is assembled by sequential activities of glycerol-3-phosphate acyltransferase (G3P-AT) and lysophosphatidic acid-acyltransferase (LPAAT). Diacylglycerol (DAG) units released from lyso-PA by phosphatidate phosphatase may be converted directly to triacylglycerol by DGAT. However, a large proportion

subunits via the thioredoxin pathway, and is subject to feedback inhibition by oleic acid (Kozaki *et al.*, 2001; Shintani and Ohlrogge, 1995). Although the β-carboxyltransferase is plastid-encoded while the remaining subunits are imported to the plastids, all four subunits are normally coordinately expressed (Ke *et al.*, 2000). Attempts to upregulate fatty acid synthesis by manipulating individual subunits of the heteromeric ACCase have had mixed results. Increased biotin carboxylase has little effect, and overexpression of BCCP actually decreased fatty acid synthesis, perhaps due to incorporation of unbiotinylated enzyme into ACCase (Shintani *et al.*, 1997; Thelen and Ohlrogge, 2002). However, Madoka *et al.* reported that transformation of tobacco with the plastidial carboxyltransferase subunit raised overall yield of seed oil by increasing seed production, although oil per seed remains constant (Madoka *et al.*, 2002). Alternatively, introduction of homomeric ACCase to rapeseed plastids increased ACCase activity and, to a lesser extent, seed oil (Roesler *et al.*, 1997).

The availability of bicarbonate and particularly of acetyl-CoA for ACCase can also impact overall fatty acid synthesis. Reduced carbonic anhydrase activity inhibited fatty acid synthesis in cotton embryos, presumably by decreasing local bicarbonate supplies (Hoang and Chapman, 2002). The sources of acetyl-CoA for ACCase probably vary between tissues and stages of development. In castor seed endosperm, malate generated by a specific phosphoenolpyruvate carboxylase isoform appears to be the major source of carbon for fatty acids (Blonde and Plaxton, 2003). In rapeseed embryos, on the other hand, malate does not contribute significantly; instead, carbon flows primarily from glycolysis, entering the plastid via transporters for glucose-6-phosphate, dihydroxyacetone phosphate, and especially phosphoenolpyruvate (Kubis and Rawsthorne, 2000; Schwender and Ohlrogge, 2002). There is also potential for increasing flow of carbon into seed oil via alternative sources of acetyl-CoA. For example, introduction of ATP:citrate lyase from rat into tobacco plastids increased total leaf fatty acids 16% (Rangasamy and Ratledge, 2000).

2.2. Fatty acid synthesis

The plastidial fatty acid synthase (FAS) is actually a complex of multiple dissociable components that uses malonyl-CoA generated by ACCase to build fatty acids, two carbons at a time. Malonyl-CoA:ACP transacylase first transfers the malonyl unit to acyl carrier protein (ACP), which holds acyl intermediates via a high energy thioester bond throughout the process of fatty acid synthesis. As diagrammed in Fig. 7.3, malonyl-ACP serves as the C_2 donor to acceptors of various lengths in condensation reactions catalyzed by 3-ketoacyl-ACP synthases (KASes). KASIII uses acetyl-CoA as the acceptor, producing acetoacetyl-ACP;

of TAG fatty acids pass through PC, which serves as a substrate for further fatty acid desaturation and other modifications. Modified fatty acids may then be transferred to TAG: (1) as part of DAG released by the reversible CDP-choline acyltransferase, (2) after return to the acyl-CoA pool, or (3) by direct transfer via PDAT. (See Page 6 in Color Section.)

KASI acetylates 4:0-ACP through 14:0-ACP; and KASII elongates a 16:0-ACP acceptor to 3-keto-18:0-ACP. After each condensation, carbon 3 of the product has a C=O group that must be reduced to CH_2 before the next condensation can occur. In the first step of this process, 3-ketoacyl-ACP reductase reduces 3-ketoacyl-ACP to 3-hydroxyacyl-ACP. 3-Hydroxyacyl-ACP dehydratase then abstracts a water molecule, producing *trans*-2-enoyl-ACP. Finally, enoyl-ACP reductase reduces the double bond to the requisite single bond (Fig. 7.3).

The end products of FAS are primarily 16:0- and 18:0-ACP. The latter product can be further modified by the stearoyl (18:0)-ACP desaturase, which catalyzes the formation of a *cis*-double bond between the C-9 and C-10 atoms of 18:0-ACP to form oleoyl (18:1Δ^9)-ACP. Unlike all other fatty acid desaturases in plants, stearoyl-ACP desaturase is a soluble enzyme which has facilitated its detailed structural characterization (Lindqvist *et al.*, 1996). The 16:0, 18:0, and 18:1Δ^9 products generated in the plastid are released from ACP for export to the cytosol by the activity of two classes of acyl-ACP thioesterases, designated FatA and FatB. FatA is most active with 18:1-ACP, whereas FatB is most active with 16:0-ACP (Salas and Ohlrogge, 2002). By the combined activities of FatA and FatB, 16:0, 18:0, and 18:1Δ^9 are made available for further modification and ultimately for storage in TAG molecules by ER-localized enzymes. The stearoyl-ACP desaturase and acyl-ACP thioesterases will be discussed further because they represent major biotechnological targets for alteration of the saturated fatty acid content of seed oils. In addition, structurally variant forms of these enzymes have arisen in seeds of certain plants and are involved in the synthesis of unusual fatty acids, many of which have potential economic value (Voelker and Kinney, 2001).

Of the FAS components, KASIII has been considered a likely gatekeeper, since the *Escherichia coli* homologue is inhibited by acyl-ACPs, the products of FAS (Heath and Rock, 1996). Similar feedback inhibition has been observed *in vitro* for the KASIII of *Cuphea lanceolata*, a plant that produces an unusual proportion of caprylic acid (8:0) (Brück *et al.*, 1996). However, Dehesh *et al.* report that over-expression of spinach KASIII in rapeseed actually reduced both FAS activity and oil content of seeds (Dehesh *et al.*, 2001). Based on elevated acetoacetyl-ACP in leaves of tobacco transformed with the same gene, as well as increased 16:0 accumulation in both organs, they propose that reduced supplies of malonyl-ACP to KASI and KASII are responsible. It should also be noted that, *in vitro*, *Cuphea* KASes can decarboxylate malonyl-ACP under conditions promoting accumulation of 3-ketoacyl-ACP (Winter *et al.*, 1997).

Reduced expression of several individual components of FAS decreases overall fatty acid synthesis in plants. Interestingly, when rapeseed 3-ketoacyl-ACP reductase mRNA and protein were decreased using antisense methods, enoyl-ACP reductase mRNA and protein were also downregulated (Slabas *et al.*, 2002). This is consistent with evidence that ratios of FAS components remain unchanged during rapeseed development (O'Hara *et al.*, 2002). Expression of FAS genes during seed maturation likewise appears coordinated with that of genes for ACCase and other enzymes related to oil production, suggesting the participation of global transcription factors comparable to the FasR factor that upregulates fatty acid synthesis in *E. coli* (Cronan and Subrahmanyam, 1998; Lee *et al.*, 2002; Ruuska *et al.*, 2002; Slabas *et al.*, 2002).

2.3. Phosphatidic acid assembly

The fatty acids released from plastids are rapidly converted to their respective acyl-CoAs by acyl-CoA synthetases, most likely those isozymes associated with the plastidial envelope (Schnurr *et al.*, 2002). Phosphatidic acid synthesis may then be initiated by transfer of an acyl group to the *sn*-1 position of glycerol-3-phosphate by membrane-bound acyl-CoA:glycerol-3-phosphate acyltransferase (GPAT) (Murata and Tasaka, 1997). Microsomal GPATs are typically capable of using a wide range of acyl-CoAs, but enzymes from some oil producing organs might be more selective. For example, a GPAT solubilized from oil palm microsomes was most active with palmitoyl (16:0)-CoA (Manaf and Harwood, 2000). Genes for ER-localized GPATs have been identified in *Arabidopsis thaliana* (Zheng *et al.*, 2003). The identification of GPATs specifically involved in the biosynthesis of TAG in seeds awaits further characterization of this seven-member gene family.

Acylation of the *sn*-2 position is subsequently catalyzed by an ER acyl-CoA: lysophosphatidic acid acyltransferases (LPAATs). In most edible oils, this position is dominated by unsaturated C_{18}-fatty acids, reflecting LPAAT discrimination against 16:0-CoA and 18:0-CoA (Brown *et al.*, 2002). Microsomal LPAAT cDNAs have been cloned from several species (Bourgis *et al.*, 1999). As will be discussed later, some plants with oils enriched in unusual fatty acids also produce functionally divergent LPAATs that accept the corresponding acyl-CoAs (Voelker and Kinney, 2001).

Although most phosphatidic acid that is a precursor to TAG is produced by ER acyltransferases, it is important to note that plastids and mitochondria also assemble phosphatidic acid. Glycerolipid backbones formed in the plastids serve primarily as precursors of phosphatidylglycerol, sulfolipid, and galactolipid, while mitochondria are the sole site of cardiolipin production. However, studies of mutants have highlighted the ability of plants to move DAG units between compartments as needed (Kunst *et al.*, 1988). In addition, genes for the acyltransferases native to any compartment have potential for seed oil modification. For example, *A. thaliana* transformed with a plastidial GPAT cDNA less its transit sequence produced about 20% more seed oil, even though the plastidial GPAT is a soluble enzyme that normally uses acyl-ACP rather than acyl-CoA (Jain *et al.*, 2000). Plastidial LPAATs, envelope-localized proteins that likewise employ acyl-ACPs as substrates, are selective for 16:0 rather than $18{:}1\Delta^{9}$ and $18{:}2\Delta^{9,12}$ (Frentzen, 1998).

2.4. Glycerolipids and fatty acid modification

Phosphatidic acid is generally metabolized by one of two enzymes. CDP-DAG synthase, an enzyme found in ER, plastids and mitochondria, generates substrate for production of phosphatidylglycerol, phosphatidylinositol, and phosphatidylserine.

The other enzyme, phosphatidate phosphatase, releases DAG, a vital precursor of PC, phosphatidylethanolamine and TAG, as well as sulfolipid and galactolipid. In some plants, microsomal phosphatidate phosphatase supplies DAG for both plastidial and microsomal glycerolipid synthesis, while in others, separate plastidial and microsomal isoforms contribute. Analysis of the phosphatase is complicated further by isozymes involved in signaling and lipid catabolism. Based on work with developing safflower seeds, Ichihara *et al.* proposed that an isoform used during oil deposition moves between a cytosolic pool and the ER, depending on cytosolic fatty acid concentrations (Ichihara *et al.*, 1990). This arrangement could allow feed-forward regulation of the TAG synthetic pathway initiated by the phosphatase.

TAG composition can be radically affected by fatty acid modifications that take place on glycerolipid substrates. As noted earlier, 18:1Δ^9 accounts for virtually all of the unsaturated fatty acid exported by a typical plastid. Production of the polyunsaturated fatty acids so common in vegetable oils involves a series of two ER-localized desaturases that act on fatty acids esterified to either *sn*-position of PC or less prominent phospholipids (Fig. 7.4 and Table 7.4). The first enzyme,

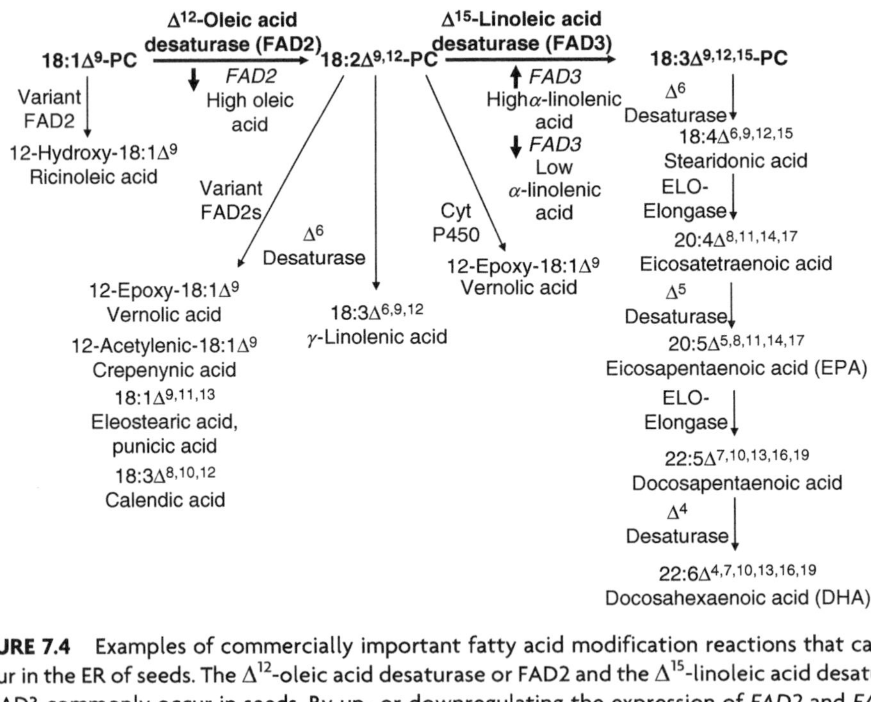

FIGURE 7.4 Examples of commercially important fatty acid modification reactions that can occur in the ER of seeds. The Δ^{12}-oleic acid desaturase or FAD2 and the Δ^{15}-linoleic acid desaturase or FAD3 commonly occur in seeds. By up- or downregulating the expression of *FAD2* and *FAD3* genes, the relative levels of vegetable oil unsaturation can be altered. Variant forms of enzymes such as FAD2, cytochrome P450 monoxygenase, and cytochrome b_5-fusion desaturases can be transgenically expressed in existing oilseeds to produce unusual fatty acids such as ricinoleic, vernolic, and GLAs. In addition, desaturases and ELO elongases from sources including mosses, fungi, and algae can be engineered into oilseed crops to produce the nutritionally important long-chain polyunsaturated fatty acids eicosapentaenoic (EPA) and docosahexaenoic (DHA) acids.

TABLE 7.4 Commonly occurring fatty acid desaturases in plants

Desaturase	Cellular location	Substrate	Product	Commercially important phenotypes
Stearoyl-ACP desaturase	Plastid	18:0-ACP	18:1Δ^9-ACP	Downregulation: increased stearic acid content
Δ^{12}-Oleic acid desaturase (FAD2)	Endoplasmic reticulum	18:1Δ^9-PC	18:2$\Delta^{9,12}$-PC	Downregulation: increased oleic acid content and reduced polyunsaturated fatty acid content
Δ^{15}-Linoleic acid desaturase (FAD3)	Endoplasmic reticulum	18:2$\Delta^{9,12}$-PC	18:3$\Delta^{9,12,15}$-PC	Downregulation: low α-linolenic acid content upregulation: increased α-linolenic acid content

The relative unsaturation of vegetable oils can be modified by up- or downregulating the expression of these fatty acid desaturases as indicated.

typically described as the Δ^{12}-oleic acid desaturase or FAD2, inserts a double bond 12 carbons from the carboxyl end of esterified 18:1Δ^9, producing 18:2$\Delta^{9,12}$ (linoleic acid). This enzyme is sometimes referred to as the ω-6 desaturase, which indicates that the double bond is inserted at the sixth carbon atom from the methyl end of the 18:1Δ^9 substrate. A more careful analysis showed that this desaturase actually references the site of double-bond insertion based on the position of the Δ^9 double bond of its monounsaturated substrate (Schwartzbeck *et al.*, 2001). The second enzyme, the Δ^{15}-linoleic acid desaturase or FAD3, converts 18:2$\Delta^{9,12}$ to 18:3$\Delta^{9,12,15}$ (α-linolenic acid). As with FAD2, this enzyme is sometimes referred to as the ω-3 desaturase, which indicates that the double bond is inserted at the third carbon atom from the methyl end of its substrate. Engeseth and Stymne found that FAD2 and FAD3 will also desaturate fatty acids that contain hydroxyl and epoxy groups (Engeseth and Stymne, 1996). When determining insertion sites for new double bonds, these enzymes appear to count the unusual functional groups as substitutes for prior double bonds.

Again, the ER enzymes have plastidial counterparts, which act primarily on glycolipid substrates. FAD2 and FAD3 and the analogous plastidial desaturases share eight conserved histidines arranged as $H(X_{3-4})H(X_{7-41})H(X_{2-3})HH(X_{61-189})H(X_{2-3})HH$, and it has been proposed that these histidines are associated with an active site di-iron cluster (Shanklin and Cahoon, 1998). The same motif occurs in enzymes catalyzing a range of fatty acyl desaturation, hydroxylation, and epoxidation reactions (Shanklin and Cahoon, 1998).

2.5. TAG synthesis and oil deposition

Acylation of the *sn*-3 position of DAG by acyl-CoA:diacylglycerol acyltranserase (DGAT) completes the synthesis of TAG. Plants, like mammals and fungi, appear to contain two very distinct families of DGAT genes. Members of the DGAT1 family are homologous to mammalian acyl CoA:cholesterol acyltransferase. However, inactivating *TAG1*, the single *A. thaliana* representative of this group, reduced DGAT activity up to 70% without an impact on sterol ester deposition (Zou *et al.*, 1999). TAG synthesis catalyzed by an *A. thaliana* DGAT2 homologue, identified based on its similarity to a fungal DGAT2, was recently confirmed in transfected insect cells (Lardizabal *et al.*, 2001).

At least one of two DGAT1 isoforms in *Brassica napus* cell suspensions was upregulated by sucrose (Nykiforuk *et al.*, 2002). This could be related to the observation that low osmotic strength inhibits TAG synthesis in wheat embryos, but that abscisic acid overcomes this inhibition (Rodriguez-Sotres and Black, 1994). Overall levels of DGAT activity appear to have an impact on levels of oil deposition, since *A. thaliana* seeds that overexpress *TAG1* displayed increased DGAT activity and seed oil (Jako *et al.*, 2001).

In yeast, a proportion of TAG is produced not by DGAT, but by phospholipid: diacylglycerol acyltransferase (PDAT), an enzyme that transfers acyl units directly from the *sn*-2 position of PC or phosphatidylethanolamine to DAG (Oelkers *et al.*, 2002). Dahlqvist *et al.* have implicated PDAT in TAG synthesis by both castor seeds and *Crepis palaestina*, plants with seed oils rich in hydroxy- and epoxy-fatty acids, respectively (Dahlqvist *et al.*, 2000). PDAT from each plant is particularly active with its characteristic oxygenated fatty acid. Since polyunsaturated fatty acids, like the oxygenated fatty acids, are formed on phospholipid substrates, PDAT activity has been proposed to account for the flow of polyunsaturates from PC to TAG observed in numerous radiolabeling studies. PDAT activity has been observed in *A. thaliana*, and several genes related to the yeast PDAT gene have been identified, although not all encode proteins with PDAT activity (Banaś *et al.*, 2000; Stymne *et al.*, 2003).

Alternative routes by which modified fatty acids could enter TAG include release of DAG from PC by the reverse reaction of CDP-choline phosphotransferase, or movement into the acyl-CoA pool via acyl-CoA:phospholipid acyltransferases or a combination of phospholipase and acyl-CoA synthase (Voelker and Kinney, 2001).

Completed TAGs are usually sequestered in 1–2 μm oil bodies bounded by a single layer of polar lipid (e.g., PC). These structures appear to arise at sites in the ER enriched in enzymes of TAG biosynthesis (Murphy, 2001). The half-unit membrane of the oil body is usually pictured as forming when TAG accumulates between the two leaflets of an ER bilayer (Huang, 1996). Oil body membranes are best known for their characteristic proteins, the oleosins and caleosins, although enzymes of TAG synthesis or catabolism have been identified in some lipid body preparations (Murphy, 2001).

3. CONTROL OF TAG COMPOSITION

As outlined, total oil deposition is the product of myriad factors, with acetyl-CoA supply and the activities of ACCase, KASIII, and acyltransferases, having prominent roles. While breeding and biotechnology continue to produce incremental improvements in yield, the most dramatic progress has been in the development of oilseed lines tailored for specific applications. Both altered proportions of common fatty acids and introduction of unusual fatty acids to crop plants have been accomplished to varying degrees.

3.1. Metabolic engineering of high oleic acid vegetable oils

The most significant achievement in the metabolic engineering of oilseed crops has been the alteration of the unsaturated fatty acid content of vegetable oils. A notable example is the development of vegetable oils with oleic acid content exceeding 70% of the total fatty acids (Kinney, 1996). Such oils have high oxidative stability (or increased shelf life) and have beneficial health properties, especially compared to ω-6 rich oils such as those obtained from soybean seeds. The high oleic acid trait has been developed in most of the major oilseed crops through either transgenic or mutagenic approaches (Auld *et al.*, 1992; Bruner *et al.*, 2001; Buhr *et al.*, 2002; Liu *et al.*, 2002; Norden *et al.*, 1987; Soldatov, 1976). In all reported cases, these oils result from the suppressed expression of FAD2, the ER Δ^{12}-oleic acid desaturase that converts monounsaturated oleic acid to polyunsaturated linoleic acid (Table 7.4 and Fig. 7.4). In the transgenic approaches, downregulation of *FAD2* gene expression has been achieved by sense and antisense suppression, or by RNA interference (RNAi) (Kinney, 1996; Liu *et al.*, 2002; Smith *et al.*, 2000). This is typically conducted using seed-specific promoters, which help to ensure that the biological and physical properties of membranes are not compromised in vegetative parts of the plant.

High oleic acid lines of most of the major oilseed crops have been developed by screening of chemically mutagenized seed populations (Auld *et al.*, 1992; Bruner *et al.*, 2001; Norden *et al.*, 1987; Soldatov, 1976). This approach has proven to be especially effective for the generation of high oleic acid lines of sunflower and peanut that also have acceptable agronomic properties. In contrast, the oleic acid content of seeds from *FAD2* mutants of crops such as soybean typically varies in response to environmental conditions, particularly temperature (Carver *et al.*, 1986; Kinney, 1994). This property has precluded commercialization of high and mid-oleic acid mutants of these crops. The environmental instability of the oleic acid content of soybean mutants is likely due to the presence of at least three *FAD2* genes, designated *GmFAD2–1a*, *GmFAD2–1b*, and *GmFAD2–2*, combined with the known influence of temperature on FAD2 activity (Cheesbrough, 1989; Heppard *et al.*, 1996; Tang *et al.*, 2005). *GmFAD2–1a* and *b* are expressed primarily in seeds, and mutations in these genes likely account for the majority of the oleic acid phenotype in high oleic acid mutants (Heppard *et al.*, 1996; Kinney, 1996). The expression levels of these genes are not significantly affected by temperature (Heppard *et al.*, 1996; Tang *et al.*, 2005). Instead, the activities of the corresponding enzymes appear to be differentially regulated through posttranslational mechanisms in response to temperature (Cheesbrough, 1989; Tang *et al.*, 2005). The GmFAD2–1a and b polypeptides, for example, display different turnover rates when expressed in heterologously in yeast at various growth temperatures (Tang *et al.*, 2005). In addition, because at least three *FAD2* genes are expressed in soybean seeds, the achievement of a high oleic phenotype would require mutations in each of these genes, including *GmFAD2–2*, which is also expressed in vegetative organs. Seedlings from such mutants would likely be poorly equipped to respond to low temperatures by increasing membrane unsaturation. Even *A. thaliana* lines with mutations in the single *FAD2* gene display reduced seed germination and seedling vigor at low temperatures (Miquel and Browse, 1994). These examples illustrate the types of difficulties that can arise with the agronomic development of mutants for genes, such as *FAD2*, that are critical to plant growth and development, as well as the difficulties associated with the breeding of phenotypes controlled by multigene families.

3.2. Metabolic engineering of high and low saturated fatty acid vegetable oils

Palmitic acid (16:0) and stearic acid (18:0) are the primary saturated fatty acid components of the seed oil of most crops. Considerable research effort has been devoted to either increasing or decreasing the content of these fatty acids in seed oils for specific commercial applications. For example, the reduction of saturated fatty acids is generally believed to result in vegetable oils with improved cardiovascular health properties. Conversely, enhancement of saturated fatty acid content results in oils with improved oxidative stability and increased melting point. The latter property is especially important for confectionary applications and margarine production. The use of conventional vegetable oils in margarine production requires chemical hydrogenation to reduce the double bonds of polyunsaturated fatty acids. The resulting oil is solid at room temperature, but contains *trans*-fatty acids that have been increasingly linked with elevated total- and low density lipoprotein (LDL)-cholesterol levels in humans (Hu *et al.*, 2001). As a result, increased emphasis has been placed on metabolic engineering of oilseeds to produce high levels of saturated fatty acids, especially stearic acid, so that the oil does not require hydrogenation for use in margarine production.

Seed oils with increased or decreased amounts of palmitic acid have been achieved through alteration of the expression levels of genes for the FatB acyl-ACP thioesterase. As described previously, this enzyme releases 16:0 and, to a lesser extent, 18:0, from ACP in plastids. By enhancing the expression of *FatB* genes using strong seed-specific promoters, oils that contain 30–40% 16:0 have been generated in seeds of a number of plants including *A. thaliana*, canola, and soybean (Dörmann *et al.*, 2000; Kinney, 1996). In contrast, 16:0 typically comprises 5–15% of the seed oil of most plant species. Downregulation of *FatB* expression in seeds has the opposite effect on the 16:0 content. Using transgenic approaches, and through the development of mutants, seed oils with as little as 2–5% 16:0 have been achieved in a variety of plants (Dörmann *et al.*, 2000; Kinney, 1996; Li *et al.*, 2002; Schnebly *et al.*, 1994). In addition, an *A. thaliana* T-DNA-insertion mutant of the *FatB1* locus was described that contained reduced 16:0 content throughout the plant, including <4% 16:0 in seeds (Bonaventure *et al.*, 2003). Interestingly, this mutant has reduced vegetative growth, indicating an essential role of 16:0 in plant growth and development.

Genetic enhancement of the stearic acid (18:0) content of seeds has been achieved by downregulating or blocking expression of the gene for the stearoyl-ACP desaturase, the enzyme that catalyzes the conversion of 18:0 into the monounsaturated 18:1Δ^9 (Table 7.4). Transgenic and mutagenic approaches have proven successful for suppressing expression of genes for the stearoyl-ACP desaturase in seeds. An early example of the transgenic production of high stearic acid seeds was achieved through antisense suppression of stearoyl-ACP desaturase genes in canola and turnip rape (Knutzon *et al.*, 1992). RNAi methods were used to generate cotton seeds with high 18:0 content. In both cases, seeds were obtained with oils containing 30–40% 18:0 (Liu *et al.*, 2002). The resulting seeds, however, germinated poorly, particularly at low temperature. Growth temperature has also been shown to have a major impact on 18:0 accumulation in seeds of a sunflower high stearic acid mutant (Fernández-Moya *et al.*, 2002). When these plants were maintained at day/night temperatures of 35/25 °C, stearic acid comprised nearly 35% of the seed oil. In contrast, when the plants were grown at day/night temperatures of 20/10 °C, levels of 18:0 in seeds decreased to ~8% of the total fatty acids, essentially the same amounts present in the nonmutant parental seeds. These results reflect the plant cell's ability to adjust fatty acid composition in order to maintain the fluidity of membranes. At low temperatures, the presence of high levels of 18:0, which has a melting point of 70 °C, likely disrupts the integrity of membranes and perhaps oil bodies and, as a result, compromises the viability of cells. It should be noted that in seeds engineered to produce high levels of 18:0, the content of this fatty acid is not only increased in the storage oil but also in microsomal membranes (Álvarez-Ortega *et al.*, 1997).

From a metabolic engineering standpoint, one would predict that an 18:0 content of 70–80% should be attainable in seeds by suppression of stearoyl-ACP desaturase expression. These levels of oleic acid are routinely obtained by suppression of *FAD2* expression in seeds of transgenic plants. Instead, stearic acid levels in transgenic or mutant seeds rarely exceed 30–40% of the total fatty acids. The likely explanation for this apparent shortfall is that seeds are not viable if stearic acid accumulates to high levels, especially under conditions of low to moderate growth temperatures. It is notable that a limited number of tropical and subtropical plants produce seed oils with >65% 18:0, suggesting that transgenic crop seeds accumulating comparable levels are a viable goal. Perhaps, a successful engineering strategy for high levels of stearic acid accumulation will include the introduction of a metabolic mechanism to exclude stearic acid from membrane lipids, coupled with the cultivation of these genetically enhanced crops in warm climates.

An alternative, but quantitatively less successful approach for producing seeds with elevated 18:0 content has involved the transgenic expression of divergent forms of the oleoyl-ACP thioesterase or FatA. The FatA enzyme typically displays a strong substrate preference for oleoyl-ACP; however, a divergent form of this enzyme with increased activity for stearoyl-ACP has been identified in seeds of *Garcinia mangostana* or mangosteen (Hawkins and Kridl, 1998). Stearic acid comprises about 55% of the oil of these seeds. Expression of the mangosteen cDNA in canola under control of a strong seed specific promoter permitted accumulation of 18:0 to ~22% of the seed oil (Hawkins and Kridl, 1998).

3.3. Metabolic engineering of high and low polyunsaturated vegetable oils

Alteration of the α-linolenic content of seed oils is an important biotechnological target. This fatty acid is a very minor component of the seed oil of a number of crops, including corn, sunflower, peanut, and canola. α-Linolenic acid, however, accounts for nearly 10% of soybean oil and over 50% of linseed (or flax) oil. The three double bonds of this fatty acid make it particularly prone to oxidation. This is an undesirable property for food processing as the oxidation products of α-linolenic result in rancidity and reduced shelf life. Conversely, the oxidative instability of α-linolenic acid is an essential property for the use of vegetable oils such as linseed oil in drying oil applications. The free radicals generated from oxidation of α-linolenic acid-rich oils result in the enhanced polymerization (or "drying") of paint, ink, and other coating materials.

The α-linolenic acid content of seed oils can be increased or decreased by altering the expression of genes for FAD3, the ER Δ^{15}-linoleic acid desaturase (Table 7.4 and Fig. 7.4). As described above, FAD3 catalyzes the conversion of linoleic acid to α-linolenic acid. Transgenic expression of the *A. thaliana FAD3* gene to high levels using a strong seed-specific promoter has been shown to increase the α-linolenic acid content to >50% of *A. thaliana* seed oil, which is comparable to the proportion found in linseed oil (Yadav *et al.*, 1993). Downregulation of *FAD3* expression in seeds can be achieved through transgenic approaches or by the generation of mutants. The development of *FAD3* mutants with good agronomic performance has been particularly effective in soybean. Mutants with as little as 1–3% α-linolenic acid in their seed oil have been reported (Ross *et al.*, 2000). These mutants do not display any significant reductions in seed yield (Ross *et al.*, 2000). It is also notable that transgenic suppression of *FAD2* genes in soybean likewise yields oils with 2–3% of α-linolenic acid. This phenotype is due to the large decrease in the linoleate substrate pool for the Δ^{15}-linoleic acid desaturase (Buhr *et al.*, 2002; Kinney, 1996).

3.4. Variant fatty acid desaturases for metabolic engineering of vegetable oil composition

3.4.1. Variant acyl-ACP desaturases

Oleic acid ($18{:}1\Delta^9$) is the primary monounsaturated fatty acid of the seed oils of most plants. However, the seed oils of a number of species contain monounsaturated fatty acids with more or fewer than 18 carbon atoms. These fatty acids can also contain a double bond other than at the Δ^9 position. Oils that contain large amounts of these novel monounsaturated fatty acids have been of biotechnological interest because they have physical or biological properties that are not found in the oils of major crop plants. For example, petroselinic acid (18:1Δ^6), a novel monounsaturated fatty acid, typically comprises >70% of the seed oil of *Apiaceae* species. Oils that contain very high levels of this fatty acid are solid at room temperature and are less susceptible to digestion by pancreatic lipases than oils enriched in oleic acid. In addition, monounsaturated fatty acids can be oxidatively cleaved to generate dicarboxylic acids for nylon production. The chain lengths of the resulting acids are determined by the position of the double bond. The cleavage of petroselinic acid (18:1Δ^6; Fig. 7.3), for example, yields adipic acid, the C6 dicarboxylic acid that is a precursor of nylon 6,6, the world's most widely manufactured nylon.

Interestingly, seeds that produce high levels of novel monounsaturated fatty acids with 18 or fewer carbon atoms have been found to contain structurally variant forms of the stearoyl (18:0)-ACP desaturase (Table 7.5) (Shanklin and Cahoon, 1998). Relative to the 18:0-ACP desaturase, these enzymes have altered specificity for the chain length of the acyl-ACP substrate or introduce double bonds at sites other than the Δ^9 position within the fatty acid chain. Variant acyl-ACP desaturases identified to date include a Δ^4-palmitoyl (16:0)-ACP desaturase associated with petroselinic acid synthesis in *Apiaceae* seeds, a Δ^6-16:0-ACP desaturase used during the synthesis of Δ^6-hexadecenoic acid (16:1Δ^6) by *Thunbergia alata* seeds, a Δ^9–16:0-ACP desaturase linked to production of palmitoleic acid (16:1Δ^9) in *Macfadyena unguis* seeds, and a Δ^9-myristoyl (14:0)-ACP desaturase from geranium trichomes that is involved in the synthesis of pest resistant anacardic acids (Cahoon and Ohlrogge, 1994; Cahoon *et al.*, 1994, 1998; Schultz *et al.*, 1996).

Attempts to produce seed oils with novel monounsaturated fatty acids by the transgenic expression of cDNAs for these enzymes have met with only marginal success (Suh *et al.*, 2002). In fact, the case of petroselinic acid highlights the complexity that can be encountered when attempting to engineer novel biosynthetic pathways in oilseed crops. The biosynthesis of this fatty acid requires at least three specialized enzymes: a Δ^4-16:0-ACP desaturase, a KAS that efficiently directs the elongation of 16:1Δ^4-ACP to 18:1Δ^6-ACP, and an acyl-ACP thioesterase that releases 18:1Δ^6 from ACP. In addition, biochemical evidence indicates the involvement of a specialized ACP in this pathway, and the sensitivity of the petroselinic acid biosynthesis to salt and detergents *in vitro* also suggests that this pathway might function as a metabolon or complex of closely associated enzymes. Perhaps as a result of this complexity, the maximum reported level of petroselinic acid accumulation in transgenic *A. thaliana* seeds is <2% of the total fatty acids (Suh *et al.*, 2002).

A more successful approach for the production of oils with novel monounsaturated fatty acids has resulted from rational modification of the substrate specificity of 18:0-ACP desaturases (Cahoon and Shanklin, 2000; Cahoon and Shanklin, 2000). The availability of a crystal structure for this enzyme has facilitated efforts to redesign its activity (Lindqvist *et al.*, 1996). By modeling the amino acid sequences of variant acyl-ACP desaturases onto the three-dimensional structure

TABLE 7.5 Examples of variant fatty acid desaturases that occur in limited number of species in the plant kingdom

Desaturase	Species	Substrate	Product	Fatty acid product name
Variant acyl-ACP desaturase				
Δ^9-Myristoyl-ACP desaturase	*Pelargonium × hortorum*	14:0-ACP	14:1Δ^9-ACP	Myristoleic acid
Δ^4-Palmitoyl-ACP desaturase	*Coriandrum sativum*	16:0-ACP	16:1Δ^4-ACP	Δ^4-Hexadecenoic acid
Δ^6-Palmitoyl-ACP desaturase	*Thunbergia alata*	16:0-ACP	16:1Δ^6-ACP	Δ^6-Hexadecenoic acid
Δ^9-Palmitoyl-ACP desaturase	*Macfadyena unguis-cati*	16:0-ACP	16:1Δ^9-ACP	Palmitoleic acid
Variant Δ^{12}-oleic acid desaturase (FAD2)				
Oleic acid hydroxylase	*Ricinus communis; Lesquerella fendleri*	18:1Δ^9-PC	12-Hydroxy-18:1Δ^9-PC	Ricinoleic acid
Linoleic acid epoxygenase	*Crepis palaestina*	18:2$\Delta^{9,12}$-PC	12-Epoxy-18:1Δ^9-PC	Vernolic acid
Linoleic acid acetylenase	*Crepis alpine*	18:2$\Delta^{9,12}$-PC	12-Acetylenic-18:1Δ^9-PC	Crepenynic acid
Linoleic acid Δ^{12}-conjugases	*Momordica charantia; Aleurites fordii; Punica granatum; Trichosanthes kirlowii*	18:2$\Delta^{9,12}$-PC	18:3$\Delta^{9,11,13}$-PC	Eleostearic acid (18:3$\Delta^{9cis,11trans,13trans}$); punicic acid (18:3$\Delta^{9cis,11cis,13trans}$)

Linolenic acid Δ^{12}-conjugase	*Impatiens balsamina*	$18{:}3\Delta^{9,12,15}$-PC	$18{:}4\Delta^{9,11,13,15}$-PC	Parinaric acid
Linoleic acid Δ^{9}-conjugase	*Calendula officinalis*	$18{:}2\Delta^{9,12}$-PC	$18{:}3\Delta^{8,10,12}$-PC	Calendic acid
Trans-Δ^{12}-linoleic acid hydroxylase/ conjugase	*Dimorphotheca sinuata*	$18{:}2\Delta^{9cis,12trans}$-PC	9-Hydroxy-$18{:}2\Delta^{10trans,12trans}$-PC	Dimorphecolic acid
Variant cytochrome b_5-fusion desaturase				
Δ^{6}-Desaturase	*Borago officinalis*	$18{:}2\Delta^{9,12}$-PC $18{:}3\Delta^{9,12,15}$-PC	$18{:}3\Delta^{6,9,12}$-PC $18{:}4\Delta^{6,9,12,15}$-PC	γ-Linolenic acid Stearidonic acid
Variant acyl-CoA desaturase				
Δ^{5}-Eicosanoic acid desaturase (Des5)	*Limnanthes douglasii*	20:0-CoA/ lipid (?)	$20{:}1\Delta^{5}$-CoA/ lipid (?)	Δ^{5}-Eicosenoic acid
Δ^{9}-Stearoyl-CoA desaturase (Des9)	*Picea glauca*	18:0-CoA	$18{:}1\Delta^{9}$-CoA	Oleic acid
Δ^{9}-Palmitic acid desaturase (ADS1, ADS2)	*Arabidopsis thaliana*	16:0-CoA/ lipid (?)	$16{:}1\Delta^{9}$-CoA/ lipid (?)	Palmitoleic acid

Genes or cDNAs for these fatty acid desaturases have been isolated from the species indicated. These genes and cDNAs can be used for the modification of the fatty acid composition of vegetable oils.

of the 18:0-ACP desaturase, it has been possible to convert the 18:0-ACP desaturase into a 16:0-ACP desaturase by substituting only one amino acid at the bottom of the substrate binding pocket. Enzymes rationally designed in this manner have been used to engineer the content of palmitoleic acid ($16{:}1\Delta^9$) and its elongation products to as high as 30% of the total fatty acids of *A. thaliana* seeds (Cahoon *et al.*, 1997). A complete understanding of how acyl-ACP desaturases position the placement of double bonds in fatty acid chains awaits further structural characterization of these enzymes.

3.4.2. Variant acyl-CoA desaturases

In contrast to plants, animals and fungi employ a membrane associated fatty acid desaturase known as Δ^9-stearoyl (18:0)-CoA desaturase for oleate synthesis. As indicated by this nomenclature, the enzyme is specific for acyl chains bound to coenzyme A, rather than to the ACP employed by soluble 18:0-ACP desaturase in plants. Genes for Δ^9–18:0-CoA desaturase-like enzymes have been identified in *A. thaliana* and other plant species (Shanklin and Cahoon, 1998; Voelker and Kinney, 2001). Two of the *Arabidopsis* genes *ADS1* and *ADS2* have been shown to encode ER-localized Δ^9 desaturases (Heilmann *et al.*, 2004). A third gene *ADS3* encodes a Δ^7 desaturase that uses 16:0 bound to the chloroplast lipid monogalatosyldiacylglycerol as its substrate (Heilmann *et al.*, 2004). Interestingly, the ADS1 and ADS2 polypeptides were converted to Δ^7 desaturases by targeting of these enzymes to chloroplasts (Heilmann *et al.*, 2004). In addition, a cDNA for a Δ^9–18:0-CoA desaturase-like enzyme from white spruce (*Picea glauca*) seeds has been recently shown to encode a Δ^9 desaturase by expression in *Saccharomyces cerevisiae* (Table 7.5). This enzyme also displays *in vitro* activity with acyl-CoA substrates (Marillia *et al.*, 2002).

An example of the use of an acyl-CoA desaturase-like enzyme for the metabolic engineering of an oilseed crop resulted from the isolation of genes from seeds of *Limnanthes douglasii* (Table 7.5) (Cahoon and Shanklin, 2000). The seed oil of this plant contains more than 60% of the unusual monounsaturated fatty acid Δ^5-eicosenoic acid ($20{:}1\Delta^5$; Table 7.3). This is an oxidatively stable oil that has superior properties for industrial lubricant applications. The cDNA for this enzyme has been identified among expressed sequence tags from *L. douglasii* seeds and shown to encode a Δ^5 desaturase by transgenic expression in soybean somatic embryos (Cahoon and Shanklin, 2000). Although this enzyme has significant activity with 16:0 and 18:0 *in vivo*, it appears to be most active with eicosanoic acid (20:0). In fact, a $20{:}1\Delta^5$ content of 12% has been achieved in soybean embryos coexpressing the Δ^5 desaturase cDNA and a cDNA for an *L. douglasii* 3-ketoacyl-CoA synthase (KCS) that initiates the elongation of 16:0-CoA to 20:0-CoA. This level of novel monounsaturated fatty acid accumulation exceeds that achieved in all previous attempts to produce such fatty acids through the transgenic expression of naturally occurring variant acyl-ACP desaturases. It is presumed that the *L. douglasii* Δ^5 desaturase uses an acyl-CoA substrate, based on results from *in vitro* assays of *Limnanthes* seed preparations (Moreau *et al.*, 1981). However, it cannot be ruled out that the substrate for the *L. douglasii* desaturase and other acyl-CoA desaturase-like enzymes in plants is, in fact, a fatty acid bound to a lipid, as is the case for FAD2 and FAD3.

3.4.3. Variant cytochrome b_5-fusion fatty acid desaturases

The most successful efforts to introduce biosynthetic pathways for novel unsaturated fatty acids in oilseed crops have involved the production of γ-linolenic acid (GLA; 18:3$\Delta^{6,9,12}$; Table 7.3) by transgenic expression of divergent forms of the sphingolipid-8-desaturase. GLA comprises up to 25% of the seed oil of many species of the families including borage (*Borago officinalis*). This fatty acid is a precursor of prostaglandins and leukotrienes, and oils enriched in GLA have a number of demonstrated therapeutic properties (Barre, 2001). As a result, GLA-enriched oils from seeds of plants such as borage are produced commercially as nutraceuticals. A cDNA for the Δ^6 desaturase associated with GLA synthesis in plants has been isolated from developing borage seeds by polymerase chain reaction (PCR) with degenerate oligonucleotides based on partially conserved amino acid sequences in membrane-associated fatty desaturases (Sayanova *et al.*, 1997). This cDNA encodes a polypeptide distantly related to FAD2 and FAD3 as well as to the functionally equivalent Δ^6 desaturase from *Synechocystis* (Table 7.5) (Reddy and Thomas, 1996; Shanklin and Cahoon, 1998). The borage polypeptide is predicted to contain not only the histidine residues characteristic of membrane-type fatty acid desaturases, but also an N-terminal cytochrome b_5 domain. Presumably this domain functions as the electron donor for the desaturase portion of the polypeptide, as cytochrome b_5 is a cofactor for ER-localized desaturases (Sayanova *et al.*, 1997). Transgenic expression of this cDNA in tobacco leaves resulted in the production of GLA via the Δ^6 desaturation of linoleic acid (Fig. 7.4), as well as stearidonic acid (18:4$\Delta^{6,9,12,15}$) comprising up to 10% of total leaf fatty acids (Sayanova *et al.*, 1997). The latter unusual polyunsaturated fatty acid arises either from the Δ^{15} desaturation of GLA by FAD3 or from the desaturation of α-linolenic acid by the Δ^6 desaturase. Stearidonic acid is found only in trace amounts in borage seeds due apparently to the lack of an active FAD3 in this tissue. Since the discovery of the borage enzyme, fatty acid desaturase-like polypeptides with close structural relationships to the Δ^6 desaturase, including the presence of the N-terminal cytochrome b_5 domain, have been shown to occur widely in plants (Sperling *et al.*, 1998). These enzymes are sphingolipid-8-desaturases associated with the synthesis of unsaturated sphingolipid long-chain bases such as sphingenine. They introduce double bonds at the sixth carbon atom from the C-2 position of long-chain base substrates (Sperling *et al.*, 1998).

A borage Δ^6 desaturase cDNA was expressed in soybean seeds under control of a strong seed-specific promoter. In the most promising transgenic line, GLA and stearidonic acid comprised 33% and 4%, respectively, of the soybean seed oil (Sato *et al.*, 2004). By comparison, the GLA content of borage oil is typically 20–25% of the total fatty acids. This is perhaps the only case to date in which the accumulation of a novel fatty acid in a transgenic seed oil exceeds that found in the naturally occurring seed oil.

Structural and functional homologues of the borage Δ^6 desaturase have been identified in other plant species as well as a wide range of organisms including the moss *Physcomitrella patens;* animals including *Caenorhabditis elegans*, mouse and human; and fungal species including *Mortierella alpina* and *Pythium irregulare* (Das *et al.*, 2000; García-Maroto *et al.*, 2002; Girke *et al.*, 1998; Hong *et al.*, 2002; Napier *et al.*, 2003). The *M. alpina* and *P. irregulare* genes have been used to engineer the production of GLA in seed oils of *Brassica* species (Das *et al.*, 2000; Hong *et al.*, 2002).

The family of cytochrome b_5-fusion desaturases has subsequently been found to include Δ^4 and Δ^5 desaturases, both of which are structurally related to the Δ^6 desaturase (Napier *et al.*, 2003). Though these enzymes occur in numerous organisms, neither fatty acid desaturase has yet been identified in plants. Genes for Δ^5 desaturases have been found in mammals, *C. elegans*, and the fungus *M. alpina* (Knutzon *et al.*, 1998; Michaelson *et al.*, 1998; Napier *et al.*, 2003). This enzyme catalyzes the introduction of the Δ^5 double bond of arachidonic acid (ARA), a C_{20} polyunsaturated fatty acid. Δ^4 Desaturase genes have been isolated from *Euglena gracilis* and *Thraustochytrium* sp. and are associated with the formation of the Δ^4 double bond of docosahexaenoic acid (DHA), a C_{22} polyunsaturated fatty acid (Meyer *et al.*, 2003; Qiu *et al.*, 2001a). Collectively, the cytochrome b_5–fusion desaturases have been referred to as "front-end" desaturases, based on their catalytic ability to position the site of double bond insertion relative to the carboxyl or "front end" of fatty acid substrates. Similar regiospecificity is displayed by acyl-CoA-type desaturases and soluble acyl-ACP desaturases (Shanklin and Cahoon, 1998). "Front-end" desaturase genes have considerable biotechnological significance for the metabolic engineering of health-promoting, very long-chain polyunsaturated fatty acids in seed oils of transgenic crops.

3.4.4. Variant FAD2s

The typical FAD2 catalyzes the synthesis of linoleic acid by the Δ^{12} desaturation of oleic acid (Table 7.4). In addition to this widely occurring enzyme, structurally divergent forms of FAD2 have been shown to function in the biosynthesis of unusual fatty acids in seeds of diverse plant species (Fig. 7.4 and Table 7.5) (Voelker and Kinney, 2001). These divergent FAD2 enzymes are of biotechnological interest because they introduce functional groups, such as hydroxy residues and epoxy rings, onto fatty acid chains. The functionalized fatty acids can be used for a variety of industrial applications. Divergent *FAD2* genes encode hydroxylases, acetylenases, and some epoxygenases, all enzymes requiring oxygen and employing similar histidine motifs and di-iron clusters (Lee *et al.*, 1998; van de Loo *et al.*, 1995). Relatively small modifications can alter the activity of these enzymes. For example, it has been possible to interconvert Δ^{12}-oleic acid desaturase and Δ^{12}-hydroxylase activities by altering only four to six amino acids (Broun *et al.*, 1998). The substrate in this case is the same, phospholipid-esterified $18{:}1\Delta^9$. Other members of the family modify preexisting double bonds. Given the right enzyme, the Δ^{12} double bond of linoleic acid ($18{:}2\Delta^{9,12}$) may be converted either to a triple bond by Δ^{12}-acetylenase, or to an epoxide ring by the related epoxygenase from *C. palaestina* (Cahoon *et al.*, 2003; Lee *et al.*, 1998). It should be noted that not all plants produce epoxy-fatty acids via a Fad2-related enzyme. A cytochrome P450 enzyme from *Euphorbia lagascae* has been shown to add oxygen across Δ^{12} double bonds (Cahoon *et al.*, 2002).

In addition, variant forms of FAD2 known as fatty acid conjugases, because they catalyze the formation of conjugated double bond systems, have been identified in seeds of a number of plant species. These enzymes use linoleic acid or α-linolenic acid as substrates. One class of fatty acid conjugase, which is found in seeds of *Calendula officinalis*, converts the *cis*-Δ^9 double bond into the two adjacent *trans*-Δ^8 and *trans*-Δ^{10} double bonds (Cahoon *et al.*, 2001; Qiu *et al.*, 2001b). A second class of conjugase converts the *cis*-Δ^{12} double bond into adjacent *trans*-Δ^{11} and *trans*-Δ^{13} or *trans*-Δ^{11} and *cis*-Δ^{13} double bonds. The second class of fatty acid conjugases is functional in seeds of species such as *Impatiens balsamina*, *Momordica charantia*, *Aleurites fordii*, and *Punica granatum* (Cahoon *et al.*, 1999; Dyer *et al.*, 2002; Hornung *et al.*, 2002). Unlike the typical Δ^{12} desaturase which abstracts hydrogens from the C-12 and C-13 atoms of the oleic acid substrate, the mechanism of fatty acid conjugases is believed to involve the removal of a hydrogen atom from the two carbon atoms that flank an existing double bond. This mechanism has been referred to as a ""1,4-desaturation" (Reed *et al.*, 2002)." Fatty acid conjugase cDNAs from *M. charantia* and *C. officinalis* have been engineered into somatic embryos of soybean to generate oils with 15–20%-conjugated trienoic (or tri-unsaturated) fatty acids (Cahoon *et al.*, 1999, 2001). Oils enriched in these fatty acids are useful as quick drying agents in coating materials, such as paint and ink, as a result of the oxidative instability of the conjugated fatty acids.

3.5. Metabolic engineering of vegetable oils with short and medium-chain fatty acids

Lauric acid (12:0) is employed extensively in the soaps and detergents industry. Although adequate supplies are available from coconut and palm oils, there has been interest in developing temperate sources of 12:0 feedstocks, and high laurate rapeseed canola was the first genetically engineered oil in commercial production. Oils enriched in shorter fatty acids such as capric acid (10:0) and caprylic acid (8:0) would also be useful in the manufacture of detergents, surfactants, and plasticizers.

Production of medium-chain fatty acids is initiated by the early release of acyl groups from ACP by medium-chain acyl-ACP thioesterases. Thioesterases specific for medium-chain fatty acids have been identified in seeds that produce high levels of 8:0, 10:0, 12:0, and 14:0 from plants as distantly related as California bay, cuphea, coconut, elm, and camphor (Voelker *et al.*, 1992, 1997). These novel thioesterases share significant amino acid sequence with FatB acyl-ACP thioesterases and have likely evolved from common predecessors. Comparison of these sequences has provided considerable insight into chain length specificity of acyl-ACP thioesterases, for example, Yuan *et al.* converted a 14:0-ACP thioesterase into a 12:0-ACP thioesterase by altering only three amino acids (Yuan *et al.*, 1995).

Transformation of rapeseed with the California bay 12:0-ACP thioesterase gene has resulted in oils with lauric acid content exceeding 50% of the total fatty acid (Voelker *et al.*, 1992). This represents one of the most significant technical achievements in the metabolic engineering of seed oil composition. Efforts to modify oil composition with thioesterases from 8:0- to 10:0-producing seeds have been less successful (Voelker and Kinney, 2001). However, cotransformation with genes for medium-chain thioesterase and KASIV, a 10:0 KAS from *Cuphea*, shifts fatty acid accumulation strongly toward shorter chain lengths (Leonard *et al.*, 1998). The KASIV gene alone does not confer deposition of medium-chain fatty acids, perhaps because it is inhibited by short and medium-chain ACPs (Dehesh *et al.*, 1998; Schütt *et al.*, 2002).

Further improvement of medium-chain production in transgenic rapeseed has been achieved by transformation with a coconut lysophosphatidic acid acyltransferase (LPAAT) gene. While native rapeseed LPAAT excludes 12:0 from the *sn*-2 position, the transformants do not (Knutzon *et al.*, 1999).

3.6. Metabolic engineering of vegetable oils with very long-chain fatty acids (VLCFAs)

Although the fatty acyl synthase cannot produce fatty acids longer than 18 carbons, extraplastidial complexes known as acyl-CoA elongases can extend fatty acids to 20, 22, or even double those numbers of carbons. While all plants require VLCFAs for production of the cuticle and extracuticular wax, some incorporate them into seed oils. For example, erucic acid ($22{:}1\Delta^{13}$) made by the elongation of oleic acid ($18{:}1\Delta^{9}$) by two C_2 units is an industrially useful fatty acid found in rapeseed oil as well as the model oilseed *A. thaliana*.

The acyl-CoA elongase performs reactions comparable to those of the fatty acyl-ACP synthase, employing KCS, 3-ketoacyl-CoA reductase, 3-hydroxyacyl dehydratase and enoyl reductase subunits. Of the elongase components, the KCS appears to be rate limiting (Ghanevati and Jaworski, 2002; Millar and Kunst, 1997). For example, the production of low erucic acid canola oil from high erucic acid varieties of rapeseed was made possible by a single amino acid change in a KCS (Katavic *et al.*, 2002). Conversely, overexpression of the *A. thaliana FAE* gene, whose KCS product initiates the elongation cycle in *A. thaliana*, results in production of fatty acids up to 22 carbons long in organs that do not normally accumulate them (Millar *et al.*, 1998). In addition, the introduction of a KCS from *L. douglasii* into soybean somatic embryos resulted in the accumulation of C_{20} and C_{22} fatty acids to amounts of >15% of the total fatty acids (Cahoon *et al.*, 2000). These fatty acids normally comprise <1% of the fatty acids of this tissue. A variety of KCS polypeptides with differing substrate specificities have been identified (Cahoon *et al.*, 2000; Lassner *et al.*, 1996; Millar *et al.*, 1999; Moon *et al.*, 2001).

Evidence indicates that KCS may not be the only enzyme class capable of initiating VLCFA synthesis in plants. For example, elongation of C_{18} Δ^{6}-polyunsaturated fatty acids in the moss *P. patens* is initiated by an enzyme with homology to a class of polypeptides termed ELOs (Zank *et al.*, 2002). These enzymes were first identified as components of the fatty acid elongation system in *S. cerevisiae*

(Toke and Martin, 1996). and have been characterized from a variety of animal and fungal species (Beaudoin *et al.*, 2000; Parker-Barnes *et al.*, 2000). Genes for ELO-like enzymes are also present in *A. thaliana* and other plants, but their functions have yet to be determined. The ELO polypeptides contain the motif HXXHH, which is similar to a sequence element found in fatty acid desaturases (Parker-Barnes *et al.*, 2000). Whether these histidines coordinate active site iron atoms, as they do in desaturases, is not known. The ELOs from organisms such as *C. elegans* and the fungus *M. alpina* participate in the synthesis of nutritionally important long-chain polyunsaturated fatty acids (Beaudoin *et al.*, 2000; Parker-Barnes *et al.*, 2000). ELO and KCS enzymes appear functionally redundant with regard to their ability to catalyze the first step in fatty acid elongation, yet these enzyme classes are structurally unrelated.

Enrichment of TAG with VLCFAs might be limited by the ability of plants to incorporate these unusual molecules into glycerolipids. For example, rapeseed, although it does accumulate $22{:}1\Delta^{13}$, is unable to incorporate this fatty acid into the *sn*-2 position. However, if the plants are transformed with a *Limnanthes alba* gene for an LPAAT utilizing $22{:}1\Delta^{13}$-CoA, some tri-$22{:}1\Delta^{13}$ is produced (Lassner *et al.*, 1995). Similarly, a yeast LPAAT, designated SLC1–1, that has specificity for C_{22} and C_{24} acyl-CoAs was shown to improve both TAG yield and proportions of VLCFAs when expressed in *B. napus* seeds (Zou *et al.*, 1997).

3.7. Metabolic engineering of nonplant pathways

The potential also exists to engineer seeds to produce oils with fatty acids not normally found in plants. Currently, few examples exist for the transgenic production of fatty acids that do not normally occur in plants. However, genes from sources other than plants have been used to alter the fatty acid composition of seed oils. Early examples include the use of genes for the yeast and mouse Δ^9-stearoyl-CoA desaturase and the *Synechocystis* Δ^6-linoleic acid desaturase to produce small amounts of novel fatty acids in plants (Moon *et al.*, 2000; Reddy and Thomas, 1996). Δ^6-linoleic acid desaturase genes from the fungi *M. alpina* and *P. irregulare* have been used to generate seed oils in *Brassica* species that contain >40% GLA (Das *et al.*, 2000; Hong *et al.*, 2002). This amount of GLA is comparable to that obtained by transgenic expression of the borage Δ^6 desaturase in soybean seeds (Sato *et al.*, 2004).

Perhaps one of the greatest challenges for metabolic engineering of seed oil composition will be the transgenic production of high levels of the long-chain ω-3 fatty acids eicosapentaenoic acid (EPA; $20{:}5\Delta^{5,8,11,14,17}$) and DHA ($22{:}6\Delta^{4,7,10,13,16,19}$) (Table 7.3). Though EPA can be found in trace amounts in the seed oil of certain gymnosperm species, this fatty acid and DHA are typically absent from seed oils but are major components of many fish and algal oils. Diets rich in the ω-3 fatty acids EPA and DHA have been linked to enhanced cardiovascular fitness (Hu *et al.*, 2001). In addition, DHA has been shown to improve brain development when supplemented in infant diets (Uauy *et al.*, 2003). DHA is present in mother's milk, but is absent from infant formula prepared from soybean and other vegetable oils. Highly refined fish oils that contain EPA and DHA can sell for >\$10/pound, whereas conventional soybean oil sells for about \$0.20/ pound. The ability to engineer EPA and DHA synthesis in oilseeds therefore offers a means for significantly increasing not only nutritional quality, but also economic value of vegetable oils.

A number of pathways might lead to the production of EPA in seeds (Napier, 2007). The most direct biosynthetic route would involve the introduction of three new enzymes: (1) a Δ^6 desaturase for the conversion of α-linolenic acid to stearidonic acid, (2) an ELO-type fatty acid elongase to initiate the elongation of stearidonic acid to ARA ($20{:}4\Delta^{8,11,14,17}$), and (3) a Δ^5 desaturase for formation of EPA from ARA (Fig. 7.4). In addition, most oilseeds contain relatively low amounts of α-linolenic acid, the initial substrate for this pathway. As a result, the production of EPA in the seeds of most crop plants would require not only the introduction of genes for the three enzymes described above but also enhanced expression of the *FAD3* gene to increase Δ^{15}-linoleic acid desaturase activity. Reports have demonstrated the feasibility of assembling the EPA biosynthetic pathway into leaves and seeds of transgenic plants (Abbadi *et al.*, 2004; Qi *et al.*, 2004; Robert *et al.*, 2005; Wu *et al.*, 2005).

The production of DHA in oilseeds requires the transgenic expression of genes for at least two additional enzymes: (1) an ELO elongase to initiate the elongation of EPA to the C_{22} fatty acid docosapentaenoic acid (DPA, $22{:}5\Delta^{7,10,13,16,19}$) and (2) a Δ^4 "front-end" desaturase to convert DPA to DHA (Fig. 7.4). Genes for Δ^4 desaturases capable of catalyzing this conversion of DPA have been isolated from *E. gracilis* and *Thraustochytrium* sp (Meyer *et al.*, 2003; Qiu *et al.*, 2001a). Indeed, production of small amounts of DHA in seeds of *Brassica juncea* was recently achieved by the transgenic expression of genes for an *Oncorhynchus mykiss* C_{18}/C_{20}-specific ELO and the *Thraustochytrium* Δ^4 desaturase along with genes for a *P. irregulare* Δ^6 desaturase, a *Phytophthora infestans* ω^3 desaturase, *Thraustochytrium* Δ^5 desaturase, acyltransferase and ELO, a *C. officinalis* Δ^{12} desaturase, and a *P. patens* ELO (Wu *et al.*, 2005). Although the introduction of this pathway into seeds of a transgenic plant via the overexpression of nine transgenes is a remarkable metabolic engineering feat, the low amounts of DHA produced (0.2% of the total fatty acids) indicate that bottlenecks exist for achieving high levels of this commercially valuable fatty acid. In this regard, biochemical studies have indicated that the desaturase and ELO elongases that are required for EPA and DHA synthesis employ fatty acid substrates esterified to different molecules (Domergue *et al.*, 2003). The Δ^5 and Δ^6 desaturases from plant, fungal, and algal species use fatty acids bound to PC as their preferred substrates. In contrast, ELO elongases accept acyl-CoAs as substrates. The need for fatty acid substrates to move between pools of PC and acyl-CoAs may limit flux through the engineered EPA and DHA biosynthetic pathways. Furthermore, polyunsaturated fatty acids (PUFAs) do not appear to be efficiently channeled into TAG and excluded from accumulating in membrane lipids of engineered seeds (Abbadi *et al.*, 2004; Wu *et al.*, 2005). Whether membrane-associated PUFAs will impair the agronomic fitness of seeds has yet to be addressed.

An alternative mechanism for EPA and DHA biosynthesis has been demonstrated in the marine bacterium *Shewanella* sp. and in the thraustochytrid *Schizochytrium* sp. These organisms contain polyketide synthases consisting of multifunctional polypeptides capable of synthesizing EPA and DHA in an anaerobic manner (Metz *et al.*, 2001). The possibility of transferring these pathways to seeds appears daunting considering that EPA synthesis in *Shewanella*, for example, results from 5 open-reading frames that code for at least 11 different protein domains. These domains include polypeptides that are related to 3-ketoacyl synthases, ACP, enoyl reductases, dehydratases, and acyltransferases.

4. SUMMARY

4.1. Alteration of seed oil content

Initial efforts to engineer oil content in crops emphasized the overexpression of enzymes responsible for incorporation of acetyl-CoA into TAG. Modest improvements have been achieved, notably by overexpression of ACCase, which catalyzes the initial step in fatty acid synthesis, and DGAT, which attaches the last fatty acid to the oil molecule (Jako *et al.*, 2001; Madoka *et al.*, 2002; Roesler *et al.*, 1997). Nevertheless, the many trials conducted to date suggest that no single enzyme in the synthetic pathway is limiting for oil accumulation in seeds. Since genes of the synthetic pathway are often coordinately expressed (Cronan and Subrahmanyam, 1998; Lee *et al.*, 2002; Ruuska *et al.*, 2002; Slabas *et al.*, 2002), it is hoped that identification of the transcription factors responsible will ultimately permit global overexpression of the pathway. In addition, it is likely that partitioning and flux of carbon into pathways such as glycolysis and the pentose phosphate cycle that generate precursors and reducing capacity for *de novo* fatty acid synthesis are major factors in determining oil production (Rangasamy and Ratledge, 2000; Rawsthorne, 2002; Schwender *et al.*, 2003).

Our understanding of carbon partitioning and flux control in seeds is currently in an early stage. Studies with isolated seed plastids have been useful for identifying cytosolic precursors of acetyl-CoA for fatty acid synthesis (Rawsthorne, 2002). The recent use of stable isotope labeling techniques coupled with nuclear magnetic resonance (NMR) and mass spectrometry has also provided useful insights into carbon flux in seeds (Schwender and Ohlrogge, 2002; Schwender *et al.*, 2003). In addition, transcriptional profiling has revealed a comprehensive view of the timing and levels of expression of genes associated with the synthesis of oil, carbohydrates, and proteins during the development of *A. thaliana* seeds (Beisson *et al.*, 2003; Ruuska *et al.*, 2002). Undoubtedly, proteomic and metabolomic analyses will yield still greater understanding of metabolic networks associated with the regulation of carbon partitioning in seeds. With these data, it should ultimately be possible to uncover the basis for differences in the relative amounts of storage compounds in seeds of different plant species. With such information, it should be possible, for example, to understand why seeds of soybean contain 18% oil and 38% protein, while seeds of peanut, which is also a legume, contain 45% oil and 23% protein.

The roles of transcription factors in the global control of seed metabolism also require extensive investigation. To date, several transcription factors associated with carbon partitioning and seed development have been identified. For example, the transcription factor WRI1 has been linked to regulation of carbohydrate metabolism in seeds (Cernac and Benning, 2004; Focks and Benning, 1998). Whether transcription factors that can be manipulated to raise oil content in oilseeds and fruits will be identified is not yet clear. However, the stunted pickle-shaped roots of the *A. thaliana pkl* mutant continue depositing the oil and protein characteristic of embryos during seed development (Ogas *et al.*, 1997). Its gene product, PICKLE, is now known to be a master regulator of embryogenic transcription factors such as LEAFY COTYLEDON 1 and 2 and FUSCA3 (Brocard-Gifford *et al.*, 2003; Ogas *et al.*, 1999; Rider *et al.*, 2003). Thus, the possibility that transcription factors can confer oil production to organs other than classical oilseeds and fruits should also be considered. Possible applications might include increasing the caloric content of vegetative organs for human and livestock nutrition or for the production of novel oils in organs such as roots that are less prone to herbivory.

4.2. Alteration of the fatty acid composition of vegetable oils

Considerable progress has been made in the genetic alteration of the relative amounts of palmitic, stearic, oleic, linoleic, and α-linolenic acids in seed oils of crop plants. These modifications have been achieved primarily by either up- or downregulating expression of genes for fatty acid desaturases or acyl-ACP thioesterases. The production of seed oils with high levels of oleic acid represents the most significant commercial achievement to date in the metabolic engineering of seed oil composition. This oil modification was achieved in numerous crop species by blocking expression of *FAD2* genes, through both transgenic and chemical mutagenic approaches. The major remaining target is development of temperate crops with seed oils that are solid at room temperature and, therefore, do not require hydrogenation for use in margarine production. Achieving this target will require the engineering of seeds to produce high levels of saturated fatty acids that are sequestered in TAG, but not accumulated in membrane lipids such as PC. The enrichment of saturated fatty acids in PC and other phospholipids likely compromises membrane integrity, especially when seeds are subjected to low germination temperatures (Álvarez-Ortega *et al.*, 1997; Knutzon *et al.*, 1992; Liu *et al.*, 2002).

Metabolic engineering of novel fatty acid synthesis and accumulation in seeds of transgenic plants has met with limited success. cDNAs for numerous divergent fatty acid modification and biosynthetic enzymes have been identified. These include variant forms of acyl-ACP desaturases, acyl-ACP thioesterases, acyl-CoA desaturases, Δ^{12}-oleic acid desaturases, fatty acid elongases, cytochrome P450s, and cytochrome b_5-fusion desaturases (Voelker and Kinney, 2001). The availability of these cDNAs offers numerous possibilities for the metabolic engineering of seeds with enhanced nutritional, industrial, and animal feed properties. The development of canola seeds with high lauric acid content for detergent applications and the development of canola and soybean seeds with high GLA content for nutraceutical applications are perhaps the most notable technical successes in this research area (Del Vecchio, 1996; Sato *et al.*, 2004).

Despite these accomplishments, most metabolic engineering efforts have resulted in the development of seeds with only low to moderate levels of unusual fatty acids. In general terms, a major limitation on unusual fatty acid accumulation in transgenic seeds appears to be the inefficient flux from the site of synthesis to the final deposition in oil bodies as a component of TAG (Cahoon *et al.*, 2007). In the case of divergent forms of FAD2, such as epoxygenases, conjugases, acetylenases, and hydroxylases, the modification reaction occurs while the fatty acid substrate is bound to the membrane lipid PC. The unusual fatty acid product then must be efficiently removed from PC and mobilized onto glycerol backbones for storage as TAG in lipid bodies. This movement or channeling of novel fatty acids between PC and TAG likely involves specialized forms of metabolic enzymes such as acyltransferases and phospholipases that are absent from seeds of transgenic plants. The lack of these specialized enzymes could result in the aberrant accumulation of novel fatty acids in membrane phospholipids in seeds of transgenic plants, as has been observed for the production of acetylenic and conjugated fatty acids (Thomaeus *et al.*, 2001; Cahoon *et al.*, 2006). The accumulation of medium-chain length fatty acids also appears to be limited by inefficient incorporation onto glycerol backbones for TAG production in seeds of transgenic plants. In this case, the synthesis of decanoic and lauric acids in plastids of transgenic *B. napus* seeds by the activity of divergent acyl-ACP thioesterases resulted in the enrichment of decanoyl-CoA and lauroyl-CoA, relative to CoA esters of common fatty acids, in acyl-CoA pools (Larson *et al.*, 2002). A similar enrichment was not observed in seeds of *Cuphea hookeriana*, which naturally accumulate high levels of these fatty acids. In addition, seeds of developing *B. napus* that have been engineered to produce decanoic (10:0) and lauric (12:0) acids contained elevated amounts of these fatty acids in phospholipids, relative to seeds that naturally accumulate these fatty acids (Wiberg *et al.*, 2000). These results suggest that specialized forms of metabolic enzymes such as acyltransferases are also important for the storage of unusual medium-chain length fatty acids generated in seed plastids. Inefficient incorporation of novel fatty acids into TAG, as evidenced by their enrichment in acyl-CoA pools, may ultimately induce β-oxidation for the breakdown of these fatty acids. Such a phenomenon has been observed in seeds that have been engineered to produce medium-chain-length fatty acids and epoxy-fatty acids (Eccleston and Ohlrogge, 1998; Moire *et al.*, 2004).

Also poorly characterized is the intracellular organization of enzymes associated with the synthesis and metabolism of unusual fatty acids. These enzymes might be closely associated in specific physical or metabolic domains. The synthesis of petroselinic acid, for example, appears to involve the association of at least three enzymes in a biosynthetic complex or metabolon, and it cannot be ruled out that specialized plastids have evolved for the synthesis of this fatty acid (Cahoon and Ohlrogge, 1994; Schultz *et al.*, 1996). In addition, it is possible that variant FAD2s (e.g., hydroxylases, epoxygenases, conjugases) (Fig. 7.4 and Table 7.5) naturally function in discrete domains of the ER that contain specialized forms of acyltransferases and phospholipases that efficiently metabolize novel fatty acids following their synthesis on PC.

In addition, aspects of gene expression and protein accumulation have not been well characterized in most attempts to produce unusual fatty acids in seeds of transgenic plants. Promoters for seed storage protein genes are typically used to mediate the expression of transgenes in these experiments. Such promoters might not provide the proper timing or levels of gene expression in the engineered seeds, particularly when compared to seeds that naturally accumulate large amounts of unusual fatty acids. It is also possible that certain enzymes associated with unusual fatty acid synthesis and metabolism are prone to high rates of turnover in transgenic plants, which may affect levels of unusual fatty acid accumulation.

Clearly, a basic understanding of the underlying enzymology, cell biology, and gene expression associated with unusual fatty acid synthesis and metabolism is essential in facilitating efforts to produce novel vegetable oils in existing crop plants, while maintaining the agronomic viability of the engineered seeds.

ACKNOWLEDGEMENTS

We thank Dr. Jan Miernyk for his insightful comments and critical reading of the text.

REFERENCES

Abbadi, A., Domergue, F., Bauer, J., Napier, J. A., Welti, R., Zähringer, U., Cirpus, P., and Heinz, E. (2004). Biosynthesis of very-long-chain fatty acids in transgenic oilseeds: Constraints on their accumulation. *Plant Cell* **16,** 2734–2748.

Álvarez-Ortega, R., Cantisán, S., Martínez-Force, E., and Garcés, R. (1997). Characterization of polar and nonpolar seed lipid classes from highly saturated fatty acid sunflower mutants. *Lipids* **32,** 833–837.

American Soybean Association (2007). Soy Stats^{TM}, A Reference Guide to Important Soybean Facts & Figures, 2007. http://www.soystats.com

Auld, D. L., Heikkinen, M. K., Erickson, D. A., Sernyk, J. L., and Romero, J. E. (1992). Rapeseed mutants with reduced levels of polyunsaturated fatty acids and increased levels of oleic acid. *Crop Sci.* **32,** 657–662.

Banaś, A., Dahlqvist, A., Ståhl, U., Lenman, M., and Stymne, S. (2000). The involvement of phospholipid: diacylglycerol acyltransferases in triacylglycerol production. *Biochem. Soc. Trans.* **28,** 703–705.

Barre, D. E. (2001). Potential of evening primrose, borage, black currant, and fungal oils in human health. *Ann. Nutr. Metab.* **45,** 47–57.

Beaudoin, F., Michaelson, L. V., Hey, S. J., Lewis, M. J., Shewry, P. R., Sayanova, O., and Napier, J. A. (2000). Heterologous reconstitution in yeast of the polyunsaturated fatty acid biosynthetic pathway. *Proc. Natl. Acad. USA* **97,** 6421–6426.

Beisson, F., Soo, A. J. K., Russka, S., Schwender, J., Pollard, M., Thelen, J. J., Paddock, T., Salas, J. J., Savage, L., Milcamps, A., Mhaske, V. B., Cho, Y., *et al.* (2003). *Arabidopsis* genes involved in acyl lipid metabolism. A 2003 census of the candidates, a study of the distribution of expressed sequence tags in organs, and a web-based database. *Plant Physiol.* **132,** 681–697.

Blonde, J. D., and Plaxton, W. C. (2003). Structural and kinetic properties of high and low molecular mass phosphoenolpyruvate carboxylase isoforms from the endosperm of developing castor oilseeds. *J. Biol. Chem.* **278,** 11867–11873.

Bonaventure, G., Salas, J. J., Pollard, M. R., and Ohlrogge, J. B. (2003). Disruption of the *FATB* gene in *Arabidopsis* demonstrates an essential role of saturated fatty acids in plant growth. *Plant Cell* **15,** 1020–1033.

Bourgis, F., Kader, J.-C., Barret, P., Renard, M., Robinson, D., Robinson, C., Delseny, M., and Roscoe, T. J. (1999). A plastidial lysophosphatidic acid acyltransferase from oilseed rape. *Plant Physiol.* **120,** 913–922.

Brocard-Gifford, I. M., Lynch, T. J., and Finkelstein, R. R. (2003). Regulatory networks in seeds integrating developmental, abscisic acid, sugar, and light signaling. *Plant Physiol.* **131,** 78–92.

Broun, P., Shanklin, J., Whittle, E., and Somerville, C. (1998). Catalytic plasticity of fatty acid modification enzymes underlying chemical diversity of plant lipids. *Science* **282,** 1315–1317.

Brown, A. P., Slabas, A. R., and Denton, H. (2002). Substrate selectivity of plant and microbial lysophosphatidic acid acyltransferases. *Phytochemistry* **61,** 493–501.

Bruner, A. C., Jung, S., Abbott, A. G., and Powell, G. L. (2001). The naturally occurring high oleate oil character in some peanut varieties results from reduced oleoyl-PC desaturase activity from mutation of aspartate 150 to asparagine. *Crop Sci.* **41,** 522–526.

Brück, F. M., Brummel, M., Schuch, R., and Spener, F. (1996). *In vitro* evidence for feed-back regulation of β-ketoacyl-acyl carrier protein synthase III in medium-chain fatty acid biosynthesis. *Planta* **198,** 271–278.

Buhr, T., Sato, S., Ebrahim, F., Xing, A., Zhou, Y., Mathiesen, M., Schweiger, B., Kinney, A., Staswick, P., and Clemente, T. (2002). Ribozyme termination of RNA transcripts down-regulate seed fatty acid genes in transgenic soybean. *Plant J.* **30,** 155–163.

Cahoon, E. B., Cranmer, A. M., Shanklin, J., and Ohlrogge, J. B. (1994). Δ^6 Hexadecenoic acid is synthesized by the activity of a soluble Δ^6 palmitoyl-acyl carrier protein desaturase in *Thunbergia alata* endosperm. *J. Biol. Chem.* **269,** 27519–27526.

Cahoon, E. B., Dietrich, C. R., Meyer, K., Damude, H. G., Dyer, J. M., and Kinney, A. J. (2006). Conjugated fatty acids accumulate to high levels in phospholipids of metabolically engineered soybean and *Arabidopsis* seeds. *Phytochemistry* **67,** 1166–1176.

Cahoon, E. B., and Ohlrogge, J. B. (1994). Metabolic evidence for the involvement of a Δ^4-palmitoyl-acyl carrier protein desaturase in petroselinic acid synthesis in coriander endosperm and transgenic tobacco cells. *Plant Physiol.* **104,** 827–837.

Cahoon, E. B., and Shanklin, J. (2000). Substrate-dependent mutant complementation to select fatty acid desaturase variants for metabolic engineering of plant seed oils. *Proc. Natl. Acad. Sci. USA* **97,** 12350–12355.

Cahoon, E. B., Shockey, J. M., Dietrich, C. R., Gidda, S. K., Mullen, R. T., and Dyer, J. M. (2007). Engineering oilseeds for sustainable production of industrial and nutritional feedstocks: Solving bottlenecks in fatty acid flux. *Curr. Opin. Plant. Biol.* **10,** 236–244.

Cahoon, E. B., Lindqvist, Y., Schneider, G., and Shanklin, J. (1997). Redesign of soluble fatty acid desaturases from plants for altered substrate specificity and double bond position. *Proc. Natl. Acad. Sci. USA* **94,** 4872–4877.

Cahoon, E. B., Shah, S., Shanklin, J., and Browse, J. (1998). A determinant of substrate specificity predicted from the acyl-acyl carrier protein desaturase of cat's claw seed. *Plant Physiol.* **117,** 593–598.

Cahoon, E. B., Carlson, T. J., Ripp, K. G., Schweiger, B. J., Cook, G. A., Hall, S. E., and Kinney, A. J. (1999). Biosynthetic origin of conjugated double bonds: Production of fatty acid components of high-value drying oils in transgenic soybean embryos. *Proc. Natl. Acad. Sci. USA* **96,** 12395–12940.

Cahoon, E. B., Marillia, E.-F., Stecca, K. L., Hall, S. E., Taylor, D. C., and Kinney, A. J. (2000). Production of fatty acid components of meadowfoam oil in somatic soybean embryos. *Plant Physiol.* **124,** 243–251.

Cahoon, E. B., Ripp, K. G., Hall, S. E., and Kinney, A. J. (2001). Formation of conjugated Δ^8, Δ^{10}-double bonds by Δ^{12}-oleic acid desaturase related enzymes: Biosynthetic origin of calendic acid. *J. Biol. Chem.* **276,** 2637–2643.

Cahoon, E. B., Ripp, K. G., Hall, S. E., and McGonigle, B. (2002). Transgenic production of epoxy fatty acids by expression of a cytochrome P450 enzyme from *Euphorbia lagascae* seed. *Plant Physiol.* **128,** 615–624.

Cahoon, E. B., Schnurr, J. A., Huffman, E. A., and Minto, R. E. (2003). Fungal responsive fatty acid acetylenases occur widely in evolutionarily distant plant families. *Plant J.* **34,** 671–683.

Carver, B. F., Burton, J. W., Carter, T. E., Jr., and Wilson, R. F. (1986). Response to environmental variation of soybean lines selected for altered unsaturated fatty acid composition. *Crop Sci.* **26,** 1176–1181.

Cernac, A., and Benning, C. (2004). *WRINKLED1* encodes an AP2/EREB domain protein involved in the control of storage compound biosynthesis in *Arabidopsis*. *Plant J.* **40,** 575–585.

Cheesbrough, T. M. (1989). Changes in the enzymes for fatty acid synthesis and desaturation during acclimation of developing soybean seeds to altered growth temperature. *Plant Physiol.* **90,** 760–764.

Cronan, J. E., Jr., and Subrahmanyam, S. (1998). FadR, transcriptional co-ordination of metabolic expediency. *Mol. Microbiol.* **29,** 937–943.

Cronan, J. E., Jr., and Waldrop, G. L. (2002). Multi-subunit acetyl-CoA carboxylases. *Prog. Lipid Res.* **41,** 407–435.

Dahlqvist, A., Ståhl, U., Lenman, M., Banas, A., Lee, M., Sandager, L., Ronne, H., and Stymne, S. (2000). Phospholipid:diacylglycerol acyltransferase: An enzyme that catalyzes the acyl-CoA-independent formation of triacylglycerol in yeast and plants. *Proc. Natl. Acad. Sci. USA* **97,** 6487–6492.

Das, T., Huang, Y.-S., and Mukerji, P. (2000). Δ6-desaturase and γ-linolenic acid biosynthesis: A biotechnology perspective. *In* "Gamma-Linolenic Acid: Recent Advances in Biotechnology and Clinical Applications" (Y.-S. Huang and V. A. Ziboh, eds.), pp. 6–23. AOCS Press, Champaign, IL.

Dehesh, K., Edwards, P., Fillatti, J., Slabaugh, M., and Byrne, J. (1998). KAS IV: A 3-ketoacyl-ACP synthase from *Cuphea sp.* is a medium chain specific condensing enzyme. *Plant J.* **15,** 383–390.

Dehesh, K., Tai, H., Edwards, P., Byrne, J., and Jaworski, J. G. (2001). Overexpression of 3-ketoacyl-acyl-carrier protein synthase IIIs in plants reduces the rate of lipid synthesis. *Plant Physiol.* **125,** 1103–1114.

Domergue, F., Abbadi, A., Ott, C., Zank, T. K., Zähringer, U., and Heinz, E. (2003). Acyl carriers used as substrates by the desaturases and elongases involved in very long-chain polyunsaturated fatty acids biosynthesis reconstituted in yeast. *J. Biol. Chem.* **278,** 35115–35126.

Dörmann, P., Voelker, T. A., and Ohlrogge, J. B. (2000). Accumulation of palmitate in *Arabidopsis* mediated by the acyl-acyl carrier protein thioesterase FATB1. *Plant Physiol.* **123,** 637–643.

Dyer, J. M., Chapital, D. C., Kuan, J. C., Mullen, R. T., Turner, C., Mckeon, T. A., and Pepperman, A. B. (2002). Molecular analysis of a bifunctional fatty acid conjugase/desaturase from tung. Implications for the evolution of plant fatty acid diversity. *Plant Physiol.* **130,** 2027–2038.

Eccleston, V. S., and Ohlrogge, J. B. (1998). Expression of lauroyl-acyl carrier protein thioesterases in *Brassica napus* seeds induces pathways for both fatty acid oxidation and biosynthesis and implies a set point for triacylglycerol accumulation. *Plant Cell* **10,** 613–621.

Engeseth, N., and Stymne, S. (1996). Desaturation of oxygenated fatty acids in *Lesquerella* and other oil seeds. *Planta* **198,** 238–245.

Fernández-Moya, V., Martínez-Force, E., and Garcés, R. (2002). Temperature effect on a high stearic acid sunflower mutant. *Phytochemistry* **59,** 33–37.

Focks, N., and Benning, C. (1998). *wrinkled1*: A novel, low-seed-oil mutant of *Arabidopsis* with a deficiency in the seed-specific regulation of carbohydrate metabolism. *Plant Physiol.* **118,** 91–101.

Frentzen, M. (1998). Acyltransferases from basic science to modified seed oils. *Fett-Lipid* **100,** 161–166.

García-Maroto, F., Garrido-Cárdenas, J. A., Rodríguez-Ruiz, J., Vilches-Ferrón, M., Adam, A. C., Polaina, J., and Alonso, D. L. (2002). Cloning and molecular characterization of the delta6-desaturase from two *Echium* plant species: Production of GLA by heterologous expression in yeast and tobacco. *Lipids* **37,** 417–426.

Ghanevati, M., and Jaworski, J. G. (2002). Engineering and mechanistic studies of the *Arabidopsis* FAE1 β-ketoacyl-CoA synthase, FAE1 KCS. *Eur. J. Biochem.* **269,** 3531–3539.

Girke, T., Schmidt, H., Zähringer, U., Reski, R., and Heinz, E. (1998). Identification of a novel Δ6-acyl-group desaturase by targeted gene disruption in *Physcomitrella patens*. *Plant J.* **15,** 39–48.

Gunstone, F. D., Harwood, J. L., and Padley, F. B. (1994). The Lipid Handbook, Second Edition, Chapman & Hall, London.

Hawkins, D. J., and Kridl, J. C. (1998). Characterization of acyl-ACP thioesterases of mangosteen (*Garcinia mangostana*) seed and high levels of stearate production in transgenic canola. *Plant J.* **13,** 743–752.

Heath, R. J., and Rock, C. O. (1996). Inhibition of β-keto-acyl carrier protein synthase III (FabH) by acyl-acyl carrier protein in *Escherichia coli.*. *J. Biol. Chem.* **271,** 10996–11000.

Heilmann, I., Pidkowich, M. S., Girke, T., and Shanklin, J. (2004). Switching desaturase enzyme specificity by alternate subcellular targeting. *Proc. Natl. Acad. Sci. USA* **101,** 10266–10271.

Heppard, E. P., Kinney, A. J., Stecca, K. L., and Miao, G.-H. (1996). Developmental and growth temperature regulation of two differential microsomal ω-6 desaturase genes in soybeans. *Plant Physiol.* **110,** 311–319.

Hoang, C. V., and Chapman, K. D. (2002). Biochemical and molecular inhibition of plastidial carbonic anhydrase reduces the incorporation of acetate into lipids in cotton embryos and tobacco cell suspensions and leaves. *Plant Physiol.* **128,** 1417–1427.

Hong, H., Datla, N., Reed, D. W., Covello, P. S., MacKenzie, S. L., and Qiu, X. (2002). High-level production of γ-linolenic acid in *Brassica juncea* using a Δ6 desaturase from *Pythium irregulare. Plant Physiol.* **129,** 354–362.

Hornung, E., Pernstich, C., and Feussner, I. (2002). Formation of conjugated Δ^{11} Δ^{13}-double bonds by Δ^{12}-linoleic acid (1,4)-acyl-lipid-desaturase in pomegranate seeds. *Eur. J. Biochem.* **269,** 4852–4859.

Hu, F. B., Manson, J. E., and Willett, W. C. (2001). Types of dietary fat and risk of coronary heart disease: A critical review. *J. Am. Coll. Nutr.* **20,** 5–19.

Huang, A. H. C. (1996). Oleosins and oil bodies in seeds and other organs. *Plant Physiol.* **110,** 1055–1061.

Ichihara, K., Murota, N., and Fujii, S. (1990). Intracellular translocation of phosphatidate phosphatase in maturing safflower seeds: a possible mechanism of feedforward control of triacylglycerol synthesis by fatty acids. *Biochim. Biophys. Acta* **1043,** 227–234.

Jain, R. K., Coffey, M., Lai, K., Kumar, A., and MacKenzie, S. L. (2000). Enhancement of seed oil content by expression of glycerol-3-phosphate acyltransferase genes. *Biochem. Soc. Trans.* **28,** 958–961.

Jako, C., Kumar, A., Wei, Y., Zou, J., Barton, D. L., Giblin, E. M., Covello, P. S., and Taylor, D. C. (2001). Seed-specific over-expression of an *Arabidopsis* cDNA encoding a diacylglycerol acyltransferase enhances seed oil content and seed weight. *Plant Physiol.* **126,** 861–874.

Jandacek, R. J. (1992). Commercial applications of fatty acid derivatives in food. *In* "Fatty Acids in Foods and Their Health Implications," (C. K. Chow, ed.), pp. 399–427. Marcel Dekker, New York.

Jones, A., Davies, H. M., and Voelker, T. A. (1995). Palmitoyl-acyl carrier protein (ACP) thioesterase and the evolutionary origin of plant acyl-ACP thioesterase. *Plant Cell* **7,** 359–371.

Katavic, V., Mietkiewska, E., Barton, D. L., Giblin, E. M., Reed, D. W., and Taylor, D. C. (2002). Restoring enzyme activity in nonfunctional low erucic acid *Brassica napus* fatty acid elongase 1 by a single amino acid substitution. *Eur. J. Biochem.* **269,** 5625–5631.

Ke, J., Wen, T.-N., Nikolau, B. J., and Wurtele, E. S. (2000). Coordinate regulation of the nuclear and plastidic genes coding for the subunits of the heteromeric acetyl-coenzyme A carboxylase. *Plant Physiol.* **122,** 1057–1071.

Kinney, A. J. (1994). Genetic modification of the storage lipids of plants. *Curr. Opin. Biotechnol.* **5,** 144–151.

Kinney, A. J. (1996). Development of genetically engineered soybean oils for food applications. *J. Food Lipids* **3,** 273–292.

Knutzon, D. S., Thompson, G. A., Radke, S. E., Johnson, W. B., Knauf, V. C., and Kridl, J. C. (1992). Modification of *Brassica* seed oil by antisense expression of a stearoyl-acyl carrier protein desaturase gene. *Proc. Natl. Acad. Sci. USA* **89,** 2624–2628.

Knutzon, D. S., Thurmond, J. M., Huang, Y. S., Chaudhary, S., Bobid, E. G., Jr., Chan, G. M., Kirchner, S. J., and Mukerji, P. (1998). Identification of Δ5-desaturase from *Mortierella alpina* by heterologous expression in Baker's yeast and canola. *J. Biol. Chem.* **273,** 29360–29366.

Knutzon, D. S., Hayes, T. R., Wyrick, A., Xiong, H., Maelor Davies, H., and Voelker, T. A. (1999). Lysophosphatidic acid acyltransferase from coconut endosperm mediates the insertion of laurate at the *sn-2* position of triacylglycerols in lauric rapeseed oil and can increase total laurate levels. *Plant Physiol.* **120,** 739–746.

Kozaki, A., Mayumi, K., and Sasaki, Y. (2001). Thiol-disulfide exchange between nuclear-encoded and chloroplast-encoded subunits of pea acetyl-CoA carboxylase. *J. Biol. Chem.* **276,** 39919–39925.

Kubis, S. E., and Rawsthorne, S. (2000). The role of plastidial transporters in developing embryos of oilseed rape (*Brassica napus* L.) for fatty acid synthesis. *Biochem. Soc. Trans.* **28,** 665–666.

Kubow, S. (1996). The influence of positional distribution of fatty acids in native, interesterified and structure-specific lipids on lipoprotein metabolism and atherogenesis. *J. Nutr. Biochem.* **7,** 530–541.

Kunst, L., Browse, J., and Somerville, C. (1988). Altered regulation of lipid biosynthesis in a mutant of *Arabidopsis* deficient in chloroplast glycerol-3-phosphate acyltransferase activity. *Proc. Natl. Acad. Sci. USA* **85,** 4143–4147.

Lardizabal, K. D., Mai, J. T., Wagner, N. W., Wyrick, A., Voelker, T., and Hawkins, D. J. (2001). DGAT2 is a new diacylglycerol acyltransferase gene family. Purification, cloning, and expression in insect cells of two polypeptides from *Mortierella ramanniana* with diacylglycerol acyltransferase activity. *J. Biol. Chem.* **276,** 38862–38869.

Larson, T. R., Edgell, T., Byrne, J., Dehesh, K., and Graham, I. A. (2002). Acyl CoA profiles of transgenic plants that accumulate medium-chain fatty acids indicate inefficient storage lipid synthesis in developing oilseeds. *Plant J.* **32,** 519–527.

Lassner, M. W., Levering, C. K., Davies, H. M., and Knutzon, D. S. (1995). Lysophosphatidic acid acyltransferase from meadowfoam mediates insertion of erucic acid at the *sn*-2 position of triacylglycerol in transgenic rapeseed oil. *Plant Physiol.* **109,** 1389–1394.

Lassner, M. W., Lardizabal, K., and Metz, J. G. (1996). A jojoba β-ketoacyl-CoA synthase cDNA complements the canola fatty acid elongation mutation in transgenic plants. *Plant Cell* **8,** 281–292.

Lee, J. M., Williams, M. E., Tingey, S. V., and Rafalski, J. A. (2002). DNA array profiling of gene expression changes during maize embryo development. *Funct. Integr. Genomics* **2,** 13–27.

Lee, M., Lenman, M., Banaś, A., Bafor, M., Singh, S., Schweizer, M., Nilsson, R., Liljenberg, C., Dahlqvist, A., Gummeson, P.-O., Sjödahl, S., Green, A., *et al.* (1998). Identification of non-heme diiron proteins that catalyze triple bond and epoxy group formation. *Science* **280,** 915–918.

Leonard, J. M., Knapp, S. J., and Slabaugh, M. B. (1998). A *Cuphea* β-ketoacyl-ACP synthase shifts the synthesis of fatty acids towards shorter chains in *Arabidopsis* seeds expressing *Cuphea* FatB thioesterases. *Plant J.* **13,** 621–628.

Li, Z., Wilson, R. F., Rayford, W. E., and Boerma, H. R. (2002). Molecular mapping genes conditioning reduced palmitic acid content in N87–2122–4 soybean. *Crop Sci.* **42,** 373–378.

Lindqvist, Y., Huang, W., Schneider, G., and Shanklin, J. (1996). Crystal structure of Δ9 stearoyl-acyl carrier protein desaturase from castor seed and its relationship to other di-iron proteins. *EMBO J.* **15,** 4081–4092.

Liu, Q., Singh, S. P., and Green, A. G. (2002). High-stearic and high-oleic cottonseed oils produced by hairpin RNA-mediated post-transcriptional gene silencing. *Plant Physiol.* **129,** 1732–1743.

Madoka, Y., Tomizawa, K.-I., Mizoi, J., Nishida, I., Nagano, Y., and Sasaki, Y. (2002). Chloroplast transformation with modified *accD* operon increases acetyl-CoA carboxylase and causes extension of leaf longevity and increase in seed yield in tobacco. *Plant Cell Physiol.* **43,** 1518–1525.

Manaf, A. M., and Harwood, J. L. (2000). Purification and characterisation of acyl-CoA: Glycerol 3-phosphate acyltransferase from oil palm (*Elaeis guineensis*) tissues. *Planta* **210,** 318–328.

Marillia, E.-F., Giblin, E. M., Covello, P. S., and Taylor, D. C. (2002). A desaturase-like protein from white spruce is a Δ^9 desaturase. *FEBS Lett.* **526,** 49–52.

Metz, J. G., Roessler, P., Facciotti, D., Levering, C., Dittrich, F., Lassner, M., Valentine, R., Lardizabal, K., Domergue, F., Yamada, A., Yazawa, K., Knauf, V., *et al.* (2001). Production of polyunsaturated fatty acids by polyketide synthases in both prokaryotes and eukaryotes. *Science* **293,** 290–293.

Meyer, A., Cirpus, P., Ott, C., Schlecker, R., Zähringer, U., and Heinz, E. (2003). Biosynthesis of docosahexaenoic acid in *Euglena gracilis*: Biochemical and molecular evidence for the involvement of a Δ4-fatty acyl group desaturase. *Biochemistry* **42,** 9779–9788.

Michaelson, L. V., Lazarus, C. M., Griffiths, G., Napier, J. A., and Stobart, A. K. (1998). Isolation of a Δ^5-fatty acid desaturase gene from *Mortierella alpina. J. Biol. Chem.* **273,** 19055–19059.

Millar, A. A., and Kunst, L. (1997). Very-long-chain fatty acid biosynthesis is controlled through the expression and specificity of the condensing enzyme. *Plant J.* **12,** 121–131.

Millar, A. A., Wrischer, M., and Kunst, L. (1998). Accumulation of very-long-chain fatty acids in membrane glycerolipids is associated with dramatic alterations in plant morphology. *Plant Cell* **10,** 1889–1902.

Millar, A. A., Clemens, S., Zachgo, S., Giblin, E. M., Taylor, D. C., and Kunst, L. (1999). *CUT1*, an *Arabidopsis* gene required for cuticular wax biosynthesis and pollen fertility, encodes a very-long-chain fatty acid condensing enzyme. *Plant Cell* **11,** 825–838.

Miquel, M. F., and Browse, J. A. (1994). High-oleate oilseeds fail to develop at low temperature. *Plant Physiol.* **106,** 421–427.

Moire, L., Rezzonico, E., Goepfert, S., and Poirier, Y. (2004). Impact of unusual fatty acid synthesis on futile cycling through β-oxidation and on gene expression in transgenic plants. *Plant Physiol.* **134,** 432–442.

Moon, H., Hazebroek, J., and Hildebrand, D. F. (2000). Changes in fatty acid composition in plant tissues expressing a mammalian Δ9 desaturase. *Lipids.* **35,** 471–479.
Moon, H., Smith, M. A., and Kunst, L. (2001). A condensing enzyme from the seeds of *Lesquerella fendleri* that specifically elongates hydroxy fatty acids. *Plant Physiol.* **127,** 1635–1643.
Moreau, R. A., Pollard, M. R., and Stumpf, P. K. (1981). Properties of a Δ5 fatty acyl-CoA desaturase in the cotyledons of developing *Limnanthes alba. Arch. Biochem. Biophys.* **209,** 376–384.
Murata, N., and Tasaka, Y. (1997). Glycerol-3-phosphate acyltransferase in plants. *Biochim. Biophys. Acta* **1348,** 10–16.
Murphy, D. J. (2001). The biogenesis and functions of lipid bodies in animals, plants and microorganisms. *Prog. Lipid Res.* **40,** 325–438.
Napier, J. A. (2007). The production of unusual fatty acids in plants. *Annu. Rev. Plant Biol.* **58,** 295–319.
Napier, J. A., Michaelson, L. V., and Sayanova, O. (2003). The role of cytochrome b_5 fusion desaturases in the synthesis of polyunsaturated fatty acids. *Prostaglandins Leukot. Essent. Fatty Acids* **68,** 135–143.
Nikolau, B. J., Ohlrogge, J. B., and Wurtele, E. S. (2003). Plant biotin-containing carboxylases. *Arch. Biochem. Biophys.* **414,** 211–222.
Norden, A. J., Gorbet, D. W., Knauft, D. A., and Young, C. T. (1987). Variability in oil quality among peanut genotypes on the Florida breeding program. *Peanut Sci.* **14,** 7–11.
Nykiforuk, C. L., Furukawa-Stoffer, T. L., Huff, P. W., Sarna, M., Laroche, A., Moloney, M. M., and Weselake, R. J. (2002). Characterization of cDNAs encoding diacylglycerol acyltransferase from cultures of *Brassica napus* and sucrose-mediated induction of enzyme biosynthesis. *Biochim. Biophys. Acta* **1580,** 95–109.
Oelkers, P., Cromley, D., Padamsee, M., Billheimer, J. T., and Sturley, S. L. (2002). The *DGA1* gene determines a second triglyceride synthetic pathway in yeast. *J. Biol. Chem.* **277,** 8877–8881.
Ogas, J., Cheng, J.-C., Sung, Z. R., and Somerville, C. (1997). Cellular differentiation regulated by gibberellin in the *Arabidopsis thaliana pickle* mutant. *Science* **277,** 91–94.
Ogas, J., Kaufmann, S., Henderson, J., and Somerville, C. (1999). PICKLE is a CHD3 chromatin-remodeling factor that regulates the transition from embryonic to vegetative development in *Arabidopsis. Proc. Natl. Acad. Sci. USA* **96,** 13839–13844.
O'Hara, P., Slabas, A. R., and Fawcett, T. (2002). Fatty acid and lipid biosynthetic genes are expressed at constant molar ratios but different absolute levels during embryogenesis. *Plant Physiol.* **129,** 310–320.
Parker-Barnes, J. M., Das, T., Bobik, E., Leonard, A. E., Thurmond, J. M., Chaung, L.-T., Huang, Y.-S., and Mukerji, P. (2000). Identification and characterization of an enzyme involved in the elongation of n-6 and n-3 polyunsaturated fatty acids. *Proc. Natl. Acad. Sci. USA* **97,** 8284–8289.
Qi, B., Fraser, T., Mugford, S., Dobson, G., Sayanova, O., Napier, J. A., Stobart, A. K., and Lazarus, C. M. (2004). Production of very long chain polyunsaturated omega-3 and omega-6 fatty acids in plants. *Nat. Biotechnol.* **22,** 739–745.
Qiu, X., Hong, H., and MacKenzie, S. L. (2001a). Identification of a Δ4 fatty acid desaturase from *Thraustochytrium* sp. involved in the biosynthesis of docosahexanoic acid by heterologous expression in *Saccharomyces cerevisiae* and *Brassica juncea. J. Biol. Chem.* **276,** 31561–31566.
Qiu, X., Reed, D. W., Hong, H., MacKenzie Rangasamy, S. L., and Covello, P. S. (2001b). Identification and analysis of a gene from *Calendula officinalis* encoding a fatty acid conjugase. *Plant Physiol.* **125,** 847–855.
Rangasamy, D., and Ratledge, C. (2000). Genetic enhancement of fatty acid synthesis by targeting rat liver ATP: citrate lyase into plastids of tobacco. *Plant Physiol.* **122,** 1231–1238.
Rawsthorne, S. (2002). Carbon flux and fatty acid synthesis in plants. *Prog. Lipid Res.* **41,** 182–196.
Reddy, A. S., and Thomas, T. L. (1996). Expression of a cyanobacterial delta 6-desaturase gene results in gamma-linolenic acid production in transgenic plants. *Nat. Biotechnol.* **14,** 639–642.
Reed, D. W., Savile, C. K., Qiu, X., Buist, P. H., and Covello, P. S. (2002). Mechanism of 1,4-dehydrogenation catalyzed by a fatty acid (1,4)-desaturase of *Calendula officinalis. Eur. J. Biochem.* **269,** 5024–5029.
Rider, S. D., Jr., Henderson, J. T., Jerome, R. E., Edenberg, H. J., Romero-Severson, J., and Ogas, J. (2003). Coordinate repression of regulators of embryonic identity by *PICKLE* during germination in *Arabidopsis. Plant J.* **35,** 33–43.

Robert, S. S., Singh, S. P., Zhou, X.-R., Petrie, J. R., Blackburn, S. I., Mansour, P. M., Nichols, P. D., Liu, Q., and Green, A. G. (2005). Metabolic engineering of *Arabidopsis* to produce nutritionally important DHA in seed oil. *Funct. Plant Biol.* **32,** 473–479.

Rodriguez-Sotres, R., and Black, M. (1994). Osmotic potential and abscisic acid regulate triacylglycerol synthesis in developing wheat embryos. *Planta* **192,** 9–15.

Roesler, K., Shintani, D., Savage, L., Boddupalli, S., and Ohlrogge, J. (1997). Targeting of the *Arabidopsis* homomeric acetyl-coenzyme A carboxylase to plastids of rapeseeds. *Plant Physiol.* **113,** 75–81.

Ross, A. J., Fehr, W. R., Welke, G. A., and Cianzio, S. R. (2000). Agronomic and seed traits of 1%-linolenate soybean genotypes. *Crop Sci.* **40,** 383–386.

Ruuska, S. A., Girke, T., Benning, C., and Ohlrogge, J. B. (2002). Contrapuntal networks of gene expression during *Arabidopsis* seed filling. *Plant Cell* **14,** 1191–1206.

Salas, J. J., and Ohlrogge, J. B. (2002). Characterization of substrate specificity of plant FatA and FatB acyl-ACP thioesterases. *Arch. Biochem. Biophys.* **403,** 25–34.

Sato, S., Xing, A., Ye, X., Schweiger, B., Kinney, A., Graef, G., and Clemente, T. (2004). Production of γ-linolenic acid and stearidonic acid in seeds of marker-free transgenic soybean. *Crop Sci.* **44,** 646–652.

Sayanova, O., Smith, M. A., Lapinskas, P., Stobart, A. K., Dobson, G., Christie, W. W., Shewry, P. R., and Napier, J. A. (1997). Expression of a borage desaturase cDNA containing an N-terminal cytochrome b_5 domain results in the accumulation of high levels of Δ^6-desaturated fatty acids in transgenic tobacco. *Proc. Natl. Acad. Sci. USA* **94,** 4211–4216.

Schnebly, S. R., Fehr, W. R., Welke, G. A., Hammond, E. G., and Duvick, D. N. (1994). Inheritance of reduced and elevated palmitate in mutant lines of soybean. *Crop Sci.* **34,** 829–833.

Schnurr, J. A., Shockey, J. M., De Boer, G.-J., and Browse, J. A. (2002). Fatty acid export from the chloroplast. Molecular characterization of a major plastidial acyl-coenzyme A synthetase from *Arabidopsis. Plant Physiol.* **129,** 1700–1709.

Schultz, D. J., Cahoon, E. B., Shanklin, J., Craig, R., Cox-Foster, D. L., Mumma, R. O., and Medford, J. I. (1996). Expression of a Δ^9 14:0-acyl carrier protein fatty acid desaturase gene is necessary for the production of ω^5 anacardic acids found in pest-resistant geranium (*Pelargonium xhortorum*). *Proc. Natl. Acad. Sci. USA* **93,** 8771–8775.

Schütt, B. S., Abbadi, A., Loddenkötter, B., Brummel, M., and Spener, F. (2002). β-ketoacyl-acyl carrier protein synthase IV: A key enzyme for regulation of medium-chain fatty acid synthesis in *Cuphea lanceolata* seeds. *Planta* **215,** 847–854.

Schwartzbeck, J. L., Jung, S., Abbott, A. G., Mosley, E., Lewis, S., Pries, G. L., and Powell, G. L. (2001). Endoplasmic oleoyl-PC desaturase references the second double bond. *Phytochemistry* **57,** 643–652.

Schwender, J., and Ohlrogge, J. B. (2002). Probing *in vivo* metabolism by stable isotope labeling of storage lipids and proteins in developing *Brassica napus* embryos. *Plant Physiol.* **130,** 347–361.

Schwender, J., Ohlrogge, J. B., and Shachar-Hill, Y. (2003). A flux model of glycolysis and the oxidative pentosephosphate pathway in developing *Brassica napus* embryos. *J. Biol. Chem.* **278,** 29442–29453.

Shanklin, J., and Cahoon, E. B. (1998). Desaturation and related modifications of fatty acids. *Annu. Rev. Plant Physiol. Plant Mol. Biol.* **49,** 611–641.

Shintani, D. K., and Ohlrogge, J. B. (1995). Feedback inhibition of fatty acid synthesis in tobacco suspension cells. *Plant J.* **7,** 577–587.

Shintani, D., Roesler, K., Shorrosh, B., Savage, L., and Ohlrogge, J. (1997). Antisense expression and overexpression of biotin carboxylase in tobacco leaves. *Plant Physiol.* **114,** 881–886.

Slabas, A. R., White, A., O'Hara, P., and Fawcett, T. (2002). Investigations into the regulation of lipid biosynthesis in *Brassica napus* using antisense down-regulation. *Biochem. Soc. Trans.* **30,** 1056–1059.

Smith, N. A., Singh, S. P., Wang, M.-B., Stoutjesdijik, P. A., Green, A. G., and Waterhouse, P. M. (2000). Totally silencing by intron-spliced hairpin RNAs. *Nature* **407,** 319–320.

Soldatov, K. I. (1976). "Chemical mutagenesis in sunflower breeding," Proceedings of the Seventh International Sunflower Association, Krasnodar, U.S.S.R., International Sunflower Association, Vlaardingen, The Netherlands, pp. 352–357.

Sperling, P., Zähringer, U., and Heinz, E. (1998). A sphingolipid desaturase from higher plants. Identification of a new cytochrome b_5 fusion protein. *J. Biol. Chem.* **273,** 28590–28596.

Ståhl, U., Carlsson, A. S., Lenman, M., Dahlqvist, A., Huang, B., Banaś, W., Banaś, A., and Stymne, S. (2004). Cloning and functional characterization of a phospholipid: diacylglycerol acyltransferase from *Arabidopsis*. *Plant Physiol.* **135,** 1324–1335.

Suh, M. C., Schultz, D. J., and Ohlrogge, J. B. (2002). What limits production of unusual monoenoic fatty acids in transgenic plants? *Planta* **215,** 584–595.

Tang, G.-Q., Novitzky, W. P., Griffin, H. C., Huber, S. C., and Dewey, R. E. (2005). Oleate desaturase enzymes of soybean: Evidence of regulation through differential stability and phosphorylation. *Plant J.* **44,** 433–446.

Thelen, J. J., and Ohlrogge, J. B. (2002). Both antisense and sense expression of biotin carboxyl carrier protein isoform 2 inactivates the plastid acetyl-coenzyme A carboxylase in *Arabidopsis thaliana. Plant J.* **32,** 419–431.

Thomaeus, S., Carlsson, A. S., and Stymne, S. (2001). Distribution of fatty acids in polar and neutral lipids during seed development in *Arabidopsis thaliana* genetically engineered to produce acetylenic, epoxy and hydroxy fatty acids. *Plant Sci.* **161,** 997–1003.

Toke, D. A., and Martin, C. E. (1996). Isolation and characterization of a gene affecting fatty acid elongation in *Saccharomyces cerevisiae. J. Biol. Chem.* **271,** 18413–18422.

Uauy, R., Hoffman, D. R., Mena, P., Llanos, A., and Birch, E. E. (2003). Term infant studies of DHA and ARA supplementation on neurodevelopment: Results of randomized controlled trials. *J. Pediatr.* **143,** S17–S25.

United States Department of Agriculture Foreign Agricultural Service (2007). Oilseeds: World Markets and Trade, Circular Series FOP 07–07, July 2007. http://www.fas.usda.gov/psdonline/circulars/oilseeds.pdf

van de Loo, F. J., Broun, P., Turner, S., and Somerville, C. (1995). An oleate 12-hydroxylase from *Ricinus communis* L. is a fatty acyl desaturase homolog. *Proc. Natl. Acad. Sci. USA* **92,** 6743–6747.

Voelker, T., and Kinney, A. J. (2001). Variations in the biosynthesis of seed-storage lipids. *Annu. Rev. Plant Physiol. Plant Mol. Biol.* **52,** 335–361.

Voelker, T. A., Worrell, A. C., Anderson, L., Bleibaum, J., Fan, C., Hawkins, D. J., Radke, S. E., and Davies, H. M. (1992). Fatty acid biosynthesis redirected to medium chains in transgenic oilseed plants. *Science* **257,** 72–74.

Voelker, T. A., Jones, A., Cranmer, A. M., Davies, H. M., and Knutzon, D. S. (1997). Broad-range and binary-range acyl-acyl-carrier protein thioesterases suggest an alternative mechanism for medium-chain production in seeds. *Plant Physiol.* **114,** 669–677.

Wiberg, E., Edwards, P., Byrne, J., Stymne, S., and Dehesh, K. (2000). The distribution of caprylate, caprate and laurate in lipids from developing and mature seeds of transgenic *Brassica napus* L. *Planta* **212,** 33–40.

Winter, E., Brummel, M., Schuch, R., and Spener, F. (1997). Decarboxylation of malonyl-(acyl carrier protein) by 3-oxoacyl-(acyl carrier protein) synthases in plant fatty acid biosynthesis. *Biochem. J.* **321,** 313–318.

Wu, G., Truska, M., Datla, N., Vrinten, P., Bauer, J., Zank, T., Cirpus, P., Heinz, E., and Qiu, X. (2005). Stepwise engineering to produce high yields of very long-chain polyunsaturated fatty acids in plants. *Nat. Biotechnol.* **23,** 1013–1017.

Yadav, N. S., Wierzbicki, A., Aegerter, M., Caster, C. S., Pérez-Grau, L., Kinney, A. J., Hitz, W. D., Booth, J. R., Jr., Schweiger, B., Stecca, K. L., Allen, S. M., Blackwell, M., *et al.* (1993). Cloning of higher plant ω-3 fatty acid desaturases. *Plant Physiol.* **103,** 467–476.

Yuan, L., Voelker, T. A., and Hawkins, D. J. (1995). Modification of the substrate specificity of an acyl-acyl carrier protein thioesterase by protein engineering. *Proc. Natl. Acad. Sci.* **92,** 10639–10643.

Zagnitko, O., Jelenska, J., Tevzadze, G., Haselkorn, R., and Gornicki, P. (2001). An isoleucine/leucine residue in the carboxyltransferase domain of acetyl-CoA carboxylase is critical for interaction with aryloxyphenoxypropionate and cyclohexanedione inhibitors. *Proc. Natl. Acad. Sci. USA* **98,** 6617–6622.

Zank, T. K., Zähringer, U., Beckman, C., Pohnert, G., Boland, W., Holtorf, H., Reski, R., Lerchl, J., and Heinz, E. (2002). Cloning and functional characterization of an enzyme involved in the elongation of Δ6-polyunsaturated fatty acids from the moss *Physcomitrella patens*. *Plant J.* **31,** 255–268.

Zheng, Z., Xia, Q., Dauk, M., Shen, W., Selvaraj, G., and Zou, J. (2003). *Arabidopsis AtGPAT1*, a member of the membrane-bound glycerol-3-phosphate acyltransferase gene family, is essential for tapetum differentiation and male fertility. *Plant Cell* **15**, 1872–1887.

Zou, J., Katavic, V., Giblin, E. M., Barton, D. L., MacKenzie, S. L., Keller, W. A., Hu, X., and Taylor, D. C. (1997). Modification of seed oil content and acyl composition in the Brassicaceae by expression of a yeast *sn*-2 acyltransferase gene. *Plant Cell* **9**, 909–923.

Zou, J., Wei, Y., Jako, C., Kumar, A., Selvaraj, G., and Taylor, D. C. (1999). The *Arabidopsis thaliana TAG1* mutant has a mutation in a diacylglycerol acyltransferase gene. *Plant J.* **19**, 645–653.

CHAPTER 8

Pathways for the Synthesis of Polyesters in Plants: Cutin, Suberin, and Polyhydroxyalkanoates

Christiane Nawrath and **Yves Poirier**

Contents

Abstract

Plants naturally produce the lipid-derived polyester cutin, which is found in the plant cuticle that is deposited at the outermost extracellular matrix of the epidermis covering nearly all aboveground tissues. Being at the interface between the cell and the external environment, cutin and the cuticle play important roles in the protection of plants from several stresses. A number of enzymes involved in the synthesis of cutin monomers have recently been identified, including several P450s and one acyl-CoA synthetase, thus representing the first steps toward the understanding of polyester formation and, potentially, polyester engineering to improve the tolerance of plants to stresses, such as drought, and for industrial applications. However, numerous processes underlying cutin synthesis, such as a controlled polymerization, still remain elusive.

Département de Biologie Moléculaire Végétale, Biophore, Université de Lausanne, CH-1015 Lausanne, Switzerland

Advances in Plant Biochemistry and Molecular Biology, Volume 1
ISSN 1755-0408, DOI: 10.1016/S1755-0408(07)01008-9

Suberin is a second polyester found in the extracellular matrix, most often synthesized in root tissues and during secondary growth. Similar to cutin, the function of suberin is to seal off the respective tissue to inhibit water loss and contribute to resistance to pathogen attack. Being the main constituent of cork, suberin is a plant polyester that has already been industrially exploited. Genetic engineering may be worth exploring in order to change the polyester properties for either different applications or to increase cork production in other species.

Polyhydroxyalkanoates (PHAs) are attractive polyesters of 3-hydroxyacids because of their properties as bioplastics and elastomers. Although PHAs are naturally found in a wide variety of bacteria, biotechnology has aimed at producing these polymers in plants as a source of cheap and renewable biodegradable plastics. Synthesis of PHA containing various monomers has been demonstrated in the cytosol, plastids, and peroxisomes of plants. Several biochemical pathways have been modified in order to achieve this, including the isoprenoid pathway, the fatty acid biosynthetic pathway, and the fatty acid β-oxidation pathway. PHA synthesis has been demonstrated in a number of plants, including monocots and dicots, and up to 40% PHA per gram dry weight has been demonstrated in *Arabidopsis thaliana*. Despite some successes, production of PHA in crop plants remains a challenging project. PHA synthesis at high level in vegetative tissues, such as leaves, is associated with chlorosis and reduced growth. The challenge for the future is to succeed in synthesis of PHA copolymers with a narrow range of monomer compositions, at levels that do not compromise plant productivity. This goal will undoubtedly require a deeper understanding of plant biochemical pathways and how carbon fluxes through these pathways can be manipulated, areas where plant "omics" can bring very valuable contributions.

Key Words: *Arabidopsis*, β-oxidation, cuticle, cutin, fatty acid, metabolic engineering, peroxisome, plastid, PHA, PHB, polyester, polyhydroxyalkanoates, polyhydroxybutyrate, suberin.

1. INTRODUCTION

Plants synthesize several classes of hydrophobic biopolyesters. Cutin and suberin, two complex lipid-based polyesters, are unique to the plant kingdom. Cutin is the main part of the cuticle (representing 40–80% of the cuticle) and evolved *circa* 400 million years ago when vascular plants established themselves on dry land and needed a barrier to protect themselves from water loss and various environmental aggressions. Although the structures of cutin and suberin are related, being primarily composed of esterified fatty acid derivatives, several features distinguish them. Notably, cutin forms a continuous layer covering the epidermal cell layer of all aerial portions of the plant, while the deposition of suberin is more diversified, encompassing both roots and aerial organs. Several reviews have been published in the past years on the structure and biochemistry of cutin and suberin (Bernards, 2002; Heredia, 2003; Kolattukudy, 2001; Nawrath, 2002). This review will particularly focus on recent insights on the complex structure and composition of cutin and suberin, as well as report on the advances that have been made to understand their biosynthesis.

The third type of polyester naturally found in plants is polyhydroxybutyrate (PHB), a polymer of 3-hydroxybutyric acid and a member of the family of polyhydroxyalkanoates (PHAs). Although the literature on PHA is primarily focused on the high-molecular-weight polyester produced in bacteria as a carbon reserve that has thermoplastic properties, a low-molecular-weight PHB is also produced in prokaryotes and eukaryotes (Reusch, 1999). This low-molecular-weight PHB, referred to as cPHB, is found in membranes associated with polyphosphate and has been detected in very small quantities in a wide spectrum of organisms, including bacteria, yeast, plants, and animal tissues (Reusch, 1999). The biochemical pathway of cPHB has not been identified and its physiological role remains uncertain, although the polyphosphate/PHB complex has been found to have ion channel properties (Reusch, 1999, 2002). In view of the paucity of information on cPHB associated with plants, this chapter will focus on the synthesis of the high-molecular-weight PHA, hereafter simply referred to as PHA, which has been produced in transgenic plants as a source of renewable and environment-friendly plastics.

Despite the interesting properties of PHAs as biodegradable thermoplastics and elastomers, use of these bacterial polyesters as substitutes for petroleum-derived plastics is limited by the expenses related to bacterial fermentation, making bacterial PHA substantially more expensive than petroleum-based polymers, such as polypropylene. It is in this context that agriculture has been regarded as a promising alternative for the production of PHAs on a large scale and at low cost (Poirier, 1999; Poirier *et al.*, 1995a). Transgenic plants producing different types of PHAs have now been demonstrated in several species and will be described in this chapter. Synthesis of PHA in crops fits into a larger concept of using plants as vectors for the renewable and sustainable synthesis of carbon building blocks that are presently largely provided by the petrochemical industry.

2. CUTIN AND SUBERIN

2.1. Functional and ultrastructural characteristics

Cutin is the main structural component of the multilayered cuticle that covers all epidermal cells of the aerial portions of plants as a continuous extracellular layer of hydrophobic material. Cutin forms, together with the intracuticular waxes in which it is embedded, the so-called cuticle proper that is overlaid by epicuticular waxes (Jeffree, 1996). Waxes are complex mixtures of long-chain fatty acids and their derivatives, but may also contain other embedded compounds, such as triterpenoids and flavonoids. The cuticle proper is linked via a so-called "cuticular layer," also containing polysaccharides, to the cell wall (Jeffree, 1996). In addition to coating the external epidermal surface, the cuticular membrane extends into the substomatal chamber (Esau, 1977). The cuticle plays an important role in protecting plants from physical, chemical, and biological aggressions, for example, ultraviolet (UV) irradiation, mechanical damage, as well as pathogen and insect attack (Kerstiens, 1996). The cuticle also covers the protoderm of the embryo, playing an important role during development in the prevention of organ fusions (Tanaka *et al.*, 2001).

Suberin is constitutively present in the secondary growth periderm of aerial tissues and in several underground tissues, for example, epidermis, hypodermis, peridermis, and the Casparian strips of the root endodermis. It may also be deposited in bundle sheets, the chalazae, and abscission zone during seed development, and in secretory organs as well as fibers. Suberin is also produced at wound sites to replace the missing cuticle (Kolattukudy, 1981). Similar to cutin, the function of suberin is to seal off the respective tissue to inhibit water loss or contribute to resistance to pathogen attack.

Despite functional, structural, and chemical similarities of suberin and cutin, both polymers are characterized by differences in their composition and location within the plant. While cutin is deposited only on the outside of the epidermal cell wall in the cuticle, suberin is deposited as part of the primary cell wall close to the plasma membrane. The ultrastructure of cutin and suberin deposition may also be different. The ultrastructure of cutin may be of amorphous, recticulate, or lamellate appearance depending on the plant tissue (Fig. 8.1). This feature is used for the classification of cutin types (Holloway, 1982). In contrast, the ultrastructure of suberin is very characteristic, having an alternation of lamellae of electron-opaque and electron-translucent materials in transmission electron microscopy (TEM) (Fig. 8.2) (Bernards, 2002; Nawrath, 2002).

2.2. Composition of cutin and suberin

Cutin and suberin consist of fatty acid derivatives, phenolic compounds, and glycerol. In most plants, cutin consists mainly of hydroxy- and epoxy-hydroxy fatty acids of 16 and 18 carbons as well as a very small portion of phenols. In contrast, suberin also contains very long-chain fatty acid derivatives, a high proportion of dicarboxylic acids, and a large fraction of phenols (Kolattukudy, 1981). The main components of suberin are largely defined to different subdomains in the polymer, a polyphenol domain that is part of the primary cell wall and an aliphatic domain close to the plasmalemma (Bernards and Lewis, 1998). It has been suggested that only the aliphatic polyester domain should be called suberin (Graça and Pereira, 1997).

FIGURE 8.1 Ultrastructure of the cuticle of the epidermis of Arabidopsis stems. Cuticle of amorphous appearance (small arrowheads) overlaying the cell wall polysaccharides. Bar = 200 nm.

Cutin can be obtained in relative pure form by separation of the cuticular membrane from the cell wall by enzyme digestion and subsequent solvent extractions to remove the wax fraction. Suberin, as part of the primary cell wall, cannot be isolated in pure form, except from cork oaks (Rocha *et al.*, 2001). The polyesters can be depolymerized by typical procedures cleaving ester bonds, for example, alkaline hydrolysis, transesterification with methanol containing boron trifluoride or sodium methoxide, as well as reductive cleavage with lithium aluminum hydride (Kolattukudy, 1981; Walton and Kolattukudy, 1972). The liberated monomers may either be first methylated or are directly converted into trimethylsilyl derivatives before subjecting them to gas chromatography/mass spectrometry (GC/MS). The monomers are identified by their characteristic fragmentation pattern (Walton and Kolattukudy, 1972).

Cutin may be formed by either hydroxylated C_{16} fatty acids (C_{16} class), or by epoxy or hydroxy C_{18} fatty acids (C_{18} class), with many cuticles having a mixed composition with different proportions of both monomer classes. The characteristic cutin monomers of the C_{16} class are 9,16- or 10,16-dihydroxypalmitic acids. Other C_{16} monomers present in cutin are palmitic acid, ω-hydroxypalmitic acid, and dihydroxypalmitic acid having the mid-chain hydroxy group at other positions. The characteristic monomers of the C_{18} cutin are 9,10,18-trihydroxystearic acid and 9,10-epoxy,18-hydroxystearic acid. Other cutin monomers of this type are stearic acid, ω-hydroxystearic acid, and some unsaturated isologs of these monomers (Kolattukudy, 1981). Minor monomers may also be other fatty acids, fatty alcohols, aldehydes, ketones, dicarboxylic acids as well as hydroxycinnamic acids. However, the cutin of *Arabidopsis* was found to be rich in dicarboxylic acids, in particular unsaturated C_{18}-dicarboxylic acids, and 2-hydroxy acids up to 26 carbons in length, revealing a monomer composition that is closer to that of suberin than to that of a canonical cutin (Bonaventure *et al.*, 2004; Franke *et al.*, 2005; Xiao *et al.*, 2004). Cutin may thus have a larger plasticity in composition within the plant kingdom than earlier expected (Nawrath, 2006). Glycerol is present in cutin to varying amounts between 1% and 14% (Graça *et al.*, 2002). Partial depolymerization by calcium oxide-catalyzed methanolysis led to the identification of 1- and 2-monoacylglyceryl esters (Graça *et al.*, 2002). Interestingly, the different types of glyceryl esters found do not always correspond to the relative proportions of the hydroxylated fatty acids present in the polyester. Some monomers also seem to be excluded from the glyceryl esters' formation, for example, epoxy fatty acids (Graça *et al.*, 2002). Thus, glycerol may contribute substantially to the three-dimensional structure of cutin, implying that the previous models based primarily on the inter-esterification of hydroxy and epoxy-hydroxy fatty acids need to be revised.

A non-hydrolysable core remains after the hydrolysis of cutin. This non-ester fraction contains a network of aliphatic compounds linked by ether bonds in which linolenic acid is preferentially incorporated (Villena *et al.*, 1999). Whether this fraction should still be called cutin or should be named cutan is still under discussion (Kolattukudy, 1996).

Suberin contains significant amounts (roughly one third) of monomeric hydroxycinnamic acids, such as ferulic, cinnamic, *p*-coumaric, or caffeic acids, in addition

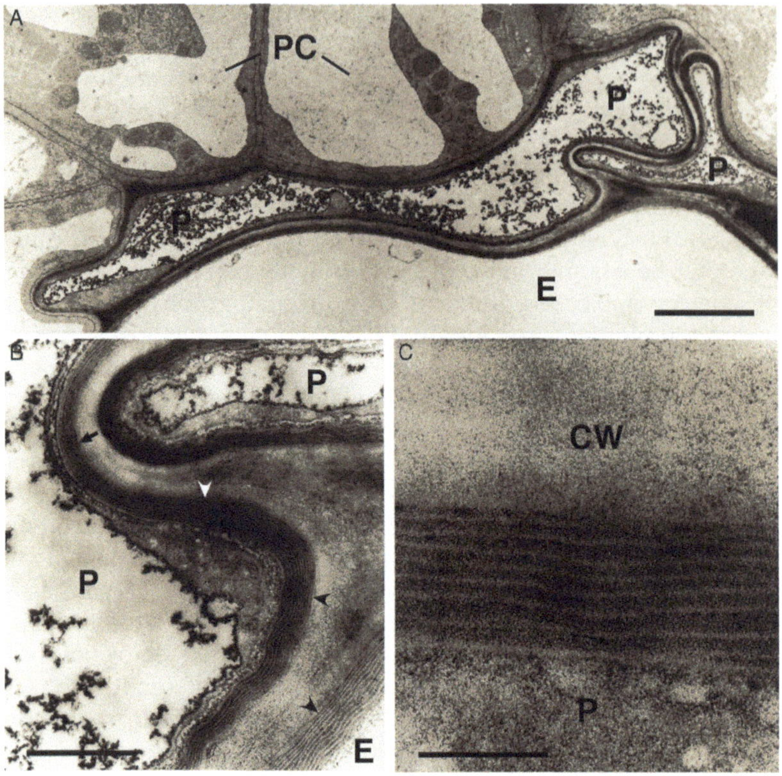

FIGURE 8.2 Ultrastructure of suberized roots tissues of Arabidopsis plants at the beginning of the secondary thickening of the root. (A) Overview of suberized endodermal and peridermal cells in the root. The suberin deposition is visible as an electron-opaque layer inside of the primary cell wall. The fully suberized peridermal cell layer typically collapses during the dehydration and embedding procedures necessary for TEM because of the low permeability of the suberized cell walls. Bar = 2.5 μm. (B) Enlargement of (A). Fine structure of suberin. The structure of the lamellae with an alternation of electron-opaque and electron-translucent layers of suberin is clearly visible when the specimen is cut perpendicularly to the suberin layers (concave arrowheads). However, the lamellate structure of suberin is barely visible when the specimen in not cut perpendicularly to the suberin layers (arrow). Bar = 500 nm. (C) Enlargement of (B). The thickness of the electron-opaque and electron-translucent layers of the suberin is very regular and characteristic for the tissue sample. Bar = 100 nm. P, peridermal cell; E, endodermal cell; PC, pericycle cell; CW, cell wall.

to aliphatic compounds and, in some species, (poly)hydroxycinnamates, like feruloyl tyramine (Bernards, 2002; Bernards and Lewis, 1998; Kolattukudy, 1981; Schreiber *et al.*, 1999). The aliphatic portion of the polymer consists of five dominant substance classes: ω-hydroxy fatty acids (C_{16}–C_{28}), α, ω-dicarboxylic acids (C_{16}–C_{26}), very-long-chain carboxylic acids, primary alcohols (C_{18}–C_{30}), and 2-hydroxy fatty acids (Kolattukudy, 1981; Schreiber *et al.*, 1999). Glycerol is a principal monomer (20%) of suberin in oak, cotton, and potato (Graça and Pereira, 2000a,b; Moire *et al.*, 1999). Partial methanolysis with calcium oxide as catalyst has identified that glycerol may be present as mono-acylglycerol esters of alkanoic acids, α,ω-diacids, and ferulic acid, as well as diglycerol esters being linked to a α,ω-diacid at both ends (Graça and Pereira, 2000b,c). Thus, the hypothesis has been proposed that glycerol and α,ω-diacids may form the backbone of the suberin polymer, implicating that suberin is a poly-(acylglycerol)-polyester (Graça and Pereira, 2000c). Glycerol may also cross-link the aromatic and aliphatic suberin components, while aliphatic and aromatic suberin monomers may only form a linear polymer on their own (Moire *et al.*, 1999). A revised model for suberin has been developed including the new compositional and structural data obtained for potato suberin (Bernards, 2002). That work also gives an excellent overview of the synthesis of the polyphenol domain of suberin, which is not subject of the present chapter (Bernards, 2002).

2.3. Biosynthesis of cutin and suberin

2.3.1. Biosynthesis of the monomers

The aliphatic monomers of cutin and suberin derive from the general fatty acid biosynthetic pathway, that is, from palmitic (16:0), stearic (18:0), and oleic (18:1) acids synthesized in the plastids of the epidermal cell.

The biosynthetic pathway leading to the characteristic cutin monomers had been largely discovered by the group of Kolattukudy in the early 70s (Kolattukudy, 1981). The major cutin monomers are synthesized by multiple hydroxylation and epoxidation reactions. These reactions are catalyzed by oxygen and NADP-dependent enzyme systems that are inhibited by CO, a typical characteristics of cytochrome P450-dependent enzymes. The research on plant cytochrome P450 has advanced much during the recent years (Kahn and Durst, 2000). Different cytochrome P450-dependent enzymes have been characterized that catalyze the internal as well as the ω-hydroxylation of fatty acids (Beneviste *et al.*, 1998; Cabello-Hurtado *et al.*, 1998; Pinot *et al.*, 1992, 1998; Tijet *et al.*, 1998). Several of these cytochrome P450-dependent monooxygenases have been cloned, including CYP86A1, CYP94A1, CYP81B1, CYP86A8 (LCR), and CYP86A2 (ATT1) (Beneviste *et al.*, 1998; Cabello-Hurtado *et al.*, 1998; Tijet *et al.*, 1998; Wellesen *et al.*, 2001; Xiao *et al.*, 2004). A function in cutin biosynthsis has been confirmed for LCR and ATT1 (Yephremov, unpublished results) (Xiao *et al.*, 2004). Mutations in ATT1 of *Arabidsopsis* lead to a 30% loss in cutin and a much looser cuticular ultrastructure (Xiao *et al.*, 2004). Alteration in the monomer composition of residual-bound lipids has been found in *lcr* plants (Yephremov, unpublished results).

A lipoxygenase/peroxygenase/epoxide hydrolase pathway has also been demonstrated for the synthesis of cutin monomers (Blée and Schuber, 1993). A peroxygenase may catalyze a hydroxyperoxide-dependent epoxidation of unsaturated fatty acids after the action of a lipoxygenase (Blée and Schuber, 1993, 1990; Hamberg and Hamberg, 1990). The *cis*-epoxy group formed by the peroxygenase may then be hydrated in the *trans* position by an epoxide hydrolase, resulting in a *threo*-diol in mid-chain position of the cutin monomers (Blée and Schuber, 1992, 1995; Pinot *et al.*, 1997; Morisseau *et al.*, 2000).

For formation of the cutin monomers, fatty acids leave the plastid after release from the fatty acid synthetase since cytochrome P450-dependent enzymes are located at the endoplasmic reticulum (ER) membrane. Precursors for the unsaturated cutin monomers of *Arabidopsis* are provided by phospholipids of the ER (Bonaventure *et al.*, 2004). Further details on the mechanism of the hydroxylation reactions have not yet been elucidated, that is, the order of hydroxylations and the substrates for the different enzymes *in vivo* or other enzymes and cofactors involved. The acyl-CoA synthetase LACS2 has been found to be involved in the synthesis of the cuticular membrane, indicating that changes in the activation status of the precursors of cutin monomers are necessary during cutin biosynthesis (Schnurr *et al.*, 2004). Recombinant LACS2 has a higher activity with 16-hydroxypalmitate than with palmitate (Schnurr *et al.*, 2004).

In *Arabidopsis*, the characterization of *HOTHEAD/ADHESION OF CALYX EDGES (HTH/ACE)* identified a gene encoding a long-chain fatty acid ω-alcohol dehydrogenase belonging to glucose-methanol-choline oxidoreductase domain-containing proteins (Krolikowski *et al.*, 2003; Kurdyukov *et al.*, 2006a). HTH/ACE catalyzes the formation of oxo-acids that are the precursors for a significant proportion of α,ω-dicarboxylic acids in *Arabidopsis* stem cutin (Kurdyukov *et al.*, 2006a).

The very-long-chain fatty acid derivatives of suberin are synthesized by fatty acid elongases that catalyze the elongation of the carbon chain of stearate to different lengths, as found in wax biosynthesis (Domergue *et al.*, 1998). Root-specific fatty acid elongases have been characterized from maize (Schreiber *et al.*, 2000). The necessary hydroxylation steps may be introduced by cytochrome P450-dependent enzymes. The formation of α,ω-dicarboxylic acids from ω-hydroxyacids is catalyzed by a ω-hydroxy fatty acid dehydrogenase (Agrawal and Kolattukudy, 1978a,b). A cytochrome P450 that oxidizes fatty acids to the corresponding ω-alcohols and subsequently to the α,ω-dicarboxylic acids was described (Le Bouquin *et al.*, 2001). Another possibility may be that HTH/ACE, or one of the closely related proteins, are involved in the formation of α,ω-dicarboxylic acids present in suberin (Kurdyukov *et al.*, 2006a). While the major types of enzymes responsible for the synthesis of aliphatic suberin monomers have been identified, none of them have been shown to be directly involved in suberin biosynthesis.

Our knowledge of cutin biosynthesis is likely to increase rapidly in the future, since genome-wide transcriptional profiling has been combined with polyester analysis in the epidermis of *Arabidopsis* stem sections, while such approaches still need to be attempted for suberized cells (Suh *et al.*, 2005).

2.3.2. Formation of the polyesters

How the cutin and suberin monomers are transported to the place of polymerization is still to be elucidated. On the other hand, an ATP-transporter involved in the transport of wax molecules across the plasmalemma has been identified by cloning the CER5 gene of *Arabidopsis* (Kunst and Samuels, 2003; Pighin *et al.*, 2004). The very-long-chain fatty acid derivatives of suberin may be transported by a similar mechanism. However, the transport mechanism of cutin monomers also remains to be discovered; cutin monomers have a shorter chain length and much lower hydrophobicity than wax molecules. For cutin formation, an additional transport through the cell wall is required, and this transport step may involve lipoproteins. This model was proposed after proteins with the activity to transport lipids *in vitro* (lipid-transfer proteins) were localized to the cell wall (Kader, 1996). Although lots of circumstantial evidence for this function of lipid-transfer proteins have been collected, the direct involvement of lipid-transfer proteins in cutin biosynthesis has not yet been substantiated (Hollenbach *et al.*, 1997; Pyee and Kolattukudy, 1995). Instead, recent work has indicated that lipid-transfer proteins act in plant defense against pathogens (Garcia-Olemedo *et al.*, 1995; Maldonado *et al.*, 2002; Molina and Garcia-Olmedo, 1997).

In order to form the three-dimensional structure of cutin and suberin, the respective monomers have to be linked together, in part by ester bonds. Classical chemical studies showed that cutin is mostly held together by primary alcohol–ester linkages between the cutin monomers with about half of the secondary hydroxyl groups involved in ester cross-links resulting in a polymeric network. The recent finding that glycerol is a substantial monomer of cutin and suberin makes it likely that the polyesters have a more complex structure whose formation remains largely to be discovered (Graça *et al.*, 2002).

Some early studies showed that the cutin monomers bound to CoA as cofactors are transferred to free hydroxyl groups present in the cutin polymer (Croteau and Kolattukudy, 1973, 1975; Kolattukudy, 1981). An hydroxyl-CoA:cutin transacylase activity has been detected in a crude extract that needs ATP for the reaction as well as cutin polymer as a primer. However, the transacylase has not been purified and no gene encoding the enzyme has been identified. A putative acyl-CoA:cutin transferase has been claimed to be purified from Agave epidermis (Reina and Heredia, 2001). After partial protein sequencing, a gene was isolated that encodes a novel small valine-rich protein with a putative HxxxE domain present in other acyltransferases (Reina and Heredia, 2001). No confirmation exists to date, however, that this protein has the proposed function.

BODYGUARD, an enzyme of the α,β hydrolase family, has been found to be critically involved in the formation of the cuticular membrane of *Arabidopsis* (Kurdyukov *et al.*, 2006b). In the *bdg* mutant, the cuticular membrane is disrupted and the outer extracellular matrix is disorganized, with polysaccharides coming to the surface and polyester also deposited within the cell wall. These structural changes were accompanied by a higher amount of residual-bound lipids of totally extracted leaves. BDG is extracellularly localized and thus functions directly in the formation/organization of the cuticular membrane (Kurdyukov *et al.*, 2006b). Since some members in the α,β hydrolase superfamily have synthase activity, it is hypothesized that BDG may also be capable of synthesizing reactions in the cuticular layer of the cell wall. Potentially, BDG may even be capable of catalyzing hydrolysis as well as synthesis, depending on the conditions in which the reaction takes place (Kurdyukov *et al.*, 2006b).

2.3.3. Mutants affected in cutin deposition

The best means for linking enzyme activities and the corresponding proteins and genes to their respective functions is by mutation. The *Sorghum bicolor bloomless* (*bm*) mutant was the first mutant identified having a thinner cuticuler membrane as well as a reduced wax deposition. The *bm* mutant exhibits a higher conductance to water vapor and an increased susceptibility to the fungal pathogen *Exserohilum turcicum* (Jenks *et al.*, 1994). Since several aspects of the cuticle were altered in *bm*, it could not be determined which cuticular component contributes which feature of the cuticle. Furthermore, *Sorghum* is not a species well suited for map-based cloning of genes.

Isolation of *Arabidopsis* mutants only affected in the deposition of either cutin or suberin by a chemical or ultrastructural screening method would be extremely work-intensive and not feasible. Thus, secondary phenotypes of plants having an altered cutin or suberin structure have to be identified in order to use the large resources available in *Arabidopsis* for this research area.

A phenotype that was at first unexpected but found to be related to cuticular changes was organ fusion (Lolle and Cheung, 1993; Lolle and Pruitt, 1999; Lolle *et al.*, 1997, 1998). Support for the idea that a disrupted cuticular membrane structure and/or less cutin lead to organ fusions was originally obtained by an indirect approach using transgenic *Arabidopsis* plants expressing and secreting a fungal cutinase and therefore degrading their own cutin (Sieber *et al.*, 2000). These transgenic plants show an altered ultrastructure and a higher permeability of the cuticle. When organs having a disrupted cuticular membrane are in close contact early during development, fusions form most likely by cross polymerization. These organ fusions are very strong so that organs do not separate during further growth, leading to distortions of the growth habit of the plant (Sieber *et al.*, 2000). A number of organ fusion mutants were shown to be altered in the cuticular polyester (Kurdyukov *et al.*, 2006a,b). Organ fusions are still used as a selection criterion for mutants having changes in cuticular structure or composition (Yephremov and Schreiber, 2005).

A very simple and much more direct way to identify mutants with an increased permeability of the cuticle is by staining of plant tissues with a dye, for example, with toluidine blue (Tanaka *et al.*, 2004). In addition, mutants with alterations in the cuticular membrane have been identified by various other phenotypes that were often not obviously associated with cuticular function, such as either altered resistance to pathogens or a number of changes in cell morphology and differentiation (Yephremov and Schreiber, 2005).

The increasing number of well-characterized *Arabidopsis* plants having alterations in the cuticular membrane enables some phenotype comparisons to be made. The organ fusion mutant *bdg* shares most of the phenotypes with cutinase-expressing plants, such as an increased permeability of the cuticle, higher wax accumulation, ectopic pollen germination, stunted growth, altered trichome formation, and increased resistance to *Botrytis cinerea* (Kurdyukov *et al.*, 2006b; Sieber *et al.*, 2000). The characteristic difference in the structure of the cuticular membranes between cutinase-expressing plants and *bdg* mutants, namely, that *bdg* accumulates, in addition, large amounts of osmophilic material deeper within the cell wall, lead to the hypothesis that BDG acts directly in the formation of the extracellular matrix, as discussed above (Kurdyukov *et al.*, 2006b).

Surprising are the differences in the phenotypes of *att1* and *lcr*, two *Arabidopsis* mutants affected in a cytochrome P450 of the same subfamily and having a higher permeability as well as ultrastructural changes in the cuticular membrane of leaves (C. Nawrath, unpublished results) (Wellesen *et al.*, 2001; Xiao *et al.*, 2004). While *lcr* shows frequently organ fusions and alterations in trichome formation, demonstrating a link between cuticle structure and the development of epidermal cells, *att1* does not show either organ fusions or any developmental disorders (Wellesen *et al.*, 2001; Xiao *et al.*, 2004). A direct link between plant disease resistance and the formation of the cuticular membrane was found in *att1* mutants. *Pseudomonas syringae* pv. *phasaelicula* expresses high levels of the type III genes when colonizing the *att1* mutant, demonstrating that ATT1 is important for the repression of bacterial virulence genes in wild-type plants (Xiao *et al.*, 2004). Therefore, *att1* mutants are more susceptible to *P. syringae* pv. *tomato* DC3000. It was speculated that the alteration of the cuticle in the substomatal chamber in which the bacteria reside is of relevance for the mechanism (Xiao *et al.*, 2004).

The organ fusion mutant *fdh* lead to the identification of a fatty acid biosynthetic enzyme with homology to condensing enzymes of which the exact substrate is unknown (Pruitt *et al.*, 2000; Yephremov *et al.*, 1999). The *fdh* mutant shows, similarly to *lcr*, *bdg*, and cutinase-expressing plants, alterations in trichome formation and ectopic pollen germination, in addition to organ fusions and a higher permeability of the cuticle (Pruitt *et al.*, 2000; Yephremov *et al.*, 1999). However, the ultrastructure of the fusion itself in *fdh* plants has a very different structure in comparison to all other characterized organ fusion mutants since the cuticular membranes are not disrupted but fuse directly to each other (C. Nawrath, unpublished results).

The analysis of the molecular basis underlaying the obvious differences in composition, structure, and function will surely be of great interest during the next years (Nawrath, 2006).

WAX2/YORE-YORE is an *Arabidopsis* protein with six-membrane spanning domains having homology to the sterol desaturase family at the N-terminus and the short-chain dehydrogenase/reductase family at the C-terminus as well as having an overall homology to CER1, a protein required for wax deposition of unknown function in *Arabidopsis* (Chen *et al.*, 2003; Kurata *et al.*, 2003). In contrast to the other mutants having a looser cuticular membrane structure and loss of cuticular membrane material, *wax2/yore-yore* has, in addition, a reduced wax deposition (Chen *et al.*, 2003; Kurata *et al.*, 2003). Other phenotypes of the *wax2/yore-yore* mutants are typical for cutin mutants, such as an increased permeability of the cuticular membrane, disorders in the development of epidermal cell types, and organ fusions (Chen *et al.*, 2003). Thus, WAX2 plays a critical role in the synthesis of both cuticular components, cutin and wax.

In addition to enzymes that are directly involved in either cutin monomer biosynthesis or polyester formation, a number of genes have been identified by mutations that are regulators of epidermal development and therefore lead to an abnormal cuticle (Aharoni *et al.*, 2004; Becraft *et al.*, 1996; Broun *et al.*, 2004; Jin *et al.*, 2000; Tanaka *et al.*, 2001, 2002; Watanabe *et al.*, 2004).

ABNORMAL LEAF SHAPE (*ALE1*) is a subtilisin-like protease that is involved in the regulation of the formation of the cuticle in embryos and juvenile plants in

Arabidopsis (Tanaka *et al.*, 2001). *ale1* mutants have a disrupted cuticular membrane in embryos, cotyledons, and juvenile leaves. The leaves of *ale1* are crinkled, often have organ fusions, and are very susceptible to low humidity, resulting in conditional lethality. Interestingly, ALE1 is expressed in certain endosperm cells adjacent to the embryo, as well as in the young embryo, and may be essential for the separation of the two entities (Tanaka *et al.*, 2001).

CRINKLY4 (*CR4*) is a receptor kinase with homology to tumor-necrosis factor receptors that is involved in proper epidermal formation that has first been identified in maize (Becraft *et al.*, 1996). *CR4* mutants have organ fusions as well as abnormal epidermal cell wall and cuticle deposition (Jin *et al.*, 2000). ACR4, the CR4 homologue in *Arabidopsis* that is expressed in the outer cell layers of embryos and mature plants, has similar function in epidermal differentiation and cuticle development as CR4 (Tanaka *et al.*, 2002; Watanabe *et al.*, 2004). ALE1 and ACR4 affect synergistically the differentiation and function of the epidermis, since *ale1/acr4* double mutants have a stronger phenotype than do both single mutants (Watanabe *et al.*, 2004).

The overexpression of *SHINE/WAX INDUCER1*, an AP2 domain transcription factor, leads to an increased wax deposition (Aharoni *et al.*, 2004; Broun *et al.*, 2004). In addition, the ultrastructure of cuticular membrane as well as permeability of the cuticle is changed. Furthermore, diverse aspects of epidermal differentiation are altered, such as epidermal cell structure, trichome shape and number, and stomatal index (Aharoni *et al.*, 2004). These diverse phenotypes make the interpretation of the physiological analyses difficult. However, the expression pattern of the different genes of the *shine* clade, whose overexpression all result in similar phenotypes, suggest diverse functions in lipid and/or cell wall metabolism, including cutin and suberin deposition (Aharoni *et al.*, 2004).

2.4. Future perspectives

During the past years, research on cutin and suberin demonstrated that the structure of these polyesters is complex and well organized. Some progress has been made to identify genes and proteins involved in cutin and suberin biosynthesis. Although mutant phenotypes indicating defects in cutin formation have been identified, no means to identify mutants in suberin formation have yet been found. However, even when mutants have been found, more work is required to link the genes identified by mutation to the exact functions of the proteins and a deeper understanding of the formation of the polyesters. In *Arabidopsis*, many resources are available that might contribute to the understanding of cutin and suberin biosynthesis in the future. In addition, analysis of cutin monomer composition have been shown to be feasible in *Arabidopsis*, an important progress for assigning functions to proteins (Bonaventure *et al.*, 2004; Franke *et al.*, 2005; Xiao *et al.*, 2004).

Meanwhile, cutin and suberin began to attract attention as biological polymers (Kolattukudy, 2001). Studies were undertaken to increase our knowledge of the physical properties of cutin and suberin (Cordeiro *et al.*, 1998; Heredia, 2003). A detailed review on the physical properties of the cutin of tomato has been published (Heredia, 2003). Results will be briefly summarized here.

Cutin is an amorphous and insoluble polymer with a molecular spacing of 0.4–0.5 nm between the polymer chains, having very low water sorption and permeability. The specific heat of cutin is higher than that of other polymers of the cell wall, possibly playing a role in thermoregulation of the plant. Most of the water diffuses as single molecules through the cuticle and not through pores (Riederer and Schreiber, 2001). These water molecules may act as a plasticizer, contributing to the molecular flexibility of the polymer(s) resulting in a viscoelastic polymer network. In this context, it may be important to consider that foliar application of chemicals may change the permeability of the polymer, possibly affecting problems related to a too rigid cutin polymer, such as cuticle cracking of fruits (Aloni *et al.*, 1998).

More research will be needed until the synthesis of these natural polyesters is understood well enough to consider engineering them to either improve their properties *in situ* in order to make plants more stress resistant, that is, reduce the cracking of the cuticle of fruits, or use them in industrial applications (Aloni *et al.*, 1998).

Until then, the natural polymers may be used for some industrial applications. For example, cuticular material containing 40–80% cutin occurs in large quantities as a valuable by-product in the waste of fruit processing. Refractory to most treatments, cutin may be recovered from waste by physical, chemical, and biological processes, and the monomers released by hydrolysis could be polymerized for various applications, for example, as either lubricants or for biomedical applications.

Cork, the bark of *Quercus suber*, contains up to 50% suberin, besides 22% lignin, 20% carbohydrates, and some additional extractable components. This polymer has already been commercially exploited for centuries. The excellent insulation property for polar liquids gives cork its special importance in the wine industry as stoppers. Cork also insulates against sound and heat and is used in insulation boards. Over 280,000 tons of raw material are used per year, from which about 20–30% is left as waste in the form of cork dust, which could potentially be useful in other applications. Recently, cork has also been tested in ink as well as a base for the synthesis of polyurethane (Cordeiro *et al.*, 1997, 2000). Cork extracts are also recognized as having antimutagenic effects (Krizkova *et al.*, 1999).

3. POLYHYDROXYALKANOATE

Synthesis of PHA in plants was first demonstrated in 1992 by the accumulation of poly(3-hydroxybutyrate) (PHB) in the cytoplasm of the cells of *Arabidopsis thaliana* (Poirier *et al.*, 1992a). Since then, a range of different PHAs has been synthesized in plants, including various copolymers such as poly(3-hydroxybutyrate-co-3-hydroxyvalerate) [P(HB-HV)] and medium-chain-length polyhydroxyalkanoates (MCL-PHAs) (Table 8.1) (Poirier, 2002). This has been achieved through the modification of various pathways localized in different subcellular compartments, such as the fatty acid and amino acid biosynthetic pathways in the

TABLE 8.1 Summary of transgenic plants producing PHA

PHA type	Species	Tissue	Organelle	PHA quantity (% dwt)	References
PHB	*Arabidopsis thaliana*	Shoot	Cytoplasm	0.1	Poirier *et al.*, 1992a
PHB	*A. thaliana*	Shoot	Plastid	40	Bohmert *et al.*, 2000; Nawrath *et al.*, 1994
PHB	Oilseed rape	Shoot	Cytoplasm	0.1	Poirier and Gruys, 2001
PHB	Oilseed rape	Seed	Plastid	8	Houmiel *et al.*, 1999
PHB	Tobacco	Shoot	Cytoplasm	0.01	Nakashita *et al.*, 1999
PHB	Tobacco	Shoot	Plastid	<1.7	Arai *et al.*, 2001; Bohmert *et al.*, 2002; Lössl *et al.*, 2003; 2005
PHB	Corn	Shoot	Plastid	6	Poirier and Gruys, 2001
PHB	Corn	Cell suspension	Peroxisome	2	Hahn *et al.*, 1999
PHB	Alfalfa	Shoot	Plastid	0.2	Saruul *et al.*, 2002
PHB	Cotton	Fiber cells	Cytoplasm	0.3	John and Keller, 1996
PHB	Potato	Shoot	Plastid	0.01	Bohmert *et al.*, 2002
PHB	Sugar beet	Hairy root	Plastid	5.5	Menzel *et al.*, 2003
PHB	Flax	Stem	Plastid	0.005	Wróbel *et al.*, 2004
P(HB-HV)	*A. thaliana*	Shoot	Plastid	1.6	Slater *et al.*, 1999
P(HB-HV)	Oilseed rape	Seed	Plastid	2.3	Slater *et al.*, 1999
P(HB-HV)	*A. thaliana*	Shoot	Cytoplasm	0.6	Matsumoto *et al.*, 2005
MCL-PHA[a]	*A. thaliana*	Whole plants	Peroxisome	0.6	Mittendorf *et al.*, 1998

(*continued*)

Table 8.1 (*continued*)

PHA type	Species	Tissue	Organelle	PHA quantity (% dwt)	References
MCL-PHA[a]	*A. thaliana*	Seed	Peroxisome	0.1	Moire *et al.*, 2004; Poirier *et al.*, 1999
MCL-PHA[b]	Potato	Cell line	Cytoplasm	1	Romano *et al.*, 2003
MCL-PHA[c]	*A. thaliana*	Whole plants	Peroxisome	0.04	Arai *et al.*, 2002
MCL-PHA[d]	Potato	Shoot	Plastid	0.03	Romano *et al.*, 2005

PHB: polyhydroxybutyrate; PHA: Polyhydroxyalkanoate; P(HB-HV): poly(3-hydroxybutyrate-co-3-hydroxyvalerate); MCL-PHA: medium-chain-length polyhydroxyalkanoate.
[a] PHA contained 3-hydroxyacid monomers from 6 to 16 carbons.
[b] PHA contained only 3-hydroxyoctanoic acid.
[c] PHA contained 3-hydroxyacid monomers from 4 to 6 carbons.
[d] PHA contained 3-hydroxyacid monomers from 6 to 12 carbons.

plastid or the fatty acid degradation pathway in the peroxisome (Poirier, 1999, 2001; Poirier *et al.*, 1995a). Although the initial driving force behind synthesis of PHA in plants has been for the biotechnological production of biodegradable polymers, PHA synthesis in plants has also been used as a useful and novel tool to study some fundamental aspects of plant metabolism.

This section will focus on the metabolic engineering of plants for PHA production. However, since PHAs are polyesters naturally synthesized in bacteria and since most of our knowledge on the biochemical pathways for PHA synthesis and degradation has been obtained from studies in bacteria, the main pathways involved in PHA synthesis in bacteria will also be briefly described.

3.1. PHA as a bacterial polyester

PHAs have been detected in over 90 genera of bacteria, including Gram-positive and Gram-negative species, as well as some cyanobacteria (Kim and Lenz, 2002; Steinbüchel and Hein, 2001). While the majority of PHAs are composed of monomers of *R*-(−)-3-hydroxyalkanoic acid ranging from 3 to 16 carbons in length (C_3–C_{16}) (Fig. 8.3), some PHAs can also incorporate 4-, 5-, or 6-hydroxy acids (Steinbüchel and Valentin, 1995). Nearly 150 different hydroxyacids have been found incorporated in bacterial PHAs, with the major diversity being found in the length and the presence of functional groups in the side–chains of the polymer (Steinbüchel and Valentin, 1995). Although some of these monomers have been found in PHA produced by bacteria in their natural environment, a larger fraction of monomers have been incorporated into PHA following growth of bacteria

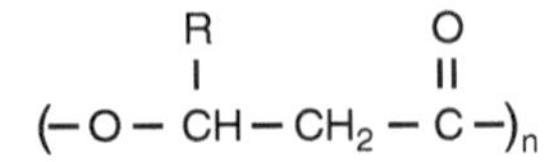

R = methyl 3-hydroxybutyrate
R = ethyl 3-hydroxyvalerate
R = hexyl 3-hydroxynonanoate
R = nonyl 3-hydroxydodecanoate

FIGURE 8.3 Chemical structure of polyhydroxyalkanoate. The monomers can range from 3 to 16 carbons in length, depending on the size of the pendant R group.

in artificial media containing exotic sources of carbon, such as fatty acids with double or triple bonds.

Bacteria synthesizing PHAs have been broadly subdivided in two groups. One group, including the bacterium *Ralstonia eutropha*, produces short-chain-length PHA (SCL-PHA) containing monomers ranging from 3 to 5 carbons in length, while a second group, including a number of Pseudomonads, synthesizes medium chain length-PHA (MCL-PHA) containing monomers ranging from 6 to 16 carbons in length. This division between SCL- and MCL-PHA is mainly determined by the substrate specificity of the PHA synthase responsible for the polymerization of the substrate *R*-3-hydroxyacyl-CoA to form PHA. This division between SCL- and MCL-PHA is however not strict, since several bacteria have been found that can synthesize a "hybrid" PHA that can include monomers from 4 to 8 carbons (Fukui and Doi, 1997).

A number of enzymes and metabolic pathways have been implicated in the synthesis of a spectrum of PHAs in bacteria. In this chapter, we wish to focus only on the pathways that have been successfully transferred in plants. The readers are referred to several excellent reviews to learn more on various aspects of bacterial PHA, including biochemical synthesis and application of PHAs (Anderson and Dawes, 1990; Braunegg *et al.*, 1998; Steinbüchel, 1991; Steinbüchel and Füchtenbusch, 1998; Steinbüchel and Lütke-Eversloh, 2003; Steinbüchel and Schlegel, 1991; Sudesh and Doi, 2000; van der Walle *et al.*, 2001).

3.2. Polyhydroxybutyrate

PHB is the most widespread and thoroughly characterized PHA found in bacteria. A large part of our knowledge on PHB biosynthesis has been obtained from *R. eutropha* (Steinbüchel and Hein, 2001). In this bacterium, PHB is synthesized from acetyl-CoA by the sequential action of three enzymes (Fig. 8.4). The first enzyme of the pathway, 3-ketothiolase, encoded by the *phbA* gene, catalyzes the reversible condensation of two acetyl-CoA moieties to form acetoacetyl-CoA. Acetoacetyl-CoA reductase, encoded by the *phbB* gene, subsequently reduces acetoacetyl-CoA to *R*-(−)-3-hydroxybutyryl-CoA, which is then polymerized to PHB by the action of a PHA synthase encoded by the *phaC* gene. The PHA synthase of *R. eutropha* has been shown to accept the *R*-isomer of 3-hydroxybutyryl-CoA but not the *S*-isomer. PHA is typically produced as a polymer of 10^3 to 10^4 monomers that accumulates as inclusions of 0.2 to 0.5 μm in diameter. In *R. eutropha*, PHB inclusions

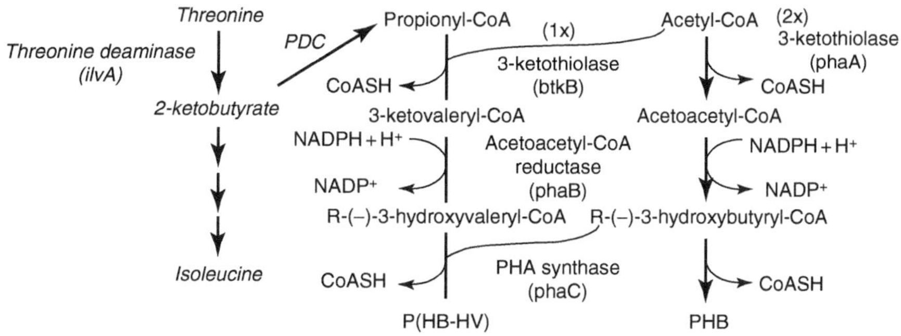

FIGURE 8.4 Pathways of PHB and P(HB-HV) synthesis. The pathways common to bacteria and transgenic plants are shown in plain letters while the pathway specific to transgenic plants is shown in italics. PDC refers to the plant endogenous pyruvate dehydrogenase complex.

can typically accumulate to 80–85% of the dry weight (dwt) when bacteria are grown in media containing excess carbon, such as glucose, but limited in one essential nutrient, such as nitrogen or phosphate (Steinbüchel and Schlegel, 1991). Under these conditions, PHB synthesis acts as a carbon reserve and an electron sink.

PHB is a highly crystalline polymer and a stiff and relatively brittle thermoplastic (de Koning, 1995). Its melting point (T_m = 175 °C) is only slightly lower than the temperature at which it starts degrading to crotonic acid, making processing difficult. These properties seriously limit its use in a wide range of commodity products. PHB has good UV light resistance but relatively poor resistance to acids and bases. The polymer is water and air impermeable as well as relatively resistant to hydrolytic degradation, making it superior to starch-derived plastics, which are moisture sensitive.

3.2.1. Synthesis of PHB in the cytoplasm

Despite its relatively poor physical properties as a thermoplastic, PHB was initially targeted for production in plants because the first bacterial PHA biosynthetic genes that were cloned were for PHB synthesis in the bacterium *R. eutropha* (Schubert *et al.*, 1988; Slater *et al.*, 1988). The cytoplasm was targeted as the first site for PHB synthesis because, in addition to containing acetyl-CoA, the building block for PHB, it also had the advantage that the bacterial enzymes could be directly expressed in this compartment without any modification of the proteins. Furthermore, an endogenous plant 3-ketothiolase is present in the cytoplasm as part of mevalonate pathway. Thus, creation of the PHB biosynthetic pathway in the cytoplasm was theoretically simpler, requiring only the expression of two additional enzymes, the reductase and synthase. The *R. eutropha phaB* and *phbC* genes, encoding, respectively, the acetoacetyl-CoA reductase and PHA synthase, were coexpressed in *A. thaliana* under the control of the cauliflower mosaic virus (CaMV) 35S promoter, allowing a relatively high expression of the enzymes in a broad range of tissues (Poirier *et al.*, 1992a). The highest amount of PHB measured in the shoots of these plants was ~0.1% dwt (Poirier *et al.*, 1992a). Detailed analysis of the PHB purified from *A. thaliana* confirmed that the polymer was isotactic poly

([*R*]-(−)-3-hydroxybutyrate) and that the thermal properties of plant PHB were similar to those of bacterial PHB (Poirier *et al.*, 1995b). Furthermore, PHB accumulated in the form of granules that had a size and appearance very similar to bacterial PHB granules (Fig. 8.5) (Poirier *et al.*, 1992a).

Plants expressing high level of acetoacetyl-CoA reductase in the cytoplasm have shown a strong reduction in growth, with the most affected plants being approximately five times smaller by fresh weight compared to wild-type plants (Poirier *et al.*, 1992b). There was an overall good correlation between the extent of the growth reduction and the level of reductase enzyme activity. While no abnormal phenotype was observed in plants expressing only the PHB synthase (and not producing PHB), combination of the acetoacetyl-CoA reductase with the PHB synthase led to a further reduction in growth compared to plants expressing only the reductase (Poirier *et al.*, 1992b). Although the reasons for the dwarf phenotype have not been unambiguously determined, it is likely that the diversion of cytoplasmic acetyl-CoA and acetoacetyl-CoA away from the endogenous isoprenoid and flavonoid pathways might lead to a depletion of essential metabolites, such as sterols, which may affect growth.

Synthesis of PHB in the cytoplasm of rape leaf cells gave results very similar to those in Arabidopsis (Poirier, 2002). Interestingly, overexpression of the bacterial 3-ketothiolase in plants expressing the reductase and PHB synthase did not lead to a significant increase in PHB accumulation, indicating that 3-ketothiolase activity was probably not limiting PHB synthesis in the cytoplasm, but that other factors, such as the low flux of acetyl-CoA, may be important.

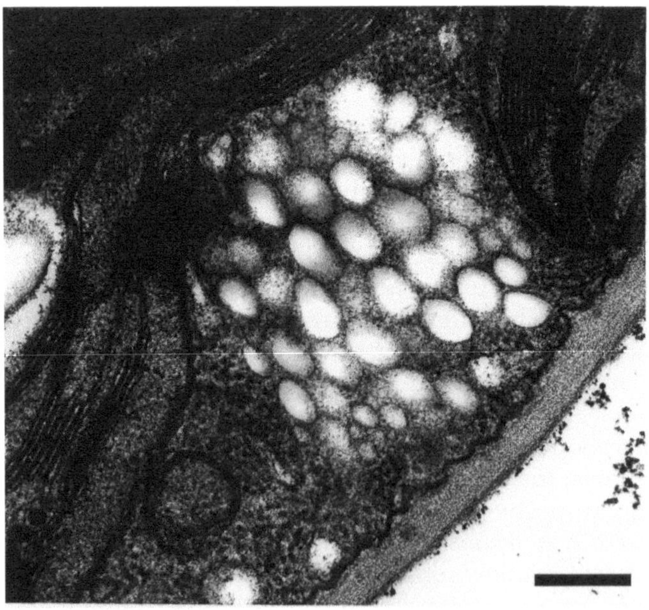

FIGURE 8.5 Accumulation of PHA inclusions in the cytoplasm of transgenic *A. thaliana* cells expressing the PHB biosynthetic pathway. Bar = 1 μm.

PHB synthesis has also been demonstrated in the cytoplasm of cotton fiber cells (John and Keller, 1996). In this approach, PHA is not produced as a source of polyester to be extracted and used in the plastic industries, but rather as an intracellular agent that modifies the heat exchange properties of the fiber. The *phaA*, *phaB*, and *phaC* genes from *R. eutropha* were expressed in transgenic cotton under the control of a fiber-specific promoter (John and Keller, 1996). PHB accumulated in the cytoplasm to 0.3% dwt of the mature fiber, a level similar to PHB production in *A. thaliana* cell cytoplasm, while no deleterious effect on fiber development was reported.

Production of PHB has been reported in leaves of *Nicotiana tabacum* through the coexpression of the *phaB* gene from *R. eutropha* and the PHA synthase from *Aeromonas caviae* (Nakashita *et al.*, 1999). Although the bacterial genes were expressed under the strong promoter CaMV35S, expression of both proteins was relatively low and the amount of PHB detected in leaves was only 10 μg/g fresh weight (fwt). Inhibition of the mevalonate pathway at the level of the 3-hydroxy-3-methylglutaryl-CoA reductase led to a twofold increase in PHB level in tobacco cell lines, indicating a link between PHB synthesis and availability of acetyl-CoA (Suzuki *et al.*, 2002). Similar levels of PHB were obtained in potato expressing the phb enzymes in the cytosol.

3.2.2. Synthesis of PHB in the plastid

The relatively limited supply of acetyl-CoA in the cytosol is thought to be responsible for the low accumulation of PHB as well as for the deleterious effects of transgene expression on plant growth observed in many plants. In this context, the plastid was viewed as a much better site for PHB synthesis, since this organelle has a larger flux of carbon through acetyl-CoA required for fatty acid biosynthesis. This is particularly true for the leucoplast of developing seeds of oil-accumulating plants, such as Arabidopsis and oilseed rape.

The phaA, phaB, and phaC proteins from *R. eutropha* were modified for plastid targeting by addition of the transit peptide of the small subunit of the ribulose bisphosphate carboxylase from pea (Nawrath *et al.*, 1994). The modified bacterial genes were first expressed individually in *A. thaliana* under the control of the constitutive CaMV35S promoter, and later the transgenes were combined through crossings. Transgenic plants expressing only the plastid-targeted reductase and PHA synthase did not produce detectable PHB, providing further evidence that plastids do not have an endogenous 3-ketothiolase activity that could support PHB synthesis (Nawrath *et al.*, 1994). However, plants expressing all three bacterial enzymes were shown to accumulate PHB inclusions exclusively in the plastids, with some organelle having a substantial portion of their volume filled with inclusions. The size and general appearance of these were similar to bacterial PHA inclusions (Nawrath *et al.*, 1994). Interestingly, the quantity of PHB in these plants was found to gradually increase over time, with fully expanded presenescing leaves typically accumulating 10 times more PHB than do young expanding leaves of the same plant. The maximal amount of PHB detected in presenescing leaves was 10 mg/g fwt, representing ~14% dwt. In contrast to PHB synthesis in the cytoplasm, expression of the PHB biosynthetic enzymes in the plastid was not accompanied by a large reduction in growth of these plants. However, leaf chlorosis was observed in plants accumulating more than 3–5% dwt. These results indicated that although the plastid can accommodate a higher production of PHB with minimal impact on plant growth compared to the cytoplasm, there was nevertheless a limit above which alteration in some of the chloroplast functions could be detected (Nawrath *et al.*, 1994).

In contrast to the individual expression of the *R. eutropha phb* genes in plants followed by stacking through crossing, an alternative strategy was devised where all three plastid-modified *phb* genes were cloned into a single binary vector. By this approach, a number of lines were identified which accumulated PHB between 3% and 40% dwt (Bohmert *et al.*, 2000). While in a line accumulating 3% dwt most of the plastids contained some PHB inclusions, all plastids of mesophyll cells were packed with inclusion in the line containing PHB to 40% dwt. Interestingly, these transgenic plants showed a negative correlation between PHB accumulation and plant growth. While plants containing 3% dwt PHB showed only a relatively small reduction in growth, plants accumulating between 30% and 40% dwt PHB were dwarfed and produced no seeds (Bohmert *et al.*, 2000). As previously observed by Nawrath and colleagues, all plants producing above 3% dwt PHB showed some chlorosis (Bohmert *et al.*, 2000; Nawrath *et al.*, 1994). Together, these experiments demonstrate that while it is possible to further increase PHB production in plastids by using new vectors, the approach of synthesizing PHB in the chloroplasts of shoots has its limits.

Since the production of PHA in the plastid typically requires the expression of several enzymes, strategies devised to simplify the number of individual genes that must be expressed could have advantages. In this respect, a novel fusion protein composed of the 3-ketothiolase and acetoacetyl-CoA reductase from *R. eutropha* was created (Kourtz *et al.*, 2005). This was a challenging project since the native thiolase and reductase enzymes act as homotetramers in bacteria. Nevertheless, one fusion protein exhibited thiolase and reductase activities in crude extracts of recombinant *Escherichia coli* that were only threefold and ninefold less than those of the individually expressed thiolase and reductase enzymes, respectively. Expression of the plastid-targeted fusion enzyme, along with the PHA synthase, resulted in plants accumulating roughly half the amount of PHB synthesized in plants expressing the individual enzymes.

As a first step to bring the technology of PHA synthesis to the field, scientists at Monsanto have demonstrated the production of PHB in the plastids of corn leaves and stalk, as well as in the leucoplast of developing seeds of *Brassica napus*. In those experiments, the same *R. eutropha* genes modified for PHB production in the plastids of *A. thaliana* were used. Levels of PHB accumulation up to 5.7% dwt were reported (Poirier and Gruys, 2001). Similar to results obtained in *A. thaliana*, there was a progressive accumulation of PHB in corn shoots with time, with older leaves having more polymer than younger leaves. Furthermore, like in *A. thaliana*, there was a correlation between leaf chlorosis and higher amount of PHB (Poirier and Gruys, 2001). Perhaps one of the most striking observations made from the experiments in corn was the fact that while the leaf mesophyll cells showed few PHB granules, the bundle sheath cells associated with the vascular tissue were packed with granules (Poirier and Gruys, 2001). This unequal distribution of PHB was not due to the promoter used, since a similar pattern was seen for plants transformed with either the CaMV35S or the chlorophyll A/B binding protein promoters, the latter promoter being known to be a strong promoter in mesophyll cells. Interestingly, a similar observation had been made by the same group for *A. thaliana* plants transformed with the *phb* genes driven by the CaMV35S promoter; that is, significantly more granules were found in cells surrounding the vascular tissue and epidermal cells compared to mesophyll cells (Poirier and Gruys, 2001). These results suggest that the availability of plastidial acetyl-CoA for PHB synthesis may be quite different in various cell types, perhaps due to metabolic channeling.

For the creation of the PHB biosynthetic pathway in developing seeds of *B. napus*, the three modified bacterial genes *phaA*, *phaB*, and *phaC* were put under the control of the fatty acid hyroxylase promoter from *Lesquerella fendeleri*, enabling strong expression to the developing seed (Houmiel *et al.*, 1999). PHB level up to 7.7% fwt of mature seeds was reported (Houmiel *et al.*, 1999). Analysis of seeds by TEM revealed that PHB accumulated exclusively within the leucoplast and that apparently every visible plastid contained the polymer. Seeds accumulating nearly 8% dwt PHB appeared normal and germinated at the same rates as nontransformed seeds (Houmiel *et al.*, 1999). These results demonstrate that at least in the range of 3–8% dwt PHB, the seed leucoplast appears a better production system than the leaf chloroplast. It is unknown at this point what is the upper limit of PHB accumulation in seeds and at what level PHB synthesis will start affecting the accumulation of lipids or proteins in the seed, two key factors that have a strong impact on the viability of this approach in the biotechnological production of PHA in oilseed crops.

Five additional crop plants have been investigated for PHA production through expression of the PHB pathway in the plastid. Transformation of alfalfa, tobacco, potato, and flax with the three *R. eutropha phb* genes modified for plastid targeting was shown to give transgenic plants producing PHB in their leaves to a maximum level of 0.18%, 0.32%, 0.009%, and 0.005% dwt (Bohmert *et al.*, 2002; Saruul *et al.*, 2002; Wróbel *et al.*, 2004). Although the reasons behind the low level of PHB accumulation in these plants compared to either Arabidopsis or corn have not been fully elucidated, it has been demonstrated that constitutive expression of the bacterial 3-ketothiolase leads to a large decrease in the recovery of transgenic plants following transformation (Bohmert *et al.*, 2002). The use of a construct where the bacterial 3-ketothiolase is expressed under the control of an inducible promoter led to an increased recovery of transgenic tobacco and potato producing PHB, although the amount of PHB produced remained relatively low at below 0.3% dwt (Bohmert *et al.*, 2002). Transformation of *in vitro* cultured hairy roots of sugar beet with the same three *R. eutropha* genes modified for plastid targeting led to significantly higher amount of PHB, with a maximum of 5.5% dwt (Menzel *et al.*, 2003). Thus, although accumulation of PHB in the plastid appears to be problematic for several plants, the success encountered with Arabidopsis, rape, corn, and roots of sugar beet indicate that there is no fundamental barrier to relatively high production of PHA in the plastids of plants.

As an alternative strategy to the transformation of the nuclear genome, transformation of the plastid genome with the *phb* gene has been examined. In theory, plastid transformation could lead to higher level of polymer production because of the much larger copy number of transplastome compared to the nuclear genome. However, transformation of tobacco plastome with the *R. eutropha* polycistronic operon containing the *phbA*, *phbB*, and *phbC* genes under the control of a bacterial promoter or of the plastid rRNA promoter (*Prrn*) has yielded plants synthesizing only low amount of PHB (<0.1% PHB dwt) (Arai *et al.*, 2001, 2004; Nakashita *et al.*, 2001). Expression of the *R. eutropha* polycistronic operon under the control of the plant *psbA* promoter and the *psbA* 5′ UTR improved PHB accumulation up to 1.7% dwt (Lössl *et al.*, 2003). In these transgenic plants, a higher level of PHB was limited to the early stage of heterotrophic *in vitro* culture and decreased through autotrophic growth despite constant transcript levels. PHB amounts were also found to be highly variable in different tissues of the same plant. Furthermore, production of PHB in transplastomic tobacco was associated with growth retardation and male sterility (Lössl *et al.*, 2003). Use of a transformation system where the plastidial polycistronic phb operon was under the control of an ethanol-inducible T7 RNA polymerase could solve the problem of growth retardation and sterility, but without further improvement in the yield of PHB (Lössl *et al.*, 2005). Although further work is required to understand the factors limiting the stable production of PHB in transplastomic tobacco, it must be stressed that accumulation of PHA in tobacco and potato, either in the cytoplasm or in the plastid, has consistently been low compared to Arabidopsis or rape. In this context, it would be very interesting to know if the application of the transplastome approach to Arabidopsis and rape would give similar or higher amount of PHB compared to nuclear transformation.

PHB synthesized in plants is not thought to be degraded, since significant hydrolysis of PHA requires the presence of specialized bacterial enzymes, the PHA depolymerases (Jendrossek, 2002). PHA in plants is thus viewed as a final and largely unrecyclable carbon sink. This opens several interesting questions about how transgenic plants accumulating PHA can cope with a new carbon sink. For example, how does PHB synthesis in the plastids affect carbon flow to other compounds synthesized in the organelle, such as starch and fatty acids? How does the plant adjust, at the metabolic and genetic levels, to accommodate for the synthesis of this new sink? Why are plants producing high amount of PHB affected in their growth? Clearly, the tools of genomics, proteomics, and metabolic profiling could provide interesting answers to these questions and give general insights on plant biochemistry that would go well beyond PHA synthesis in plants.

In a first small-scale study of metabolite profiling, over 60 metabolites were measured in transgenic *A. thaliana* lines producing high amount of PHB (Bohmert *et al.*, 2000). Surprisingly, no changes in fatty acids were observed. There was, however, a correlation between an increase in PHB with a decrease in levels of isocitrate and fumarate, indicating a reduction in tricarboxylic acid cycle activity, leading perhaps to a reduction in pools of acetyl-CoA that may result in growth retardation. There was also a positive correlation between PHB accumulation and levels of several sugars such as mannitol, glucose, fructose, and sucrose. Together, these data indicate that a high amount of accumulation of PHB in chloroplasts has a negative and complex effect on plant metabolism that go beyond the chloroplast. At the gene expression level, no correlation could be found between level of expression of the three *phb* genes and PHB accumulation, leaving unresolved the question of what limits PHB synthesis in the plastids.

3.2.3. Synthesis of PHB in the peroxisome

Acetyl-CoA is found not only in the cytoplasm and plastids but also in the mitochondria and peroxisomes, being primarily implicated in these organelles in the tricarboxylic acid and β-oxidation cycles, respectively. Although no conclusive demonstration of PHB in plant mitochondria has been reported, synthesis of PHB in the peroxisome was described in transgenic Black Mexican sweet corn suspension cell cultures (Hahn *et al.*, 1999). In these experiments, the *phaA*, *phaB*, and *phaC* genes from *R. eutropha* were modified in order to add a peroxisomal targeting signal at the carboxy terminal end of each protein. Biolistic transformation of maize suspension culture with a mixture of all three genes led to the isolation of transformants expressing all three enzyme activities and accumulating PHB up to 2% dwt (Hahn *et al.*, 1999). As no transgenic plants have been obtained from these transformed cells, it is difficult at this point to evaluate the potential effects of PHB synthesis in peroxisome on growth and metabolism.

3.3. Poly(3-hydroxybutyrate-co-3-hydroxyvalerate)

Because PHB homopolymer has relatively poor physical properties, extensive efforts have been invested on the synthesis of SCL-PHA copolymers that have better properties. Incorporation of either 3- or 5-carbon monomers into a polymer composed mainly of 3-hydroxybutyrate leads to a decrease in the crystallinity and melting point compared to PHB homopolymer (de Koning, 1995). The copolymer P(HB-HV) is, thus, less stiff and tougher than PHB, as well as easier to process, making it a good target for commercial application (de Koning, 1995). In *R. eutropha*, addition of either propionic acid or valeric acid to the growth media containing glucose leads to the production of a random copolymer composed of 3-hydroxybutyrate and 3-hydroxyvalerate P(HB-HV) (Steinbüchel and Schlegel, 1991). The biochemical pathway of P(HB-HV) synthesis from propionic acid is shown in Fig. 8.4. In *R. eutropha*, condensation of propionyl-CoA with acetyl-CoA is mediated by a distinct 3-ketothiolase, named btkB, which has a higher specificity for propionyl-CoA than do the 3-ketothiolase encoded by the *phaA* gene (Slater *et al.*, 1998). Reduction of 3-ketovaleryl-CoA to *R*-3-hydroxyvaleryl-CoA and subsequent polymerization to form P(HB-HV) are catalyzed by the same enzymes involved in PHB synthesis, namely, the acetoacetyl-CoA reductase and PHA synthase.

3.3.1. Synthesis of poly(3-hydroxybutyrate-co-3-hydroxyvalerate) in the cytosol

As described in a previous section, expression of the *R. eutropha* acetoacetyl-CoA reductase and PHB synthase in the cytosol of *A. thaliana* plants led to the accumulation of only 0.1% of the homopolymer PHB (Poirier *et al.*, 1992a). However, expression of the same reductase along with the PHA synthase from *A. caviae* led to the accumulation of a similar amount of a PHA copolymer containing mostly 3-hydroxybutyrate with 0.2–0.8 mol% of 3-hydroxyvalerate (Matsumoto *et al.*, 2005). The PHA synthase of *A. caviae* has been previously shown to have unique substrate specificity, being capable of producing a PHA copolymer composed of monomers ranging from 4 to 6 carbons (Fukui and Doi, 1997). Although several potential pathways could provide either the propionyl-CoA or 3-hydroxyvaleryl-CoA thought to be required for the synthesis of P(HB-HV), including amino acid synthesis or degradation, as well as β-oxidation of odd-chain fatty acids (see below for further details), it is not known which of these pathways provides the substrate for copolymer synthesis in the cytosol. Interestingly, use of an *in vitro* mutated *A. caviae* PHA synthase having higher catalytic activity led to an approximate fivefold increase in PHA accumulation in the cytoplasm, indicating that the improvement of enzymatic properties though mutagenesis is a valuable approach to increase the amount of PHA produced in plants.

3.3.2. Synthesis of poly(3-hydroxybutyrate-co-3-hydroxyvalerate) in the plastid

Because of the improved properties of P(HB-HV) copolymers over PHB, bacterial production of P(HB-HV), also known under the trade name Biopol™, has been central to the marketing and commercial production of PHA. It was therefore natural that after the demonstration of high-level PHB synthesis in the plastids, efforts would be focused on the synthesis of PHA copolymers, such as P(HB-HV).

Since synthesis of P(HB-HV) in bacteria relies on the production of propionyl-CoA, it was necessary to create an endogenous pool of propionyl-CoA in plants that could be used by the PHA pathway. Furthermore, since the plastid was shown to be the best subcellular compartment for the synthesis of PHB from acetyl-CoA, it was also chosen as the site for P(HB-HV) synthesis from acetyl-CoA and propionyl-CoA. Although several metabolic pathways exist in prokaryotes and eukaryotes that can generate propionyl-CoA, the simplest strategy adopted was the conversion of 2-ketobutyrate to propionyl-CoA by the pyruvate dehydrogenase complex (PDC), an enzyme naturally located in the plastid (Slater *et al.*, 1999). Although PDC normally decarboxylates pyruvate to give acetyl-CoA, the same enzyme can also decarboxylate 2-ketobutyrate, albeit at low efficiency, to give propionyl-CoA. Since 2-ketobutyrate is also found in the plastid as an intermediate in the synthesis of isoleucine from threonine, both the substrate and the enzyme complex required for the generation of propionyl-CoA are present in this organelle. However, since PDC would have to compete for the 2-ketobutyrate with the acetolactate synthase, an enzyme involved in isoleucine biosynthesis, the quantity of 2-ketobutyrate present in the plastid was enhanced through the expression of the *E. coli ilvA* gene, which encodes a threonine deaminase (Slater *et al.*, 1999).

The genes encoding the *E. coli ilvA*, the *R. eutropha phaB*, and *phaC*, as well as the *bktB* gene from *R. eutropha* encoding a 3-thiolase having high affinity for both acetyl-CoA and propionyl-CoA, were all modified by adding a plastid leader sequence to the enzymes (Slater *et al.*, 1999). All genes were expressed under the control of the CaMV35S promoter. Constitutive expression of the ilvA protein along with bktB, phaB, and phaC proteins in the plastids of *A. thaliana* led to the synthesis of P(HB-HV) in the range of 0.1–1.6% dwt, with the fraction of HV units being between 2 and 17 mol% (Slater *et al.*, 1999).

Expression of the P(HB-HV) pathway in the leucoplast of *B. napus* seeds has also been achieved by putting the bacterial genes under the control of the seed-specific promoter from the *Lesquerella* hydroxylase gene. In these experiments, an isoleucine-insensitive mutant of the *ilvA* gene was coexpressed along with the *bktB*, *phaA*, and *phaC* genes, and all four genes were inserted in a single multigene vector. P(HB-HV) synthesis in the range of 0.7–2.3% dwt was reported with an HV content of 2.3–6.4 mol% (Slater *et al.*, 1999). Interestingly, there was an inverse relationship between the amount of PHA and the proportion of the HV monomer, indicating a "bottleneck" in providing 3-hydroxyvaleryl-CoA to the PHA synthase. This bottleneck is thought to be caused by the inefficiency of the PDC in converting 2-ketobutyrate to propionyl-CoA.

3.4. Medium-chain-length polyhydroxyalkanaote

MCL-PHAs are typically described as elastomers, although their actual physical properties are very diverse, ranging from soft plastic to glue and rubber, and are primarily dependent on the monomer composition (de Koning, 1995). Monomers present in MCL-PHA may contain a wide spectrum of functional groups, including unsaturated bonds and halogenated groups (Steinbüchel and Valentin, 1995).

There are two main routes for the synthesis of MCL-PHA in bacteria (Fig. 8.6) (Steinbüchel and Füchtenbusch, 1998; Steinbüchel and Hein, 2001). The first is the

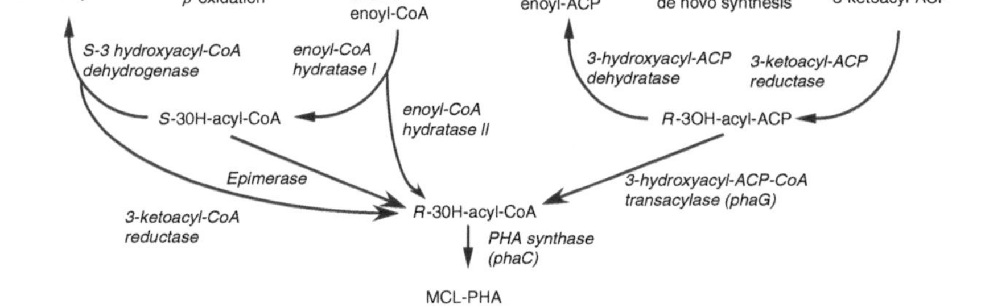

FIGURE 8.6 Pathways for MCL-PHA synthesis. Synthesis of MCL-PHA in bacteria can be accomplished either through the use of intermediates of the fatty acid β-oxidation cycle (left) or of the *de novo* fatty acid biosynthetic pathway (right).

synthesis of PHA using intermediates of fatty acid β-oxidation. This pathway is found in several bacteria, such as *Pseudomonas oleovorans* and *Pseudomonas fragii*, which can synthesize MCL-PHA from either alkanoic acids or fatty acids. In these bacteria, the monomer composition of the PHA produced is directly influenced by the carbon source added to the growth media. Typically, the PHA is composed of monomers that are $2n$ ($n \geq 0$) carbons shorter than the substrates added to the media. For example, growth of *P. oleovorans* on octanoate (C_8) generates a PHA copolymer containing C_8 and C_6 monomers, whereas growth on dodecanoate (C_{12}) generates a PHA containing C_{12}, C_{10}, C_8, and C_6 monomers (Lageveen *et al.*, 1995). Alkanoic acids present in the media are transported into the cell where they are first converted to CoA esters before being directed to the β-oxidation pathway where a number of 3-hydroxyacyl-CoA intermediates can be generated. Since the PHA synthase accepts only the *R*-isomer of 3-hydroxyacyl-CoA and the bacterial β-oxidation of saturated fatty acids generates only the *S*-isomer of 3-hydroxyacyl-CoA, bacteria must have enzymes capable of generating *R*-3-hydroxyacyl-CoA. One potential enzyme is a 3-hydroxyacyl-CoA epimerase, mediating the reversible conversion of the *S*- and *R*-isomers of 3-hydroxyacyl-CoA, although no protein or gene encoding such activity has yet been unambiguously identified (Yang *et al.*, 1986). In contrast, monofunctional enoyl-CoA hydratase II enzymes, converting directly enoyl-CoA to *R*-3-hydroxyacyl-CoA, have been identified in several bacteria, including *Aeromonas caviae* (Fukui *et al.*, 1998; Reiser *et al.*, 2000; Tsuge *et al.*, 2000). Finally, it is speculated that a 3-ketoacyl-CoA reductase that could specifically generate *R*-3-hydroxyacyl-CoA may exist in bacteria, although such an enzyme has not yet been unambiguously identified. It has, however, been shown that the enzyme 3-ketoacyl-acyl carrier protein (ACP) reductase, participating normally in the fatty acid biosynthetic pathway, may also act on 3-ketoacyl-CoA to generate *R*-3-hydroxyacyl-CoA, and thus contribute to MCL-PHA synthesis (Taguchi *et al.*, 1999).

The second route for MCL-PHA in bacteria is through the use of intermediates of fatty acid biosynthesis (Fig. 8.6). This pathway is also found in numerous Pseudomonads. In contrast to *P. oleovorans* and *P. fragii*, which can synthesize MCL-PHA only from related alkanoic acids present in the growth media, *Pseudomonas aeruginosa* and *Pseudomonas putida* can synthesize a similar type of MCL-PHA when grown on unrelated substrates, such as glucose (Huijberts *et al.*, 1992; Steinbüchel and Lütke-Eversloh, 2003). In these bacteria, MCL-PHA is formed from the 3-hydroxyacyl-ACP intermediates of the *de novo* fatty acid biosynthetic pathway. PhaG is a key enzyme in this pathway, having a 3-hydroxyacyl-CoA-ACP transferase activity responsible for converting the *R*-3-hydroxyacyl-ACP intermediate of the fatty acid biosynthetic pathway to *R*-3-hydroxyacyl-CoA, the substrate for the PHA synthase (Rehm *et al.*, 1998).

3.4.1. Synthesis of MCL-PHA in plants

The first approach used to synthesize MCL-PHA in plants was to divert the 3-hydroxyacyl-CoA intermediates of the β-oxidation of endogenous fatty acids. Since in plants β-oxidation occurs in the peroxisomes, PHA biosynthetic proteins needed to be targeted to this organelle. The phaC1 synthase from *P. aeruginosa* was thus modified at the carboxy end by the addition of peroxisomal targeting signal. The modified *phaC1* gene was expressed under the control of the CaMV35S promoter and transformed into *A. thaliana* (Mittendorf *et al.*, 1998). Appropriate targeting of the PHA synthase in plant peroxisomes was demonstrated by immunolocalization. TEM also showed the presence of typical PHA inclusions within the peroxisomes. The monomer composition of the MCL-PHA produced in plants reflected well the broad substrate specificity of the PHA synthase of *P. aeruginosa*. Thus, peroxisomal PHA was composed of over 14 different monomers, including saturated and unsaturated monomers ranging from 6 to 16 carbons (Mittendorf *et al.*, 1998). The majority of 3-hydroxyacids found in plant MCL-PHA could be clearly linked to the corresponding 3-hydroxyacyl-CoA generated by the β-oxidation of saturated and unsaturated fatty acids. The production of peroxisomal MCL-PHA was relatively low, with a maximal level of 0.4% dwt in 7-day-old germinating seedlings. In leaves, PHA level decreased to ~0.02% dwt. Interestingly, a two- to threefold increase in PHA was observed during leaf senescence. These data support the link between β-oxidation and PHA synthesis, since this pathway, in association with the glyoxylate cycle, is most active during germination and senescence where they are involved in the conversion of fatty acids to carbohydrates. In contrast to PHB synthesis in the cytoplasm and plastid, no negative effects of peroxisomal MCL-PHA accumulation on plant growth or seed germination were observed (Mittendorf *et al.*, 1998).

Similar to the PHA synthase from *R. eutropha*, the PHA synthase of *P. aeruginosa* is thought to accept only the *R*-isomer of 3-hydroxyacyl-CoAs. The wide range of monomers found in plant MCL-PHA suggests that, as with bacteria, plants also have enzymes capable of converting the β-oxidation intermediates *S*-3-hydroxyacyl-CoA to the *R*-isomer. Such enzymes could be either the 3-hydroxyacyl-CoA epimerase present on the plant MFP or an enoyl-CoA hydratase II activity that is specific for the generation of *R*-3-hydroxyacyl-CoA from *trans*-2-enoyl-CoA. A third route for the synthesis of a narrow range of *R*-3-hydroxyacyl-CoA is the hydration of *cis*-2-enoyl-CoA by the enoyl-CoA hydratase I activity of the MFP. The substrate *cis*-2-enoyl-CoA is derived from the β-oxidation of unsaturated fatty acids having a *cis* double bond at an even position, such as that found in linoleic and linolenic acids (Poirier, 2002).

Growth of transgenic plants in liquid media supplemented with detergents containing various fatty acids was used to study how to influence the quantity and monomer composition of PHA produced from β-oxidation. Addition of external fatty acids to plants resulted in both an increased accumulation of MCL-PHA and a shift in the monomer composition that reflected the intermediates generated by the β-oxidation of the external fatty acids (Mittendorf *et al.*, 1999). For example, addition of the detergent polyoxyethylenesorbitan esterified to lauric acid (Tween-20) to the media resulted in an eight- to tenfold increase in the amount of PHA synthesized in 14-day-old plants compared to plants growing in the same media without detergent. The monomer composition of the MCL-PHA synthesized media containing Tween-20 showed a large increase in the proportion of saturated even-chain monomers with $\leq$12 carbons, and a corresponding decrease in the proportion of all unsaturated monomers. This shift in monomer composition is accounted by the fact that β-oxidation of lauric acid, a 12 carbon saturated fatty acid, gives saturated 3-hydroxyacyl-CoA intermediates of 12 carbons and lower. Further experiments have shown that addition of either tride-, tridecenoic acid (C13:1 Δ12), or 8-methyl-nonanoic acid in the plant growth media resulted in the production of MCL-PHA containing mainly saturated odd-chain, unsaturated odd-chain, or branched-chain 3-hydroxyacid monomers, respectively (Mittendorf *et al.*, 1999). These results demonstrated that the plant β-oxidation cycle was capable of generating a large spectrum of monomers that can be included in MCL-PHA even from fatty acids that are not present in significant quantities in plants. Furthermore, "feeding" experiments with these unusual fatty acids demonstrated that all 3-hydroxyacids between 6 and 16 carbons that could be generated by the β-oxidation cycle (via the 3-hydroxyacyl-CoA intermediate) were found in the MCL-PHA. These results supported the concept that the monomer composition of PHA could be used as a tool to study the degradation pathway of fatty acids, including unsaturated fatty acids.

As an alternative to the addition of external fatty acids, modulation of the monomer composition of MCL-PHA synthesized in peroxisomes was also achieved by modifying the endogenous fatty acid biosynthetic pathway (Mittendorf *et al.*, 1999). The first example of this approach was the expression of the peroxisomal PHA synthase in a mutant of *A. thaliana* deficient in the synthesis of triunsaturated fatty acids. MCL-PHA produced from this mutant was almost completely deficient in all 3-hydroxyacids derived from the degradation of triunsaturated fatty acids, including triunsaturated monomers (Mittendorf *et al.*, 1999). Since numerous fatty acids desaturases have now been cloned and expressed in transgenic plants to control the number and position of unsaturated bonds in fatty acids, this approach could be extended to further modulate the proportion of a number of 3-hydroxyacid monomers in PHAs.

The second approach used to influence the quantity and monomer composition of MCL-PHA was the coexpression of a medium-chain thioesterase in the plastid with a PHA synthase in the peroxisome. Studies on transgenic plants expressing a laurate acyl-ACP thioesterase in the plastid of either leaves or seeds of rape revealed the presence of a futile cycling of lauric acid whereas a substantial portion of the unusual fatty acid was degraded through peroxisomal β-oxidation instead of accumulating in lipids (Eccleston and Ohlrogge, 1998; Eccleston *et al.*, 1996). These studies on lauric acid-producing rapeseed indicated that expression of a thioesterase might be a way of increasing the carbon flux toward β-oxidation and peroxisomal PHA biosynthesis. This hypothesis was tested in *A. thaliana* by combining the constitutive expression of the peroxisomal PHA synthase with the caproyl-ACP thioesterase from *Cuphea lanceolata* in the plastid (Mittendorf *et al.*, 1999). Expression of both enzymes led to a seven- to eightfold increase in the amount of MCL-PHA synthesized in plant shoots as compared to transgenics expressing only the PHA synthase. Furthermore, the composition of the MCL-PHA in the thioesterase/PHA synthase double transgenic plant was shifted toward saturated 3-hydroxyacid monomers containing

10 or fewer carbons. This shift is in agreement with an increase in the flux of decanoic acid toward β-oxidation triggered by the expression of the caproyl-ACP thioesterase (Mittendorf *et al.*, 1999). Interestingly, constitutive expression of the related lauroyl-ACP thioesterase in *A. thaliana* was shown not to lead to an increase in the genes or enzymes involved in β-oxidation (Hooks *et al.*, 1999).

The relation between fatty acid futile cycling and peroxisomal PHA synthesis was further extended to the developing seeds (Poirier *et al.*, 1999). Synthesis of MCL-PHA has been demonstrated in seeds of *A. thaliana* by expressing the peroxisomal PHA synthase gene under the control of the seed-specific napin promoter. In such transgenic plants, MCL-PHAs accumulated to 0.006% dwt in mature seeds and the monomer composition was relatively similar to the PHA synthesized in germinating seedlings. Expression of both the PHA synthase and caproyl-ACP thioesterase in the leucoplasts of developing seeds resulted in a nearly 20-fold increase in seed PHA, reaching 0.1% dwt in mature seeds. Furthermore, as found with the expression of these two enzymes in whole plants, coexpression in seeds resulted in a large increase in the proportion of 3-hydroxyacid monomers containing 10 or fewer carbons in PHA. These data clearly indicate that even though expression of the caproyl-ACP thioesterase in seeds leads to the accumulation of medium-chain fatty acids in triacylglycerides, there are still a significant proportion of these fatty acids that are channeled toward β-oxidation. This flux toward the β-oxidation cycle is thought to be quite significant, considering that there is only a fourfold difference between the maximal amount of PHA synthesized in germinating seedlings (0.4% dwt), where β-oxidation is thought to be maximal, and the PHA synthesized in the developing seeds expressing the thioesterase (0.1% dwt), where metabolism should be mainly devoted to the synthesis of fatty acid instead of degradation.

Synthesis of MCL-PHA in the peroxisomes of developing seeds has also demonstrated the presence of an increased cycling of fatty acids toward β-oxidation in plants deficient in the enzyme diacylglycerol acyltransferase (DAGAT) (Poirier *et al.*, 1999). The *tag1* mutant of *A. thaliana* was shown to be deficient in DAGAT activity in developing seeds, resulting in a decreased accumulation of triacylglycerides and corresponding increase in diacylglycerides and free fatty acids in mature seeds (Katavic *et al.*, 1995). It was hypothesized that the imbalance created between the capacity of the plastid to synthesize fatty acids and the capacity of the lipid biosynthetic machinery of the ER to include these fatty acids into triacylglycerides might have two basic consequences: either fatty acid biosynthesis would be reduced (feedback inhibited) in order to match it with triacylglyceride biosynthesis or excess fatty acids that cannot be included in triacylglycerides would be channeled toward β-oxidation. Expression of the peroxisomal PHA synthase in the *tag1* mutant resulted in a tenfold increase in the amount of MCL-PHA accumulating in mature seeds compared to expression of the transgene in wild-type plants (Poirier *et al.*, 1999). Although these results do not address whether fatty acid biosynthesis is decreased in the *tag1* mutant, they nevertheless clearly indicate that a decrease in triacylglyceride biosynthesis results in an increase in the flux of fatty acids toward β-oxidation. Thus, carbon flux to the β-oxidation cycle can be modulated to a great extent and appears to play an important role in lipid homeostasis in plants even in tissues that are primarily devoted to lipid biosynthesis, such as the developing seeds.

Analysis of futile cycling of fatty acids in developing seeds has been extended to transgenic plants accumulating the unusual fatty acids, ricinoleic acid and vernolic acid (Moire *et al.*, 2004). *A. thaliana* expressing either the *Ricinus communis* oleate 12-hydroxylase or the *Crepis palaestina* linoleate 12-epoxygenase under the control of the napin promoter was shown to accumulate approximately twofold more MCL-PHA in developing seeds compared to control. Although relatively small compared to the increase in PHA observed in transgenic plants expressing the *C. lanceolata* caproyl-ACP thioesterase, the twofold increase in MCL-PHA was quite significant considering that the steady level of either hydroxy or epoxy fatty acids accumulated in transgenic seeds represented only 6.3 mol% or 3.1 mol%, respectively. Thus, clearly, a larger proportion of unusual fatty acids were being degraded via peroxisomal β-oxidation in developing seeds compared to the common fatty acids. Interestingly, microarray analysis of nearly 200 genes involved in fatty acid biosynthesis and degradation, including the genes encoding enzymes of the β-oxidation cycle, revealed no changes in gene expression in transgenic developing seeds expressing either *C. lanceolata* caproyl-ACP thioesterase, *R. communis* oleate 12-hydroxylase, or *C. palaestina* linoleate 12-epoxygenase (Moire *et al.*, 2004). These results indicated that analysis of peroxisomal PHA is a better indicator of the flux of fatty acid through β-oxidation than the expression profile of genes involved in lipid metabolism.

Synthesis of a "hybrid" PHA copolymer has been reported in *A. thaliana* expressing a PHA synthase from *A. caviae* modified at the carboxy terminal end for targeting to the peroxisome (Arai *et al.*, 2002). Expression of this PHA synthase under the control of the CaMV35S promoter leads to the accumulation of a PHA containing even-chain and odd-chain monomers ranging from 4 to 6 carbons. The maximal amount of PHA accumulated in leaves and seeds was 0.04% and 0.0032% dwt, respectively. Growth of transgenic plants in media containing Tween-20 increased the total amount of PHA synthesized without affecting appreciably the monomer composition (Arai *et al.*, 2002). The incorporation of ~25 mol% of 3-hydroxyvalerate into PHA raises the interesting question of the source of the odd-chain monomer. Although odd-chain monomers have been detected in MCL-PHA synthesized from the expression of the *P. aeruginosa* PHA synthase in the peroxisome, the amount of odd-chain monomers was very low (<1 mol%). It is possible that an α-oxidation pathway could generate odd-chain intermediate from even-chain fatty acids and that this pathway is more active toward shorter chain intermediates (i.e., 6 carbon fatty acids). Although a gene involved in α-oxidation has been identified, the corresponding protein has not been linked to the peroxisome (Hamberg *et al.*, 1999). Thus, despite evidence of a complete α-oxidation pathway in plants, the link between this pathway and the peroxisome needs to be established. PHA thus offers potentially a unique handle to study α-oxidation in plants.

In the bacterial pathway of MCL-PHA synthesis from intermediates of fatty acid biosynthesis, the enzyme phaG plays a key role, catalyzing the conversion of *R*-3-hydroxyacyl-ACP to *R*-3-hydroxyacyl-CoA, the latter being the substrate for the PHA synthase (Rehm *et al.*, 1998). The identification and cloning of the *P. putida phaG* gene opened the possibility of synthesizing PHA copolymers in the plastids of plants from intermediates of fatty acid biosynthesis. Unfortunately, constitutive expression in the plastid of *A. thaliana* of only the phaG enzyme led to a marked deleterious effect on plant growth, the plants being dwarfed with crinkly leaves and the seed set being strongly reduced (V. Mittendorf, unpublished results). The reason for this phenotype is not known but is thought to be perhaps due to interference of the transacylase with fatty acid biosynthesis. If this is the case, it would be interesting to know why this does not occur in bacteria expressing phaG. Coexpression in the plastid of the *P. aeruginosa* PHA synthase along with phaG did not conclusively lead to PHA accumulation in Arabidopsis (V. Mittendorf, unpublished results). Analogous experiments in potato led to similar conclusions, although evidence for the synthesis of a very small amount of a hydrophobic polymer that could be MCL-PHAs was provided (Romano *et al.*, 2005). Thus, despite the obvious advantages of the plastid as a location for the production of PHB and P(HB-HV), the synthesis in this organelle of PHA copolymer using fatty acid biosynthetic intermediates appears problematic at present.

The synthesis of MCL-PHA in potato cell lines has been demonstrated through expression of the PHA synthase from *P. oleovorans* in the cytoplasm (Romano *et al.*, 2003). PHA could be detected only after "feeding" the cell lines with 3-hydroxyoctanoic acid, with the PHA containing only the 8 carbon monomer. These results indicate that while no endogenous 3-hydroxyacyl-CoA could be detected in the cytoplasm, an acyl-CoA synthetase activity capable of converting 3-hydroxyoctanoic acid (that originally comes from the external media) to the corresponding 3-hydroxyacyl-CoA was present. The amount of PHA detected reached up to 1% dwt.

3.5. Future perspectives

A spectrum of PHAs has now been successfully synthesized in plants by using various metabolic pathways. These ranges from the stiff and brittle PHB to the more flexible P(HB-HV) plastic and MCL-PHA elastomers and glues. Experiments have shown that in some case very high amount of polymer can be produced at, however, a considerable metabolic cost. The challenge for the future is to succeed in the accumulation of adequate amounts of PHA ($\geq$15% dwt) without affecting yield. For some agricultural production strategies, it will also be necessary to succeed in harvesting PHA without affecting the recovery of other plant products, such as oils, protein, or starch. This is important since in contrast to the production of PHA by bacterial fermentation, where the system is designed to produce mainly PHA with little residual waste, a large-scale agricultural production of PHA may be viable only through the recovery of not only PHA but also all other valuable components of the crop. For example, in the case of an oil crop such as *B. napus*, one must be able to recover PHA and the oil, as well as still being able to use the de-lipidized protein-rich meal for animal feed. In the case of a carbohydrate-producing crop such as either sugar beet or sugarcane, both sucrose and PHA would have to be recovered. An alternative strategy could be used whereby crop plants would be grown only for biomass and PHA production. An example would be the synthesis of PHA in switchgrass, where the residual biomass remaining after PHA extraction could be used for energy production. We know thus far that PHB can be produced in the seed of rape to 8% dwt without obvious deleterious effects on plant growth and germination (Houmiel *et al.*, 1999). Thus, the goal of producing adequate level of PHA in crops without yield penalty appears realistic.

The success of using transgenic plants as a source of novel material will depend not only on the production levels achieved but also on whether the polymers can be extracted efficiently, economically, and ecologically from crops. Although a number of strategies have been described in the literature for the extraction of PHA, further work is required to validate these extraction processes in the context of large-scale production in plants (Poirier, 2001).

REFERENCES

Agrawal, V. P., and Kolattukudy, P. E. (1978a). Purification and characterization of a wound-induced omega-hydroxy fatty acid: NADP oxidoreductase from potato tuber disks. *Arch. Biochem. Biophys.* **191,** 452–465.

Agrawal, V. P., and Kolattukudy, P. E. (1978b). Mechanism of action of a wound-induced omega-hydroxy fatty acid: NADP oxidoreductase isolated from potato tubers (*Solanum tuberosum* L.). *Arch. Biochem. Biophys.* **191,** 466–478.

Aharoni, A., Dixit, S., Jetter, R., Thoenes, E., Van Arkel, G., and Pereira, A. (2004). The shine clade of AP2 domain transcription factors activates wax biosynthesis, alters cuticular properties, and confers drought resistance when overexpressed in Arabidopsis. *Plant Cell* **16,** 2463–2480.

Aloni, B., Karni, L., Rylski, I., Cohen, Y., Lee, Y., Fuchs, M., Moreshet, S., and Yao, C. (1998). Cuticular cracking in pepper fruit. Effects of night temperature and humidity. *J. Hortic. Sci.* **73,** 743–749.

Anderson, A. J., and Dawes, E. A. (1990). Occurrence, metabolism, metabolic role, and industrial uses of bacterial polyhydroxyalkanoates. *Microbiol. Rev.* **54,** 450–472.

Arai, Y., Nakashita, H., Doi, Y., and Yamaguchi, I. (2001). Plastid targeting of polyhydroxybutyrate biosynthetic pathway in tobacco. *Plant Biotechnol.* **18,** 289–293.

Arai, Y., Nakashita, H., Suzuki, Y., Kobayashi, Y., Shimizu, T., Yasuda, M., Doi, Y., and Yamaguchi, I. (2002). Synthesis of a novel class of polyhydroxyalkanoates in *Arabidopsis* peroxisomes, and their use in monitoring short-chain-length intermediates of β-oxidation. *Plant Cell Physiol.* **43,** 555–562.

Arai, Y., Shikanai, T., Doi, Y., Yoshida, S., Yamaguchi, I., and Nakashita, H. (2004). Production of polyhydroxybutyrate by polycistronic expression of bacterial genes in tobacco plastid. *Plant Cell Physiol.* **45,** 1176–1184.

Becraft, P. W., Stinard, P. S., and Mccarty, D. R. (1996). CRINKLY4: A TNKR-like receptor kinase involved in epidermal maize differentiation. *Science* **273,** 1406–1409.

Beneviste, I., Tijet, N., Adas, F., Philipps, G., Salaun, J. P., and Durst, F. (1998). CYP86A1 from *Arabidopsis thaliana* encodes a cytochrome P450-dependent fatty acid omega-hydroxylase. *Biochem. Biophys. Res. Commun.* **243,** 688–693.

Bernards, M. A. (2002). Demystifying suberin. *Can. J. Bot.* **80,** 227–240.

Bernards, M. A., and Lewis, N. G. (1998). The macromolecular aromatic domain in suberized tissue: A changing paradigm. *Phytochemistry* **47,** 915–933.

Blée, E., and Schuber, F. (1990). Efficient epoxidation of unsaturated fatty acids by a hydroperoxide-dependent oxygenase. *J. Biol. Chem.* **265,** 12887–12894.

Blée, E., and Schuber, F. (1992). Occurrence of fatty acid epoxide hydrolases in soybean (*Glycine max.*). *Biochem. J.* **282,** 711–714.

Blée, E., and Schuber, F. (1993). Biosynthesis of cutin monomers: Involvement of a lipoxygenase/peroxygenase pathway. *Plant J.* **4,** 113–123.

Blée, E., and Schuber, F. (1995). Stereocontrolled hydrolysis of the linoleic acid monoepoxide regioisomers catalyzed by soybean epoxide hydrolase. *Eur. J. Biochem.* **230,** 229–234.

Bohmert, K., Balbo, I., Kopka, J., Mittendorf, V., Nawrath, C., Poirier, Y., Tischendorf, G., Trethewey, R. N., and Willmitzer, L. (2000). Transgenic Arabidopsis plants can accumulate polyhydroxybutyrate to up 4% of their fresh weight. *Planta* **211,** 841–845.

Bohmert, K., Balbo, I., Steinbüchel, A., Tischendorf, G., and Willmitzer, L. (2002). Constitutive expression of the β-ketothiolase gene in transgenic plants. A major obstacle for obtaining polyhydroxybutyrate-producing plants. *Plant Physiol.* **128,** 1282–1290.

Bonaventure, G., Beisson, F., Ohlrogge, J., and Pollard, M. (2004). Analysis of the aliphatic monomer composition of polyesters associated with *Arabidopsis* epidermis: Occurrence of octadeca-*cis*-6, *cis*-9-diene-1,18-dioate as the major component. *Plant J.* **40,** 920–930.

Braunegg, G., Lefebvre, G., and Genser, K. F. (1998). Polyhydroxyalkanoates, biopolyesters from renewable resources: Physiological and engineering aspects. *J. Biotechnol.* **65,** 127–161.

Broun, P., Poindexter, P., Osborne, E., Jiang, C. Z., and Riechmann, J. L. (2004). WIN1, a transcriptional activator of epidermal wax accumulation in *Arabidopsis*. *Proc. Natl. Acad. Sci. USA* **101,** 4706–4711.

Cabello-Hurtado, F., Batard, Y., Salaün, J.-P., Durst, F., Pinot, F., and Werck-Reichart, D. (1998). Cloning, expression in yeast, and functional characterization of CYP81B1, a plant cytochrome P450 that catalyzes in-chain hydroxylation of fatty acids. *J. Biol. Chem.* **273,** 7260–7267.

Chen, X., Goodwin, S. M., Boroff, V. L., Liu, X., and Jenks, M. A. (2003). Cloning and characterization of the *wax2* gene of Arabidopsis involved in cuticle membrane and wax production. *Plant Cell* **15,** 1170–1185.

Cordeiro, N., Belgacem, M. N., Gandini, A., and Neto, C. P. (1997). Urethanes and polyurethanes from suberin. 1. Kinetic study. *Ind. Crop Prod.* **6,** 163–167.

Cordeiro, N., Belgacam, N. M., Gandini, A., and Pascal Neto, C. (1998). Cork suberin as a new source of chemical: 2. Crystallinity, thermal and rheological properties. *Bioresour. Technol.* **63,** 153–158.

Cordeiro, N., Blayo, A., Belgacem, N. M., Gandini, A., Pascoal Neto, C., and Lenest, J.-F. (2000). Cork suberin as an additive in offset lithographic printing. *Ind. Crop Prod.* **11,** 63–71.

Croteau, R., and Kolattukudy, P. E. (1973). Enzymatic biosynthesis of a hydroxy fatty acid polymer, cutin, by a particulate preparation from *Vicia faba* epidermis. *Biochem. Biophys. Res. Commun.* **52,** 863–869.

Croteau, R., and Kolattukudy, P. E. (1975). Biosynthesis of hydroxy fatty acid polymers. Enzymatic epoxidation of 18-hydroxyoleic acid to 18-hydroxy-*cis*-9,10-epoxystearic acid by a particular preparation from spinach (*Spinacea oleracea*). *Arch. Biochem. Biophys.* **170,** 61–72.

de Koning, G. (1995). Physical properties of bacterial poly((R)3-hydroxyalkanoates). *Can. J. Microbiol.* **41**(Suppl. 1), 303–309.

Domergue, F., Bessoule, J.-J., Moreau, P., Lessire, R., and Cassagne, C. (1998). Recent advances in plant fatty acid elongation. *In* "Plant Lipid Biosynthesis" (J. L. Harwood, ed.), pp. 185–220. Cambridge University Press, Cambridge.

Eccleston, V. S., and Ohlrogge, J. B. (1998). Expression of lauroyl-acyl carrier protein thioesterase in *Brassica napus* seeds induces pathways for both fatty acid oxidation and biosynthesis and implies a set point for triacylglycerol accumulation. *Plant Cell* **10,** 613–621.

Eccleston, V. S., Cranmer, A. M., Voelker, T. A., and Ohlrogge, J. B. (1996). Medium-chain fatty acid biosynthesis and utilization in *Brassica napus* plants expressing lauroyl-acyl carrier protein thioesterase. *Planta* **198,** 46–53.

Esau, K. (1977). "Anatony of Seed Plants," 2nd Edn. Wiley, New York.

Franke, R., Briesen, I., Wojciechowski, T., Faust, A., Yephremov, A., Nawrath, C., and Schreiber, L. (2005). Apoplastic polyesters in *Arabidopsis* surface tissues—A typical suberin and a particular cutin. *Phytochemistry* **66,** 2643–2658.

Fukui, T., and Doi, Y. (1997). Cloning and analysis of the poly(3-hydroxybutyrate-*co*-3-hydroxyhexanoate) biosynthesis genes of *Aeromonas caviae*. *J. Bacteriol.* **179,** 4821–4830.

Fukui, T., Shiomi, N., and Doi, Y. (1998). Expression and characterization of (R)-specific enoyl coenzyme a hydratase involved in polyhydroxyalkanoate biosynthesis by *Aeromonas caviae*. *J. Bacteriol.* **180,** 667–673.

Garcia-Olemedo, F., Molina, A., Segura, A., and Moreno, M. (1995). The defensive role of nonspecific lipid-transfer proteins in plants. *Trends Microbiol.* **3,** 72–74.

Graça, J., and Pereira, H. (1997). Cork suberin: A glyceryl-based polyester. *Holzforschung* **51,** 225–234.
Graça, J., and Pereira, H. (2000a). Methanolysis of bark suberins: Analysis of glycerol and acid monomers. *Phytochem. Anal.* **11,** 45–51.
Graça, J., and Pereira, H. (2000b). Suberin in potato periderm: Glycerol, long-chain monomers, and glyceryl and feruloyl dimers. *J. Agric. Food Chem.* **48,** 5476–5483.
Graça, J., and Pereira, H. (2000c). Diglycerol alkenedioates in suberin: Building units of a poly(acylglycerol) polyester. *Biomacromolecules* **1,** 519–522.
Graça, J., Schreiber, L., Rodrigues, J., and Pereira, H. (2002). Glycerol and glyceryl esters of ω-hydroxyacids in cutins. *Phytochemistry* **61,** 205–215.
Hahn, J. J., Eschenlauer, A. C., Sleytr, U. B., Somers, D. A., and Srienc, F. (1999). Peroxisomes as sites for synthesis of polyhydroxyalkanoates in transgenic plants. *Biotechnol. Prog.* **15,** 1053–1057.
Hamberg, M., and Hamberg, G. (1990). Hydroperoxide-dependent epoxidation of unsaturated fatty acids in the broad bean (*Vicia faba* L.). *Arch. Biochem. Biophys.* **283,** 409–416.
Hamberg, M., Sanz, A., and Castresana, C. (1999). Alpha-oxidation of fatty acids in higher plants: Identification of a pathogen-inducible oxygenase (PIOX) as an alpha-dioxygenase and biosynthesis of 2-hydroperoxylinolenic acid. *J. Biol. Chem.* **274,** 24503–24513.
Heredia, A. (2003). Biophysical and biochemical characteristics of cutin, a plant barrier biopolymer. *Biochim. Biophys. Acta* **1620,** 1–7.
Hollenbach, B., Schreiber, L., Hartung, W., and Dietz, K.-J. (1997). Cadmium leads to stimulated expression of the lipid transfer protein genes in barley: Implications for the involvement of lipid transfer proteins in wax assembly. *Planta* **203,** 9–19.
Holloway, P. J. (1982). Structure and histochemistry of plant cuticular membranes: An overview. *In* "The Plant Cuticle" (D. F. Cutler, K. L. Alvin, and C. E. Price, eds.), pp. 1–32. Academic Press, London.
Hooks, M. A., Fleming, Y., Larson, T. R., and Graham, I. A. (1999). No induction of β-oxidation in leaves of *Arabidopsis* that over-produce lauric acid. *Planta* **207,** 385–392.
Houmiel, K. L., Slater, S., Broyles, D., Casagrande, L., Colburn, S., Gonzalez, K., Mitsky, T. A., Reiser, S. E., Shah, D., Taylor, N. B., Tran, M., Valentin, H. E., *et al.* (1999). Poly(beta-hydroxybutyrate) production in oilseed leucoplasts of *Brassica napus*. *Planta* **209,** 547–550.
Huijberts, G. N. M., Eggink, G., De Waard, P., Huisman, G. W., and Witholt, B. (1992). *Pseudomonas putida* KT2442 cultivated on glucose accumulates poly(3-hydroxyalkanoates) consisting of saturated and unsaturated monomers. *Appl. Environ. Microbiol.* **58,** 536–544.
Jeffree, C. E. (1996). Structure and ontogeny of plant cuticles. *In* "Plant Cuticles: An Integrated Functional Approach" (G. Kertiens, ed.), pp. 33–82. BIOS Scientific Publishers Limited, Oxford.
Jendrossek, D. (2002). Microbial degradation of polyesters. *Adv. Biochem Eng. Biotechnol.* **71,** 293–325.
Jenks, M., Joly, R. J., Peters, P. J., Rich, P. J., Axtell, J. D., and Ashworth, E. N. (1994). Chemically induced cuticle mutation affecting epidermal conductance to water vapor and disease susceptibility in *Sorghum bicolor* (L.) Moench. *Plant Physiol.* **105,** 1239–1245.
Jin, P., Guo, T., and Becraft, P. W. (2000). The maize CR4 receptor-like kinase mediates a growth factor-like differentiation response. *Genesis* **27,** 104–116.
John, M. E., and Keller, G. (1996). Metabolic pathway engineering in cotton: Biosynthesis of polyhydroxybutyrate in fiber cells. *Proc. Natl. Acad. Sci. USA* **93,** 12768–12773.
Kader, J.-C. (1996). Lipid-transfer proteins in plants. *Annu. Rev. Plant Physiol. Plant Mol. Biol.* **47,** 627–654.
Kahn, R. A., and Durst, F. (2000). Function and evolution of plant cytochrome P450. *In* "Evolution of Metabolic Pathways" (J. T. Romeo, R. Ibrahim, L. Varin, and V. de Luca, eds.), pp. 151–189, Recent Advances in Phytochemistry **34,** Elsevier Science Ltd, London.
Katavic, V., Reed, D. W., Taylor, D. C., Giblin, E. M., Barton, D. L., Zou, J., Mackenzie, S. L., Covello, P. S., and Kunst, L. (1995). Alteration of seed fatty acid composition by an ethyl methanesulfonate-induced mutation in *Arabidopsis thaliana* affecting diacylglycerol acyltransferase activity. *Plant Physiol.* **108,** 399–409.
Kerstiens, G. (1996). Diffusion of water vapour and gases across cuticles and through stomatal pores presumed closed. *In* "Plant Cuticles: An Integrated Functional Approach" (G. Kertiens, ed.), pp. 121–134. BIOS Scientific Publishers Limited, Oxford.

Kim, Y. B., and Lenz, R. W. (2002). Polyesters from microorganisms. *Adv. Biochem. Eng. Biotechnol.* **71,** 51–79.

Kolattukudy, P. E. (1981). Structure, biosynthesis, and biodegradation of cutin and suberin. *Annu. Rev. Plant. Physiol.* **32,** 539–567.

Kolattukudy, P. E. (1996). Biosynthetic pathways of cutin and waxes, and their sensitivity to environmental stresses. *In* "Plant Cuticles: An Integrated Functional Approach" (G. Kerstiens, ed.), pp. 83–108. BIOS Scientific Publishers Limited, Oxford.

Kolattukudy, P. E. (2001). Polyesters in higher plants. *Adv. Biochem. Eng. Biotechnol.* **71,** 1–49.

Kourtz, L., Dillon, K., Daughtry, S., Madison, L. L., Peoples, O., and Snell, K. D. (2005). A novel thiolase-reductase gene fusion promotes the production of polyhydroxybutyrate in Arabidopsis. *Plant Biotechnol. J.* **3,** 435–447.

Krizkova, L., Lopes, M. H., Polonyi, J., Belicova, A., Dobias, J., and Ebringer, L. (1999). Antimutagenicity of a suberin extract from *Quercus suber* cork. *Mutat. Res.* **446,** 225–230.

Krolikowski, K. A., Victor, J. L., Wagler, T. N., Lolle, S. J., and Pruitt, R. E. (2003). Isolation and characterization of the Arabidopsis organ fusion gene HOTHEAD. *Plant J.* **35,** 501–511.

Kunst, L., and Samuels, A. L. (2003). Biosynthesis and secretion of plant cuticular wax. *Prog. Lipid Res.* **42,** 51–80.

Kurata, T., Kawabata-Awai, C., Sakuradani, E., Shimizu, S., Okada, K., and Wada, T. (2003). The *YORE-YORE* gene regulates multiple aspects of epidermal cell differentiation in. *Arabidopsis. Plant J.* **36,** 55–66.

Kurdyukov, S., Faust, A., Trenkamp, S., Bär, S., Franke, R., Efremova, N., Tietjen, K., Schreiber, L., Saedler, H., and Yephremov, A. (2006a). Genetic and biochemical evidence for involvement of α,ω-dicarboxylic acids in the formation of extracellular matrix. *Planta* **224,** 315–329.

Kurdyukov, S., Faust, A., Nawrath, C., Bär, S., Voisin, D., Franke, R., Schreiber, L., Saedler, H., Métraux, J.-P., and Yephremov, A. (2006b). The epidermis-specific extracellular α/β hydrolase BODYGUARD, an Arabidopsis homologue of fungal cutinases, controls cuticle development and morphogenesis in plants. *Plant Cell* **18,** 321–339.

Lageveen, R. G., Huisman, G. W., Preusting, H., Ketelaar, P., Eggink, G., and Witholt, B. (1995). Formation of polyesters by *Pseudomonas oleovorans*: Effect of substrates on formation and composition of poly-(R)-3-hydroxyalkanoates and poly-(R)-3-hydroxyalkenoates. *Appl. Environ. Microbiol.* **54,** 2924–2932.

Le Bouquin, R., Skarbs, M., Kahn, R., Beneviste, I., Salaün, J.-P., Schreiber, L., Durst, F., and Pinot, F. (2001). CYP94A5, a new cytochrome P450 from *Nicotiana tabacum* is able to catalyze the oxidation of fatty acids to the omega-alcohol and to the corresponding diacid. *Eur. J. Biochem.* **268,** 3083–3090.

Lolle, S. J., and Cheung, A. Y. (1993). Promiscuous germination and growth of wildtype pollen from *Arabidopsis* and related species on the shoot of the *Arabidopsis* mutant, *fiddlehead. Dev. Biol.* **155,** 250–258.

Lolle, S. J., and Pruitt, R. E. (1999). Epidermal cell interactions: A case for local talk. *Trends Plant Sci.* **4,** 14–20.

Lolle, S. J., Berlyn, G. P., Engstrom, E. M., Krolikowski, K. A., Reiter, W. D., and Pruitt, R. E. (1997). Developmental regulation of cell interactions in the *Arabidopsis fiddlehead1* mutant: A role for the epidermal cell wall and cuticle. *Dev. Biol.* **189,** 311–321.

Lolle, S. J., Hsu, W., and Pruitt, R. E. (1998). Genetic analysis of organ fusion in *Arabiopsis thaliana. Genetics* **149,** 607–619.

Lössl, A., Eibl, C., Harloff, H.-J., Jung, C., and Koop, H.-U. (2003). Polyester synthesis in transplastomic tobacco (*Nicotiana tabacum* L.): Significant contents of polyhydroxybutyrate are associated with growth reduction. *Plant Cell Rep.* **21,** 891–899.

Lössl, A., Bohmert, K., Harloff, H., Eibl, C., Mühlbauer, S., and Koop, H.-U. (2005). Inducible trans-activation of plastid transgenes: Expression of the *R. eutropha* phb operon in transplastomic tobacco. *Plant Cell Physiol.* **46,** 1462–1471.

Maldonado, A. M., Doerner, P., Dixon, R. A., Lamb, C. J., and Cameron, R. K. (2002). A putative lipid tranfer protein involved in systemic resistance signaling in Arabidopsis. *Nature* **419,** 399–403.

Matsumoto, K., Nagao, R., Murata, T., Arai, Y., Kichise, T., Nakashita, H., Taguchi, S., Shimada, H., and Doi, Y. (2005). Enhancement of poly(3-hydroxybutyrate-co-3-hydroxyvalerate) production in the

transgenic *Arabidopsis thaliana* by the in vitro evolved highly active mutants of polyhydroxyalkanoate (PHA) synthase from *Aeromonas caviae*. *Biomacromolecules* **6,** 2126–2130.
Menzel, G., Harloff, H.-J., and Jung, H. -C. (2003). Expression of bacterial poly(3-hydroxybutyrate) synthesis genes in hairy roots of sugar beet (*Beta vulgaris* L.). *Appl. Microbiol. Biotechnol.* **60,** 571–576.
Mittendorf, V., Robertson, E. J., Leech, R. M., Krüger, N., Steinbüchel, A., and Poirier, Y. (1998). Synthesis of medium-chain-length polyhydroxyalkanoates in *Arabidopsis thaliana* using intermediates of peroxisomal fatty acid β-oxidation. *Proc. Natl. Acad. Sci. USA* **95,** 13397–13402.
Mittendorf, V., Bongcam, V., Allenbach, L., Coullerez, G., Martini, N., and Poirier, Y. (1999). Polyhydroxyalkanoate synthesis in transgenic plants as a new tool to study carbon flow through β-oxidation. *Plant J.* **20,** 45–55.
Moire, L., Schmutz, A., Buchala, A., Yan, B., Stark, R., and Ryser, U. (1999). Glycerol is a suberin monomer. New experimental evidence for an old hypothesis. *Plant Physiol.* **119,** 1137–1146.
Moire, L., Rezzonico, E., Goepfert, S., and Poirier, Y. (2004). Impact of unusual fatty acid synthesis on futile cycling through β-oxidation and on gene expression in transgenic plants. *Plant Physiol.* **134,** 432–442.
Molina, A., and Garcia-Olmedo, F. (1997). Enhanced tolerance to bacterial pathogens caused by the transgenic expression of barley lipid transfer protein LTP2. *Plant J.* **12,** 669–675.
Morisseau, C., Beetham, J. K., Pinot, F., Debernard, S., Newman, J. W., and Hammock, B. D. (2000). Cress and potato soluble epoxide hydrolases: Purification, biochemical characterization, and comparison to mammalian enzymes. *Arch. Biochem. Biophys.* **378,** 321–332.
Nakashita, H., Arai, Y., Yoshioka, K., Fukui, T., Doi, Y., Usami, R., Horikoshi, K., and Yamaguchi, I. (1999). Production of biodegradable polyester by a transgenic tobacco. *Biosci. Biotechnol. Biochem.* **63,** 870–874.
Nakashita, H., Arai, Y., Shikanai, T., Doi, Y., and Yamaguchi, I. (2001). Introduction of bacterial metabolism into higher plants by polycistronic transgene expression. *Biosci. Biotechnol. Biochem.* **65,** 1688–1691.
Nawrath, C. (2002). The biopolymers cutin and suberin. *In* "The Arabidopsis Book" (C. Somerville and E. Meyerowitz, eds.), pp. 1–15. American Society of Plant Physiologists, Rockville.
Nawrath, C. (2006). Unraveling the complex network of cuticular structure and function. *Curr. Opin. Plant Biol.* **9,** 281–287.
Nawrath, C., Poirier, Y., and Somerville, C. R. (1994). Targeting of the polyhydroxybutyrate biosynthetic pathway to the plastids of *Arabidopsis thaliana* results in high-levels of polymer accumulation. *Proc. Natl. Acad. Sci. USA* **91,** 12760–12764.
Pighin, J. A., Zheng, H. Q., Balakshin, L. J., Goodman, I. P., Western, T. L., Jetter, R., Kunst, L., and Samuels, A. L. (2004). Plant cuticular lipid export requires an ABC transporter. *Science* **306,** 702–704.
Pinot, F., Salaün, J.-P., Bosch, H., Lesot, A., Mioskowski, C., and Durst, F. (1992). Omega-hydroxylation of Z-9-octadecenoic, Z-9,10-epoxistearic and 9,10-dihydroxystearic acids by microsomal cytochrome P450 systems from *Vicia sativa*. *Biochem. Biophys. Res. Commun.* **184,** 183–193.
Pinot, F., Bosch, H., Salaün, J. P., Durst, F., Mioskowski, C., and Hammock, B. D. (1997). Epoxide hydrolase activities in the microsomes and the soluble fraction from *Vicia sativa* seedlings. *Plant Physiol. Biochem.* **35,** 103–110.
Pinot, F., Beneviste, I., Salaün, J. P., and Durst, F. (1998). Methyl jasmonate induces lauric acid omega-hydroxylase activity and accumulation of CYP94A1 transcripts but does not affect epoxide hydrolase activities in *Vicia sativa* seedlings. *Plant Physiol.* **118,** 1481–1486.
Poirier, Y. (1999). Production of new polymeric compounds in plants. *Curr. Opin. Biotechnol.* **10,** 181–185.
Poirier, Y. (2001). Production of polyesters in transgenic plants. *Adv. Biochem. Eng. Biotechnol.* **71,** 209–240.
Poirier, Y. (2002). Poylhydroxyalkanoate synthesis in plants as a tool for biotechnology and basic studies of lipid metabolism. *Prog. Lipid Res.* **41,** 131–155.
Poirier, Y., and Gruys, K. J. (2001). Production of PHAs in transgenic plants. *In* "Biopolyesters" (Y. Doi and A. Steinbüchel, eds.), pp. 401–435. Wiley-VCH, Weinheim.
Poirier, Y., Dennis, D. E., Klomparens, K., and Somerville, C. (1992a). Polyhydroxybutyrate, a biodegradable thermoplastic, produced in transgenic plants. *Science* **256,** 520–523.

Poirier, Y., Dennis, D., Klomparens, K., Nawrath, C., and Somerville, C. (1992b). Perspectives on the production of polyhydroxyalkanoates in plants. *FEMS Microbiol. Rev.* **103,** 237–246.

Poirier, Y., Nawrath, C., and Somerville, C. (1995a). Production of polyhydroxyalkanoates, a family of biodegradable plastics and elastomers, in bacteria and plants. *Biotechnology* **13,** 142–150.

Poirier, Y., Schechtman, L. A., Satkowski, M. M., Noda, I., and Somerville, C. (1995b). Synthesis of high molecular weight poly([R]-(−)-3-hydroxybutyrate) in transgenic *Arabidopsis thaliana* plant cells. *Int. J. Biol. Macromol.* **17,** 7–12.

Poirier, Y., Ventre, G., and Caldelari, D. (1999). Increased flow of fatty acids towards β-oxidation in developing seeds of *Arabidopsis thaliana* deficient in diacylglycerol acyltransferase activity or synthesizing medium-chain fatty acids. *Plant Physiol.* **121,** 1359–1366.

Pruitt, R. E., Vielle-Calzada, J.-P., Ploense, S. E., Grossniklaus, U., and Lolle, S. J. (2000). *FIDDLEHEAD,* a gene required to suppress epidermal cell interactions in Arabidopsis, encodes a putative lipid biosynthetic enzyme. *Proc. Natl. Acad. Sci.* **97,** 1311–1316.

Pyee, J., and Kolattukudy, P. E. (1995). The gene for the major cuticular wax-associated protein and three homologous genes from broccoli (*Brassica oleracea*) and their expression patterns. *Plant J.* **7,** 49–59.

Rehm, B. H., Krüger, N., and Steinbüchel, A. (1998). A new metabolic link between fatty acid de novo synthesis and polyhydroxyalkanoic acid synthesis. The PHAG gene from *Pseudomonas putida* KT2440 encodes a 3-hydroxyacyl-acyl carrier protein-coenzyme a transferase. *J. Biol. Chem.* **273,** 24044–24051.

Reina, J. J., and Heredia, A. (2001). Plant cutin biosynthesis: The involvement of a new acyltransferase. *Trends Plant Sci.* **6,** 296.

Reiser, S. E., Mitsky, T. A., and Gruys, K. J. (2000). Characterization and cloning of an (R)-specific trans-2,3-enoylacyl-CoA hydratase from *Rhodospirillum rubrum* and use of this enzyme for PHA production in. *Escherichia coli. Appl. Microbiol. Biotechnol.* **53,** 209–218.

Reusch, R. N. (1999). Polyphosphate/poly-(R)-3-hydroxybutyrate) ion channels in cell membranes. *Prog. Mol. Subcell. Biol.* **23,** 151–182.

Reusch, R. N., Huang, R., and Bramble, L. L. (2002). Poly-3-hydroxybutyrate/polyphosphate complexes form voltage-activated Ca^{2+} channels in the plasma membranes of *Escherichia coli. Biophys. J.* **527,** 319–322.

Riederer, M., and Schreiber, L. (2001). Protecting against water loss: Analysis of the barrier properties of plant cuticles. *J. Exp. Bot.* **52,** 2023–2032.

Rocha, S. M., Goodfellow, B. J., Delgadillo, I., Neto, C. P., and Gil, A. M. (2001). Enzymatic isolation and structural characterization of polymeric suberin of cork from *Quercus suber* L. *Int. J. Biol. Macromol.* **28,** 107–119.

Romano, A., Vreugdenhil, D., Jamar, D., van der Plas, L. H. W., De Roo, G., Witholt, B., Eggink, G., and Mooibroek, H. (2003). Evidence of medium-chain-length polyhydroxyoctanoate accumulation in transgenic potato lines expressing the *Pseudomonas oleovorans* Pha-C1 polymerase in the cytoplasm. *Biochem. Eng. J.* **3728,** 1–9.

Romano, A., van der Plas, L. H. W., Witholt, B., Eggink, G., and Mooibroek, H. (2005). Expression of poly-3-(R)-hydroxyalkanoate (PHA) polymerase and acyl-CoA-transacylase in plastids of transgenic potato leads to the synthesis of a hydrophobic polymer, presumably medium-chain-length PHAs. *Planta* **220,** 455–464.

Saruul, P., Srienc, F., Somers, D. A., and Samac, D. A. (2002). Production of a biodegradable plastic polymer, poly-*beta*-hydroxybutyrate, in transgenic alfalfa. *Crop Sci.* **42,** 919–927.

Schnurr, J., Shockey, J., and Browse, J. (2004). The acyl-CoA synthetase encoded by *lacs2* is essential for normal cuticle development in Arabidopsis. *Plant Cell* **16,** 629–642.

Schreiber, L., Hartmann, K., Skrabs, M., and Zeier, J. (1999). Apoplastic barriers in roots: Chemical composition of endodermal and hypodermal cell walls. *J. Exp. Bot.* **50,** 1267–1280.

Schreiber, L., Skrabs, M., Hartmann, K., Becker, D., Cassagne, C., and Lessire, R. (2000). Biochemical and molecular characterization of corn (*Zea mays* L.) root elongases. *Biochem. Soc. Trans.* **28,** 647–649.

Schubert, P., Steinbüchel, A., and Schlegel, H. G. (1988). Cloning of the *Alcaligenes eutrophus* genes for synthesis of poly-β-hydroxybutyric acid (PHB) and synthesis of PHB in *Escherichia coli. J. Bacteriol.* **170,** 5837–5847.

Sieber, P., Schorderet, M., Ryser, U., Buchala, A., Kolattukudy, P. E., Métraux, J.-P., and Nawrath, C. (2000). Transgenic Arabidopsis plants expressing a fungal cutinase show alterations in the structure and properties of the cuticle and postgenital organs. *Plant Cell* **12,** 721–737.

Slater, S. C., Voige, W. H., and Dennis, D. E. (1988). Cloning and expression in *Escherichia coli* of the *Alcaligenes eutrophus* H16 poly-β-hydroxybutyrate biosynthetic pathway. *J. Bacteriol.* **170,** 4431–4436.

Slater, S., Houmiel, K. L., Tran, M., Mitsky, T. A., Taylor, N. B., Padgette, S. R., and Gruys, K. J. (1998). Multiple β-ketothiolases mediate poly(β-hydroxyalkanoate) copolymer synthesis in *Ralstonia eutropha. J. Bacteriol.* **180,** 1979–1987.

Slater, S., Mitsky, T. A., Houmiel, K. L., Hao, M., Reiser, S. E., Taylor, N. B., Tran, M., Valentin, H. E., Rodriguez, D. J., Stone, D. A., Padgette, S. R., Kishore, G., *et al.* (1999). Metabolic engineering of *Arabidopsis* and *Brassica* for poly(3-hydroxybutyrate-co-3-hydroxyvalerate) copolymer production. *Nat. Biotechnol.* **17,** 1011–1016.

Steinbüchel, A. (1991). Polyhydroxyalkanoic acids. *In* "Novel Biomaterials from Biological Sources" (D. Byrom, ed.), pp. 123–126. MacMillan, New York.

Steinbüchel, A., and Füchtenbusch, B. (1998). Bacterial and other biological systems for polyester production. *Trends Biotechnol.* **16,** 419–427.

Steinbüchel, A., and Hein, S. (2001). Biochemical and molecular basis of microbial synthesis of polyhydroxyalkanoates in microorganisms. *Adv. Biochem. Eng. Biotechnol.* **71,** 81–123.

Steinbüchel, A., and Lütke-Eversloh, T. (2003). Metabolic engineering and pathway construction for biotechnological production of relevant polyhydroxyalkanoates in microorganisms. *Biochem. Eng. J.* **3734,** 1–16.

Steinbüchel, A., and Schlegel, H. G. (1991). Physiology and molecular genetics of poly(β-hydroxyalkanoic acid) synthesis in *Alcaligenes eutrophus. Mol. Microbiol.* **5,** 535–542.

Steinbüchel, A., and Valentin, H. E. (1995). Diversity of bacterial polyhydroxyalkanoic acids. *FEMS Microbiol. Lett.* **128,** 219–228.

Sudesh, K., and Doi, Y. (2000). Synthesis, structure and properties of polyhydroxyalkanoates: Biological polyesters. *Prog. Polym. Sci.* **25,** 1503–1555.

Suh, M. C., Samuels, A. L., Jetter, R., Kunst, L., Pollard, M., Ohlrogge, J., and Reisson, F. (2005). Cuticular lipid composition, surface structure, and gene expression in *Arabidopsis* stem epidermis. *Plant Physiol.* **139,** 1649–1665.

Suzuki, Y., Kurano, K., Arai, Y., Nakashita, H., Doi, Y., Usami, R., Horikoshi, K., and Yamaguchi, I. (2002). Enzyme inhibitors to increase poly-3-hydroxybutyrate production by transgenic tobacco. *Biosci. Biotechnol. Biochem.* **66,** 2537–2542.

Taguchi, K., Aoyagi, Y., Matsusaki, H., Fukui, T., and Doi, Y. (1999). Co-expression of 3-ketoacyl-ACP reductase and polyhydroxyalkanoate synthase genes induces PHA production in *Escherichia coli* HB101 strain. *FEMS Microbiol. Lett.* **176,** 183–190.

Tanaka, H., Onouchi, H., Kondo, M., Hara-Nishimura, I., Nishimura, M., Machida, C., and Machida, Y. (2001). A subtilisin-like serine protease is required for epidermal surface formation in *Arabidopsis* embryos and juvenile plants. *Development* **128,** 4681–4689.

Tanaka, H., Watanabe, M., Watanabe, D., Tanaka, T., Machida, C., and Machida, Y. (2002). ACR4, a putative receptor kinase gen of *Arabidopsis thaliana,* that is expressed in the outer cell layers of embryos and plants, is involved in proper embryogenesis. *Plant Cell Physiol.* **43,** 419–428.

Tanaka, T., Tanaka, H., Machida, C., Watanabe, M., and Machida, Y. (2004). A new method for rapid visualization of defects in leaf cuticle reveals five intrinsic patterns of surface defects in *Arabidopsis. Plant J.* **37,** 139–146.

Tijet, N., Helvig, C., Pinot, F., Le Bouquin, R., Lesot, A., Durst, F., Salaün, J. P., and Beneviste, I. (1998). Functional expression in yeast and characterization of a clofibrate-inducible plant cytochrome P450 (CYP94A1) involved in cutin monomer synthesis. *Biochem. J.* **332,** 583–589.

Tsuge, T., Fukui, T., Matsusaki, H., Taguchi, S., Kobayashi, G., Ishizaki, A., and Doi, Y. (2000). Molecular cloning of two (R)-specific enoyl-CoA hydratase genes from *Pseudomonas aeruginosa* and their use for polyhydroxyalkanoate synthesis. *FEMS Microbiol. Lett.* **184,** 193–198.

van der Walle, G. A. M., de Koning, G. J. M., Weusthuis, R. A., and Eggink, G. (2001). Properties, modifications and applications of biopolyesters. *Adv. Biochem. Eng. Biotechnol.* **71,** 263–291.

Villena, J. F., Dominguez, E., Stewart, D., and Heredia, A. (1999). Characterization and biosynthesis of non-degradable polymers in plant cuticles. *Planta* **208,** 181–187.

Walton, T. J., and Kolattukudy, P. E. (1972). Determination of the structures of cutin monomers by a novel depolymerization procedure and combined gas-chromatography and mass spectrometry. *Biochemistry* **11,** 1885–1897.

Watanabe, M., Tanaka, H., Watanabe, D., Machida, C., and Machida, Y. (2004). The ACR4 receptor-like kinase is required for surface formation of epidermis-related tissues in *Arabidopsis thaliana. Plant J.* **39,** 298–308.

Wellesen, K., Durst, F., Pinot, F., Beneviste, I., Nettesheim, K., Wisman, E., Steiner-Lange, S., Saedler, H., and Yephremov, A. (2001). Functional analysis of the *LACERATA* gene of *Arabidopsis* provides evidence for different roles of fatty acid omega-hydroxylation in development. *Proc. Natl. Acad. Sci.* **98,** 10694–10699.

Wróbel, M., Zebrowski, J., and Szopa, J. (2004). Polyhydroxybutyrate synthesis in transgenic flax. *J. Biotechnol.* **107,** 41–54.

Xiao, F., Goodwin, S. M., Xiao, Y., Sun, Z., Baker, D., Tang, X., Jenks, M. A., and Zhou, J.-M. (2004). Arabidopsis CYP86A2 represses *Pseudomonas syringae* type III genes and is required for cuticle development. *EMBO J.* **23,** 2903–2913.

Yang, S. Y., Cuebas, D., and Schultz, H. (1986). 3-Hydroxyacyl-CoA epimerase of rat liver peroxisomes and *Escherichia coli* function as auxiliary enzymes in the β-oxidation of polyunsaturated fatty acids. *J. Biol. Chem.* **261,** 12238–12243.

Yephremov, A., and Schreiber, L. (2005). The dark side of the cell wall: Molecular genetics of plant cuticle. *Plant Biosyst.* **139,** 74–79.

Yephremov, A., Wisman, E., Huijser, P., Huijser, C., Wellesen, K., and Saedler, H. (1999). Characterization of the *FIDDLEHEAD* gene of Arabidopsis reveals a link between adhesion response and cell differentiation in the epidermis. *Plant Cell* **11,** 2187–2201.

CHAPTER 9

Plant Sterol Methyltransferases: Phytosterolomic Analysis, Enzymology, and Bioengineering Strategies

Wenxu Zhou,* **Henry T. Nguyen,**[†] and **W. David Nes***

Contents

Abstract

So far as is known, the biosynthesis of phytosterols is a ubiquitous property of plants and microbes, whereas insects fail to synthesize the sterol nucleus. All of them require phytosterols to grow and mature. In recent years, the availability of stable isotopes, modern instrumentation such as high-field nuclear magnetic resonance spectroscopy and molecular biology have offered new insights into the evolutionary development of sterol structure–enzyme function relationships, pathway sequencing, chemical ecology, and in the case of bioengineering provides a mechanism to generate value-added traits. Here we describe results obtained with these techniques in relation to the critical enzyme that controls the pattern of C-24 side-chain diversity, the (*S*)-adenosyl-L-methionine: Δ^{24}-sterol methyltransferase (SMT).

Key Words: Sterol methyltransferase, Sitosterol, Cholesterol, Sterol biosynthesis, Phytosterol biosynthesis.

* Department of Chemistry and Biochemistry, Texas Tech University, Lubbock, Texas 79409

† Division of Plant Sciences, National Center for Soybean Biotechnology, University of Missouri-Columbia, Columbia, Missouri 65211

Advances in Plant Biochemistry and Molecular Biology, Volume 1
ISSN 1755-0408, DOI: 10.1016/S1755-0408(07)01009-0

1. INTRODUCTION

The vast majority of naturally occurring sterols contain an alkyl group at carbon-24 in the side-chain. 24-Alkyl sterols are collectively known as phytosterols and possess chemical structures similar to that of cholesterol (cholest-5-en-3β-ol). Sitosterol is the major C_{29}-Δ^5-phytosterol and ergosterol is the major C_{28}-Δ^5-phytosterol of plants and fungi, respectively. Cholesterol is the major C_{27}-Δ^5-zoosterol of animals and insects (Fig. 9.1). Pure sitosterol resembles cholesterol, white in color and waxy in nature. The physical resemblance of the two sterols is mirrored in the similarities of the biosynthetic pathways and functions of these compounds. A striking feature of phytosterol synthesis is that the pathway is directed to form membrane components, in a similar fashion as for cholesterol production and processing in animal systems. However, the subtle differences in the structures of the sterol side chains, 24-methyl (or ethyl) group compared to a 24-hydrogen atom, make for plant-specific functions of sterols.

Phytosterols make up greater than 80% of the total sterol content of the vegetative parts of plants and accumulate to 500–3,000 fg/cell or 1 mg/blade, mostly as free sterol (Nes, 1990; Nes and McKean, 1977). High phytosterol levels are found esterified to fatty acids in seed oils and in leaves of aging plants (Nes, 1990; Wojciechowski, 1991). In humans, cholesterol is present in cells about 500–3,000 fg/cell, similar to the sterol content of plant cells, and about 10 mg/liter in the serum (Nes, 1990; Nes *et al.*, 2000). Alternatively, under normal physiological conditions phytosterols are found at 800–1,000 times lower concentration than that of endogenous cholesterol in the serum. When the concentration of phytosterol is maintained at a high level, ca. 300 mg up to 5 g/day, by a diet rich in vegetables and fruits or through vegetable oil spreads, such as Benecol or Take Control, there are important benefits. For example, phytosterols have been shown experimentally to inhibit colon, breast, and prostate cancers; induce anti-inflammatory effects; and reduce cholesterol levels (Awad and Fink, 2000; Moreau *et al.*, 2002; Tapiero *et al.*, 2003). The exact mechanism by which sitosterol offers protection from cancer is not known and several theories have been advanced as reviewed (Ling and Jones, 1995).

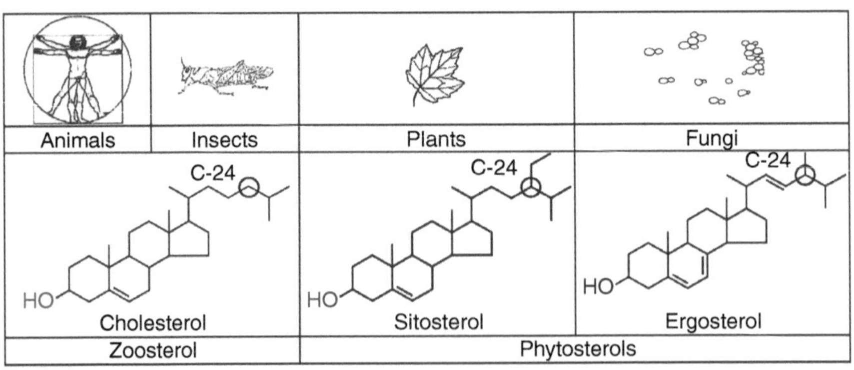

FIGURE 9.1 Major sterols found in living systems. (See Page 7 in Color Section.)

On the other hand, it is generally assumed that cholesterol reduction in the blood stream results directly from inhibition of food-based cholesterol absorption through displacement of cholesterol from micelles (Akihisa *et al.*, 2000; Moreau *et al.*, 2002). For these reasons, phytosterols are considered nutraceuticals and the engineering of plants to increase the phytosterol content as a nonpharmacological approach to prevent certain diseases is underway. It is worth pointing out that there is a portion of the literature that continues to use beta (β)-sitosterol; the beta in this case does not refer to the stereochemistry of the molecule but is used to distinguish the compound from α- and γ-sitosterol. The notation is dropped in common usage (Nes, 2000).

In contrast to other plant lipids, such as fatty acids defined on the basis of their physical properties, phytosterols are defined by their common chemical structure and biosynthetic reasoning related to the cyclization of squalene oxide (Nes and McKean, 1977). Phytosterols, which are insoluble in water and can be extracted from cells by nonpolar organic solvents (such as hexane or chloroform), are characterized by a cyclopentanoperhydrophenanthrene structured nucleus, a flexible side-chain of 9 or 10 carbon atoms and equatorial attachments of a polar head (3β-hydroxyl group) and nonpolar tail (17β-side-chain). The three-dimensional shape is established by the alternating all *trans–anti* stereochemistry of the ring system and the 20*R*-configuration that directs the side-chain to a "right-handed" conformation (Nes *et al.*, 1984; Parker and Nes, 1992; Fig. 9.2). When the side-chain of, for example, sitosterol is oriented to the "right," the sterol has the appropriate

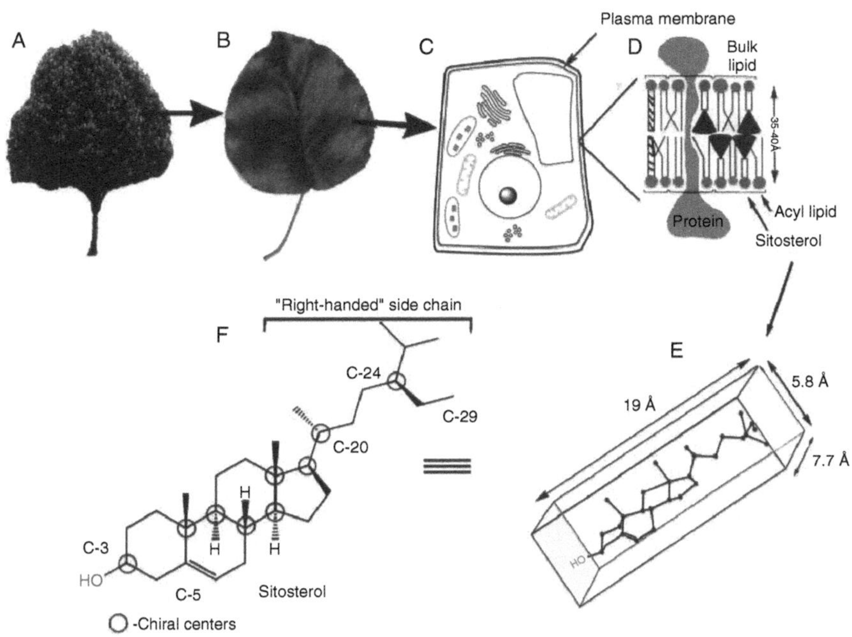

FIGURE 9.2 Sitosterol distribution in the intact plant: (A) whole plant; (B) leaf; (C) cell structure; (D) membrane lipid leaflet; (E) conformational perspective of sterol; (F) structure and stereochemistry in sitosterol. (See Page 7 in Color Section.)

length—ca. 19 Å to fit the monolayer—that is, one-half of the lipid leaflet structure. The combination of asymmetry and electronics in the sitosterol structure gives rise to an amphipathic molecule, basically flat with a length suitable for the sterol to insert the membrane. The presence of 24-ethyl sterols in plant membranes correlates with their efficiency in developing interactions with plant phospholipids to affect membrane fluidity (Mckersie and Thompson, 1979; Schuler *et al.*, 1991).

The C-24 alkylated family of phytosterols contains a greater number of individual compounds in plants than fungi or protozoa. The variant structures and health benefits of generating a modified phytosterol composition provide the basis for investigations of phytosterol profiling and metabolic engineering. Studies on testing inhibitors of sterol methyltransferase (SMT) action in cultured cells and genetic manipulation of the phytosterol pathway in several plants have revealed a physiological requirement for a 24-alkyl substituted steroid in plant growth, the central position of C-methylation in the plant sterol pathway and the connection of the SMT to manufacturing value-added traits (Chappell *et al.*, 1995; Harker *et al.*, 2003; Nes *et al.*, 1991c; Rahier *et al.*, 1980). Since the biosynthesis and functions of sterols in plants are covered in recent review articles (Benveniste, 2004; Clouse, 2002; Lindsey *et al.*, 2003; Nes, 2003; Schaller, 2004), we have included only background material pertinent to the phytosterol pathway as influenced by the SMT. Initially, the pathways of sterol biosynthesis will be described. After a brief coverage of phytosterolomics, with emphasis on structure determination, we review the enzymology and evolution of the SMT and conclude with the current state of metabolic engineering sterol pathways in plants. We have not attempted to be comprehensive in our presentation of the literature; rather the focus is on illustrative examples that may serve to indicate future opportunities to engineer plants with unusual sterol profiles or traits.

2. PATHWAYS OF PHYTOSTEROL BIOSYNTHESIS

The pathway to isoprenoids (five carbon units similar to isoprene) and phytosterols in plants begins with CO_2 fixation and sugar formation (Fig. 9.3). Sugar can be converted to acetate which can be converted to isoprenoids and phytosterols in the cytosol. In producing phytosterols, acetate is converted to mevalonic acid (MVA). The MVA is phosphorylated and the carboxyl carbon is lost as CO_2 to produce Δ^3-isopentyl diphosphate (IPP). However, the sugar can be converted to isoprenoids in the plastids without necessarily involving the intermediacy of acetate via the mevalonate-independent pathway [=1-deoxy xylulose-5-phosphate (DXP) pathway]. The DXP synthase is considered rate-limiting in this pathway. Carbons from Δ^3-IPP can flow to the sterol or fatty acid pathways (via the MVA shunt) (Nes and Bach, 1985). HMGCoA-reductase (HMGR) is considered to be the rate-limiting enzyme of the acetate–mevalonate pathway to isoprenoids. Six of the isoprenoid units are joined to produce squalene. The C-30 olefin is converted to squalene oxide which is cyclized to 24-desalkyl sterols. The extra "methyl or ethyl" group at C-24 in the side-chain of phytosterols is added

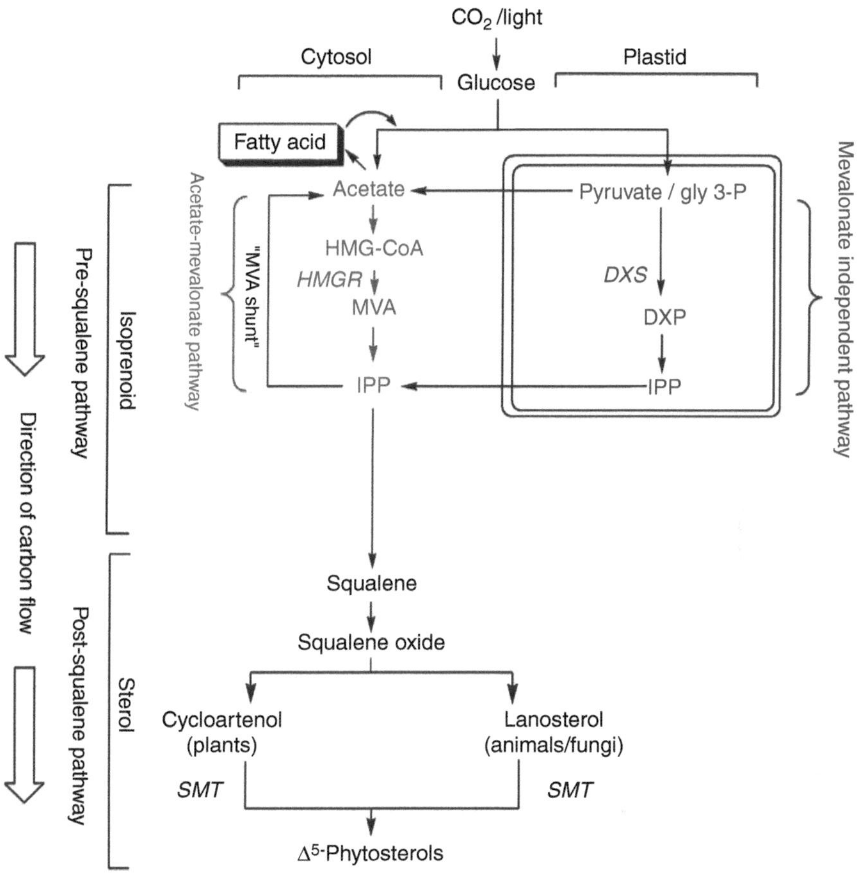

FIGURE 9.3 Stages in the isoprenoid–phytosterol pathway and compartmentation of acetate–mevalonate pathways. (See Page 8 in Color Section.)

after formation of the first tetracycle by the action of SMT. In the phytosterol pathway, SMT is considered to be a rate-limiting enzyme.

Biosynthetic tracer studies have been carried out on almost all classes of phytosterols, and in many instances cell-free preparations capable of converting Δ^{24}-sterols to methylated products are available. In almost all instances, the C-methylation processes are consistent with an ionic mechanism hypothesized by Castle *et al.* (1963), attesting to a common set of reactions that evolved for this class of enzyme. Detailed information about isoprenoid-sterol biosynthesis and the pathways involved in sterol side-chain construction has been obtained recently using ^{13}C and ^{2}H-labeled compounds (Kresge *et al.*, 2005; Seo *et al.*, 1988, 1990). Isotopic labeling experiments with stable isotopes are often used to avoid radioactive measurements and laborious chemical degradation of the sterol side-chain (Nes and Le, 1990). The sites of labeling in the end product are evident through peak enhancements in the ^{13}C NMR spectra of the ^{13}C-labeled sterols and reveal the route of carbon flux from the isoprenoid pathway to phytosterols.

Plant biochemists have shown that two distinct pathways to sterols exist in the cell and they are compartmentalized so that the acetate–mevalonate pathway is cytosolic and the mevalonate-independent pathway is plastidial (Arigoni *et al.*, 1997; De-Eknamkul and Potduang, 2003; Laule *et al.*, 2003; Lichtenthaler *et al.*, 1997; Nes *et al.*, 1992; Umlauf *et al.*, 2004; Zhou and Nes, 2000). Some typical experiments to determine the pathway of phytosterol synthesis involve treating plants with [1-^{13}C]glucose. The sugar is converted to [3-^{13}C]pyruvate which is converted to [2-^{13}C]acetate and the C-2 unit is converted in several reactions to [2,4,5-$^{13}C_3$]Δ^3 IPP; or the [3-^{13}C]pyruvate is converted to [3-^{13}C]glyceraldehyde 3-phosphate and this intermediate is transformed to [1,5-$^{13}C_2$]Δ^3-IPP. The labeling pattern of, for example, ergosterol derived from administering cells [1-^{13}C]glucose has been found to be different in different organisms whether the Δ^3-IPP is formed from the glucose breakdown product of [2-^{13}C]acetate or originates directly from the intermediate DXP (Zhou and Nes, 2000; Lichtenthaler *et al.*, 1997). When the plastid-derived intermediates are labeled with [1-^{13}C]glucose, the labeling pattern of the biosynthetically formed sterol molecule is predicted to be [2,6,11,12,16,18,19,23,27-$^{13}C_9$]ergosterol. A spectrum of ergosterol, biosynthesized from [1-^{13}C]glucose using the alga *Prototheca wickerhamii* shows enhanced peaks corresponding to nine carbon atoms (earlier we incorrectly identified the enhanced peak corresponding to C-11 and C-21 to be from [1-^{13}C]glucose (Zhou and Nes, 2000); actually the enhanced peak is due to C-11 only; Fig. 9.5; spectrum c), suggesting the alga operates the mevalonate-independent pathway to sterols. Alternatively, the labeling pattern of the products of the acetate-mevalonate pathway is predicted to be [1,3,5,7,9,13,15,17,18,19,21,22,24,26,27-$^{13}C_{15}$]sitosterol (De-Eknamkul and Potduang, 2003; Umlauf *et al.*, 2004). Since neither C-28 nor C-29 will be labeled by these intermediates, the labeling patterns of ergosterol and sitosterol assayed with [1-^{13}C]glucose will be the same (Fig. 9.4). Evidence to confirm the traditional isoprenoid–sterol pathway was obtained by administering [2-^{13}C] MVA or [5-^{13}C] MVA to cultured sunflower cells and noting the number and position of the enhanced peaks in the ^{13}C NMR spectrum (Fig. 9.5; spectra a and b) (Nes *et al.*, 1992).

As shown in Fig. 9.5, C-26 and C-27 of the sterol side-chain are chemically equivalent yet they are biosynthetically distinct. Seo *et al.* (1990) have shown that C-26 is derived from C-6 of MVA and C-27 is derived from C-2 of MVA. Appropriate rotations in the structures of these compounds generate a view of the sterol side-chain in equivalent conformation so that the stereochemistry at C-25 can be rationalized and the existence of stereodifferentiated enzymes determined from tracer studies with ^{13}C-labeled substrates. Nes *et al.* (1992) examined the incorporation of [2-^{13}C]MVA to sitosterol and found that C-26 is labeled and Seo *et al.* labeled C-26 in ergosterol with [2-$^{13}CH_3$]acetate (Zhou and Nes, 2000). To confirm that the stereochemistry at C-25 is the same in sitosterol and ergosterol, acceptors [27-^{13}C]zymosterol, [27-^{13}C]lanosterol, and [27-^{13}C]cycloartenol were prepared and assayed with cell-free preparations from plants, fungi, and algae (Guo *et al.*, 1996; Mangla and Nes, 2000; Nes *et al.*, 1998b; Zhou *et al.*, 1996). The enzymatically formed side-chains of fecosterol, 24(28)-methylene lanosterol, and cyclolaudenol contained the C-25*R*-configuration as determined by ^{13}C NMR

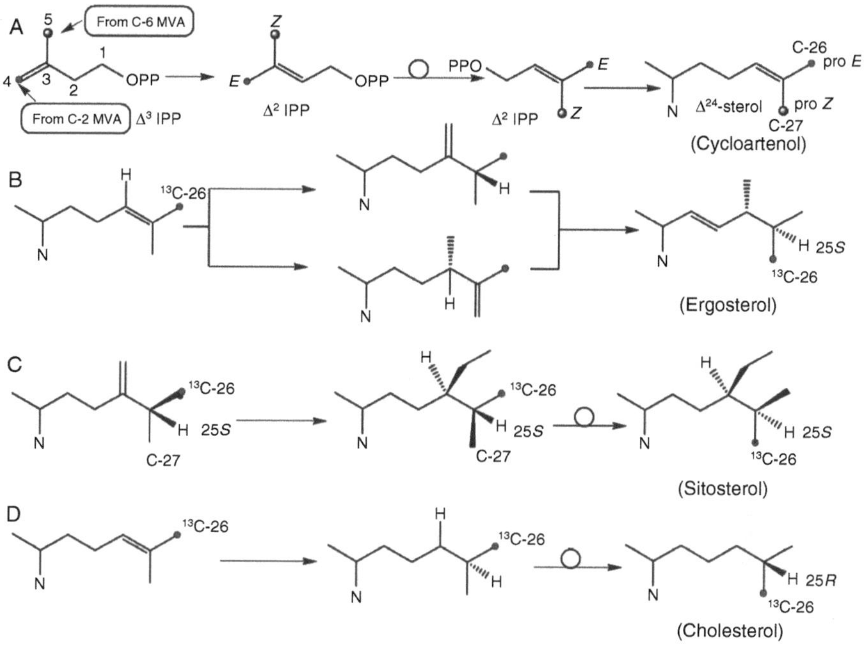

FIGURE 9.4 Determination of labeling patterns in isopentyl diphosphate and sterols synthesized by different pathways. (See Page 9 in Color Section.)

spectroscopic analysis, thereby showing the *ProZ* C-27 of the Δ^{24}-intermediate generates the *R*-C-27 methyl group of the phytosterol.

Incubation of [1-^{13}C]glucose with cultured cells or intact plants can generate different ^{13}C-labeled species of phytosterol and the purity of the labeling pattern of the sterol can indicate the degree to which the acetate–mevalonate pathway or mevalonate-independent pathway operates and the extent of cross talk between the pathways (Arigoni *et al.*, 1997; Laule *et al.*, 2003). These studies indicate that the acetate–mevalonate pathway (=isoprenoid pathway) is preferred in vascular plants. This observation is further substantiated by physiology experiments with inhibitors of HMGR such as mevinolin added to radish seedlings (Bach and Lichtenthaler, 1983) or fosmidomycin, an inhibitor of 1-deoxy-D-xylulose-5-phosphate reductoisomerase, added to tobacco cells (Sauvaire *et al.*, 1997), and by flux studies using tracer amounts of radiolabeled acetate and high concentrations of mevalonate to inhibit sterol synthesis in sorghum seedlings (Hemmerlin *et al.*, 2003).

In the past several years, there have been several remarkable advances in the study of phytosterol enzymes to support the hypothesis that multiple phytosterol pathways exist in nature. The start and direction of the pathway is established by whether squalene oxide cyclizes to cycloartenol (plants and algae) or lanosterol (animals and fungi) (Fig. 9.6; Nes and McKean, 1977). Thereafter, the C-24 methylation pathways provide direction to phytosterol synthesis; "primitive" plants catalyze the $\Delta^{25(27)}$-route whereas advanced plants and fungi catalyze

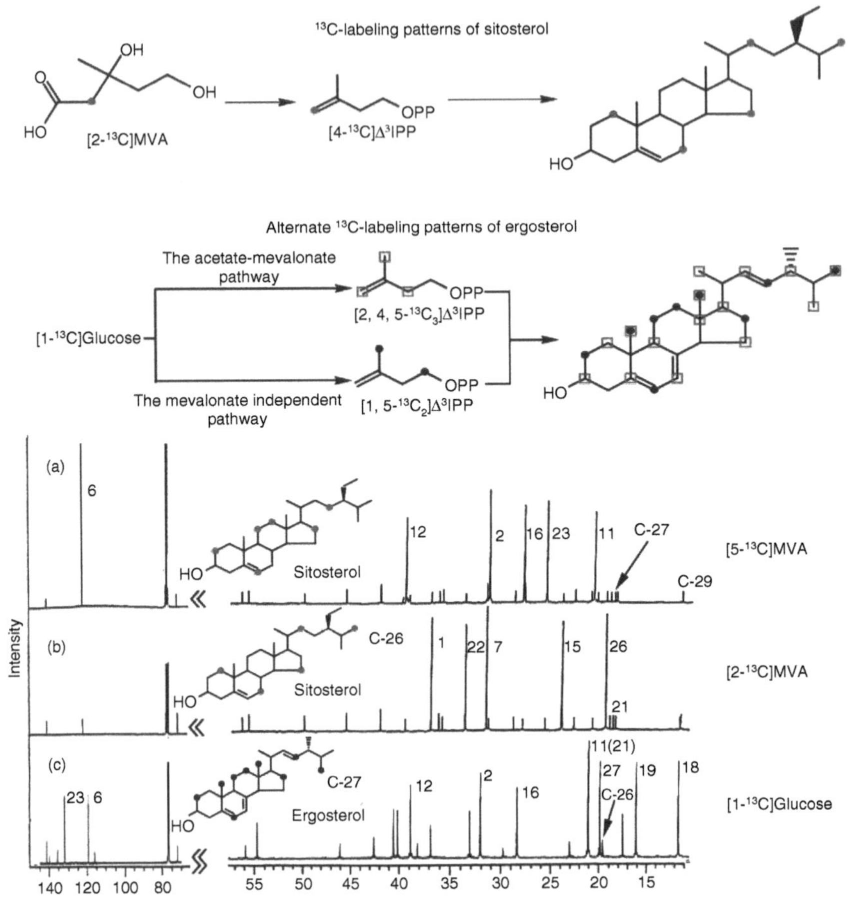

FIGURE 9.5 Stereochemistry of phytosterols at C-25 after ^{13}C labeling of the Pro*E* C-26 of Δ^{24}-sterols. (See Page 10 in Color Section.)

the $\Delta^{24(28)}$-route (Goad *et al.*, 1974; Nes *et al.*, 1977). Mass spectroscopy (MS) can be used as a first screen to determine whether ergosterol is formed by either a fungal or algal route. For example, administering [$^{2}H_3$]methionine to a yeast sterol auxotroph GL7 cultured on lanosterol led to the biosynthesis of [28-^{2}H]ergosterol (Zhou *et al.*, 1996). The mass spectrum of the deuterated ergosterol was found to be two mass units higher than the control species, consistent with methylation at C-24 proceeding by a $\Delta^{24(28)}$-methylene intermediate. The site of introduction of methyl in the sterol side-chain was determined by inspection of the ^{1}H NMR spectra of ergosterol and the corresponding deuterium-labeled ergosterol (Fig. 9.7, Panels B and A). The only signal in the ^{1}H NMR spectrum affected by the incorporation of deuterium in the molecule corresponds to C-28 which is lowered compared to the control. In related experiments with a cell-free preparation of SMT from yeast assayed with [$^{2}H_3$-*methyl*]AdoMet and zymosterol, the enzymatic product [28-^{2}H]fecosterol contained two extra deuterium atoms as

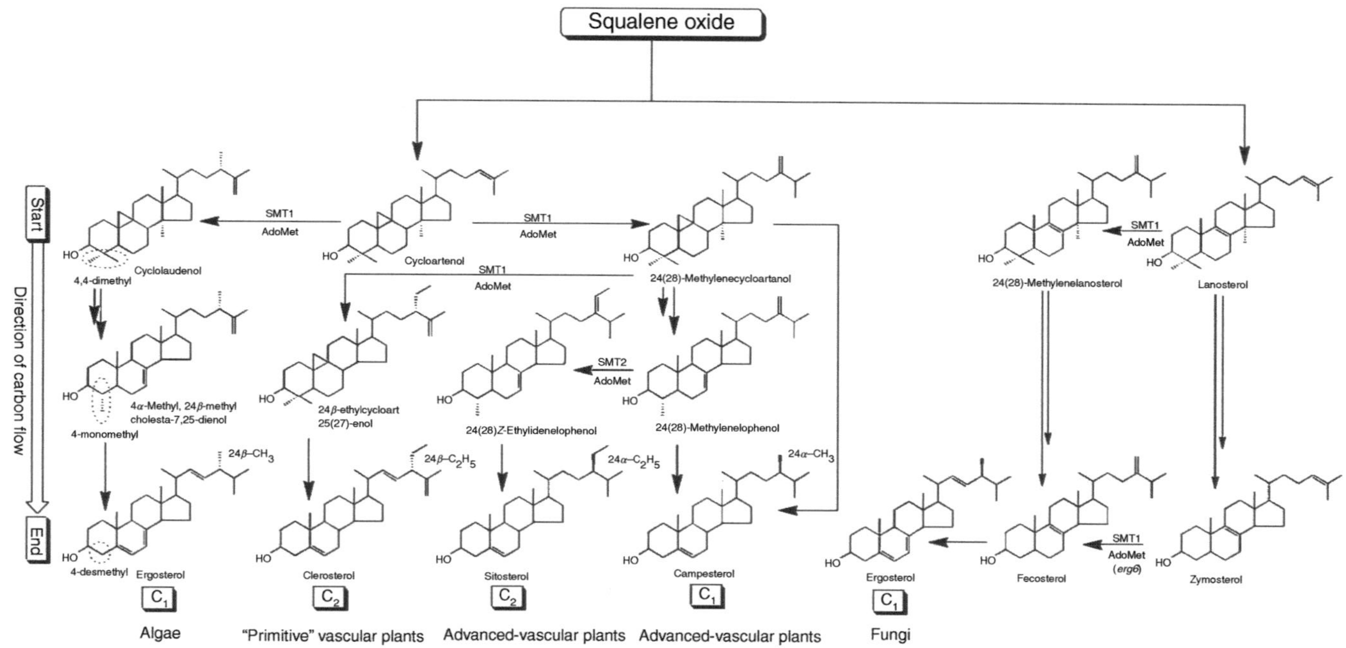

FIGURE 9.6 Hypothetical pathways to Δ^5-phytosterols: C_1 and C_2 refer to the CH_3 and C_2H_5 groups attached to C-24 of the sterol side-chain. Circled C-4 is to emphasize C-4 methyl group removal; 4,4-dimethyl sterol to 4-monomethyl sterol to 4-desmethyl sterol.

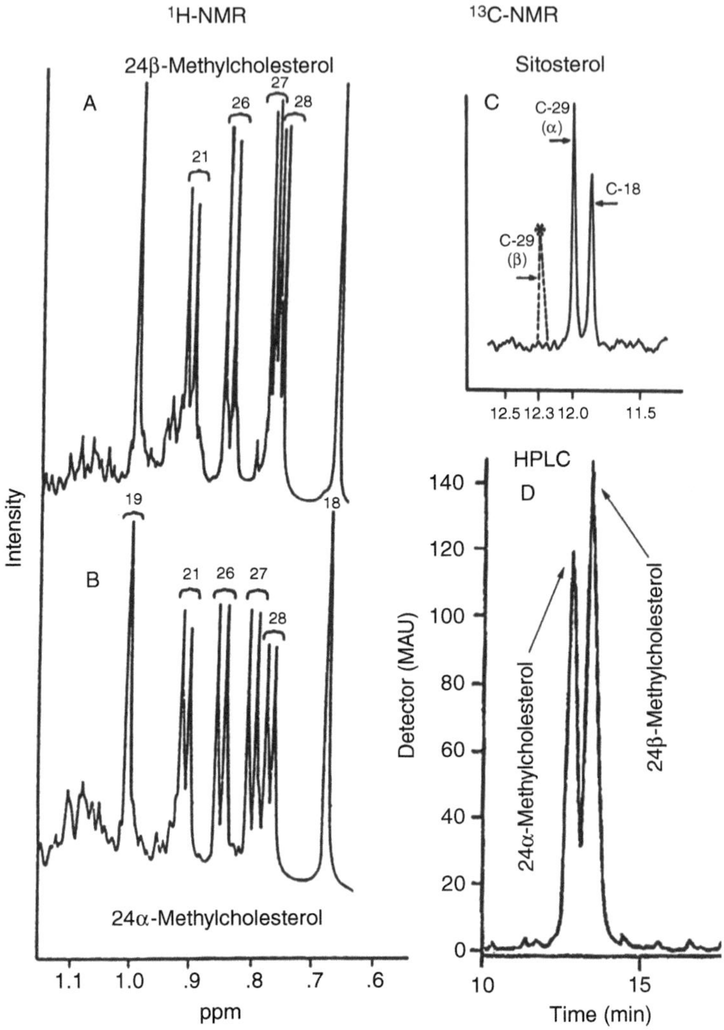

FIGURE 9.7 Spectroscopic and chromatographic analysis of sterols.

determined by MS; in the ^{1}H NMR spectrum, the olefin peak at ca. 4.65 ppm (not shown) corresponding to C-28 was missing but it was present in the spectrum of unlabeled fecosterol (Fig. 9.7; Nes *et al.*, 1998b). These findings show methyl from AdoMet is added to C-24 of the sterol side-chain and becomes C-28 of yeast ergosterol via a pathway that involves methylation of zymosterol to produce the $\Delta^{24(28)}$-olefin fecosterol, as expected (Goad *et al.*, 1974). In the pathway to algal ergosterol, activity assay with [$^{2}H_3$-*methyl*]AdoMet and cycloartenol produces an intermediate molecule cyclolaudenol that mass spectroscopy reveals contains three rather than two deuterium atoms (Zhou *et al.*, 1996). Thus, the methylation route to fungal ergosterol proceeds by a $\Delta^{24(28)}$-route whereas the algal route proceeds by the $\Delta^{25(27)}$-route.

Phytosterols will incorporate either four (α-configuration) or five (β-configuration) deuterium atoms in the 24-ethyl group after treatment with [$^2H_3$-methionine] (Goad *et al.*, 1974), but in these instances the sterols can be derived from the same SMT by a similar C-24-methylation mechanism (Kaneshiro *et al.*, 2002; Nes *et al.*, 2003). To address whether the mechanism of sterol C-methylation leading to vascular plant campesterol (24α-methyl group) is the same as the one leading to fungal ergosterol (24β-methyl group), Nes and coworkers recovered 27-^{13}C-24(28)-methylene lanosterol from activity assay of 27-^{13}C-lanosterol prepared in a cell-free corn system (Guo *et al.*, 1996). The ^{13}C-labeled compound was administered to a yeast sterol auxotroph GL7 which converted the dietary supplement to [27-^{13}C] ergosterol (Zhou *et al.*, 1996). The ^{13}C NMR spectrum of either the [27-^{13}C]ergosterol derived by the plant- or fungal-generated [27-^{13}C]-labeled intermediates were identical, indicating the phytosterol C-methylation pathway in the two organisms is similar and involves a 1,2-hydride shift from the *Re*-face of the original substrate undergoing methylation. Using cloned SMTs from a mutant yeast (Y81W) and wild-type soybean and Arabidopsis, a single enzyme was found to catalyze both the first and second C_1-transfer activities, and in the case of the second C_1-transfer activity the stereochemistry of the 24-ethyl group (β) in the product was shown to be opposite to that of the 24-ethyl group in sitosterol (α) (Fig. 9.6). Inevitably, the conclusion that different phytosterol pathways are present in plants and fungi stems from the fact that different enzymes control the stereochemistry of the methylated products and that the substrate affinities and catalytic competence of SMTs can be different in different organisms.

3. PHYTOSTEROLOMICS

The study of metabolic networks and feedback response to developmental and environmental conditions involves the profiling of natural products. In this context, phytosterolomic research can be defined as the complete identification and quantification of all sterols present in a specific biological sample. Phytosterol fingerprinting is another aspect of phytosterolomics where the unusual sterol composition of a pathogen can serve as a signature lipid for disease such as for immunocompromised patients harboring pneumocystis (Kaneshiro *et al.*, 2002; Zhou *et al.*, 2002). Phytosterol profiling involves techniques such as thin layer chromatography (TLC) and high-performance liquid chromatography (LC) mass spectroscopy (LC/MS) and gas chromatography mass spectroscopy (GC/MS) where individual components are separated and tentatively identified and quantified by one set of techniques with final characterization of structure determined by ^{1}H or ^{13}C NMR spectroscopy (Kalinowska *et al.*, 1990; Nes *et al.*, 1998a, 1999, 2003). HPLC equipped with a diode array detector to monitor UV is useful in the identification of the type and amount of conjugation in the sterol molecule (Nes *et al.*, 1985; Norton and Nes, 1991; Xu *et al.*, 1988). Separation factors in GLC and HPLC based on the retention of cholesterol relative to the retention time of sterol standards have been reported to enable the prediction of the identity of an unknown sterol (Xu *et al.*, 1988). High-throughput screening of phytosterol mixtures by conventional chromatography and spectroscopy has not yet progressed to the point where a single method can establish the identities of a complicated set of structurally similar phytosterols. Although the sterol patterns of many plant extracts deduced by GLC are rather simple, usually consisting of three or four major sterols, the sterol composition of specific plant parts or the type of sterol in primitive versus advanced plants can be strikingly different (Nes, 1990).

The sterol side-chain can be modified by the addition of one or two supernumerary carbon atoms at C-24 with either α- or β-chirality. Establishing the stereochemistry of the 24-alkyl group of individual sterols is a formidable task and can be achieved by identifying specific side-chain carbons by high-field NMR (Fig. 9.7, Panels A–C). The assignments of the signals from the ^{1}H or ^{13}C NMR spectra of ^{1}H- or ^{13}C-labeled sterols are diagnostic for the origin of the methyl group at C-24 and the biosynthetic stereospecificity of phytosterol side-chains (Guo *et al.*, 1996, 1995; Popják *et al.*, 1977). The relative proportion of α- to β-isomer of a 24α/β-methyl cholesterol mixture can also be quantified by reversed-phase HPLC (Guo *et al.*, 1995) (Fig. 9.7, Panel D) (Parker and Nes, 1992). The ratio of these diastereoisomers can change as plants evolve from less to more advanced (Nes *et al.*, 1977).

A more exacting analysis of sterol structure and stereochemistry involves the use of several spectroscopic techniques, including high-field ^{1}H and/or ^{13}C-NMR, X-ray crystallography, as well as molecular modeling (Nes *et al.*, 1998a). An example where the use of several techniques has played a major role in our understanding of conformational analysis of plant sterols is the characterization of cycloartenol (with a 9β,19-cyclopropane ring) and lanosterol (with a $\Delta^{8(9)}$-bond). There is a hypothesis that the structural isomers are shaped differently, bent versus flat (Bloch, 1983; Goodwin, 1981; Rahier *et al.*, 1984). The putative bent 9β,19-cyclosterol was considered to be an unique structural trait of intermediate plant sterols, and bent compounds were acceptable only to either specific plant enzymes or membrane systems. The rationale for 9β,19-cyclosterols to be bent is based on the *syn–cis* configuration at the A/B and B/C ring junctions, resulting in an unfavorable interaction between the 9,10-bridgehead and the 8β-hydrogen atom at C-8. Furthermore, on the basis of manipulation of ball and stick models, it was proposed that ring B in 9β,19-cyclosterols becomes a boat and the A/B/C rings orient in the chair-boat-chair conformation. However, we discovered for the first time excellent agreement between a set of crystallographically observed cycloartenol-type structures and their solution conformations deduced from 2D-NMR spectroscopic analysis, analysis of the NOE networks (Fig. 9.8) and MM/MD calculations to be pseudoplanar (A/B/C rings are chair/half-chair/twist-chair conformer), and the sterol side-chain to orient to the "right" (Nes *et al.*, 1984, 1991a). These results with sitosterol, lanosterol, and related compounds reveal that the compounds possess similar three-dimensional shapes (Nes *et al.*, 1991b) and differ in the tilt of the C-3 group and C-17(20) side-chain (Fig. 9.9), structural features that can be responsible for the different activities of sterols (Nes *et al.*, 1991b, 1993).

Using the modern methods of sterol analysis, it has been possible to show that phytosterols appear in all cells at all stages of development, but the type and

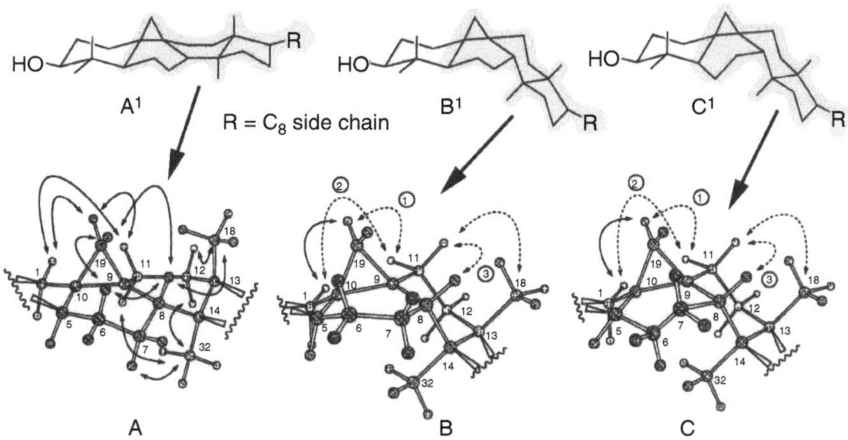

FIGURE 9.8 NOE networks that indicate whether cycloartenol can orient into a flat or bent shape. Adapted from Nes *et al.* (1998b). (See Page 11 in Color Section.)

amount of sterol is greatly influenced by ontogeny and speciation, and consequently seems to be a carefully regulated process (Nes, 1990). In spite of findings that show variability of the sterol side C-24 alkyl structure occurs widely, suggesting some sort of association between sterol structure and plant biology, relatively little is known about individual sterols or sterol sets and their role in plant physiology. As many as 60 sterols have been reported in a vascular plant, *Zea mays* (Guo *et al.*, 1995), which is about one-third to one-half the number of phytosterols found in living systems; 39 sterols have also been identified in an ascomycetous fungus, *Gibberella fujikuroi* (Nes *et al.*, 1989a).

The ratio of 4,4-dimethyl and 4-desmethyl sterol intermediate to 4-desmethyl sterol end product, easily determined by TLC, can be as much as 1:3 in pollen, 1:3 in the seed, and 1:9 in vegetative parts (Heupel *et al.*, 1986; Marshall *et al.*, 2001; Nes and Schmidt, 1988; Nes *et al.*, 1991c). These differences in intermediate to end product ratios suggest that the phytosterol pathway is undergoing significant changes during development. Consistent with this observation, as plants mature from seed to flower the amounts of total sterol in individual parts and various cell types differ dramatically. For example, in sunflower the sterol levels are reported to be (Nes, 1990, 1991b,c) in: a seed, 83 μg; the 4-day dark grown sprout, 15 μg; the primary leaf, 15 μg; mature leaf, 100 μg; immature green flower bud, 35 μg; disk flower, 11 μg; cultured cells, 3,000 fg/cell; and the ray mesophyll cells released by macerase digestion, 500 fg/cell. Pollen grains were found to be the richest source of sterol, 585,000 fg/grain.

Sterols accumulate in vegetative parts mostly as 24-ethylsterols (e.g., sitosterol and stigmasterol [22-dehydrositosterol]) whereas pollen can accumulate significant amounts of 24-methylene sterols and 24-desalkylsterols (4,4-dinorcycloartenol; 24-dehydropollinastanol) (Guo *et al.*, 1995; Nes and Schmidt, 1988). However,

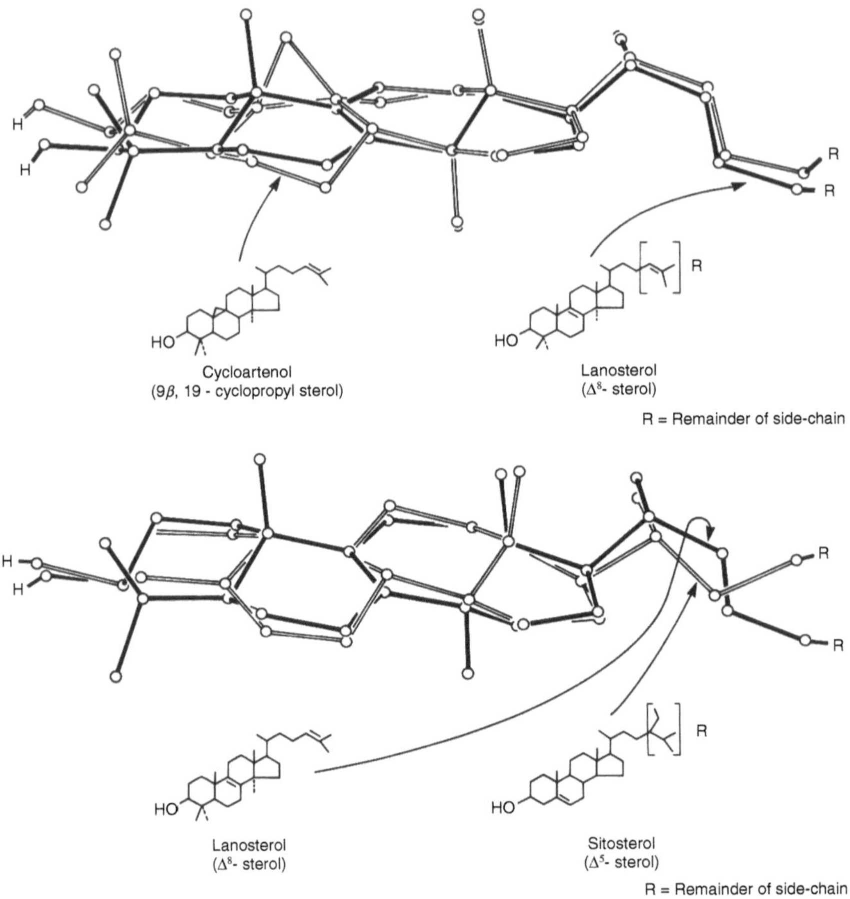

FIGURE 9.9 Partial X-ray crystallographic structures of cycloartenol, lanosterol, and sitosterol. Adapted from Nes *et al.* (1991b).

the sterols of some pollen and seeds can be mostly sitosterol and related phytosterols. Stem trichomes separated from sunflower had no detectable sterol within a GLC limit of detection set at <0.1% total sterol. Alternatively, the leaf wax contains cholesterol, as reported for sorghum leaves (Nes, 1990). Cycloartenol, usually an intermediate that accumulates to trace levels in the sterol mixture, can accumulate to as much as 25% in aging tomato leaves (unpublished data) and similarly can be a major sterol in soybean seeds (Marshall *et al.*, 2001). The inability to detect cycloartenol in some instances may be due to the manner in which sterol was prepared for analysis. For instance, in many early studies sterol purification involved digitonin precipitation that fails to precipitate 4,4-dimethyl sterols along with the major phytosterols; therefore only the Δ^5-sterol content in the tissues was determined.

In contrast to other lipids, a correlation exists between the amount of sterol and seed size. Thus, as the seed increases in size the amount of sterol increases

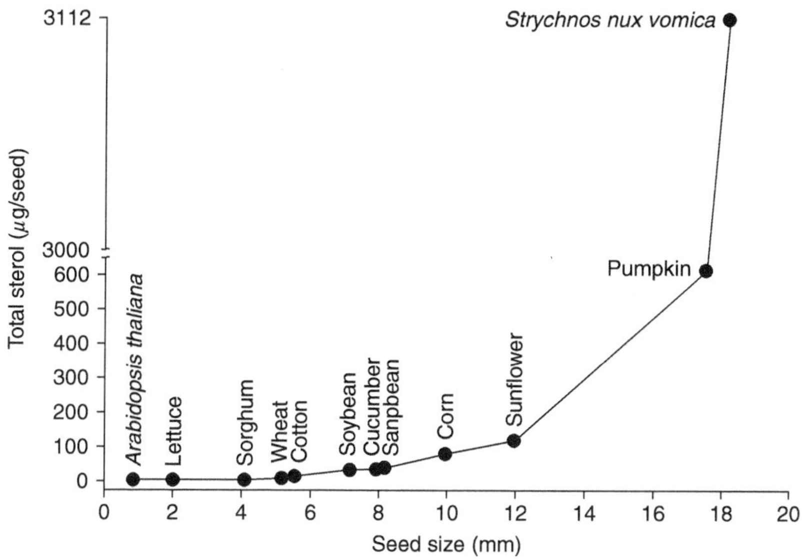

FIGURE 9.10 Correlation of the sterol content with the increase in size of seeds. From unpublished data.

from as little as a few micrograms per seed in Arabidopsis to approximately 3,000 μg/seed in *Strychnos nux vomica* (Fig. 9.10). In similar fashion, the amount of sterol in vegetative plant parts (e.g., leaves) increases as the blade size increases during growth and as the plant height approaches maturity (Fig. 9.11). It would appear that the free sterol content of a system is roughly related to the cell number; therefore the amount of total sterol of a system can be an approximate measure of the total number of cells. Thus, Arabidopsis seeds will contain a smaller number of cells than Strychnos seeds.

The amount of total phytosterol in seedlings has been correlated to the level of SMT activity during plant maturation (Fig. 9.12). Transcript expression levels of SMT from Arabidopsis and soybean also vary during plant maturation (Carland *et al.*, 2002; Diener *et al.*, 2000; Shi *et al.*, 1996). For example, young roots, leaves, and stems had higher levels of steady state transcript levels as compared with mature leaves, suggesting a high rate of sterol biosynthesis in growing vegetative tissues. Growing apical meristems appear to be a major site for sterol biosynthesis (Devarenne *et al.*, 2002). In plants, phytosterol synthesis is measurably a slow event, ca. 1.0 pmol/h/100 shoots, consistent with their primary function as architectural components of membranes, and cycloartenol is turned over rapidly at 24 pmol/h/mg protein consistent with its role as an intermediate (Guo *et al.*, 1995). Thus, to meet the continued needs of campesterol and sitosterol formation, carbon from the isoprenoid pathway must be continually made available to cycloartenol synthesis.

The phytosterol composition can change in regard to the structure of the sterol side chains with plant development. A "switching" mechanism in the biosynthesis of phytosterols has been shown in the cucurbits to produce different

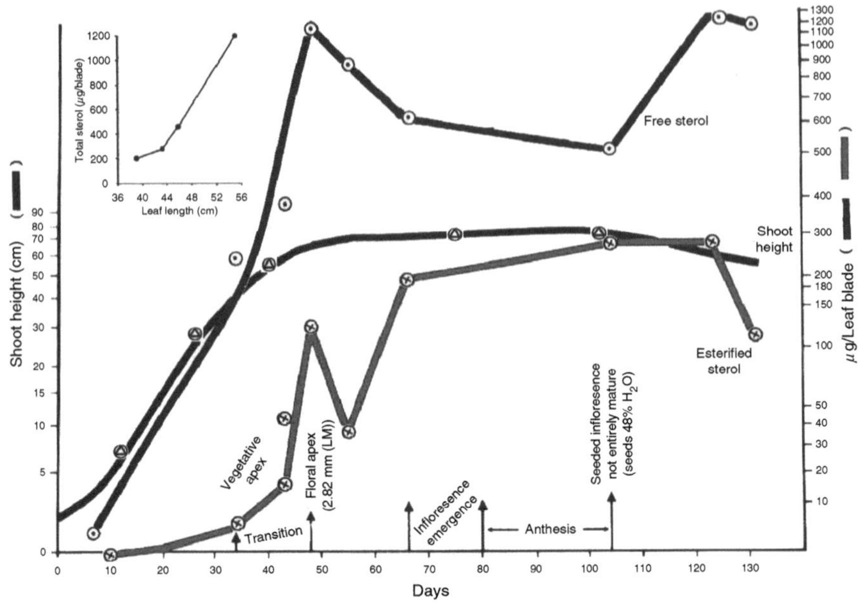

FIGURE 9.11 Correlation of the sterol content with the increase in shoot height of sorghum. Inset is a correlation of the increase in sterol content with maturation of the leaf blade, measured as changes in leaf length. Adapted from Heupel *et al.* (1986). (See Page 11 in Color Section.)

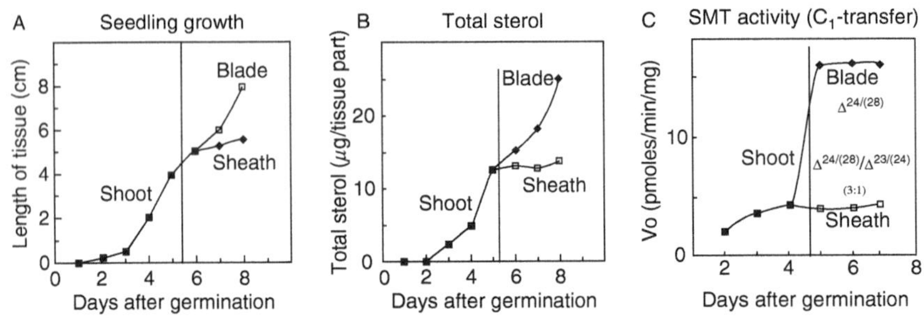

FIGURE 9.12 Correlation of the rate of phytosterol synthesis with plant growth and SMT activity. Data adapted from Guo *et al.* (1995).

24-alkyl sterol stereoisomers with development (Kalinowska *et al.*, 1990). The seeds of squash and pumpkin synthesize 24β-ethyl sterols whereas the seedlings of these plants synthesize 24α-ethyl sterols (Nes, 1987a). Similarly, the ratio of 24α-ethyl sterol to 24β-ethyl sterol can switch in Kalanchoe; roots (total sterol about 285 μg/fwt) contain exclusively the α-isomer and leaves (total sterol 232 μg/fwt) contain the β-isomer. Flowers (total sterol 160 μg/fwt) have an equal mixture of 24α/β-ethyl sterols. These profiling results strongly suggest that the level and perhaps type of SMT expressed in individual cells and cell types during ontogeny is regulated differentially and is an important determinant of phytosterol diversity.

The principal approach adopted seeks to derive developmental and evolutionary understanding of sterol synthesis by combining the sterol and genetic composition of plants with sterol functions and the morphological evidence of plants at different time points. These studies indicate that sterol biosynthesis can proceed in plants by a cycloartenol pathway and in animals and fungi by a lanosterol pathway (Goodwin, 1981; Nes *et al.*, 1990). In addition, the size and direction of the 24-alkyl group of the sterol side-chain can be an indicator of an early or late stage of development and whether the plants are less or more advanced. Primitive organisms synthesize phytosterols with a 24β-methyl group and vascular plants synthesize sterols with a 24α-ethyl group (Kalinowska *et al.*, 1990). Since C-methylation is an energy expensive process, it seems unlikely that the cell would commit energy to produce the side-chains of ergosterol or sitosterol unless the methylation event was functionally significant. Although few pure enzymes of sterol catalysis have been examined in any detail, it does seem safe to suggest that a relatively small number of SMT enzymes determine the basic structural character of the phytosterol side chains produced in a given species.

The type and amount of phytosterol can change during the life history of plants and fungi yet there is a sterol homeostasis maintained at the cell level throughout development. The question arises as to why certain sterols accumulate and what biology maintains their balance in the cell. In other words, is one sterol as good as another? Can a single sterol play multiple roles? Is there something special about a particular sterol cocktail? One way to get at this problem would be to determine what sterols actually do, to find a way to quantify the value of individual features, and then to determine the extent to which deviations in sterol homeostasis affect function. The occurrence of intermediates is also problematic since different pathways can lead to the same end product. In the first case, the biosynthetic sequence determines the structure of the end product. In the second case, a choice presumably is made in a way that has no impact on the structure of the end product. Nes discussed the possible multiple roles of sterols in plants and fungi (Nes, 1980) and the difference between functional and phylogenetic control of biosynthesis (Parker and Nes, 1992). Evidence in support of phylogenetic control is the cycloartenol-lanosterol bifurcation whereby either precursor can give rise to ergosterol. In contrast, functional control is related to the structure and function of enzymes that act on sterols that, once impaired, interrupt flux to end products resulting in aberrant morphology or cell death. In support of the latter hypothesis is recent work from animal and plant studies that show mutations in the terminal segments of the cholesterol and sitosterol pathway lead to malformations (Clouse, 2002; Herman, 2003).

The hypothesis that sterols play multiple roles in plants and fungi has its origins with the work of Clark and Bloch (1959) who studied the insect nutritional requirements of sterol. Insects cannot make their own sterol due to a block in the pathway before squalene oxide cyclization, therefore they are dependent on an exogenous or dietary source of sterol to satisfy their structural and physiological needs for them (Clark and Bloch, 1959). Clark and Bloch discovered a sterol function other than the bulk membrane role for sterols on the basis of structural and quantitative requirements. Thus, feeding different amounts of a mixture of cholesterol and cholestanol to the insect resulted in the finding that cholestanol can spare (partly replace) cholesterol. The function of cholestanol was considered to be that of a regulatory molecule, not necessarily as a precursor to a hormone, for example, ecdysteroid. Fungi have been found to utilize sterols in multiple roles as well and various actions similar to sparing cholesterol in insects have been described with the yeast ergosterol (Nes, 1987b). The requirement for sparing amounts of a 24-ethyl sterol was demonstrated using cultured celery cells (Haughan *et al.*, 1987). Support for the involvement of the different phytosterols in different physiological roles has been obtained with the fungus *G. fujikuroi* where there is delayed expression of certain sterolic enzymes that generate different sterol compositions (Nes and Heupel, 1986) and mutation studies in plant sterol synthesis that show changes in sterol composition affects morphology (Lindsey *et al.*, 2003; Schaller, 2004).

4. ENZYMOLOGY AND EVOLUTION OF THE SMT

SMTs are a family of AdoMet-dependent C-methyltransferases that use two substrates, AdoMet as a methyl source and sterols with a Δ^{24}-bond as the acceptor molecules, for the transmethylating reaction yielding AdoHcy and phytosterols with single or double methylation at C-24. AdoMet-dependent methylations are important in generating phytosterols as primary metabolites, and other AdoMet-dependent methylations contribute to generating many secondary products, including phenylpropanoids, flavonoids, and alkaloids (Nes *et al.*, 1986; Ounaroon *et al.*, 2002; Schubert *et al.*, 2003; Zubieta *et al.*, 2002). The primary structures and enzyme kinetics as well as the requirements for the substrate may be quite different among the methyltransferases. However, a related evolution may be inferred to the generation of these seemingly different plant enzymes in which the active center evolved a common core of structurally similar amino acid residues for interaction with AdoMet.

SMT-catalyzed reactions are remarkable in that they convert lipophilic compounds to methylated olefins in a single step. The details of these processes have intrigued scientists for half a century. A comparison of biomimetic studies of uncatalyzed versus SMT-catalyzed reactions reveal that the elementary chemical steps required for C-methylation of an olefin can take place to produce multiple products in a nonezymatic reaction following predictable chemical principles (Julia and Marazano, 1985; Venkatramesh *et al.*, 1996). Mechanistically, the key to the C-methylation reaction is the positively charged sulfonium center in AdoMet which renders the methyl group electrophilic and susceptible to the relatively nucleophilic Δ^{24}-double bond. The ensuing reaction proceeds by a reorganization of at least three bonds: (1) cleavage of the C–S bond in the AdoMet donor, (2) formation of the C-24(28)-bond between the donor and the acceptor, and (3) loss of a proton from either the donor or acceptor. SMT catalysis proceeds stereo- and regiospecifically to generate distinct product sets by variations of a similar ionic mechanism (Fig. 9.13). Together these enzymes are capable of converting sterol precursor into more than 100 distinct phytosterols in plants (Nes and McKean, 1977).

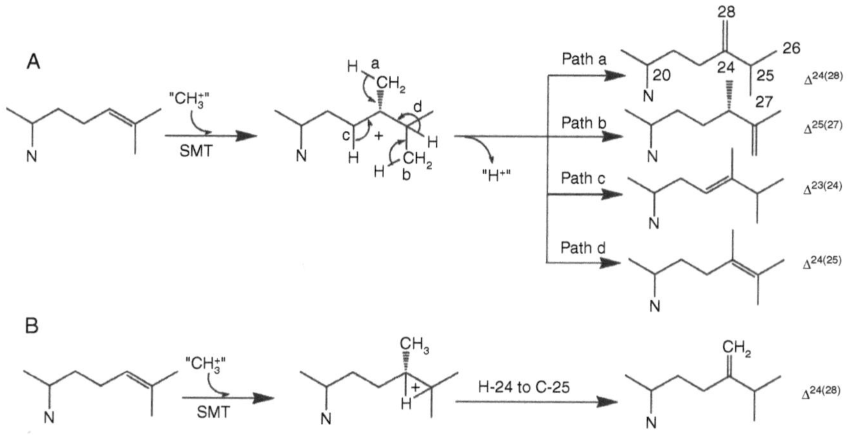

FIGURE 9.13 Possible C-methylation mechanisms for producing C-28 olefins. (A) Stepwise or carbocationic pathway; (B) nonstop or concerted pathway.

SMTs are membrane-bound enzymes which share a high degree of similarity in primary structure as revealed by a comparison of their amino sequences reported in the GenBank. Hydropathy analyses of SMTs from different organisms indicate that they are similar and moderately hydrophobic with no membrane spanning domains (Fig. 9.14). Several cloned SMT enzymes from plants, fungi, and protozoa have been overexpressed in *Escherichia coli* (Nes *et al.*, 1998b, 2003; Zhou *et al.*, 2006). The open reading frame of these catalysts will code for predicted proteins of 336–383 amino acids with a molecular mass that ranges from 38.5 to 43.3 kDa. In all cases studied, the purified protein possesses a tetrameric subunit organization that ranges from 160 to 172 kDa. Equilibrium dialysis and Scatchard plotting of K_d measurements indicate that the SMT has a single binding site for sterol and AdoMet. Catalytic constants for the native substrates for these enzymes are generally K_m ca. 30 μM and k_{cat} ca. 0.01 s^{-1}. The secondary structure of several SMT1 enzymes, as determined by circular dichroism, was found to be similar (Nes *et al.*, 2004; Zhou and Nes, 2003). In the case of the yeast SMT, the following population of structures was recorded: 43% α-helix, 29% β-sheet, 7% turn, and 21% random coil (Zhou and Nes, 2003). These findings suggest the conformational features of the native SMTs will be similar. Although X-ray crystallographic analysis of heavy atom-labeled crystalline SMT bound to a suicide substrate could provide much helpful information about structure–function relationships, crystallization of the SMT has proven to be a formidable task and no three-dimensional structure of the enzyme is available to date. Despite this fact, sufficient activity assays of substrate and transition state analogs with different SMTs are available to speculate a *steric–electric* model of SMT catalysis (Fig. 9.15) (Parker and Nes, 1992).

The model predicts that four domains of the sterol molecule are critical to productive binding and catalysis. Domain 1 recognizes an equatorial C_3–OH group that is nucleophilic to bind to a polar amino acid; Domain 2 recognizes

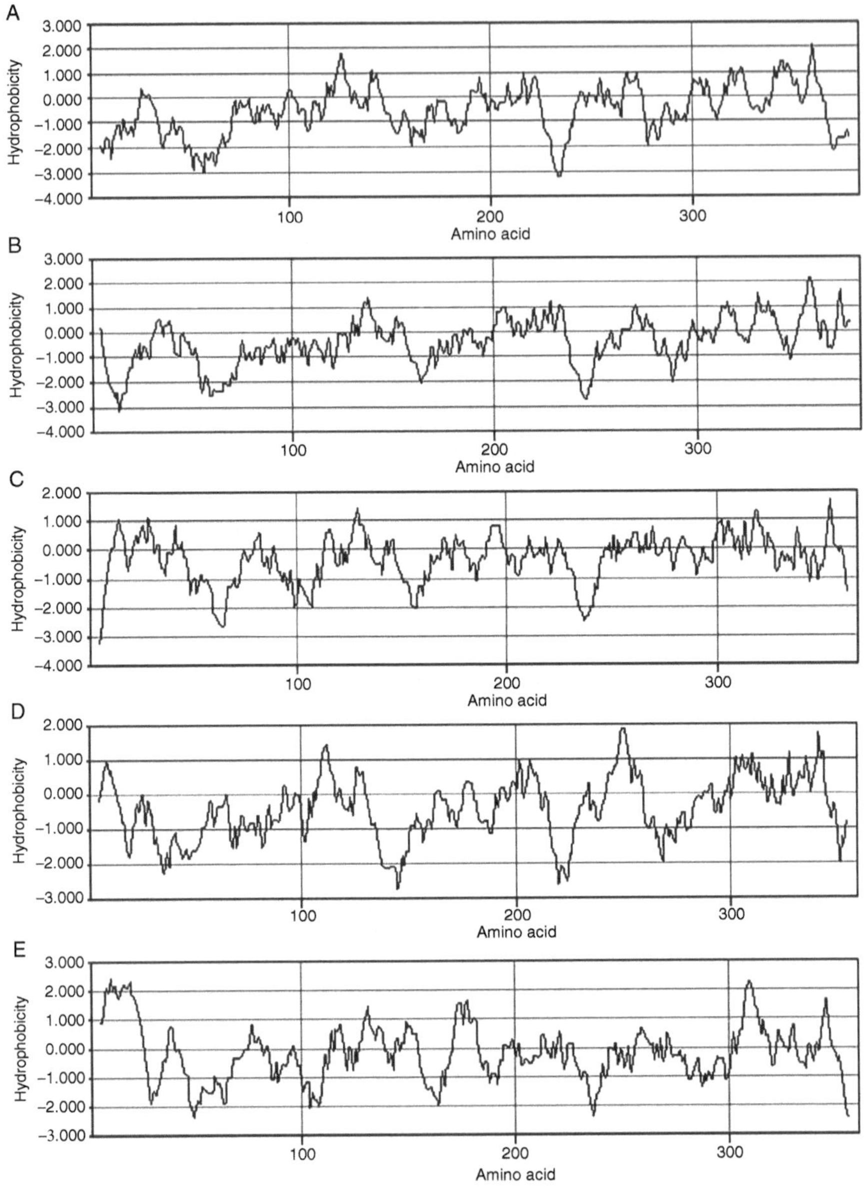

FIGURE 9.14 Hydropathy plots corresponding to SMTs for (A) *S. cerevisiae*; (B) *Gibberella fujikuroi*; (C) *Glycine max*; (D) *Trypanosoma brucei*; (E) *Arabidopsis thaliana.*

a double bond in the nucleus, preferable at $\Delta^{8(9)}$ (or a 9β,19 cyclopropane group), to lock the nucleus in a pseudoplanar conformation thereby generating a flat shape and interacts with the angular methyl groups at C-10 and C-13 to secure the sterol in the hydrophobic cleft; Domain 3 recognizes a 20*R*-configuration to direct the side-chain into a "right-handed" conformation thereby positioning

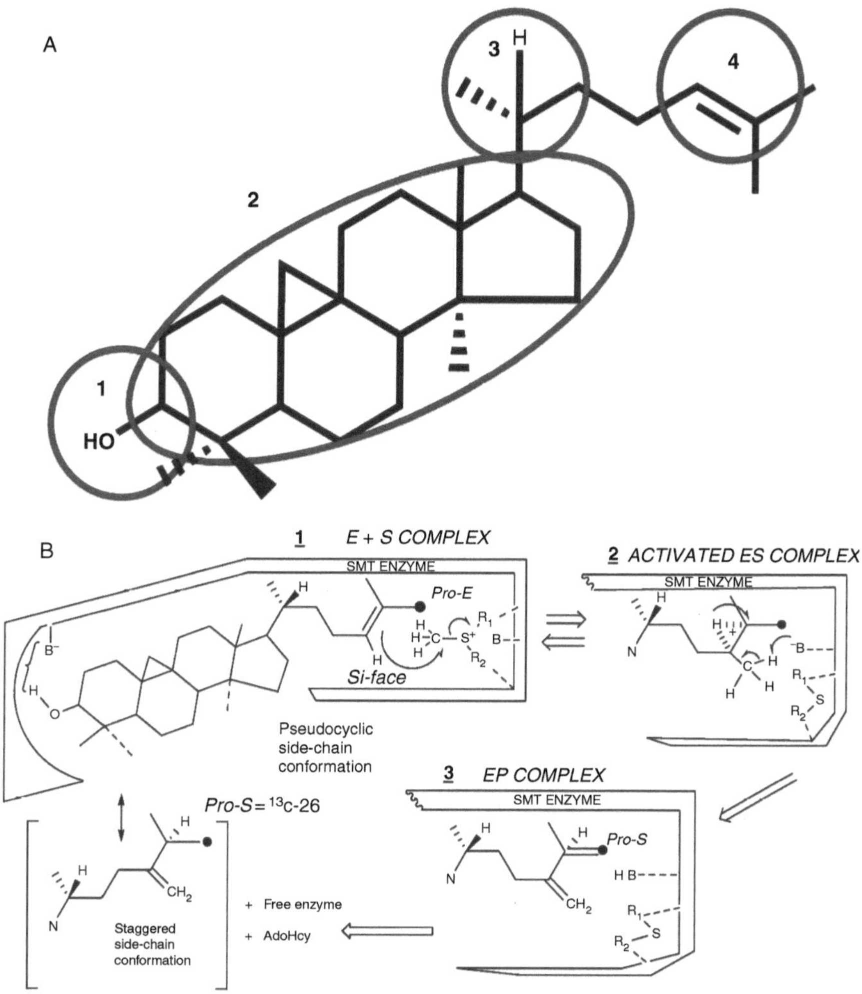

FIGURE 9.15 Postulated domains of sterol molecule (Top, A) recognized by the SMT and the *steric–electric plug* model of SMT catalysis (Bottom, B). (See Page 12 in Color Section.)

C-22–C-26 to a catalytically acceptable conformation; and Domain 4 recognizes a side-chain that contains the terminal C-25-C-26–C-27-isopropyl group and Δ^{24}-bond to anchor the side-chain near AdoMet. The model also predicts a conformational change in the enzyme during catalysis to allow for the different kinetics of the reaction (Parker and Nes, 1992).

For the first C_1-activity catalyzed by the yeast SMT, the methyl transfer reaction proceeds by a simple S_N2 reaction involving a dative bond formed between C-24 and C-28 and a bridged carbenium ion formed opposite to the side of the original double bond facing the AdoMet. A 1,2-hydride shift of H-24 to C-25 proceeds because both C-24 and C-25 have become tertiary sites, equally stabilized by hyperconjugation. The hydride shift of H-24 to C-25 and methyl transfer (to C-24 from AdoMet) are concerted; only in this way will there be no formation of a discrete C-24 or C-25 cation and channeling will be restricted to formation of a single product (Nes *et al.*, 1998b; Nes *et al.*, 2003).

The plant SMT1, in contrast to the yeast SMT1, can catalyze both the first and second C_1-transfer reaction. The first C_1-activity acts mechanistically similar to the yeast SMT1 to produce a $\Delta^{24(28)}$-product. However, the second C_1-activity can generate a mixture of 24-ethyl sterols by an ionic mechanism involving a C-24 isofucosterol cation and a reversible 1,2-hydride shift of H-24 to C-25 that leads to deprotonation and formation of different olefins corresponding to the side-chains of fucosterol ($\Delta^{24(28)}E$), isofucosterol ($\Delta^{24(28)}Z$), and clerosterol ($\Delta^{25(27)}$ 24β-ethyl) (Nes *et al.*, 2003). The chemical mechanisms associated with the first and second C_1-activities differ in the order the two bonds are cleaved in the $\Delta^{24(25)}$- and $\Delta^{24(28)}$-substrates. In the carbocation mechanism, the bond to the leaving group is broken first whereas in the concerted mechanism, both bonds are cleaved simultaneously without intervention of an intermediate. Thus, the second C_1-activity can operate a stepwise mechanism whereas the first C_1-activity can operate a nonstop mechanism. In similar fashion, the first C_1-transfer can proceed by a stepwise mechanism to produce multiple products, as recently found for the cloned *Trypanosoma brucei* SMT that converts the $\Delta^{24(25)}$- sterol to a mixture of $\Delta^{24(28)}$-, $\Delta^{25(27)}$-, and $\Delta^{24(25)}$-sterols (Zhou *et al.*, 2006). The operation of path a in the yeast SMT compared to the operation of paths a and b versus path d in the *T. brucei* SMT (Fig. 9.13) suggests that the yeast SMT can operate a 1-base mechanism whereas the *T. brucei* SMT can operate a 2-base mechanism for the coupled methylation–deprotonation reaction.

Studies on a set of cloned wild-type and mutant yeast SMT (Nes *et al.*, 1999; Nes *et al.*, 2002; Zhou and Nes, 2003) indicate that the conserved acidic amino acids at D125 and D152 form a wall of the AdoMet binding site, perhaps hydrogen bonding to the methionine and ribose moieties of the substrate; D276 and E195 are positioned directly or by way of a water bridge to the proximal (C_3-hydroxyl group of the sterol) and distal (Δ^{24}-bond of the sterol) nucleophilic segments of the acceptor, respectively. E195 may also interact with the positive charge on the sulfur residue of AdoMet in which case it may serve as a counterion to AdoMet. Alternatively, through cation-π interactions, a neighboring aromatic amino acid may serve as the counterion to AdoMet. H90, positioned above (*Si*-face of the 24,25-double bond) the substrate double bond in the same plane as AdoMet, may serve as the base involved with C-28 deprotonation that leads CH_3 to CH_2 production. The residue at Y81 may lie on the *Re*-face of the substrate double bond and act during the methylation-deprotonation reaction to stabilize the high-energy intermediate(s) formed during the reaction progress. Negatively charged ions included in the SMTs are considered to be arranged in the active site so as to stabilize the intermediary carbocations formed during catalysis through cation-π interactions (Nes *et al.*, 2004), thereby restricting channeling and leading to an acceleration of the C-methylation reaction. Homology modeling and the enzymatic studies with SMT predict a spatial arrangement of the secondary structural elements in relation to sterol and AdoMet substrates as shown in Fig. 9.16.

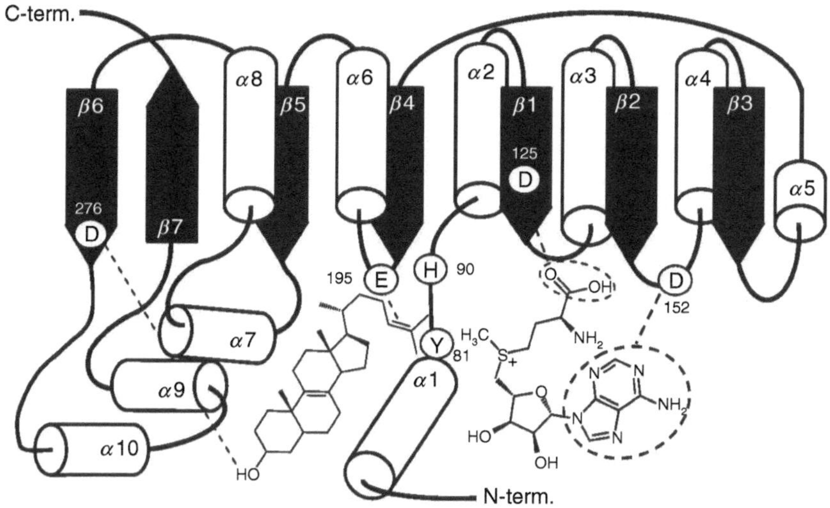

FIGURE 9.16 Schematic representation of the methyltransferase fold of the SMT; spatial arrangement of the secondary structure elements in relation to sterol and AdoMet substrates. Adapted from Nes *et al*., 2004.

Recent developments in the cloning and purification of SMTs and studies performed in our laboratory (Mangla and Nes, 2000; Nes, 2000; Nes *et al.*, 2003; Zhou *et al.*, 2006) and that of Benventiste (2004) have led the enzyme commission (EC) to reclassify the SMT family according to substrate preference into three classes: the fungal SMT1 prefers zymosterol [$SMT1_{ZY}$], EC 2.1.1.41; the plant SMT1 prefers cycloartenol [$SMT1_{CA}$], EC 2.1.1.142; and the plant SMT2 prefers 24(28)-methylene lophenol [$SMT2_{ML}$], EC 2.1.1.143. SMT1 and SMT2 catalyze the first and second methylation activities, respectively. If two enzymes belong to the same class in this classification, they are considered to have similar chemical functions. To date, the EC does not recognize the ability of SMTs to catalyze different product distributions as a measure of function. Moreover, there must be additional SMTs with substrate preferences not heretofore recognized by the EC. For example, fungi exist that accumulate lanosterol rather than zymosterol when treated with an inhibitor of SMT (Nes *et al.*, 2002), suggesting these catalysts, referred to as $SMT1_{LA}$, prefer lanosterol to zymosterol.

The steric–electric plug recognizes structural complementarity between the SMT and the substrate molecules. However, the substrate affinity and enzymatic product are not always either obvious or predictable for an unknown SMT. For example, the plant $SMT1_{CA}$ from algae and the protozoan $SMT1_{ZY}$ can generate similar 24-methyl-$\Delta^{25(27)}$-olefins as the major product. Alternatively, the fungal $SMT1_{ZY}$ from yeast and the plant $SMT1_{CA}$ from soybean can generate similar 24 (28)-methyl(ene)-$\Delta^{24(28)}$-olefins. The first C_1-activity of plant SMT2 catalyzes cycloartenol to a single methylated product whereas the second C_1-activity catalyzes 24(28)-methylene lophenol to three methylated products. The ability of the plant SMT2 to catalyze substrates of different features to different product sets was shown to result from the specificity in molecular recognition of the Δ^{24}-sterol structure. Kinetically, the SMT1 from fungi and plants differ. In the case of the yeast SMT1, the mechanism is random and for the plant SMT1 ordered so that AdoMet is the leading substrate and must bind before sterol to the enzyme (Nes, 2000; Nes *et al.*, 2003). Most likely, the different kinetic mechanisms relate to the significance that the plant SMT1 has activities for both the first and second C_1-transfer reaction and an ordered mechanism enables sequential C_1-transfer of different acceptors in the active site.

SMT1 and SMT2 activities vary with respect to one another in plants (Nes, 2000; Nes *et al.*, 1989b; Wentzinger *et al.*, 2002). For example, the proportion of SMT1 to SMT2 activity measured as their catalytic competence (V_{max}/K_m) is similar in cultured cells of tobacco during active cell proliferation. Alternatively, as the plant matures and cell proliferation is arrested, the ratio of the two activities change dramatically and can favor the activity of the second C_1-transfer reaction. These findings are in agreement with the suggestion of Benveniste and coworkers (Arnqvist *et al.*, 2003; Benveniste, 2004) of the possible importance of SMT2 as a branch point enzyme in phytosterol synthesis. The activity of SMT1 to control carbon flux into the phytosterol pathway and to cholesterol (Diener *et al.*, 2000; Parker and Nes, 1992) is evidenced in the high specific activity of cholesterol synthesis in mature leaves (Nes, 1990) and in potato plants overexpressing a foreign SMT1 (Fonteneau *et al.*, 1977). Given the regulatory role of SMT1 and SMT2 to balance the ratio of C-8 to C-9 to C-10 sterols suggests that the induced change in the ratio of SMT1 to SMT2 activities during development is largely controlled at the level of transcription. However, effectors, such as sitosterol or ATP, may also increase or decrease carbon flux through a branch point in phytosterol synthesis by modulating SMT activity (Nes, 2000).

AdoMet-dependent methylation has been found to be the target of functional convergence (Schubert *et al.*, 2003) generating a subfamily of SMTs that clearly evolved their substrate specificity for sterol independently of related AdoMet-dependent methyltransferases. Genetic and bioinformatic research of the derived amino acid sequences of over 60 SMTs deposited in the GenBank and other databases, indicate the primary structure encodes about 380 ($\pm$20) amino acid long proteins with an amino acid sequence relatedness which allows subdivision of the SMT gene family into five subfamilies, designated *SMTa* through *SMTe*, each distinguished by sharing a minimum of 40% identity among members (Fig. 9.17). Certain sequence motifs shared by all SMTs suggest that they share a common evolutionary origin. For example, three conserved domains in the primary structure of SMTs, which we have previously designated as Regions I, II, and III, are hydrophobic and always found in the same order on the polypeptide chain and are separated by comparable intervals (Fig. 9.18). Chemical affinity labeling and site-directed mutagenesis experiments have shown that Regions I and III correspond to a sterol binding site and in the yeast SMT maps to Y81EYGWGSSFHF and Y192AIEATCHAP (Nes *et al.*, 1999; Sinha, 2004; Zhou and Nes, 2003). Photoaffinity labeling and site-directed mutagenesis experiments show that Region II corresponds to a AdoMet binding site and in the yeast SMT maps to L24DVGCGVGGP (Schaeffer *et al.*, 2000). That these enzymes are so similar in much of their primary and presumably three-dimensional structures

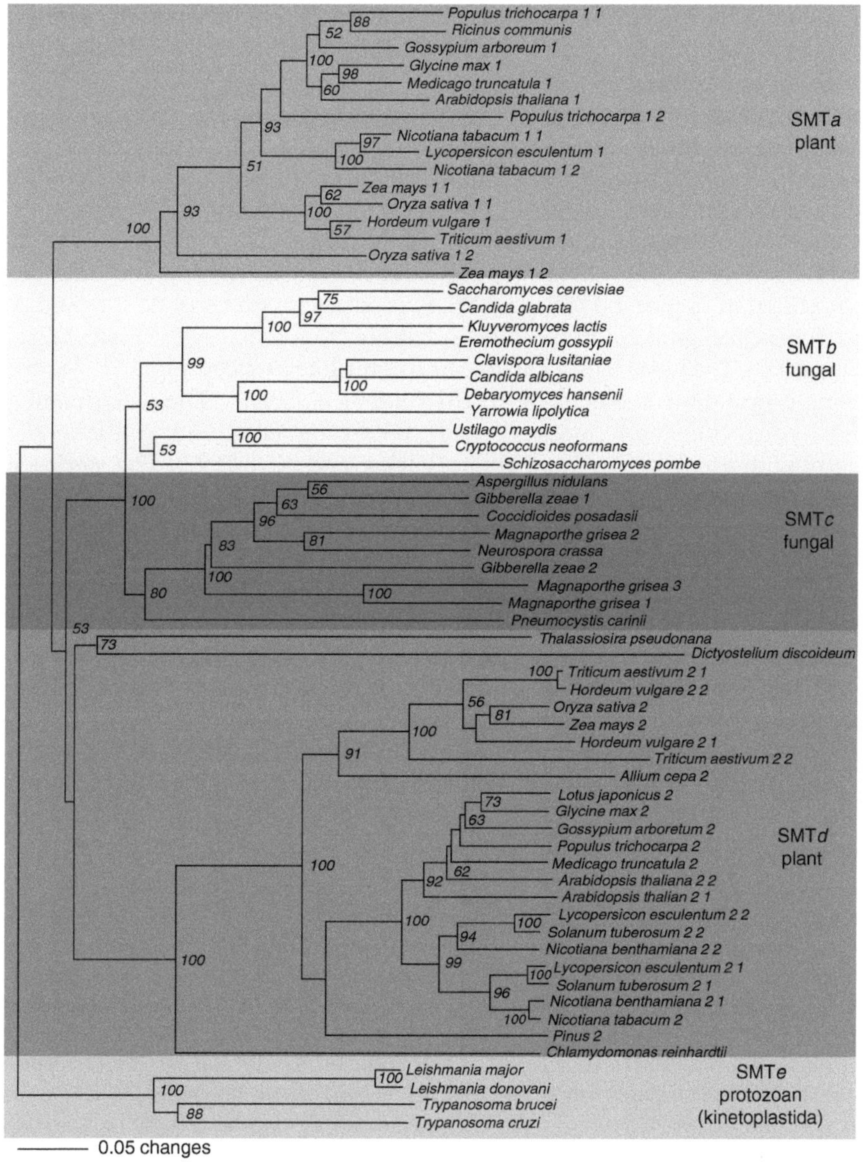

FIGURE 9.17 Rooted phylogenetic tree of eukaryotic SMT was created with PAUP using the neighbor-joining method with kinetoplastida SMTs as the out group. The scale bar represents a distance of 0.05 substitutions per site. Numbers are the percentage bootstrap values for 1,000 replicates. SMTa through SMTe designate SMT subfamilies defined by a minimum of 35% identity between members at the amino acid level. Accession number of SMT sequences obtained from NCBI (http://www.ncbi.nlm.nih.gov): *Ricinus communis*, T10173; *Glycine max* 1, T06780; *Arabidopsis thaliana* 1, AGG28462; *Nicotiana tabacum* 1-1, AAC43951; *Nicotiana tabacum* 1-2, AAC35787; *Zea mays* 1-1, TO4138; *Oryza sativa* 1-1, AAC34988; *Triticum aestivum* 1, ABB49388; *Oryza sativa* 1-2, AAP21419; *Saccharmyces cerevisiae*, NP_013706; *Candida glabrata*, CAG59930; *Kluyveromyces lactis*, AAS52116; *Clavispora lusitaniae*, CAO21936; *Candida albicans*, O74198; *Debaryomyces*

in spite of taxonomic differences in the species from which they are derived suggests that they arose by divergent evolution from an ancestral gene prior to the origin of eukaryotes.

To understand further the importance of conserved amino acids in the primary structure and the molecular interactions between sterol and enzyme to correctly position and discriminate the nucleophile that serves to undergo C-methylation to a single product, mutagenesis and activity assay with carefully designed mechanism-based inactivators were undertaken of highly conserved amino acid residues in the SMT from yeast (Fig. 9.19). We have reported that several mutants of yeast SMT modified in Region I, D79L and Y81F, produce mixtures of $\Delta^{24(28)}$- and $\Delta^{25(27)}$-olefins depending on the nature of Δ^{24}-substrate (Nes *et al.*, 1999; Sinha, 2004; Zhou and Nes, 2003). The yeast SMT1 can become plant-like in accepting $\Delta^{24(28)}$-substrates by a single mutation in Region I at Y81 (Nes *et al.*, 1999). The suicide substrate, 26,27-dehydrozymosterol, assayed with wild-type enzyme produces a novel Δ^{23}-sterol with an elongated side-chain (Parker and Nes, 1992). The steric–electric plug model permits differences in product specificity among SMT enzymes to have arisen through point mutations which change either the shape of the catalytic site and/or positions of the crucial functional groups. The D79L and Y81F mutants are capable of overcoming the topological impediment to generate multiple products, suggesting similar binding segments in the active center are present in all SMTs.

Testing with different sterol substrates of either heterologously expressed or wild-type enzymes indicates that substrate specificities have evolved differently in plants compared to the fungi and protozoa. All plant SMT1s reveal a strict substrate specificity accepting cycloartenol whereas the fungal SMT1 and protozoan SMT1 accept either zymosterol or lanosterol. A major difference between

---

FIGURE 9.17 *(continued)*
hansenii, CAG87427; *Yarrowia lipolytica*, CAG77980; *Ustilago maydis*, EAK84412; *Schizosaccharomyces pombe*, CAB16897; *Gibberella zeae* 1, XP_382959; *Magnaporthe grisa* 2, EAA48309; *Neurospora crassa*, CAB97289; *Gibberella zeae* 2, XP_355916; *Magnaporthe grisa* 3, EAA50587; *Magnaporthe grisa* 1, EAA47049; *Pneumocystis carinii*, AKK54439; *Oryza sativa* 2, ACC34989; *Arabidopsis thaliana* 2-1, ABB62809; *Arabidopsis thaliana* 2-2, CAA61966; *Nicotiana tabacum*, TO3848; *Leishmania donovani*, AAR92098; *Trypanosoma cruzi*, TIGR_5693. Gene indices of SMT obtained from TIGR (http://www.tigr.org): *Populus trichocarpa* 1-1, TC35619; *Medicago trucatula*, TC86500; *Gassypium arboretum* 1, TC20798; *Populus trichocarpa* 1-1, TC36329; *Hordeum vulgare* 1, TC 110279; *Zea mays* 1-2, TC234797; *Cryptococcus neoformans*, TC4573; *Aspergillus nidulans*, TC6295; *Coccidioides posadasii*, TC4513; *Triticum aestivum* 2-1, TC165856; *Hordeum vulgare* 2–2, TC121611; *Zea mays*, TC224796; *Hordeum vulgare* 2-1, TC123636; *Triticum aestivum* 2-2, TC172448; *Allium cape* 2, TC2207; *Lotus japonicus*, TC7995; *Glycine max* 2, TC189052; *Gossypium arboretum* 2, TC 21049; *Populus trichocarpa* 2, TC36329; *Medicago truncatula* 2, TC77751; *Lycopersicon esculentum* 2-2, TC126730; *Solanum tuberosum* 2-2, TC126449; *Nicotiana benthamiana* 2-2, TC7251; *Lycopersicon esculentum* 2-1, TC124648; *Solanum tuberosum* 2-1, TC112127; *Nicotiana benthamiana* 2-1, TC8230; *Pinus* 2, TC52326; *Chlamydomonas reinhardtii*, TC29837. Gene identification number of SMT from The Wellcome Trust Sanger Institute (http://www.sanger.ac.uk): *Leishmanis major*, LM3731Bb05.p1c; *Trypanosoma brucei*, TB10.1520. The SMT gene of *Dictyostelium discoideum* was from IMB (http://genome.imb-jena.de) and gene identification number is pcr25kl1p3887. SMT gene of *Thalassiosira pseudonana* is identified from GRI (http://genome.jgi-psf.org) in scaffold_30, 94659:96009. (See Page 13 in Color Section.)

FIGURE 9.18 Alignment of representative deduced amino acid sequences of SMT from fungi, protozoa, and plants that represent SMT1 and SMT2 isoforms. Conserved regions corresponding to sterol (Regions I and III) and AdoMet binding sites (Region II) are boxed. (See Page 14 in Color Section.)

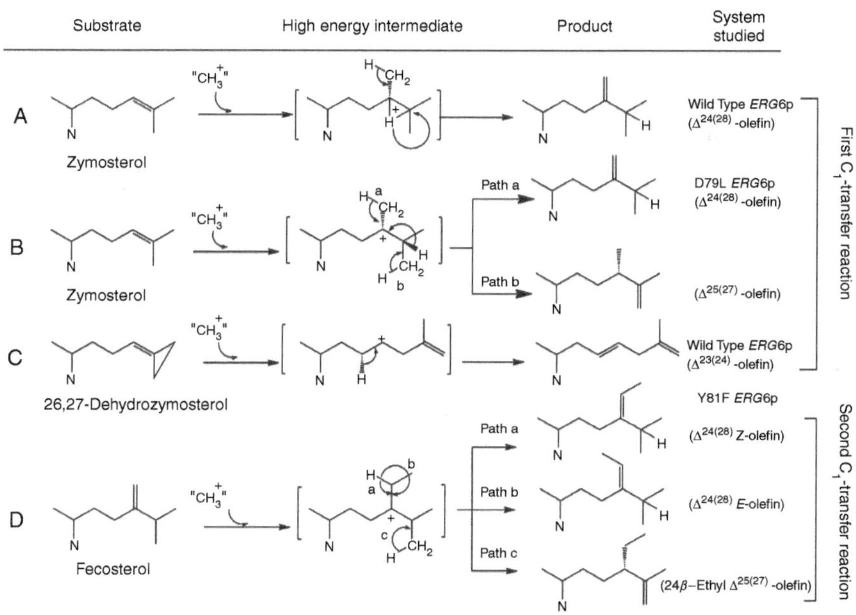

FIGURE 9.19 Catalytic competence of native and mutant yeast SMTs tested with different substrates.

zymosterol and lanosterol is the presence of the geminal methyl group at C-4 in the lanosterol structure. The 4,4-dimethyl group can sterically interfere with the hydrogen bonding ability of the C_3-hydroxyl group thereby affecting substrate affinity. Whereas SMT1 enzymes can be distinguished on their recognition of the nucleus structure, SMT1 and SMT2 enzymes can be distinguished on their recognition of the side-chain functional group $\Delta^{24(25)}$- versus $\Delta^{24(28)}$-substrate. To account for the different substrate acceptability, different arrangements of similar amino acid residues may have evolved in the active site of SMTs to interact with the nucleophilic groups at C-3 and C-24.

No conclusive discussion for the order of SMT evolution could be drawn from the unrooted phyogenetic tree (Fig. 9.17), based on the order of intermediates and the positioning of SMT isoforms in phytosterol synthesis. However, plant SMT2s which utilize 24(28)-methylene lophenol as the preferred substrate, but maintain vestige substrate specificity for cycloartenol, must have evolved from a progenitor plant SMT1 perhaps by duplication followed by mutation and divergence. An important question that warrants further study is whether cycloartenol served as a template providing functional constraints in the design and genetic formation of the plant SMT active site during either evolution or random mutation with natural selection provided the guiding principle of enzyme redesign.

A similarity in function (catalytic competence) among SMTs is reflected in the enzyme ability to catalyze a sterol acceptor to a methylated product. However, enzyme function of individual SMT isoforms in either the same plant or those formed in different phyla can be different in terms of substrate acceptability, kinetics, product outcome, and selectivity to inhibitors or effectors of SMT action. Although it may be possible to use the conserved Regions I and II of cDNA corresponding to SMT for homology-based cloning strategies, there is insufficient information for predictions of catalysis of new SMTs that might be cloned from, for example prokaryotes, since none of the putative prokaryote SMTs in the GenBank contain Region I. The trace level of phytosterol in cyanobacteria, suggesting they are contaminants (Marshall, 2007), and the likelihood SMTs are not synthesized by photosynthetic bacteria make the timing of SMT production and hence phytosterol accumulation during the course of evolution unclear.

5. BIOENGINEERING STRATEGIES FOR GENERATING PLANTS WITH MODIFIED STEROL COMPOSITIONS

Plant metabolic engineering the isoprenoid–phytosterol pathway to understand sterol biosynthesis and function has been underway for about 10 years. The enzymes catalyzing the committed step in the isoprenoid and phytosterol pathways are usually the most important control sites, and in plants they are the HMGR and SMT enzymes, respectively (Bach, 1995; Holmberg *et al.*, 2003; Nes, 2000; Volkman, 2005). Phytosterol synthesis is likely controlled by allosteric interactions involving end products of the pathway and nonsteroidal effectors, changes in the amount of the SMT isoforms, compartmentation and interactions between metabolically distinctive organs. The identification and biochemical characterization of a number of Arabidopsis mutant lines (EMS mutants,

T-DNA/Transposon insertion lines, transgenic plants) with an altered sterol profile has enabled researchers in defining the distinct functional metabolic units in the pathway. These mutants also point to the essential roles of sterols in regulating plant development and morphogenesis (Benveniste, 2004). As the genes encoding more enzymes of sterol synthesis are identified and manipulated, it is becoming apparent that phytosterol homeostasis, carbon flux, and growth are intimately tied to a variety of cellular functions and signaling pathways; therefore careful metabolic manipulation of the pathways will be required to generate value-added traits. From the biotechnological perspective the engineering of SMT activity has yet to lead to a desired trait that can be commercialized. However, considerable literature exists on the transgenic alteration of SMTs in plants, which indicates agronomically important applications will be forthcoming.

Studies on the results of overexpression and underexpression of the SMT isoforms in tobacco, tomato, potato, and Arabidopsis (Table 9.1) have recently confirmed that SMT1 catalyzes the first step in the cycloartenol-sitosterol (Arnqvist *et al.*, 2003; Fonteneau *et al.*, 1977; Holmberg *et al.*, 2002; Schaeffer *et al.*, 2001; Schaller *et al.*, 2001; Sitbon and Jonsson, 2001) and that SMT2 can regulate the levels of 24-methyl sterol to 24-ethyl sterol in the plant (Arnqvist *et al.*, 2003). SMT1 from plants is feedback inhibited by sitosterol but not by either ergosterol or cholesterol (Nes, 2000). Alternatively, ATP serves as an activator of SMT1 (Nes, 2000). Sitosterol inhibits the SMT1 in a competitive manner by decreasing its affinity for substrates without affecting its V_{max}. Sitosterol can also inhibit SMT2 activity but with significantly less effectiveness ca. K_i 100 μM versus 300 μM, respectively (Parker and Nes, 1992). Pulse-chase experiments using radioactively labeled intermediates (Bush and Grunwald, 1973; Heupel *et al.*, 1987; Nes, 1997; Nes, W. D. and Nguyen, H. T., unpublished data) and microarray analysis (Ledford *et al.*, 2004) have shown that light stimulates phytosterol synthesis and the expression of SMT. Three strategies have been employed to genetically modify the phytosterol composition in plants. These strategies are designed to either interrupt or enhance carbon flux at the stage of SMT1 or SMT2 activity or to elaborate improved enzymes aimed at reshaping enzyme specificities and mechanisms (Fig. 9.20).

The first approach to modify the phytosterol composition involves mutation of a gene encoding SMT (Diener *et al.*, 2000). Single-enzyme mutation can result in an inability to synthesize the enzyme in active form. Such a defect leads to a block in the metabolic pathway at the point where the enzyme acts and the enzyme's substrate accumulates. In some cases the functional consequences of the mutation have been investigated (Benveniste, 2004; Diener *et al.*, 2000). The *cvp1* Arabidopsis plants defective in SMT2 activity accumulate campesterol, much the same way *erg6* yeast mutants defective in SMT1 activity accumulate zymosterol (Fig. 9.20, Panel I) (Ledford *et al.*, 2004). The *smt1–3/cph* Arabidopsis plants defective in SMT1 activity accumulate cholesterol rather than cycloartenol. The accumulation of cholesterol (nonalkylated sterol) in significant amounts suggests enzymatic reactions which normally do not recognize cycloartenol can process the intermediate to a Δ^5-sterol in the transgenic plants. The induced mutations at the stage of either SMT1 or SMT2 activity whereby a 24-desalkyl (C-8-sterol side-chain) sterol accumulates or a 24-methyl (C-9-sterol side-chain) sterol accumulates agrees with the order of intermediates positioned in the kinetically favored pathway of

TABLE 9.1 Sterol content of transgenic plants engineered for modified phytosterol compositions

	Constructions		Affect on SMT expression	Sterol content of transgenic plant	References
	cDNA to	Plant system			
1	Nt SMT1	Tobacco	Cosuppression SMT1	Elevated C8 side-chain level; decrease C10 side-chain level	Arnqvist *et al.*, 2003
			Overexpression SMT1	Elevated C9 side-chain level	
2	Gm SMT1	Tobacco	Overexpression SMT1	Elevated C9 side-chain level	Fonteneau *et al.*, 1977
3	$ScSMT1_{Y81W}$	Tobacco	Co-suppression SMT1	Decrease C8 side-chain level; elevated C10 side-chain level	Schaller *et al.*, 2001
4	$ScSMT1_{FSM}$	Tomato	Cosuppression SMT1	Elevated C8 side-chain level; decrease C10 side-chain level	Schaller *et al.*, 2001
5	AtSMT2	Tobacco	Cosuppression SMT1		Schaeffer *et al.*, 2001
6	$AtSMT2_{anti}$	Arabidopsis	Cosuppression SMT2	Elevated C9 side-chain level	Schaeffer *et al.*, 2001
			Overexpression SMT2	Decrease C9 side-chain level	Sitbon and Jonsson, 2001
7	Nt $SMT1_{anti}$	Tobacco	Antisense gene expression SMT1	Elevated C8 side-chain level; decrease C10 side-chain level	Holmberg *et al.*, 2002

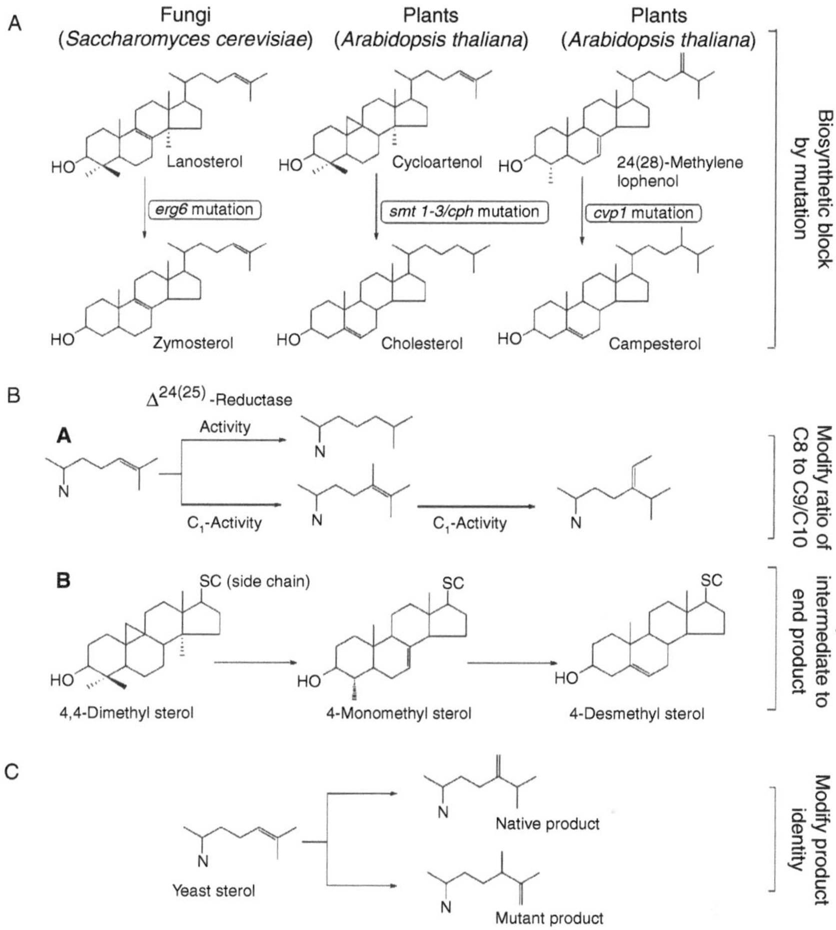

FIGURE 9.20 Different strategies to engineer plants with modified sterol compositions. (A) Engineer change in sterol composition through mutation; (B) engineer change in sterol composition through antisense/cosuppression technology; (C) engineer change through introduction of a foreign SMT that generates novel products.

phytosterol biosynthesis. Mutant plants having a modified cholesterol to phytosterol ratio are stunted consistent with the requirement for a fixed phytosterol homeostasis (Fig. 9.21). In related work, inhibitors of sterol biosynthesis administered to cultured plant cells (Nes *et al.*, 1991c) revealed a sequence of reactions in the phytosterol pathway that posits SMT1 as a critical slow step. These transition state analogs possess similar binding properties to either SMT1 or SMT2 (Nes *et al.*, 2003). Therefore, both SMT activities are impaired equally *in vivo* as well as *in vitro*. Studies in the design of inhibitors of SMT, activity have not yet progressed to the stage where one compound can inhibit an individual SMT, although mechanism-based inhibitors prepared in our laboratory are tailored with the expectation to be SMT specific (Zhou *et al.*, 2004).

FIGURE 9.21 Comparison of *smt1* (*Ac*-mutagenized plant that generates high levels of cholesterol) and *SMT1* (wild-type that generates high levels of sitosterol) plants. (A) Rosette of 2-week plants; (B) rosette of 2-week-old *smt1*-*smt1*–3 plant; (C) mature (5-week) SMT1 and *smt1*–3 plant. Adapted from Diener *et al.* (2000). (See Page 15 in Color Section.)

The second approach to modify the phytosterol composition is to engineer plants with an SMT construct in the sense or antisense direction either of which can lead to cosuppression of native SMT synthesis (Fig. 9.21, Panel B). Phytosterol regulation can favor different branch point enzymes to generate a sterol mixture containing side-chains of varying degrees of C-24 alkylation. If carbon flux in phytosterol synthesis passes along the same set of tracks separated by SMT1 and SMT2, equilibrium considerations will dictate that either slowing or increasing the traffic in one direction by modifying SMT expression will lead to a change in the ratio of C-8- to C-9- to C-10-sterol side-chains (Table 9.1). The simplest approach to disturb the steady state concentration of SMT is to engineer either a sense or antisense construct of the native SMT1 to a plant. Transgenic tobacco plants with modified expression of the native SMT1 gene were prepared by introducing sense and antisense expression cassettes of cDNA NtSMT1–1 to tobacco plants (Arnqvist *et al.*, 2003; Holmberg *et al.*, 2002; Schaeffer *et al.*, 2001; Sitbon and Jonsson, 2001). The resulting plants possessed variations in the cycloartenol proportions and a concomitant effect on the proportion of 24-ethyl sterols. In these plants, the total amount of sterol remained relatively unchanged, consistent with our hypothesis that plants will tend to maintain a sterol homeostasis to the extent possible (Nes, 1990). Arabidopsis plants showing cosuppression of SMT2–1 were characterized by high campesterol levels and depletion of sitosterol (Sitbon and Jonsson, 2001). Pleiotropic effects on development such as reduced growth appear to result from changes in expression of SMT1 and SMT2 (Benveniste, 2004; Lindsey *et al.*, 2003; Schaller, 2004). In a related study, an Arabidopsis frill1 (fri1) mutant was generated that had a mutation in SMT2 and an altered C_1/C_2-methyl sterol composition. Petal morphogenesis was found to be impaired by the change in phytosterol homeostasis (Hase *et al.*, 2005). Overexpression of soybean SMT1 in transgenic potato plants results in a marked reduction of cholesterol and glycoalkaloids (Sitbon and Jonsson, 2001), consistent with the role of SMT1 to control the level of C-8- to C-9/C-10-sterols.

The reports published thus far on engineering modified sterol pathways in plants revolve around engineering a plant SMT back into plants. An alternative strategy adopted in our laboratory (Nes and Nguyen unpublished data) is to engineer a fungal SMT into plants. We were concerned initially whether plants would either express or otherwise tolerate a fungal SMT. This could be due to either genetic mechanism or due to a failure of the enzyme to catalyze plant substrates which normally are unacceptable to the plant SMT. According to structure–activity tests with the yeast SMT, neither cycloartenol nor 24(28)-methylene lophenol will bind productively (Zhou and Nes, 2003), although zymosterol, the optimal substrate for the yeast SMT1, is a good substrate for SMT2. These findings led to the hypothesis that engineering a yeast SMT to plants will promote the underexpression of the native SMT1 by cosuppression and interrupt carbon flux to SMT2 by providing a foreign SMT that can compete for substrates targeted for SMT catalysis. If substrates normally converted by the plant SMT isoforms were acceptable to the yeast SMT1, then it might be possible to engineer plants with a yeast mutant SMT1 (or some related SMT) capable to generate novel products (Fig. 9.21, Panel C).

To study our hypothesis further, two different *ERG6* SMT constructs were developed. The first *ERG6* SMT construct was developed as a consequence of PCR modification which gave rise to a mutation that introduced a frameshift in the gene toward the N-terminus (Nes, unpublished data). This *Saccharomyces cerevisiae* SMT1 containing a *f*rameshift *m*utation is referred to as ScSMT1-FSM. A second *ERG6* SMT construct was developed from site-directed mutagenesis of an amino acid at position-81 which corresponds to the sterol binding site (Nes *et al.*, 1999). This *S. cerevisiae* SMT1 containing a tyrosine to phenylalanine mutation at position-81, referred to as ScSMT1-Y81F, was chosen for the engineering studies because it has altered substrate specificity and product sets that make it plant-like. Tomato plants harboring the ScSMT1-FSM transgene contained a high level of cycloartenol and a corresponding decreased level of 24-ethyl sterols (Table 9.2). Tobacco plants harboring the Y81F yeast mutant contained decreased levels of cholesterol and increased levels of 24-ethyl sterols. It is expected that

TABLE 9.2 Sterol composition of 6-week-old tomato and tobacco transformants expressing a mutant yeast SMT1[a]

Analysis (as % of total sterol)[b]		Tomato[c]: control		Tomato: $ScSMT1_{FSM}$		Tobacco: control		Tobacco[d]: $ScSMT1_{Y81W}$	
Cycloartenol	C8 side-chain	39	(48)	56	(62)	4	(21)	1	(12)
Cholesterol		9		6		10		10	
Campesterol	C9 side-chain	5	(5)	6	(6)	14	(4)	12	(12)
Stigmasterol	C10 side-chain	14	(47)	10	(32)	48	(75)	44	(66)
Sitosterol		15		11		17		10	
Isofucosterol		16		11		10		2	
Ratio C8/C9/C10		1:tr:1		2:tr:1		1:0.5:5		1:1:7	

[a] To obtain transgenic plants with altered sterol profiles, a DNA fragment containing the open reading frame of SMT *ERG6* gene of yeast was modified by PCR to include restriction sites for *NcoI* on either ends of the open reading frame. This PCR product gave rise to a mutation which introduced a frame shift in the gene. This mutation (Frame Shift Mutation, FSM) made from the *ERG6* gene introduced into the plant untranslatable but capable of inhibiting the endogenous SMT via cosuppression mechanism (Schaller *et al.*, 2001). The Y81W yeast mutant was prepared by site-directed mutagenesis as described by Nes *et al.* (1999).

[b] Sterol side-chain ratios are compared as (C8)-24-desmethylsterol; (C9)-24-methyl(ene)sterols, and (C10)-24-ethyl (idene)sterols.

[c] Nontransgenic segregated gene as control.

[d] Y81W *ERG6* mutant engineered in the sense direction. Conformation of the expression of yeast SMT mutant was by Northern blot analysis, Western blot analysis with an antibody against yeast SMT, and activity assay of leaf microsomes with cycloartenol and zymosterol.

overexpressing a foreign SMT1 in plants will increase the overall SMT1 activity (resulting from the combination of native protein and transgene protein) thereby increasing flux after the formation of 24(28)-methylene cycloartenol, which appears to be the case for either plant or fungal SMTs engineered into plants. The ScSMT1-FSM was not expressed in tomato whereas the Y81F yeast mutant was constitutively expressed, as determined by several techniques including activity assay, Northern blot analysis, and immunochemistry using the yeast SMT antibody (unpublished data). The overexpressed yeast SMT1-Y81F did not appear to compete for endogenous substrates of tobacco since the plant did not make any acceptors suitable for the fungal SMT catalysis.

Because it is generally undesirable to alter the balance of 4,4-dimethyl sterol intermediate to 4-desmethyl sterol end products, except according to the normal developmental program, single-enzyme manipulations designed to either increase the amount of intermediates or change the ratio of C-24-alkylated sterols in the sterol mixture (Fig. 9.21, Panel B), are limited to upregulating HMGR activity and downregulating SMT activity (Bach, 1995; Holmberg *et al.*, 2003; Guo *et al.*, 1995). Generally, changes in the intermediate to end product ratio result in cycloartenol accumulating as sterol ester in lipid droplets whereas the modified ratio of 24-methyl to 24-ethyl Δ^5-sterols results in little change in the total cellular sterol thereby maintaining sterol balance in the membrane.

The third approach being developed in our laboratory is to engineer a combination of transgenes in the antisense and sense directions to crop plants. We

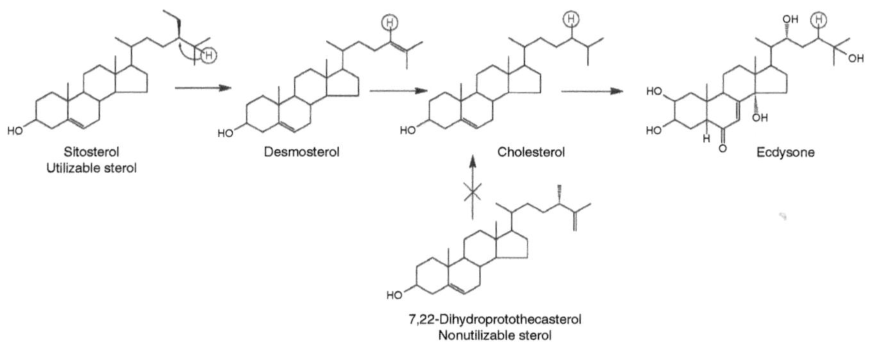

FIGURE 9.22 Conversion of sitosterol to cholesterol and ecdysone by phytophagous insects. The hydrogen circled in red migrates from C-25 in sitosterol to C-24 in cholesterol during the C-24-dealkylation reaction. The methyl group at C-24 in 7,22-dihydroprotothecasterol remains intact for mechanistic reasons. (See Page 15 in Color Section.)

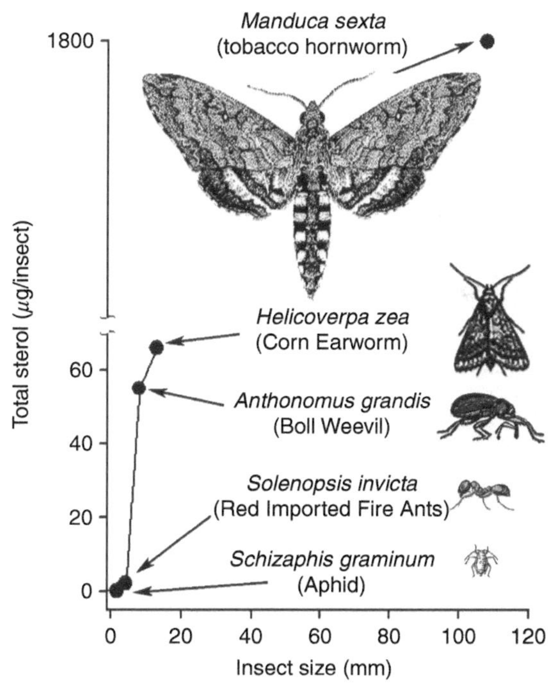

FIGURE 9.23 Correlation of sterol content with insect size. Adapted from Behmer and Nes, 2003.

expect to eliminate the expression of the native SMT1 thereby permitting a foreign SMT to be expressed that possesses an unusual catalytic competence (Fig. 9.21, Panel C). This strategy is based on the observation that insects do not synthesize their own sterol and require cholesterol to synthesize ecdysteroid involved in the molting process (Behmer and Nes, 2003). Phytophagous insects convert sitosterol, obtained from the host plant, to cholesterol (Fig. 9.22) (Behmer and Nes, 2003; Nes *et al.*, 1997). Insects, like plants, have different sterol requirements depending on size and with respect to their phylogeny (Fig. 9.23). In the C-24-dealkylation pathway, the hydrogen atom at C-25 migrates to C-24 during the elimination of the 24-ethyl group. Thus, phytosterols with a $\Delta^{25(27)}$-bond will not undergo side-chain metabolism by the insect; therefore these sterols will be nonutilizable nutrients for growth (Svoboda *et al.*, 1995 Nes *et al.*, 1997). In order to develop insect-resistant plants, a strategy we considered is to redesign the yeast SMT1 to affect its catalytic competence in such a way to bind cycloartenol, to remove the affinity for effectors (sitosterol) that might interfere with SMT activity, and to modify the active site to promote channeling to generate $\Delta^{25(27)}$-olefins. It is clear that we can tailor SMTs to produce new substrate affinities and products. The opportunities to generate transgenic plants that appear similar to wild types with modified sterol compositions are limitless and soon the commercial benefits of this evolving bioengineering technology directed at phytosterols will be realized.

ACKNOWLEDGEMENT

This work was supported by the National Science Foundation (MCB 0417436 and MCB 0115401) and the Welch Foundation (D-1276) to W.D.N.

REFERENCES

Akihisa, T. K., Yasukawa, M., Yamaura, M., Ukiya, Y., Kimura, N., Shimizu, N., and Arai, K. (2000). Triterpene alcohol and sterol ferulates from rice bran and their anti-inflammatory effects. *J. Agric. Food Chem.* **48,** 2313–2319.

Arigoni, D., Sagner, S., Latzel, C., Eisenreich, W., Bacher, A., and Zenk, M. H. (1997). Terpenoid biosynthesis from 1-deoxy-D-xylulose in higher plants by intramolecular skeletal rearrangement. *Proc. Natl. Acad. Sci.* **94,** 10600–10605.

Arnqvist, L., Dutta, P. C., Jonsson, L., and Sitbon, F. (2003). Reduction of cholesterol and glycoalkaloid levels in transgenic potato plants by overexpression of a typ1 sterol methyltransferase cDNA. *Plant Physiol.* **131,** 1792–1799.

Awad, A. B., and Fink, C. S. (2000). Phytosterols as anticancer dietary components: Evidence and mechanism of action. *J. Nutr.* **130,** 2127–2130.

Bach, T. J. (1995). Some new aspects of isoprenoid biosynthesis in plants—a review. *Lipids* **30,** 191–202.

Bach, T. J., and Lichtenthaler, H. K. (1983). Inhibition by mevinolin of plant growth, sterol formation and pigment accumulation. *Physiol. Plant* **59,** 50–60.

Behmer, S. T., and Nes, W. D. (2003). Insect sterol nutrition and physiology: A global overview. *Adv. Insect Physiol.* **31,** 1–72.

Benveniste, P. (2004). Biosynthesis and accumulation of sterols. *Annu. Rev. Plant Biol.* **55,** 429–457.

Bloch, K. E. (1983). Sterol structure-membrane function. *CRC Crit. Rev. Biochem.* **14,** 47–82.

Bush, P. B., and Grunwald, C. (1973). Effect of light on mevalonic acid incorporation into the phytosterols of *Nicotiana tabacum L.* seedlings. *Plant Physiol.* **51,** 110–114.

Carland, F. M., Fujiko, S., Takatsuto, S., Yoshida, S., and Nelson, T. (2002). The identification of *CVP*1 reveals a role for sterols in vascular patterning. *The Plant Cell* **14,** 2045–2058.

Castle, M., Blondin, G., and Nes, W. R. (1963). Evidence for the origin of the ethyl group of β-sitosterol. *J. Am. Chem. Soc.* **85,** 3306–3308.

Chappell, J., Wolf, F., Proulx, J., Cuellar, R., and Saunders, C. (1995). Is the reaction catalyzed by 3-hydroxy-3-methylglutaryl coenzyme A reductase a rate-limiting step for isoprenoid biosynthesis in plants? *Plant Physiol.* **109,** 1337–1343.

Clark, A. J., and Bloch, K. (1959). Function of sterols in *Dermetes vulpinus*. *J. Biol. Chem.* **234,** 2583–2588.

Clouse, S. D. (2002). *Arabidopsis* mutants reveal multiple roles for sterols in plant development. *The Plant Cell* **14,** 1995–2000.

De-Eknamkul, W., and Potduang, B. (2003). Biosynthesis of β-sitosterol and stigmasterol in *Croton sublyratus* proceeds via a mixed origin of isoprene units. *Phytochemistry* **62,** 389–398.
Devarenne, T. P., Ghosh, A., and Chappell, J. (2002). Regulation of squalene synthase, a key enzyme of sterol biosynthesis in tobacco. *Plant Physiol.* **129,** 1095–1106.
Diener, A. C., Li, H., Zhou, W., Whoriskey, W. J., Nes, W. D., and Fink, G. R. (2000). Sterol methyltransferase 1 controls the level of cholesterol in plants. *The Plant Cell* **12,** 853–870.
Fonteneau, P., Hartmann, M. A., and Benveniste, P. (1977). A 24-methylene lophenol C-28 methyltransferase from suspension cultures of bramble cells. *Plant Sci. Lett.* **10,** 147–155.
Goad, L. J., Lenton, J. R., Knapp, F. F., and Goodwin, T. W. (1974). Phytosterol side-chain biosynthesis. *Lipids* **9,** 582–594.
Goodwin, T. W. (1981). Biosynthesis of plant sterols and other triterpenoids. *In* "Biosynthesis of Isoprenoid Compounds" (J. W. Porter and S. L. Spurgeon, eds.), Vol. 1, pp. 444–480. Wiley and Sons, New York.
Guo, D., Venkatramesh, M., and Nes, W. D. (1995). Developmental regulation of sterol biosynthesis in *Zea mays*. *Lipids* **30,** 203–219.
Guo, D., Jia, Z., and Nes, W. D. (1996). Stereochemistry of hydrogen migration from C-24 to C-25 during phytosterol biomethylation. *J. Am. Chem. Soc.* **118,** 8507–8508.
Harker, M., Hellyer, A., Clayton, J. C., Duvoix, A., Lanot, A., and Stafford, R. (2003). Co-ordinate regulation of sterol biosynthesis activity during accumulation of sterols in developing rape and tobacco seeds. *Planta* **216,** 707–715.
Hase, Y., Fujioka, S., Yoshida, S., Sun, G., Umeda, M., and Tanaka, A. (2005). Ectopic endoreduplication caused by sterol alteration results in serrated petals in *Arabidopsis*. *J. Exp. Bot.* **56,** 1263–1268.
Haughan, P. A., Lenton, J. R., and Goad, L. J. (1987). Paclobutrazol inhibition of sterol biosynthesis in cell suspension culture and evidence of an essential role for 24-ethylsterol in plant cell division. *Biochem. Biophys. Res. Commun.* **146,** 510–516.
Hemmerlin, A., Hoeffler, J.-F., Meyer, O., Tritsch, D., Kagan, I. A., Grosdemange-Billiard, C., Rohmer, M., and Bach, T. J. (2003). Cross-talk between the cytosolic mevalonate and the plastidial methylerythritol phosphate pathways in tobacco bright yellow-2 cells. *J. Biol. Chem.* **278,** 26666–26676.
Herman, G. E. (2003). Disorders of cholesterol biosynthesis: Prototypic metabolic malformation syndromes. *Human Mole. Gene* **2,** 75–88.
Heupel, R. C., Sauvaire, Y., Le, P. H., Parish, E. J., and Nes, W. D. (1986). Sterol composition and biosynthesis in sorghum: Importance to developmental regulation. *Lipids* **21,** 69–76.
Heupel, R. D., Nes, W. D., and Verbeke, J. A. (1987). Developmental regulation of sterol and pentacyclic triterpenoid biosynthesis and composition: A correlation with sorghum floral initiation. *In* "The Metabolism, Structure, and Function of Plant Lipids" (P. K. Stumpf, J. B. Mudd, and W. D. Nes, eds.), pp. 53–56. Plenum Press, New York.
Holmberg, N., Harker, M., Gibbard, C. L., Wallce, A. D., Clayton, J. C., Rawlins, S., Hellyer, A., and Safford, R. (2002). Sterol methyltransferase type 1 controls the flux of carbon into sterol biosynthesis in tobacco seed. *Plant Physiol.* **130,** 303–311.
Holmberg, N., Harker, M., Wallace, A. D., Clayton, A. D., Gibbard, C. L., and Safford, R. (2003). Co-expression of *N*-terminus truncated 3-hydroxy-3-methylglutaryl CoA reductase and C24-methyltransferase type 1 in transgenic tobacco enhances carbon flux towards end-product sterols. *Plant J.* **36,** 12–20.
Julia, M., and Marazano, C. (1985). Biomimetic methyltransfer to olefins. *Tetrahedron* **41,** 3717–3724.
Kalinowska, M., Nes, W. R., Crumley, F. G., and Nes, W. D. (1990). Stereochemical differences in the anatomical distribution of C-24 alkylated sterols in *Kalanchoe diagremontiana*. *Phytochemistry* **29,** 3427–3434.
Kaneshiro, E. S., Rosenfeld, J. A., Basselin-Eiweida, M., Stringer, J. R., Keeley, S. P., Smulian, A. G., and Giner, J.-L. (2002). The *Pneumocystis carinii* drug target S-adenosyl-methionine: Sterol C-24 methyl transferase has a unique substrate preference. *Mol. Microbiol.* **44,** 989–999.
Kresge, N., Simoni, R. D., and Hill, R. L. (2005). The biosynthetic pathway for cholesterol: Konrad Bloch. *J. Biol. Chem.* **280,** 7–10.
Laule, O., Furholz, A., Chang, H.-S., Zhu, T., Wang, X., Heifetz, P. B., Gruissem, W., and Lange, B. M. (2003). Cross-talk between cytosolic and plastidial pathways of isoprenoid biosynthesis in *Arabidopsis thaliana*. *Proc. Natl. Acad. Sci.* **100,** 6866–6871.

Ledford, H. K., Baroli, I., Shin, J. W., Fisher, B. B., Eggen, R. I. L., and Niyogi, K. K. (2004). Comparative profiling of lipid-soluble antioxidants and transcript reveals two phases of photo-oxidative stress in a xanthophyll-deficient mutant of *Chlamydomonas reinhardtii*. *Mol. Gen. Genomics* **272,** 47–479.

Lichtenthaler, H. K., Schwender, J., Disch, A., and Rohmer, M. (1997). Biosynthesis of isoprenoids in higher plant chloroplasts proceeds via a mevalonate-independent pathway. *FEBS Lett.* **400,** 271–274.

Lindsey, K., Pullen, M. L., and Topping, J. F. (2003). Importance of plants sterols in pattern formation and hormone signaling. *Trends Plant Sci.* **8,** 521–525.

Ling, W. H., and Jones, P. J. (1995). Dietary phytosterols: A review of metabolism, benefits and side effects. *Life Sci.* **57,** 195–206.

Mangla, A. T., and Nes, W. D. (2000). Sterol C-methyl transferase from *Prototheca wickerhamii*: Mechanism, sterol specificity and inhibition. *Bioorg. Med. Chem.* **8,** 925–936.

Marshall, J. A. (2007). Studies on the enzymology of sterol methyltransferase from *Saccharomyces cervevisiae*. *Dissertation* pp. 1–95. Texas Tech University.

Marshall, J. A., Dennis, A. L., Haynes, A., Kumazawa, T., and Nes, W. D. (2001). Sterol composition and utilization of soybean sterols by *Phytophthora sojae*. *Phytochemistry* **58,** 423–428.

Mckersie, B. D., and Thompson, J. E. (1979). Influence of plant sterols on the phase properties of phospholipid bilayers. *Plant Physiol.* **63,** 802–806.

Moreau, R. A., Whitaker, B. D., and Hicks, K. B. (2002). Phytosterols, phytostanols and their conjugates in foods: Structural diversity, quantitative analysis, and health-promoting uses. *Prog. Lipid Res.* **41,** 457–500.

Nes, W. D. (1990). Control of sterol biosynthesis and its importance to developmental regulation and evolution. *Rec. Adv. Phytochem.* **24,** 283–327.

Nes, W. D. (1997). Development of Transgenic Plants with Modified Sterol Compositions Patent Application Ser. No. 08/998,339.

Nes, W. D. (2000). Sterol methyltransferase: Enzymology and inhibition. *Biochim. Biophys. Acta* **1529,** 63–88.

Nes, W. D. (2003). Enzyme mechanisms for sterol C-methylations. *Phytochemistry* **64,** 75–95.

Nes, W. D., and Bach, T. J. (1985). Evidence for a mevalonate shunt in a tracheophyte. *Proc. R. Soc., Lond., B* **224,** 425–444.

Nes, W. D., and Heupel, R. C. (1986). Physiological requirement for biosynthesis of multiple 24β-methylsterols in *Gibberella fujikuroi*. *Arch. Biochem. Biophys.* **244,** 211–217.

Nes, W. D., and Le, P. H. (1990). Evidence for separate intermediates in the biosynthesis of multiple 24β-methylsterol end products by *Gibberella fujikuroi*. *Biochim. Biophys. Acta* **1042,** 119–125.

Nes, W. D., and Nguyen, H. T. (unpublished data).

Nes, W. D., and Schmidt, J. O. (1988). Isolation of 25(27)-dehydrolanost-8-enol from *Cereus giganteus* and its biosynthetic importance. *Phytochemistry* **27,** 1705–1708.

Nes, W. D., Wong, R. Y., Benson, M., Landrey, J. R., and Nes, W. R. (1984). Rotational isomerism about the 17(20)-bond of steroids and euphoids as shown by the crystal structures of euphol and tirucallol. *Proc. Natl. Acad. Sci.* **81,** 5896–5900.

Nes, W. D., Heupel, R. C., and Le, P. H. (1985). Biosynthesis of ergosta-6(7), 8(14), 22(23)-trien-3β-ol by *Gibberella fujikuroi*: Its importance to ergosterol's metabolic pathway. *J. Chem. Chem. Commun.* **8,** 1431–1433.

Nes, W. D., Hanners, P. K., and Parish, E. J. (1986). Control of fungal sterol C-24 transalkylation. Importance to developmental regulation. *Biochem. Biophys. Res. Commun.* **139,** 410–415.

Nes, W. D., Xu, S., and Haddon, W. F. (1989a). Evidence for similarities and differences in the biosynthesis of fungal sterols. *Steroids* **53,** 533–558.

Nes, W. D., Xu, S., and Parish, E. J. (1989b). Metabolism of 24(*R*,*S*), 25-epiminolanosterol to 25-aminolanosterol and lanosterol by *Gibberella fujikuroi*. *Arch. Biochem. Biophys.* **272,** 323–331.

Nes, W. D., Norton, R. A., Crumley, F. G., Madigan, S. J., and Katz, E. R. (1990). Sterol phylogenesis and algal evolution. *Proc. Natl. Acad. Sci.* **87,** 7565–7569.

Nes, W. D., Wong, R. Y., Benson, M., and Akihisa, T. (1991a). Conformational analysis of 10-α cucurbitadienol. *J. Chem. Soc. Chem. Commun.* **18,** 1272–1274.

Nes, W. D., Janssen, G. G., and Bergenstrahle, A. (1991b). Structural requirements for transformation of substrates by the (S)-adenosyl-L-methionine: $\Delta^{24(25)}$-sterol methyl transferase. *J. Biol. Chem.* **266,** 15202–15212.

Nes, W. D., Janssen, G. G., Norton, R. A., Kalinowska, M., Crumley, F. G., Tal, B., Bergenstrahle, A., and Jonsson, L. (1991c). Regulation of sterol biosynthesis in sunflower by 24(*R,S*)-25-epiminolanosterol, a novel C-24 methyl transferase inhibitor. *Biochem. Biophys. Res. Commun.* **177,** 566–574.

Nes, W. D., Norton, R. A., and Benson, M. (1992). Carbon-13 NMR studies on sitosterol biosynthesized from [^{13}C]mevalonates. *Phytochemistry* **31,** 805–816.

Nes, W. D., Janssen, G. G., Crumley, F. G., Kalinowska, M., and Akihisha, T. (1993). The structural requirements of sterols for membrane function in *Saccharomyces cerevisiae*. *Arch. Biochem. Biophys.* **300,** 724–733.

Nes, W. D., Lopez, M., Zhou, W., Dowd, P. F., and Norton, R. A. (1997). Sterol utilization and metabolism by *Heliothis zea*. *Lipids* **32,** 1317–1323.

Nes, W. D., Koike, K., Jia, Z., Sakamoto, Y., Satou, T., Nikaido, T., and Griffin, J. F. (1998a). 9β,19-Cyclosterol analysis by ^{1}H- and ^{13}C-NMR, crystallographic observations and molecular mechanics calculations. *J. Am. Chem. Soc.* **120,** 5970–5980.

Nes, W. D., McCourt, B. S., Zhou, W., Ma, J., Marshall, J. A., Peek, L. A., and Brennan, M. (1998b). Overexpression, purification, and stereochemical studies of the recombinant (S)-adenosyl-L-methionine: $\Delta^{24(25)}$- to $\Delta^{24(28)}$-sterol methyl transferase enzyme from *Saccharomyces cerevisiae*. *Arch. Biochem. Biophys.* **353,** 297–311.

Nes, W. D., McCourt, B. S., Marshall, J. A., Ma, J., Dennis, A. L., Lopez, M., Li, H., and He, L. (1999). Site-directed mutagenesis of the sterol methyltransferase active site from *Saccharomyces cerevisiae* results in formation of 24-ethyl sterols. *J. Org. Chem.* **64,** 1535–1542.

Nes, W. D., Lukyanenko, Y. O., Jia, Z., Quideau, S., Howald, W. N., Pratum, T. K., West, R. R., and Hutson, J. C. (2000). Identification of the lipophilic factor produced by macrophages that stimulates steroidogenesis. *Endocrinology* **142,** 953–958.

Nes, W. D., Marshall, J. A., Jia, Z., Jaradat, T. T., Song, Z., and Jayasimha, P. (2002). Active site mapping and substrate channeling in the sterol methyltransferase pathway. *J. Biol. Chem.* **277,** 42549–42556.

Nes, W. D., Song, Z., Dennis, A. L., Zhou, W., Nam, J., and Miller, M. (2003). Biosynthesis of phytosterols: Kinetic mechanism for the enzymatic *C*-methylation of sterols. *J. Biol. Chem.* **278,** 34505–34516.

Nes, W. D., Jayasimha, P., Zhou, W., Ragu, K., Jin, C., Jaradat, T. T., Shaw, R. W., and Bujinicki, J. M. (2004). Sterol methyltransferase. Functional analysis of highly conserved residues by site-directed mutagenesis. *Biochemistry* **43,** 569–576.

Nes, W. R. (1980). Function as an evolutionary determinant of biosynthesis. *In* "Biogenesis and Function of Plant Lipids" (P. Mazilak, P. Benveniste, C. Costes, and R. Douce, eds.), pp. 387–394. Elsevier, North Holland Biomedical Press.

Nes, W. R. (1987a). Multiple roles for plant sterols. *In* "The Metabolism, Structure and Function of Plant Lipids" (P. K. Stumpf, J. B. Mudd, and W. D. Nes, eds.), pp. 3–9. Plenum Press, New York.

Nes, W. R. (1987b). Structure-function relationships for sterols in *Saccharomyces cerevisiae*. *ACS Symp. Ser.* **325,** 252–267.

Nes, W. R., and McKean, M. L. (1977). "Biochemistry of Steroids and Other Isopentenoids," pp. 412–452. University Park Press, Baltimore, MD.

Nes, W. R., Krevitz, K., Joseph, J., Nes, W. D., Harris, B., Gibbons, G. F., and Patterson, G. W. (1977). The phylogenetic distribution of sterols in tracheophytes. *Lipids* **12,** 511–527.

Norton, R. A., and Nes, W. D. (1991). Identification of ergosta-6(8), 8(14), 25(27)-trien-3β-ol and ergosta-5(6), 7(8), 25(27)-trien-3β-ol; two new steroidal trienes synthesized by *Prototheca wickerhamii*. *Lipids* **26,** 247–249.

Ounaroon, A., Decker, G., Schmidt, J., Lottspeich, F., and Kutchan, T. M. (2002). (*R,S*)-Retuculine 7-*O*-methyltransferase and (*R,S*)-norcoclaurine 6-*O*-methyltransferase of *Papaver somniferum*—cDNA cloning and characterization of methyl transfer enzymes of alkaloid biosynthesis in opium poppy. *Plant J.* **36,** 808–819.

Parker, S. R., and Nes, W. D. (1992). Regulation of sterol biosynthesis and its phylogenetic implications. *In* "Regulation of Isopentenoid Metabolism" (W. D. Nes, E. J. Parish, and J. M. Trzaskos, eds.), Vol. 497, pp. 110–145. American Chemical Society Symposium Series, Washington, DC.

Parks, L. W., Crowley, J. H., Leak, F. W., Smith, S. J., and Tomeo, M. E. (1997). Use of sterol mutants as probes for sterol functions in the yeast *Saccharomyces cerevisiae*. *In* "Biochemistry and Function of Sterols" (E. J. Parish and W. D. Nes, eds.), pp. 257–262. CRC Press, Boca Raton.

Popják, G., Edmond, J., Anet, F. A., and Easton, N. R., Jr. (1977). Carbon-13 NMR studies on cholesterol biosynthesized from [^{13}C]mevalonates. *J. Amer. Chem. Soc.* **99,** 931–935.

Rahier, A., Narula, A. S., Benveniste, P., and Schmitt, P. (1980). 25-Azacycloartanol, a potent inhibitor of S-adenosyl-methionine sterol-C24 and C28 methyltransferase in higher plants. *Biochem. Biophys. Res. Commun.* **92,** 20–25.

Rahier, A., Genot, J.-C., Benveniste, P., and Narula, A. S. (1984). Inhibition of S-adenosyl-L-methionine sterol C-24 methyl transferase by analogues of a carbocationic high energy intermediate. *J. Biol. Chem.* **259,** 15213–15215.

Sauvaire, Y., Tal, B., Heupel, R. C., England, R., Hanners, P. K., Nes, W. D., and Mudd, J. B. (1997). A comparison of sterol and long chain fatty acid biosynthesis in *Sorghum bicolor*. *In* "The Metabolism, Structure and Function of Plant Lipids" (P. K. Stumpf, J. B. Mudd, and W. D. Nes, eds.), pp. 107–110. Plenum Press, New York.

Schaeffer, A., Bouvier-Nave, P., Benveniste, P., and Schaller, H. (2000). Plant sterol C-24-methyltransferases: Different profiles of tobacco transformed with SMT1 and SMT2. *Lipids* **35,** 263–269.

Schaeffer, A., Bronner, R., Benveniste, P., and Schaller, H. (2001). The ratio of campesterol to sitosterol that modulates growth in Arabidopsis is controlled by *STEROL METHYLTRANSFERASE 2;1*. *Plant J.* **25,** 605–615.

Schaller, H. (2004). New aspects of sterol biosynthesis in growth and development of higher plants. *Plant Physiol. Biochem.* **42,** 465–476.

Schaller, H., Bouvier-Nave, P., and Benveniste, P. (2001). Overexpression of an *Arabidopsis* cDNA encoding a sterol-C-241-methyltranferase in tobacco modifies the ratio of 4-methyl cholesterol to sitosterol and is associated with growth reduction. *Plant Physiol.* **118,** 461–469.

Schubert, H. L., Blumenthal, R. M., and Cheng, X. (2003). Many paths to methyltransfer: A chronicle of convergence. *Trends Biochem. Sci.* **28,** 329–335.

Schuler, I., Milon, A., Nakatani, Y., Ourisson, G., Albrecht, A.-M., Benveniste, P., and Hartmann, M.-A. (1991). Differential effects of plant sterols on water permeability and on acyl chain ordering of soybean phosphatidylcholine bilayers. *Proc. Natl. Acad. Sci.* **88,** 6926–6930.

Seo, S., Uomori, A., Yashimura, Y., Takeda, K. J., Seto, H., Ebizuka, Y., Noguchi, H., and Sankawa, U. (1988). Biosynthesis of sitosterol, cycloartenol, and 24-methylene cycloartanol in tissue cultures of higher plants and ergosterol in yeast from [1,2–$^{13}C_2$]- and [2–$^{13}C_2H_3$]-acetate and [5–$^{13}C^2H_2$]mevalonic acid. *J. Chem. Soc. Perkin I* **8,** 2407–2413.

Seo, S., Uomori, A., Yoshimura, Y., Seto, H., Ebizuka, Y., Noguchi, H., Sankawa, U., and Takeda, K. (1990). Biosynthesis of isofucosterol from [2–$^{13}C_2H_3$] and [1,2–^{13}C]acetate in tissue cultures of *Physalis peruviana*-the stereochemistry of the hydride shift from C-24 to C-25. *J. Chem. Soc. Perkin I* **1,** 105–109.

Shi, J., Gonzales, R. A., and Bhattacharyya, M. K. (1996). Identification and characterization of an S-adenosyl-L-methionine: Δ^{24}-sterol C-methyl transferase cDNA from soybean. *J. Biol. Chem.* **271,** 9384–9389.

Sinha, A. (2004). Protein engineering soybean sterol methyltransferase leads to altered substrate binding and catalysis, pp. 1–63. Texas Tech University, *Master Thesis.*

Sitbon, F., and Jonsson, L. (2001). Sterol composition and growth of transgenic tobacco plants expressing type-1 and type 2 sterol methyltransferases. *Planta* **212,** 568–572.

Svoboda, J. A., Ross, S. R., and Nes, W. D. (1995). Comparative studies of metabolism of 4-desmethyl, 4-monomethyl, and 4,4-dimethyl sterols in *Manduca sexta*. *Lipids* **30,** 91–94.

Tapiero, H., Townsend, D. M., and Tew, K. D. (2003). Phytosterols in the prevention of human pathologies. *Biomed. Pharmacother.* **57,** 321–325.

Umlauf, D., Zapp, J., Becker, H., and Adam, K. P. (2004). Biosynthesis of irregular monoterpene artemisia ketone, the sesquiterpene germacrene D and other isoprenoids in *Tanacetum vulgare* L. (*Asteracease*). *Phytochemistry* **65,** 2463–2470.

Venkatramesh, M., Guo, D., Jia, Z., and Nes, W. D. (1996). Mechanism and structural requirements for transformation of substrates by the (S)-adenosyl-L methionine: $\Delta^{24(25)}$-sterol methyl transferase from *Saccharomyces cerevisiae*. *Biochim. Biophys. Acta* **1299,** 313–324.

Volkman, J. K. (2005). Sterols and other triterpenoids: Source specificity and evolution of biosynthetic pathways. *Org. Geochem.* **36,** 139–159.

Wentzinger, L. F., Bach, T. J., and Hartmann, M.-A. (2002). Inhibition of squalene synthase and squalene epoxidase in tobacco cells triggers an upregulation of 3-hydroxy-3-methylglutaryl coenzyme A reductase. *Plant Physiol.* **130,** 1–13.

Wojciechowski, Z. A. (1991). Biochemistry of phytosterol conjugates. *In* ''Physiology and Biochemistry of Sterols'' (G. W. Patterson and W. D. Nes, eds.), pp. 361–395. Amer. Oil Chem. Soc. Press, Champaign.

Xu, S., Norton, R. A., Crumley, F. G., and Nes, W. D. (1988). Comparison of the chromatographic properties of sterols, select additional steroids and triterpenoids: Gravity-flow liquid chromatography, thin-layer chromatography, gas-liquid chromatography, and high performance liquid chromatography. *J. Chromatogr.* **452,** 377–398.

Zhou, W., and Nes, W. D. (2000). Stereochemistry of hydrogen introduction at C-25 in ergosterol synthesized by the mevalonate-independent pathway. *Tetrahedron Lett.* **41,** 2791–2795.

Zhou, W., and Nes, W. D. (2003). Sterol methyltransferase2: Purification, properties and inhibition. *Arch. Biochem. Biophys.* **420,** 18–34.

Zhou, W., Guo, D., and Nes, W. D. (1996). Stereochemistry of hydrogen migration from C-24 to C-25 during biomethylation in ergosterol biosynthesis. *Tetrahedron Lett.* **37,** 1339–1342.

Zhou, W., Minh, T. T., Collins, M. S., Cushion, M. T., and Nes, W. D. (2002). Evidence for multiple sterol methyltransferase pathways in *Pneumocystis carinii. Lipids* **37,** 1177–1186.

Zhou, W., Song, Z., Kanagasabai, R., Liu, J., Jayasimha, P., Sinha, A., Veeramachanemi, P., Miller, M. B., and Nes, W. D. (2004). Mechanism-based enzyme inactivators of phytosterol biosynthesis—a review. *Molecules* **9,** 185–203.

Zhou, W., Lepesheva, G. I., Waterman, M. R., and Nes, W. D. (2006). Mechanistic analysis of a multiple product sterol methyltransferase implicated in ergosterol biosynthesis in *Trypanosoma brucei. J. Biol. Chem.* **281,** 6290–6296.

Zubieta, C., Kota, P., Ferrer, J.-L., Dixon, R. A., and Noel, J. (2002). Structural basis for modulation of lignin monomer methylation by caffeic acid/5-hydroxyferulic acid 3/5-*O*-methyltransferase. *The Plant Cell* **14,** 1265–1277.

CHAPTER 10

Engineering Plant Alkaloid Biosynthetic Pathways: Progress and Prospects

Toni M. Kutchan,[*,†] **Susanne Frick,**[*,†] and **Marion Weid**[†]

Contents

[*] Donald Danforth Plant Science Center, St. Louis, Missouri 63132

[†] Leibniz Institut für Pflanzenbiochemie, Weinberg 3, 06120 Halle/Saale, Germany

Advances in Plant Biochemistry and Molecular Biology, Volume 1

ISSN 1755-0408, DOI: 10.1016/S1755-0408(07)01010-7

Abstract

With the successful application of molecular genetic methods to the plant alkaloid field, we now have sophisticated tools at our disposal to study regulation of enzymatic biosynthesis, as well as determining the cellular and subcellular localization of these enzymes. The availability of ever-increasing numbers of recombinant enzymes has enabled thorough analyses of selected alkaloid biosynthetic enzymes at the biochemical and structural levels. We are just beginning to use this knowledge to metabolically engineer alkaloid metabolism in plants and in *in vitro* cultures. Multicellular compartmentation of alkaloid pathways must be considered if meaningful metabolic engineering experiments are to be designed; for example, we will need to use promoters that drive transgene expression in the correct cell types. Regulation of these pathways at the gene and enzyme level is complex and there is still much to be learned about metabolite levels, multienzyme complexes, and pathway interconnections, as we systematically overexpress and suppress gene transcription. Today, pathway engineering in plants remains highly variable. When we perturb cellular physiology, metabolite homeostasis and intra- and intercellular partitioning can be affected in unpredictable ways. Predictive metabolic engineering to generate plants with tailored alkaloid profiles for basic research and for commercial production is clearly a challenge for the future.

Key Words: Tryptophan decarboxylase, Strictosidine synthase, Geraniol 10-hydroxylase, Secologanin synthase, Strictosidine glucosidase, Tabersonine 16-hydroxylase, Desacetoxyvindoline 4-hydroxlyase, Deacetylvindoline 4-*O*-acetyltransferase, Polyneuridine aldehyde esterase, Vinorine synthase, Tyrosine/dopa decarboxylase, (*R,S*)-Norcoclaurine 6-*O*-methyltransferase, (*R,S*)-Coclaurine, *N*-Methyltransferase, (*S*)-*N*-Methylcoclaurine 3′-hydroxylase, (*R,S*)-3′-Hydroxy-*N*-methylcoclaurine 4′-*O*-methyltransferase, (*R,S*)-Reticuline 7-*O*-methyltransferase, Salutaridinol 7-*O*-acetyltransferase, Codeinone reductase, Berberine bridge enzyme, (*S*)-Scoulerine 9-*O*-methyltransferase, (*S*)-Canadine synthase, Major latex protein, Putrescine *N*-methyltransferase, Tropinone reductase I, Tropinone reductase II, Hyoscyamine 6β-hydroxylase, Vindoline, Ajmaline, Morphine, Sanguinarine, Laudanine, Hyoscyamine, Scopolamine, Calistegin, Cocaine, Berberine, Dopamine, Strictosidine.

1. INTRODUCTION

Twenty-five years have passed since Volume 7 "Secondary Plants Products" was published in the series *The Biochemistry of Plants* (Waller and Dermer, 1981). Chapter 12 in that volume, authored by George Waller and Otis Dermer, presented the current day knowledge of the enzymology of alkaloid metabolism in plants. Scant information was available on the enzymology of plant alkaloid formation at the time. Most biosynthetic pathways were inferred from the results of "feeding" experiments with radiolabeled putative precursor molecules, but did not involve a study of the enzyme catalysts. Most of the enzymes associated with alkaloid biosynthesis up until 1979 were decarboxylases, aminotransferases, amine oxidases, and phenol oxidases of relatively broad substrate specificity. Examples of enzyme activities in cell-free extracts known then that, in retrospect, were clearly dedicated to alkaloid biosynthesis were the berberine bridge enzyme (BBE) described by Rink and Böhm in Halle in 1975 (Rink and Böhm, 1975); the cytochrome *P450*-dependent geraniol 10-hydroxylase by Coscia and coworkers in St. Louis in 1976 (Madyastha *et al.*, 1976, 1977); the strictosidine synthase described by Stöckigt and Zenk in Bochum in 1977 (Stöckigt and Zenk, 1977a,b); and the strictosidine β-D-glucosidase by Treimer and Zenk in 1978 (Treimer and Zenk, 1978).

Immediately after Volume 7 was published in 1981, rapid advances were made in the ability to detect and purify enzymes involved in plant alkaloid biosynthesis. This was in large part due to the development of plant cell suspension cultures and root cultures that produced alkaloids in the tens to hundreds of milligrams per liter range (Zenk, 1991). Methods of protein purification were greatly improved using these systems. These protocols could then be applied back to the native plant systems to identify even more enzymes. In total, in the past 20 years, approximately 80–100 new enzymes of alkaloid biosynthesis have been identified [summarized in part in Kutchan (1998)].

Central to the advances of the past two decades is also the development of plant molecular genetic methods. Concurrent to the identification and purification of new enzymes of alkaloid biosynthesis, the cDNA cloning and functional expression techniques of plant molecular biology were being developed. In 1988, the first cDNA of alkaloid formation, encoding strictosidine synthase, was isolated from *Rauwolfia serpentina* (Kutchan *et al.*, 1988). One year later, it was functionally expressed in *Escherichia coli* (Kutchan, 1989), thus opening a new period in the study of alkaloid biosynthesis. Other clones have rapidly followed and today the isolation and expression of alkaloid biosynthetic genes, although not yet routine, proceed at an accelerated pace. The full exploitation of these genes lies in the ability to ultimately engineer alkaloid pathways to generate plants with tailored alkaloid profiles for basic research and for commercial production. We do not yet fully understand how alkaloid accumulation is regulated in plants nor do we have a full complement of plant transformation and regeneration protocols necessary to producing transgenics of all species that may be interesting or useful.

In retrospect, many aspects of the study of alkaloid metabolism have advanced tremendously during the past 20 years, but several of the insightful statements of Waller and Dermer in 1981 remain true today (Waller and Dermer, 1981). To paraphrase their words, alkaloids play an important role in the ecology and physiology of the plants in which they occur. Although more easily described than precisely defined, there are several major incentives for studying alkaloids: they are the oldest drugs and still have a significant use in modern medicine, organic chemists are fascinated by the variety and complexity of structures presented, plant physiologists seek to learn the function of these compounds in plants, and pharmaceutical industry searches to improve methods of production. The main difference in the alkaloid field today compared to 1981 is the tools that are available to aid researchers in analyzing structure, biosynthesis, regulation, function, and production. Our understanding in 2006 of selected medicinally relevant alkaloid pathways and our ability to manipulate them is as follows.

2. MONOTERPENOID INDOLE ALKALOIDS

The Madagascar periwinkle *Catharanthus roseus* contains nearly 200 monoterpenoid indole alkaloids. It is the source of the chemotherapeutic dimeric alkaloid vinblastine. Vinblastine is formed when the Iboga-type indole alkaloid catharanthine is oxidatively coupled to the *Aspidosperma*-type indole alkaloid vindoline. Many research groups worldwide have investigated alkaloid formation in *C. roseus*, such that our understanding of this particular system is relatively advanced. Success in the isolation of cDNAs encoding both enzymes involved in the biosynthesis of strictosidine aglycone from primary metabolites and enzymes involved in the formation of vindoline from tabersonine warrants presentation of our current comprehension of alkaloid biosynthesis in *C. roseus*.

2.1. Monoterpenoid indole alkaloid biosynthesis

On the pathway leading from L-tryptophan and geraniol to the central monoterpenoid indole alkaloid intermediate 3α(*S*)-strictosidine aglycone, the cDNAs encoding five biosynthetic enzymes have been described (Fig. 10.1). These are *tydc* (tryptophan decarboxylase) (De Luca *et al.*, 1989), *g10h* or *cyp76b6* (geraniol 10-hydroxylase) (Collu *et al.*, 2001), *cyp72a1* (secologanin synthase) (Irmler *et al.*, 2000), *str1* (strictosidine synthase) (Kutchan *et al.*, 1988, 1989; Mcnight *et al.*, 1990), and *sgd* (strictosidine glucosidase) (Geerlings *et al.*, 2000; Gerasimenko *et al.*, 2002).

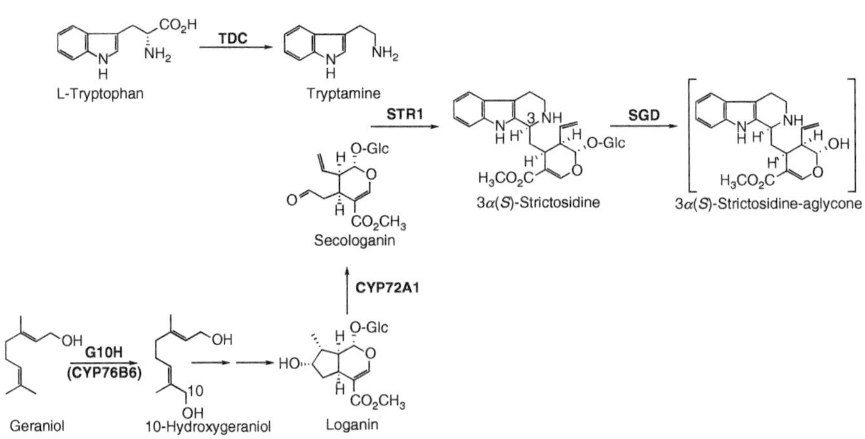

FIGURE 10.1 Schematic representation of the biosynthetic pathway leading from L-tryptophan and geraniol to the hypothetical intermediate 3α(*S*)-strictosidine-aglycone. TDC, tryptophan decarboxylase; STR1, strictosidine synthase; G10H (CYP76B6), geraniol 10-hydroxylase; CYP72A1, secologanin synthase; SGD, strictosidine glucosidase.

3α(*S*)-Strictosidine aglycone is an unstable intermediate that can be transformed into a number of chemical structures, which then lead into specific monoterpenoid indole alkaloid pathways. Along the biosynthetic pathway that leads from tabersonine to vindoline, three cDNAs encoding biosynthetic enzymes have been identified (Fig. 10.2). The cytochrome *P450*-dependent monooxygenase gene *cyp71d12* encodes tabersonine 16-hydroxylase (Schröder *et al.*, 1999; St-Pierre and De Luca, 1995). Desacetoxyvindoline is hydroxylated at the 4-position by the oxoglutarate-dependent dioxygenase desacetoxyvindoline–4-hydroxylase (encoded by *d4h*) (De Carolis and De Luca, 1993; Vazquez-Flota *et al.*, 1997). Finally, deacetylvindoline is acetylated to vindoline by acetylcoenzyme A: deacetylvindoline 4-*O*-acetyltransferase, the gene product of *dat* (Fahn *et al.*, 1985; Power *et al.*, 1990; St-Pierre *et al.*, 1998).

Although not specific to monoterpenoid indole alkaloid biosynthesis, the nonmevalonate pathway has been shown to be the biosynthetic route to loganin and secologanin (Contin *et al.*, 1998; Eichinger *et al.*, 1999), and three cDNAs encoding 1-deoxy-D-xylulose 5-phosphate synthase, 1-deoxy-D-xylulose 5-phosphate reductoisomerase, and 2C-methyl-D-erythritol 2,4-cyclodiphosphate synthase of this pathway have been isolated from *C. roseus* (Chahed *et al.*, 2000; Veau *et al.*, 2000).

Progress has also been made in identifying cDNAs encoding enzymes of monoterpenoid indole alkaloid biosynthesis in *R. serpentina* (Fig. 10.3). Two cDNAs involved in the transformation of the sarpagan alkaloid polyneuridine aldehyde into the ajmalan-type alkaloid ajmaline have been characterized: *pnae* encoding polyneuridine aldehyde esterase (Dogru *et al.*, 2000; Pfitzner and Stöckigt, 1983) that converts polyneuridine aldehyde into epivellosimine which

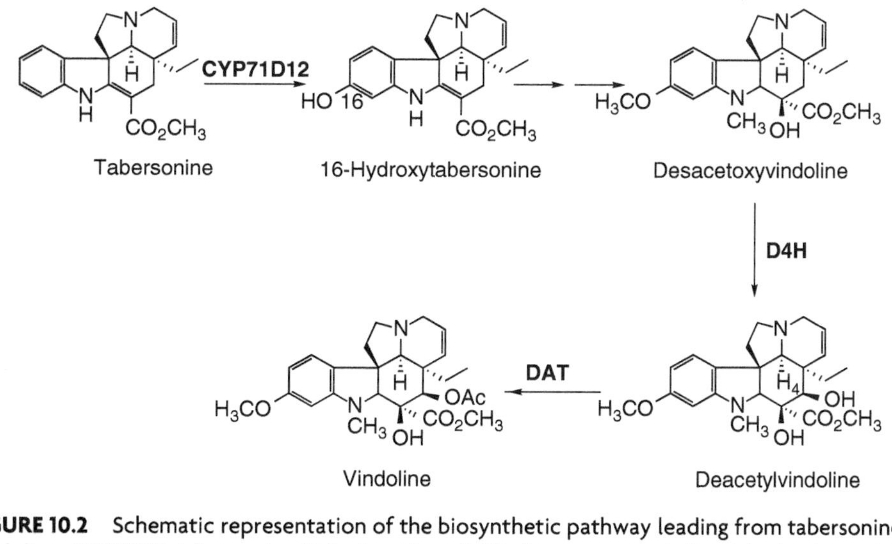

FIGURE 10.2 Schematic representation of the biosynthetic pathway leading from tabersonine to vindoline. CYP71D12, tabersonine 16-hydroxylase; D4H, desacetoxyvindoline 4-hydroxlyase; DAT, deacetylvindoline 4-*O*-acetyltransferase.

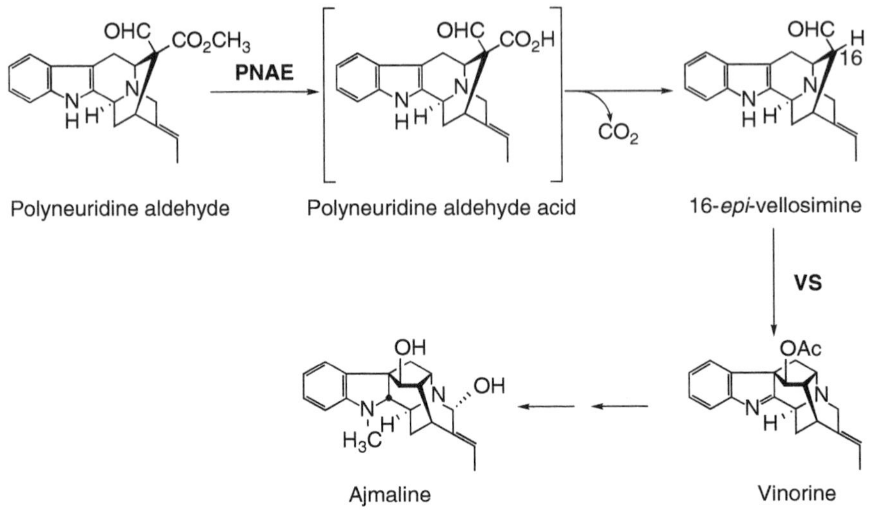

FIGURE 10.3 Schematic representation of the biosynthetic pathway leading from polyneuridine aldehyde to ajmaline. PNAE, polyneuridine aldehyde esterase; VS, vinorine synthase.

is then rearranged and acetylated to vinorine by vinorine synthase encoded by the *vs* gene (Bayer *et al.*, 2004).

With the current collection of cDNAs encoding enzymes of monoterpenoid indole alkaloid biosynthesis, progress has also been made with respect to our understanding of the regulation of this biosynthesis. The first topic to be covered here is the cellular localization of monoterpenoid indole alkaloid biosynthesis.

2.2. Cell-specific expression of monoterpenoid indole alkaloid biosynthetic genes

The differentiation of plant cells into specialized structures can result in a biochemical specialization essential for the biosynthesis and accumulation of selected classes of alkaloids. The lack of cytodifferentiation is considered a possible reason for the failure of plant cell cultures to accumulate alkaloids such as the dimeric monoterpenoid indole alkaloid vinblastine. This chemotherapeutic alkaloid is formed by the oxidative coupling of vindoline and catharanthine. *C. roseus* cell cultures accumulate catharanthine, but not vindoline. At least three enzymes of vindoline biosynthesis are absent from cell culture (reviewed in Kutchan, 1998). This could be due to the absence of the correct differentiated cell types in culture.

A detailed study of the expression of several genes of vindoline biosynthesis in developing *C. roseus* leaves using *in situ* hybridization and immunolocalization provides the first clear insight into the spatial distribution of monoterpenoid indole alkaloid biosynthesis (St-Pierre *et al.*, 1999). *In situ* hybridization of two genes, *tdc* and *str1*, occurring early in the vindoline biosynthetic pathway and two genes, *d4h* and *dat*, occurring late in the vindoline biosynthesis revealed that multiple cell types are involved. *C. roseus* produces more than 180 monoterpenoid indole alkaloids all of which are derived from the central intermediate 3α(*S*)-strictosidine. The enzymes tryptophan decarboxylase and strictosidine synthase lead to this central intermediate and are, therefore, involved in the biosynthesis of all of the *C. roseus* alkaloids. The transcripts of *tdc* and *str1* were found in the epidermis of developing leaves (Fig. 10.4A–D, K, and L) (Irmler *et al.*, 2000; St-Pierre *et al.*, 1999). In contrast, transcripts of the vindoline-specific biosynthetic genes *d4h* and *dat* localized to laticifer and idioblasts of developing leaves (Fig. 10.4E–H, M, and N) (Irmler *et al.*, 2000; St-Pierre *et al.*, 1999). In addition, transcript of the secologanin biosynthetic enzyme secologanin synthase (*cyp72a1*) localized to epidermis of developing leaves is consistent with this tissue as the site of formation of the central intermediate 3α(*S*)-strictosidine (Fig. 10.4I and J; Irmler *et al.*, 2000). Immunolocalization studies corroborated these results

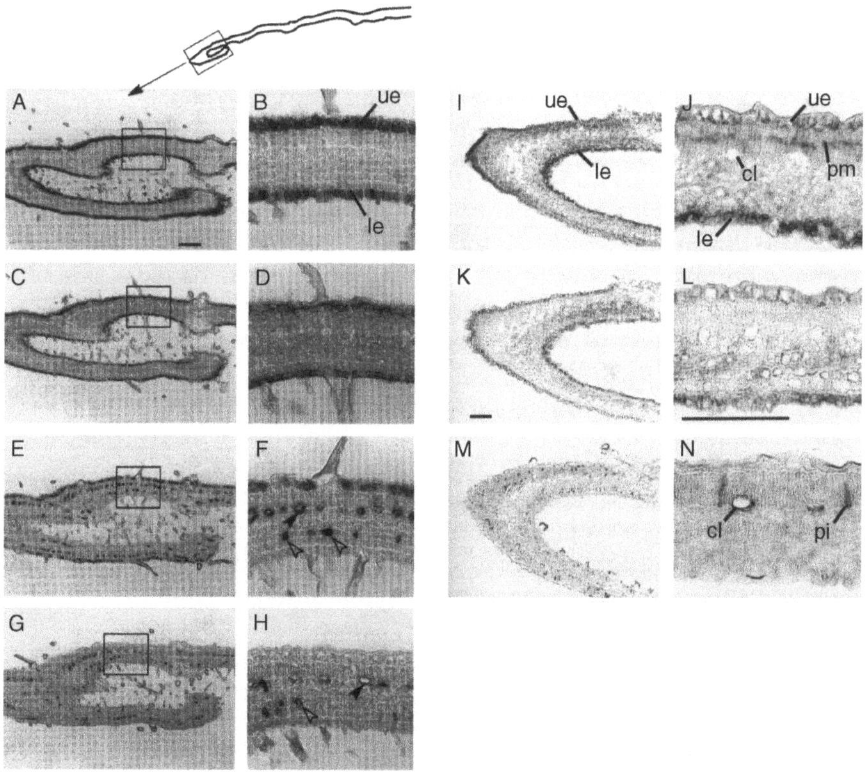

FIGURE 10.4 Localization of *cyp72a1, tdc, str1, d4h,* and *dat* transcripts in developing *C. roseus* leaves (Irmler *et al.*, 2000; St-Pierre *et al.*, 1999). Panels (A, B, K, L) *tdc*; panels (C, D) *str1*; panels (E, F) *d4h*; panels (G, H, M, N) *dat*; panels (I, J) *cyp72a1*. cl, cross-connecting laticifer cells; le, lower epidermis; pi, idioblast cells associated with palisade mesophyll cells; pm, palisade mesophyll cells; si, idioblast cells associated with spongy mesophyll cells; ue, upper epidermis; *tdc*, tryptophan decarboxylase; *str1*, strictosidine synthase; *cyp72a1*, secologanin synthase; *d4h*, desacetoxyvindoline 4-hydroxlyase; *dat*, deacetylvindoline 4-*O*-acetyltransferase. Bar shown in panels (A, K), 100 μm; in panel (L), 50 μm. Solid arrows in panels (F, H), laticifer cells; open arrows, idioblast cells. (See Page 16 in Color Section.)

(Irmler *et al.*, 2000; St-Pierre *et al.*, 1999). Taken together, interpretation of these results implies a translocation of a pathway intermediate. They also suggest that a central alkaloid pathway may occur in one cell type and branches in the pathway take place in various other cell types. Cellular localization and intermediate transport can therefore be one level of regulation of alkaloid biosynthesis.

Transcript of *tdc* and *str1* are also found in protoderm and cortical cells around the apical meristem of root tips (St-Pierre *et al.*, 1999). Likewise, the transcript of a gene involved in the biosynthesis of the root-specific monoterpenoid indole alkaloid minovincinine was detected in root tissue. RNA *in situ* hybridization studies located minovincinine 19-hydroxy-*O*-acetyltransferase gene expression within the cortex and epidermis of tissues near the root tip (Laflamme *et al.*, 2001).

Multicellular compartmentation of alkaloid biosynthesis should be a central consideration in the metabolic engineering of alkaloid pathways. Promoters should be chosen that would direct transgene expression to the cell types in which the appropriate biosynthetic intermediates are expected to occur or in which biosynthetic genes are expressed.

2.3. Genetic engineering of monoterpenoid indole alkaloid biosynthetic pathways

Early attempts at engineering plants with cDNAs from monoterpenoid indole alkaloid pathways were limited to those plant species for which transformation and regeneration protocols were available. Later attempts have been made with *C. roseus* cell cultures. Although *C. roseus* can be transformed with *Agrobacterium*, there are not yet reports of regeneration of rooted plants. Selected examples of each are discussed in the following.

One of the first introductions of a monoterpenoid indole alkaloid cDNA was *tdc* from *C. roseus* into tobacco (Songstad *et al.*, 1990, 1991). Transgene expression was driven by the CaMV 35S promoter. Young, fully expanded leaves of the transgenic tobacco plants had up to 45 times greater tryptophan decarboxylase activity than did control plants. Tryptamine accumulation in the plants was proportional to the tryptophan decarboxylase specific activity in protein extracts. Transgenic plants accumuated up to 1 mg tryptamine per gram fresh weight, but were unaffected in their indole-3-acetic acid levels.

Transgenic tobacco plants containing the *C. roseus tdc* gene and accumulating the protoalkaloid tryptamine have been analyzed for their effects on insect feeding (Thomas *et al.*, 1995). An advantage of this type of system is that it is possible to test plants in parallel that differ only in their protoalkaloid content. In addition, tryptamine was found in phloem extracts. The sweet potato whitefly *Bemisia tabaci* was allowed to feed and reproduce on transgenic and control tobacco plants. The sweet potato whitefly, like the aphid, pierces phloem cells with a stylet, thereby obtaining nutrients from the vascular system of the host plant. Sweet potato whitefly pupae emergence was reduced by as much as 97% in transgenic plants as compared to control plants. The mechanism of this action of tryptamine on whitefly reproduction is not yet understood, but poses an interesting possible use for tryptamine in insect control.

Tobacco plants expressing *tdc* from *C. roseus*, *tydc* (tyrosine decarboxylase) from *Papaver somniferum* or both transgenes were analyzed for effects on metabolism and visible phenotype (Guillet *et al.*, 2000). Expression of these transgenes *tdc* and *tydc* should create an artificial metabolic sink for the aromatic amino acids tryptophan and tyrosine, respectively. The results obtained are complex, but may be summarized as follows. Reduction of the tryptophan and tyrosine pools affected the phenylalanine pool in a light-dependent manner. Also perturbed were pool sizes of the nonaromatic amino acids methionine, valine, and leucine. This depletion of amino acids correlated with increases in activity of enzymes of the shikimate and phenylpropanoid pathways in older, light-treated seedlings. In addition, nicotine and chlorogenic acid levels were also increased.

Expression of the *C. roseus tdc* cDNA in *Brassica napus* (canola) created an artificial sink for tryptophan and resulted in reduced levels of indole glucosinolates (Chavadej *et al.*, 1994). Transgene expression driven by the CaMV 35S promoter accumulated tryptamine while lower levels of tryptophan-derived indole glucosinolates accumulated in all plant parts compared to control plants. Significant to this particular study, seeds from transgenic plants contained as little as 3% indole glucosinate compared to control plant seeds. This potentially yielded a more palatable protein fodder that is produced after extraction of the oil from seed material.

The final nonnative plant example considered here is expression of the *C. roseus tdc* cDNA in potato tuber under transcriptional control of the CaMV 35S promoter (Yao *et al.*, 1995). Again, an artificial metabolic sink was created for tryptophan. Transgene expression altered the balance of substrate and product pools in the shikimate and phenylpropanoid pathways. Transgenic tubers accumulated tryptamine and contained decreased levels of tryptophan, phenylalanine, and phenylalanine-derived phenolic compounds compared to control tubers. Wound-induced accumulation of chlorogenic acid was reduced as was the accumulation of soluble and cell wall-bound phenolics. The transgenic tubers were also more susceptible to infection by *Phytophthora infestans* possibly due to the modified cell wall.

Attempts at engineering monoterpenoid indole alkaloid profiles in *C. roseus* have also recently been made. *Str1* and/or *tdc* cDNAs were introduced into leaves from 6- to 8-week-old *C. roseus* seedlings by *Agrobacterium tumefaciens*-mediated transformation (Canel *et al.*, 1998). Transgenic cell cultures showed that CaMV 35S-driven *str1* demonstrated tenfold higher strictosidine synthase activity than untransformed cultured cells and higher levels of strictosidine, ajmalicine, catharanthine, serpentine, and tabersonine. Alkaloid production in these cell lines was, however, found to be unstable. Overexpression of *tdc* was apparently not necessary for alkaloid overproduction, but rather was detrimental to normal growth of the transgenic cultures. Feeding of a transgenic cell line of *C. roseus* that contained an *str1* transgene with the monoterpenoid indole alkaloid biosynthetic precursors tryptamine and/or loganin indicated that utilization of tryptamine for alkaloid biosynthesis increases flux through the indole pathway (Whitmer *et al.*, 1998). Overexpression of *tdc* is not necessary for high rates of tryptamine biosynthesis, and addition of tryptamine to the cultures did not increase overall levels of alkaloid. Addition of either loganin or loganin and tryptamine to the cell cultures resulted in increased accumulation of monoterpenoid indole alkaloids, suggesting that flux through the secoiridoid biosynthetic pathway is rate limiting. Rate limitation of the secologanin rather than the tryptamine pathway has also been demonstrated with *C. roseus* hairy root cultures (Morgan and Shanks, 2000). Methodologies for controlled expression in *C. roseus* using a glucocorticoid-inducible promoter system (Hughes *et al.*, 2002) as well as models for monitoring metabolic flux in plant metabolism using HPLC and NMR spectroscopy have been developed (Morgan and Shanks, 2002; Rijhwani *et al.*, 1999).

As a final point for consideration, transcription factors could also prove useful tools for engineering plant alkaloid accumulation where gene transcription is rate limiting. Analysis of the promoter regions of the *C. roseus tdc* and *str1* genes led to the identification of *cis* elements and *trans* factors that affect gene expression (Menke *et al.*, 1999; Van Der Fits and Memelink, 2000, 2001). Ectopic expression of one of these *trans* factor-encoding cDNAs, *ORCA3*, in cultured cells of *C. roseus* increased expression of *tdc*, *str1*, and *d4h*, but did not lead to an increase in accumulation of monoterpenoid indole alkaloids. Although ORCA3 acted pleiotropically on the transcription of several alkaloid biosynthetic genes, it did not activate all of the biosynthetic genes indicating that additional regulatory factors could be involved. Methods for identifying alkaloid gene regulatory factors and their potential use in metabolic engineering of alkaloid pathways have been reviewed and will not be discussed further here (Gantet and Memelink, 2002; Van Der Fits *et al.*, 2001).

3. TETRAHYDROBENZYLISOQUINOLINE ALKALOIDS

The opium poppy *P. somniferum* contains more than 80 tetrahydrobenzylisoquinoline-derived alkaloids. It is the source of the narcotic analgesics codeine and morphine. *P. somniferum* is one of mankind's oldest medicinal plants, and the plant as we know it today is the result of centuries of breeding. Our biochemical and molecular genetic knowledge of *P. somniferum* is relatively advanced. We understand how morphine is enzymatically formed and cDNAs encoding the enzymes occurring mainly downstream of formation of the first alkaloid in the pathway, (*S*)-norcoclaurine, have been isolated (Kutchan, 1998). Our current understanding of alkaloid biosynthesis in *P. somniferum* is as follows.

3.1. Tetrahydrobenzylisoquinoline alkaloid biosynthesis

On the pathway leading from L-tyrosine to the first tetrahydrobenzylisoquinoline alkaloidal intermediate (*S*)-norcoclaurine, cDNAs encoding tyrosine/dopa decarboxylases (*tydc*) have been isolated (Fig. 10.5) (Facchini and De Luca, 1994). The transformation of (*S*)-norcoclaurine to the central isoquinoline alkaloid biosynthetic intermediate (*S*)-reticuline is quite well understood at both the enzyme

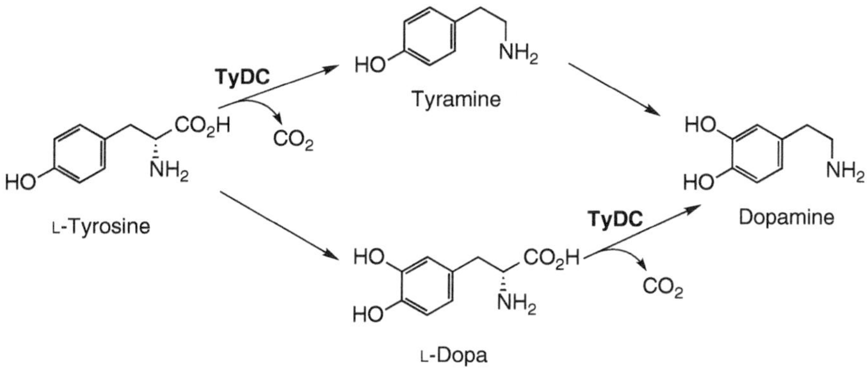

FIGURE 10.5 Schematic representation of the biosynthetic grid leading from L-tyrosine to dopamine. TyDC, tyrosine/dopa decarboxylase.

and gene level (Fig. 10.6). (*S*)-Norcoclaurine is *O*-methylated at the 6-position by (*R,S*)-norcoclaurine 6-*O*-methyltransferase (6-OMT) (Rüffer *et al.*, 1983). cDNAs encoding this enzyme have been isolated from *Thalictrum tuberosum*, *Coptis japonica*, and *P. somniferum* (Frick and Kutchan, 1999; Morishige *et al.*, 2000; Ounaroon *et al.*, 2003). (*S*)-Coclaurine is next *N*-methylated by (*R,S*)-coclaurine *N*-methyltransferase (Frenzel and Zenk, 1990a). This cDNA has been characterized from *C. japonica* (Choi *et al.*, 2002) and *P. somniferum* (S. Haase, J. Ziegler, S. Frick, and T. M. Kutchan, unpublished data). (*S*)-*N*-Methylcoclaurine is hydroxylated by the cytochrome *P450* dependent monooxygenase CYP80B1 (*S*)-*N*-methylcoclaurine 3′-hydroxlyase (Pauli and Kutchan, 1998). The cDNA encoding this cytochrome *P450* has been isolated from the California poppy *Eschscholzia californica* and from *P. somniferum* (Huang and Kutchan, 2000; Pauli and Kutchan, 1998). (*S*)-3′-Hydroxy-*N*-methylcoclaurine is methylated to (*S*)-reticuline by (*R,S*)-3′-hydroxy-*N*-methylcoclaurine 4′-*O*-methyltransferase (4′-OMT) (Frenzel and Zenk, 1990b). The cDNA *4′-omt* has been isolated from *C. japonica* (Morishige *et al.*, 2000) and from *P. somniferum* (Ziegler *et al.*, 2005).

(*S*)-Reticuline is the chemical chameleon of isoquinoline alkaloid biosynthesis, which can lead to a plethora of alkaloidal structures. In *P. somniferum*, (*R,S*)-reticuline can be methylated by (*R,S*)-reticuline 7-*O*-methyltransferase, for which the cDNA *7-omt* has been described, to the tetrahydrobenzylisoquinoline laudanine (Fig. 10.6 Ounaroon *et al.*, 2003). Along the pathway in which (*S*)-reticuline is specifically converted to morphine, cDNAs encoding two biosynthetic enzymes have been identified (Fig. 10.7). Salutaridinol 7-*O*-acetyltransferase, encoded by *SalAT*, transfers an acetyl moiety from acetyl-CoA to the 7-hydroxyl group of salutaridinol (Grothe *et al.*, 2001; Lenz and Zenk, 1995a). Codeinone reductase is encoded by *cor1* and catalyzes the penultimate step in morphine biosynthesis, the NADPH-dependent reduction of the keto moiety of codeinone to the 6-hydroxyl group of codeine (Lenz and Zenk, 1995b; Unterlinner *et al.*, 1999).

In *P. somniferum* and *E. californica*, the *N*-methyl group of (*S*)-reticuline can be oxidatively cyclized by the BBE to the bridge carbon, C-8, of (*S*)-scoulerine

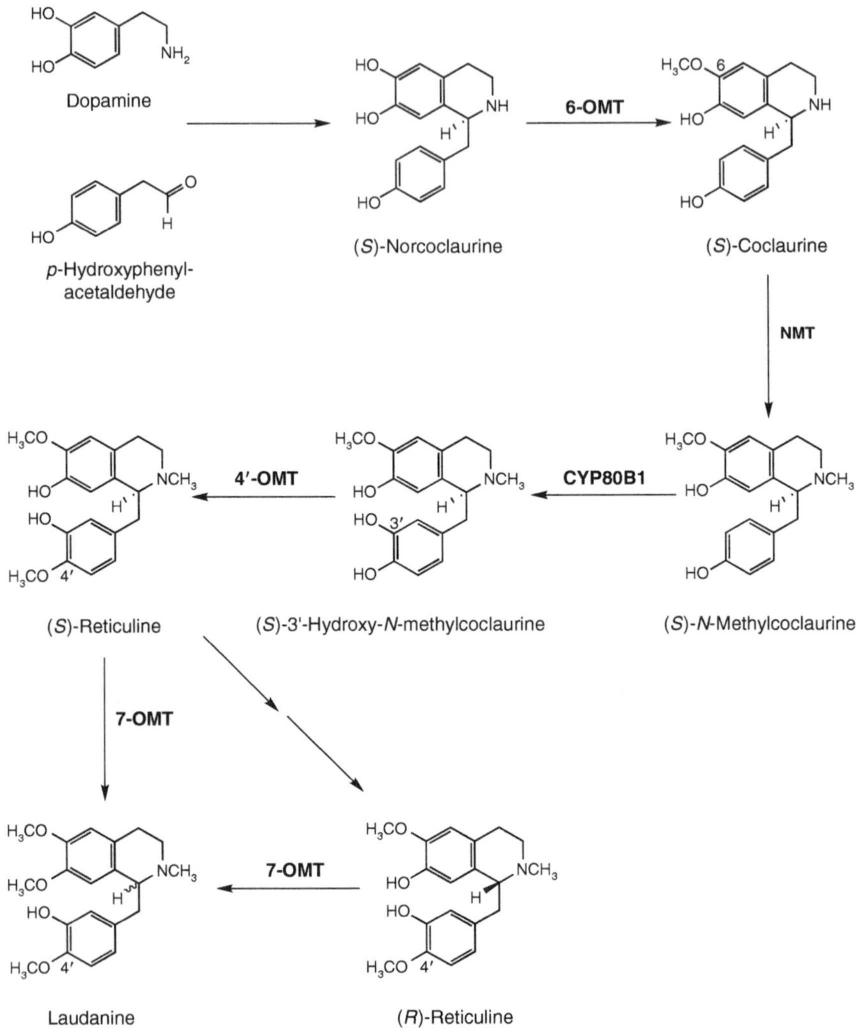

FIGURE 10.6 Schematic representation of the biosynthetic pathway leading from dopamine and p-hydroxyphenylacetaldehyde to laudanine. 6-OMT, (R,S)-norcoclaurine 6-O-methyltransferase; NMT, (R,S)-coclaurine, N-methyltransferase; CYP80B1, (S)-N-methylcoclaurine 3′-hydroxylase; 4′-OMT, (R,S)-3′-hydroxy-N-methylcoclaurine 4′-O-methyltransferase; 7-OMT, (R,S)-reticuline 7-O-methyltransferase.

(Fig. 10.8; Rink and Böhm, 1975; Steffens *et al.*, 1985). (*S*)-Scoulerine is then further converted in these plants to antimicrobial benzo[*c*]phenathridine alkaloids, such as sanguinarine. cDNAs encoding the BBE have been isolated from *E. californica*, *P. somniferum*, and *Berberis stolonifera* (Chou and Kutchan, 1998; Dittrich and Kutchan, 1991; Facchini *et al.*, 1996; Huang and Kutchan, 2000). (*S*)-Reticuline is converted via (*S*)-scoulerine to berberine alkaloids in *Berberis* and *Coptis*

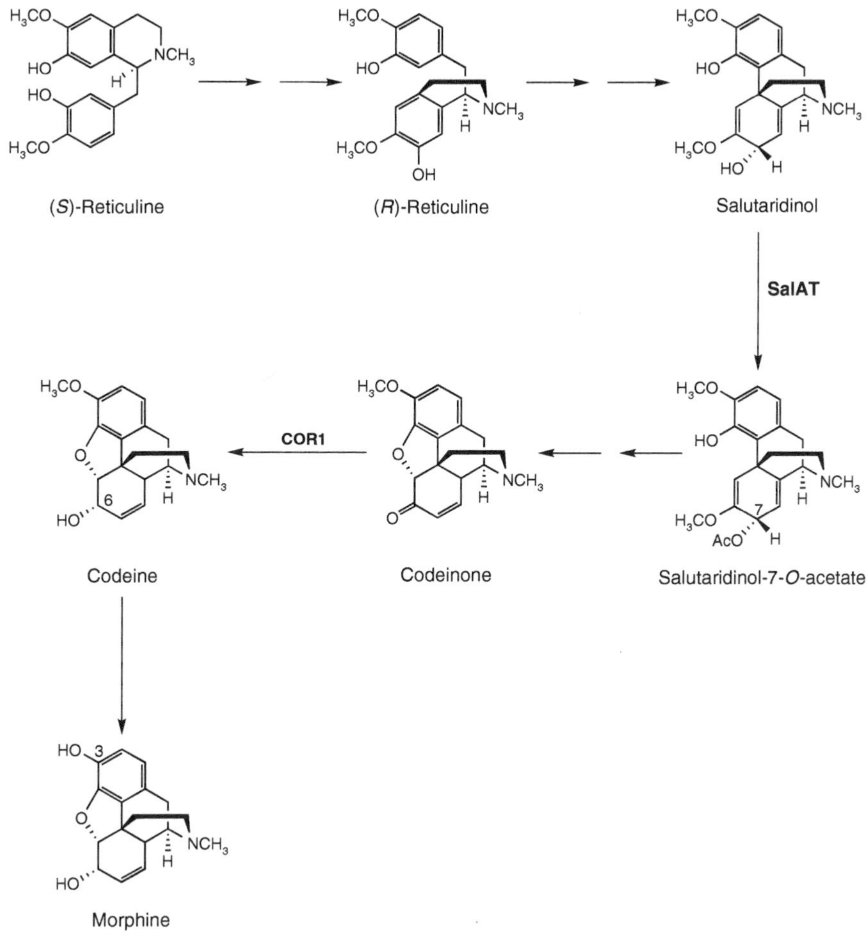

FIGURE 10.7 Schematic representation of the biosynthetic pathway leading from (*S*)-reticuline to morphine. SalAT, salutaridinol 7-*O*-acetyltransferase; COR1, codeinone reductase.

species. Along the biosynthetic pathway to berberine, two cDNAs have been identified from *C. japonica*. (*S*)-Scoulerine is methylated by (*S*)-scoulerine 9-*O*-methyltransferase (9-OMT) (Muemmler *et al.*, 1985; Takeshita *et al.*, 1995) to (*S*)-tetrahydrocolumbamine which is subsequently acted upon by CYP719 (Bauer and Zenk, 1991; Ikezawa *et al.*, 2003; Rueffer and Zenk, 1994), a cytochrome *P450*-dependent enzyme that catalyzes formation of the methylenedioxy bridge of (*S*)-canadine.

With the current collection of cDNAs encoding enzymes of tetrahydrobenzylisoquinoline alkaloid biosynthesis, some progress has also been made with respect to our understanding of the spatial regulation of this biosynthesis. The cellular localization of tetrahydrobenzylisoquinoline alkaloid biosynthesis will next be considered.

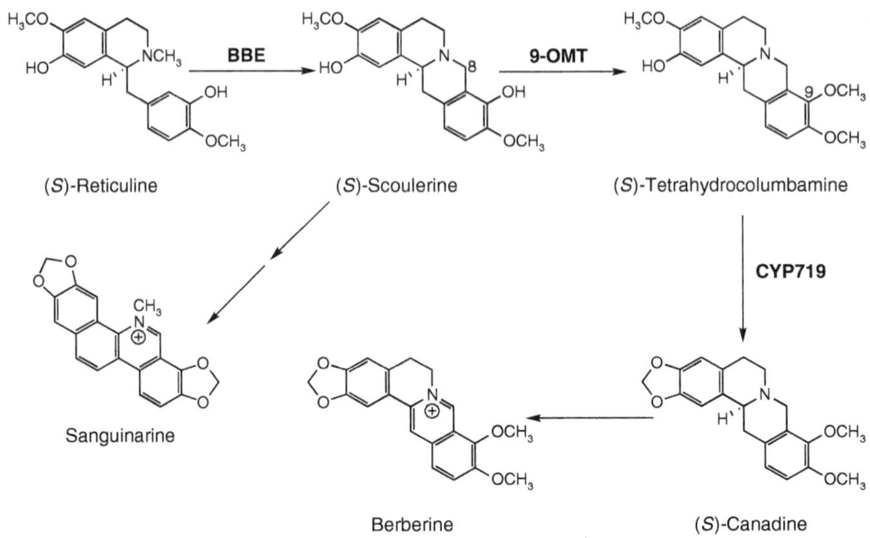

FIGURE 10.8 Schematic representation of the biosynthetic pathway leading from (*S*)-reticuline to sanguinarine and berberine. BBE, berberine bridge enzyme; 9-OMT, (*S*)-scoulerine 9-*O*-methyltransferase; CYP719, (*S*)-canadine synthase.

3.2. Cell-specific expression of tetrahydrobenzylisoquinoline alkaloid biosynthetic genes

The opium poppy contains specialized internal secretory cells called laticifers. In the aerial parts of the plant, the laticifer cells are anastomosed, forming a reticulated network. Laticifers are found associated with the vascular bundle in all plant parts. Morphine is found both in roots and in aerial plant parts and specifically accumulates in vesicles in laticifers. The benzo[*c*]phenathridine sanguinarine is found in root tissue. In plant cell cultures of *P. somniferum*, accumulation of sanguinarine can be elicited by addition of methyl jasmonate (Huang and Kutchan, 2000), but conditions have not been found under which morphine accumulates. The reason for the absence of morphine in cell culture is not completely clear, since all of the enzymes for which *in vitro* assays have been developed are also found in cell culture extracts. With availability of several biosynthetic cDNAs from *P. somniferum*, information as to the localization of selected biosynthetic enzymes and, therefore, the spatial distribution of alkaloid biosynthesis becomes clearer.

Tyrosine/dopa decarboxylase participates in the very early stages of tetrahydrobenzylisoquinoline alkaloid biosynthesis. In *P. somniferum*, this enzyme is encoded by a multigene family, which is classified into two groups *tydc1* and *tydc2* (Facchini and De Luca, 1994). From *in situ* hybridization experiments, transcript of *tydc1* was more abundant than *tydc2* in roots, while *tydc2* transcript was more abundant than *tydc1* transcript in stem (Facchini and De Luca, 1995). *Tydc* transcript was detected in the metaphloem and protoxylem of vascular bundles in aerial plant parts (Fig. 10.9A and C). This localization is consistent with latex as the site of morphinan alkaloid accumulation.

Immunolocalization of five enzymes of alkaloid biosynthesis, two of which occur late in the morphine-specific pathway, has been carried out with *P. somniferum*. Morphine biosynthesis is localized to the vascular bundle in capsule, stem, and root, and involves two different cell types—paranchyma associated with phloem and laticifer cells (Weid *et al.*, 2004). Whereas 4′-OMT and SalAT were detected in

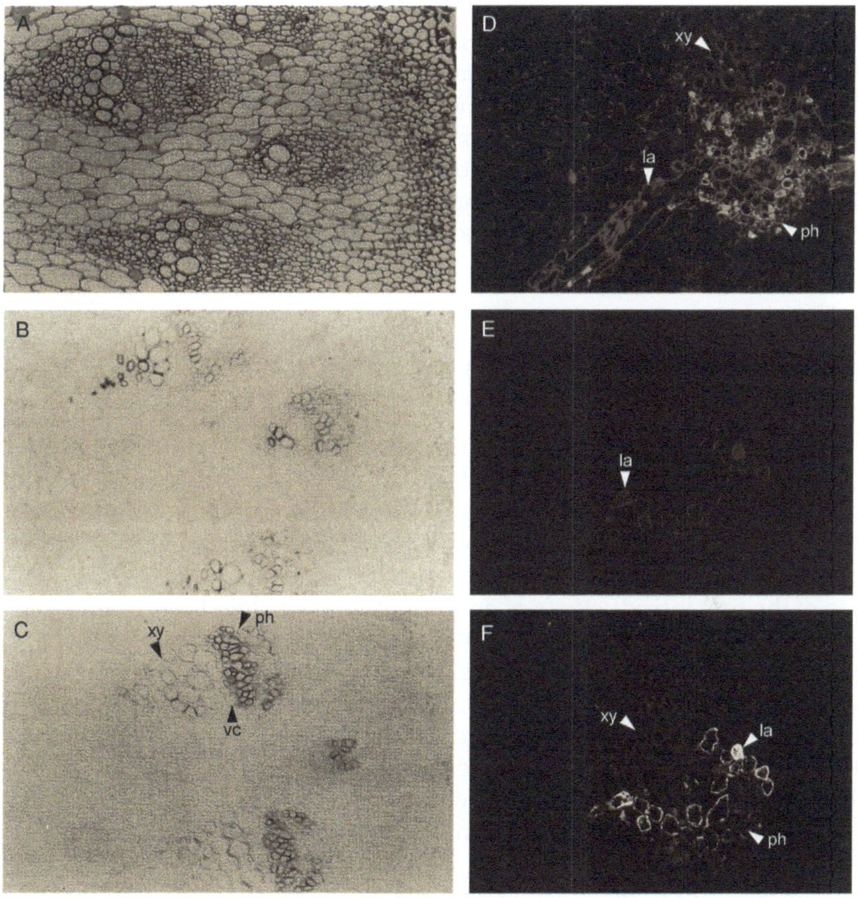

FIGURE 10.9 Localization of *tydc* (Facchini and De Luca, 1995), SalAT, COR1, and MLP (Weid *et al.*, 2004) in stem of *P. somniferum*. Panel (A) root cross-section stained with aniline safranine and astra blue; (B) *in situ* hybridization of *tydc1*; (C) *in situ* hybridization of *tydc2*; (D) immunolocalization of MLP to laticifers (red fluorescence) and SalAT to phloem parenchyma (green fluorescence); (E) immunolocalization of MLP to laticifers (red fluorescence); and (F) coimmunolocalization of MLP and COR1 to laticifers (yellow fluorescence). Green fluorescence indicates COR1 is present also in phloem parenchyma. xy, xylem; ph, phloem; vc, vascular cambium; la, laticifers; *tydc*, tyrosine/dopa decarboxylase; SalAT, salutaridinol 7-*O*-acetyltransferase; COR1, codeinone reductase; MLP, major latex protein. (See Page 17 in Color Section.)

phloem parenchyma cells, COR1 was abundantly colocalized with major latex proteins to laticifers (Fig. 10.9D and F). It appears that at a late stage in morphine formation, at the level of salutaridinol-7-*O*-acetate or thebaine, biosynthesis moves out of the phloem parenchyma into the laticifers, the ultimate site of thebaine, codeine, and morphine accumulation. As for vindoline biosynthesis in *C. roseus* (St-Pierre *et al.*, 1999), more than one cell type is implied in morphine biosynthesis in *P. somniferum*. This spatial distribution of the biosynthetic pathway infers that transport processes are integral to alkaloid formation. Again, cellular localization and intermediate transport could be one level of regulation of alkaloid biosynthesis in *P. somniferum*.

Due to commercial importance, *P. somniferum* is an alkaloid-producing plant of choice for metabolic engineering. Promoters will need to be chosen, however, that will direct transgene expression to the cell types in *P. somniferum* in which the appropriate biosynthetic gene transcripts are expected to occur. As model systems, plant cell cultures of a multitude of isoquinoline alkaloid-producing species can be used for metabolic engineering experiments, bypassing in some instances the complications that arise from multicellular compartmentation in differentiated plants.

3.3. Genetic engineering of tetrahydrobenzylisoquinoline alkaloid biosynthetic pathways

Attempts at engineering plant cell cultures and differentiated plants that produce isoquinoline alkaloids using cDNAs from tetrahydrobenzylisoquinoline-derived alkaloid biosynthesis have been hampered by the difficulties associated with establishing transformation and regeneration protocols for each new species. Just the same, there are selected successes to be reported.

One of the first introductions of an isoquinoline alkaloid biosynthetic gene into an isoquinoline alkaloid-producing plant was the *Agrobacterium*-mediated transformation of the *C. japonica 9-omt* under transcriptional of the CaMV 35S promoter into *C. japonica* cell cultures (Sato *et al.*, 2001). Ectopic expression of *9-omt* in a high berberine-producing cell culture resulted in a 15% increase in berberine and columbamine. Likewise, the same construct was introduced into *E. californica* seedling segments from which transgenic cell cultures were derived. *E. californica* produces benzo[*c*]phenanthridine alkaloids rather than berberine alkaloids. Introduction of the *9-omt* cDNA resulted in the accumulation of columbamine (not normally present in *E. californica*) and a reduction in the level of the *E. californica* native alkaloid sanguinarine. Ectopic expression of *9-omt* successfully introduced a new branch point in the *E. californica* isoquinoline alkaloid pathway and redirected (*S*)-scoulerine away from benzo[*c*]phenanthridine alkaloid biosynthesis into berberine alkaloid formation.

Root cultures of *E. californica* have been engineered with *bbe1* and *cyp80b1* under transcriptional control of the CaMV 35S promoter by *Agrobacterium rhizogenes*-mediated transformation (Park *et al.*, 2002, 2003). Transgenic root cultures containing either antisense *bbe1* or antisense *cyp80b1* accumulated lower levels of benzo[*c*]phenanthridine alkaloids compared to controls, whereas transgenic root cultures that overexpressed *bbe1* showed increased accumulation of benzo[*c*]phenanthridine alkaloids compared to controls. These types of experiments demonstrate that manipulation of alkaloid biosynthetic pathways is possible, but illustrate that intact transgenic plants are difficult to obtain.

There are several reports in the literature of the transformation and regeneration of *P. somniferum*. The earliest reports of regeneration of *P. somniferum* plants from cell culture preceeded the transformation work (Wakhlu and Bajwa, 1986). Transformation of *P. somniferum* with *A. rhizogenes* appeared shortly thereafter (Williams and Ellis, 1993; Yoshimatsu and Shimomura, 1992). The first report of *Agrobacterium*-mediated transformation of *P. somniferum* cell suspension cultures in which an *Arabidopsis thaliana sam1* trangene was introduced occurred in 1997 (Belny *et al.*, 1997). The first report of *Agrobacterium*-mediated transformation with subsequent regeneration of *P. somniferum* appeared in 1999 (Larkin *et al.*, 1999). A second report appeared in 2000 (Park and Facchini, 2000). Attempts at metabolic engineering of alkaloid biosynthesis in *P. somniferum* are currently being made. The known genes of tetrahydrobenzylisoquinoline-derived alkaloid formation in *P. somniferum* have been reintroduced into a highly inbred, elite Tasmanian variety in the sense and antisense orientation. Several hundred transgenic plants have been produced by *Agrobacterium*-mediated transformation (S. Frick, P. J. Larkin, and T. M. Kutchan, unpublished data). The first set of transgenic plants for which the analyses are complete are those that contained the *bbe1* gene in an antisense orientation under transcriptional control of the S4S4 promoter (Frick *et al.*, 2004). These experiments were designed to reduce flow of the central intermediate (*S*)-reticuline into the sanguinarine pathway. A complex picture of ratios of tetrahydrobenzylisoquinoline intermediate alkaloids emerged from these experiments, but morphinan levels did not increase (Fig. 10.10). Preliminary analyses of plants that contain various other transgenes indicate that it is, however, possible to manipulate morphine levels. *P. somniferum* has tremendous potential for alkaloid engineering because the plant is the commercial source of pharmaceutically important morphinan alkaloids and the transformation and regeneration, although still a slow process, has been established.

4. TROPANE ALKALOIDS

Solanaceous plants are the main source of the tropane class of alkaloids, notably the anticholinergic hyoscyamine and scopolamine. These plants have traditionally been used for their medicinal, hallucinogenic, and poisonous properties. Biosynthetically related to tropane alkaloids, nicotine is also found in the Solanaceae. The biosynthesis of nicotine and tropane alkaloids has not yet been completely elucidated. Our biochemical and molecular genetic understanding of nicotine/tropane alkaloid biosynthesis is as follows.

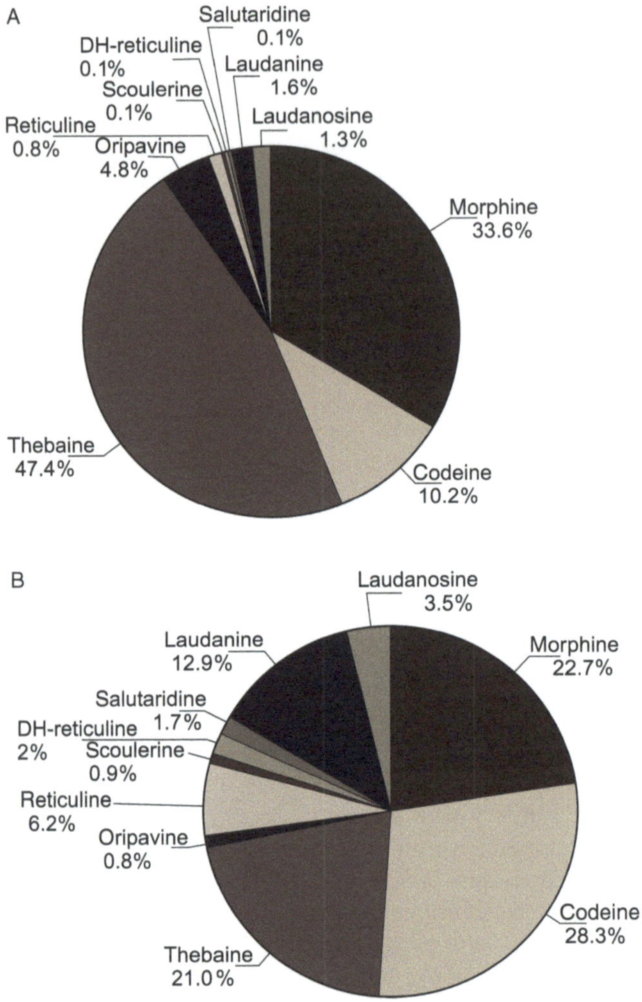

FIGURE 10.10 Benzylisoquinoline alkaloid quantitation in latex from (A) untransformed *P. somniferum* and (B) transgenic *anti-bbe1* T_2 plants (Frick *et al.*, 2004). The data are presented in percentages (μg alkaloid/100 μg soluble protein in latex). The values given in (A) are the mean of individual analyses of 29 plants. (See Page 18 in Color Section.)

4.1. Tropane alkaloid biosynthesis

The biosynthetic building blocks leading to the tropane alkaloids are the amino acids L-arginine and L-phenylalanine (Fig. 10.11). On the pathway leading from L-arginine to the *N*-methylpyrrolinium ion, the enzyme putrescine *N*-methyltransferase (PMT) has been well characterized at the enzymatic and molecular genetic levels (Hibi *et al.*, 1992). Several *pmt* genes have been isolated from *Nicotiana tabacum, Atropa belladonna, Hyoscyamus niger, N. sylvestris*, and *N. attenuate* (Hibi *et al.*, 1994; Shoji *et al.*, 2000; Suzuki *et al.*, 1999a; Winz and Baldwin, 2001). Further along the

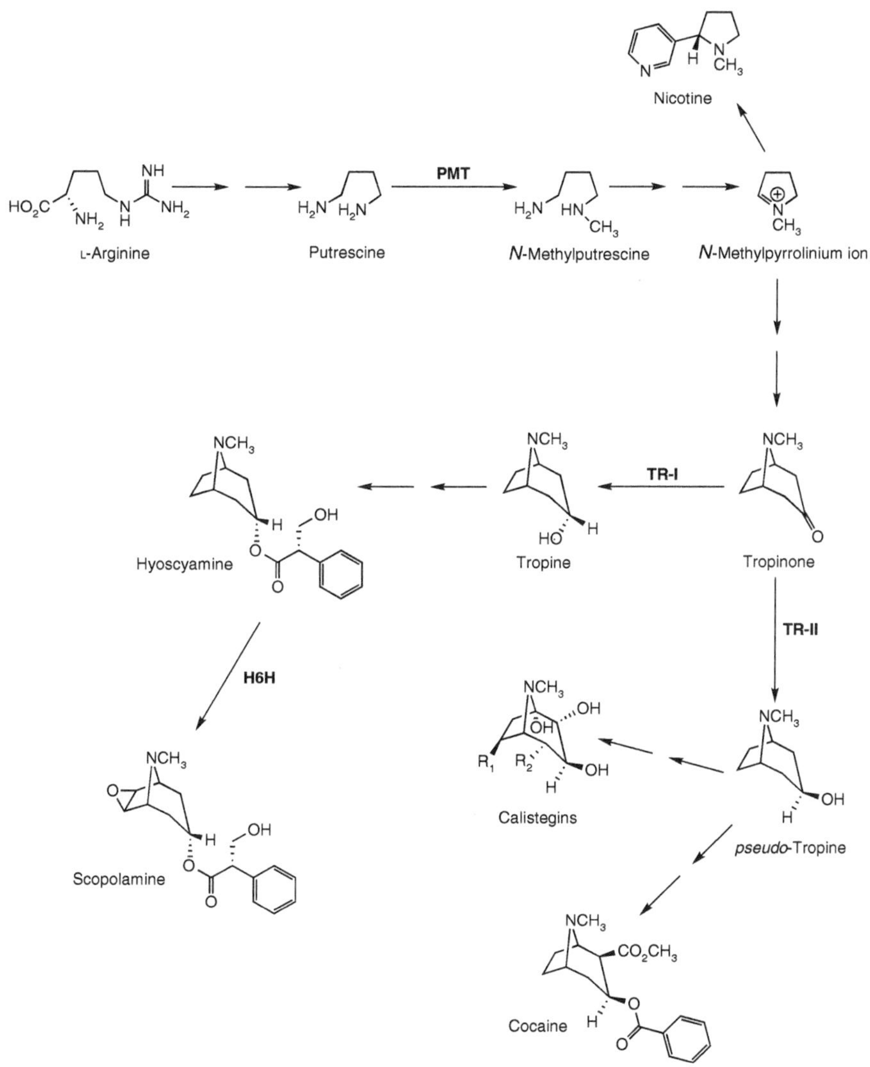

FIGURE 10.11 Schematic representation of the biosynthetic pathway leading from L-arginine to nicotine, scopolamine, calistegins, and cocaine. PMT, putrescine N-methyltransferase; TR-I, tropinone reductase I; TR-II, tropinone reductase II; H6H, hyoscyamine 6β-hydroxylase.

biosynthetic pathway to tropane alkaloids, tropinone is reduced by tropinone reductase I (TR-I) to tropine. TR-I specifically reduces the tropinone 3-keto moiety to the 3α-hydroxyl group of tropine, the biosynthetic precursor of hyoscyamine and scopolamine (Hashimoto *et al.*, 1992; Koelen and Gross, 1982). Tropinone reductase II (TR-II) reduces the 3-keto group of tropinone to the 3β-hydroxy moiety of pseudotropine, which serves as precursor to the calistegins (Dräger and Schaal, 1994). The gene *tr-I* has been characterized from *Datura stramonium* and *H. niger* (Nakajima *et al.*, 1993, 1999). The gene *tr-II* is known from *D. stramonium*,

H. niger, and *Solanum tuberosum* (Keiner *et al.*, 2002; Nakajima *et al.*, 1993, 1999). The final gene from the scopolamine biosynthetic pathway that has been identified is *h6h* encoding hyoscyamine 6β-hydroxylase (Hashimoto and Yamada, 1986). This 2-oxoglutarate-dependent dioxygenase is bifunctional, catalyzing both the monooxygenation of hyoscyamine to 6β-hydroxyhyoscyamine and the subsequent epoxidation to scopolamine. This gene has been characterized from *H. niger* and *A. belladonna* (Matsuda *et al.*, 1991; Suzuki *et al.*, 1999b).

With the cloning of the genes *pmt*, *tr-I*, *tr-II*, and *h6h*, our understanding of the cellular localization of solanaceous alkaloid biosynthesis has also advanced. The current view is discussed in the following sections.

4.2. Cell-specific expression of tropane alkaloid biosynthetic genes

The main site of tropane alkaloid biosynthesis was demonstrated to be in the roots by classical grafting experiments in which scions from tropane alkaloid-producing species were grafted onto rootstock from nonproducing species. The resultant plants did not accumulate alkaloids. The reciprocal graft experiments yielded plants that accumulated alkaloids (cited in Hashimoto *et al.*, 1991). These experimental results suggest that tropane alkaloids are synthesized in roots and transported to aerial parts of the plant. With these early experiments in mind, histochemical localization of *pmt*, TR-I, TR-II, and H6H has been carried out.

TR-I and H6H, specific to scopolamine biosynthesis, and TR-II, specific to calistegin biosynthesis, have been localized to the *H. niger* root (Hashimoto *et al.*, 1991; Nakajima and Hashimoto, 1999). Accumulation was highest in lateral roots. TR-I and TR-II accumulated with cell-specific patterns that differed from those of H6H, implying a transport of an alkaloid biosynthetic intermediate that must occur somewhere between tropine and hyoscyamine (Fig. 10.12). Using GUS fusions, *A. belladonna pmt* promoter activity has been histochemically localized to root pericycle and *h6h* promoter activity was found both in root pericycle and tapetum and pollen grains (Suzuki *et al.*, 1999a,b) The cell-specific expression of these promoters appears to be species specific (Kanegae *et al.*, 1994).

Once again, multiple cell types are involved in an alkaloid biosynthetic pathway and transport of biosynthetic intermediates must be involved, thereby introducing another possible level of regulation.

4.3. Genetic engineering of tropane alkaloid biosynthetic pathways

The earliest examples of engineering an alkaloid pathway in a medicinal plant/plant tissue culture come from the tropane alkaloid field. The commercial source of scopolamine is *Duboisia*, originally cultivated in Australia. *A. belladonna* accumulates hyoscyamine rather than scopolamine as the major alkaloid. The cDNA encoding H6H from *H. niger* was introduced into *A. belladonna* using either *A. tumefaciens*- or *A. rhizogenes*-mediated transformation. Elevated levels of scopolamine accumulated in the transgenic plants (up to 1.2% dry weight compared to trace levels in control plants) and 0.3% in hairy roots compared to 0.03% in control roots (Yun *et al.*, 1992). Since these original experiments, the tobacco *pmt*

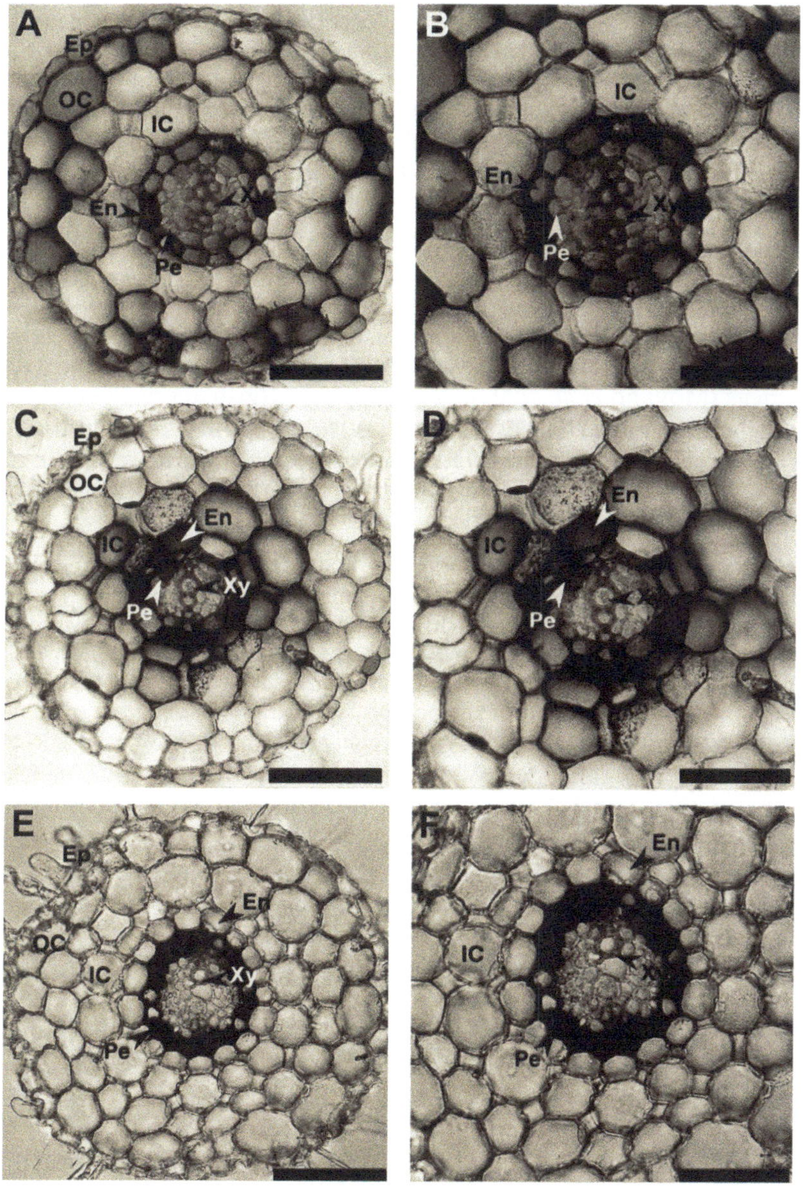

FIGURE 10.12 Immunohistochemical localization of TR-I, TR-II, and H6H in root cross-sections of *H. niger* (Nakajima and Hashimoto, 1999). Panels (A, B) TR-I; panels (C, D) TR-II; panels (E, F) H6H. Ep, epidermis; OC, outer cortex; IC, inner cortex; En, endodermis; Pe, pericycle; Xy, xylem. (See Page 19 in Color Section.)

has been expressed under transcriptional control of the dual CaMV 35S promoter in *A. belladonna* (Sato *et al.*, 2001). Regardless of the transgene transcript level, the plants were phenotypically normal and had levels of hyoscyamine comparable to control plants. The alkaloid profiles of transgenic hairy root lines were also quantitatively and qualitatively similar to wild type. In contrast, *N. sylvestris* overexpressing the tobacco *pmt* showed increases in leaf nicotine of up to 40% compared to wild type (Sato *et al.*, 2001). Cosuppressed plants accumulated only 2% of wild-type nicotine levels and contained elevated levels of putrescine and spermidine. Cosuppressed plants also showed phenotypic abnormalities.

The tobacco *pmt* under transcritional control of the CaMV 35S promoter has also been introduced into a scopolamine-rich *Duboisia* hybrid (Moyano *et al.*, 2002). The *N*-methylputrescine levels of the resultant hairy roots increased two- to fourfold compared to wild type, but levels of tropane alkaloids were not increased. When the tobacco *pmt* gene was introduced into *Datura metel* and *Hyoscyamus muticus*, the resultant hairy roots showed improved hyoscyamine and scopolamine production in *D. metel*, but only hyoscyamine content increased in *H. muticus* (Moyano *et al.*, 2003). A similar *pmt*-construct has been introduced into *A. belladonna* (Rothe *et al.*, 2003). Transgenic plants and derived root cultures showed alkaloid profiles similar to those of controls. In summary, the expression level of *pmt* is apparently not rate limiting in all tropane alkaloid-producing species.

Efforts have also been made to establish particle bombardment-mediated transformation of *H. muticus* to facilitate future metabolic engineering of this species (Zeef *et al.*, 2000). Importantly in these experiments, fertile, mature plants were recovered that could be self-pollinated.

5. SUMMARY

To return to Chapter 12 of Volume 7 "Secondary Plants Products" in the series *The Biochemistry of Plants* (Waller and Dermer, 1981) authored by George Waller and Otis Dermer, the authors concluded at the time that the important characteristics of alkaloid biosynthetic enzymes should be catalytic properties, regulation, intracellular localization, and tissue distribution. The authors were arguably not satisfied with the progress that had been made in these areas up until 1979. Since that time, astonishingly much progress has been made in our understanding the nature of the enzymes that synthesize alkaloids in plants. With the advent of the application of molecular genetic methods to the field, we have sophisticated tools at our disposal to study the regulation of enzyme biosynthesis as well as the cellular and subcellular localization. We still purify enzymes with ever more refined instrumentation, but we can also identify biosynthetic enzymes with genetic approaches. The first reports of successful crystallization of enzymes of alkaloid biosynthesis, a requisite to X-ray crystallographic structure determination, have now appeared (Ma *et al.*, 2004a,b). Notably, the first alkaloid biosynthetic enzyme for which a crystal structure has been determined is also the first one for which a cDNA was isolated, strictosidine synthase from *R. serpentina* (Ma *et al.*, 2006). A topic not touched upon at all by Waller and Dermer was the possibility to alter alkaloid metabolism in plants and plant tissue and cell cultures. In 2006, we are just beginning to metabolically engineer alkaloid metabolism in plants and in *in vitro* culture. Multicellular compartmentation of alkaloid pathways must be considered if meaningful metabolic engineering experiments are to be designed. We will need to use promoters that drive transgene expression in the correct cell types. Regulation of these pathways at the gene and enzyme level is complex, and there is still much to be learned about metabolite levels and pathway interconnections as we systematically overexpress and suppress gene transcription. Today, pathway engineering in plants remains highly variable. When we perturb cellular physiology, metabolite homeostasis and intra- and intercellular partitioning can be affected in unpredictable ways. Another aspect that needs attention is the development of efficient transformation and regeneration protocols for alkaloid-producing plants that do not belong to the Solanaceae. All told, there is still much to be achieved.

ACKNOWLEDGEMENTS

Our results reported in this chapter were from research supported by the Deutsche Forschungsgemeinschaft, Bonn and Fonds der Chemischen Industrie, Frankfurt.

REFERENCES

Bauer, W., and Zenk, M. H. (1991). Two methylenedioxy bridge-forming cytochrome P-450-dependent enzymes are involved in (*S*)-stylopine biosynthesis. *Phytochemistry* **30,** 2953–2961.

Bayer, A., Ma, X., and Stöckigt, J. (2004). Acetyltransfer in natural product biosynthesis—functional cloning and molecular analysis of vinorine synthase. *Bioorg. Med. Chem.* **12,** 2787–2795.

Belny, M., Hérouart, D., Thomasset, B., David, H., Jacquin-Dubreuil, A., and David, A. (1997). Transformation of *Papaver somniferum* cell suspension cultures with *sam1* from *A. thaliana* results in cell lines of different *S*-adenosyl-l-methionine synthetase activity. *Physiol. Plant.* **99,** 233–240.

Canel, C., Lopes-Cardoso, M. I., Whitmer, S., Van Der Fits, L., Pasquali, G., Van Der Heijden, R., Hoge, J. H. C., and Verpoorte, R. (1998). Effects of over-expression of strictosidine synthase and tryptophan decarboxylase on alkaloid production by cell cultures of *Catharanthus roseus*. *Planta* **205,** 414–419.

Chahed, K., Oudin, A., Guivare'h, N., Hamdi, S., Chénieux, J. C., Rideau, M., and Clastre, M. (2000). 1-Deoxy-D-xylulose 5-phosphate synthase from periwinkle: cDNA identification and induced gene expression in terpenoid indole alkaloid-producing cells. *Plant Physiol. Biochem.* **38,** 559–566.

Chavadej, S., Brisson, N., Mcneil, J. N., and De Luca, V. (1994). Redirection of tryptophan leads to production of low indole glucosinolate canola. *Proc. Natl. Acad. Sci. USA* **91,** 2166–2170.

Choi, K.-B., Morishige, T., Shitan, N., Yazaki, K., and Sato, F. (2002). Molecular cloning and characterization of coclaurine *N*-methyltransferase from cultured cells of *Coptis japonica*. *J. Biol. Chem.* **277,** 830–835.

Chou, W.-M., and Kutchan, T. M. (1998). Enzymatic oxidations in the biosynthesis of complex alkaloids. *Plant J.* **15,** 289–300.

Collu, G., Unver, N., Peltenburg-Looman, A. M. G., Van Der Heijden, R., Verpoorte, R., and Memelink, J. (2001). Geraniol 10-hydroxylase, a cytochrome P450 enzyme involved in terpenoid indole alkaloid biosynthesis. *FEBS Lett.* **508,** 215–220.

Contin, A., Van Der Heijden, R., Lefeber, A. W. M., and Verpoorte, R. (1998). The iridoid glucoside secologanin is derived from the novel triose phosphate/pyruvate pathway in a *Catharanthus roseus* cell culture. *FEBS Lett.* **434,** 413–416.

De Carolis, E., and De Luca, V. (1993). Purification, characterization, and kinetic analysis of a 2-oxoglutarate-dependent dioxygenase involved in vindoline biosynthesis from *Catharanthus roseus*. *J. Biol. Chem.* **268,** 5505–5511.

De Luca, V., Marineau, C., and Brisson, N. (1989). Molecular cloning and analysis of cDNA encoding a plant tryptophan decarboxylase: Comparison with animal dopa decarboxylases. *Proc. Natl. Acad. Sci. USA* **86,** 2582–2586.

Dittrich, H., and Kutchan, T. M. (1991). Molecular cloning, expression and induction of berberine bridge enzyme, an enzyme essential to the formation of benzophenanthridine alkaloids in the response of plants to pathogenic attack. *Proc. Natl. Acad. Sci. USA* **88,** 9969–9973.

Dogru, E., Warzecha, H., Seibel, F., Haebel, S., Lottspeich, F., and Stöckigt, J. (2000). The gene encoding polyneuridine aldehyde esterase of monoterpenoid indole alkaloid biosynthesis in plants is an ortholog of the alpha/betahydrolase super family. *Eur. J. Biochem.* **267,** 1392–1406.

Dräger, B., and Schaal, A. (1994). Tropinone reduction in *Atropa belladonna* root cultures. *Phytochemistry* **35,** 1441–1447.

Eichinger, D., Bacher, A., Zenk, M. H., and Eisenreich, W. (1999). Analysis of metabolic pathways via quantitative prediction of isotope labeling patterns: A retrobiosynthetic ^{13}C NMR study on the monoterpene loganin. *Phytochemistry* **51,** 223–236.

Facchini, P. J., and De Luca, V. (1994). Differential and tissue-specific expression of a gene family for tyrosine/dopa decarboxylase from opium poppy. *J. Biol. Chem.* **269,** 26684–26690.

Facchini, P. J., and De Luca, V. (1995). Phloem-specific expression of tyrosine/dopa decarboxylase genes and the biosynthesis of isoquinoline alkaloids in opium poppy. *Plant Cell* **7,** 1811–1821.

Facchini, P. J., Penzes, C., Johnson, A. G., and Bull, D. (1996). Molecular characterization of berberine bridge enzyme genes from opium poppy. *Plant Physiol.* **112,** 1669–1677.

Fahn, W., Gundlach, H., Deus-Neumann, B., and Stöckigt, J. (1985). Late enzymes of vindoline biosynthesis. Acetyl-CoA: 17-*O*-deacetylvindoline 17-*O*-acetyltransferase. *Plant Cell Rep.* **4,** 333–336.

Frenzel, T., and Zenk, M. H. (1990a). Purification and characterization of three isoforms of *S*-adenosyl-L-methionine: (*R,S*)-tetrahydrobenzylisoquinoline-*N*-methyltransferase from *Berberis koetineana* cell cultures. *Phytochemistry* **29,** 3491–3497.

Frenzel, T., and Zenk, M. H. (1990b). *S*-adenosyl-L-methionine: 3′-hydroxy-*N*-methyl-(*S*)-coclaurine 4′-*O*-methyltransferase, a regio- and stereoselective enzyme of the (*S*)-reticuline pathway. *Phytochemistry* **29,** 3505–3511.

Frick, S., and Kutchan, T. M. (1999). Molecular cloning and functional expression of *O*-methyltransferases common to isoquinoline alkaloid and phenylpropanoid biosynthesis. *Plant J.* **17,** 329–339.

Frick, S., Chitty, J. A., Kramell, R., Schmidt, J., Allen, R. S., Larkin, P. J., and Kutchan, T. M. (2004). Transformation of the opium poppy (*Papaver somniferum* L.) with anti-sense berberine bridge enzyme 1 (*anti-bbe1)* via somatic embryogenesis results in an altered ratio of alkaloids in latex but not in roots. *Transgenic Res.* **13,** 607–613.

Gantet, P., and Memelink, J. (2002). Transcription factors: Tools to engineer the production of pharmacologically active plant metabolites. *Trends Pharmacol. Sci.* **23,** 563–569.

Geerlings, A., Ibanez, M. M., Memelink, J., Van Der Heijden, R., and Verpoorte, R. (2000). Molecular cloning and analysis of strictosidine β-D-glucosidase, an enzyme in terpenoid indole alkaloid biosynthesis in *Catharanthus roseus*. *J. Biol. Chem.* **275,** 3051–3056.

Gerasimenko, I., Sheludko, Y., Ma, X., and Stöckigt, J. (2002). Heterologous expression of a *Rauvolfia* cDNA encoding strictosidine glucosidase, a biosynthetic key to over 2000 monoterpenoid indole alkaloids. *Eur. J. Biochem.* **269,** 2204–2213.

Grothe, T., Lenz, R., and Kutchan, T. M. (2001). Molecular characterization of the salutaridinol 7-*O*-acetyltransferase involved in morphine biosynthesis in opium poppy *Papaver somniferum*. *J. Biol. Chem.* **276,** 30717–30723.

Guillet, G., Poupart, J., Basurco, J., and De Luca, V. (2000). Expression of tryptophan decarboxylase and tyrosine decarboxylase genes in tobacco results in altered biochemical and physiological phenotypes. *Plant Physiol.* **122,** 933–943.

Hashimoto, T., and Yamada, Y. (1986). Hyoscyamine 6β-hydroxylase, a 2-oxoglutarate-dependent dioxygenase, in alkaloid-producing root cultures. *Plant Physiol.* **81,** 619–625.

Hashimoto, T., Hayashi, A., Amano, Y., Kohno, J., Iwanari, H., Usuda, S., and Yamada, Y. (1991). Hyoscyamine 6β-hydroxylase, an enzyme involved in tropane alkaloid biosynthesis, is localized to the pericycle of the root. *J. Biol. Chem.* **266,** 4648–4653.

Hashimoto, T., Nakajima, K., Ongena, G., and Yamada, Y. (1992). Two tropinone reductases with distinct stereospecificities from cultured roots of *Hyoscyamus niger*. *Plant Physiol.* **100,** 836–845.

Hibi, N., Fujita, T., Hatano, M., Hashimoto, T., and Yamada, Y. (1992). Putrescine *N*-methyltransferase in cultured roots of *Hyoscyamus albus*. *Plant Physiol.* **100,** 826–835.

Hibi, N., Higashiguchi, S., Hashimoto, T., and Yamada, Y. (1994). Gene expression in tobacco low-nicotine mutants. *Plant Cell* **6,** 723–735.

Huang, F.-C., and Kutchan, T. M. (2000). Distribution of morphinan and benzophenanthridine alkaloid gene transcript accumulation in the opium poppy *Papaver somniferum*. *Phytochemistry* **53,** 555–564.

Hughes, E. H., Hong, S.-B., Shanks, J. V., San, K.-Y., and Gibson, S. I. (2002). Characterization of an inducible promoter system in *Catharanthus roseus* hairy roots. *Biotechnol. Prog.* **18,** 1183–1186.

Ikezawa, N., Tanaka, M., Nagayoshi, M., Shinkyo, R., Sakaki, T., Inouye, K., and Sato, F. (2003). Molecular cloning and characterization of CYP719, a methylenedioxy bridge-forming enzyme that belongs to a novel P450 family, from cultured *Coptis japonica* cells. *J. Biol. Chem.* **278,** 38557–38565.

Irmler, S., Schröder, G., St-Pierre, B., Crouch, N. P., Hotze, M., Schmidt, J., Strack, D., Matern, U., and Schröder, J. (2000). Indole alkaloid biosynthesis in *Catharanthus roseus*: New enzyme activities and identification of cytochrome P450 CYP72A1 as secologanin synthase. *Plant J.* **24,** 797–804.

Kanegae, T., Kajiya, H., Amano, Y., Hashimoto, T., and Yamada, Y. (1994). Species-dependent expression of the hyoscyamine 6β-hydroxylase gene in the pericycle. *Plant Physiol.* **105,** 483–490.

Keiner, R., Kaiser, H., Nakajima, K., Hashimoto, T., and Dräger, B. (2002). Molecular cloning, expression and characterization of tropinone reductase II, an enzyme of the SDR family in *Solanum tuberosum* (L.). *Plant Mol. Biol.* **48,** 299–308.

Koelen, K. J., and Gross, G. G. (1982). Partial purification and properties of tropine dehydrogenase from root cultures of *Datura stramonium*. *Planta Med.* **44,** 227–230.

Kutchan, T. M. (1989). Expression of enzymatically active cloned strictosidine synthase from the higher plant *Rauwolfia serpentina* in *Escherichia coli.*. *FEBS Lett.* **257,** 127–130.

Kutchan, T. M. (1998). Molecular genetics of plant alkaloid biosynthesis. *In* "The Alkaloids—Chemistry and Biology" (G. A. Cordell, ed.), Vol. 50, pp. 257–316. Academic Press, San Diego.

Kutchan, T. M., Hampp, N., Lottspeich, F., Beyreuther, K., and Zenk, M. H. (1988). The cDNA clone for strictosidine synthase from *Rauwolfia serpentina*: DNA sequence determination and expression in *Escherichia coli*. *FEBS Lett.* **237,** 40–44.

Laflamme, P., St-Pierre, B., and De Luca, V. (2001). Molecular and biochemical analysis of a Madagascar periwinkle root-specific minovincinine 19-hydroxy-*O*-acetyltransferase. *Plant Physiol.* **125,** 189–198.

Larkin, P. J., Chitty, J. A., and Brettell, R. I. S. (1999). "Methods for Plant Transformation and Regeneration." International Patent Publication Number WO 99/34663.

Lenz, R., and Zenk, M. H. (1995a). Acetyl coenzyme A: Salutaridinol 7-*O*-acetyltransferase from *Papaver somniferum* plant cell cultures. *J. Biol. Chem.* **270,** 31091–31096.

Lenz, R., and Zenk, M. H. (1995b). Purification and properties of codeinone reductase (NADPH) from *Papaver somniferum* cell cultures and differentiated plants. *Eur. J. Biochem.* **233,** 132–139.

Ma, X.-Y., Koepke, J., Linhard, V., Bayer, A., Fritzsch, G., Zhang, B., Michel, H., and Stöckigt, J. (2004a). Vinorine synthase from *Rauvolfia*—the first example of crystallization and preliminary x-ray diffraction analysis of an enzyme of the BAHD super family. *Biochim. Biophys. Acta* **1701,** 129–132.

Ma, X.-Y., Koepke, J., Fritzsch, G., Diem, R., Kutchan, T. M., Michel, H., and Stöckigt, J. (2004b). Crystallization and preliminary crystallographic analysis of strictosidine synthase from *Rauvolfia*—the first member of a novel enzyme family. *Biochim. Biophys. Acta* **1702,** 121–124.

Ma, X.-Y., Panjikar, S., Koepke, J., Loris, E., and Stöckigt, J. (2006). The structure of *Rauvolfia serpentina* strictosidine synthase is a novel six-bladed *beta*-propeller fold in plant proteins. *Plant Cell* **18,** 907–920.

Madyastha, K. M., Meehan, T. D., and Coscia, C. J. (1976). Characterization of a cytochrome P-450 dependent monoterpene hydroxylase from the higher plant *Vinca rosea*. *Biochemistry* **15,** 1097–1102.

Madyastha, K. M., Ridgway, J. E., Dwyer, J. G., and Coscia, C. J. (1977). Subcellular localization of a cytochrome P-450-dependent monogenase in vesicles of the higher plant *Catharanthus roseus*. *J. Cell Biol.* **72,** 302–313.

Matsuda, J., Okabe, S., Hashimoto, T., and Yamada, Y. (1991). Molecular cloning of hyoscyamine 6β-hydroxylase, a 2-oxoglutarate-dependent dioxygenase, from cultured roots of *Hyoscyamus niger*. *J. Biol. Chem.* **266,** 9460–9464.

Mcnight, T. D., Roessner, C. A., Devagupta, R., Scott, A. I., and Nessler, C. L. (1990). Nucleotide sequence of a cDNA encoding the vacuolar protein strictosidine synthase from *Catharanthus roseus*. *Nucleic Acids Res.* **18,** 4939.

Menke, F. L., Champion, A., Kijne, J. W., and Memelink, J. (1999). A novel jasmonate- and elicitor-responsive element in the periwinkle secondary metabolite biosynthetic gene *Str* interacts with a jasmonate- and elicitor-inducible AP2-domain transcription factor, ORCA2. *EMBO J.* **18,** 4455–4463.

Morgan, J. A., and Shanks, J. V. (2000). Determination of metabolic rate-limitations by precursor feeding in *Catharanthus roseus* hairy root cultures. *J. Biotechnol.* **79,** 137–145.

Morgan, J. A., and Shanks, J. V. (2002). Quantitation of metabolic flux in plant secondary metabolism by a biogenetic organizational approach. *Metab. Eng.* **4,** 257–262.

Morishige, T., Tsujita, T., Yamada, Y., and Sato, F. (2000). Molecular characterization of the *S*-adenosyl-L-methionine: 3′-hydroxy-*N*-methylcoclaurine 4′-*O*-methyltransferase involved in isoquinoline alkaloid biosynthesis in *Coptis japonica*. *J. Biol. Chem.* **275,** 23398–23405.

Moyano, E., Fornalé, S., Palazón, J., Cusidó, R. M., Bagni, N., and Piñol, M. T. (2002). Alkaloid production in *Duboisia* hybrid hairy root cultures overexpressing the *pmt* gene. *Phytochemistry* **59,** 697–702.

Moyano, E., Jouhikainen, K., Tammela, P., Palazón, J., Cusidó, R. M., Piñol, M. T., Teeri, T. H., and Oksman-Caldentey, K.-M. (2003). Effect of *pmt* gene overexpression on tropane alkaloid production in transformed root cultures of *Datura metel* and *Hyoscyamus muticus*. *J. Exp. Bot.* **54,** 203–211.

Muemmler, S., Rueffer, M., Nagakura, N., and Zenk, M. H. (1985). *S*-Adenosyl-L-methionine: (*S*)-scoulerine 9-*O*-methyltransferase, a highly stereo- and regiospecific enzyme in tetrahydroprotoberberine biosynthesis. *Plant Cell Rep.* **4,** 36–39.

Nakajima, K., and Hashimoto, T. (1999). Two tropinone reductases, that catalyze opposite stereospecific reductions in tropane alkaloid biosynthesis, are localized in plant root with different cell-specific patterns. *Plant Cell Physiol.* **40,** 1099–1107.

Nakajima, K., Hashimoto, T., and Yamada, Y. (1993). Two tropinone reductases with different stereospecificities are short-chain dehydrogenases evolved from a common ancestor. *Proc. Natl. Acad. Sci. USA* **90,** 9591–9595.

Nakajima, K., Oshita, Y., Kaya, M., Yamada, Y., and Hashimoto, T. (1999). Structures and expression patterns of two tropinone reductase genes from *Hyoscyamus niger*. *Biosci. Biotechnol. Biochem.* **63,** 1756–1764.

Ounaroon, A., Decker, G., Schmidt, J., Lottspeich, F., and Kutchan, T. M. (2003). (*R,S*)-Reticuline 7-*O*-methyltransferase and (*R,S*)-norcoclaurine 6-*O*-methyltransferase of *Papaver somniferum*—cDNA cloning and characterization of methyl transfer enzymes of alkaloid biosynthesis in opium poppy. *Plant J.* **36,** 808–819.

Park, S.-U., and Facchini, P. J. (2000). *Agrobacterium*-mediated transformation of opium poppy, *Papaver somniferum*, via shoot organogenesis. *J. Plant Physiol.* **157,** 207–214.

Park, S. U., Yu, M., and Facchini, P. J. (2002). Antisense RNA-mediated suppression of benzophenanthridine alkaloid biosynthesis in transgenic cultures of California poppy. *Plant Physiol.* **128,** 696–706.

Park, S. U., Yu, M., and Facchini, P. J. (2003). Modulation of berberine bridge enzyme levels in transgenic root cultures of California poppy alters accumulation of benzophenanthridine alkaloids. *Plant Mol. Biol.* **51,** 153–164.

Pauli, H. H., and Kutchan, T. M. (1998). Molecular cloning and functional heterologous expression of two alleles encoding (*S*)-*N*-methylcoclaurine 3′-hydroxylase (CYP80B1), a new methyl jasmonate-inducible cytochrome P-450-dependent monooxygenase of benzylisoquinoline alkaloid biosynthesis. *Plant J.* **13,** 793–801.

Pfitzner, A., and Stöckigt, J. (1983). Characterization of polyneuridine aldehyde esterase, a key enzyme in the biosynthesis of sapargine/ajmaline type alkaloids. *Planta Med.* **48,** 221–227.

Power, R., Kurz, W. G. W., and De Luca, V. (1990). Purification and characterization of acetylcoenzyme A: Deacetylvindoline 4-*O*-acetyltransferase from *Catharanthus roseus*. *Arch. Biochem. Biophys.* **279,** 370–376.

Rijhwani, S. K., Ho, C.-H., and Shanks, J. V. (1999). *In vivo* ^{31}P and multilabel ^{13}C NMR measurements for evaluation of plant metabolic pathways. *Metab. Eng.* **1,** 12–25.

Rink, E., and Böhm, H. (1975). Conversion of reticuline into scoulerine by a cell free preparation from *Macleaya microcarpa* cell suspension cultures. *FEBS Lett.* **49,** 396–399.

Rothe, G., Hachiya, A., Yamada, Y., Hashimoto, T., and Dräger, B. (2003). Alkaloids in plants and root cultures of *Atropa belladonna* overexpressing putrescine *N*-methyltransferase. *J. Exp. Bot.* **54,** 2065–2070.

Rueffer, M., and Zenk, M. H. (1994). Canadine synthase from *Thalictrum tuberosum* cell cultures catalyses the formation of the methylenedioxy bridge in berberine biosynthesis. *Phytochemistry* **36,** 1219–1223.

Rüffer, M., Nagakura, N., and Zenk, M. H. (1983). Partial purification and properties of *S*-adenosylmethionine: (*R,S*)-norlaudanosoline 6-*O*-methyltransferase from *Argemone platyceras* cell cultures. *Planta med.* **49,** 131–137.

Sato, F., Hashimoto, T., Hachiya, A., Tamura, K.-I., Choi, K.-B., Morishige, T., Fujimoto, H., and Yamada, Y. (2001). Metabolic engineering of plant alkaloid biosynthesis. *Proc. Natl. Acad. Sci. USA* **98,** 367–372.

Schröder, G., Unterbusch, E., Kaltenbach, M., Schmidt, J., Strack, D., De Luca, V., and Schröder, J. (1999). Light-induced cytochrome P450-dependent enzyme in indole alkaloid biosynthesis: Tabersonine 16-hydroxylase. *FEBS Lett.* **458,** 97–102.

Shoji, T., Yamada, Y., and Hashimoto, T. (2000). Jasmonate induction of putrscine *N*-methyltransferase genes in the root of *Nicotiana sylvestris*. *Plant Cell Physiol.* **41,** 831–839.

Songstad, D. D., De Luca, V., Brisson, N., Kurz, W. G. W., and Nessler, C. L. (1990). High levels of tryptamine accumulation in transgenic tobacco expressing tryptophan decarboxylase. *Plant Physiol.* **94,** 1410–1413.

Songstad, D. D., Kurz, W. G. W., and Nessler, C. (1991). Tyramine accumulation in *Nicotiana tabacum* transformed with a chimeric tryptophan decarboxylase gene. *Phytochemistry* **30,** 3245–3246.

Steffens, P., Nagakura, N., and Zenk, M. H. (1985). Purification and characterization of the berberine bridge enzyme from *Berberis beaniana* cell cultures. *Phytochemistry* **24,** 2577–2583.

Stöckigt, J., and Zenk, M. H. (1977a). Isovincoside (strictosidine), the key intermediate in the enzymatic formation of indole alkaloids. *FEBS Lett.* **79,** 233–237.

Stöckigt, J., and Zenk, M. H. (1977b). Strictosidine (isovincoside): The key intermediate in the biosynthesis of monoterpenoid indole alkaloids. *J. Chem. Soc. Chem. Commun.* 646–648.

St-Pierre, B., and De Luca, V. (1995). A cytochrome P-450 monooxygenase catalyzes the first step in the conversion of tabersonine to vindoline in *Catharanthus roseus*. *Plant Physiol.* **109,** 131–139.

St-Pierre, B., Laflamme, P., Alarco, A.-M., and De Luca, V. (1998). The terminal *O*-acetyltransferase involved in vindoline biosynthesis defines a new class of proteins responsible for coenzyme A-dependent acyl transfer. *Plant J.* **14,** 703–713.

St-Pierre, B., Vazquez-Flota, A., and De Luca, V. (1999). Multicellular compartmentation of *Catharanthus roseus* alkaloid biosynthesis predicts intercellular translocation of a pathway intermediate. *Plant Cell* **11,** 887–900.

Suzuki, K.-I., Yamada, Y., and Hashimoto, T. (1999a). Expression of *Atropa belladonna* putrescine *N*-methyltransferase gene in root pericycle. *Plant Cell Physiol.* **40,** 289–297.

Suzuki, K.-I., Yun, D.-J., Chen, X.-Y., Yamada, Y., and Hashimoto, T. (1999b). An *Atropa belladonna* hyoscyamine 6β-hydoxylase gene is differentially expressed in the root pericycle and anthers. *Plant Mol. Biol.* **40,** 141–152.

Takeshita, N., Fujiwara, H., Mimura, H., Fitchen, J. H., Yamada, Y., and Sato, F. (1995). Molecular cloning and characterization of *S*-adenosyl-L-methionine: Scoulerine 9-*O*-methyltransferase from cultured cells of *Coptis japonica*. *Plant Cell Physiol.* **36,** 29–36.

Thomas, J. C., Adams, D. G., Nessler, C. L., Brown, J. K., and Bohnert, H. J. (1995). Tryptophan decarboxylase, tryptamine, and reproduction of whitefly. *Plant Physiol.* **109,** 717–720.

Treimer, J. F., and Zenk, M. H. (1978). Enzymic synthesis of corynanthe-type alkaloids in cell cultures of *Catharanthus roseus*: Quantitation by radioimmunoassay. *Phytochemistry* **17,** 227–231.

Unterlinner, B., Lenz, R., and Kutchan, T. M. (1999). Molecular cloning and functional expression of codeinone reductase—the penultimate enzyme in morphine biosynthesis in the opium poppy *Papaver somniferum*. *Plant J.* **18,** 465–475.

Van Der Fits, L., and Memelink, J. (2000). ORCA3, a jasmonate-responsive transcriptional regulator of plant primary and secondary metabolism. *Science* **289,** 295–297.

Van Der Fits, L., and Memelink, J. (2001). The jasmonate-inducible AP2/ERF-domain transcription factor ORCA3 activates gene expression via interaction with a jasmonate-responsive promoter element. *Plant J.* **25,** 43–53.

Van Der Fits, L., Hilliou, F., and Memelink, J. (2001). T-DNA activation tagging as a tool to isolate regulators of a metabolic pathway from a genetically non-tractable plant species. *Transgenic Res.* **10,** 513–521.

Vazquez-Flota, F., De Carolis, E., Alarco, A.-M., and De Luca, V. (1997). Molecular cloning and characterization of desacetoxyvindoline-4-hydroxylase, a 2-oxoglutarate-dependent dioxygenase involved in the biosynthesis of vindoline in *Catharanthus roseus* (L.) G. Don. *Plant Mol. Biol.* **34,** 935–948.

Veau, B., Courtois, M., Oudin, A., Chénieux, J.-C., Rideau, M., and Clastre, M. (2000). Cloning and expression of cDNAs encoding two enzymes of the MEP pathway in *Catharanthus roseus. Biochim. Biophys. Acta* **1517,** 159–163.

Wakhlu, A. K., and Bajwa, P. S. (1986). Regeneration of uniform plants from somatic embryos of *Papaver somniferum* (opium poppy). *Phytomorphology* **36,** 101–105.

Waller, G. R., and Dermer, O. C. (1981). Enzymology of alkaloid metabolism in plants and microorganisms. *In* "The Biochemistry of Plants—A Comprehensive Treatise, Volume 7, Secondary Plant Products" (E. E. Conn, ed.), pp. 317–402. Academic Press, New York.

Weid, M., Ziegler, J., and Kutchan, T. M. (2004). The roles of latex and the vascular bundle in morphine biosynthesis in the opium poppy *Papaver somniferum. Proc. Natl. Acad. Sci. USA* **101,** 13957–13962.

Whitmer, S., Canel, C., Hallard, D., Gonçalves, C., and Verpoorte, R. (1998). Influence of precursor availability on alkaloid accumulation by transgenic cell line of *Catharanthus roseus. Plant Physiol.* **116,** 853–857.

Williams, R. D., and Ellis, B. E. (1993). Alkaloids from *Agrobacterium rhizogenes*-transformed *Papaver somniferum* cultures. *Phytochemistry* **32,** 719–723.

Winz, R. A., and Baldwin, I. T. (2001). Molecular interactions between the specialist herbivore *Manduca sexta* (Lepidoptera, Sphingidae) and its natural host *Nicotiana attenuata*. IV. Insect-induced ethylene reduces jasmonate-induced nicotine accumulation by regulating putrescine *N*-methyltransferase transcripts. *Plant Physiol.* **125,** 2189–2202.

Yao, K., De Luca, V., and Brisson, N. (1995). Creation of a metabolic sink for tryptophan alters the phenylpropanoid pathway and the susceptibility of potato to *Phytophthora infestans. Plant Cell* **7,** 1787–1799.

Yoshimatsu, K., and Shimomura, K. (1992). Transformation of opium poppy (*Papaver somniferum* L.) with *Agrobacterium rhizogenes* MAFF 03–01724. *Plant Cell Rep.* **11,** 132–136.

Yun, D.-J., Hashimoto, T., and Yamada, Y. (1992). Metabolic engineering of medicinal plants: *Atropa belladonna* with an improved alkaloid composition. *Proc. Natl. Acad. Sci. USA* **89,** 11799–11803.

Zeef, L. A., Christou, P., and Leech, M. J. (2000). Transformation of the tropane alkaloid-producing medicinal plant *Hyoscyamus muticus* by particle bombardment. *Transgenic Res.* **9,** 163–168.

Zenk, M. H. (1991). Chasing the enzymes of secondary metabolism: Plant cell cultures as a pot of gold. *Phytochemistry* **30,** 3861–3863.

Ziegler, J., Diaz-Chávez, M. L., Kramell, R., Ammer, C., and Kutchan, T. M. (2005). Comparative macroarray analysis of morphine containing *Papaver somniferum* and eight morphine free *Papaver* species identifies an *O*-methyltransferase involved in benzylisoquinoline biosynthesis. *Planta* **222,** 458–471.

CHAPTER 11

Engineering Formation of Medicinal Compounds in Cell Cultures

Fumihiko Sato* and **Yasuyuki Yamada**†

Contents

Abstract

Higher plants are rich sources of medicinal compounds. Many medicinal plants, however, are still harvested in the wild due to technical difficulties of cultivation, as well as for economic reasons. The increased demand and

* Department of Plant Gene and Totipotency, Graduate School of Biostudies, Kyoto University, Kyoto 606-8502, Japan
† Graduate School of Biological Sciences, Nara Institute of Science and Technology, 8916-5 Takayama, Ikoma, Nara 630-0192, Japan

Advances in Plant Biochemistry and Molecular Biology, Volume 1
ISSN 1755-0408, DOI: 10.1016/S1755-0408(07)01011-9

drastic reduction in plant availability increase the pressure to produce medicinal compounds via alternative ways, especially using cell/tissue cultures and transgenic plants. Furthermore, the demands for quality materials have also increased. Before 1970, the reported yields from cell cultures were generally lower than those in plants. However, several cell cultures can have considerable productivity, and in some cases their production exceeds that present in intact plants. The current advancements in understanding and manipulating the molecular and cellular biology of secondary metabolism provide a basis for optimism regarding the commercial production of secondary products in cell/organ cultures and/or transgenic plants. We summarize recent progress in alkaloid research as a principal model and discuss the factors that control productivity and quality, focusing on organ differentiation, genetic instability, rate-limiting enzyme steps in metabolism, transcriptional regulators, transport, and storage.

Key Words: Secondary metabolite, Plant cell culture, Metabolic engineering, RNA interference (RNAi), Expressed sequence tag (EST), Organ differentiation, Compartmentalization, Cell-specific gene expression, Biosynthetic pathway, Isoquinoline alkaloid, Terpenoid indole alkaloid, Tropane alkaloid.

Abbreviations: TYDC, Tyrosine/dopa decarboxylase; NCS, Norcoclaurine synthase; SAM, *S*-adenosyl methionine; 6OMT, Norcoclaurine 6-*O*-methyltransferase; 4′OMT, 3′ hydroxy *N*-methylcoclaurine 4′-*O*-methyltransferase; BBE, Berberine bridge enzyme; CDS, Canadine synthase; SOMT, Scoulerine 9-*O*-methyltransferase; SAT, Acetylcoenzyme A:salutaridinol-7-*O*-acetyltransferase; COR, Codeinone reductase; STR, Strictosidine synthase; TDC, Tryptophan decarboxylase; G10H, Geraniol 10-hydroxylase; CYP72A1, Secologanin synthase; SGD, Strictosidine glucosidase; T16H, Tabersonine 16-hydroxylase; PCR, Polymerase chain reaction; CPR, Cytochrome *P450* reductase; D4H, Desacetoxyvindoline 4-hydroxylase; DAT, Acetylcoenzyme A: deacetylvindoline 4-*O*-acetyltransferase; ODC, Ornithine decarboxylase; ADC, Arginine decarboxylase; PMT, Putrescine *N*-methyltransferase; GUS, β-glucuronidase; TR-I/II, Tropinone reductase I/II; H6H, Hyoscyamine 6β-hydroxylase; MeJA, Methyl jasmonate; SA, Salicylic acid; EST, expressed sequence tag; AFLP, amplified fragment length polymorphism.

1. INTRODUCTION

Higher plants produce a wide range of chemicals. More than 25,000 terpenoids, about 12,000 alkaloids, and 8000 phenolic substances have been identified thus far (Croteau *et al.*, 2000). These chemicals serve in a variety of functions in plants. They defend against herbivores and pathogens, aid in interplant competition, attract beneficial organisms such as pollinators, and have protective effects with regard to abiotic stresses such as UV exposure, temperature changes, water status, and mineral nutrients. In addition, many secondary metabolites produced in plants are used by humans as spices, dyes, fragrances, flavoring agents, or pharmaceuticals. Many of these chemicals also promote human health and enrich our lives in many different ways.

The use of plant metabolites as natural medicines has a long history that can be traced back more than 3500 years, when Egyptian, Sumerian, Ayurvedic, and Chinese medicines were developed (Askin *et al.*, 2002). Morphine was isolated as the principal active ingredient in opium just 200 years ago (1806), and the pain-killing and fever-reducing aspirin (acetyl salicylic acid) became available via chemical synthesis, instead of from willow bark extract, only about 100 years ago (1897). Following rapid progress in chemical synthesis, the microbial production of antibiotics, as well as biotransformation approaches, has increased the supply of modern medicines. Today, however, a large proportion (more than 25%) of products beneficial to humans are still derived from natural sources, especially from plants (Askin *et al.*, 2002; Briskin, 2000; Fig. 11.1).

Many medicinal plants are still harvested in the wild due to the technical difficulties of cultivation, as well as for economic reasons. This harvesting of medicinal plants along with human disturbance of the natural environment increasingly raises concerns about diminishing biodiversity. For example, *Taxus brevifolia* (yew), which is used for the production of paclitaxel (Taxol^{TM}), a potent antineoplastic agent, is an endangered species on the west coast of the United States. The increased demand and drastic reduction in plant availability increase the pressure to produce medicinal compounds in alternative ways, especially via cell/tissue cultures and transgenic plants since plant cells have a high potential for totipotency. Furthermore, the demand for quality materials has also increased

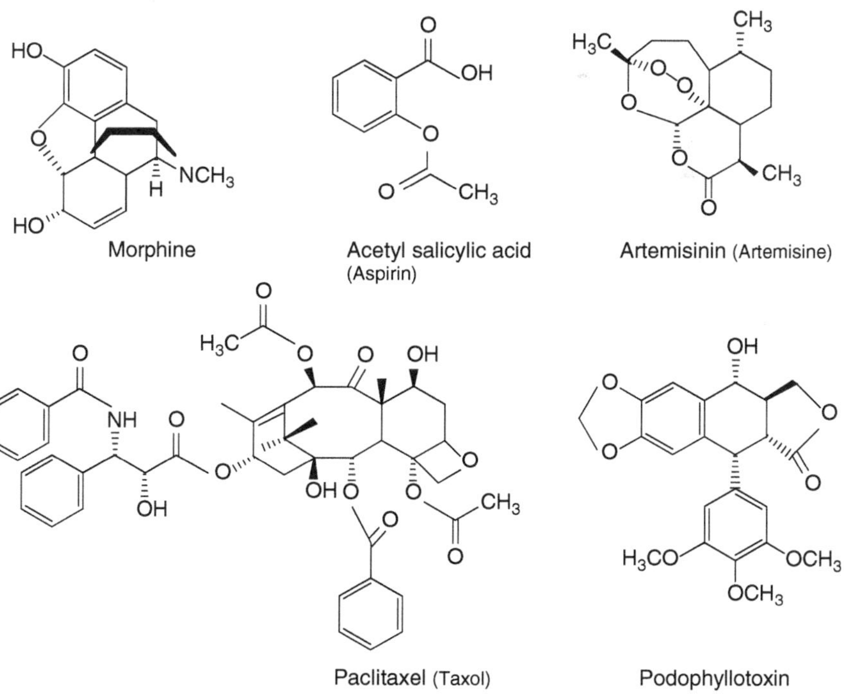

FIGURE 11.1 Chemical structures of some important medicinal compounds.

since variations in medicinal plant quality and incorrect plant identification occasionally cause tragic consequences. Thus, the application of biotechnological approaches to produce secondary metabolites is an attractive alternative for their production, particularly in transgenic cell cultures.

In vitro cell culture systems have several advantages, whose benefits have been discussed in depth previously (Dougall, 1981). In summary, the desired metabolites are produced in a controlled environment, independent of climatic changes and soil conditions, and free of microbes and insects. In addition, materials of definitive quality can be produced, and whose productivity can be improved through cell selection and optimization of culture conditions. The established cell culture systems are also useful for understanding the synthesis of natural products since the results obtained are generally uniform and reproducible. The products are obtained under controlled conditions and are available in large quantities from cells that are less structurally organized than those of higher plants. On the other hand, cell culture systems also have several disadvantages; for example, the labor required to maintain cell culture systems by serial passages, and the possibility that cell properties may change during serial cultures over long periods of time or that cultures may be lost due to contamination.

Before 1970, the reported yields from cell cultures were generally lower than those in plants. However, as shown in Table 11.1, several cell culture systems can have considerable productivity, and in some cases their production exceeds than that found in the intact plant. Also some plant cell cultures still produce little, if any, of the desired compounds. These facts highlight the difficulties faced in producing useful metabolites in economically viable amounts. Accordingly, the current advancements in the molecular and cellular biology of secondary metabolism provide a basis for optimism regarding the commercial production of secondary products in cell/organ cultures and/or transgenic plants (Facchini and St-Pierre, 2005; Kutchan, 2005a,b; Zhao *et al.*, 2005).

Since the biochemistry and cell biology of secondary metabolism represent the foundation of attempts directed toward biotechnological improvement, we first review studies that focus on alkaloids for illustrative purposes only. Simplest applications are biochemical conversions of chemicals that are readily available using isolated native, or recombinant, enzymes and cells as biocatalysts. More complicated metabolic engineering attempts require additional biochemical and cell biological information to optimize the conditions for production. The first section thus addresses the biochemistry and cell biology and then is directly related to the application of metabolic engineering.

2. BIOCHEMISTRY AND CELL BIOLOGY OF SECONDARY METABOLITES

Many secondary metabolites in plants promote human health not only as pharmaceuticals but also as dietary supplements and functional foods (Askin *et al.*, 2002; Briskin, 2000). Plant-produced recombinant proteins are also important as botanical therapeutics (Askin *et al.*, 2002). As suggested by their diverse functions,

TABLE 11.1 Compounds produced commercially and potential candidates

Products	Plant species	Contents in original plants	Contents in cells	Remarks
Berberine	*Coptis japonica* L.	2–4% DW	10% DW	Cellular selection, optimization of medium and high-density culture, berberine 3.5 g/liter/14 days (Sato and Yamada, 1984; Matsubara *et al.*, 1989).
Ginsenosides	*Panax ginseng* C. A. Meyer	4.5% DW	27% DW	Selection and optimization of culture (cited from Misawa, 1994).
Paclitaxel	*Taxus canadensis*	0.015% DW	23.4 mg/liter/day	Elicitation (non-elicitation; 23 mg/liter/28 days) (Ketchum *et al.*, 2003).
Polysaccharide	*Polianthes tuberosa* L.	n.d.	3.5 g/liter/month	Production in intact plants has not been investigated. Selection and optimization of culture (Honda *et al.*, 1996).
Scopolamine	*Duboisia myoporoides*	0.15% DW	3.2% DW	Selection, optimization of culture, and high-density culture (120 g DW/liter) of root with two-stage culture method, 2.5 g/liter/21 days (Yukimune *et al.*, 1994).
Shikonin	*Lithospermum erythrorizon*	1.5% DW	20% DW	Selection and optimization of two-stage culture, shikonin 3.6 g/g DW inoculum/23 days (Fujita *et al.*, 1986).

DW: dry weight.

secondary metabolites are structurally divergent (see Croteau *et al.*, 2000 or other chapters) and are usually classified into several major groups. Terpenoids and steroids, which are formed from isoprenoid moieties, include the antineoplastic diterpene paclitaxel, from *T. brevifolia* Nutt; the antimalarial sesquiterpene artemisinin, isolated from *Artemisia annua* L.; diosgenin, a precursor for oral contraceptives and hormonal drugs, from *Dioscorea* spp.; and the cardiotonic steroidal glycoside digoxin, from *Digitalis* spp. Phenylpropanoids, derived from the shikimate pathway, include the antineoplastic lignan, podophyllotoxin, from *Podophyllum peltatum* L. Alkaloids, which are a diverse family of alkaline nitrogen-containing compounds, include many biologically active chemicals such as caffeine, coniine, morphine, nicotine, and strychnine. We focus here on the alkaloids, justified by their relatively high beneficial biological activities, and especially the isoquinoline alkaloids, the terpenoid indole alkaloids, and the tropane alkaloids, which are currently the subjects of intensive study (De Luca and Laflamme, 2001; Facchini, 2001; Facchini and St-Pierre, 2005; Hashimoto and Yamada, 2003; Kutchan, 2005a,b; Verpoorte and Memelink, 2002; Zhao *et al.*, 2005). Excellent reviews of the potential of lignans in cell cultures (Petersen and Alfermann, 2001), as well as monoterpenes (Lange and Croteau, 1999) and taxol (Jennewein and Croteau, 2001), have already been published.

2.1. Isoquinoline alkaloid biosynthesis

Isoquinoline alkaloids are a large and diverse group of alkaloids with ~2500 defined structures. They include the analgesic morphine from *Papaver somniferum* L., the antigout colchicine from *Colchicum autumnale* L., the emetic and antiamoebic emetine from *Cephaelis ipecacuanha* (Brot.) A. Rich., the skeletal muscle relaxant tubocurarine from *Strychonos toxifera* Bentham, and the antimicrobial compounds berberine and sanguinarine from divergent plant species including *Berberis* spp. and *Sanguinaria* spp., many of which are used as pharmaceuticals.

Isoquinoline alkaloid biosynthesis begins with the conversion of tyrosine to both dopamine and 4-hydroxyphenylacetaldehyde by decarboxylation, *ortho*-hydroxylation, and deamination (Fig. 11.2; Facchini, 2001). Among these early steps, only tyrosine/dopa decarboxylase (TYDC; an aromatic L-amino acid decarboxylase), which converts tyrosine and dopa to their corresponding amines, has been purified and characterized. This small family of genes (~15 genes) was isolated from opium poppy (*P. somniferum*) and each subfamily has been shown to have distinct developmental and inducible expression patterns (Facchini and De Luca, 1994, 1995; Park *et al.*, 1999). Members of the TYDC gene family are classified into two groups (TYDC1 and TYDC2) that are differentially expressed in opium poppy. In the mature plant, TYDC2-like transcripts are predominant in the stem and are also present in roots, whereas TYDC1-like transcripts are abundant only in roots. The localization of TYDC transcripts in the phloem is consistent with the expected developmental origin of laticifers, which are specialized internal secretory cells that accompany vascular tissues in all organs of select species and contain alkaloid-rich latex in aerial organs (Facchini and De Luca, 1995).

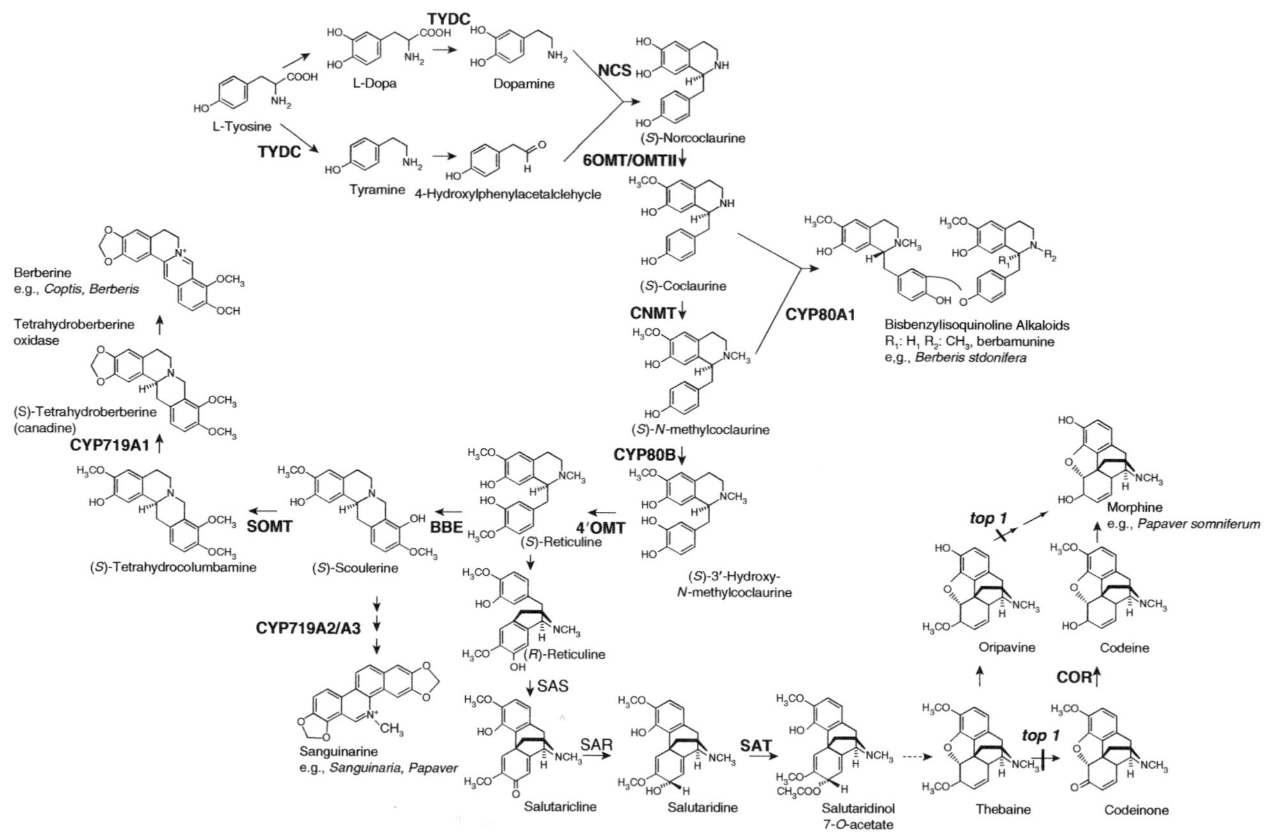

FIGURE 11.2 Biosynthetic pathways to various isoquinoline alkaloids. Unbroken arrows indicate single enzymatic conversions and broken arrows indicate multiple enzymatic steps. Enzymes for which the corresponding genes have been cloned are indicated in bold. TYDC, tyrosine/dopa decarboxylase; NCS, norcoclaurine synthase; 6OMT, norcoclaurine 6-*O*-methyltransferase; CNMT, coclaurine *N*-methyltransferase; CYP80B1, *N*-methylcoclaurine 3′-hydroxylase; 4′OMT, 3′ hydroxy *N*-methylcoclaurine 4′-*O*-methyltransferase; BBE, berberine bridge enzyme; CYP719A1, canadine synthase (methylenedioxy bridge-forming enzyme); SOMT, scoulerine 9-*O*-methyltransferase; SAS, salutaridine synthase; SAR, salutaridine reductase; SAT, acetylcoenzyme A:salutaridinol-7-*O*-acetyltransferase; COR, codeinone reductase; CYP80A1, berbamunine synthase.

Dopamine and 4-hydroxyphenylacetaldehyde are condensed by norcoclaurine synthase (NCS) to yield (*S*)-norcoclaurine, which is the central precursor to all isoquinoline alkaloids. Recently, NCS has been purified and characterized (Samanani and Facchini, 2002) from cultured *Thalictrum flavum* spp., and a TfNCS cDNA belonging to PR10 family was isolated from *T. flavum* (Samanani *et al.*, 2004), whereas a novel dioxygenase-like protein (CjNCS) from cultured *Coptis japonica* cells was also shown to catalyze this NCS reaction (Minami *et al.*, 2007). The presence of TYDC (Facchini, 2001) and TfNCS (Samanani *et al.*, 2004) and CjNCS homologues (Minami *et al.*, 2007) in *Arabidopsis* or rice suggests that these genes either have other basic biological roles or that the isoquinoline biosynthesis pathway is relatively universal in the plant kingdom, although sequence homology of TfNCS or CjNCS with *Arabidopsis* or rice homologues were relatively low (less than 20% in amino acid basis) (Liscombe *et al.*, 2005; Minami *et al.*, 2007) and no isoquinoline alkaloid has been found in *Arabidopsis* or rice. Strictosidine synthase (STR) (the key reaction in terpenoid indole alkaloid)-like genes has also been found in animals and *Arabidopsis* (De Luca and Laflamme, 2001).

(*S*)-Norcoclaurine is sequentially converted to coclaurine by *S*-adenosyl methionine (SAM)-dependent norcoclaurine 6-*O*-methyltransferase (6OMT) (Morishige *et al.*, 2000), to *N*-methylcoclaurine by coclaurine *N*-methyltransferase (Choi *et al.*, 2002), to 3′-hydroxy-*N*-methyl coclaurine by P450 hydroxylase (Pauli and Kutchan, 1998), and then to (*S*)-reticuline by 3′-hydroxy *N*-methylcoclaurine 4′-*O*-methyltransferase (4′OMT; see Fig. 11.2; Morishige *et al.*, 2000). All of the cDNAs for these reactions have been isolated and functional recombinant proteins subsequently produced. Detailed biochemical studies using recombinant enzymes have shown their strict reaction specificities, and these enzymes regulate biosynthesis sequentially and in a coordinated manner. For example, CNMT prefers coclaurine than 6-*O*-methylnorlaudanosoline and 4′OMT prefers an *N*-methylated substrate, which suggest that the pathway in Fig. 11.2 is preferable to a sequence of *N*-methylation, hydroxylation, and 4′-*O*-methylation. On the other hand, *Thalictrum* cells may show some variation since *Thalictrum* *O*-methyltransferases can form heterodimers and exhibit broad substrate specificity (Frick and Kutchan, 1999). Current data also indicate that all of these enzymes, except the membrane-bound P450 CYP80B1, are located in the cytosol.

While dimeric bisbenzylisoquinoline alkaloids, such as berbamunine and tubocurarine, are produced from the intermediates of the (*S*)-reticuline pathway by the action of a phenol coupling P450-dependent oxidase (berbamunine synthase, CYP80A1) (Kraus and Kutchan, 1995), reticuline is the central intermediate in branch pathways that lead to benzophenanthridine alkaloids (e.g., sanguinarine and marcarpine), protoberberine alkaloids (e.g., berberine and palmatine), and morphinan alkaloids (e.g., morphine and codeine) (Fig. 11.2). Many of the enzymes involved in these branch pathways have been purified and the corresponding cDNAs have been cloned.

The first committed step in the biosynthesis of benzophenanthridine, a protoberberine alkaloid, involves conversion of the *N*-methyl group of (*S*)-reticuline into the methylene bridge moiety of (*S*)-scoulerine by the berberine bridge enzyme (BBE)

(Dittrich and Kutchan, 1991). This unique enzyme is soluble but localized in vesicles (Bock *et al.*, 2002). Immunocytological staining of *P. somniferum* tissue with antibodies against BBE led to a characteristic labeling of electron-dense aggregates in idioblasts that are not connected to the laticifer system, which demonstrates that benzophenanthridine and morphine biosyntheses show strict cytological separation within this plant (Bock *et al.*, 2002).

In benzophenanthridine alkaloid biosynthesis, (*S*)-scoulerine can be converted to (*S*)-stylopine by two P450-dependent oxidases, (*S*)-cheilanthifoline and (*S*)-stylopine synthase, which result in the formation of two methylenedioxy groups (not shown) (see Facchini, 2001; Ikezawa *et al.*, 2007). On the other hand, in protoberberine biosynthesis, (*S*)-scoulerine is converted to (*S*) tetrahydrocolumbamine by the SAM-dependent scoulerine 9-*O*-methyltransferase (SOMT) (Takeshita *et al.*, 1995) and then to tetrahydroberberine (canadine) by a P450-dependent canadine synthase (CDS or CYP719A1) (Ikezawa *et al.*, 2003). The isolation and characterization of these enzymes and the corresponding cDNAs have confirmed that berberine biosynthesis proceeds via canadine and not via columbamine. Again, the enzyme substrate specificity shows a clear preference for this pathway. While the hydrophobic N-terminal region of SOMT suggests that this enzyme may be targeted to the membrane fraction, its localization in both the cytosol (Muemmler *et al.*, 1985) and within the lumen of alkaloid-specific vesicles (Galneder *et al.*, 1988) has been reported. Note that the CYP719A1 family as well as CYP80 were not found in *Arabidopsis* and rice and that these members of the cytochrome *P450* superfamily are unique for benzylisoquinoline alkaloid biosynthesis (Nelson *et al.*, 2004).

In morphinan alkaloid biosynthesis, (*S*)-reticuline is converted to its (*R*)-enantiomer via the stereospecific reduction of 1,2-dehydroreticuline with NADPH-dependent cytosolic 1,2-dehydroreticuline reductase. Subsequent intramolecular carbon–carbon phenol coupling of (*R*)-reticuline by a P450-dependent salutaridine synthase results in the formation of salutaridine. Salutaridine: NADPH 7-oxidoreductase then reduces salutaridine to (7*S*)-salutaridinol. Transformation of salutaridinol into thebaine involves the closure of an oxide bridge between C-4 and C-5 by acetylcoenzyme A:salutaridinol-7-*O*-acetyltransferase (SAT) (Grothe *et al.*, 2001). Furthermore, thebaine can be converted to codeinone and then reduced to codeine by cytosolic NADPH-dependent codeinone reductase (COR) (Unterlinner *et al.*, 1999). Finally, codeine is demethylated to give morphine. Interestingly, COR genes have been found in some *Papaver* spp. that do not produce morphine (Unterlinner *et al.*, 1999), whereas SAT transcript was detected in *Papaver* spp. that accumulate alkaloids with a morphinan nucleus, consistent with the expected distribution (Grothe *et al.*, 2001; Unterlinner *et al.*, 1999). The recent isolation of the *top1* mutant from poppy, and the demonstration of the activity of the protein, has illustrated that thebaine can be demethylated in two steps either through codeinone or oripavine to morphine (Millgate *et al.*, 2004).

Northern blot analysis using the eight available genes in morphinan alkaloid biosynthesis showed that while all of the transcripts are detected in every organ, the highest levels are seen in stems and flower buds and the lowest levels are seen in leaves (Millgate *et al.*, 2004; Unterlinner *et al.*, 1999). The accumulation of each transcript, with the exception of COR, was markedly induced in response to treatment with an elicitor or wounding of cultured cells. All known enzymes in the morphine pathway have been detected in cultured cells derived from the fruit capsule (Facchini and Park, 2003; Grothe *et al.*, 2001; Huang and Kutchan, 2000; Unterlinner *et al.*, 1999). Conversely, different cell type-specific localizations of biosynthetic enzymes have been reported in the capsule and stem of intact opium poppy plants. *In situ* localization of alkaloid biosynthetic gene transcripts indicated seven biosynthetic enzymes: 6OMT, CNMT, CYP80B, 4′OMT and BBE involved in reticuline biosynthesis, and SAT and COR in morphine pathway. These proteins have apparently been localized to sieve elements in opium poppy and the corresponding gene transcripts to adjacent phloem companion cells (Bird *et al.*, 2003; Facchini and St-Pierre, 2005). In contrast, a different immunocytochemical analysis clearly showed that 4′OMT and SAT were located in phloem parenchyma cells in vascular bundles. COR, catalyzing the penultimate step in morphine biosynthesis, was localized to laticifers, the site of morphinan alkaloid accumulation (Weid *et al.*, 2004). Although this discrepancy in the cell type-specific localization of enzymes remains to be clarified, it is noteworthy that both studies showed different localizations of the biosynthetic gene transcripts and their corresponding enzymes.

Cell type-specific expression has recently been reported in protoberberine alkaloid biosynthesis in *T. flavum* subsp. (Samanani *et al.*, 2005). While gene transcripts for biosynthetic enzyme were most abundant in rhizomes, they were also detected at lower levels in roots and other organs. Further *in situ* RNA hybridization analysis revealed that all transcripts were mainly localized to immature endodermis cells, the pericycle of roots, and restricted to the protoderm of leaf primordia in rhizomes. These data and analysis of alkaloid accumulation clearly indicated that distinct and different cell types are involved in the biosynthesis and accumulation of benzylisoquinoline alkaloids in *T. flavum* and *P. somniferum*.

2.2. Terpenoid indole alkaloid biosynthesis

Terpenoid indole alkaloids consist of about 3000 compounds, including the antineoplastic vinblastine from Madagascar periwinkle (*Catharanthus roseus*), camptothecin from *Camptotheca acuminata*, and the antimalarial quinine from *Cinchona* spp. The central intermediate in the biosynthesis of terpenoid indole alkaloids is strictosidine, which is produced from tryptamine and the iridoid glucoside secologanin by STR (Fig. 11.3). Tryptamine is produced by tryptophan decarboxylase (TDC). While a single gene in *C. roseus* responds in both developmental and inducible expression (De Luca *et al.*, 1989; Goddijn *et al.*, 1992), two genes in *C. acuminata* showed different expression profiles, suggesting that one is involved in developmentally controlled camptothecin production in shoot apex and bark, while the other is involved in an inducible defense mechanism (Lopez-Meyer and Nessler, 1997).

Two steps in secologanin biosynthesis are catalyzed by P450-dependent enzymes: geraniol 10-hydroxylase (G10H) converts geraniol to 10-hydroxygeraniol (Collu *et al.*, 2001), while secologanin synthase (CYP72A1) converts loganin to secologanin and shows epidermis-specific expression in immature leaves of *C. roseus* (Irmler *et al.*, 2000). The supply of terpenoid precursors should be rate-limiting in

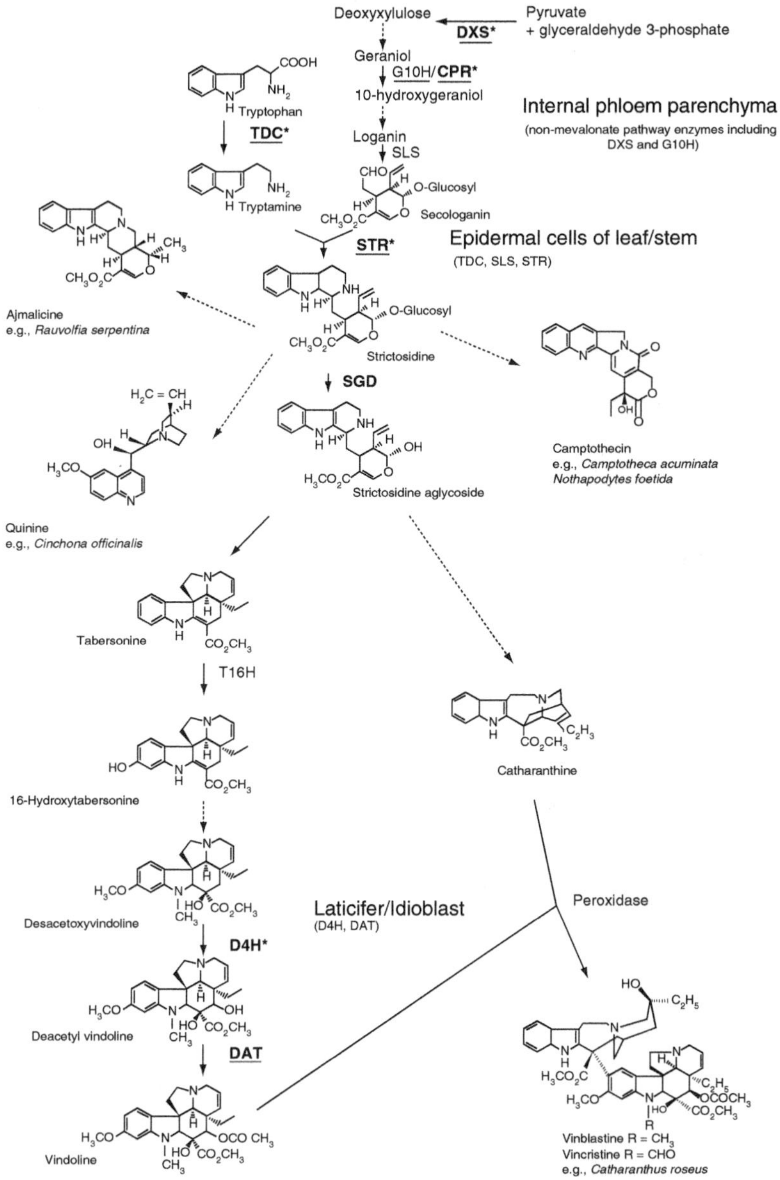

FIGURE 11.3 Biosynthetic pathways to various terpenoid indole alkaloids. Unbroken arrows indicate single enzymatic conversions and dotted arrows indicate multiple enzymatic steps. Enzymes for which the corresponding genes have been cloned are indicated. JA-inducible genes are indicated in bold. Underlining indicates that the corresponding genes have been tested for being an ORCA3 target gene in *C. roseus* cells. Enzyme-encoding genes regulated by ORCA3 are asterisked (Vazquez-Flota *et al.*, 2000). DXS, D-1-deoxyxylulose 5-phosphate synthase; STR, strictosidine synthase; TDC, tryptophan decarboxylase; G10H, geraniol 10-hydroxylase; CYP72A1, secologanin synthase; SGD, strictosidine glucosidase; T16H, tabersonine 16-hydroxylase; CPR, cytochrome *P450* reductase; D4H, desacetoxyvindoline 4-hydroxylase; DAT, acetylcoenzyme A: deacetylvindoline 4-*O*-acetyltransferase.

terpenoid indole alkaloid biosynthesis. The addition of secologanin or loganin to *C. roseus* cell culture increases alkaloid accumulation (Whitmer *et al.*, 1998), and the level of G10H activity is also positively correlated with the accumulation of alkaloids (Facchini, 2001). However, secologanin is inefficiently used in strictosidine synthesis when added to the medium since exogenous secologanin appears to be compartmentalized differently from endogenous secologanin. This result also suggests that the proper subcellular localization of biosynthetic enzymes and substrates is important for the efficient biosynthesis of metabolites. These isoprenoid precursors are also known to be derived from a nonmevalonate pathway (Contin *et al.*, 1998).

STR is a key enzyme in terpenoid indole alkaloid biosynthesis and cDNAs have been isolated from *Rauvolfia serpentina* (Kutchan *et al.*, 1988) and *C. roseus* (Mcknight *et al.*, 1990). STR is one of the most investigated biosynthetic genes in secondary metabolism. Strictosidine is deglucosylated by strictosidine glucosidase (SGD) (Geerlings *et al.*, 2000) and then converted via several unstable intermediates. While there is limited information available on the pathway to catharanthine, vindoline biosynthesis has been relatively well characterized, although the production of vindoline in cultured cells is limited.

The first of six steps in the conversion of tabersonine to vindoline consists of hydroxylation at the C-16 position by tabersonine 16-hydroxylase (T16H), a P450-dependent monooxygenase. While several P450 sequences were amplified by polymerase chain reaction (PCR), the active principle of CYP71D12 was finally identified to be T16H using translationally fused protein expressed in *Escherichia coli* (Schroeder *et al.*, 1999). Interestingly, while *C. roseus* has a single copy of the TDC, STR, and cytochrome *P450* reductase (CPR) genes, it has at least two T16H genes. The 16-hydroxylation of tabersonine is followed by 16-*O*-methylation by a cytosolic SAM: 16 hydroxyltabersonine *O*-methyltransferase (St-Pierre and De Luca, 1995), hydration of the 2,3-double bond by an as yet uncharacterized enzyme, and *N*-methylation of the indole-ring nitrogen by a thylakoid-associated SAM: 2,3-dihydro-3-hydroxytabersonine-*N*-methyltransferase.

The penultimate step in vindoline biosynthesis is catalyzed by a cytosolic 2-oxoglutarate-dependent dioxygenase that hydroxylates the C-4 position of desacetoxyvindoline 4-hydroxylase (D4H) (Vazquez-Flota *et al.*, 1997), and the final step is catalyzed by the cytosolic acetylcoenzyme A: deacetylvindoline 4-*O*-acetyltransferase (DAT) (St-Pierre *et al.*, 1998). The expression of T16H, D4H, and DAT in developing *C. roseus* seedlings is light regulated. Although D4H and DAT activities are detected exclusively under conditions that result in vindoline biosynthesis, T16H is expressed at low levels in *C. roseus* cell cultures that do not accumulate vindoline (St-Pierre and De Luca, 1995). The expression of D4H appears to be under complex, multilevel developmental and light regulation.

A series of experiments with leaves of *C. roseus* (Burlat *et al.*, 2004; St-Pierre *et al.*, 1999) showed that at least three cell types are involved in vindoline biosynthesis. The nonmevalonate pathway genes (DXS, 1-deoxy-D-xylulose 5-phosphate synthase, 1-deoxy-D-xylulose 5-phosphate reductoisomerase, and 2C-methyl-D-erythriotol 2,4-cyclodiphosphate synthase) as well as G10H were found to be expressed in internal phloem parenchyma of the young aerial organs

(Burlat *et al.*, 2004). Other early stage enzymes in the biosynthesis of strictosidine, such as TDC, SLS, and STR, were expressed specifically in the upper and lower epidermis of young leaves, stem, and flower buds. Late-stage enzymes in vindoline biosynthesis, such as D4H and DAT, were localized in laticifer and idioblast cells, which show greater yellow autofluorescence with few chloroplasts, compared to the surrounding red-autofluorescent mesophyll cells. Light, which is not required for the formation of these cell types, is required for activation of the localized expression of the late stages of vindoline biosynthesis (Vazquez-Flota *et al.*, 2000).

Vindoline biosynthesis is restricted to the aboveground organs, and the pathway beyond tabersonine is not expressed in tissue cultures (Vazquez-Flota *et al.*, 2002), whereas catharanthine accumulates in cultured cells as well as etiolated seedlings. These facts, along with the recovery of vindoline biosynthesis in regenerated shoots, suggest that the biosynthesis of catharanthine and vindoline is differentially regulated and that vindoline biosynthesis is under more rigid tissue-, developmental-, and environmental-specific control than that of catharanthine (St-Pierre *et al.*, 1999). These results raise the possibility that these cell cultures lack the cell types required to accommodate the late stages of vindoline biosynthesis.

Until recently, characterization of terpenoid indole alkaloid biosynthesis has mainly been carried out with *C. roseus*, but the recent establishment of hairy root cultures of *Ophiorrhiza pumila* (Rubiaceae) that showed high camptothecin production provided another useful experimental system. Computer-aided atomic reconstruction of metabolism and tracer experiments with [1–^{13}C] glucose indicated that camptothecin is formed by the combined activities of the 2C-methyl-D-erythritol 4-phosphate pathway and the shikimate pathway (Yamazaki *et al.*, 2004).

2.3. Tropane alkaloid and nicotine biosynthesis

Tropane alkaloids are mainly found in the Solanaceae and include the anticholinergic drugs atropine, hyoscyamine, and scopolamine and the narcotic cocaine. *N*-methylputrescine, the central precursor in tropane alkaloid biosynthesis, is also an intermediate in the nicotine pathway. *N*-methylputrescine is produced by the decarboxylation of ornithine or arginine by ornithine decarboxylase (ODC) or arginine decarboxylase (ADC), respectively. Tropane alkaloids and nicotine biosynthesis are also closely related to polyamine metabolism (Fig. 11.4). The first committed step in tropane/nicotine alkaloid biosynthesis is catalyzed by the SAM-dependent putrescine *N*-methyltransferase (PMT) (Hibi *et al.*, 1994), which is highly homologous to spermidine synthase. Methylputrescine is subsequently deaminated by a diamine oxidase, and spontaneous cyclization then forms the reactive *N*-methyl-Δ^1-pyrrolinium cation. The latter is thought to provide a precursor of the tropane ring or nicotinic acid to form nicotine, although details are not available. PMT in nicotine biosynthesis is expressed specifically in the cortex and endodermis of tobacco root tips, whereas strong expression is seen in the xylem parenchyma and outer cortex cells in more differentiated parts of the root (Hibi *et al.*, 1994).

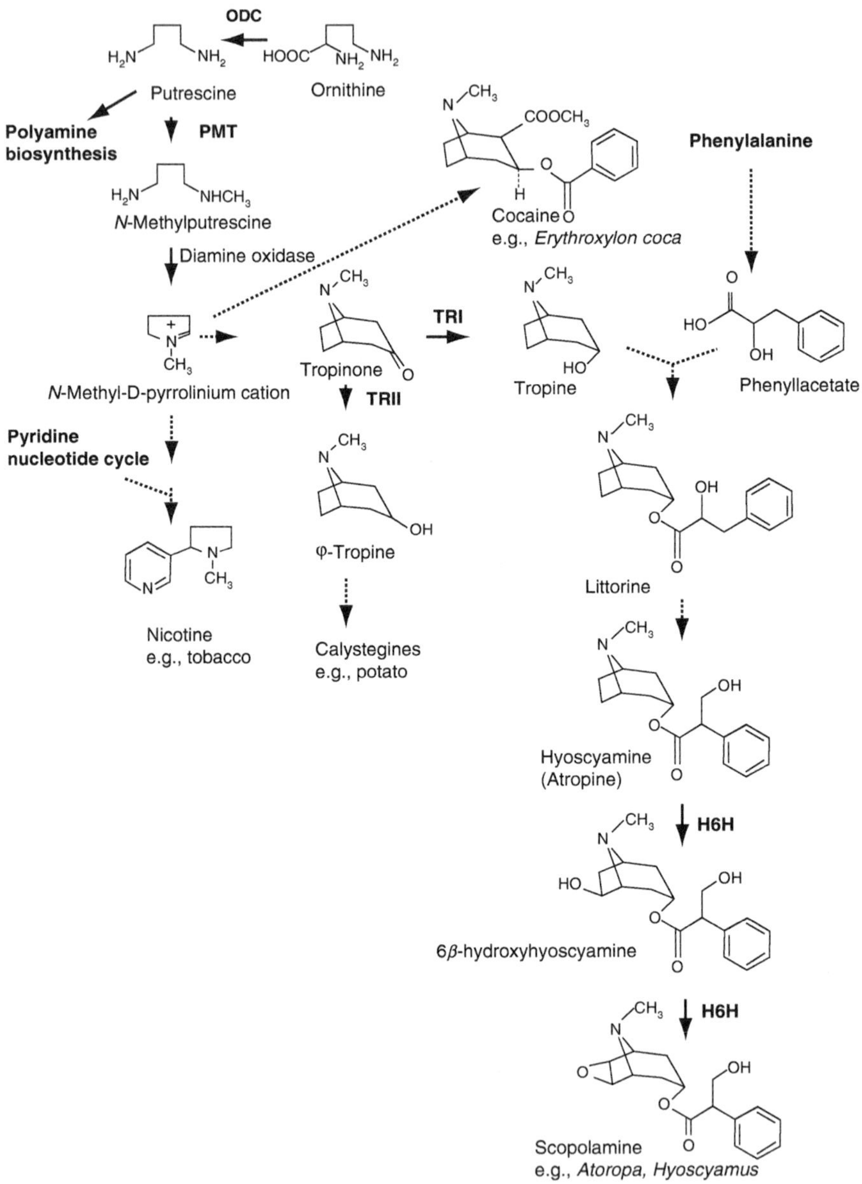

FIGURE 11.4 Biosynthetic pathways to tropane alkaloids, related compounds, and nicotine. Unbroken arrows indicate single enzymatic conversions and broken arrows indicate multiple enzymatic steps. Enzymes for which the corresponding genes have been cloned are indicated in bold. ODC, ornithine decarboxylase; PMT, putrescine *N*-methyltransferase; TR-I/II, tropinone reductase I/II; H6H, hyoscyamine 6β-hydroxylase.

PMT genes have also been isolated from tropane alkaloid-producing *Hyoscymus niger* (*HnPMT*) and *Atropa belladonna* (*AbPMT*). PMT promoter and β-glucuronidase (GUS) fusion gene showed that AbPMT is expressed specifically in root pericycle cells (Suzuki *et al.*, 1999a). While tropane alkaloids and nicotine are mainly synthesized in the root and transported to aerial parts where they accumulate in vacuoles to high levels, the biosynthesis of these alkaloids might nevertheless be differentially regulated (see below).

Tropinone is located at a branch point in tropane alkaloid synthesis. Two related dehydrogenases, tropinone reductase I (TR-I) and tropinone reductase II (TR-II), stereospecifically reduce the 3-keto group of tropinone to the 3α- and 3β-groups of tropine and ψ-tropine, respectively. cDNA clones for TR-I and TR-II have been isolated from *Datura stramonium* (Nakajima *et al.*, 1993). A further analysis of their localization suggested that TR-I and TR-II were localized differently and might have different functions (Nakajima and Hashimoto, 1999). Nortropane polyhydroxylated alkaloids, calistegines, are also assumed to originate from ψ-tropine. They have been isolated from different species in the Solanaceae (Scholl *et al.*, 2003, and references cited therein). Calistegines show glycosidase-inhibiting activities and are considered nutritional mediators in the rhizosphere (Tepfer *et al.*, 1988).

Hyoscyamine is produced by condensation of tropine and the phenylalanine-derived intermediate (*R*)-phenyllactate. Hyoscyamine can be converted to its epoxide scopolamine via 6β-hydroxylhyoscyamine by a 2-oxoglutarate-dependent dioxygenase, hyoscyamine 6β-hydroxylase (H6H) (Matsuda *et al.*, 1991). H6H localizes in the pericycle in branch roots of several scopolamine-producing Solanaceae plants (Hashimoto *et al.*, 1991). Histochemical analysis using *H. niger* and *A. belladonna* H6H promoter::GUS fusion gene also showed that cell-specific expression of the H6H gene is controlled by (unknown) genetic regulation specific to scopolamine-producing plants but is absent in tobacco that does not produce scopolamine (Kanegae *et al.*, 1994; Suzuki *et al.*, 1999b).

In *Nicotiana sylvestris*, a set of nicotine biosynthesis genes was activated by the exogenous application of methyl jasmonate (MeJA), but this activation was effectively suppressed by simultaneous treatment with ethylene (Shoji *et al.*, 2000), even though ethylene and JA are generally considered to act synergistically. In contrast, treatment of *A. belladonna* roots with MeJA did not lead to upregulated expression of AbPMT genes (Suzuki *et al.*, 1999a). The different responses of tropane alkaloids and nicotine biosynthesis to JA and ethylene suggest that these biosyntheses might be under the control of different genetic regulation systems (Shoji *et al.*, 2000).

3. CELL CULTURE AND METABOLITE PRODUCTION

When we cultivate plant cells *in vitro*, there are different types of cultures with distinct metabolite productivities. One type produces metabolites in undifferentiated cells cultured *in vitro*, while another produces metabolites only under differentiated conditions, for example, shoot or root cultures. In some cases, cells lose their biosynthetic potential even after redifferentiation, that is, showing the complete loss of their natural biosynthetic potential.

The first type of cell culture is in high demand because cells are easily cultivated under *in vitro* conditions. Plant cells often do not produce desired secondary metabolites, although they have the potential to regenerate whole plants from single cells. For example, plant growth regulators, such as auxins and cytokinins, can regulate morphological differentiation, but the chemical regulation of functional differentiation in secondary metabolism without morphological differentiation is rather limited. For example, morphinan alkaloids are not produced in cell cultures of *P. somniferum* without organ differentiation (Facchini and Park, 2003; Grothe *et al.*, 2001; Huang and Kutchan, 2000; Unterlinner *et al.*, 1999), whereas several cell cultures, for example, *P. somniferum* and *C. japonica*, produce large quantities of structurally related isoquinoline alkaloids, sanguinarine and berberine, respectively (Facchini and Park, 2003; Huang and Kutchan, 2000; Sato and Yamada, 1984). Interestingly, morphine, sanguinarine, and berberine are derived from tyrosine through the same intermediate, reticuline (Fig. 11.2), indicating that early steps in metabolic pathway do not determine the end-product in cell culture.

Biochemical, molecular biological, and cell biological studies of biosynthetic enzymes have gradually revealed the mechanisms of regulation. For example, all enzymes examined were highly expressed in cultured *C. japonica* cells, which show a high production of berberine (Ikezawa *et al.*, 2003; Sato *et al.*, unpublished data), and in *P. somniferum* cells, which do not produce morphinan alkaloids under undifferentiated conditions (Facchini and Park, 2003; Grothe *et al.*, 2001; Huang and Kutchan, 2000; Unterlinner *et al.*, 1999). However, biosynthetic enzymes in sanguinarine biosynthesis in roots have been localized to the immature endodermis and the protodermis of leaf primordia in the rhizome of *Thalictrum* (Samanani *et al.*, 2005). Similarly, the enzymes in morphinan alkaloid biosynthesis were localized in different cell types (see above); localization of 4′OMT and SAT were both in phloem parenchyma cells, but the enzyme catalyzing the penultimate step, COR, in morphine biosynthesis was located in the laticifers (Kutchan, 2005a; Weid *et al.*, 2004). These results suggest that secondary metabolite production in cell cultures is regulated in a complicated manner. The coordinated expression of biosynthetic genes and their enzymes at a high level seems to be an essential requirement for high production of metabolites.

Many medicinal compounds are used as chemical defense agents in whole plants. These metabolites often have activities as phytoalexins which are induced in response to fungal attack, that is, to act as endogenous chemical weapons for defense in plants. For example, berberine and sanguinarine are antibacterial and antifungal agents (Schmeller *et al.*, 1997). These chemicals are also produced in cells/tissues cultured *in vitro*, even though they are cultivated under aseptic conditions without infection by microbes or attack by animals. The high expression of pathogenesis-related protein genes, as well as of cell proliferation-related genes in cultured cells, has been indicated by protein analysis and expressed sequence tag (EST)- and microarray analyses, indicating that the cell cultures exhibit stress responses (Sasaki *et al.*, 1994; Takeda *et al.*, 1990; Sato *et al.*, unpublished data).

JA (or MeJA) (Fig. 11.5) is an active component in stress responses, especially in elicitation induced by microbial cell walls, heavy metals, and so on (Gundlach *et al.*, 1992; Zhao *et al.*, 2005). Among signal mediators, JA has a pronounced effect in the production of several secondary metabolites, including paclitaxel (Yukimune *et al.*, 1996). However, the application of JA alone is not sufficient to induce the entire biosynthetic pathway for indole alkaloids (Eilert *et al.*, 1987; Van Der Fits and Memelink, 2000): the application of JA induced only a set of biosynthetic genes (Fig. 11.3). Some high-metabolite-producing cells do not respond to JA, suggesting that the signal transduction system and/or the expression of some downstream biosynthetic gene(s) may be highly activated during cell selection (Sato *et al.*, unpublished data). While many of the biosynthetic genes in secondary metabolism respond to JA, PMT in tropane alkaloid synthesis does not (Suzuki *et al.*, 1999a).

Regarding defense responses, we have identified several mediators for such signals other than JA, including salicylic acid (SA) and ethylene. Recent advances in the molecular biology of signal transduction have shown that the overall mechanism of regulation of the expression of defense genes is more complicated and divergent among plant species than expected (Vom Endt *et al.*, 2002; Zhao *et al.*, 2005): for example, the "ethylene response factor 1" in *Arabidopsis* acts downstream of the intersection between the ethylene and JA pathways, suggesting that these signals are somehow integrated (Lorenzo *et al.*, 2003; Vom Endt *et al.*, 2002). Although a general antagonism between JA and SA and synergistic interaction between JA and ethylene have been noted in plant–pathogen interactions, it is still too early to make any definite conclusions about this topic.

3.1. Establishment of high-metabolite-producing lines

While some metabolites can be produced in cell culture systems, or can be induced by signal mediators, the establishment of a high-yield culture, which is essential for commercialization, requires both high productivity and stability. In this regard, the starting materials should show suitable genetic diversity, especially since correlations between the productivity in the originator plant and in cultured cells have not been established (Suzuki *et al.*, 1987). Thus, empirical trials are needed to establish cell lines with suitable metabolite productivity. For example, *H. niger* is a good plant species for producing tropane alkaloids in cell culture, whereas the productivity compared to the intact plant is lower in *A. belladonna* and *D. stramonium* cell cultures (Hashimoto *et al.*, 1986).

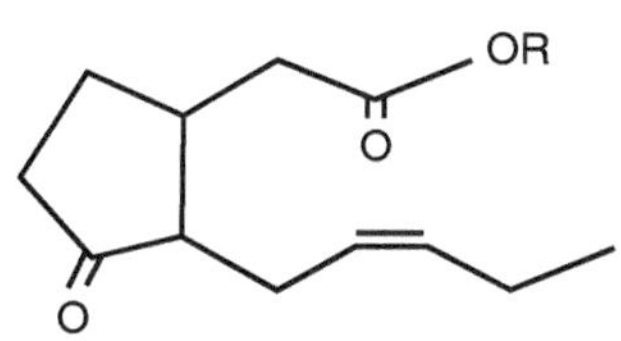

FIGURE 11.5 Active jasmonic acid structure. JA; R=H; MeJA, R=CH_3.

The culture medium and culture conditions, including temperature, illumination, aeration, and so on, are also important factors. Other factors such as shearing stress, gas composition, and cell density are also important for expanding the scale of the culture (Bisaria and Panda, 1991; Matsubara *et al.*, 1989; Taticek *et al.*, 1994). For example, berberine production requires higher aeration because of the greater need for oxygen in the different biosynthetic steps (Sato and Yamada, 1984). Similarly, illumination clearly enhances terpenoid indole alkaloid production in *Catharanthus* cells (Kutchan *et al.*, 1988). The optimization of culture conditions for production is not yet well understood; a more detailed characterization of the biosynthetic pathways and regulation mechanisms may help to guide such optimization.

Furthermore, the selection of high- and stable-metabolite-producing cells is crucial. Since plant cells change their ploidy condition during development and cell culture induces spontaneous mutation (Galbraith *et al.*, 1991; Hirochika *et al.*, 1996; Phillips *et al.*, 1994), the selection of cells should be effective for establishing a high-betabolite-producing line (Sato and Yamada, 1984), while continuous cell culture also leads to instability (see below) (Deus-Neumann and Zenk, 1984).

The continuous maintenance of high- and stable-metabolite-producing cell lines is also important because subtle changes in the culture conditions and the transfer method can easily change the cell phenotype. One endogenous factor that influences productivity is the effect of metabolites that are produced on cell viability. Since many metabolites are biologically active, that is, cytotoxic, the high accumulation of these chemicals can induce cell death or growth retardation; for example, berberine inhibits the growth of nonberberine-producing plant cells (Sakai *et al.*, 2002). The avoidance of cytotoxicity by the ectopic expression of an ABC transporter clearly indicated that metabolites themselves can be toxic if the cells lack a detoxification/segregation machinery or tolerance system (Goossens *et al.*, 2003a). While the essential factors needed to stabilize a cell culture have not yet been identified, our experience suggests that slow-growing cells are more stable. This might be due to the presence of slow cell division and low mutation frequency, as well as less competition for primary metabolites between cell growth and secondary metabolism; a negative correlation is often observed between growth and alkaloid yield (Hashimoto *et al.*, 1986). Alternatively, it seems appropriate to optimize the two-step growth and production system that has been employed for cell growth and shikonin production in *Lithospermum erythrorizon* (Fujita *et al.*, 1986).

3.2. Organ differentiation and secondary plant products

For the second case where metabolites are produced during organ differentiation, organ culture is a suitable alternative even on a large scale (Curtis, 1993; Kusakari *et al.*, 2000). Hairy roots, transformed with *Agrobacterium rhizogenes*, have also been found to be suitable for the production of secondary metabolites due to their stability and high productivity in hormone-free culture conditions (Oksman-Caldentey, 2002; Shanks and Morgan, 1999), and several medicinal plant species have been successfully transformed in this manner. The selection of high-productivity root lines based on somaclonal variation also offers an interesting option for enhancing productivity. While this type of metabolite-producing culture provides stable material and an interesting field of study, how hairy root formation affects metabolite production is not clear. The hairy root system could have a more complicated level of regulation due to tissue organization and the effect of integrated transgenes (see below).

In higher plants and other complex organisms, certain pathways of secondary metabolism can depend on the general development of the organism, including organ, tissue, and particular specialized cell development (Wiermann, 1981). The biosynthesis and accumulation of several secondary compounds occur during defined developmental stages in an organism. Compounds are not necessarily synthesized in organs and tissues with high levels of accumulation; for example, tropane alkaloids in *Atropa* and nicotine alkaloids in tobacco are produced in root and transported to aerial parts (Yun *et al.*, 1992).

One of the most well-known examples of cell differentiation regarding secondary metabolites is the glandular trichome for essential oils in mint and related species (Kutchan, 2005a). These glandular cells show a striking differentiation of tubular smooth ER, and the biosynthetic characteristics of glandular trichomes have been examined by EST analysis (Lange *et al.*, 2000). Highly cytotoxic monoterpenoids require this specific structure for biosynthesis and accumulation. While it is not clear whether the process of the formation of glandular trichomes is similar to that of nonglandular trichomes, the successful trichome formation seen on all epidermal surfaces by the constitutive overexpression of transcriptional factor (GL1 and maize R) genes may provide insight into cellular differentiation and metabolite production (Lange and Croteau, 1999).

The extent to which secondary metabolism depends on the development of specific cellular structures is not yet clear. Poppy plants have idioblast cells specialized for the storage of secondary products and laticifers for excretion (Bird *et al.*, 2003; Facchini and St-Pierre, 2005; Kutchan, 2005a; Weid *et al.*, 2004). Terpenoid indole and tropane alkaloids also need similar cellular collaboration (Burlat *et al.*, 2004; Facchini and St-Pierre, 2005; Kutchan, 2005a; St-Pierre *et al.*, 1999): nonmevalonate pathway enzymes and G10H to produce 10-hydroxygeraniol are localized in internal phloem parenchyma cells, TDC and STR, in the early pathway of terpenoid indole alkaloid biosynthesis, are in epidermal cells, while D4H and DAT, in the late pathway, are in idioblast cells of aerial organs. In tropane alkaloid biosynthesis, the entry enzyme, PMT, is expressed in the cortex and endodermis, whereas the last step of the pathway, H6H, is specifically expressed in pericycle cells in root (Fig. 11.6). These results indicate that intermediates should drive the site of the primary reaction to that of the end reaction. The morphological differentiation of cells would be needed for functional differentiation of each specific reaction in metabolism. However, the molecular basis of the link between function and morphological differentiation is not yet clear. Indeed, this observation can explain why the production of some metabolites requires organ differentiation. For example, the overexpression of a key enzyme (H6H) in tropane alkaloid biosynthesis can improve the production of scopolamine in cultured cells (Yun *et al.*, 1992).

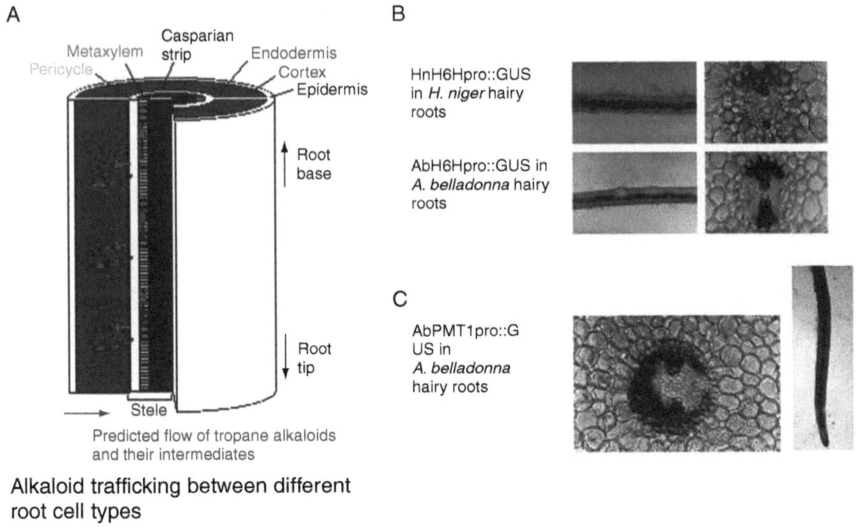

FIGURE 11.6 Cell-specific gene expression in tropane alkaloids. (A) A model for alkaloid trafficking between different cell types in root. (B) Pericycle cell-specific gene expression of the H6H gene. (C) Cell-specific gene expression of the PMT gene. (See Page 20 in Color Section.)

3.3. Genetic instability of productivity

A loss of productivity is a critical factor in industrial applications. The continuous maintenance of cultured cells will lead to the accumulation of mutations, as previously mentioned (Hirochika *et al.*, 1996; Phillips *et al.*, 1994). Currently, the only means to avoid the mutational loss of productivity is cryopreservation, although some empirical trials have shown that repeated selection can overcome this problem based on the establishment of cell lines that are more stable for production (Sato and Yamada, 1984). In fact, selected *C. japonica* cells showed the full chromosome number of 18, while other cell lines showed quite different numbers of chromosomes (Sato *et al.*, unpublished data; Yamada and Mino, 1986). Additional studies are needed to determine how to control spontaneous mutation/genetic and epigenetic modification during culture.

While the molecular investigation of the loss of biosynthetic activity is rather limited, the recovery of biosynthetic activity after the regeneration of whole plants is common as is a loss of productivity that accompanies a phenotype abnormality. Some secondary metabolites might be involved in the normal development of plants; for example, the cosuppression of PMT in tobacco induced an abnormal morphology (Sato *et al.*, 2001). The silencing of obtusifoliol-14α-demethylase (CYP51) in sterol biosynthesis also reduced the growth of transgenic tobacco (Burger *et al.*, 2003).

Fluctuation of biosynthetic activity is more frequently observed, especially during continuous subculture. For example, the accumulation of alkaloids by highly productive transgenic lines showed considerable instability and was strongly influenced by culture conditions, such as the hormonal composition of the medium and the availability of precursors (Whitmer *et al.*, 2003). High transgene-encoded TDC activity was not only unnecessary for increased productivity but also detrimental to the normal growth of the cultures (Canel *et al.*, 1998). Since primary and secondary metabolism compete for the pool of substrates, it is quite likely that rapidly growing cells consume substrates so that none are available for production of secondary metabolites, for example, rapid growth but poor alkaloid production. The optimization of cell growth and metabolite production are important issues for industrialization. We point in particular to two-step culture methods that promote growth and metabolite production separately as essential to overcome these constraints (Fujita *et al.*, 1986).

Another reason for fluctuation is the active channeling (catabolism) of metabolites. We sometimes encounter changes in metabolites within cells, which suggests that metabolism is not static. There is often a difference in the metabolite profile between intact tissue and cultured cells (Fu, 1998; Ketchum *et al.*, 2003). A recent metabolite profile analysis of alkaloids in camptothecin-producing plants showed that anthraquinones, which can be considered phytoalexins, were present in the extracts of hairy roots and calli but not in the differentiated plant of *O. pumila* (Yamazaki *et al.*, 2003). This is just one example of a difference in composition between differentiated plants and hairy roots callus tissues. The poor productivity of opium poppy cells for morphinan alkaloids might be due to the active channeling of intermediates into benzophenanthridine alkaloids rather than to morphinan alkaloids. Similarly, first attempts to reduce the pathway to benzophenanthridine alkaloid with antisense BBE gene in opium poppy plants have been unsuccessful or, at least, ambiguous (Frick *et al.*, 2004). A detailed molecular characterization of metabolic flow should be helpful and clarify mechanisms that lead to the loss of productivity.

The long-term instability of alkaloid production is currently unavoidable, even when stable transformed cells are maintained in selective medium containing selective agents and GUS activity is monitored. For example, the activities of exogenous STR and TDC varied greatly over time, occasionally falling to the levels seen in nontransgenic cultures, indicating that indirect selection, such as antibiotic resistance, is insufficient to maintain the concerted expression of a secondary metabolite pathway necessary for high productivity (Whitmer *et al.*, 2003).

4. BEYOND THE OBSTACLES: MOLECULAR BIOLOGICAL APPROACHES TO IMPROVE PRODUCTIVITY OF SECONDARY METABOLITES IN PLANT CELLS

After the establishment of genetic engineering techniques, the modification of metabolic pathways and genetic regulation of secondary metabolism became feasible. In fact, the industrial production of secondary metabolites is more feasible if we can improve productivity, either by reducing the culture period or increasing the level of metabolites (Misawa, 1994). Furthermore, it may be possible to improve the quality of metabolites by reducing undesired pathways, or introducing new pathways to produce novel compounds, or by completely

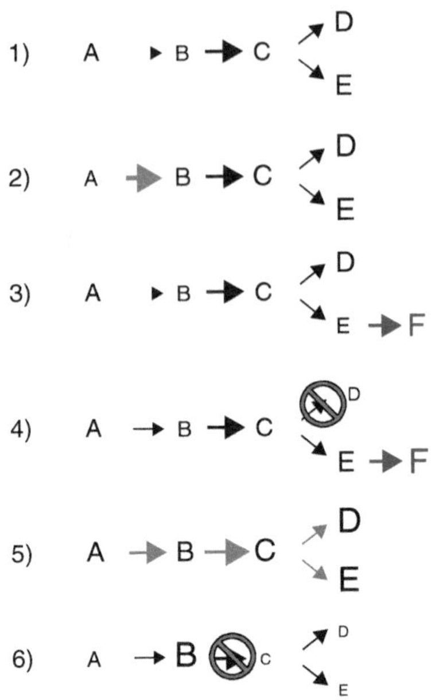

FIGURE 11.7 Strategies for overcoming the limitations in the production of metabolites in plant cells. (1) Preexisting pathway, (2) overexpression of rate-limiting (early) step to clarify the regulation and to overproduce metabolites, (3) creation of a new branch pathway to produce novel compound F and to increase the sink strength, (4) reducing an undesired pathway to enhance the metabolic flow to a desired path, (5) overexpression of transcriptional factor(s) to activate the entire biosynthetic pathway simultaneously, (6) mutation/downregulation of gene expression of biosynthetic enzyme to accumulate the intermediate of pathway. (See Page 21 in Color Section.)

blocking a pathway to accumulate intermediates. Such metabolic engineering would modify plant cells to become "green chemical factories" to obtain the desired compounds. In the sections below, we discuss how genetic engineering could be useful for removing obstacles for the production of metabolites in cell culture or intact plants; general strategies are shown in Fig. 11.7 (also see Croteau *et al.*, 2000; Verpoorte and Memelink, 2002). One promising result suggests that transformation itself might be effective for overcoming these obstacles. *C. roseus* cells transformed with *Agrobacterium tumefaciens* or *A. rhizogenes* gave a stable production of vindoline, which was not found in cultured cells (O'Keefe *et al.*, 1997). This metabolic activation in transformed cells suggests the possibility that the cell type-specific expression required for alkaloid biosynthesis may not be absolutely necessary under all circumstances.

4.1. Overcoming rate-limiting processes in the pathway

If we can identify the rate-limiting step and isolate the target gene, overexpression of the rate-limiting enzyme should be useful for increasing the accumulation of the desired compounds. The pioneering work by Yun *et al.* (1992) is an example of such a successful application in scopolamine biosynthesis in transgenic *Atropa* with *Hyoscyamus* H6H. The crucial point is identification of the rate-limiting step. While the early step is speculated to be rate-limiting for the whole pathway, simple overexpression of the early step enzyme into secondary metabolism is usually not sufficient for inducing the biosynthesis of the desired end-product due to a lack of sufficient induction of late-step enzymes. For example, the genes encoding TDC and STR have been extensively studied in *C. roseus* cell cultures. Overexpression of TDC resulted in higher levels of the immediate product tryptamine, but not increased levels of alkaloids; for STR, higher levels of alkaloids were noted (Canel *et al.*, 1998). Administering such cell lines with tryptophan and terpenoid intermediates showed that they have the capacity for high alkaloid production, indicating that the terpenoid branch of the pathway is limiting (Canel *et al.*, 1998; Whitmer *et al.*, 1998). Such studies indicate that there are multiple rate-limiting processes. Accumulating results suggest that overall and integrated regulation of the biosynthetic pathway is crucial.

Other studies have also been aimed at increasing the flux through the biosynthetic pathway for both the tropane alkaloids and nicotine (Sato *et al.*, 2001). Thus, the tobacco PMT gene was overexpressed in *A. belladonna* and *N. sylvestris*. Although a modest increase in PMT activity of up to 3.3-fold was found in transgenic *A. belladonna* plants, no increase in alkaloid levels was observed and only the levels of methyl putrescine was increased. In some transgenic *N. sylvestris* plants, PMT activity was increased four- to eightfold, whereas cosuppression was noted in other plants. The PMT-overexpressing transgenic lines showed a 40% increase in nicotine levels, whereas it was only 3% of that in the wild type in the case of cosuppression (Sato *et al.*, 2001). A certain step in a pathway might appear to be rate-limiting, but overexpression of the encoding gene will, in most cases, immediately reveal new rate-limiting steps (Verpoorte and Memelink, 2002). These results are in contrast to those seen in overexpression of the gene encoding deoxyxylulose phosphate reductoisomerase in mint, which resulted in plants that had a normal phenotype and an almost 50% increase in essential oil (monoterpenoid) production (Mahmoud and Croteau, 2002, 2001).

4.2. Transcriptional regulation and overall activation

Due to the difficulty of studying biosynthetic pathways in secondary metabolism, researchers have in general isolated few biosynthetic genes and characterized their promoter sequences, except for anthocyanin biosynthesis (Winkel-Shirley, 2001). The molecular characterization of anthocyanin biosynthesis is a successful application of transcriptional regulation to metabolite production; that is, introduction of transcriptional regulator-R and C1 in a heterologous system (Lloyd *et al.*, 1992) and recent successes with activation tagging methods (Borevitz *et al.*, 2000) show the high potential and bright future of these approaches. For example, the *pap1* general transcriptional factor gene in anthocyanin biosynthesis was isolated from *Arabidopsis* based on the visible inspection of highly pigmented plants after activation tagging (Borevitz *et al.*, 2000). Additional successful isolation of ORCA3, a transcriptional factor with a JA-responsive AP2 domain, in indole alkaloid biosynthesis in *Catharanthus* using activation tagging demonstrated the effectiveness of this approach, while the selection methodology had to be further developed due to the colorless nature of target compounds (Van Der Fits and Memelink, 2000).

ORCA expression is induced by JA. Ectopic expression of ORCA3 in cultured cells from *C. roseus* increases the expression of the terpenoid indole alkaloid biosynthetic genes TDC, STR, CPR, and D4H but does not regulate the genes encoding G10H and DAT (Fig. 11.3). Transgenic cells that overexpressed ORCA3 accumulated significantly more tryptophan and tryptamine. However, since no terpenoid indole alkaloids were detected, the terpenoid branch of the pathway remains limiting for terpenoid indole alkaloid production. Although ORCA3 plays an important role in regulating terpenoid indole alkaloid biosynthesis, it is not sufficient to regulate the complete pathway. This indicates that other transcription factors are also involved. The use of an enhancer domain of the STR promoter as bait in a yeast-one-hybrid screen resulted in the isolation of CrBPF1, an MYB-like transcription factor (Van Der Fits *et al.*, 2000). CrBPF1 expression is induced by elicitor but not JA. In addition, the STR promoter contains a promoter element that is conserved in plants, called the G-box, which is located adjacent to the JERE element. A yeast-one-hybrid screen using the G-box as bait isolated G-box-binding factors (crGBF) of the basic leucine zipper class and MYC-type bHLH transcription factors (CrMYC) (Chatel *et al.*, 2003). CrGBFs have been shown to repress STR expression (Siberil *et al.*, 2001), whereas this factor is not sufficient to control the overall gene expression for indole alkaloid biosynthesis. As discussed above, regulation of the whole metabolic pathway would be more complicated since the spatial and developmental integration of metabolism are needed.

4.3. Qualitative control of metabolites and the isolation of desired biosynthetic genes

For the successful industrial application of secondary metabolite production in plant cells, both the quantity and quality of metabolites must be improved, that is, control of the metabolite profile is important. Creation of a new branch to produce novel compounds by the introduction of a novel gene is a positive approach to modify this profile. To achieve this end, there are at least two possible approaches. One is to use biotechnological methods to produce the desired compounds, that is, heterologous expression of a complete designed biosynthetic pathway in simple and rapidly growing microorganism systems and/or their use in biotransformation (Rathbone and Bruce, 2002). The recombinant enzymes would provide novel catalytic activities, whereas low activity in intact cells often makes such evaluation difficult. Microbial enzymes have been shown to be useful for the biotransformation of chemicals. Plant enzymes are much more likely to be biocatalysts, although they have rather high substrate specificity, for example, a P450, CYP719A1 (Ikezawa *et al.*, 2003).

Another approach is to introduce a new branch pathway into a preexisting biosynthetic pathway (Sato *et al.*, 2001). A crucial point in creating a new branch in a preexisting pathway is the substrate-affinity/reaction specificity of introduced enzyme(s). Complete modification of the alkaloid profile with the introduction of *C. japonica* SOMT cDNA into *Eschscholzia californica* cells suggested that *C. japonica* SOMT was superior to *E. californica* chelanthifoline synthase (Sato *et al.*, 2001); that is, the alkaloid profile changed from sanguinarine (benzophenanthridine-type) to columbamine (berberine-type) (Fig. 11.2). However, plant cells do not always innately show a new pathway; that is, a newly introduced pathway can provide the substrate for further enzymatic conversion and produce novel compounds that are not detected in wild-type cells. This result clearly suggests the potential of plant cells for the production of divergent chemicals after the introduction of a new branch in a pathway. This experimental result also suggests how plant cells can obtain a divergent array of metabolites using preexisting pathways. In either case, it would be very important to isolate genes with novel and desired functions.

4.3.1. Isolation of specific genes

To modify a given pathway, the availability of biosynthetic enzymes is important. While biosynthetic enzymes with adequate substrate and reaction specificity, or rational protein engineering, are needed, our knowledge on these subjects is still limited (Zubieta *et al.*, 2003). Thus, the isolation of specific genes from specialized cells would be more practical, although it is questionable how we can acquire good materials to isolate such desired enzymes, cDNAs and genes. While biosynthetic genes in microorganisms are usually clustered, the genes for secondary metabolites are not (Hauschild *et al.*, 1998). Previously selected high-metabolite-producing cells would be good candidates for isolating such enzymes and/or genes (Hashimoto and Yamada, 2003). For example, berberine-producing *C. japonica* cells have been shown to be very useful for isolating biosynthetic genes, as seen in oil gland cells (Lange *et al.*, 2000; Morishige *et al.*, 2002). Similarly, the comparison of morphine producing and nonproducing *Papaver* spp. has been used to identify the gene(s) specific for alkaloid production (Ziegler *et al.*, 2005). In either case, the further characterization of ESTs by microarray or Northern analysis should be effective for evaluating the functional linkages of genes. Another possibility is the use of compounds to induce metabolism. The cDNA-amplified fragment length polymorphism (AFLP) method is an attractive alternative for identifying genes involved in plant secondary metabolism, in combination with targeted metabolite analysis (Goossens *et al.*, 2003b). Alternatively, if we could isolate a general transcription factor(s) that targets secondary metabolism, the use of transgenic cells as the starting material could become possible. One example is provided by *Arabidopsis thaliana* overexpressing the PAP1 gene encoding an MYB transcriptional factor. This line was used to identify a novel gene involved in flavonoid biosynthesis using the integration of metabolomics and transcriptomics (Tohge *et al.*, 2005). A proteomics approach can also be useful for examining proteins by two-dimensional gel electrophoresis and internal peptide microsequencing. Using this approach, a representative enzyme in morphine biosynthesis could be detected within the serum fraction of latex of opium poppy (Decker *et al.*,

2000). As shown above, a recent trend is to avoid biochemical purification and to directly isolate candidate clones by a combination of expression pattern analysis and homology-based screening (Hashimoto and Yamada, 2003). However, our understanding of the resulting data is limited due to a fundamental lack of biochemical and physiological knowledge about network organization in plants, although the development of metabolomic methods and tools is progressing rapidly (Tohge *et al.*, 2005; Weckwerth and Fiehn, 2002).

4.3.2. Quality control of metabolites

The simple composition of metabolites is desirable for industrial production. The simplest approach to producing desired compounds is to reconstruct entire biosynthetic processes *in vitro*, as mentioned above. So far, a considerable number of genes involved in alkaloid biosynthesis have been cloned and expressed in *E. coli* or insect cells (De Luca and Laflamme, 2001; Facchini, 2001; Hashimoto and Yamada, 2003; Rathbone and Bruce, 2002; Verpoorte and Memelink, 2002). While microbial cells have less capacity to store metabolites, heterologous systems could be useful for bioconversion.

A more promising approach is the downregulation of gene expression for an undesired pathway. Downregulation using antisense material and cosuppression have been shown to be effective since mint plants transformed with the antisense version of menthofuran synthase cDNA produced less than half of the undesired monoterpene oil component than did wild type (Mahmoud and Croteau, 2001). Suppression of a P450 hydroxylase gene in plant trichome glands has also been associated with the accumulation of cembratriene-ol and enhanced resistance against aphids (Wang *et al.*, 2001).

The recent development of the RNA interference (RNAi) method using double-stranded (ds)RNA-induced posttranscriptional and/or transcriptional silencing is a more efficient method for modifying a pathway to shut down (Wang and Waterhouse, 2002; Waterhouse and Helliwell, 2003; Wesley *et al.*, 2001). While transgenic *E. californica* cells with antisense BBE RNA lost alkaloid productivity with no accumulation of intermediate reticuline (Park and Facchini, 2000; Park *et al.*, 2002), our experiment with the dsRNA expression vector for BBE (BBEir) evidently induced reticuline accumulation with the reduction of end-products (Fujii *et al.*, 2007). This result clearly shows that this new tool is useful for metabolic engineering and that plant cells are capable of storing intermediates even when the metabolic pathway has been altered. A note of caution about the selection of target sequences must be included because short stretches of oligonucleotides that perfectly match a gene sequence may effectively lead to silencing of homologous genes (Ishihara *et al.*, 2005). However, the silencing of undesired genes and the elimination of certain proteins can also provide a way to alter metabolic pathways. Accumulation of reticuline in transgenic opium poppy with a COR RNAi vector could lead to silencing effects, while disruption of enzyme complex for morphine alkaloid biosynthesis after reticuline also might be involved in the accumulation of reticuline (Allen *et al.*, 2004).

4.4. Accumulation and storage

Since the secondary metabolites produced in plant cells have high biological activity, they are potentially toxic to the plant cells themselves. However, metabolite-producing cells can grow without serious disruption of their basic metabolism (Sakai *et al.*, 2002). The regulatory mechanisms of alkaloid accumulation and detoxification in plant cells have been less well studied. Functional analysis of yeast PDR5 (pleiotrophic drug resistance-type ATP-binding cassette transporter) genes in transgenic tobacco (*Nicotiana tabacum* L.) BY2 cell lines has suggested that these genes can be used to stimulate the secretion of secondary metabolites in plant cell cultures (Goossens *et al.*, 2003a). Recent identification of the function of CjMDR (a multidrug resistance gene) isolated from *C. japonica* cells for uptake in rhizome may lead to a better understanding of the mechanism of accumulation and detoxification of these biologically active metabolites (Shitan *et al.*, 2003).

5. FUTURE PERSPECTIVES

The most common problems for the commercial application of metabolite production in cell culture systems are low productivity and the high cost associated with the maintenance of cell cultures, compared to harvesting in the field; for example, morphine is not produced in cultured *Papaver* cells and only low levels of vinblastine are produced in cultured *Catharanthus* cells. However, molecular engineering is a powerful tool and may be able to overcome these problems. First, other aspects of cell culture will need to be clarified for production. For example, while it is clear that selected cultured *Coptis* cells are superior to naturally grown *Coptis* rhizomes with regard to berberine production and uniform metabolite composition, these cultured cells are not accepted as being equivalent to natural rhizome. Since it has often been reported that *in vitro*-cultured cells/tissues have different metabolite profiles than intact tissues (Fu, 1998; Yamazaki *et al.*, 2003), the safety of these materials for direct use as natural medicines has been questioned. Thus, cultured cells require additional certification to be considered equivalent to *Coptis* rhizome. On the other hand, the costs of tedious downstream processing to purify berberine from cultured *Coptis* cells cannot compete with that obtained by wild harvest from other sources, if metabolic engineering is not applicable. The size of the anticipated market is another important factor for commercialization.

So far, three metabolites produced in cultured cells/roots have been reported to be successfully commercialized (Table 11.1). One is cultured ginseng, the extract of which has been approved as a food additive with a market of $3 million in 1995 (Fu, 1998). The other two products are used in cosmetics: polysaccharides produced by cultured *Polyanthes tuberosa* cells (Honda *et al.*, 1996) and Saiko extracts from cultured Saiko (*Bupleurum falcatum* L.) roots (Kusakari *et al.*, 2000). In the latter cases, researchers found that polysaccharides and saikosaponins had novel activities regarding skin protection and developed their use as cosmetics.

These successes clearly indicate that the discovery of novel product activities is essential for commercial development, except for a few chemicals such as paclitaxel, which have a high price and high demand. Functional evaluation of the metabolic profile in plant cells is crucial for the future engineering of medicinal compounds.

6. SUMMARY

Many secondary metabolites that have been isolated from higher plants are used as important natural resources for pharmaceuticals. Since plant cells have high totipotency, empirical trials to produce these secondary metabolites in *in vitro* cell and/or tissue culture have been carried out. Over the past 30 years, studies of these cell culture systems have provided a basis for understanding the basic mechanism of biosynthesis, whereas practical applications are still limited due to relatively low productivity. Recent advances in molecular biology in plant sciences, for example, the comprehensive analysis of expressed genes in biosynthesis-specialized cells and integrated analyses of expression profiles, as well as the development of metabolic profiling analysis, have also considerably stimulated the development of metabolic engineering in plants, even in secondary metabolism. The identification of many biosynthetic genes and the characterization of the spatial and developmental regulation of their expression have clarified their importance in the biosynthesis of secondary metabolites and revealed bottlenecks for their production in cell culture. The molecular engineering of secondary metabolites may lead to a new era for the production of medicinal compounds in cell/tissue cultures as well as in transgenic plants to improve the production and quality of metabolites. The forthcoming decades should be an exciting time for basic and applied sciences regarding secondary metabolite production in plant cells.

ACKNOWLEDGEMENTS

We appreciate Dr. T. Hashimoto for his critical reading and his courtesy for providing Fig. 11.6. The research is supported in part by Research for the Future Program Grant JSPS-RFTF00L01606 from the Japan Society for the Promotion of Science (to F.S.).

REFERENCES

Allen, R. S., Millgate, A. G., Chitty, J. A., Thisleton, J., Miller, J. A., Fist, A. J., Gerlach, W. L., and Larkin, P. J. (2004). RNAi-mediated replacement of morphine with the nonnarcotic alkaloid reticuline in opium poppy. *Nat. Biotechnol.* **22,** 1559–1566.

Askin, I., Ribnicky, D. M., Komarnytsky, S., Ilic, N., Poulev, A., Borisjuk, N., Brinker, A., Moreno, D. A., Ripoll, C., Yakoby, N., O'Neal, J. M., Cornwell, T., *et al.* (2002). Plants and human health in the twenty-first century. *Trends Biotechnol.* **20,** 522–531.

Bird, D. A., Franceschi, V. R., and Facchini, P. J. (2003). A tale of three cell-types: Alkaloid biosynthesis is localized to sieve elements in opium poppy. *Plant Cell* **15,** 2626–2635.

Bisaria, V., and Panda, A. (1991). Large-scale plant cell culture: Methods, applications and products. *Curr. Opin. Biotechnol.* **2,** 370–374.

Bock, A., Wanner, G., and Zenk, M. H. (2002). Immunocytological localization of two enzymes involved in berberine biosynthesis. *Planta* **216,** 57–63.

Borevitz, J. O., Xia, Y. J., Blount, J., Dixon, R. A., and Lamb, C. (2000). Activation tagging identifies a conserved MYB regulator of phenylpropanoid biosynthesis. *Plant Cell* **12,** 2383–2393.

Briskin, D. P. (2000). Medicinal plants and phytomedicines. Linking plant biochemistry and physiology to human health. *Plant Physiol.* **124,** 507–514.

Burger, C., Rondet, S., Benveniste, P., and Schaller, H. (2003). Virus-induced silencing of sterol biosynthetic genes: Identification of a *Nicotiana tabacum* L. obtusifoliol-14 alpha-demethylase (CYP51) by genetic manipulation of the sterol biosynthetic pathway in *Nicotiana benthamiana* L. *J. Exp. Bot.* **54,** 1675–1683.

Burlat, V., Oudin, A., Courtois, M., Rideau, M., and St.-Pierre, B. (2004). Co-expression of three MEP pathway genes and geraniol 10-hydroxylase in internal phloem parenchyma of *Catharanthus roseus* implicates multicellular translocation of intermediates during the biosynthesis of monoterpene indole alkaloids and isoprenoid-derived primary metabolites. *Plant J.* **38,** 131–141.

Canel, C., Lopes-Cardoso, M. I., Whitmer, S., Van der Fits, L., Pasquali, G., Van Der Heijden, R., Hoge, J. H., and Verpoorte, R. (1998). Effects of over-expression of strictosidine synthase and tryptophan decarboxylase on alkaloid production by cell cultures of *Catharanthus roseus. Planta* **205,** 414–419.

Chatel, G., Montiel, G., Pre, M., Memelink, J., Thiersault, M., Saint-Pierre, B., Doireau, P., and Gantet, P. (2003). CrMYC1, a *Catharanthus roseus* elicitor- and jasmonate-responsive bHLH factor that binds the G-box element of the strictosidine synthase gene promoter. *J. Exp. Bot.* **54,** 2587–2588.

Choi, K. B., Morishige, T., Shitan, N., Yazaki, K., and Sato, F. (2002). Molecular cloning and characterization of coclaurine *N*-methyltransferase from cultured cells of *Coptis japonica. J. Biol. Chem.* **277,** 830–835.

Collu, G., Unver, N., Peltenburg-Looman, A. M., Van Der Heijden, R., Verpoorte, R., and Memelink, J. (2001). Geraniol 10-hydroxylase, a cytochrome P450 enzyme involved in terpenoid indole alkaloid biosynthesis. *FEBS Lett.* **508,** 215–220.

Contin, A., Van Der Heijden, R., Lefeber, A. W., and Verpoorte, R. (1998). The iridoid glucoside secologanin is derived from the novel triose phosphate/pyruvate pathway in a *Catharanthus roseus* cell culture. *FEBS Lett.* **434,** 413–416.

Croteau, R., Kutchan, T. M., and Lewis, N. G. (2000). Natural products (Secondary metabolites). *In* "Biochemistry & Molecular Biology of Plants" (B. B. Buchanan, W. Gruissem, and R. L. Jones, eds.), pp. 1250–1318. American Society of Plant Physiologists, Maryland.

Curtis, W. R. (1993). Cultivation of roots in bioreactors. *Curr. Opin. Biotechnol.* **4,** 205–210.

Decker, G., Wanner, G., Zenk, M. H., and Lottspeich, F. (2000). Characterization of proteins in latex of the opium poppy (*Papaver somniferum*) using two-dimensional gel electrophoresis and microsequencing. *Electrophoresis* **21,** 3500–3516.

De Luca, V., and Laflamme, P. (2001). The expanding universe of alkaloid biosynthesis. *Curr. Opin. Plant Biol.* **4,** 225–233.

De Luca, V., Marineau, C., and Brisson, N. (1989). Molecular cloning and analysis of cDNA encoding a plant tryptophan decarboxylase: Comparison with animal dopa decarboxylases. *Proc. Natl. Acad. Sci. USA* **86,** 2582–2586.

Deus-Neumann, B., and Zenk, M. H. (1984). Instability of indole alkaloid production in *Catharanthus roseus* cell suspension cultures. *Planta Med.* **50,** 427–431.

Dittrich, H., and Kutchan, T. M. (1991). Molecular cloning, expression, and induction of berberine bridge enzyme, an enzyme essential to the formation of benzophenanthridine alkaloids in the response of plants to pathogen attack. *Proc. Natl. Acad. Sci. USA* **88,** 9969–9973.

Dougall, D. K. (1981). Tissue culture and the study of secondary (natural) products. *In* "The Biochemistry of Plants" (E. E. Conn, ed.), Vol. 7, pp. 21–34. Academic Press, New York.

Eilert, U., De Luca, V., Constabel, F., and Kurz, W. G. (1987). Elicitor-mediated induction of tryptophan decarboxylase and strictosidine synthase activities in cell suspension cultures of *Catharanthus roseus. Arch. Biochem. Biophys.* **254,** 491–497.

Facchini, P. (2001). Alkaloid biosynthesis in plants: Biochemistry, cell biology, molecular regulation, and metabolic engineering applications. *Annu. Rev. Plant Physiol. Plant Mol. Biol.* **52,** 29–66.

Facchini, P. J., and De Luca, V. (1994). Differential and tissue-specific expression of a gene family for tyrosine/dopa decarboxylase in opium poppy. *J. Biol. Chem.* **269,** 26684–26690.
Facchini, P. J., and De Luca, V. (1995). Phloem-specific expression of tyrosine/dopa decarboxylase genes and the biosynthesis of isoquinoline alkaloids in opium poppy. *Plant Cell* **7,** 1811–1821.
Facchini, P. J., and Park, S. U. (2003). Developmental and inducible accumulation of gene transcripts involved in alkaloid biosynthesis in opium poppy. *Phytochemistry* **64,** 177–186.
Facchini, P. J., and St-Pierre, B. (2005). Synthesis and trafficking of alkaloid biosynthetic enzymes. *Curr. Opin. Plant Biol.* **8,** 657–666.
Frick, S., and Kutchan, T. M. (1999). Molecular cloning and functional expression of O-methyltransferases common to isoquinoline alkaloid and phenylpropanoid biosynthesis. *Plant J.* **17,** 329–339.
Frick, S., Chitty, J. A., Karamell, R., Schumidt, J., Allen, R. S., Larkin, P. J., and Kutchan, T. M. (2004). Transformation of opium poppy (*Papaver somniferum* L.) with antisense berberine bridge enzyme gene (*anti-bbe*) via somatic embryogenesis results in an altered ratio of alkaloids in latex but not in roots. *Transgenic Res.* **13,** 607–613.
Fu, T.-J. (1998). Safety considerations for food ingredients produced by plant cell and tissue culture. *Chemtech* **28,** 40–46. http://pubs.acs.org/hotartcl/chemtech/98/
Fujii, N., Inui, T., Iwasa, K., Morishige, T., and Sato, F. (2007). Knockdown of berberine bridge enzyme by RNAi accumulates (S)-reticuline and activates a silent pathway in cultured California poppy cells. *Transgenic Research* **16,** 363–375.
Fujita, Y., Suga, C., Matsubara, K., and Hara, Y. (1986). Production of shikonin derivatives by plant-cell cultures. *Nippon Nogeikagaku Kaishi* **60,** 849–854 (in Japanese).
Galbraith, D. W., Harkins, K. R., and Knapp, S. (1991). Systemic endopolyploidy in *Arabidopsis thaliana. Plant Physiol.* **96,** 985–989.
Galneder, E., Rueffer, M., Wanner, G., Tabata, M., and Zenk, M. H. (1988). Alternative final steps in berberine biosynthesis in *Coptis japonica* cell cultures. *Plant Cell Rep.* **7,** 1–4.
Geerlings, A., Martinez-Lozano, M., Memelink, J., Van Der Heijden, R., and Verpoorte, R. (2000). Molecular cloning and analysis of strictosidine β-D-glucosidase, an enzyme in terpenoid indole alkaloid biosynthesis in *Catharanthus roseus. J. Biol. Chem.* **275,** 3051–3056.
Goddijn, O. J., De Kam, R. J., Zanetti, A., Schilperoort, R. A., and Hoge, J. H. (1992). Auxin rapidly down-regulates transcription of the tryptophan decarboxylase gene from *Catharanthus roseus. Plant Mol. Biol.* **18,** 1113–1120.
Goossens, A., Haekkinen, S. T., Laakso, I., Oksman-Caldentey, O.-M., and Inze, D. (2003a). Secretion of secondary metabolites by ATP-binding cassette transporters in plant cell suspension cultures. *Plant Physiol.* **131,** 1161–1164.
Goossens, A., Haekkinen, S. T., Laakso, I., Seppaenen-Laakso, T., Biondi, S., De Sutter, V., Lammertyn, F., Nuutila, A. M., Soederlund, H., Zabeau, M., Inze, D., and Oksman-Caldentey, K.-M. (2003b). A functional genomics approach toward the understanding of secondary metabolism in plant cells. *Proc. Natl. Acad. Sci. USA* **100,** 8595–8600.
Grothe, T., Lenz, R., and Kutchan, T. M. (2001). Molecular characterization of the salutaridinol-7-O-acetyltransferase involved in morphine biosynthesis in opium poppy *Papaver somniferum. J. Biol. Chem.* **276,** 30717–30723.
Gundlach, H., Mueller, M., Kutchan, T. M., and Zenk, M. H. (1992). Jasmonic acid is a signal transducer in elicitor-induced plant cell cultures. *Proc. Natl. Acad. Sci. USA* **89,** 2389–2393.
Hashimoto, T., and Yamada, Y. (2003). New genes in alkaloid metabolism and transport. *Curr. Opin. Biotechnol.* **14,** 163–168.
Hashimoto, T., Yukimune, Y., and Yamada, Y. (1986). Tropane alkaloid production in *Hyoscyamus* root cultures. *J. Plant Physiol.* **124,** 61–76.
Hashimoto, T., Hayashi, A., Amano, Y., Kohno, J., Iwanari, H., Usuda, S., and Yamada, Y. (1991). Hyoscyamine 6 β-hydroxylase, an enzyme involved in tropane alkaloid biosynthesis, is localized at the pericycle of the root. *J. Biol. Chem.* **266,** 4648–4653.
Hauschild, K., Pauli, H. H., and Kutchan, T. M. (1998). Isolation and analysis of a gene *bbe1* encoding the berberine bridge enzyme from the California poppy *Eschscholzia californica. Plant Mol. Biol.* **36,** 473–478.
Hibi, N., Higashiguchi, S., Hashimoto, T., and Yamada, Y. (1994). Gene expression in tobacco low-nicotine mutants. *Plant Cell* **6,** 723–735.

Hirochika, H., Sugimoto, K., Otsuki, Y., Tsugawa, H., and Kanda, M. (1996). Retrotransposons of rice involved in mutations induced by tissue culture. *Proc. Natl. Acad. Sci. USA* **93,** 7783–7788.
Honda, Y., Inaoka, H., Takei, A., Sugimura, Y., and Otsuji, K. (1996). Extracellular polysaccharides produced by tuberose callus. *Phytochemistry* **41,** 1517–1521.
Huang, F.-C., and Kutchan, T. M. (2000). Distribution of morphinan and benzo[*c*]phenanthridine alkaloid gene transcript accumulation in *Papaver somniferum*. *Phytochemistry* **53,** 555–564.
Ikezawa, N., Tanaka, M., Nagayoshi, M., Shinkyo, R., Sakaki, T., Inouye, K., and Sato, F. (2003). Molecular cloning and characterization of CYP719, a methylenedioxy bridge-forming enzyme that belongs to a novel P450 family, from cultured *Coptis japonica* cells. *J. Biol. Chem.* **278,** 38557–38565.
Ikezawa, N., Iwasa, K., and Sato, F. (2007). Molecular cloning and characterization of methylenedioxy bridge-forming enzymes involved in stylopine biosynthesis in *Eschscholzia californica*. *FEBS J.* **274,** 1019–1035.
Irmler, S., Schroeder, G., St-Pierre, B., Crouch, N. P., Hotze, M., Schmidt, J., Strack, D., Matern, U., and Schroeder, J. (2000). Indole alkaloid biosynthesis in *Catharanthus roseus*: New enzyme activities and identification of cytochrome P450 CYP72A1 as secologanin synthase. *Plant J.* **24,** 797–804.
Ishihara, S., Yamamoto, Y., Ifuku, K., and Sato, F. (2005). Functional analysis of four members of the PsbP family in photosystem II in *Nicotiana tabacum* using differential RNA interference. *Plant Cell Physiol.* **46,** 1885–1893.
Jennewein, S., and Croteau, R. (2001). Taxol: Biosynthesis, molecular genetics, and biotechnological applications. *Appl. Microbiol. Biotechnol.* **57,** 13–19.
Kanegae, T., Kajiya, H., Amano, Y., Hashimoto, T., and Yamada, Y. (1994). Species-dependent expression of the hyoscyamine 6-β-hydroxylase gene in the pericycle. *Plant Physiol.* **105,** 483–490.
Ketchum, R. E., Rithner, C. D., Qiu, D., Kim, Y. S., Williams, R. M., and Croteau, R. B. (2003). *Taxus* metabolomics: Methyl jasmonate preferentially induces production of taxoids oxygenated at C-13 in *Taxus x media* cell cultures. *Phytochemistry* **62,** 901–909.
Kraus, P. F. X., and Kutchan, T. M. (1995). Molecular-cloning and heterologous expression of a cDNA-encoding berbamunine synthase, a C-O-phenol-coupling cytochrome-P450 from the higher-plant *Berberis-stolonifera*. *Proc. Natl. Acad. Sci. USA* **92**(6), 2071–2075.
Kusakari, K., Yokoyama, M., and Inomata, S. (2000). Enhanced production of saikosaponins by root culture of *Bupleurum falcatum* L. using two-step control of sugar concentration. *Plant Cell Rep.* **19,** 1115–1120.
Kutchan, T. M. (2005a). A role for intra-and intercellular translocation in natural product biosynthesis. *Curr. Opin. Plant Biol.* **8,** 292–300.
Kutchan, T. M. (2005b). Predictive metabolic engineering in plants: Still full of surprises. *Trends Biotechnol.* **23,** 381–383.
Kutchan, T. M., Hampp, N., Lottspeich, F., Beyreuther, K., and Zenk, M. H. (1988). The cDNA clone for strictosidine synthase from *Rauvolfia serpentina*: DNA sequence determination and expression in *Escherichia coli*. *FEBS Lett.* **237,** 40–44.
Lange, B. M., and Croteau, R. (1999). Genetic engineering of essential oil production in mint. *Curr. Opin. Plant Biol.* **2,** 139–144.
Lange, B. M., Wildung, M. R., Stauber, E. J., Sanchez, C., Pouchnik, D., and Croteau, R. (2000). Probing essential oil biosynthesis and secretion by functional evaluation of expressed sequence tags from mint glandular trichomes. *Proc. Natl. Acad. Sci. USA* **97,** 2934–2939.
Liscombe, D. K., Macleod, B. P., Loukanina, N., Nandi, O. I., and Facchini, P. J. (2005). Evidence for the monophyletic evolution of benzylisoquinoline alkaloid biosynthesis in angiosperms. *Phytochemistry* **66,** 1374–1393.
Lloyd, A. M., Walbot, V., and Davis, R. W. (1992). *Arabidopsis* and *Nicotiana* anthocyanin production activated by maize regulator-R and regulator-C1. *Science* **258,** 1773–1775.
Lopez-Meyer, M., and Nessler, C. L. (1997). Tryptophan decarboxylase is encoded by two autonomously regulated genes in *Camptotheca acuminata* which are differentially expressed during development and stress. *Plant J.* **11,** 1167–1175.
Lorenzo, O., Piqueras, R., Sanchez-Serrano, J. J., and Solano, R. (2003). Ethylene response factor1 integrates signals from ethylene and jasmonate pathways in plant defense. *Plant Cell* **15,** 165–178.

Mahmoud, S. S., and Croteau, R. B. (2001). Metabolic engineering of essential oil yield and composition in mint by altering expression of deoxyxylulose phosphate reductoisomerase and menthofuran synthase. *Proc. Natl. Acad. Sci. USA* **98,** 8915–8920.

Mahmoud, S. S., and Croteau, R. B. (2002). Strategies for transgenic manipulation of monoterpene biosynthesis in plants. *Trends Plant Sci.* **7,** 366–373.

Matsubara, K., Kitani, S., Yoshioka, T., Morimoto, T., and Fujita, Y. (1989). High-density culture of *Coptis japonica* cells increases berberine production. *J. Chem. Technol. Biotechnol.* **46,** 61–69.

Matsuda, J., Okabe, S., Hashimoto, T., and Yamada, Y. (1991). Molecular cloning of hyoscyamine 6 β-hydroxylase, a 2-oxoglutarate-dependent dioxygenase, from cultured roots of *Hyoscyamus niger*. *J. Biol. Chem.* **266,** 9460–9464.

Mcknight, T. D., Toessner, C. A., Devagupta, R., Scott, A. I., and Nessler, C. L. (1990). Nucleotide sequence of a cDNA encoding the vacuolar protein strictosidine synthase from *Catharanthus roseus*. *Nucleic Acids Res.* **18,** 4939.

Millgate, A. G., Pogson, B. J., Wilson, I. W., Kutchan, T. M., Zenk, M. H., Gerlach, W. L., Fist, A. J., and Larkin, P. J. (2004). Morphine-pathway block in *top1* poppies. *Nature* **431,** 413–414.

Minami, H., Dubouzet, E., Iwasa, K., and Sato, F. (2007). Functional analysis of norcoclaurine synthase in *Coptis japonica, J. Biol. Chem.* **282,** 6274–6282.

Misawa, M. (1994). Plant tissue culture: An alternative for production of useful metabolite. *FAO AGRICULTURAL SERVICES BULLETIN* No. 108, M-06, ISBN 92-5-103391-9, http://www.fao.org/docrep/t0831e/t0831e00.htm#con.

Morishige, T., Tsujita, T., Yamada, Y., and Sato, F. (2000). Molecular characterization of the *S*-adenosyl-L-methionine: 3′-hydroxyl-*N*-methylcoclaurine 4′-*O*-methyltransferase involved in isoquinoline alkaloid biosynthesis in *Coptis japonica*. *J. Biol. Chem.* **275,** 23398–23405.

Morishige, T., Dubouzet, E., Choi, K. B., Yazaki, K., and Sato, F. (2002). Molecular cloning of columbamine *O*-methyltransferase from cultured *Coptis japonica* cells. *Eur. J. Biochem.* **269,** 5659–5667.

Muemmler, S., Rueffer, M., Nagakura, N., and Zenk, M. H. (1985). *S*-Adenosyl-L-methionine: (*S*)-scoulerine 9-*O*- methyltransferase, a highly stereo- and regio-specific enzyme in tertrahydroprotoberberine biosynthesis. *Plant Cell Rep.* **4,** 36–39.

Nakajima, K., and Hashimoto, T. (1999). Two tropinone reductases, that catalyze opposite stereospecific reductions in tropane alkaloid biosynthesis, are localized in plant root with different cell-specific patterns. *Plant Cell Physiol.* **40,** 1099–1107.

Nakajima, K., Hashimoto, T., and Yamada, Y. (1993). Two tropinone reductases with different stereospecificities are short-chain dehydrogenases evolved from a common ancestor. *Proc. Natl. Acad. Sci. USA* **90,** 9591–9595.

Nelson, D. R., Schuler, M. A., Paquette, S. M., Werck-Reichhart, D., and Bak, S. (2004). Comparative genomics of rice and arabidopsis. Analysis of 727 cytochrome P450 genes and psuedogenes from a monocot and a dicot. *Plant Physiol.* **135,** 756–772.

O'Keefe, B. R., Mahady, G. B., Gills, J. J., Beecher, C. W. W., and Schilling, A. B. (1997). Stable vindoline production in transformed cell cultures of *Catharanthus roseus*. *J. Nat. Prod.* **60,** 261–264.

Oksman-Caldentey, K. M. (2002). *Agrobacterium rhizogenes*-mediated transformation: Root cultures as a source of alkaloids. *Planta Med.* **68,** 859–868.

Park, S. U., and Facchini, P. J. (2000). *Agrobacterium rhizogenes*- mediated transformation of opium poppy, *Papaver somniferum* l., and California poppy, *Eschscholzia californica* cham., root cultures. *J. Exp. Bot.* **51,** 1005–1016.

Park, S. U., Johnson, A. G., Penzes-Yost, C., and Facchini, P. J. (1999). Analysis of promoters from tyrosine/dihydroxyphenylalanine decarboxylase and berberine bridge enzyme genes involved in benzylisoquinoline alkaloid biosynthesis in opium. *Plant Mol. Biol.* **40,** 121–131.

Park, S. U., Yu, M., and Facchini, P. J. (2002). Antisense RNA-mediated suppression of benzophenanthridine alkaloid biosynthesis in transgenic cell cultures of California poppy. *Plant Physiol.* **128,** 696–706.

Pauli, H. H., and Kutchan, T. M. (1998). Molecular cloning and functional heterologous expression of two alleles encoding (*S*)-*N*-methylcoclaurine 3′-hydroxylase (CYP80B1), a new methyl jasmonate-inducible cytochrome P-450-dependent mono-oxygenase of benzylisoquinoline alkaloid biosynthesis. *Plant J.* **13,** 793–801.

Petersen, M., and Alfermann, A. W. (2001). The production of cytotoxic lignans by plant cell cultures. *Appl. Microbiol. Biotechnol.* **55,** 135–142.

Phillips, R. L., Kaeppler, S. M., and Olhoft, P. (1994). Genetic instability of plant tissue cultures: Breakdown of normal controls. *Proc. Natl. Acad. Sci. USA* **91,** 5222–5226.

Rathbone, D. A., and Bruce, N. C. (2002). Microbial transformation of alkaloids. *Curr. Opin. Microbiol.* **5,** 274–281.

Sakai, K., Shitan, N., Sato, F., Ueda, K., and Yazaki, K. (2002). Characterization of berberine transport into *Coptis japonica* cells and the involvement of ABC protein. *J. Exp. Bot.* **53,** 1879–1886.

Samanani, N., and Facchini, P. J. (2002). Purification and characterization of norcoclaurine synthase. *J. Biol. Chem.* **277,** 33878–33883.

Samanani, N., Liscombe, D. K., and Facchini, P. J. (2004). Molecular cloning and characterization of norcoclaurine synthase, an enzyme catalyzing the first committed step in benzylisoquinoline alkaloid biosynthesis. *Plant J.* **40,** 302–313.

Samanani, N., Park, S.-U., and Facchini, P. J. (2005). Cell type-specific localization of transcripts encoding nine consecutive enzymes involved in protoberberine alkaloid biosynthesis. *Plant Cell* **17,** 915–926.

Sasaki, T., Song, J., Koga-Ban, Y., Matsui, E., Fang, F., Higo, H., Nagasaki, H., Hori, M., Miya, M., Murayama-Kayano, E., Takiguchi, T., Takasuga, A., *et al.* (1994). Toward cataloguing all rice genes: Large-scale sequencing of randomly chosen rice cDNAs from a callus cDNA library. *Plant J.* **6,** 615–624.

Sato, F., and Yamada, Y. (1984). High berberine-producing cultures of *Coptis japonica* cells. *Phytochemistry* **23,** 281–285.

Sato, F., Hashimoto, T., Hachiya, A., Tamura, K., Choi, K. B., Morishige, T., Fujimoto, H., and Yamada, Y. (2001). Metabolic engineering of plant alkaloid biosynthesis. *Proc. Natl. Acad. Sci. USA* **98,** 367–372.

Schmeller, T., Latz-Bruning, B., and Wink, M. (1997). Biochemical activities of berberine, palmatine and sanguinarine mediating chemical defence against microorganisms and herbivores. *Phytochemistry* **44,** 257–266.

Scholl, Y., Schneider, B., and Drager, B. (2003). Biosynthesis of calystegines: ^{15}N NMR and kinetics of formation in root cultures of *Calystegia sepium*. *Phytochemistry* **62,** 325–332.

Schroeder, G., Unterbusch, E., Kaltenbach, M., Schmidt, J., Strack, D., De Luca, V., and Schroder, J. (1999). Light-induced cytochrome P450-dependent enzyme in indole alkaloid biosynthesis: Tabersonine 16-hydroxylase. *FEBS Lett.* **458,** 97–102.

Shanks, J. V., and Morgan, J. (1999). Plant ''hairy root'' culture. *Curr. Opin. Biotechnol.* **10,** 151–155.

Shitan, N., Bazin, I., Dan, K., Obata, K., Kigawa, K., Ueda, K., Sato, F., Forestier, C., and Yazaki, K. (2003). Involvement of CjMDR1, a plant multidrug-resistance-type ATP-binding cassette protein, in alkaloid transport in *Coptis japonica*. *Proc. Natl. Acad. Sci. USA* **100,** 751–756.

Shoji, T., Nakajima, K., and Hashimoto, T. (2000). Ethylene suppresses jasmonate-induced gene expression nicotine biosynthesis. *Plant Cell Physiol.* **41,** 1072–1076.

Siberil, Y., Benhamron, S., Memelink, J., Giglioli-Guivarch, N., Thiersault, M., Boisson, B., Doireau, P., and Gantet, P. (2001). *Catharanthus roseus* G-box binding factors 1 and 2 act as repressors of strictosidine synthase gene expression in cell cultures. *Plant Mol. Biol.* **45,** 477–488.

St-Pierre, B., and De Luca, V. (1995). A cytochrome P-450 monooxygenase catalyzes the first step in the conversion of tabersonine to vindoline in *Catharanthus roseus*. *Plant Physiol.* **109,** 131–139.

St-Pierre, B., Laflamme, P., Alarco, A. M., and De Luca, V. (1998). The terminal *O*-acetyltransferase involved in vindoline biosynthesis defines a new class of proteins responsible for coenzyme A-dependent acyltransferase. *Plant J.* **14,** 703–713.

St-Pierre, B., Vazquez-Flota, F. A., and De Luca, V. (1999). Multicellular compartmentation of *Catharanthus roseus* alkaloid biosynthesis predicts intercellular translocation of a pathway intermediate. *Plant Cell* **11,** 887–900.

Suzuki, K., Yamada, Y., and Hashimoto, T. (1999a). Expression of *Atropa belladonna* putrescine *N*-methyltransferase gene in root pericycle. *Plant Cell Physiol.* **40,** 289–297.

Suzuki, K., Yun, D. Y., Chen, X. Y., Yamada, Y., and Hashimoto, T. (1999b). An *Atropa belladonna* hyoscyamine 6 β-hydroxylase gene is differentially expressed in the root pericycle and anthers. *Plant Mol. Biol.* **40,** 141–152.

Suzuki, M., Nakagawa, K., Fukui, H., and Tabata, M. (1987). Relationship of berberine-producing capability between *Thalictrum* plants and their tissue cultures. *Plant Cell Rep.* **6,** 260–263.

Takeda, S., Sato, F., Ida, K., and Yamada, Y. (1990). Characterization of polypeptides that accumulate in cultured *Nicotiana tabacum* cells. *Plant Cell Physiol.* **31**(2), 215–221.

Takeshita, N., Fujiwara, H., Mimura, H., Fitchen, J. H., Yamada, Y., and Sato, F. (1995). Molecular cloning and characterization of *S*-adenosyl-L-methionine: Scoulerine-9-*O*-methyltransferase from cultured cells of *Coptis japonica*. *Plant Cell Physiol.* **36,** 29–36.

Taticek, R. A., Lee, C. W., and Shuler, M. L. (1994). Large-scale insect and plant cell culture. *Curr. Opin. Biotechnol.* **5,** 165–174.

Tepfer, D. A., Goldman, A., Pamboukdjian, N., Maille, M., Lepingle, A., Chevalier, D., Denarie, J., and Rosenberg, C. (1988). A plasmid of *Rhizobium meliloti* 41 encodes catabolism of two compounds from root exudate of *Calystegia sepium*. *J. Bacteriol.* **170,** 1153–1161.

Tohge, T., Nishiyama, Y., Hirai, M. Y., Yano, M., Nakajima, J., Awazuhara, M., Inoue, E., Takahashi, H., Goodenovwe, D. B., Kitayama, M., Noji, M., Yamazaki, M., *et al.* (2005). Functional genomics by integrated analysis of metabolome and transcriptome of Arabidopsis plants over-expressing an MYB transcription factor. *Plant J.* **42,** 218–235.

Unterlinner, B., Lenz, R., and Kutchan, T. M. (1999). Molecular cloning and functional expression of codeinone reductase: The penultimate enzyme in morphine biosynthesis in the opium poppy *Papaver somniferum*. *Plant J.* **18,** 465–475.

Van Der Fits, L., and Memelink, J. (2000). ORCA3, a jasmonate-responsive transcriptional regulator of plant primary and secondary metabolism. *Science* **289,** 295–297.

Van Der Fits, L., Zhang, H., Menke, F. L. H., Deneka, M., and Memelink, J. (2000). A *Catharanthus roseus* BPF-1 homologue interacts with an elicitor-responsive region of the secondary metabolite biosynthetic gene *Str* and is induced by elicitor via a JA-independent signal transduction pathway. *Plant Mol. Biol.* **44,** 675–685.

Vazquez-Flota, F. A., De Carolis, E., Alarco, A. M., and De Luca, V. (1997). Molecular cloning and characterization of deacetoxyvindoline 4-hydroxylase, a 2-oxoglutarate-dependent dioxygenase involved in the biosynthesis of vindoline in *Catharanthus roseus* (I.) G. Don. *Plant Mol. Biol.* **34,** 935–948.

Vazquez-Flota, F., St-Pierre, B., and De Luca, V. (2000). Light activation of vindoline biosynthesis does not require cytomorphogenesis in *Catharanthus roseus* seedlings. *Phytochemistry* **55,** 531–536.

Vazquez-Flota, F., De Luca, V., Carrillo-Pech, M., Canto-Flick, A., and De Lourdes Miranda-Ham, M. (2002). Vindoline biosynthesis is transcriptionally blocked in *Catharanthus roseus* cell suspension cultures. *Mol. Biotechnol.* **22,** 1–8.

Verpoorte, R., and Memelink, J. (2002). Engineering secondary metabolite production in plants. *Curr. Opin. Biotechnol.* **13,** 181–187.

Vom Endt, D., Kijne, J. W., and Memelink, J. (2002). Transcriptional factors controlling plant secondary metabolism: What regulates the regulators? *Phytochemistry* **61,** 107–114.

Wang, E., Wang, R., Deparasis, J., Loughrin, H. H., Gan, S., and Wagner, G. J. (2001). Suppression of a P450 hydroxylase gene in plant trichome glands enhances natural-product-based aphid resistance. *Nat. Biotechnol.* **19,** 371–374.

Wang, M. B., and Waterhouse, P. M. (2002). Application of gene silencing in plants. *Curr. Opin. Plant Biol.* **5**(2), 146–150.

Waterhouse, P. M., and Helliwell, C. A. (2003). Exploring plant genomes by RNA-induced gene silencing. *Nat. Rev. Genet.* **4,** 29–38.

Weckwerth, W., and Fiehn, O. (2002). Can we discover novel pathways using metabolomic analysis? *Curr. Opin. Biotechnol.* **13,** 156–160.

Weid, M., Ziegler, J., and Kutchan, T. M. (2004). The roles of latex and the vascular bundle in morphine biosynthesis in the opium poppy, *Papaver somniferum*. *Proc. Natl. Acad. Sci. USA* **101,** 13957–13962.

Wesley, S. V., Helliwell, C. A., Smith, N. A., Wang, M. B., Rouse, D. T., Liu, Q., Gooding, P. S., Sing, S. P., Abbott, D., Stoutjesdijk, P. A., Robinson, S. P., Gleave, A. P., *et al.* (2001). Construct design for efficient, effective and high-throughput gene silencing in plants. *Plant J.* **27,** 581–590.

Whitmer, S., Canel, C., Hallard, D., Goncalves, C., and Verpoorte, R. (1998). Influence of precursor availability on alkaloid accumulation by transgenic cell line of *Catharanthus roseus*. *Plant Physiol.* **116,** 853–857.

Whitmer, S., Canel, C., Van Der Heijden, R., and Verpoorte, R. (2003). Long-term instability of alkaloid production by stably transformed cell lines of *Catharanthus roseus*. *Plant Cell Tissue Organ Cult.* **74,** 73–80.

Wiermann, R. (1981). Secondary plant products and cell and tissue differentiation. *In* "The Biochemistry of Plants" (E. E. Conn, ed.), Vol. 7, pp. 85–116. Academic Press, New York.

Winkel-Shirley, B. (2001). Flavonoid biosynthesis. A colorful model for genetics, biochemistry, cell biology, and biotechnology. *Plant Physiol.* **126**(2), 485–493.

Yamada, Y., and Mino, M. (1986). Instability of chromosomes and alkaloid content in cell lines derived from single protoplasts of cultured *Coptis japonica* cells. *Curr. Top. Dev. Biol.* **20,** 409–417.

Yamazaki, Y., Urano, A., Sudo, H., Kitajima, M., Takayama, H., Yamazaki, M., Aimi, N., and Saito, K. (2003). Metabolite profiling of alkaloids and strictosidine synthase activity in camptothecin-producing plants. *Phytochemistry* **62,** 461–470.

Yamazaki, Y., Kitajima, M., Arita, M., Takayama, H., Sudo, H., Yamazaki, M., Aimi, N., and Saito, K. (2004). Biosynthesis of camptothecin. In silico and in vivo tracer study from [1–^{13}C] glucose. *Plant Physiol.* **134,** 161–170.

Yukimune, Y., Tabata, H., Hara, Y., and Yamada, Y. (1994). Scopolamine yield in cultured roots of *Duboisia myoporoides* improved by a novel 2-stage culture method. *Biosci. Biotechnol. Biochem.* **58,** 1820–1823.

Yukimune, Y., Tabata, H., Higashi, Y., and Hara, Y. (1996). Methyljasmonate-induced overproduction of paclitaxel and baccatin III in Taxus cell suspension cultures. *Nat. Biotechnol.* **14,** 1129–1132.

Yun, D. J., Hashimoto, T., and Yamada, Y. (1992). Metabolic engineering of medicinal-plants—transgenic *Atropa belladonna* with an improved alkaloid composition. *Proc. Natl. Acad. Sci. USA* **89,** 11799–11803.

Zhao, J., Davis, L. C., and Verpoorte, R. (2005). Elicitor signal transduction leading to production of plant secondary metabolites. *Biotechnol. Adv.* **23,** 283–333.

Ziegler, J., Diaz-Chavez, M. L., Karamell, R., Ammer, C., and Kutchan, T. M. (2005). Comparative macroarray analysis of morphine containing *Papaver somniferum* and eight morphine free *Papaver* species identifies an O-methyltransferase involved in benzylisoquinoline biosynthesis. *Planta* **222,** 458–471.

Zubieta, C., Ross, J. R., Koschesk, I. P., Yang, Y., Pichersky, E., and Noel, J. P. (2003). Structural basis for substrate recognition in the salicylic acid carboxyl methyltransferase family. *Plant Cell* **15,** 1704–1716.

CHAPTER 12

Genetic Engineering for Salinity Stress Tolerance

Ray A. Bressan,* Hans J. Bohnert,† and P. Michael Hasegawa*

Contents

Abstract

Multiple biotic and abiotic environmental factors may constitute stresses that affect plant growth and yield in crop species. With a focus on ionic stress exerted by the presence of sodium, and the associated water deficit, recent

* Department of Horticulture and Landscape Architecture, Purdue University, Horticulture Building 1165, West Lafayette, Indiana 47907-1165
† Departments of Plant Biology and of Crop Sciences, University of Illinois at Urbana-Champaign, Urbana, Illinois 61801

Advances in Plant Biochemistry and Molecular Biology, Volume 1
ISSN 1755-0408, DOI: 10.1016/S1755-0408(07)01012-0

advances in our understanding are reviewed. Established physiological, biochemical, and genetic approaches are made more meaningful by the inclusion of genomics-type tools, which have been most helpful by making available a global view of transcriptome responses to salinity stress, and by providing lines from the global mutagenesis of model species, in particular for *Arabidopsis thaliana*. Many of the genetic elements that assure ion homeostasis and ion transport have become known, as have several elements that control ion homeostasis. Genes that respond to salinity stress have been identified through mutant screens, from comparative functional studies that relied on known physiological and phenotypic parameters. Until now, the resulting concepts and strategies for engineering salinity stress tolerance in their majority targeted single genes in biochemical pathways, which represent end points of response cascades, but engineering of upstream master switches that regulate the activity of many downstream genes and proteins is increasingly attempted. The rapidly growing body of results on (salinity) stress sensing and signaling promises to lead to the identification of those genes that are of superior significance in salt stress response pathways, and abiotic stresses in general.

Key Words: Abiotic stresses, Ion homeostasis, Osmotic adjustment, Stress engineering, Abscisic acid, Biochemical pathways, Stress sensing, Stress signaling pathways.

1. SALINITY STRESS ENGINEERING

Apart from satisfying scientific curiosity, the value that can be associated with knowledge about plant reactions to high salinity, leading to tolerance or explaining sensitivity, can be significant. One incentive is the agronomical value that crops might acquire if they could be made salt tolerant. Such modification, if accomplished without major yield penalties, would provide food security through yield stability in areas where crop production predominantly satisfies daily sustenance. Also, if more fresh water could be made available for growing urban populations, instead of its present primary use in agriculture, human lives could improve in many countries. On the other hand, the engineering of truly halophytic crops might generate incentives to utilize land that has not previously been under the plow, which could then endanger saline wetlands, estuaries, or semideserts that provide refuge for endemic organisms. Also, using saline groundwater for irrigation possibly represents a short-lived achievement, necessitating leaching of land by fresh water. In their majority, present crops are glycophytes depending on fresh water to approach their yield potential; their metabolism and growth are, however, affected at low concentrations of sodium, 50–100 mM NaCl, equivalent to ~20% of seawater strength. While it seems possible to engineer salinity tolerance at this level or at even somewhat higher concentrations of NaCl (Apse and Blumwald, 2002; Ward *et al.*, 2003), productivity might still be compromised. Results are already available from a number of experiments that used a controlled environment to gauge differences between a progenitor species and its engineered lines (Apse *et al.*, 1999; Ellul *et al.*, 2003; Garg *et al.*, 2002; Jang *et al.*, 2003; Kasuga *et al.*, 1999; Lee *et al.*, 2004; Mckersie *et al.*, 1996; Mittova *et al.*, 2003; Quintero *et al.*, 1996; Romero *et al.*, 1997; Roxas *et al.*, 1997; Shi *et al.*, 2003; Singla-Pareek *et al.*, 2003; Sulpice *et al.*, 2003; Urano *et al.*, 2004; Van Camp *et al.*, 1996; Wang *et al.*, 2003; Wu *et al.*, 2004; Zhifang and Loescher, 2003). The results with single gene engineering attempts, exemplified by these references that mostly targeted biochemical and physiological characters during the last decade, have identified a number of genes whose altered expression affects tolerance or sensitivity but agronomical benefits have yet to be documented. The characters that emerged as providing increased tolerance tend to support membrane and protein integrity, the synthesis of carbohydrates and N-containing compounds, energy provision, detoxification reactions, and a variety of transport proteins that establish ion and metabolite homeostasis.

New approaches have become possible as information emerged about the nature of the genetic elements that control expression of stress alleviating biochemical determinants, and about how hormonal changes elicit stress response pathways (Mukhopadhyay *et al.*, 2004; Nagaoka and Takano, 2003; Novillo *et al.*, 2004; Perruc *et al.*, 2004; Sakamoto *et al.*, 2004; Teige *et al.*, 2004; Villalobos *et al.*, 2004; Winicov and Bastola, 1999; Zhang *et al.*, 2004). This new frontier and orientation of abiotic stress research has profited from, and in many cases has been initiated by, work on the presently most tractable genetic model, *Arabidopsis thaliana* (Hasegawa *et al.*, 2000a, b; Ishitani *et al.*, 1997; Serrano and Rodriguez-Navarro, 2002; Shinozaki and Yamaguchi-Shinozaki, 1999; Shinozaki *et al.*, 2003; Tester and Davenport, 2003; Ward *et al.*, 2003; Xiong *et al.*, 2002a; Zhu, 2002). In this chapter, we will review the classical approaches and then outline recent developments with special emphasis on results that have become possible with the advent of a new age: genome sequences, transcript profiles, protein dynamics, and, especially, a large number of phenotyped and functionally identified mutant lines for *Arabidopsis*, as well as a growing number of other plants, including crop species.

2. THE CONTEXT OF SALINITY STRESS

The considerable increase in worldwide crop production that occurred during the green revolution did not result in substantially greater land use but focused on adapting germplasms to respond to altered farm management practices (Trewavas, 2001). However, even with better adjustment of crops and increased production efficiency, the actual yield is less than the crop genetic potential. By now, the increased population in developing countries puts even more constraints on production as urban populations compete with agriculture for fresh water. In parts of the world, this has necessitated the use of less suitable irrigation water, often water of low quality that is unsuitable for high-yield agriculture with existing crops. This in turn makes it paramount to find ways that realize the genetic potential and yield capacity of crop genomes, even under moderate stresses, or to enhance them by transgenic means.

Significant among the abiotic stresses is salinity, which not only constraints crop production in a particular growing cycle but also leads to steady deterioration of soils and irrigation water that compounds the effects of salinity on subsequent crop generations. In many countries around the globe where water is already scarce and droughts are recurring, soil salinity is a major constraint to crop productivity that negatively affects much of the cultivated land and substantially reduces yield (Flowers and Yeo, 1995; Läuchli and Epstein, 1990; Maas, 1990; Munns, 1993). These facts, and the prospect of erratic rainfall patterns that could increase in the future, have led to efforts to improve yield stability of present-day crops by focusing on abiotic stress factors (Flowers, 2004; Flowers and Yeo, 1995; Serrano, 1996). Crop improvement strategies seek to develop more salt-adapted or -adaptable germplasms by utilizing molecular genetic approaches and resources developed during the mid-1980s, notably marker-assisted breeding techniques, exploration of halophytic species, biotechnology, and genomics (Apse and Blumwald, 2002; Garciadeblas *et al.*, 2003; Hasegawa *et al.*, 2000b; Koyama *et al.*, 2001; Loudet *et al.*, 2003; Ribaut and Hoisington, 1998; Tuberosa *et al.*, 2002; Zhu, 2002). Table 12.1 lists strategies that have been suggested, initially with respect to different classical breeding approaches, and more recently, including genome-anchored methods that can significantly enhance breeding because physical maps, mapped and phenotyped mutants, and genome sequences provide precision.

High salt in the root environment, salinity that typically appears in the form of increased NaCl in the soil, is a combination of ionic and hyperosmotic imbalance and secondary effects including pathologies that inhibit growth and can affect development or cause cell death (Hasegawa *et al.*, 2000a; Zhu, 2001, 2002). Ionic and osmotic stress signals are sensed and decoded by all plants via distinct and interconnecting signal pathways that are response relays for the control of unique and stress-specific programs. These pathways are response relays that control genetic programs and coordinate determinants and processes required for adaptation. Both stress conditions constitute environmental perturbations that modulate normal cellular or developmental programs (Zhu, 2002, 2003). Consequently, salt adaptation involves determinants that establish ion homeostasis and/or osmolyte biosynthesis, termed osmotic adjustment. In cases where the severity of a stress condition exceeds the capacity of a species, ecotype or line to acclimate, this then precludes cell division, expansion and normal development and may result in death (Apse and Blumwald, 2002; Blumwald, 2000; Hasegawa *et al.*, 2000b; Zhu, 2001, 2002). The capacity of species to adapt to salt stress distinguishes glycophytic species with a reduced capacity from halophytes that are to various degrees able to adapt well, or even grow better at slightly increased levels of sodium, and many of the latter may in fact use NaCl as a "cheap" osmoticum (Adams *et al.*, 1998).

TABLE 12.1 Strategies for enhancing salinity stress tolerance in crop species[a]

Approach	Description	References
Yield	Disregarding tolerance/ sensitivity—yield alone is important	Condon *et al.*, 2004
Mutation breeding	Designing new phenotypes by generating mutants	Ramage, 1980
Screen within phenotype	Utilizing genetic variation within a species	Norlyn, 1979
Wide crosses	Introducing germplasm from wild species into crop species	Rush and Epstein, 1981
Develop halophytes	Generate new crops from halophytic species	O'Leary *et al.*, 1985
Transgenic modification	Altered metabolic (or signaling) pathways based on perceived functional knowledge	Cushman and Bohnert, 2000; Hill *et al.*, 2004
Chromosome typing	Map- or (genome-) sequence-based QTL analysis	Lin *et al.*, 2004; Price *et al.*, 2002; Quesada *et al.*, 2002
Genomics	Transcript/protein/metabolite profiling of model and crop species, and their stress-tolerant relatives	Bennetzen, 2002; Bressan *et al.*, 2002; Cushman and Bohnert, 2000; Zhu, 2002
Systems biology	Combining characters from mutation screening, genome sequencing, reverse and forward genetics, and computational tools into QTL-characterized backgrounds	Henikoff *et al.*, 2004; Provart and Mccourt, 2004; Tani *et al.*, 2004

[a] Expanded from Bohnert and Bressan (2001) and Flowers and Yeo (1995).

Evolutionary adaptations have resulted in species that exhibit different competence to tolerate or resist high salt and complete their life cycles. Although glycophytes and halophytes differ substantially in their capacity to tolerate salt, the cytosolic and organellar machineries of the two plant categories seem to be equally sensitive to Na^+ and Cl^- (Flowers, 2004; Greenway and Osmond, 1972; Hasegawa *et al.*, 2000b; Jacoby, 1999; Serrano, 1996). Consequently, adaptation by plants in both groups requires cellular responses that attenuate the osmotic and ionic components of salt stress. The options are limited. They involve NaCl exclusion or compartmentalizing Na^+ and Cl^- into an "inert" compartment, vacuole, or tissue. Other response mechanisms, such as avoidance reactions, in essence induced dormancy, are not considered here. Simultaneously, mechanisms that confine salt must be accompanied by the accumulation of solutes that are compatible with cellular metabolism. Such osmolytes must increase in cytosol and organelles to achieve osmotic adjustment (Blumwald, 2000; Hasegawa *et al.*, 2000a,b; Zhu, 2002, 2003). Both osmotic adjustment and the confinement of sodium in (pre)vacuoles have been shown in the single cell model, *Saccharomyces cerevisiae* (Gaxiola *et al.*, 1999; Hohmann, 2002). That is, the mechanisms by which all plants achieve osmotic and ionic equilibria are mediated by orthologous mechanisms based on conserved biochemical and/or physiological functions that are inherently necessary for essential plant processes (Hasegawa *et al.*, 2000a; Serrano *et al.*, 1999; Van Camp, 2005; Van Camp *et al.*, 1996; Zhu, 2000, 2001). This statement has been substantiated by the genomic DNA sequences of two glycophytes, *Arabidopsis thaliana* and *Oryza sativa*, which seem to include all components that have been researched as essential or necessary for plants to cope with salt stress in different model species and crops (*Arabidopsis* Genome Initiative, 2000; Goff *et al.*, 2002; Yu *et al.*, 2002).

What then, if the important stress tolerance components are ubiquitous, distinguishes glycophytes and halophytes? To solve this conundrum, research is directed into several areas. One is to determine if halophytic versions of salt adaptation determinants have greater innate operational capacity to facilitate survival, growth, and development in saline environments, that is, if halophytic versions of genes may represent an allele that encodes a more effective protein that functions in the presence of high salt (Waditee *et al.*, 2002). An example supporting such a view may be the case of L-*myo*-inositol-1-phosphate synthase that distinguishes rice (*O. sativa*) from a wild relative (*Porteresia coarctata*). In *Porteresia*, the homodimeric enzyme retains its aggregation state in high salt, while the rice protein disintegrates into enzymatically inactive monomers at much lower salt concentrations. This may be due to a domain that discriminate the two forms of the enzyme, and, indeed, overexpression of the *Porteresia* enzyme enhances salt tolerance (Majee *et al.*, 2004).

Alternatively, halophytes may control universal determinants in a manner that imparts to the species a preadapted state or a faster and superior ''adaptive response capacity'' when the saline environment becomes increasingly severe. A point in favor of such a scenario may be studies targeting the *Arabidopsis* relative *Thellungiella halophila* (salt cress), which is salt tolerant. Preliminary transcript profiling and analysis of expressed sequence tags (ESTs) seems to indicate that the salt cress constitutively shows high nonstress activities for a range of genes/transcripts, and that induction of these transcripts is initiated at a higher stress level than in *Arabidopsis* (Inan *et al.*, 2004; Taji *et al.*, 2004). The fact that (eu) halophytes show increased growth at moderate concentrations of NaCl, higher than in fresh water, might be causally related to the high constitutive expression of stress response pathways.

Third, outlining a related hypothesis, it must also be considered that some halophytes have evolved specialized adaptations (e.g., salt glands for excretion or bladder cells for the storage of NaCl). It is therefore possible that such species also possess other unique determinants with specialized function to mediate adaptation that are missing from the genomes of glycophytes. Such uniqueness will only be revealed when we have identified the relevant genes in halophytic models. A variation of this theme is chromosomal context and genome size. The fusion of genomes during speciation or endoreduplication events that have been documented for many species in diverse plant families may have resulted in duplicated genes, paralogues of ubiquitous genes that further evolved in some species in response to changing environments (Arango *et al.*, 2003). Such genes could, although they might not be superior in their biochemical function to those in glycophytes, impart higher tolerance to a species by their expression at constitutively higher levels, in a stress-inducible manner, in different compartments, or by being connected to altered or novel regulatory circuits. We will briefly review salt stress tolerance mechanisms and transgenic approaches that have begun to engineer ionic and osmotic tolerance mechanisms into model species. Subsequently, we will place emphasis on the regulatory circuits that control mechanisms of tolerance acquisition (Schachtman, 2000; Zhu, 2002).

3. ION HOMEOSTASIS

High concentration of NaCl is the dominant salt stress globally because of its abundance in the environment (Hasegawa *et al.*, 2000b). Although both Na^+ and Cl^- can be cytotoxic, best understood are effects of Na^+ toxicity and the mechanisms by which homeostasis is established (Blumwald *et al.*, 2000; Hasegawa *et al.*, 2000b; Niu *et al.*, 1995; Zhu, 2003). The capacity to maintain a high cytoplasmic K^+ to Na^+ ratio is essential. Apparently, halophytes have an inherently greater capacity to maintain this balance in saline environments (Flowers and Yeo, 1992; Flowers *et al.*, 1986). Ca^{2+} facilitates K^+ relative to Na^+ selective uptake through mechanisms that control the uptake of both ions (Epstein, 1961; Läuchli, 1996; Läuchli and Epstein, 1990).

Vacuolar compartmentalization of Na^+ is a salt-adaptive mechanism used by all plants and is a conserved process in organisms as taxonomically distant as yeasts (Blumwald *et al.*, 2000; Flowers and Yeo, 1995; Flowers *et al.*, 1986; Glenn *et al.*, 1999; Hasegawa *et al.*, 2000b; Pardo and Quintero, 2002). This process not only mitigates against toxic accumulation of ions in the cytoplasm, but it also is physiologically crucial for osmotic adjustment in saline environments, which is necessary for cell volume regulation and development (Hasegawa *et al.*, 2000b). Although increased Na^+ compartmentalization in the vacuole enhances salt tolerance (Apse *et al.*, 1999; Blumwald *et al.*, 2000; Hasegawa *et al.*, 2000b; Zhang and Blumwald, 2001; Zhang *et al.*, 2001). Na^+ sequestration in this compartment is critically dependent on the regulation of net uptake at the plasma membrane (Hasegawa *et al.*, 2000b; Ros *et al.*, 1998). Functional disruption of the plasma membrane Na^+ efflux system in *Arabidopsis* or yeast results in toxic levels of Na^+ to accumulate in the cytosol, even when the vacuolar compartmentalization machinery is functional (Garciadeblas *et al.*, 1993; Munns, 2002; Zhu, 2000).

Once loaded into the root xylem from the soil solution, ions move to the shoot in the transpirational stream (Flowers and Yeo, 1992; Läuchli, 1996; Maathuis and Sanders, 1996; Munns *et al.*, 2002). Controlling Na^+ content in the root xylem thus regulates ion content in the shoot and leaf apoplast to a level where intracellular compartmentalizing processes allow cells to be ion sinks (Hasegawa *et al.*, 2000a,b; Maathuis and Sanders, 1996; Munns *et al.*, 2002). Ion homeostasis *in planta* requires coordination of cellular processes with those that function in intercellular, tissue, and organ ion regulation (Flowers *et al.*, 1986; Hasegawa *et al.*, 2000b; Maathuis and Sanders, 1996; Munns *et al.*, 2002). For example, vacuolar compartmentalization allows interconnected root cells, from the epidermis to the xylem parenchyma, to function as at least temporary Na^+ and Cl^- sinks that substantially reduce ion content in the transpiration stream moving into the shoots.

3.1. Ion transport

Loss- or gain-of-function experimentation has identified many determinants that control and mediate Na^+ and K^+ uptake, as well as homeostasis *in planta* (Amtmann and Sanders, 1999; Hasegawa *et al.*, 2000b; Horie and Schroeder, 2004; Maathuis *et al.*, 1996; Mäser *et al.*, 2002; Niu *et al.*, 1995; Schachtman, 2000; Sondergaard *et al.*, 2004; Tester and Davenport, 2003; Zhu, 2003). Illustrated in Fig. 12.1A are known or suspected cellular Na^+ uptake systems, as well as other facilitators of homeostasis such as aquaporins and the proton transporters that establish membrane potentials in different compartments (Gaxiola *et al.*, 2002). Included are channels and transporters responsible for K^+ and Na^+ flux and H^+ pumps that generate the requisite electrochemical potential necessary to facilitate channel function or secondary-active transport (Arango *et al.*, 2003; Borsani *et al.*, 2001; Horie and Schroeder, 2004; Schachtman, 2000; Tester and Davenport, 2003; Vitart *et al.*, 2001; Ward *et al.*, 2003; Zhu, 2003). Other, yet unidentified, genetic loci are involved in the control of Na^+ and K^+ homeostasis, including those that regulate Ca^{2+} homeostasis (Nublat *et al.*, 2001; Rus *et al.*, 2001). Focal are Na^+ transport systems that control not only intracellular distribution of Na^+ but also homeostasis between tissues and organs (Fig. 12.1B) (Berthomieu *et al.*, 2003; Horie and Schroeder, 2004; Laurie *et al.*, 2002; Rus *et al.*, 2004; Tester and Davenport, 2003; Ward *et al.*, 2003; Yokoi *et al.*, 2002). Possibly most important is the salt overly sensitive (SOS) pathway described from *Arabidopsis*, described in detail below (Aharon *et al.*, 2003; Gaxiola *et al.*, 2001; Quintero *et al.*, 2002; Su *et al.*, 2001; Talke *et al.*, 2003). One of the three proteins in this pathway, SOS1, controls Na^+ uptake into the root xylem. It is hypothesized that high-affinity K^+ transporter 1 (HKT1), originally described as a K^+ transporter, acts as a Na^+ transporter that recirculates the ion from the shoot to the root when Na^+ is in excess (Horie and Schroeder, 2004; Munns, 2002; Rus *et al.*, 2004; Shi *et al.*, 2003; Yokoi *et al.*, 2002). Phenotypic suppression of *sos1–1* NaCl sensitivity by dysfunctional *hkt1* alleles is genetic evidence that these two transport systems have opposing functions in Na^+ homeostasis (Laurie *et al.*, 2002). Thus, SOS1 and HKT1 may function in concert to regulate Na^+ content in the shoot. However, *AtHKT1* overexpression does not increase salt tolerance of *Arabidopsis* as would be

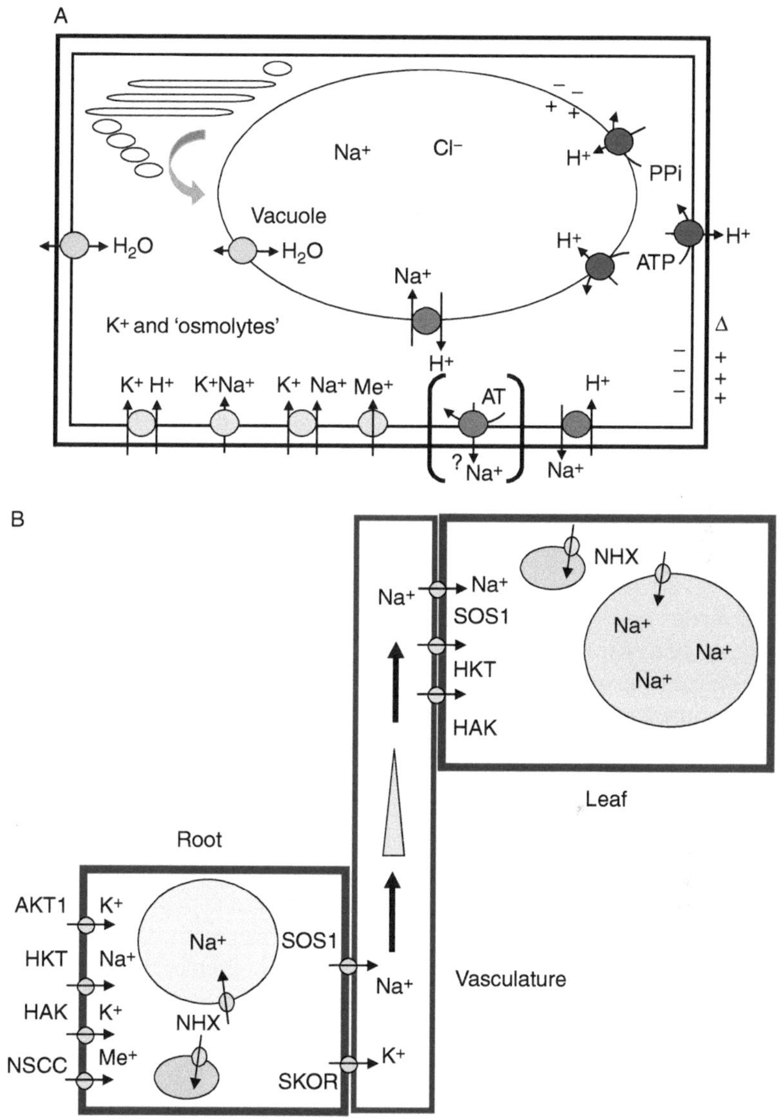

FIGURE 12.1 Transporters in plant cells and transport of sodium in plants. (A) Complexity of transporters and facilitators involved in ion homeostasis. A cellular view of ion and water facilitators located in the plasma membrane or tonoplast membrane, although the location of a number of these proteins in other membranes (e.g., the prevacuolar complex, plastids, mitochondria, or the endomembrane system) is not excluded, and has rarely been investigated. White symbols—proton ATPases and inorganic pyrophosphatases; dark grey symbols—sodium/proton antiporters (SOS1 and NHXn); light grey symbols—different K^+ (or alkali ion) transporters (HKT, KAT, HAK, CNGC, NSCC); stippled symbol—aquaporins (and/or low molecular weight metabolite facilitators); dotted symbols—the presence of yeast ENAx-like (bracketed) sodium ATPases has not been observed in plants. However, the transfer of the yeast ENA1 gene into tobacco suspension culture cells resulting in altered salt stress tolerance has been reported

expected (Laurie *et al.*, 2002). Most likely the endomembrane cation/H^+ transporters, exemplified by the NHX family, are principally responsible for vacuolar compartmentalization of Na^+ in plant cells (Aharon *et al.*, 2003; Munns, 2002; Rigas *et al.*, 2001). However, NHX gene copy number and their expression under nonsaline conditions might indicate that the various NHX forms could have very different, additional functions or that their function under saline conditions becomes modified.

Na^+ detrimentally affects K^+ acquisition. K^+ is an essential macronutrient that functions in critical processes ranging from charge balance, osmotic adjustment, and enzyme catalysis to growth and development (Elumalai *et al.*, 2002; Maathuis *et al.*, 1996; Rains and Epstein, 1965). Na^+ disturbs intracellular K^+ homeostasis possibly because it can compete for binding sites on enzymes and transport proteins (Cramer *et al.*, 1987; Epstein, 1961; Niu *et al.*, 1995; Tester and Davenport, 2003). Ca^{2+} enhances K^+/Na^+ selective accumulation because it facilitates K^+ uptake (Epstein, 1961; Hasegawa *et al.*, 2000b; Li *et al.*, 1998).

3.2. Control of ion homeostasis

Na^+ and K^+ homeostasis control occurs through direct modulation of transport protein properties resulting from the physiological and biochemical status of cells or as an output of a responsive signal relay system(s) that mediates transcriptional or posttranscriptional regulation (Hasegawa *et al.*, 2000a; Zhu, 2002, 2003). Conductance through transport systems is affected directly by pH, membrane potential, hyper- or hypoosmolarity, Ca^{2+}, cyclic nucleotides, phosphorylation/dephosphorylation, as well as protein interaction and modification (Cherel *et al.*, 2002; Demidchik *et al.*, 2002; Qui *et al.*, 2002; Quintero *et al.*, 2002; Talke *et al.*, 2003; Tester and Davenport, 2003; Vera-Estrella *et al.*, 1999). There is substantial evidence for transcriptional regulation by salinity or osmotic stresses of plasma membrane and vacuolar H^+ pumps, as well, additional posttranslational regulation of the resident proton pumps by phosphorylation with participation of 14:3:3 proteins (Dambly and Boutry, 2001; Hasegawa *et al.*, 2000b; Shi *et al.*, 2000). Also, transcription of genes encoding Na^+ and K^+ channels and transporters are modulated by salt and osmotic stresses (Aharon *et al.*, 2003; Pilot *et al.*, 2003; Rains and Epstein, 1967; Shi and Zhu, 2002; Su *et al.*, 2001, 2002, 2003; Tester and Davenport, 2003; Zhu, 2003). *SOS1* mRNA stability is enhanced by salt stress. NaCl regulates Ca^{2+} modulation of Na^+ uptake, attributable to the inhibition of

FIGURE 12.1 *(continued)*
(Marin *et al.*, 2003). (B) Long-distance K^+ and Na^+ transport. The distribution of sodium in plants is controlled by a limited number of potassium (or alkali ion) transporters and sodium/proton antiporters (Tester and Davenport, 2003). The transport of sodium and redistribution into sinks, such as specialized cells, inert tissues, or vacuoles, generates a gradient that can be maintained as long as spaces are available into which sodium can be caged, and as long as water movement can be maintained. In at least some halophytes, the gradient can have a reversed orientation, with sodium accumulation in distal plant parts, thus generating an osmotic potential that draws water into aerial tissues.

unidirectional Na^+ influx across the plasma membrane through nonselective cation channels (NSCCs) (Davenport and Tester, 2000; Demidchik and Tester, 2002; Epstein, 1961; Essah *et al.*, 2003; Roberts and Tester, 1997). Although unconfirmed, prediction is that these transport proteins are cyclic nucleotide-gated channels (CNGC) whose gating properties are controlled by numerous effectors, including pharmacological agents known to inhibit similar channels in animal models, and whose Na^+ conductance is blocked by Ca^{2+} (Cheng *et al.*, 2003; Hirschi, 2004; Tester and Davenport, 2003). Genetic modulation of intracellular Ca^{2+} levels affects numerous ion homeostatic processes (Matsumoto *et al.*, 2002).

Models for osmotic and ionic stress signal recognition assume that distinct sensors activate discrete signal transduction pathways that effect transcriptional and/or posttranscriptional control over determinants that mediate ion homeostasis, osmotic regulation, and obviate or attenuate stress pathologies (Cheng *et al.*, 2003; Shinozaki *et al.*, 2003; Zhu, 2002, 2003). Ca^{2+} and abscisic acid (ABA) are focal regulatory intermediates in hypersaline stress signaling that controls adaptation. Since it is the agricultural goal to enhance yield stability in saline environments, hypersaline stress effects on metabolic, cell division, growth, and development programs combinatorially increases the genetic determinants that become involved.

The search continues for the osmotic and/or the Na^+ sensors. As indicated, hyper- or hypoosmolarity might affect directly the gating of ion channels that initiate a signal transduction pathway (Zhu, 2003). Hyperosmotically induced Ca^{2+} transients activate signaling through calcineurin resulting in the transcription of *ENA1* (encoding a plasma membrane Na^+/H^+ antiporter) increasing salt tolerance of yeast cells and in transformed plant cells (Marin *et al.*, 2003; Nakayama *et al.*, 2004). Alternatively, electrophoretic ion flux dissipates the membrane potential resulting in a cascade that controls osmotic or ion homeostasis. AtHK1 is implicated as a two component histidine kinase (Hik) that functions as a phospho-relay transducer controlling ABA-independent or dependent signaling that activates expression of genes encoding putative tolerance determinants (Shinozaki *et al.*, 2003). *Synechocystis* sp. PCC 6803 Hiks, Hik16, Hik33, Hik34, and Hik41 implicated in salt and ion perception, while K^+ and ionic strength activate autophosphorylation of KdpD, the putative turgor sensor expression of an operon encoding for a high affinity K^+ transport system (Chinnusamy *et al.*, 2004; Jung *et al.*, 2000).

The Ca^{2+}-activated SOS signal transduction and response pathway facilitates Na^+ and perhaps K^+ homeostasis *in planta* (Zhu, 2003). It is presumed that hypersaline stress induces a Ca^{2+} transient that is decoded by components of the SOS pathway to facilitate Na^+ homeostasis (Shinozaki *et al.*, 2003; Zhu, 2003). SOS3 recognizes Ca^{2+} signals and binds the divalent cation. SOS3 then activates the serine/threonine kinase SOS2, which phosphorylates the plasma membrane localized SOS1 to induce its Na^+/H^+ antiporter capacity. SOS1 has been suggested to be a Na^+ sensor but the determinant(s) responsible for the salt-induced Ca^{2+} transient is not yet identified.

The *sos* mutants (*sos1*, *sos2*, and *sos3*) are NaCl sensitive and exhibit K^+ deficiency. The later phenotype indicates that the SOS signaling pathway has a positive regulatory effect on K^+ acquisition, although neither the regulatory components nor the regulated determinants have been identified. SOS1 does not possess innate K^+ transport capacity. Pretreatment of seedlings with 50 mM NaCl reduces K^+ permeability of root cell plasma membranes isolated from *sos1-1* but not of wild-type seedlings (Qiu *et al.*, 2003). Reduced K^+ uptake into (and consequential K^+ deficiency) *sos1* plants is attributed to elevated cytosolic Na^+ levels that occur because of the defective Na^+ efflux system (Guo *et al.*, 2001; Qiu *et al.*, 2003; Talke *et al.*, 2003). *htkt1* Mutations suppress K^+ deficiency of *sos1*, *2* or *3* plants by reducing the intracellular Na^+ accumulation (Rus *et al.*, 2004). Mutations to other loci that suppress Na^+ accumulation also attenuate the K^+ deficient phenotype of *sos3–1* plants.

4. STRATEGIES TO IMPROVE SALT TOLERANCE BY MODULATING ION HOMEOSTASIS

Discoveries about the identity and function of salt stress signaling components and the transport proteins that mediate Na^+ homeostasis make it possible to propose strategies for the biotechnological improvement of crops with high probability to increase yield stability in saline environments encountered in cultivated agriculture. Strategies include regulating salt stress signal pathway(s) to be constitutively active or more responsive to stress or modulating effector activity or efficacy. Constitutive activation of the SOS pathway is achievable by modifying the SOS2 kinase through deletion of its autoinhibitory domain or site-specific modifications to the catalytic region, or by ectopic inducible coexpression of SOS3/SOS2 (Gaxiola *et al.*, 2001; Quintero *et al.*, 2002; Talke *et al.*, 2003). Presumably, the SOS signal pathway modulation can be engineered to enhance the salt adaptation capacity of plants. Another approach is the coordinate control of net Na^+ flux across the plasma membrane and vacuolar compartmentalization. Other yet unidentified salt adaptation determinants that are outputs of the SOS pathway could also be positively affected (Zhu, 2003).

Obviously, overexpression of the putative sodium/proton antiporter AtNHX1 also enhances plant salt tolerance, possibly by increasing vacuolar Na^+ compartmentalization that minimizes the toxic accumulation of the ion in the cytosol and facilitates growth in the saline environment (Apse *et al.*, 1999; Zhang and Blumwald, 2001; Zhang *et al.*, 2001). The authors report that in the NHX overexpressing tomato plants, Na^+ is not accumulating in all organs, which might indicate that the altered Na^+ flux could initiate yet other mechanisms, possibly influencing other ion transporters. Furthermore, overexpression of SOS1 increases salt tolerance of *Arabidopsis* (Shi *et al.*, 2003). These results indicate that regulating net Na^+ influx across the plasma membrane together with enhancing the capacity for vacuolar compartmentalization should substantially facilitate Na^+ homeostasis and salt tolerance. With current understanding, this is achievable by modulating the expression or activity of SOS1 (Na^+ efflux) and/or HKT1 (Na^+ influx) at the plasma membrane, as well as modulating the activity of the vacuolar Na^+/H^+ antiporter and/or H^+ pump (Rubio *et al.*, 1995; Vitart *et al.*, 2001; Zhu, 2003).

Future molecular genetic resources for bioengineering of salt tolerance include alleles that encode transport determinants with greater capacity to mediate Na^+ homeostasis. Halophytes are a potential germplasm resource. Alternatively, new alleles may be generated by directed molecular evolution. For example, mutant variant forms of HKT1 transport more K^+ at the expense of Na^+ and render greater salt tolerance (Cheeseman, 1988; Rubio *et al.*, 1999) Promoters that direct the tissue- and/or inducer-specific regulation of target genes can condition the expression of the signal intermediates and the effectors. Thus regulation of the numerous salt tolerance determinants can be coordinated for an effective plant response but much of the cost associated with salt tolerance in nature might be minimized because some essential evolutionary necessities can be compensated for by agricultural practices. Again, halophytes may be natural resources or synthetic promoters will be constructed. Apart from the SOS cascade, it is presumed that yet other salt adaptation signal regulatory pathways exist that await discovery and dissection. Conceivably of equally critical importance are growth and development pathways that perceive and interact with salinity perception and response, and are modulated by salt to affect yield stability. In this respect, the redistribution of sodium along the xylem stream from root to reproductive organs could become an avenue for additional intervention. The capacity to short-circuit effects of increasing Na^+ on growth and development will require a clear understanding about how and why environmental perturbation has such a negative impact on processes required for crop production.

5. STRATEGIES TO IMPROVE SALT TOLERANCE BY MODULATING METABOLIC ADJUSTMENTS

Many plant species or ecotypes of species, most conspicuously salt-tolerant species, alter the intracellular osmotic potential when salt stressed, and this has given rise to the concept of ''osmotic adjustment,'' which, although not unchallenged, has become a widely accepted concept (Bohnert and Shen, 1999; Rontein *et al.*, 2002). The increase of metabolically inert compounds would provide both a sink for unutilized products of primary metabolism and assure continued water influx into the plant. The topic has received much attention and has been reviewed extensively (Apse and Blumwald, 2002; Hasegawa *et al.*, 2000b; Hill *et al.*, 2004; Hohmann, 2002; Knight *et al.*, 1997; Verslues and Bray, 2004). The concept is based on many observations demonstrating the accumulation of a number of metabolites that are normally present in low concentrations in response to an osmotic or ionic imbalance. As the determinants that control entire (metabolic) pathways have emerged, engineering has moved to altering expression of transcription factors (TFs), and components of calcium-dependent responses that elicit phospho-relay regulatory systems that respond to signals caused by various abiotic stresses (Table 12.2) (Himmelbach *et al.*, 2003; Qi and Spalding, 2004; Serrano *et al.*, 1999; Shinozaki *et al.*, 2003; Teige *et al.*, 2004; Zhu, 2003).

TABLE 12.2 Transgenic approaches to engineering salt stress tolerance

Gene	Protein	Source	Cellular role(s)	Transgenic host	Comments	References
TFs						
alfin1[a]	Zn-finger family	*M. sativa*	TF	*M. sativa*	Overexpressors show salinity tolerance	Winicov and Bastola, 1999
tsi1[a]	Tobacco stress-induced gene 1	*N. tabacum*	TF	*N. tabacum*	Tolerance to salinity and salicylic acid; transgene has homology to ethylene-responsive element binding protein (EREBP)/AP2	Park *et al.*, 2001
MYB10[a]	Myb-family	*Craterostigma plantagineum*	TF	*A. thaliana*	General stress tolerance; glucose insensitivity	Villalobos *et al.*, 2004
OSISAP1[a]	Zinc-finger family	*O. sativa*	TF	*N. tabacum*	Improved growth under several stress conditions	Mukhopadhyay *et al.*, 2004
Components of signal transduction or ROS detoxification						
cnb1[a]	Calcineurin B1	*S. cerevisiae*	Ca^{2+} binding	*N. tabacum*	Coexpression of catalytic and regulatory subunits; strong tolerance to salinity	Pardo *et al.*, 1998
Oscdpk7[a]	Ca-dependent protein kinase	*O. sativa*	Protein kinase (PK)	*O. sativa*	Induction of stress-responsive genes in response to salinity, drought and cold	Saijo *et al.*, 2000
GST/GPX[a]	Glutathione-*S*-transferase/ glutathione peroxidase	*E. coli*	Detoxification of herbicides and toxic substances	*N. tabacum*	Overexpressors of GST/GPX show stimulated seedling growth under chilling and salt stress	Roxas *et al.*, 1997
Mn-SOD[b]	Mn-superoxide dismutase		Dismutation of ROS in mitochondria	*M. sativa*	Transformants showed significantly higher survival under water stress and freezing	Mckersie *et al.*, 1996

Osmolyte production (detoxification or, possibly, acting as secondary signal molecules)[c]						
Bet[a]	Choline dehydrogenase	*E. coli*	Glycinebetaine biosynthesis	*Synecho-coccus* sp.	Transformants with enhanced survival of Rubisco in plants under salt stress	Nomura *et al.*, 1995
betB[a]	Betaine aldehyde dehydrogenase	*E. coli*	Glycinebetaine biosynthesis	*N. tabacum*	Transformed plants with better growth in osmotic stress conditions	Holmstrom *et al.*, 2000
codA[b]	Choline oxidase A	*Arthrobacter globiformis*	Glycinebetaine biosynthesis	*A. thaliana*	Transformants tolerant to salt and cold	Hayashi *et al.*, 1997
ectA, ectB, ectC[a]	L-2,4-Diamino-butyric acid acetyltransferase, L-2,4-diamino butyric acid transaminase, L-ectoine synthase	*Halomonas elongata*	Ectoine biosynthesis	*N. tabacum*	Transformants with increased tolerance to hyperosmotic stress	Nakayama *et al.*, 2000
Imt1[a]	Myo-inositol-*O*-methyltransferase	*M. crystallinum*	D-Ononitol biosynthesis	*N. tabacum*	Transformants better adapted to water and salt stress	Sheveleva *et al.*, 1997
mtlD[a]	Mannitol-1 phosphate dehydrogenase	*E. coli*	Mannitol metabolism	*N. tabacum*	Transformants with better growth under salt stress	Tarczynski *et al.*, 1993
mtlD[b]	Mannitol-1 phosphate dehydrogenase	*E. coli*	Mannitol metabolism	*N. tabacum*	Transformants more tolerant to salt and oxidative stress	Shen *et al.*, 1997
P5CS[a]	*O*-pyrroline 5-carboxylate synthase	*V. acontitifolia*	Proline biosynthesis	*N. tabacum*	Transformants accumulated twofold more proline than wild type; more tolerant to water stress	Hong *et al.*, 2000; Kishor *et al.*, 1995
ProDH[a]	Proline dehydrogenase	*A. thaliana*	Proline biosynthesis	*A. thaliana*	Antisense transgenics more tolerant to freezing and high salinity than wild type	Nanjo *et al.*, 1999

(*continued*)

TABLE 12.2 *(continued)*

Gene	Protein	Source	Cellular role(s)	Transgenic host	Comments	References
Tsp1[a]	Trehalose 6-phosphate synthase	*S. cerevisiae*	Trehalose biosynthesis	*N. tabacum*	Transformants with trehalose accumulation and improved drought tolerance	Romero *et al.*, 1997
Transport proteins						
AtNhx1[d]	Na^+/H^+ antiporter	*A. thaliana*	Vacuolar antiporter	*A. thaliana*	Transformants with sustained growth and development in soil water with high sodium chloride	Apse *et al.*, 1999
AtNhx1[a]	Na^+/H^+ antiporter	*A. thaliana*	Vacuolar antiporter	*L. esculentum*	Transformants with sustained growth in high salt (200 mM) without Na^+ accumulation in fruits	Zhang and Blumwald, 2001; Zhang *et al.*, 2001
Hal1[a]	Protein involved in regulation of K^+ transport and Na^+ extrusion	*S. cerevisiae*	Regulation of K^+ transport	*L. esculentum*	Transformants with higher level of salt tolerance; transgenics able to retain more K^+ than control under salt stress	Ellul *et al.*, 2003; Gisbert *et al.*, 2000
Others						
AtNced3	9-*Cis*-epoxy carotenoid synthesis	*A. thaliana*	ABA biosynthesis	*A. thaliana*	Transformants with increased endogenous ABA; enhanced levels of drought/ABA-inducible genes; reduced transcription rates; improved drought tolerance	Iuchi *et al.*, 2000

AtRabG3e[a]	Vesicle trafficking protein	*A. thaliana*	Vesicle traffic	*A. thaliana*	Induction of salt and drought tolerance	Mazel *et al.*, 2004
Bip[a]	Binding protein	*G. max*	Molecular chaperone; unfolded protein response (UPR)	*N. tabacum*	Transformants with higher tolerance to water stress	Alvim *et al.*, 2001
Gly1[a]	Glyoxylase-1	*Brassica juncea*	Converts 2-oxaldehydes into 2-hydroxy acids	*N. tabacum*	Overexpressors with tolerance to methylglyoxal and high salt	Veena *et al.*, 1999
Gpd[a]	Nicotinamide adenine dinucleotide (NAD^+)-dependent glyceraldehydes 3-phosphate dehydrogenase	*Pleurotus sajor-caju*	Glycolytic pathway	*S. tuberosum*	Transformants with salt stress tolerance	Jeong *et al.*, 2001

[a] CaMV35S promoter used in transgene expression.
[b] Modified CaMV35S promoter used.
[c] Metabolic disturbance caused by the overexpression of (foreign) enzymes may constitute signal inducing stress responses, indicating the existence of yet unknown signal perception modules (Eastmond and Graham, 2003).
[d] Other promoters used.

A comprehensive listing of engineering studies can be found at: www.plantstress.com/Files%5CAbiotic-stress_gene.htm.

A number of tagged, phenotyped, and identified mutants, either using a "root bending" or a "luciferase" screen (Hasegawa *et al.*, 2000; Ishitani *et al.*, 1997) are available at: http://www.life.uiuc.edu/bohnert/arizonamut.html and at: http://www.life.uiuc.edu/bohnert/purduemut.html.

5.1. Osmotic adjustments and controlling factors

Among the accumulators are amino acids, predominantly proline, sugars and sugar alcohols, such as sucrose or mannitol, trehalose, mannitol/sorbitol, or inositol derivatives, and more complex carbohydrates, such as fructans and raffinose-related compounds. This list is likely to expand in the future as other models are investigated. In a drought-adapted watermelon, for example, citrulline, an intermediate in the urea cycle, has been detected as a drastically accumulating metabolite during drought stress. Citrulline function may be in radical oxygen scavenging (Akashi *et al.*, 2001, 2004). A number of experiments have been reported that attempted to engineer osmolyte accumulation into glycophytic plants that show only marginal accumulation of metabolites with the intention to improve tolerance. Table 12.2 provides a selection of such studies, also including experiments to transgenically engineer ionic stress tolerance and to engineer regulatory circuits. The range included exemplifies the major categories of genes or cDNAs that have been used in engineering: osmolytes and other protectants [chaperones, late-embryogenesis-abundant (LEA) and heat-shock proteins (HSPs)], transporters and pumps, scavengers of radicals, adjustments in hormone biosynthesis, and regulatory genes, outlined in detail at: www.plantstress.com/Files%5CAbiotic-stress_gene.htm. Under strictly controlled growth conditions, it has been shown in many of these experiments that the plants exhibited showed increased osmotic, ionic, or temperature tolerance (Table 12.2). Often the actual increase or accumulation did not amount to concentrations found in the natural models gave rise to the "compatible osmolyte" concept.

These experiments demonstrated several other aspects as well. First, osmotic/ionic abiotic stress tolerance seems to be controlled by different mechanisms depending on age or developmental stage, that is, seedling, vegetative, and reproductive stages, each seem to require stage-specific regulation of tolerance and protective determinants (Rontein *et al.*, 2002). In one of the first attempts at engineering, for example, salinity stress tolerance (mannitol accumulation) was observed only when the transgenic tobacco plants received stress during early vegetative growth (Tarczynski *et al.*, 1993). Second, the common use of strong, constitutively expressed regulatory elements, while potentially leading to high (enzyme) expression, product accumulation, and vegetative tolerance, is or can be nonphysiological. This was demonstrated by the high accumulation of D-ononitol and mannitol in transgenic tobacco that protected the plants at vegetative growth stages (salinity and drought), but prevented normal seed formation due to the interference of the accumulating metabolites, both nonutilizable metabolic end-products in tobacco, with sucrose unloading in the developing seeds (Sheveleva *et al.*, 2000). Figure 12.2A gives a schematic rendition of biochemical and physiological mechanisms, structures, and determinants that have recognized as stress relevant from experiments conducted during the last decade.

Third, synthesis leading to accumulation is not necessarily the major purpose of a purported osmolyte, while the accumulation may be a pathological side effect. One example is yeast that increases the pathway leading to increased synthesis of trehalose (Hohmann, 2002). However, yeast also increases levels of enzymes that degrade trehalose, indicating that flux through the pathway (consuming reducing power) may be more important than simply making more (Fig. 12.2B).

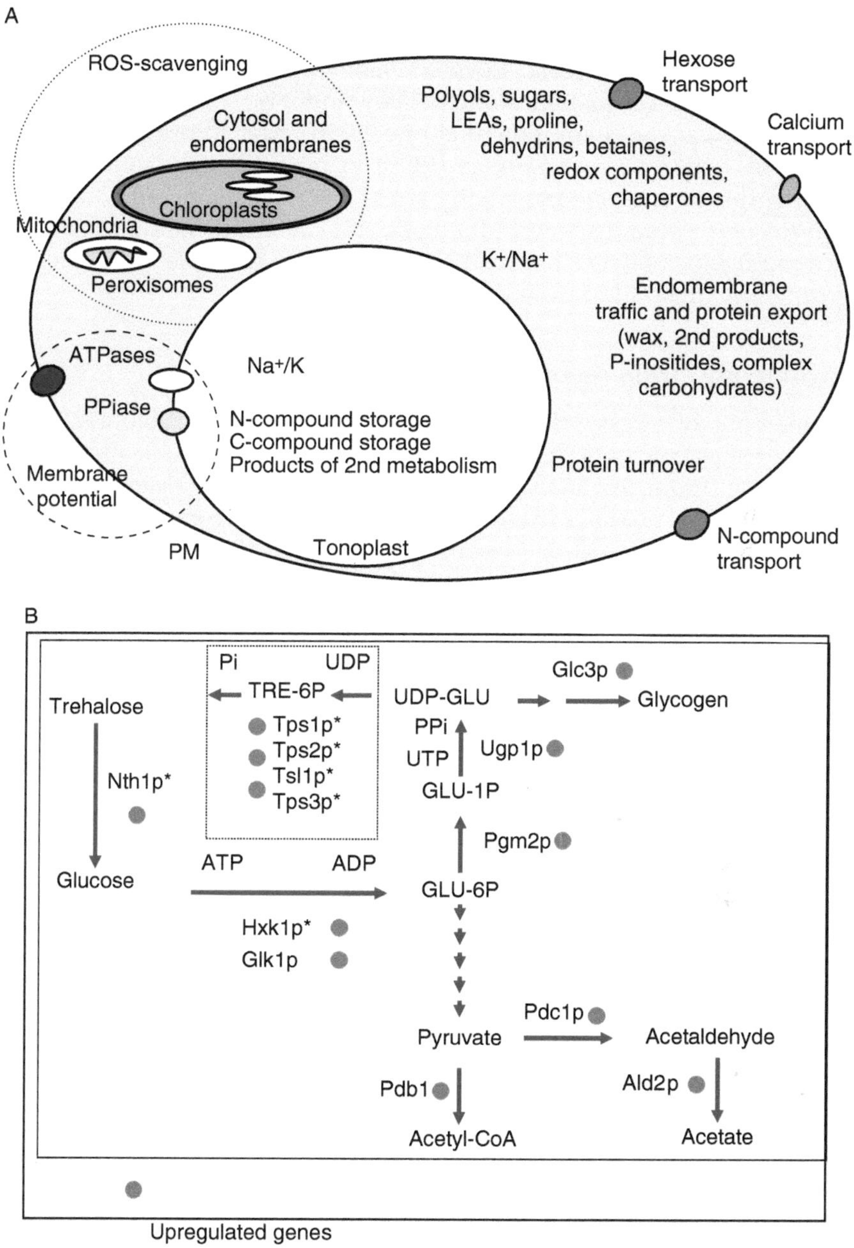

FIGURE 12.2 Biochemical and metabolic determinants of salinity stress tolerance. (A) A schematic representation of cellular mechanisms depicted as functional categories (e.g., maintenance of membrane potential, ROS-scavenging, altered membrane traffic, or protein turnover), and identification of major metabolites and protein families that constitute cellular defenses against ionic stress (modified after Hasegawa *et al.*, 2000b). Included are functions such as protein turnover, membrane structure, and vesicular traffic reorientations. Not included are molecular functions that also play important roles: chromatin remodeling, transcription/splicing, RNA transport, or regulation of translation. (B) Metabolic reactions leading to trehalose synthesis and

This behavior has been shown in several studies (Yale and Bohnert, 2001). Only at very high salt concentrations, which arrests or slows growth, will trehalose accumulate drastically in yeast. Such a view might then lead to a different interpretation for the compatible solutes. They might be seen as metabolic valves that adjust or lower the redox potential of cells in order to prevent or minimize production of radical oxygen species (ROS) in mitochondria and chloroplasts. Finally, attempts at engineering salinity stress tolerance have only begun to pay attention to compartment-specific strategies and the appreciation of stress tolerance as a multigenic network has not been tackled, for example, by the accumulation of different stress alleviating determinants.

5.2. Engineering stress response control determinants

The engineering of stress response control determinants (Table 12.2) is more recent. Few transgenic experiments have been reported, specifically for salinity stress tolerance. From the analysis of tagged mutants in *Arabidopsis*, one might deduce a number of possibly fruitful avenues, in the categories of: TFs, mitogen-activated protein kinases (MAPKs), calcium-regulated signaling components (for example, the SOS pathway genes), or site-directed mutants of these components that alter activity or activation of particular pathways.

However, one aspect that has to be considered in all engineering attempts is the effect of abiotic stresses on growth. Yeast and plants have been shown to include a signaling link that connects stress, the stress response, and growth retardation (Bressan *et al.*, 2002; Hohmann, 2002; Zhu, 2001, 2002, 2003). As shown in Fig. 12.3, first outlined by Zhu (2001), abiotic stresses lead to type-specific responses (SOS in the clearly documented case of salinity stress), but there are also general responses that extend beyond the specific elicitor. Most likely, these are responses to metabolic or hormonal deviations from a species-specific checkpoint or injury caused by the deviation in the form of, for example, ROS. It will require much work to address the dichotomy between protection and growth in the future.

5.3. How to analyze transgenic lines resulting from (salinity) stress engineering

Apart from satisfying scientific curiosity, what is the value that can be associated with knowledge about plant reactions, leading to tolerance or sensitivity, to high salinity? The major incentive is the agronomical value that crops might acquire if they could be made tolerant to high salinity because most crops are glycophytes,

FIGURE 12.2 (*continued*)
reutilization in *S. cerevisiae* stressed by addition of 1 M NaCl (90 min). Upregulated transcripts are indicated by filled circles, and the presence of a stress response element (STRE) in promoters of individual genes is indicated (*). Although trehalose accumulates long term, the trehalose-cleaving enzyme, trehalase, is also upregulated (Yale and Bohnert, 2001).

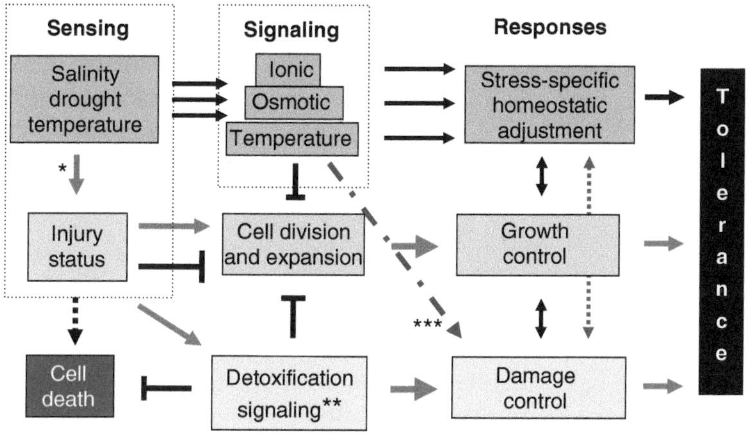

FIGURE 12.3 Processes and mechanisms connecting abiotic stresses, injury, defense, and tolerance. Reactions to various abiotic stresses are seen as an integrated network of sensing, signaling, and downstream response pathways. Reactions are elicited that are stress factor specific. General responses are those that originate from the perception of metabolic imbalance and/or cellular injury. This leads to altered regulation of sets of genes that overlap several environmental stress conditions. Repression of cell cycle, growth, and development is seen as a consequence of additional hormone-regulated pathways (modified) (Bohnert and Bressan, 2001; Zhu, 2001). *Protein unfolding, membrane leakage, water/ion imbalance, ROS production. **General pathway distinct from the stress in functional categories of genes that are affected, and are not specific for a particular stress. Different stresses can elicit distinct isoforms (orthologues and paralogues) of genes in the same functional categories. ***Overlapping pathways that link specific stresses with individual genes that may be coregulated, or hyperregulated, by the specific condition and the general response.

whose metabolism and growth are affected at low concentrations of sodium, well below 100 mM that would not pose problems for halophytes, in the soil. It seems possible to engineer tolerance at this level and at even higher NaCl concentrations, yet productivity might well be compromised (Apse *et al.*, 1999; Kasuga *et al.*, 1999; Mckersie *et al.*, 1996; Romero *et al.*, 1997; Roxas *et al.*, 1997; Tarczynski *et al.*, 1993; Van Camp *et al.*, 1996; Zhifang and Loescher, 2003). Providing additional credibility to claims of engineered tolerance to a particular stress condition, it seems appropriate to establish rules about the experimental design of abiotic stress engineering, and how the results should be reported. The text box added below includes a suggestion for such rules. They were established by the attendees of a conference on salinity stress responses in plants in 2001. These recommendations (compiled by A. D. Hanson, University of Florida) describe what the participants considered essential and sufficient experimental process for the analysis and description of the effect of single transgenes on engineering or altering complex genetic and physiological traits, such as tolerance to salinity stress, drought, low temperature, or freezing (Serrano and Rodriguez-Navarro, 2002).

Essential Process for Evaluating Transgene Impact

Genetic engineering of salt tolerance is becoming a mature field, so it seems appropriate to define a minimum set of criteria for establishing unambiguously that transgenic plants do indeed show tolerance that is attributable to the transgene.

Do:

- Establish that the stress-tolerant phenotype of interest is shown by a fair-sized population of primary transformants (e.g., 10 independently generated transgenic lines) that have been shown to express the transgene. This avoids being misled by insertional mutagenesis, positional effects, and somaclonal variation. A range of expression levels may be more informative than selecting the highest expressers, especially in studies involving metabolic traits.
- Likewise, show that a comparably sized population of control plants harboring the empty vector does not show the stress-tolerant phenotype. Another desirable type of control (so far not used) is using an inactivated transgene (e.g., lacking its catalytically essential residues). This type of control is unsuitable for oligomeric proteins, for example, enzymes in pathways for which channeling of substrates is shown or suspected, due to dominant negative effects.
- For progeny of primary transformants, work with single-insert lines. Either backcross the primary transformants to the wild type and compare the hemizygotes (% of progeny) to the azygotes, or self the primary transformants and compare the homozygotes and azygotes (each % of progeny). This controls for somaclonal effects.
- Carry out stress tests using the established, statistically sound procedures that are mandatory in the agronomy literature. Blind designs (in which the experimenter is unaware of the plants' identities) are advisable.
- For *Arabidopsis*, deposit viable seeds of transgenic lines in a stock center.
- Discourage institutional publicity officers or the press from extrapolating from small advances in basic research to major benefits to agriculture.

Do Not:

- Use only wild type as a control.
- Use the stress-tolerant phenotype of interest as the first criterion for screening transgenic lines to be studied (and do not discard others).

6. PLANT SIGNAL TRANSDUCTION FOR ADAPTATION TO SALINITY

We have earlier discussed the SOS pathway under the aspect of controlling sodium homeostasis, but detection of the pathway provides an example of another kind because SOS genes were first found in a salt-sensitive, glycophytic species. Indeed, it has become clear that the cells of virtually all plants possess the capability to sense and respond to a saline environment. Tolerance is, to a species specific and widely varying degree, genetically possible (Serrano and Rodriguez-Navarro, 2002). Although the salt tolerance of many halophytes is constitutive, the tolerance of others is induced by the salinity level of the environment (facultative halophytes) but functions that determine halophytism are ubiquitous. From many physiological and biochemical studies, it has also become clear that the plant adaptive response to salinity involves four basic co-coordinated adjustments, outlined in Fig. 12.3: ion homeostasis, osmotic compensation (water homeostasis), injury repair or avoidance, and growth reduction (Zhu, 2001). Yet, the mechanisms and precise genetic components involved in adaptation to salinity have remained a mystery for long. Our understanding of how plants perceive the salinity of their environment and adjust appropriately has improved tremendously with the introduction and use of *Arabidopsis* as a model system. Although *Arabidopsis* is not salt tolerant in the sense of halophytic tolerance, screening for mutants with lower tolerance than the wild type in *Arabidopsis* in the mid 1990s has been successful and much work has confirmed that *Arabidopsis* does have genetic components that control the ability to survive and grow in a salinized environment (Ishitani *et al.*, 1997; Warren *et al.*, 1996; Werner and Finkelstein, 1995; Xiong *et al.*, 2002b). Our present understanding of plant signal transduction for adaptation to salinity has become possible based on common ancestry of land plants. The approach has been successful because the advantages of well-developed genetic and molecular tools developed for *Arabidopsis* could be exploited.

6.1. The SOS signal pathway controls adaptation to hypersalinity

In *Arabidopsis*, the Ca^{2+}-dependent SOS signaling pathway transduces salt stress to activate the plasma membrane Na^+/H^+ antiporter (SOS1), which mediates Na^+ efflux and homeostasis necessary for salt adaptation (Guo *et al.*, 2001; Shi *et al.*, 2003; Zhu, 2002). Current evidence indicates that the myristoylated Ca^{2+}-binding protein SOS3 activates the SOS2 serine/threonine kinase and recruits it to the plasma membrane (Quintero *et al.*, 2002; Talke *et al.*, 2003; Zhu, 2003). An NaCl-induced Ca^{2+} transient is transduced by SOS3 leading to the activation of SOS2 (Verslues and Bray, 2004; Zhu, 2003). The SOS3–SOS2 protein kinase complex phosphorylates SOS1 resulting in Na^+/H^+ antiport activity (Quintero *et al.*, 2002; Talke *et al.*, 2003). The SOS3–SOS2 complex also activates *SOS1* expression, perhaps through processes that affect *SOS1* mRNA stability (Shi *et al.*, 2003; Su *et al.*, 2001; Zhu, 2003). The Ca^{2+} sensor SCaBP5 (SOS3 family) and its interacting kinase PKS3 (SOS2 family) are components of a negative regulatory circuit that controls ABA signaling through ABI1/2. This regulatory circuit presumably is a negative controller of ABA-induced Ca^{2+} channel gating and is necessary for Ca^{2+} oscillations that activate the SOS pathway and other signaling required for salt adaptation (Gaxiola *et al.*, 2001). The sensor and remaining Ca^{2+} perturbation components of the SOS pathway, and few downstream targets of the SOS pathway, are yet to be identified. SOS1 has been implicated to be the Na^+ sensor (Zhu, 2003). Besides SOS1, only the VSP2 gene has been implicated as a SOS pathway target for transcriptional control. Importantly, the affect of the SOS pathway on transcription of SOS1 and VSP2 indicates that an unknown TF that interacts with a promoter element(s) on SOS1 and VSP2 is also activated by the SOS pathway.

6.2. What do we know about stress sensors in plants?

Salinity stress shares some important physical–chemical characteristics with other abiotic stresses such as desiccation and cold stress. All of these stresses impose an osmotic gradient on plant cells that impedes the movement of water into or retention inside the plasma membrane. Also, these stresses all mediate gene expression changes and other responses through a transient Ca^{++} influx causing elevated cytosolic Ca^{++} levels (Xiong *et al.*, 2002a). As such, genes that control the impact of the response of plants to one of these stresses also affect responses to other factors (Qi and Spalding, 2004). Although we will not discuss here genes that have been identified in screens for response to desiccation or cold stress, it is relevant to note that in all of these screens relatively few genes encoding putative stress sensing (receptor) proteins have been identified. It is interesting to note also that an important class of environmental sensors, two component Hiks, has been found to be involved in thermosensing in prokaryotic cyanobacteria and in *Bacillus subtilis* (Suzuki *et al.*, 2000; Urao *et al.*, 1999). The well-studied yeast model system has also been utilized to identify the important salinity/osmotic sensor SLN1. Shinozaki *et al.* have identified in *Arabidopsis* an Hik, AtHK1, which is able to complement the yeast SLN1 mutant, and is thereby implicated as an osmosensor in plants (Urao *et al.*, 1999). The inability to detect mutations in abiotic stress sensors, at first thought, appears puzzling since these signaling components are the first molecules involved in the plant's adaptive responses to stress and should therefore have large, easily detectable, influences on adaptive phenotypes such as reducing injury and death, and reduced growth. However, it may be just for this reason of high importance to survival under stress that sensor signal components would likely be highly redundant. This may not only result from the occurrence of more than one gene with the same sensing function but more likely derives from the overlapping function of sensors with some but not complete specificity for particular environmental cues.

6.3. SOS independent pathways and protein kinase systems

The most highly developed model for NaCl (osmotic) signaling and response systems is the common baker's yeast, *S. cerevisiae*. Apart from the two-component Hik receptor SLN1, an Src-homology 3 (SH3) domain protein is also involved in osmosensing (Hohmann, 2002). These sensors control both the ion homeostasis and osmotic adjustment responses to NaCl stress in yeast, through the activation of MAPK signal cascades. These are also commonly involved in environmental signal responses of animals' systems (Chang and Karin, 2001). Several MAPKs of plants have been shown to be activated by osmotic or NaCl stress (Jonak *et al.*, 2002). In addition, the transcript levels of MAPK genes have been found to be elevated by osmotic stresses (Ichimura *et al.*, 2000). Although there is considerable likelihood that some of these MAPK-encoding genes are involved in controlling adaptive responses of plants to NaCl/osmotic stress, very few genetic studies have been conducted to determine the effects of loss of function or overexpression of such genes on stress tolerance phenotypes. Their definitive roles in stress adaptation therefore, remain elusive. Several other protein kinases besides MAPKs have also been implicated in control of stress response in plants. Especially interesting is a 42 kDa nonspecific protein kinase belonging to the sucrose nonfermenting (SNF1) kinase group that is activated by NaCl (Mikolajczyk *et al.*, 2000). This group of kinases is apparently independent of the SOS pathway, and being unrelated to the MAPK family, suggests that plants may have a novel signal system to respond to NaCl stress.

7. ABA IS A MAJOR MEDIATOR OF PLANT STRESS RESPONSE SIGNALING

For many years, investigators have observed that the plant hormone ABA accumulates in plant tissues, especially leaves, in response to osmotic-based stresses, including NaCl stress (Himmelbach *et al.*, 2003; Zhu, 2002). Although other plant hormones including ethylene, salicylic acid, and jasmonic acid may also participate in various stress responses and even have interactive roles, ABA has remained the most important plant hormone controlling response and adaptation to abiotic stress.

A major mechanism by which ABA controls the plant adaptive response to osmotic/NaCl stress is thought to involve the alteration of gene expression, and many osmotic stress-responsive genes whose expression is mediated by ABA have been identified (Hoth *et al.*, 2002; Seki *et al.*, 2002). Confirmation of the control of gene expression in adaptation to stress has been provided by studies that show the effects of genes controlling ABA sensitivity on gene expression (Xiong and Zhu, 2003). The ABA-insensitive mutants *abi1* and *abi2* dramatically affect the ability of ABA to induce gene expression changes. Although important gene expression changes induced by stress are induced by ABA, it is now well accepted that stress-induced gene expression changes in plants are rather complex; they can be both dependent and independent of ABA mediation (Grillo *et al.*, 1995; Ingram and Bartels, 1996; Kurkela and Franck, 1990; Nordin *et al.*, 1991).

In addition to genes that affect ABA sensitivity of target gene expression, genes that encode proteins controlling ABA biosynthesis have been identified and found to be themselves controlled by stress exposure (Xiong and Zhu, 2003). Therefore, the long-standing observation of a stress-induced increase in tissue

ABA levels appears to be at least partially the result of transcriptional regulation. The main biosynthetic pathway for ABA is initiated in plants by the conversion of zeaxanthin to violaxanthin by the enzyme zeaxanthin epoxidase (ZEP) (Audran *et al.*, 1998). An important and perhaps rate limiting step is the oxidative cleavage of neoxanthin by 9-*cis* epoxycarotenoid deoxygenase (NCED). These enzymes and the ABA aldehyde oxidase (AAO) and the molybdenum cofactor sulfurylase (MCSU) have all been shown to be transcriptionally upregulated by drought or salt stress (Audran *et al.*, 1998; Iuchi *et al.*, 2000; Seo *et al.*, 2000; Xiong *et al.*, 2001a, 2002a). Since, genes encoding enzymes for several steps in the ABA biosynthetic pathway may be also transcriptionally activated by ABA itself, through a possible Ca^{++}/phosphorylation signal cascade, ABA may participate in a feed-forward loop to amplify ABA-mediated stress responses (Xiong *et al.*, 2001a,b, 2002a,b; Zhao *et al.*, 2001).

Besides controlling phenotype by way of modulating the activity of TFs such as AB1–3, AB1–4, and AB1–5, it is clear that ABA is involved in controlling many aspects of accessing the mediation between the environment and appropriate response information encoded within plant genomes (Hoth *et al.*, 2002). The SAD1 gene describes a mutant with phenotypic alterations involving ABA that are manifested through processes at the posttranscriptional or RNA metabolism level of control and several other genes in this function have been described (ABH1, CPL3, CPL1, HYL-1) (Hoth *et al.*, 2002; Hugouvieux *et al.*, 2002; Koiwa *et al.*, 2004; Xiong *et al.*, 2001a, 2002a). The SAD1 gene encodes an SM-like SnRNA protein that when mutated conveys supersensitive responses to drought and ABA (Xiong *et al.*, 2001a, 2002a). The HYL1 gene also mediates ABA responses by causing hyperaccumulation of AB15 mRNA and protein [HYL1 encodes a double stranded (dsRNA) bind protein and probably affects ABA] responses by processing or stabilization of RNA (Lu *et al.*, 2002). The Fiery-2 and AtCPL1 are alleles of the gene that encodes RNA polymerase II carboxy-terminal domain phosphatase-like proteins that also control RNA polymerase II activity to specifically mediate the efficiency of ABA/stress responses (Koiwa *et al.*, 2002; Xiong *et al.*, 2002b). In addition, mRNA cap structure and transcript maturation are also involved in ABA control of stress responses as revealed by the ABA hypersensitive 1-cap-binding protein2 (Hugouvieux *et al.*, 2001, 2002). It is likely that several more genes involving ABA-mediated phenotypes will be discovered that control many aspects of transcript processing and translation. Also, given the rapidly rising importance of unusual epigenetic control processes such as miRNA, SiRNA, DNA, and chromatin protein modifications (e.g., methylation or acetylation), it is becoming clear that there are epigenetic coding systems in plants that encode dynamic information guiding development and environmental responses beyond that directed by the encoded information of the DNA sequence and the central dogma of transcription and translation (Wada *et al.*, 2004). These information-encoding systems will certainly play important roles in the function of plant stress responses including those mediated by ABA (Himmelbach *et al.*, 2003).

It is clear now that there are many signaling components that control the adaptive responses of plants to salt/osmotic stress. Furthermore, the identification of a number of these components has clearly shown that stress-responsive signal pathways overlap between many different stresses (Hasegawa *et al.*, 2000a; Xiong *et al.*, 2002b). This is a clear indication that we will need to identify many more components of stress signaling in order to understand their interrelationships and how they cooperate to mediate responses to multiple stresses that often occur together in the environment. Several important technical advances are allowing the opportunity to accomplish this formidable task.

The completion of the sequencing of the *Arabidopsis* genome has paved the way for both forward and reverse genetic screens aimed at the identification of mutants with a myriad of phenotypes including many with altered stress tolerances. In addition, the collection of tagged insertion mutants available from the SALK Center that represent mutations in nearly all expressed genes of *Arabidopsis* has so simplified the task of identifying which gene is responsible for any particular mutant phenotype, that functional genomics analysis has become a tool now available to investigators with less experience in molecular genetics. In fact, the tool of insertion tagging has been surprisingly successful and is now rapidly allowing further dissection of signal pathways through the production of many more mutant phenotypes including second site mutations that suppress or enhance original phenotypes (Rus *et al.*, 2004). The emerging new phenotypes and the genes responsible for them will greatly increase our understanding of the genetic basis of adaptation of plants to stress environments.

8. SUMMARY

The present focus on genomics-type plant biology has been ushered in by the generation of the *Arabidopsis* sequencing project initiated in the end of 1980s, mirrored on the coincidental focus on bacterial, yeast, and animal, human in particular, genome sequencing projects. Genome sequences have become essential requisites for anchoring ESTs and expression profiles, and even more significantly, for determining which protein and pathways are present in an organism. The recognition of syntenic relationships between species is increasingly exploited for comparative genomics analyses (Bennetzen, 2002). Similar comprehensive data collection methods have emerged for proteins and metabolites, with improvements in tools and technologies continuous or accelerating. Sequences and dynamic expression profiles are only a starting point: the number of predicted reading frames is continuously increasing as predictive bioinformatics tools improve, as additional reading frames, dismissed or not recognized in the past, are confirmed by their presence, and as the siRNAs, RNA genes that selectively silence particular transcripts, are now being added as novel, important components of gene expression regulation. The number of splicing variants, leading to different protein sequences from one gene, can be expected to increase as well. Within this multitude of genes will be many functions relevant for ion and metabolic homeostasis under saline conditions, as well as specialized pathways for other abiotic stresses; functions that underlie the multigenic trait that is stress tolerance or resistance. In *S. cerevisiae*, the model that has most crucially contributed to our understanding of salt stress tolerance, more than 500 genes confer a "severe salt phenotype" to the cells when deleted (http://www.yeastgenome.org/cache/genome-wide-analysis.html) (Hohmann, 2002; Serrano *et al.*, 1999). More than 400 of these genes have homologues in *Arabidopsis*. Even when trivial causes, for example, the deletion of a ribosomal protein gene or an essential RNA polymerase subunit, are excluded, the genes that are essential for yeast cell survival identify many different functional categories, most likely in any cell.

The most promising way forward will, most likely, be to identify the stress-relevant genes in model species through mutagenesis and forward screens and tilling methods (Henikoff *et al.*, 2004; Tani *et al.*, 2004). This strategy will be especially useful when the population of tagged mutants carries a reporter gene that reports altered responses to stress (Ishitani *et al.*, 1997). A second opportunity is to become more aware of evolutionarily related naturally stress tolerant species that are relatives of established glycophytic model species. In comparisons of gene and protein expression patterns, and by determining divergent gene numbers (paralogues of ubiquitous genes), we can learn about the underlying functions that determine different plant life styles (Bressan *et al.*, 2002; Inan *et al.*, 2004; Taji *et al.*, 2004).

Additionally, the immediate future will be characterized by high throughput localization studies for all or most of the salinity stress-related transcripts and proteins, using, for example, cell ablation techniques combined with microarray analysis, and cellular and subcellular painting of transcripts and proteins by high-throughput *in situ* and real-time, *in vivo* fluorescence detection and localization methods. Eventually, we will have a virtual representation of all transcripts, proteins, and major metabolites during the life of a number of model species from seed to seed under optimal conditions, and when challenged by abiotic stresses.

This information when combined with classical and marker-assisted breeding and correlating quantitative trati loci (QTL) regions with genome information may enable us to generate stress tolerant species and lines of crops that rely on the immense genetic variability that exists in plants (Koyama *et al.*, 2001; Loudet *et al.*, 2003; Tuberosa *et al.*, 2002).

ACKNOWLEDGEMENTS

We thank the many students and collaborators over many years, too numerous to list, whose dedicated work supported conclusions that we were able to summarize. Especially, we thank Albino Maggio, Federika Consiglio, and Jae Cheol Jeong for contributions to this chapter. We also thank the many colleagues whose critique and discussion we have taken to heart and enjoyed, while we could not cite all the literature that the salt stress community has accumulated. Work in our laboratories has been supported by the Rockefeller Foundation, the US Department of Agriculture (NRI), the US Department of Energy (Biological Energy Program), and in several programs—in particular the Plant Genome Program (DBI9813360 and DBI022905)—by the US National Science Foundation.

REFERENCES

Adams, P., Nelson, D. E., Yamada, S., Chmara, W., Jensen, R. G., Bohnert, H. J., and Griffiths, H. (1998). Growth and development of *Mesembryanthemum crystallinum* (Aizoaceae). *New Phytol.* **138,** 171–190.

Aharon, G. S., Apse, M. P., Duan, S., Hua, X., and Blumwald, E. (2003). Characterization of a family of vacuolar Na^+/H^+ antiporters in *Arabidopsis thaliana*. *Plant Soil* **253,** 245–256.

Akashi, K., Miyake, C., and Yokota, A. (2001). Citrulline, a novel compatible solute in drought-tolerant wild watermelon leaves, is an efficient hydroxyl radical scavenger. *FEBS Lett.* **508,** 438–442.

Akashi, K., Nishimura, N., Ishida, Y., and Yokota, A. (2004). Potent hydroxyl radical-scavenging activity of drought-induced type-2 metallothionein in wild watermelon. *Biochem. Biophys. Res. Commun.* **323,** 72–78.

Alvim, F. C., Carolino, S. M., Cascardo, J. C., Nunes, C. C., Martinez, C. A., Otoni, W. C., and Fontes, E. P. (2001). Enhanced accumulation of BiP in transgenic plants confers tolerance to water stress. *Plant Physiol.* **126,** 1042–1054.

Amtmann, A., and Sanders, D. (1999). Mechanisms of Na^+ uptake by plant cells. *Adv. Bot. Res.* **29,** 75–112.

Apse, M. P., and Blumwald, E. (2002). Engineering salt tolerance in plants. *Curr. Opin. Biotech.* **13,** 146–150.

Apse, M. P., Aharon, G. S., Snedden, W. A., and Blumwald, E. (1999). Salt tolerance conferred by overexpression of a vacuolar Na^+/H^+ antiport in *Arabidopsis*. *Science* **285,** 1256–1258.

Arabidopsis Genome Initiative (2000). Analysis of the genome sequence of the flowering plant *Arabidopsis thaliana*. *Nature* **408,** 796–815.

Arango, M., Gévaudant, F., Oufattole, M., and Boutry, M. (2003). The plasma membrane proton pump ATPase, the significance of gene subfamilies. *Planta* **216,** 355–365.

Audran, C., Borel, C., Frey, A., Sotta, B., Meyer, C., Simonneau, T., and Marion-Poll, A. (1998). Expression studies of the zeaxanthin epoxidase gene in *Nicotiana plumbaginifolia*. *Plant Physiol.* **118,** 1021–1028.

Bennetzen, J. (2002). The rice genome. Opening the door to comparative plant biology. *Science* **296,** 60–63.

Berthomieu, P., Conéjéro, G., Nublat, A., Brackenbury, W. J., Lambert, C., Savio, C., Uozumi, N., Oiki, S., Yamada, K., Cellier, F., Gosti, F., Simonneau, T., *et al.* (2003). Functional analysis of *AtHKT1* in *Arabidopsis* shows that Na^+ recirculation by the phloem is crucial for salt tolerance. *EMBO J.* **22,** 2004–2014.

Blumwald, E. (2000). Sodium transport and salt tolerance in plants. *Curr. Opin. Cell Biol.* **12,** 431–434.

Blumwald, E., Aharon, G. S., and Apse, M. P. (2000). Sodium transport in plant cells. *Biochim. Biophys. Acta* **1465,** 140–151.

Bohnert, H. J., and Bressan, R. A. (2001). Abiotic stresses, plant reactions, and approaches towards improving stress tolerance. *In* "Crop Science, Progress and Prospects" (J. Nössberger, H. H. Geiger, and P. C. Struik, eds.), pp. 81–100. CABI International, Wallingford, UK.

Bohnert, H. J., and Shen, B. (1999). Transformation and compatible solutes. *Sci. Horticult.* **78,** 237–260.

Borsani, O., Cuartero, J., Fernández, J. A., Valpuesta, V., and Botella, M. A. (2001). Identification of two loci in tomato reveals distinct mechanisms for salt tolerance. *Plant Cell* **13,** 873–887.

Bressan, R. A., Zhang, C., Zhang, H., Hasegawa, P. M., Bohnert, H. J., and Zhu, J. K. (2002). Learning from the *Arabidopsis* experience. The next gene search paradigm. *Plant Physiol.* **127,** 1354–1360.

Chang, L., and Karin, M. (2001). Mammalian MAP kinase signaling cascades. *Nature* **410,** 37–40.

Cheeseman, J. M. (1988). Mechanisms of salinity tolerance in plants. *Plant Physiol.* **87,** 547–550.

Cheng, N.-H., Pittman, J. K., Barkla, B. J., Shigaki, T., and Hirschi, K. D. (2003). The *Arabidopsis cax1* mutant exhibits impaired ion homeostasis, development, and hormonal responses and reveals interplay among vacuolar transporters. *Plant Cell* **15,** 347–364.

Cherel, I., Michard, E., Platet, N., Mouline, K., Alcon, C., Sentenac, H., and Thibaud, J. B. (2002). Physical and functional interaction of the *Arabidopsis* K^+ channel AKT2 and phosphatase AtPP2CA. *Plant Cell* **14,** 1133–1146.

Chinnusamy, V., Schumaker, K., and Zhu, J.-K. (2004). Molecular genetic perspectives on cross-talk and specificity in abiotic stress signaling in plants. *J. Exp. Bot.* **55,** 225–236.

Condon, A. G., Richards, R. A., Rebetzke, G. J., and Farquhar, G. D. (2004). Breeding for high water-use efficiency. *J. Exp. Bot.* **55,** 2447–2460.

Cramer, G. R., Lynch, J. L., Laüchli, A., and Epstein, E. (1987). Influx of Na^+, K^+, and Ca^{2+} into roots of salt-stressed cotton seedling. Effects of supplemental Ca^{2+}. *Plant Physiol.* **83,** 510–516.

Cushman, J. C., and Bohnert, H. J. (2000). Genomic approaches to plant stress tolerance. *Curr. Opin. Plant Biol.* **3,** 117–124.

Dambly, S., and Boutry, M. (2001). The two major plant plasma membrane H^+ -ATPase display different regulatory properties. *J. Biol. Chem.* **276,** 7017–7022.
Davenport, R. J., and Tester, M. (2000). A weakly voltage-dependent, nonselective cation channel mediates toxic sodium influx in wheat. *Plant Physiol.* **122,** 823–834.
Demidchik, V., and Tester, M. (2002). Sodium fluxes through nonselective cation channels in the plasma membrane of protoplasts from *Arabidopsis* roots. *Plant Physiol.* **128,** 379–387.
Demidchik, V., Davenport, R. J., and Tester, M. (2002). Nonselective cation channels in plants. *Annu. Rev. Plant Biol.* **53,** 67–107.
Eastmond, P. J., and Graham, I. A. (2003). Trehalose metabolism: A regulatory role for trehalose-6-phosphate? *Curr. Opin. Plant Biol.* **6,** 231–235.
Ellul, P., Rios, G., Atares, A., Roig, L. A., Serrano, R., and Moreno, V. (2003). The expression of the *Saccharomyces cerevisiae* HAL1 gene increases salt tolerance in transgenic watermelon [*Citrullus lanatus* (Thunb.) Matsun. & Nakai.]. *Theor. Appl. Genet.* **107,** 462–469.
Elumalai, R. P., Nagpal, P., and Reed, J. W. (2002). A mutation in the *Arabidopsis KT2/KUP2* potassium transporter gene affects shoot cell expansion. *Plant Cell* **14,** 119–131.
Epstein, E. (1961). The essential role of calcium in selective cation transport by plant cells. *Plant Physiol.* **36,** 437–444.
Essah, P. A., Davenport, R., and Tester, M. (2003). Sodium influx and accumulation in *Arabidopsis. Plant Physiol.* **133,** 307–318.
Flowers, T. J. (2004). Improving crop salt tolerance. *J. Exp. Bot.* **55,** 307–319.
Flowers, T. J., and Yeo, A. R. (1992). "Solute Transport in Plants," pp. 1–176. Blackie Academic & Professional, London.
Flowers, T. J., and Yeo, A. R. (1995). Breeding for salinity resistance in crop plants, where next? *Aust. J. Plant Physiol.* **22,** 875–884.
Flowers, T. J., Hajibagheri, M. A., and Clipson, N. J. W. (1986). Halophytes. *Quart. Rev. Biol.* **61,** 313–337.
Garciadeblas, B., Rubio, F., Quintero, F. J., Banuelos, M. A., Haro, R., and Rodriguez-Navarro, A. (1993). Differential expression of two genes encoding isoforms of the ATPase involved in sodium efflux in *Saccharomyces cerevisiae. Mol. Gen. Genet.* **236,** 363–368.
Garciadeblas, B., Senn, M. E., Banuelos, M. A., and Rodriguez-Navarro, A. (2003). Sodium transport and HKT transporters, the rice model. *Plant J.* **34,** 788–801.
Garg, A. K., Kim, J. K., Owens, T. G., Ranwala, A. P., Choi, Y. D., Kochian, L. V., and Wu, R. J. (2002). Trehalose accumulation in rice plants confers high tolerance levels to different abiotic stresses. *Proc. Natl. Acad. Sci. USA* **99,** 15898–15903.
Gaxiola, R. A., Rao, R., Sherman, A., Grisafi, P., Alper, S. L., and Fink, G. R. (1999). The *Arabidopsis thaliana* proton transporters, AtNhx1 and Avp1, can function in cation detoxification in yeast. *Proc. Natl. Acad. Sci. USA* **96,** 1480–1485.
Gaxiola, R. A., Li, J., Undurrage, S., Dang, L. M., Allen, G. J., Alper, S. L., and Fink, G. R. (2001). Drought- and salt-tolerant plants result from overexpression of the AVP1 H^+ -pump. *Proc. Natl. Acad. Sci. USA* **98,** 11444–11449.
Gaxiola, R. A., Fink, G. R., and Hirschi, K. D. (2002). Genetic manipulation of vacuolar proton pumps and transporters. *Plant Physiol.* **129,** 967–973.
Gisbert, C., Rus, A. M., Bolarin, M. C., Lopez-Coronado, J. M., Arrillaga, I., Montesinos, C., Caro, M., Serrano, R., and Moreno, V. (2000). The yeast HAL1 gene improves salt tolerance of transgenic tomato. *Plant Physiol.* **123,** 393–402.
Glenn, E., Brown, J. J., and Blumwald, E. (1999). Salt tolerance and crop potential of halophytes. *Crit. Rev. Plant Sci.* **18,** 227–255.
Goff, S. A., Ricke, D., Lan, T. H., Presting, G., Wang, R., Dunn, M., Glazebrook, J., Sessions, A., Oeller, P., Varma, H., Hadley, D., Hutchison, D., *et al.* (2002). A draft sequence of the rice genome (*Oryza sativa* L. ssp. *japonica*). *Science* **296,** 92–100.
Greenway, H., and Osmond, C. B. (1972). Salt responses of enzymes from species differing in salt tolerance. *Plant Physiol.* **49,** 256–259.
Grillo, S., Leone, A., Xu, Y., Tucci, M., Francione, R., Hasegawa, P. M., Monti, L., and Bressan, R. A. (1995). Control of osmotin gene expression by ABA and osmotic stress in vegetative tissues of wild-type and ABA-deficient mutants of tomato. *Physiol. Plant* **93,** 498–504.

Guo, Y., Halfter, U., Ishitani, M., and Zhu, J. K. (2001). Molecular characterization of functional domains in the protein kinase SOS2 that is required for plant salt tolerance. *Plant Cell* **13,** 1383–1400.
Hasegawa, P. M., Bressan, R. A., and Pardo, J. M. (2000a). The dawn of plant salt to tolerance genetics. *Trends Plant Sci.* **5,** 317–319.
Hasegawa, P. M., Bressan, R. A., Zhu, J.-K., and Bohnert, H. J. (2000b). Plant cellular and molecular responses to high salinity. *Annu. Rev. Plant Physiol. Plant Mol. Biol.* **51,** 463–499.
Hayashi, H., Alia, Mustardy, L., Deshnium, P., Ida, M., and Murata, N. (1997). Transformation of *Arabidopsis thaliana* with the coda gene for choline oxidase; accumulation of glycinebetaine and enhanced tolerance to salt and cold stress. *Plant J.* **12,** 133–142.
Henikoff, S., Till, B. J., and Comai, L. (2004). Tilling. Traditional mutagenesis meets functional genomics. *Plant Physiol.* **135,** 630–636.
Hill, A. E., Shachar-Hill, B., and Shachar-Hill, Y. (2004). What are aquaporins for? *J. Membr. Biol.* **197,** 1–32.
Himmelbach, A., Yang, Y., and Grill, E. (2003). Relay and control of abscisic acid signaling. *Curr. Opin. Plant Biol.* **6,** 470–479.
Hirschi, K. D. (2004). The calcium conundrum. Both versatile nutrient and specific signals. *Plant Physiol.* **136,** 2438–2442.
Hohmann, S. (2002). Osmotic stress signaling and osmoadaptation in yeasts. *Microbiol. Mol. Biol. Rev.* **66,** 300–372.
Holmstrom, K. O., Somersalo, S., Mandal, A., Palva, T. E., and Welin, B. (2000). Improved tolerance to salinity and low temperature in transgenic tobacco producing glycine betaine. *J. Exp. Bot.* **51,** 177–185.
Hong, Z., Lakkineni, K., Zhang, Z., and Verma, D. P. (2000). Removal of feedback inhibition of delta(1)-pyrroline-5-carboxylate synthetase results in increased proline accumulation and protection of plants from osmotic stress. *Plant Physiol.* **122,** 1129–1136.
Horie, T., and Schroeder, J. I. (2004). Sodium transporters in plants. Diverse genes and physiological functions. *Plant Physiol.* **136,** 2457–2462.
Hoth, S., Morgante, M., Sanchez, J. P., Hanafey, M. K., Tingey, S. V., and Chua, N. H. (2002). Genome-wide gene expression profiling in *Arabidopsis thaliana* reveals new targets of abscisic acid and largely impaired gene regulation in the abi1–1 mutant. *J. Cell Sci.* **115,** 4891–4900.
Hugouvieux, V., Kwan, J. M., and Schroeder, J. I. (2001). An mRNA binding protein, ABH1, modulates early abscisic acid signal transduction in *Arabidopsis*. *Cell* **106,** 477–487.
Hugouvieux, V., Murata, Y., Young, J. J., Kwak, J. M., Mackesy, D. Z., and Schroeder, J. I. (2002). Localization, ion channel regulation, and genetic interactions during abscisic acid signaling of the nuclear mRNA cap-binding protein, ABH1. *Plant Physiol.* **130,** 1276–1287.
Ichimura, K., Mizoguchi, T., Yoshida, R., Yuasa, T., and Shinozaki, K. (2000). Various abiotic stresses rapidly activate *Arabidopsis* MAP kinases ATMPK4 and ATMPK6. *Plant J.* **24,** 655–665.
Inan, G., Zhang, Q., Li, P., Wang, Z., Cao, Z., Zhang, H., Zhang, C., Quist, T. M., Goodwin, S. M., Zhu, J., Shi, H., Damsz, B., *et al.* (2004). Salt cress. A halophyte and cryophyte *Arabidopsis* relative model system and its applicability to molecular genetic analyses of growth and development of extremophiles. *Plant Physiol.* **135,** 1718–1737.
Ingram, J., and Bartels, D. (1996). The molecular basis of dehydration tolerance in plants. *Annu. Rev. Plant Physiol. Plant Mol. Biol.* **48,** 377–403.
Ishitani, M., Xiong, L., Stevenson, B., and Zhu, J. K. (1997). Genetic analysis of osmotic and cold stress signal transduction in *Arabidopsis*, interactions and convergence of abscisic acid-dependent and abscisic acid-independent pathways. *Plant Cell* **9,** 1935–1949.
Iuchi, S., Kobayashi, M., Yamaguchi-Shinozaki, K., and Shinozaki, K. (2000). A stress-inducible gene for 9-cis-epoxycarotenoid dioxygenase involved in abscisic acid biosynthesis under water stress in drought-tolerant cowpea. *Plant Physiol.* **123,** 553–562.
Jacoby, B. (1999). Nutrient uptake in plants. *In* "Handbook of Plant and Crop Stress" (M. Pessarakli, ed.), pp. 1–23. Dekker, New York.
Jang, I. C., Oh, S. J., Seo, J. S., Choi, W. B., Song, S. I., Kim, C. H., Kim, Y. S., Seo, H. S., Choi, Y. D., Nahm, B. H., and Kim, J. K. (2003). Expression of a bifunctional fusion of the *Escherichia coli* genes for trehalose-6-phosphate synthase and trehalose-6-phosphate phosphatase in transgenic rice plants increases trehalose accumulation and abiotic stress tolerance without stunting growth. *Plant Physiol.* **131,** 516–524.

Jeong, M. J., Park, S. C., and Byun, M. O. (2001). Improvement of salt tolerance in transgenic potato plants by glyceraldehyde-3-phosphate dehydrogenase gene transfer. *Mol. Cell* **12,** 185–189.

Jonak, C., Ökrész, L., Bögre, L., and Hirt, H. (2002). Complexity, cross talk and integration of plant MAP kinase signaling. *Curr. Opin. Plant Biol.* **5,** 415–424.

Jung, K., Veen, M., and Altendorf, K. (2000). K^+ and ionic strength directly influence the autophosphorylation activity of the putative turgor sensor KdpD of *Escherichia coli*. *J. Biol. Chem.* **275,** 40142–40147.

Kasuga, M., Liu, Q., Miura, S., Yamaguchi-Shinozaki, K., and Shinozaki, K. (1999). Improving plant drought, salt, and freezing tolerance by gene transfer of a single stress-inducible transcription factor. *Nat. Biotechnol.* **17,** 287–291.

Kishor, P., Hong, Z., Miao, G. H., Hu, C., and Verma, D. (1995). Overexpression of [delta]-pyrroline-5-carboxylate synthetase increases proline production and confers osmotolerance in transgenic plants. *Plant Physiol.* **108,** 1387–1394.

Knight, H., Trewavas, A. J., and Knight, M. R. (1997). Calcium signaling in *Arabidopsis thaliana* responding to drought and salinity. *Plant J.* **12,** 1067–1078.

Koiwa, H., Barb, A. W., Xiong, L., Li, F., Mcculy, M. G., Lee, B. H., Sokolchik, I., Zhu, J., Gong, Z., and Reddy, M. (2002). C-terminal domain phosphatase-like family members (AtCPLs) differentially regulate *Arabidopsis thaliana* abiotic stress signaling, growth, and development. *Proc. Natl. Acad. Sci. USA* **99,** 10893–10898.

Koiwa, H., Hausmann, S., Bang, W. Y., Ueda, A., Kondo, N., Hiraguri, A., Fukuhara, T., Bahk, J. D., Yun, D. J., Bressan, R. A., Hasegawa, P. M., and Shuman, S. (2004). *Arabidopsis* C-terminal domain phosphatase-like 1 and 2 are essential Ser-5-specific C-terminal domain phosphatases. *Proc. Natl. Acad. Sci. USA* **101,** 14539–14544.

Koyama, M. L., Levesley, A., Koebner, R. M. D., Flowers, T. J., and Yeo, A. R. (2001). Quantitative trait loci for component physiological traits determining salt tolerance in rice. *Plant Physiol.* **125,** 406–422.

Kurkela, S., and Franck, M. B. (1990). Cloning and characterization of a cold- and ABA-inducible Arabidopsis gene. *Plant Mol. Biol.* **15,** 137–144.

Läuchli, A. (1996). Salinity—Potassium interaction in crop plants. *In* "Frontiers in Potassium Nutrition, New Perspectives on the Effects of Potassium on Physiology of Plants" (D. M. Oosterhuis, and G. A. Berkowitz, eds.), pp. 71–76. Proceeding of a Symposium, published by the Potash & Phosphate Institute/Potash & Phosphate Institute of Canada.

Läuchli, A., and Epstein, E. (1990). Plant responses to saline and sodic conditions. *In* "Salinity Assessment and Management" (K. K. Tanji, ed.), pp. 113–137. Amer. Soc. Civil Eng., New York.

Laurie, S., Feeney, K. A., Maathuis, F. J. M., Heard, P. J., Brown, S. J., and Leigh, R. A. (2002). A role for HKT1 in sodium uptake by wheat roots. *Plant J.* **32,** 139–149.

Lee, E. K., Kwon, M., Ko, J. H., Yi, H., Hwang, M. G., Chang, S., and Cho, M. H. (2004). Binding of sulfonylurea by AtMRP5, an *Arabidopsis* multidrug resistance-related protein that functions in salt tolerance. *Plant Physiol.* **134,** 528–538.

Li, J., Lee, Y. R. J., and Assmann, S. M. (1998). Guard cells possess a calcium-dependent protein kinase that phosphorylates the KAT1 potassium channel. *Plant Physiol.* **116,** 785–795.

Lin, H. X., Zhu, M. Z., Yano, M., Gao, J. P., Liang, Z. W., Su, W. A., Hu, X. H., Ren, Z. H., and Chao, D. Y. (2004). QTLs for Na^+ and K^+ uptake of the shoots and roots controlling rice salt tolerance. *Theor. Appl. Genet.* **108,** 253–260.

Loudet, O., Chaillou, S., Krapp, A., and Daniel-Vedele, F. (2003). Quantitative trait loci analysis of water and anion contents in interaction with nitrogen availability in *Arabidopsis thaliana*. *Genetics* **163,** 711–722.

Lu, C., Han, M. H., Guevara-Garcia, A., and Fedoroff, N. V. (2002). Mitogen-activated protein kinase signaling in postgermination arrest of development by abscisic acid. *Proc. Natl. Acad. Sci. USA* **99,** 15812–15817.

Maas, E. V. (1990). Crop salt tolerance. *In* "Agricultural Salinity Assessment and Management" (K. K. Tanji, ed.), pp. 262–304. Amer. Soc. Civil Eng., New York.

Maathuis, F. J. M., and Sanders, D. (1996). Mechanisms of potassium absorption by higher plant roots. *Physiol. Plant.* **96,** 158–168.

Maathuis, F. J. M., Verlin, D., Smith, F. A., Sanders, D., Fernández, J. A., and Walker, N. A. (1996). The physiological relevance of Na^+-coupled K^+-transport. *Plant Physiol.* **112,** 1609–1616.

Majee, M., Maitra, S., Dastidar, K. G., Pattnaik, S., Chatterjee, A., Hait, N. C., Das, K. P., and An Majumder, A. L. (2004). A novel salt-tolerant L-myo-inositol-1-phosphate synthase from *Porteresia coarctata* (Roxb.) Tateoka, a halophytic wild rice, molecular cloning, bacterial overexpression, characterization, and functional introgression into tobacco conferring salt tolerance phenotype. *J. Biol. Chem.* **279,** 28539–28552.

Marin, K., Suzuki, I., Yamaguchi, K., Ribbeck, K., Yamamoto, H., Kanesaki, Y., Hagemann, M., and Murata, N. (2003). Identification of histidine kinases that act as sensors in the perception of salt stress in *Synechocystis* sp. PCC 6803. *Proc. Natl. Acad. Sci. USA* **100,** 9061–9066.

Mäser, P., Gierth, M., and Schroeder, J. I. (2002). Molecular mechanisms of potassium and sodium uptake in plants. *Plant Soil* **247,** 43–54.

Matsumoto, T. K., Ellsmore, A. J., Cessna, S. G., Low, P. S., Pardo, J. M., Bressan, R. A., and Hasegawa, P. M. (2002). An osmotically induced cytosolic Ca^{2+} transient activates calcineurin signaling to mediate ion homeostasis and salt tolerance of *Saccharomyces cerevisiae. J. Biol. Chem.* **277,** 33075–33080.

Mazel, A., Leshem, Y., Tiwari, B. S., and Levine, A. (2004). Induction of salt and osmotic stress tolerance by overexpression of an intracellular vesicle trafficking protein AtRab7 (AtRabG3e). *Plant Physiol.* **134,** 118–128.

Mckersie, B. D., Bowley, S. R., Harjanto, E., and Leprince, O. (1996). Water-deficit tolerance and field performance of transgenic alfalfa overexpressing superoxide dismutase. *Plant Physiol.* **111,** 1177–1181.

Mikolajczyk, M., Awotunde, O. S., Muszyska, G., Klessig, D. F., and Dobrowolska, G. (2000). Osmotic stress induces rapid activation of a salicylic acid-induced protein kinase and a homolog of protein kinase ASK1 in tobacco cells. *Plant Cell* **12,** 165–178.

Mittova, V., Theodoulou, F. L., Kiddle, G., Gomez, L., Volokita, M., Tal, M., Foyer, C. H., and Guy, M. (2003). Coordinate induction of glutathione biosynthesis and glutathione-metabolizing enzymes is correlated with salt tolerance in tomato. *FEBS Lett.* **554,** 417–421.

Mukhopadhyay, A., Vij, S., and Tyagi, A. K. (2004). Overexpression of a zinc-finger protein gene from rice confers tolerance to cold, dehydration, and salt stress in transgenic tobacco. *Proc. Natl. Acad. Sci. USA* **101,** 6309–6314.

Munns, R. (1993). Physiological processes limiting plant growth in saline soils, some dogmas and hypotheses. *Plant Cell Environ.* **16,** 15–24.

Munns, R. (2002). Comparative physiology of salt and water stress. *Plant Cell Environ.* **25,** 239–250.

Munns, R., Husain, S., Rivelli, A. R., James, R. A., Condon, A. G., Lindsay, M. P., Lagudah, E. S., Schachtman, D., and Hare, R. A. (2002). Avenues for increasing salt tolerance of crops, and the role of physiologically-based selection traits. *Plant Soil* **247,** 93–105.

Nagaoka, S., and Takano, T. (2003). Salt tolerance-related protein STO binds to a Myb transcription factor homologue and confers salt tolerance in *Arabidopsis. J. Exp. Bot.* **54,** 2231–2237.

Nakayama, H., Yoshida, K., Ono, H., Murooka, Y., and Shinmyo, A. (2000). Ectoine, the compatible solute of *Halomonas elongata,* confers hyperosmotic tolerance in cultured tobacco cells. *Plant Physiol.* **122,** 1239–1247.

Nakayama, H., Yoshida, K., and Shinmyo, A. (2004). Yeast plasma membrane Ena1p ATPase alters alkali-cation homeostasis and confers increased salt tolerance in tobacco cultured cells. *Biotechnol. Bioeng.* **85,** 776–789.

Nanjo, T., Kobayashi, M., Yoshiba, Y., Kakubari, Y., Yamaguchi-Shinozaki, K., and Shinozaki, K. (1999). Antisense suppression of praline degradation improves tolerance to freezing and salinity in *Arabidopsis thaliana. FEBS Lett.* **461,** 205–210.

Niu, X., Bressan, R. A., Hasegawa, P. M., and Pardo, J. M. (1995). Ion homeostasis in NaCl stress environments. *Plant Physiol.* **109,** 735–742.

Nomura, M., Ishitani, M., Takabe, T., Rai, A. K., and Takabe, T. (1995). Synechococcus sp. PCC7942 transformed with *Escherichia coli bet* genes produces glycine betaine from choline and acquires resistance to salt stress. *Plant Physiol.* **107,** 703–708.

Nordin, K., Heino, P., and Palva, E. T. (1991). Separate signal pathways regulate the expression of a low-temperature-induced gene in *Arabidopsis thaliana* (L.) Heynh. *Plant Mol. Biol.* **16,** 1061–1071.

Norlyn, J. D. (1979). Breeding salt-tolerant crop plants. *Basic Life Sci.* **14,** 293–309.

Novillo, F., Alonso, J. M., Ecker, J. R., and Salinas, J. (2004). CBF2/DREB1C is a negative regulator of CBF1/DREB1B and CBF3/DREB1A expression and plays a central role in stress tolerance in *Arabidopsis*. *Proc. Natl. Acad. Sci. USA* **101,** 3985–3990.

Nublat, A., Desplans, J., Casse, F., and Berthomieu, P. (2001). *sas1*, an *Arabidopsis* mutant overaccumulating sodium in the shoot, shows deficiency in the control of the root radial transport of sodium. *Plant Cell* **13,** 125–137.

O'Leary, J. W., Glenn, E. P., and Watson, M. C. (1985). Agricultural production of halophytes irrigated with seawater. *Plant Soil* **89,** 311–321.

Pardo, J. M., and Quintero, F. J. (2002). Plants and sodium ions, keeping company with the enemy. *Genome Biol.* **3,** 1017.1–1017.4.

Pardo, J. M., Reddy, M. P., Yang, S., Maggio, A., Huh, G.-H., Matsumoto, T., Coca, M. A., Paino-D'Urzo, M., Koiwa, H., Yun, D.-J., Watad, A. A., Bressan, R. A., *et al.* (1998). Stress signaling through Ca^{2+}/calmodulin-dependent protein phosphatase calcineurin mediates salt adaptation in plants. *Proc. Natl. Acad. Sci. USA* **95,** 9681–9686.

Park, J. M., Park, C. J., Lee, S. B., Ham, B. K., Shin, R., and Paek, K. H. (2001). Overexpression of the tobacco Tsi1 gene encoding an EREBP/AP2-type transcription factor enhances resistance against pathogen attack and osmotic stress in tobacco. *Plant Cell* **13,** 1035–1046.

Perruc, E., Charpenteau, M., Ramirez, B. C., Jauneau, A., Galaud, J. P., Ranjeva, R., and Ranty, B. (2004). A novel calmodulin-binding protein functions as a negative regulator of osmotic stress tolerance in *Arabidopsis thaliana* seedlings. *Plant J.* **38,** 410–420.

Pilot, G., Gaymard, F., Mouline, K., Cherel, I., and Sentenac, H. (2003). Regulated expression of *Arabidopsis* shaker K^+ channel genes involved in K^+ uptake and distribution in the plant. *Plant Mol. Biol.* **51,** 773–787.

Price, A. H., Cairns, J. E., Horton, P., Jones, H. G., and Griffiths, H. (2002). Linking drought-resistance mechanisms to drought avoidance in upland rice using a QTL approach, progress and new opportunities to integrate stomatal and mesophyll responses. *J. Exp. Bot.* **53,** 989–1004.

Provart, N. J., and Mccourt, P. (2004). Systems approaches to understanding cell signaling and gene regulation. *Curr. Opin. Plant Biol.* **7,** 605–609.

Qi, Z., and Spalding, E. P. (2004). Protection of plasma membrane K^+ transport by the salt overly sensitive1 Na^+ -H^+ antiporter during salinity stress. *Plant Physiol.* **136,** 2548–2555.

Qiu, Q.-S., Guo, Y., Dietrich, M. A., Schumaker, K. S., and Zhu, J.-K. (2002). Regulation of SOS1, a plasma membrane Na^+/H^+ exchanger in *Arabidopsis thaliana*, by SOS2 and SOS3. *Proc. Natl. Acad. Sci. USA* **99,** 8436–8441.

Qiu, Q.-S., Barkla, B. J., Vera-Estrella, R., Zhu, J.-K., and Schumaker, K. S. (2003). Na^+/H^+ exchange activity in the plasma membrane of *Arabidopsis*. *Plant Physiol.* **132,** 1041–1052.

Quesada, V., Garcia-Martinez, S., Piqueras, P., Ponce, M. R., and Micol, J. L. (2002). Genetic architecture of NaCl tolerance in *Arabidopsis*. *Plant Physiol.* **130,** 951–963.

Quintero, F. J., Garciadeblas, B., and Rodriguez-Navarro, A. (1996). The SAL1 gene of *Arabidopsis*, encoding an enzyme with 3′(2′),5′-bisphosphate nucleotidase and inositol polyphosphate 1-phosphatase activities, increases salt tolerance in yeast. *Plant Cell* **8,** 529–537.

Quintero, F. J., Ohta, M., Shi, H., Zhu, J.-K., and Pardo, J. M. (2002). Reconstitution in yeast of the *Arabidopsis* SOS signaling pathway for Na^+ homeostasis. *Proc. Natl. Acad. Sci. USA* **99,** 9061–9066.

Rains, D. W., and Epstein, E. (1965). Transport of sodium in plant tissue. *Science* **148,** 1611.

Rains, D. W., and Epstein, E. (1967). Sodium absorption by barley roots: Role of the dual mechanisms of alkali cation transport. *Plant Physiol.* **42,** 314–318.

Ramage, R. (1980). Genetic methods to breed salt tolerance in plants. *In* "Genetic Engineering of Osmoregulation Impact on Plant Productivity for Food, Chemicals and Energy" (D. Rains, R. Valentine, and A. Hollaender, eds.), pp. 311–318. Plenum Press, New York.

Ribaut, J.-M., and Hoisington, D. (1998). Marker-assisted selection, new tools and strategies. *Trends Plant Sci.* **3,** 236–239.

Rigas, S., Debrosses, G., Haralampidis, K., Vicente-Agullo, F., Feldmann, K. A., Grabov, A., Dolan, L., and Hatzopoulos, P. (2001). *TRH1* encodes a potassium transporter required for tip growth in *Arabidopsis* root hairs. *Plant Cell* **13,** 139–151.

Roberts, S. K., and Tester, M. (1997). A patch clamp study of Na^+ transport in maize roots. *J. Exp. Bot.* **48,** 431–440.

Romero, C., Belles, J. M., Vaya, J. L., Serrano, R., and Culiañez-Macià, F. A. (1997). Expression of the yeast trehalose-6-phosphate synthase gene in transgenic tobacco plants, pleiotropic phenotypes include drought tolerance. *Planta* **201,** 293–297.

Rontein, D., Basset, G., and Hanson, A. D. (2002). Metabolic engineering of osmoprotectant accumulation in plants. *Metab. Eng.* **4,** 49–56.

Ros, R., Montesinos, C., Rimon, A., Padan, E., and Serrano, R. (1998). Altered Na^+ and Li^+ homeostasis in *Saccharoyces cerevisiae* expressing the bacterial cation antiporter NhaA. *J. Bacteriol.* **180,** 3131–3136.

Roxas, V. R., Smigh, R. K., Jr., Allen, E. R., and Allen, R. D. (1997). Overexpression of glutathione-*S*-transferase/glutathione peroxidase enhances the growth of transgenic tobacco seedlings during stress. *Nat. Biotechnol.* **15,** 988–991.

Rubio, F., Gassmann, W., and Schroeder, J. I. (1995). Sodium-driven potassium uptake by the plant transporter HKT1 and mutations conferring salt tolerance. *Science* **270,** 1660–1663.

Rubio, F., Schwarz, M., Gassmann, W., and Schroeder, J. I. (1999). Genetic selection of mutations in the high affinity K^+ transporter HKT1 that defines functions of a loop site for reduced Na^+ permeability and increased Na^+ tolerance. *J. Biol. Chem.* **274,** 6839–6847.

Rus, A., Yokoi, S., Sharkhuu, A., Reddy, M., Lee, B.-H., Matsumoto, T. K., Koiwa, H., Zhu, J.-K., Bressan, R. A., and Hasegawa, P. M. (2001). AtHK1 is a salt tolerance determinant that controls Na^+ entry into plant roots. *Proc. Natl. Acad. Sci. USA* **98,** 14150–14155.

Rus, A., Lee, B.-H., Munoz-Mayor, A., Sharkhuu, A., Miura, K., Zhu, J.-K., Bressan, R. A., and Hasegawa, P. M. (2004). AtHKT1 facilitates Na^+ homeostasis and K^+ nutrition *in planta*. *Plant Physiol.* **136,** 2500–2511.

Rush, P. W., and Epstein, E. (1981). Breeding and selection for salt tolerance by the incorporation of wild germplasm into a domestic tomato. *J. Am. Soc. Hort. Sci.* **106,** 699–704.

Saijo, Y., Hata, S., Kyozuka, J., Shimamoto, K., and Izui, K. (2000). Over-expression of a single Ca^{2+} -dependent protein kinase confers both cold and salt/drought tolerance on rice plants. *Plant J.* **23,** 319–327.

Sakamoto, H., Maruyama, K., Sakuma, Y., Meshi, T., Iwabuchi, M., Shinozaki, K., and Yamaguchi-Shinozaki, K. (2004). *Arabidopsis* Cys2/His2-type zinc-finger proteins function as transcription repressors under drought, cold, and high-salinity stress conditions. *Plant Physiol.* **136,** 2734–2746.

Schachtman, D. P. (2000). Molecular insights into the structure and function of plant K^+ transport mechanisms. *Biochim. Biophys. Acta* **1465,** 127–139.

Seki, M., Ishida, J., Narusaka, M., Fujita, M., Nanjo, T., Umezawa, T., Kamiya, A., Nakajima, M., Enju, A., and Sakurai, T. (2002). Monitoring the expression pattern of around 7,000 *Arabidopsis* genes under ABA treatments using a full-length cDNA microarray. *Funct. Integr. Genom.* **2,** 282–291.

Seo, M., Koiwai, H., Akaba, S., Komano, T., Oritani, T., Kamiya, Y., and Koshiba, T. (2000). Abscisic aldehyde oxidase in leaves of *Arabidopsis thaliana*. *Plant J.* **23,** 481–488.

Serrano, R. (1996). Salt tolerance in plants and microorganisms, toxicity targets and defense responses. *Int. Rev. Cytol.* **165,** 1–52.

Serrano, R., and Rodriguez-Navarro, P. L. (2002). Plants, genes and ions. Workshop on the molecular basis of ionic homeostasis and salt tolerance in plants. *EMBO Rep.* **3,** 116–119.

Serrano, R., Culiañz-Maciá, A., and Moreno, V. (1999). Genetic engineering of salt and drought tolerance with yeast regulatory genes. *Sci. Hortic.* **78,** 261–269.

Shen, B., Jensen, R. G., and Bohnert, H. J. (1997). Mannitol protects against oxidation by hydroxyl radicals. *Plant Physiol.* **115,** 527–532.

Sheveleva, E., Chmara, W., Bohnert, H. J., and Jensen, R. G. (1997). Increased salt and drought tolerance by D-ononitol production in transgenic *Nicotiana tabacum* L. *Plant Physiol.* **115,** 1211–1219.

Sheveleva, E., Jensen, R. G., and Bohnert, H. J. (2000). Disturbance in the allocation of carbohydrates to regenerative organs in transgenic *Nicotiana tabacum* L. *J. Exp. Bot.* **51,** 115–122.

Shi, H., and Zhu, J.-K. (2002). Regulation of expression of the vacuolar Na^+/H^+ antiporter gene AtNHX1 by salt stress and ABA. *Plant Mol. Biol.* **50,** 543–550.

Shi, H., Ishitani, M., Kim, C., and Zhu, J.-K. (2000). The *Arabidopsis thaliana* salt tolerance gene *SOS1* encodes a putative Na^+/H^+ antiporter. *Proc. Natl. Acad. Sci. USA* **97,** 6896–6901.

Shi, H., Lee, B.-H., Wu, S.-J., and Zhu, J.-K (2003). Overexpression of a plasma membrane Na^+/H^+ antiporter gene improves salt tolerance in *Arabidopsis thaliana*. *Nat. Biotechnol.* **21,** 81–85.

Shinozaki, K., and Yamaguchi-Shinozaki, K. (1999). Molecular responses to drought stress. *In* "Molecular Responses to Cold, Drought, Heat and Salt Stress in Higher Plants" (K. Shinozaki and K. Shinozaki-Yamaguchi, eds.), pp. 11–28. R. G. Landes, Austin, Texas.

Shinozaki, K., Yamaguchi-Shinozaki, K., and Seki, M. (2003). Regulatory network of gene expression in the drought and cold stress responses. *Curr. Opin. Plant Biol.* **6,** 410–417.

Singla-Pareek, S. L., Reddy, M. K., and Sopory, S. K. (2003). Genetic engineering of the glyoxalase pathway in tobacco leads to enhanced salinity tolerance. *Proc. Natl. Acad. Sci. USA* **100,** 14672–14677.

Sondergaard, T. E., Schulz, A., and Palmgren, M. G. (2004). Energization of transport processes in plants. Roles of the plasma membrane H^+-ATPase. *Plant Physiol.* **136,** 2475–2482.

Su, H., Golldack, D., Katsuhara, M., Zhao, C., and Bohnert, H. J. (2001). Expression and stress-dependent induction of potassium channel transcripts in the common ice plant. *Plant Mol. Biol.* **51,** 773–787.

Su, H., Golldack, D., Zhao, C., and Bohnert, H. J. (2002). The expression of HAK-type K(+) transporters is regulated in response to salinity stress in common ice plant. *Plant Physiol.* **129,** 1482–1493.

Su, H., Balderas, E., Vera-Estrella, R., Golldack, D., Quigley, F., Zhao, C., Pantoya, O., and Bohnert, H. J. (2003). Expression of the cation transporter McHKT1 in a halophyte. *Plant Mol. Biol.* **52,** 967–980.

Sulpice, R., Tsukaya, H., Nonaka, H., Mustardy, L., Chen, T. H., and Murata, N. (2003). Enhanced formation of flowers in salt-stressed *Arabidopsis* after genetic engineering of the synthesis of glycine betaine. *Plant J.* **36,** 165–176.

Suzuki, I., Los, D. A., Kanesaki, Y., Mikami, K., and Murata, N. (2000). The pathway for perception and transduction of low-temperature signals in *Synechocystis*. *EMBO J.* **19,** 1327–1334.

Taji, T., Seki, M., Satou, M., Sakurai, T., Kobayashi, M., Ishiyama, K., Narusaka, Y., Narusaka, M., Zhu, J. K., and Shinozaki, K. (2004). Comparative genomics in salt tolerance between *Arabidopsis* and Arabidopsis-related halophyte salt cress using *Arabidopsis* microarray. *Plant Physiol.* **135,** 1697–1709.

Talke, I. N., Blaudez, D., Maathuis, F. J. M., and Sanders, D. (2003). CNGCs, prime targets of plant cyclic nucleotide signaling? *Trends Plant Sci.* **8,** 286–293.

Tani, H., Chen, X., Nurmberg, P., Grant, J. J., Santamaria, M., Chini, A., Gilroy, E., Birch, P. R., and Loake, G. J. (2004). Activation tagging in plants, a tool for gene discovery. *Funct. Integr. Genom.* **4,** 258–266.

Tarczynski, M., Jensen, R. G., and Bohnert, H. J. (1993). Stress protection of transgenic tobacco by production of the osmolyte mannitol. *Science* **259,** 508–510.

Teige, M., Scheikl, E., Eulgem, T., Doczi, R., Ichimura, K., Shinozaki, K., Dangl, J. L., and Hirt, H. (2004). The MKK2 pathway mediates cold and salt stress signaling in *Arabidopsis*. *Mol. Cell.* **15,** 141–152.

Tester, M., and Davenport, R. (2003). Na^+ tolerance and Na^+ transport in higher plants. *Ann. Bot. (Lond.)* **91,** 503–527.

Trewavas, A. J. (2001). The population/biodiversity paradox. Agricultural efficiency to save wilderness. *Plant Physiol.* **125,** 174–179.

Tuberosa, R., Salvi, S., Sanguineti, M. C., Landi, P., MacCaferri, M., and Conti, S. (2002). Mapping QTLs regulating morpho-physiological traits and yield: Case studies, shortcomings and perspectives in drought-stressed maize. *Ann. Bot. (Lond.)* **89,** 941–963.

Urano, K., Yoshiba, Y., Nanjo, T., Ito, T., Yamaguchi-Shinozaki, K., and Shinozaki, K. (2004). *Arabidopsis* stress-inducible gene for arginine decarboxylase AtADC2 is required for accumulation of putrescine in salt tolerance. *Biochem. Biophys. Res. Commun.* **313,** 369–375.

Urao, T., Yakubov, B., Satoh, R., Yamaguchi-Shinozaki, K., Seki, M., Hirayama, T., and Shinozaki, K. (1999). A transmembrane hybrid-type histidine kinase in *Arabidopsis* functions as an osmosensor. *Plant Cell* **11,** 1743–1754.

Van Camp, W. (2005). Yield enhancement genes: Seeds for growth. *Curr. Opin. Biotechnol.* **16,** 147–153.

Van Camp, W., Capiau, K., Van Montagu, M., Inze, D., and Slooten, L. (1996). Enhancement of oxidative stress tolerance in transgenic tobacco plants overproducing Fe-superoxide dismutase in chloroplasts. *Plant Physiol.* **112,** 1703–1714.

Veena, Reddy, V. S., and Sopory, S. K. (1999). Glyoxalase I from *Brassica juncea*, molecular cloning, regulation and its over-expression confer tolerance in transgenic tobacco under stress. *Plant J.* **17,** 385–395.

Vera-Estrella, R., Barkla, B. J., Bohnert, H. J., and Pantoja, O. (1999). Salt-stress in *Mesembryanthemum crystallinum* suspensions activates adaptive mechanisms similar to those observed in the whole plant. *Planta* **207,** 426–435.

Verslues, P. E., and Bray, E. A. (2004). LWR1 and LWR2 are required for osmoregulation and osmotic adjustment in *Arabidopsis*. *Plant Physiol.* **136,** 2831–2842.

Villalobos, M. A., Bartels, D., and Iturriaga, G. (2004). Stress tolerance and glucose insensitive phenotypes in *Arabidopsis* overexpressing the CpMYB10 transcription factor gene. *Plant Physiol.* **135,** 309–324.

Vitart, V., Baxter, I., Doerner, P., and Harper, J. F. (2001). Evidence for a role in growth and salt resistance of a plasma membrane H^+-ATPase in the root endodermis. *Plant J.* **27,** 191–201.

Wada, Y., Miyamoto, K., Kusano, T., and Sano, H. (2004). Association between up-regulation of stress-responsive genes and hypomethylation of genomic DNA in tobacco plants. *Mol. Genet. Genom.* **271,** 658–666.

Waditee, R., Hibino, T., Nakamura, T., Incharoensakdi, A., and Takabe, T. (2002). Overexpression of a Na^+/H^+ antiporter confers salt tolerance on a freshwater cyanobacterium, making it capable of growth in sea water. *Proc. Natl. Acad. Sci. USA* **99,** 4109–4114.

Wang, W., Vinocur, B., and Altman, A. (2003). Plant responses to drought, salinity and extreme temperatures, towards genetic engineering for stress tolerance. *Planta* **218,** 1–14.

Ward, J. M., Hirschi, K. D., and Sze, H. (2003). Plants pass the salt. *Trends Plant Sci.* **8,** 200–201.

Warren, G., Mckown, R., Martin, A. L., and Teutonico, V. (1996). Isolation of mutations affecting the development of freezing tolerance in *Arabidopsis thaliana* (L.) Heynh. *Plant Physiol.* **111,** 1011–1019.

Werner, J. E., and Finkelstein, R. R. (1995). *Arabidopsis* mutants with reduced response to NaCl and osmotic stress. *Physiol. Plant.* **93,** 659–666.

Winicov, I., and Bastola, D. R. (1999). Transgenic overexpression of the transcription factor A1fin1 enhances expression of the endogenous MsPRP2 gene in alfalfa and improves salinity tolerance of the plants. *Plant Physiol.* **120,** 473–480.

Wu, C. A., Yang, G. D., Meng, Q. W., and Zheng, C. C. (2004). The cotton GhNHX1 gene encoding a novel putative tonoplast Na(+)/H(+) antiporter plays an important role in salt stress. *Plant Cell Physiol.* **45,** 600–607.

Xiong, L., and Zhu, J. K. (2003). Regulation of abscisic acid biosynthesis. *Plant Physiol.* **133,** 29–36.

Xiong, L., Gong, Z., Rock, C. D., Subramanian, S., Guo, Y., Xu, W., Galbraith, D., and Zhu, J.-K. (2001a). Modulation of abscisic acid signal transduction and biosynthesis by an Sm-like protein in *Arabidopsis*. *Dev. Cell* **1,** 771–781.

Xiong, L., Ishitani, M., Lee, H., and Zhu, J. K. (2001b). *The Arabidopsis* LOS5/ABA3 locus encodes a molybdenum cofactor sulfurase and modulates cold stress- and osmotic stress-responsive gene expression. *Plant Cell* **13,** 2063–2083.

Xiong, L., Lee, H., Ishitani, M., Tanaka, Y., Stevenson, B., Koiwa, H., Bressan, R. A., Hasegawa, P. M., and Zhu, J.-K. (2002a). Repression of stress-responsive genes by FIERY2, a novel transcriptional regulator in *Arabidopsis*. *Proc. Natl. Acad. Sci. USA* **99,** 10899–10904.

Xiong, L., Schumaker, K. S., and Zhu, J.-K. (2002b). Cell signaling during cold, drought, and salt stress. *Plant Cell* **14,** S165–S183.

Yale, J., and Bohnert, H. J. (2001). Transcript expression in *Saccharomyces cerevisiae* at high salinity. *J. Biol. Chem.* **276,** 15996–16007.

Yokoi, S., Quintero, F. J., Cubero, B., Ruiz, M. T., Bressan, R. A., Hasegawa, P. M., and Pardo, J. M. (2002). Differential expression and function of *Arabidopsis thaliana* NHX Na^+/H^+ antiporters in the salt stress response. *Plant J.* **30,** 529–539.

Yu, J., Hu, S., Wang, J., Wong, G. K., Li, S., Liu, B., Deng, Y., Dai, L., Zhou, Y., Zhang, X., Cao, M., Liu, J., *et al.* (2002). A draft sequence of the rice genome (*Oryza sativa* L. ssp. *indica*). *Science* **296,** 79–92.

Zhang, H. X., and Blumwald, E. (2001). Transgenic salt-tolerant tomato plants accumulate salt in foliage but not in fruit. *Nat. Biotech.* **19,** 765–768.

Zhang, H. X., Hodson, J. N., Williams, J. P., and Blumwald, E. (2001). Engineering salt-tolerant *Brassica* plants, characterization of yield and seed oil quality in transgenic plants with increased sodium accumulation. *Proc. Natl. Acad. Sci. USA* **98,** 12832–12836.

Zhang, H., Huang, Z., Xie, B., Chen, Q., Tian, X., Zhang, X., Zhang, H., Lu, X., Huang, D., and Huang, R. (2004). The ethylene-, jasmonate-, abscisic acid- and NaCl-responsive tomato transcription factor

JERF1 modulates expression of GCC box-containing genes and salt tolerance in tobacco. *Planta* **220,** 262–270.

Zhao, Z., Chen, G., and Zhang, C. (2001). Interaction between reactive oxygen species and nitric oxide in drought-induced abscisic acid synthesis in root tips of wheat seedlings. *Aust. J. Plant Physiol.* **28,** 1055–1061.

Zhifang, G., and Loescher, W. H. (2003). Expression of a celery mannose 6-phosphate reductase in *Arabidopsis thaliana* enhances salt tolerance and induces biosynthesis of both mannitol and a glucosyl-mannitol dimmer. *Plant Cell Environ.* **26,** 275–283.

Zhu, J.-K. (2000). Genetic analysis of plant salt tolerance using *Arabidopsis. Plant Physiol.* **124,** 941–948.

Zhu, J.-K. (2001). Plant salt tolerance. *Trends Plant Sci.* **6,** 66–71.

Zhu, J.-K. (2002). Salt and drought stress signal transduction in plants. *Annu. Rev. Plant Biol.* **53,** 247–273.

Zhu, J.-K. (2003). Regulation of ion homeostasis under salt stress. *Curr. Opin. Plant Biol.* **6,** 441–445.

CHAPTER 13

Metabolic Engineering of Plant Allyl/Propenyl Phenol and Lignin Pathways: Future Potential for Biofuels/Bioenergy, Polymer Intermediates, and Specialty Chemicals?

Daniel G. Vassão, Laurence B. Davin, and **Norman G. Lewis**

Contents

Abstract Exciting recent developments in the enzymology and molecular biology of plant phenylpropanoids offer numerous opportunities to re-engineer the composition of plant biomass. Two main targets of such modifications are

Institute of Biological Chemistry, Washington State University, Pullman, Washington 99164

Advances in Plant Biochemistry and Molecular Biology, Volume 1
ISSN 1755-0408, DOI: 10.1016/S1755-0408(07)01013-2

the optimized production of valuable compounds and reductions in the levels of less desirable products, such as the structural biopolymeric lignins. For example, the amounts of lignin biopolymers in (woody) species might be reduced, with carbon flow concurrently redirected toward production of related nonpolymeric phenylpropanoids, such as the more valuable allyl/propenyl phenols (e.g., eugenol, chavicol).

Lignins are monolignol-derived polymeric end-products of the phenylpropanoid pathway (originating from the amino acids phenylalanine and tyrosine). In general, lignins represent a formidable technical challenge, particularly due to their intractable nature, for improved plant biomass utilization, for example, when considering the use of woody biomass for bioethanol production, as well as for wood, pulp, and paper manufacture.

Other species-specific outcomes of the phenylpropanoid pathway, however, include metabolites such as lignans, flavonoids, and allyl/propenyl phenols. The recent discovery of the biochemical pathway resulting in the production of the more valuable liquid allyl/propenyl phenols (e.g., eugenol, chavicol, estragole, and anethole), important components of plant spice aromas and flavors, presents one potential approach to the engineering of plant metabolism in new directions. These compounds are synthesized from monolignols in two consecutive enzymatic reactions: (1) acylation of the terminal (C-9) oxygen of the monolignol forming an ester and (2) regiospecific, NAD(P)H-dependent reduction of the phenylpropanoid side chain with displacement of the carboxylate ester as leaving group. The proteins involved in the latter step are homologous to well-characterized phenylpropanoid reductases (pinoresinol-lariciresinol, isoflavone, phenylcoumaran-benzylic ether, and leucoanthocyanidin reductases), with similar catalytic mechanisms being operative. The proteins (and corresponding genes) involved in these transformations have been isolated and characterized and offer the potential of engineering plants to partially redirect carbon flow from lignin (or lignans) into these liquid volatile compounds in oilseeds, leafy or heartwood-forming tissues, or woody stems.

The emerging knowledge could also potentially facilitate wood processing in pulp/paper industries and offer sources of renewable plant-derived biofuels, intermediate chemicals in polymer industries, or specialty chemicals in perfume and flavor industries.

Key Words: Acetyltransferase, Acyltransferase, Allylphenol, Anethole, *Arabidopsis thaliana*, *Asparagus officinalis*, Bacterial cell culture, Basil, Biocatalysis, Biodiesel, Bioethanol, Biofuel, Biomass, Biosynthesis, C3H, C4H, CAD, Catalytic hydrogenation, Cationic polymerization, CCOMT, CCR, Cellulose, Chavicol, Chavicol synthase, Cinnamate 4-hydroxylase, Cinnamoyl CoA oxidoreductase, Cinnamyl alcohol dehydrogenase, COMT, Coniferyl acetate, Coniferyl alcohol, Corn, Creosote bush, *Cryptomeria japonica*, Deoxygenation, Double-bond reductase, Essential oil, Estragole, Eugenol, Eugenol synthase, F5H, Ferulate 5-hydroxylase, Flavor, Flavoring substances, Flax, Flaxseed, Furanocoumaran, Heartwood, Heat of combustion, Hinokiresinol, Hydrogenation, Hydroxycinnamoyl CoA ligase, Hydroxycinnamoyl shikimate/quinate transferase, IFR, Isoeugenol, Isoeugenol synthase, Isoflavone reductase, *Larrea tridentata*, Leucoanthocyanidin reductase, Lignan, Lignification, Lignin, Lignin challenge, Lignin downregulation, Lignin mutant, Lignin problem, Lignin reduction, Lumber, Metabolic engineering, Metabolic optimization, Methylchavicol, Methyleugenol, Methyltransferase, *Miscanthus*, Monolignol, Natural flavors, Natural food additives, NDGA, Nordihydroguaiaretic acid, *Ocimum basilicum*, Oilseed, *p*-anol, *p*-coumarate 3-hydroxylase, *p*-coumaryl acetate, *p*-coumaryl alcohol, *p*-coumaryl coumarate, PAL, PCBER, Petunia, *Petunia hybrida*, Phenylalanine, Phenylalanine ammonia lyase, Phenylcoumaran-benzylic ether reductase, Phenylpropanoid, Phenylpropanoid pathway, Pinoresinol, Pinoresinol-lariciresinol reductase, *Pinus taeda*, Plant cell culture, Plant metabolism, Plicatic acid, PLR, Polymer, Polymer intermediate, Propenylbenzene, Propenylphenol, Pulp/paper manufacture, Quinone methide, Reductase, Regiospecific reduction, Secoisolariciresinol, Spice, Stereospecific reduction, Styrene polymer, Switchgrass, *Syzygium aromaticum*, TAL, *Thuja plicata*, Vanillin, Vegetable oil, Wood.

1. INTRODUCTION

1.1. The challenge for humanity: Renewable, sustainable sources of bioenergy/biofuels, intermediate chemicals, and specialty chemical bioproducts

Humanity, as we know it, currently faces enormous political and scientific challenges in identifying and securing stable future sources of renewable energy in an environmentally acceptable and sustainable manner (i.e., leading to so-called biofuels/bioenergy). Similar concerns/considerations also apply to the continued future supply of other key petrochemical intermediates, such as monomers needed for industrial polymer production (e.g., polystyrenes and polyethylenes), as well as stable sources of key specialty chemicals (e.g., flavor and fragrance chemicals). This is, however, by no means a new scientific problem. It reflects instead one that has been difficult to solve over a period spanning more than three decades until now, and has been again brought to the forefront by the most recent biofuels/bioenergy crisis. Thus, there is an urgent need for highly creative and sound technological solutions for renewable (plant) resource utilization. To date, the difficulties in plant biomass utilization have centered on the recalcitrance of the various lignocellulosic matrices present in (woody) plants. This is largely, but not exclusively, due to the so-called lignin problem or challenge.

Regarding possible alternative sources of sustainable biofuels/bioenergy and various forms of bioproducts, it is generally recognized that truly novel (bio) technological solutions must be found and/or developed. Indeed, the U.S. Department of Energy was recently directed by the U.S. Congress to identify technology to produce 60 billion gallons of bioethanol annually by 2030, in order to replace some of the petroleum-derived gasoline. This, in turn, would require approximately 1.3 billion tons of lignified (woody) biomass annually in the United States alone, with the main plant species currently considered being corn, poplar, wheat, and more recently switchgrass (U.S. Department of Energy's Genomics, 2006). The enormous scale of the proposed cultivation of plant biomass for biofuel/bioenergy emphasizes, by itself, the increasing need to identify at least one possible means of securing sustainable bioenergy/biofuel supplies over the long term. However, little detailed thought has apparently been given to the potential ramifications of utilizing large swaths of forestry, agricultural, or marginal land for this purpose, as well as the ethical and/or practical issues that may arise.

Such measures, however, address only the biofuels need. An additional challenge generally overlooked at present is the ability to produce sufficient levels of industrial polymers, such as polyethylenes, polystyrenes, and other products which historically have come from the petrochemical industry. Today, about 12% of all the petroleum resources are used for nonfuel/nonenergy purposes, including polymer and other specialty chemical applications. There is also the need to obtain stable supplies of key specialty chemicals, such as flavors and fragrances, which at present are produced in regions of varying political stability and can also be subject to seasonal (climatic) variations; such factors often result in unpredictable market prices for these commodities.

1.2. Lignified biomass utilization: The lignin challenge

The major potential source of renewable energy/biofuels is that from plant biomass, for example through fermentation of polymeric carbohydrates to provide bioethanol. In this context, bioethanol production levels in the United States have steadily grown over the last decade, from approximately 1.4 to 4.26 billion gallons between 1995 and 2005 (Henniges and Zeddies, 2006), with this predominantly being obtained from the partial fermentation of corn. Yet this represents only approximately 4% of the current U.S. annual gasoline consumption (~100 billion gallons) and 7% of that needed (60 billion gallons) by 2030. There are two major scientific hurdles, however, that have not been technically overcome for the facile utilization of this and other plant renewable resources, both of which involve the polymeric lignins.

The first results from their intractable nature, since lignin removal has long been a limitation in the processing of wood both for pulp/paper manufacture and for forage digestibility by ruminants. This is largely due to the lack of isolated enzymes and/or proteins that can efficiently degrade lignin macromolecules, in contrast to reports in the 1980s that indicated that this problem had been solved (Glenn *et al.*, 1983; Kirk *et al.*, 1986; Tien and Kirk, 1983; Tien and Tu, 1987). That is, nearly 20 years ago, it was reported that several productive routes for lignin removal from wood had been both discovered and attained via utilization of lignin-degrading enzymes in fungi/bacteria, and where three candidates ultimately emerged (lignin peroxidase, manganese peroxidase, and laccase). However, this "lignin peroxidase" or "ligninase" (Tien and Tu, 1987) was assayed initially only with an aqueous acetone extract of spruce wood (Tien and Kirk, 1983), which does not actually extract the lignins from wood. Twenty years later, none of these enzymes is (routinely) utilized in biotechnological applications for lignin removal/separation, and their roles in enzymatic lignin biodegradation are still in question, as we had noted earlier (Sarkanen *et al.*, 1991). Today, more than 50 million tons of lignin-derived substances are generated annually as by-products of pulp/paper manufacture within the United States alone (Committee on Biobased Industrial Products; Board on Biology; Commission on Life Sciences; National Research Council, 2000). Interestingly, other possibilities now perhaps considered as being more likely to be useful are putative true lignin depolymerases targeting specific interunit linkages in lignin macromolecules (Chen *et al.*, 2001).

From a structural perspective, the lignins, nature's second most abundant organic substances after cellulose, are amorphous cell wall polymers that make up approximately 20–30% of all plant stem biomass (Lewis and Yamamoto, 1990; Lewis *et al.*, 1999). More specifically, vascular plant species have different lignin contents, with values ranging from approximately 30% in conifers (softwoods) to lower amounts (~20–25%) in hardwoods (such as poplar) and herbaceous species, to even smaller levels in various "primitive" plant species. The physiological roles of lignins are to engender structural support to the vascular apparatus, thereby enabling such organisms to stand upright, as well as providing conduits for water and nutrient transport, and to provide physical barriers against opportunistic pathogens. It is currently not known, however, what actual (i.e., minimal) lignin contents and/or compositions are needed for a particular plant to avoid any deleterious effects for growth/development/stem structural integrity, etc.

The second technological hurdle is that lignins cannot readily be converted into either ethanol and/or other liquid/gaseous fuels using currently available fermentation processes. Indeed, the polymeric lignins themselves are a formidable physical barrier to an efficient fermentation of carbohydrate biomass for ethanol generation, and thus their presence represents a critical problem in making these technologies more economical. Therefore, an approach whereby the carbon allocated toward lignification is redirected, resulting in inherently useful and/or more easily tractable materials, could potentially facilitate the generation of biofuels from the remaining plant biomass. One such strategy would be the generation *in planta* of allyl/propenyl phenols, such as eugenol and chavicol. In addition, these liquid/combustible phenolic products could themselves be potentially utilized for (nonethanol) biofuel/bioenergy purposes.

2. LIGNIN FORMATION AND MANIPULATION

Outcomes of the phenylpropanoid (C_6C_3) pathway (Fig. 13.1) include not only the lignins in woody/nonwoody vascular plants but also, to varying extents, lignans, flavonoids, coumarins, anthocyanins, as well as allyl and propenyl phenols in different species. This pathway has attracted much attention since the 1960s, and the various biochemical steps—including the historical development of this field—were comprehensively discussed (Lewis *et al.*, 1999), as were the trends observed in various genetic manipulations of the monolignol pathway and the corresponding downstream effects on lignification (Anterola and Lewis, 2002).

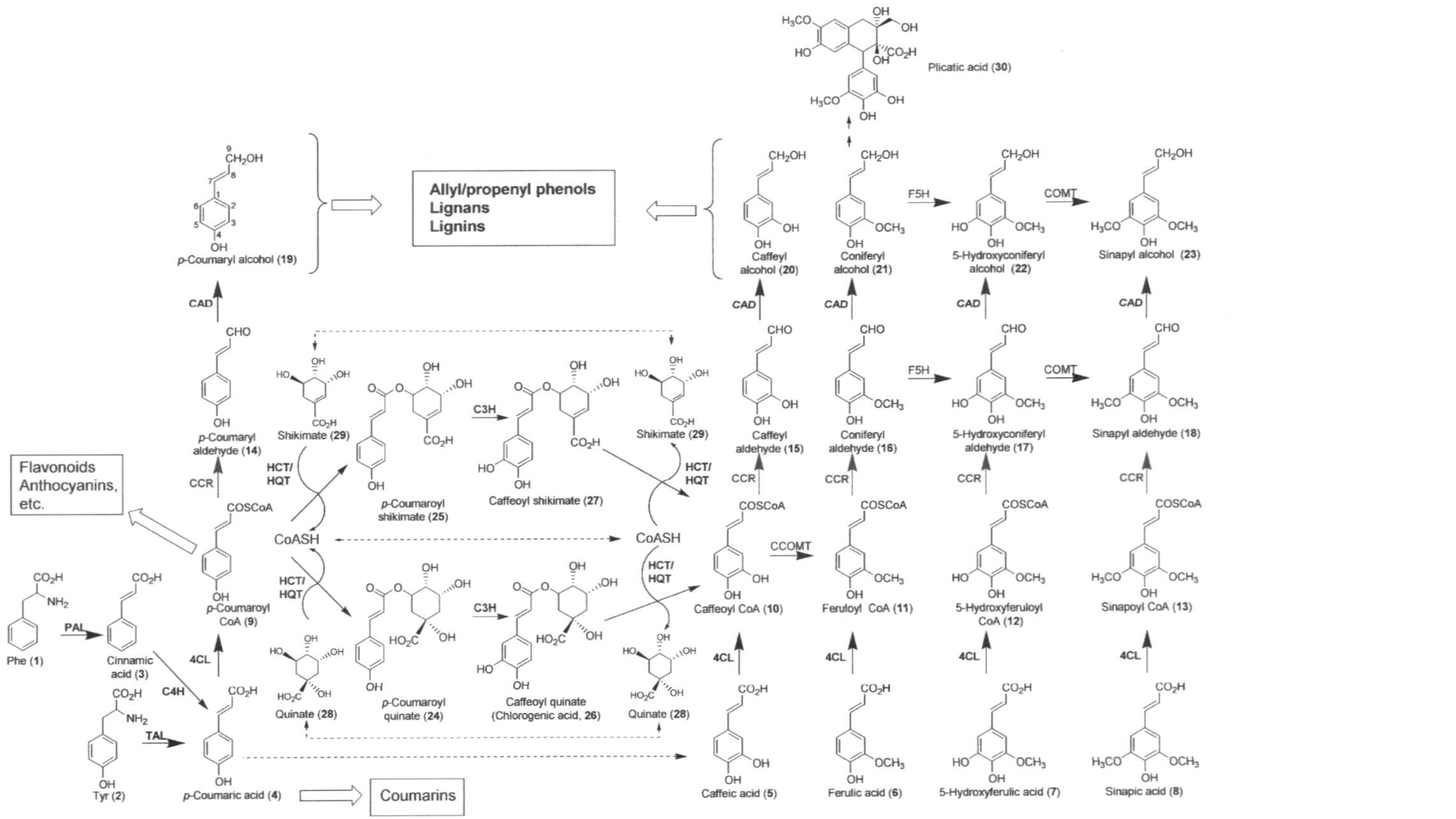

FIGURE 13.1 Current view of the phenylpropanoid pathway. 4CL, hydroxycinnamoyl CoA ligases; C3H, *p*-coumarate 3-hydroxylase; C4H, cinnamate 4-hydroxylase; CAD, cinnamyl alcohol dehydrogenases; CCOMT, hydroxycinnamoyl CoA *O*-methyltransferases; CCR, cinnamoyl CoA oxidoreductases; COMT, caffeic acid *O*-methyltransferase; F5H, ferulate 5-hydroxylase; HCT/HQT, hydroxycinnamoyl shikimate/quinate transferase; PAL, phenylalanine ammonia lyase; TAL, tyrosine ammonia lyase. (See Page 22 in Color Section.)

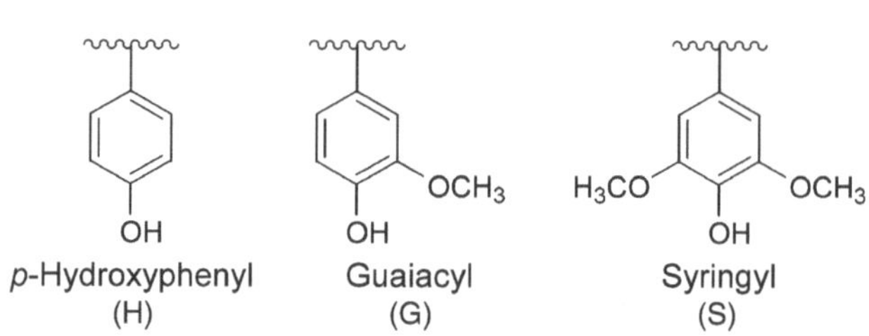

FIGURE 13.2 Aromatic (H, G, and S) residues in plant lignins.

A common theme that has emerged is the action of several enzymes of the phenylpropanoid pathway (e.g., hydroxylases, *O*-methyltransferases) not on free hydroxycinnamic acids, as previously believed, but on hydroxycinnamic esters, alcohols, and aldehydes. [For needed context, lignin assembly mainly utilizes three monolignols as monomers, namely the phenylpropanoid pathway-derived *p*-coumaryl (**19**), coniferyl (**21**), and sinapyl (**23**) alcohols (Fig. 13.1), giving rise to H (hydroxyphenyl), G (guaiacyl), and S (syringyl) moieties (''units''), respectively (Fig. 13.2). The H and G units are present in the lignin of gymnosperms and ''primitive'' plants, with angiosperms also containing S components, the latter being associated with fiber/vessel formation (Lewis *et al.*, 1999)].

2.1. Biosynthesis of monolignols

The phenylpropanoid pathway (Fig. 13.1) entry point is generally considered to be the amino acid phenylalanine (**1**, Phe), through action of phenylalanine ammonia lyase (PAL, EC 4.3.1.5) forming *trans*-cinnamic acid (**3**). Some monocots (e.g., maize), though, are also able to utilize the *p*-hydroxylated amino acid tyrosine (**2**, Tyr) as a partial entry point to this pathway through tyrosine ammonia lyase (TAL). The mechanism of phenylalanine (**1**) formation *in planta* has been a matter of debate, with arogenate decarboxylation/dehydration appearing to be the preferred biosynthetic route (Cho *et al.*, 2007; Jung *et al.*, 1986). The subsequent deamination reaction, catalyzed by PAL [forming cinnamic acid (**3**) and stoichiometric amounts of ammonium ion], was discovered by Eric Conn and his group (Koukol and Conn, 1961) and is probably the most extensively studied reaction in plant secondary metabolism (Lewis *et al.*, 1999).

The aromatic amino acid ammonia lyases (PAL, TAL, and HAL, histidine ammonia lyase) all possess the interesting internal cofactor 3,5-dihydro-5-methylidene-4*H*-imidazol-4-one (MIO) formed by spontaneous dehydration and cyclization of a conserved tripeptide sequence (Ala-Ser-Gly) (Baedeker and Schulz, 2002; Calabrese *et al.*, 2004; Schwede *et al.*, 1999), which acts as a strong electrophile during catalysis. The mechanism of ammonia elimination is still controversial, with the two most likely possibilities currently being (1) a E1cb-like mechanism where MIO forms a bond to the amine group of the amino acid (Fig. 13.3A) or (2) a Friedel-Crafts-like mechanism where MIO reacts with the aromatic ring of the substrate, which therefore loses aromaticity and increases the acidity of the nearby β-proton (Fig. 13.3B) (Calabrese *et al.*, 2004; Louie *et al.*, 2006;

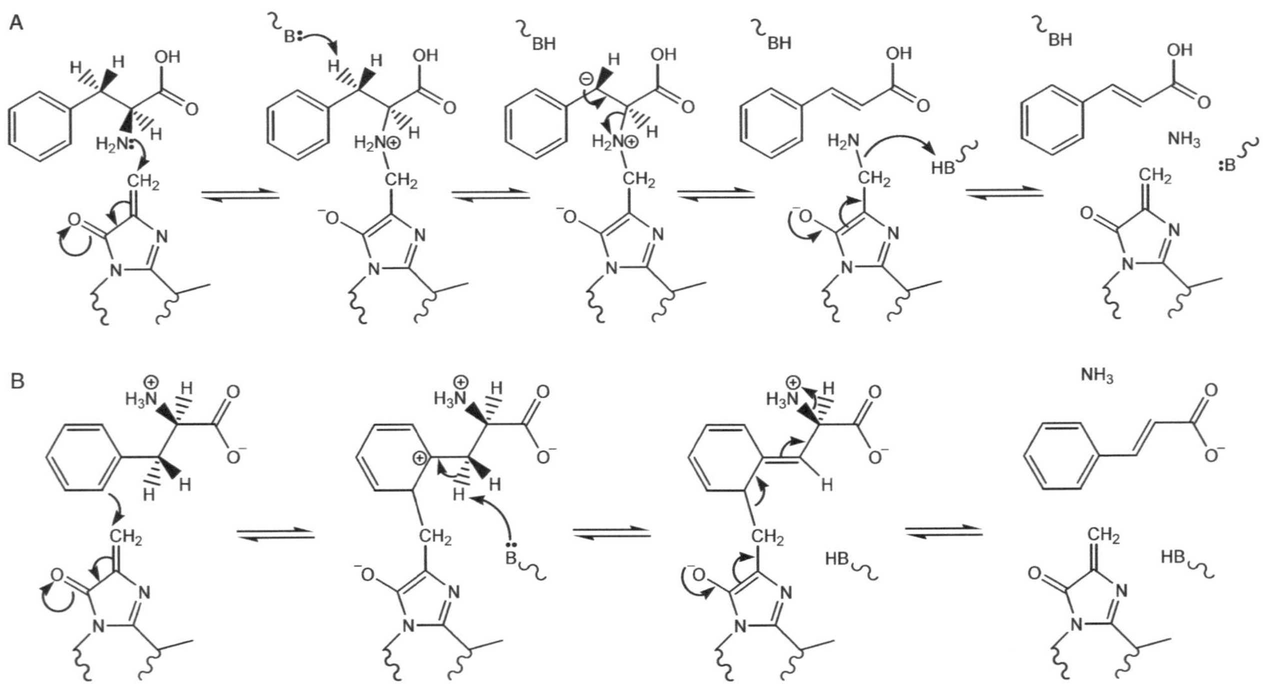

FIGURE 13.3 Potential mechanisms for PAL catalysis: (A) E1cb-like mechanism and (B) Friedel-Crafts-like mechanism. Source: Redrawn from Calabrese *et al.* (2004).

Poppe and Rétey, 2005). Recent studies of TAL from *Rhodobacter sphaeroides* have determined that Tyr (**2**) preference (i.e., TAL activity) is largely dictated by one specific TAL-conserved active site residue, H89, which is thought to be hydrogen-bonded to the phenolic moiety of the substrate Tyr (**2**). When H89 is replaced with a Phe residue (characteristic of PAL), substrate selectivity is switched and the mutant H89F TAL prefers Phe (**1**) (i.e., PAL activity). When the corresponding inverse mutation was made in PAL from *Arabidopsis thaliana*, the mutant (F144H) PAL enzyme lost PAL activity and instead gained TAL activity (Louie *et al.*, 2006; Watts *et al.*, 2006).

Although initially regarded as a rate-determining reaction in the pathway (Camm and Towers, 1973; Rubery and Fosket, 1969), PAL does not seem to serve as a major carbon allocation regulator, with flux apparently being determined by the availability of Phe (**1**) and by downstream enzymes in the pathway, especially C4H/C3H (Anterola *et al.*, 1999, 2002). In *A. thaliana*, a four-membered *PAL* multigene family (*AtPAL1–4*) has been characterized, with Tyr (**2**) serving, as expected, as a very poor substrate. Kinetic parameters for AtPAL1, 2, and 4 with Phe (**1**) were similar, with K_m values of ~60–70 μM and V_{max} ~5–10 pkat/μg protein, whereas AtPAL3 was barely active (K_m = 2.56 mM, V_{max} = 0.4 pkat/μg protein) (Cochrane *et al.*, 2004). Rohde *et al.* (2004) have also analyzed the effects of *PAL* downregulation in *A. thaliana*. In particular, *pal1* and *pal2* mutants apparently presented no clear phenotypic alterations, with lignin levels of approximately 60–70% of wild type (WT), perhaps as a result of compensatory upregulation of the remaining *PAL* genes in each mutant (i.e., *AtPAL2/4* were upregulated in *pal1* mutants, *AtPAL1/4* in *pal2* mutants). Double *pal1 pal2* mutants, on the other hand, exhibited approximately 25% of WT PAL activity (resulting mainly from the presence of AtPAL4), were sterile, and had estimated lignin levels of approximately 30–35% of WT, with this lignin having its S/G ratio apparently nearly doubled relative to WT.

Trans-cinnamic acid (**3**) is next *para*-hydroxylated by cinnamate 4-hydroxylase (C4H, EC 1.14.13.11), a membrane-associated cytochrome P-450 hydroxylase acting along with an associated NADPH-dependent reductase (Lewis *et al.*, 1999), to afford *p*-coumaric acid (**4**). Subsequent ring hydroxylation reactions [i.e., C3H-catalyzed 3-hydroxylation leading to formation of the catechol-like substructure of caffeic acid (**5**) and derivatives, as well as F5H-catalyzed 5-hydroxylation leading to the 5-hydroxyguaiacyl substructure present in 5-hydroxyconiferyl alcohol (**22**)] are also performed by cytochrome P-450 hydroxylases. Therefore, these enzymes (C4H, C3H, and F5H) are often grouped together as a single class. In *A. thaliana*, each of these enzymes is encoded by either one- (C4H and C3H) or two-membered (F5H) gene families, with locus numbers At2g30490 (C4H), At2g40890 (C3H), and At4g36220/At5g04330 (F5H), respectively. As of yet, no X-ray crystal structures for any of these plant cytochrome P-450 enzymes have been reported, although they play important (i.e., possibly regulatory, in the case of C4H and C3H) roles in monolignol biosynthesis and lignification.

Interestingly, 3-hydroxylation of the *p*-coumaric acid (**4**) moiety apparently occurs only after esterification, first to its CoA form (see below), then to shikimate and/or quinate esters; that is, *p*-coumarate 3-hydroxylase (C3H) acts on *p*-coumaroyl quinate (**24**) and/or shikimate (**25**) to form the corresponding caffeic acid derivative (**26** and/or **27**, respectively). Recognition that the *p*-coumaroyl esters serve as substrates for cytochrome P-450 hydroxylation was established using parsley (*Petroselinum crispum*) cell suspension cultures (Heller and Kühnl, 1985); more recent studies identified the gene and expressed functional recombinant protein (Schoch *et al.*, 2001). In this regard, transesterases (HCT/HQT) that transfer hydroxycinnamoyl groups from the corresponding CoA esters to shikimate (**29**) and quinate (**28**), respectively, were also recently reported (Hoffmann *et al.*, 2003; Niggeweg *et al.*, 2004). HCT, which prefers shikimate (**29**) as an acyl acceptor, has been reported to be present in stem vascular tissues in tobacco, silencing of which in *A. thaliana* led to dwarfed plants. In HCT-silenced *Nicotiana benthamiana* plants, the corresponding lignin reportedly decreased and presented lower S and increased H contents, respectively, that is, in agreement with its presumed participation in monolignol biosynthesis (Hoffmann *et al.*, 2004).

The final ring hydroxylation catalyzed by the cytochrome P-450 F5H was first reported to act on ferulic acid (**6**) (Grand, 1984); however, findings by Fukushima and colleagues indicated that the hydroxylation step preferentially involved coniferyl alcohol (**21**) (Chen *et al.*, 1999a,b). The corresponding *F5H* (*CYP84*) gene was subsequently isolated from *A. thaliana* (Meyer *et al.*, 1996), and further analysis of the recombinant CYP84 protein indicated a preference for both coniferyl alcohol (**21**) and coniferyl aldehyde (**16**) rather than ferulic acid (**6**) as substrates [K_m values of 3 and 1 μM, V_{max} of 6 and 5 pkat/mg protein for **21** and **16**, respectively, while $K_m = 1$ mM and $V_{max} = 4$ pkat/mg protein for the latter (**6**)] (Humphreys *et al.*, 1999), which is in agreement with Fukushima's seminal findings.

Through further modification of the aromatic ring, the catechol-like substructure of caffeic acid (**5**) [or an ester derivative thereof, e.g., caffeoyl CoA (**10**)], caffeyl aldehyde (**15**), and/or caffeyl alcohol (**20**) can serve as substrate for SAM (*S*-adenosyl methionine)-dependent *O*-methyltransferases (OMTs). In this way, the guaiacol substructure of ferulic acid (**6**) [or the respective ester derivative, e.g., feruloyl CoA (**11**)], coniferyl aldehyde (**16**), and/or coniferyl alcohol (**21**), respectively, can be formed. This *O*-methylation was originally reported to be performed by the so-called caffeic acid *O*-methyltransferases (COMTs) but was later shown to much more likely utilize caffeoyl CoA (**10**) as substrate i.e., caffeoyl CoA *O*–methyltransferase (CCOMT) (Pakusch *et al.*, 1989; Schmitt *et al.*, 1991; Ye *et al.*, 1994). Analogously, the 5-hydroxyguaiacol substructure resulting from F5H-mediated hydroxylation is *O*-methylated to afford a syringyl moiety. This is catalyzed by an OMT originally thought to act on caffeic acid (**5**) as primary substrate (COMT), and although the enzyme has now been established to be a 5-hydroxyguaiacyl *O*-methyltransferase (Anterola and Lewis, 2002; Atanassova *et al.*, 1995), the original "COMT" denomination still largely remains in use.

In *A. thaliana*, 22 genes have been putatively annotated as either COMT (EC 2.1.1.68, 17 putative genes) or CCOMT (EC 2.1.1.104, 5 putative genes). Biochemical analysis of 12 putative AtCOMT proteins bearing highest homology to the previously characterized AtCOMT1 indicated that only the latter is able to efficiently *O*-methylate a variety of potential substrates, with the remaining 11 not able to act on any of the phenolic compounds tested (Zhang *et al.*, manuscript in preparation). Caffeyl (**15**) and 5-hydroxyconiferyl (**17**) aldehydes are the preferred substrates, having k_{cat}/K_m values of 78,400 and 159,000 M^{-1} s^{-1}, respectively (Table 13.1). This is in accordance with previous analyses (see Anterola and Lewis, 2002), where the largest effect of COMT downregulation was the decrease of S contents in lignin (whereas G levels and total lignin amounts apparently remained relatively unchanged), with the concomitant incorporation of 5-hydroxyconiferyl alcohol (**22**) into the mutated lignin resulting in weakened and discolored stems. Interestingly, AtCOMT1 has also been described as an enzyme involved in quercetin to isorhamnetin formation (Muzac *et al.*, 2000); however, the kinetic data indicate that quercetin is a poorer substrate by an order of magnitude (Table 13.1) (Zhang *et al.*, manuscript in preparation).

Correspondingly, work on the five putative AtCCOMT enzymes has thus far focused on AtCCOMT1 and 2. AtCCOMT1 uses caffeoyl CoA (**10**) and 5-hydroxyferuloyl CoA (**12**) as preferred substrates (Table 13.2), together with quercetin, although with efficacies 10–100-fold lower than those observed for AtCOMT1 acting on the corresponding substrates [caffeyl aldehyde (**15**) and 5-hydroxyconiferyl aldehyde (**17**), respectively]. The corresponding acids, as expected, did not serve efficiently as substrates. AtCCOMT2, by contrast, more efficiently methylated quercetin, rather than the possible monolignol pathway intermediates, thus suggesting a role in flavonoid metabolism. Interestingly, AtCCOMT2-catalyzed methylation occurred at both positions 4 and 5 with 5-hydroxyferuloyl CoA (**12**) and 5-hydroxyconiferyl aldehyde (**17**) (Takahashi *et al.*, manuscript in preparation), indicating that it has no role in monolignol formation, where position 4 remains unmodified. In summary, while the participation of AtCOMT1 in monolignol biosynthesis in *A. thaliana* appears to be quite well established (i.e., S monomer formation), the particular roles of the multiple AtCCOMT isoforms, and the particular substrates they act upon, are still a matter of debate and thus remain to be further determined.

The transformations occurring on the phenolic ring (e.g., hydroxylations, *O*-methylations) are accompanied by side-chain modification of the aforementioned

TABLE 13.1 Kinetic parameters of AtCOMT1 with various phenylpropanoids and quercetin (Zhang *et al.*, manuscript in preparation)

	K_m (μM)	V_{max} (pkat/μg prot.)	k_{cat} (s^{-1})	k_{cat}/K_m (M^{-1} s^{-1})
Caffeic acid (**5**)	24.2	14.6	0.62	26,000
5-OH Ferulic acid (**7**)	32.0	30.1	1.30	40,500
Caffeyl aldehyde (**15**)	19.7	35.9	1.55	78,400
5-OH Coniferyl aldehyde (**17**)	17.9	66.2	2.85	159,000
Caffeyl alcohol (**20**)	51.5	35.3	1.52	29,500
5-OH Coniferyl alcohol (**22**)	31.6	51.9	2.23	70,700
Quercetin	23.7	3.8	0.16	6900

TABLE 13.2 Kinetic parameters of AtCCOMT1 and 2 with various phenylpropanoids and quercetin (Takahashi *et al.*, manuscript in preparation)

	K_m (μM)	V_{max} (pkat /μg prot.)	k_{cat} (s^{-1})	k_{cat}/K_m (M^{-1} s^{-1})
AtCCOMT1				
Caffeoyl CoA (**10**)	40.7	7.47	0.25	6140
5-OH Feruloyl CoA (**12**)	67.6	3.13	0.094	1630
Caffeyl aldehyde (**15**)	76.4	0.43	0.013	200
5-OH Coniferyl aldehyde (**17**)	23.1	0.55	0.017	820
Quercetin	25.0	1.03	0.081	1400
AtCCOMT2				
Caffeoyl CoA (**10**)	282.0	0.72	0.019	80
Caffeyl aldehyde (**15**)	20.1	0.32	0.0096	480
Quercetin	10.3	1.58	0.047	4560

cinnamic acid derivatives (**4–8**), first generating CoA esters (**9–13**) through the two-step activity of hydroxycinnamoyl CoA ligases (4CL, EC 6.2.1.12). These utilize ATP to form an AMP-acyl intermediate that stays enzyme-bound, with AMP then enzymatically displaced by CoA. In *A. thaliana*, 14 genes had been originally annotated as putative *4CLs*, but through further analysis and biochemical characterization, only 4 genes were shown to encode functional 4CL enzymes, albeit with very distinct kinetic properties [using all six possible substrates, cinnamic (**3**), *p*-coumaric (**4**), caffeic (**5**), ferulic (**6**), 5-hydroxyferulic (**7**), and sinapic (**8**) acids]. At4CL1 is by far the most active, with k_{cat}/K_m being ~15–600 times higher than that for others (Table 13.3) (Costa *et al.*, 2005).

The hydroxycinnamoyl CoA esters (**9–13**) thus formed serve as substrates for reduction by NADPH-dependent cinnamoyl CoA oxidoreductases (CCR, EC 1.2.1.44) generating the corresponding aldehydes (**14–18**), with 11 genes being putatively annotated as *CCRs* in *A. thaliana*. To date, only three of the putative AtCCRs have been biochemically characterized, namely, AtCCR1, AtCCR2 (Lauvergeat *et al.*, 2001; Patten *et al.*, 2005), and AtCCR8. The most active of these is AtCCR1, which most effectively utilized feruloyl CoA (**11**) and 5-hydroxyferuloyl CoA (**12**) as substrates (Table 13.4) (Patten *et al.*, 2005). This is in harmony with the corresponding aldehydes [coniferyl (**16**) and 5-hydroxyconiferyl (**17**) aldehydes] serving as substrates for subsequent monolignol formation. AtCCR2, by contrast, was a less effective catalyst, but again slightly preferred compounds **11**/**12** as substrates (Table 13.4). Interestingly, AtCCR8 displays both CCR and CAD catalytic properties, being able to slowly convert cinnamyl, *p*-coumaryl (**14**), caffeyl (**15**), and sinapyl (**18**) aldehydes into the corresponding monolignols; however, the physiological significance of these latter observations is still unclear, given the substrate versatility of many proteins.

The hydroxycinnamyl aldehydes (**14–18**) are then further reduced (by cinnamyl alcohol dehydrogenases, CAD, EC 1.1.1.195) to afford the monolignols **19–23**

TABLE 13.3 Kinetic parameters of At4CL1–3 and 5 (Costa *et al.*, 2005)

Substrate (acid)	K_m (μM)	V_{max} (pkat/μg prot.)	k_{cat} (s^{-1})	k_{cat}/K_m (M^{-1} s^{-1})
At4CL1				
Cinnamic (**3**)	135 ± 4.9	9.2 ± 0.10	0.59	4400
p-Coumaric (**4**)	6 ± 0.4	61.3 ± 0.95	3.96	666,000
Caffeic (**5**)	2 ± 0.1	42.0 ± 0.25	2.71	1,214,500
Ferulic (**6**)	16 ± 1.3	39.5 ± 1.10	2.55	154,600
5-OH Ferulic (**7**)	21 ± 1.8	10.2 ± 0.31	0.66	31,000
Sinapic (**8**)	No conversion			
At4CL1 native				
Cinnamic (**3**)	Not determined			
p-Coumaric (**4**)	6 ± 0.5	8.7 ± 0.14	0.53	91,000
Caffeic (**5**)	2 ± 0.2	3.9 ± 0.07	0.24	97,400
Ferulic (**6**)	17 ± 1.0	6.9 ± 0.13	0.42	25,000
5-OH Ferulic (**7**)	15 ± 2.9	1.1 ± 0.01	0.06	4260
Sinapic (**8**)	No conversion			
At4CL2				
Cinnamic (**3**)	181 ± 4.3	3.2 ± 0.03	0.21	1140
p-Coumaric (**4**)	45 ± 2.4	46.6 ± 1.14	3.00	66,600
Caffeic (**5**)	4 ± 0.4	30.4 ± 0.54	1.96	492,400
Ferulic (**6**)	686 ± 32.8	9.9 ± 0.22	0.64	930
5-OH Ferulic (**7**)	142 ± 7.9	7.0 ± 0.12	0.45	3200
Sinapic (**8**)	No conversion			
At4CL3				
Cinnamic (**3**)	42 ± 1.8	4.0 ± 0.06	0.26	6200
p-Coumaric (**4**)	4 ± 0.2	12.9 ± 0.16	0.84	227,900
Caffeic (**5**)	13 ± 1.0	7.0 ± 0.21	0.45	33,600
Ferulic (**6**)	10 ± 0.7	6.4 ± 0.11	0.41	41,000
5-OH Ferulic (**7**)	37 ± 2.3	0.9 ± 0.02	0.06	1600
Sinapic (**8**)	No conversion			
At4CL5				
Cinnamic (**3**)	No conversion			
p-Coumaric (**4**)	25 ± 1.7	3.1 ± 0.07	0.20	8000
Caffeic (**5**)	7 ± 0.9	2.9 ± 0.09	0.19	26,000
Ferulic (**6**)	6 ± 0.9	0.9 ± 0.03	0.06	9300
5-OH Ferulic (**7**)	6 ± 0.8	2.5 ± 0.05	0.17	26,500
Sinapic (**8**)	10 ± 1.1	1.1 ± 0.04	0.08	7900

(Lewis *et al.*, 1999). These NADPH-dependent CADs thus catalyze the final step in monolignol biosynthesis and have been studied in some detail in *A. thaliana* and other plants. In *A. thaliana, in silico* analyses had indicated the existence of a 17-membered CAD family, but further biochemical examination established that 2 of those (AtCAD4/5) were catalytically the most active (Kim *et al.*, 2004).

TABLE 13.4 Kinetic parameters for AtCCR1 (Patten *et al.*, 2005) and AtCCR2

	K_m (μM)	V_{max} (pkat/ μg prot.)	k_{cat} (s^{-1})	k_{cat}/K_m (M^{-1} s^{-1})
AtCCR1				
p-Coumaroyl CoA (**9**)	13.7	2.6	0.10	7170
Caffeoyl CoA (**10**)	7.1	3.4	0.13	17,870
Feruloyl CoA (**11**)	3.0	41.4	1.55	518,890
5-OH Feruloyl CoA (**12**)	2.9	37.0	1.39	485,770
Sinapoyl CoA (**13**)	11.9	4.0	1.10	91,470
AtCCR2				
p-Coumaroyl CoA (**9**)	20.8	10.0	0.37	17,630
Caffeoyl CoA (**10**)	11.2	7.4	0.27	24,260
Feruloyl CoA (**11**)	6.4	14.8	0.54	84,700
5-OH Feruloyl CoA (**12**)	6.5	10.8	0.40	61,275
Sinapoyl CoA (**13**)	16.7	24.0	0.88	52,820

Moreover, none of the isoforms displayed any strong sinapyl aldehyde (**18**) preference, contrary to reports of a putative sinapyl alcohol dehydrogenase (SAD) specific for sinapyl aldehyde (**18**) (Li *et al.*, 2001). X-ray crystal structures for AtCAD4 and 5 have been recently determined (Fig. 13.4), with each being a dimer containing two zinc ions per subunit, one participating in catalysis and the other playing a structural role (Youn *et al.*, 2006a). An *A. thaliana cad4 cad5* double mutant has also been obtained (Jourdes *et al.*, 2007; Sibout *et al.*, 2005) and presents a visible prostrate phenotype (i.e., cannot stand upright, see Fig. 13.5A) after approximately 4 weeks growth, resulting from the large decrease in amounts of lignin proper. Furthermore, this double mutant produces only about approximately 10% of WT lignin amounts, based on monolignol-derived thioacidolysis cleavage data, with small amounts of a poly-*p*-hydroxycinnamyl aldehyde also being formed. Interestingly, deposition of the latter is prematurely aborted relative to lignin proper (Jourdes *et al.*, 2007). This further emphasizes the stringency of the lignification process, where three of the CAD monolignol products (**19**, **21**, and **23**) have evolutionarily become the highly conserved substrates for subsequent controlled one electron (radical) oxidation and polymerization of the H, G, and S "building blocks" of lignin (Fig. 13.2) (Anterola and Lewis, 2002). That is, strong evolutionary pressure resulted in lignins being formed from monolignols (as well as, to a much lesser extent, monolignol esters in grasses).

2.2. The challenge of lignin manipulation: Plant growth/development versus stem structural integrity

There are compelling and long-standing reasons to identify novel ways to more effectively either utilize the lignin biopolymers or manipulate the amounts or forms of carbon allocated to the lignin-forming pathway, for example, to produce more desirable bioproducts in commercially cultivated plant species. Indeed, a number of biotechnological manipulations of both lignin contents and

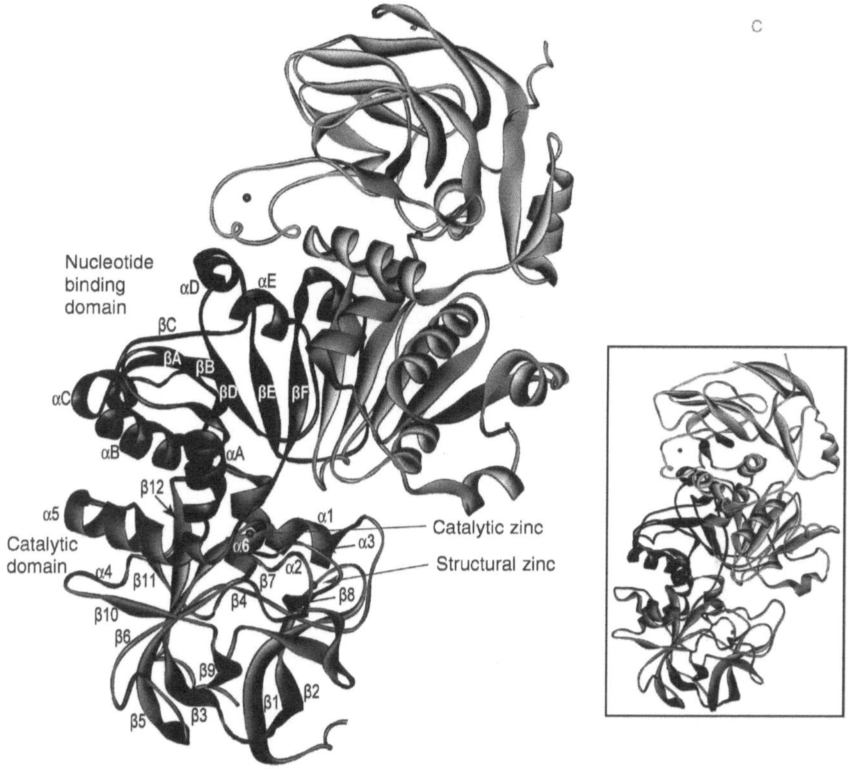

FIGURE 13.4 Schematic representation of the crystal structure of the AtCAD5 homodimer and energy-minimized model of AtCAD4 (inset). Source: Reprinted from Youn *et al.* (2006a). (See Page 23 in Color Section.)

compositions in various plant species have already been carried out; that is, various transgenic/mutant lines have been successfully obtained using standard transformation procedures (see Anterola and Lewis, 2002, for examples and references therein). Generally, though, the effects of drastically reducing lignin contents in both woody and nonwoody vascular plants result in a significant impairment/weakening of the vascular apparatus, for example, collapsed vessels (for a discussion and examples, see Anterola and Lewis, 2002). Such defects potentially lead to severe drawbacks in growing biotechnologically modified plant lines commercially, as this can lead to, for example, premature lodging, weakening of plant stems, and dwarfing during growth/development.

To put the utility of employing mutant and/or genetically modified lines into sharper focus, the following examples should illustrate why this issue deserves attention. It is often overlooked that many lignin mutants have been described over a period spanning nearly a century, particularly the brown midrib mutants (see Anterola and Lewis, 2002). All had significant deleterious effects on vascular tissue integrity, and none, to our knowledge, has yet found commercial application. Three of these are COMT, CAD, and cinnamoyl CoA oxidoreductase (CCR)

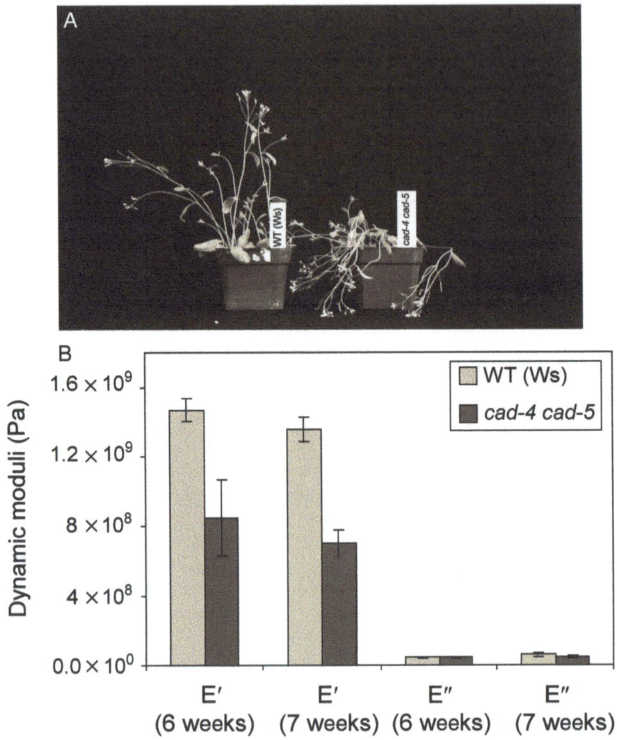

FIGURE 13.5 Effects of knocking out *Atcad4* and *cad5* (*cad-c cad-d*) in *Arabidopsis thaliana* (ecotype Wassilewskija). (A) Phenotypical differences between 4-week-old wild type (Ws) and lignin-deficient *cad-4 cad-5* double mutant plants. (B) Tensile storage and loss moduli of WT and lignin-deficient (~90%) *cad-4 cad-5* double mutant lines. Source: Redrawn from Jourdes *et al.* (2007). (See Page 24 in Color Section.)

mutants (see Fig. 13.1 for biochemical pathway steps). The first, due to introduction of a 5-hydroxyconiferyl alcohol (**22**) monomer in lignin, results in brittle stems, which are more susceptible to lodging (Anterola and Lewis, 2002). The CAD double mutation also results in a generally weakened vasculature. The CAD double mutant in *A. thaliana* has a greatly compromised ability to form monolignols **19, 21,** and **23** (by ~90–94%), with only very small amounts of lignin proper being formed (~10% of the natural levels). The resulting stems (of the CAD double mutant) are thus unable to stand upright (Fig. 13.5A) and their tensile modulus is greatly compromised (~50%) when tested in the tension mode (Fig. 13.5B) (Jourdes *et al.*, 2007). Such a phenotype may be a disadvantage for either large-scale commercial cultivation, harvesting, or processing due to the weakened vascular apparatus. It is also unknown whether such modifications may also impact/decrease resistance to opportunistic pathogens.

CCR mutation in *A. thaliana* also resulted in a severely dwarfed phenotype and a delayed but coherent lignification program (Fig. 13.6A–C) (Laskar *et al.*, 2006; Patten *et al.*, 2005), whereas in tobacco, it resulted in a compromised vasculature and dwarfing as well (Piquemal *et al.*, 1998). In an analogous manner, 4-coumarate CoA ligase (4CL), PAL, and C4H downregulation resulted in a significant loss of vascular integrity (see Anterola and Lewis, 2002) and/or other effects, such as dwarfing, due (mainly) to reduced lignin levels. Other concerns about deleterious effects on vascular integrity hold also for C3H downregulation (Patten *et al.*, 2007).

These examples underscore the central question as to what extent lignin compositions/contents can actually be manipulated, without introducing structural defects prohibiting field applications of the resulting plant cultivars in, for example, bioethanol/biofuel/bioproduct generation. Another possible concern is that a weaker vascular apparatus may result in plants more susceptible to opportunistic pathogens. In short, it is becoming increasingly evident that a judicious balance must be maintained in growing vascular plants for commercial purposes and in reducing/modifying lignin contents/compositions. However, what flexibility exists in modifying lignin amounts/composition to avoid such adverse growth/developmental effects has not yet been determined.

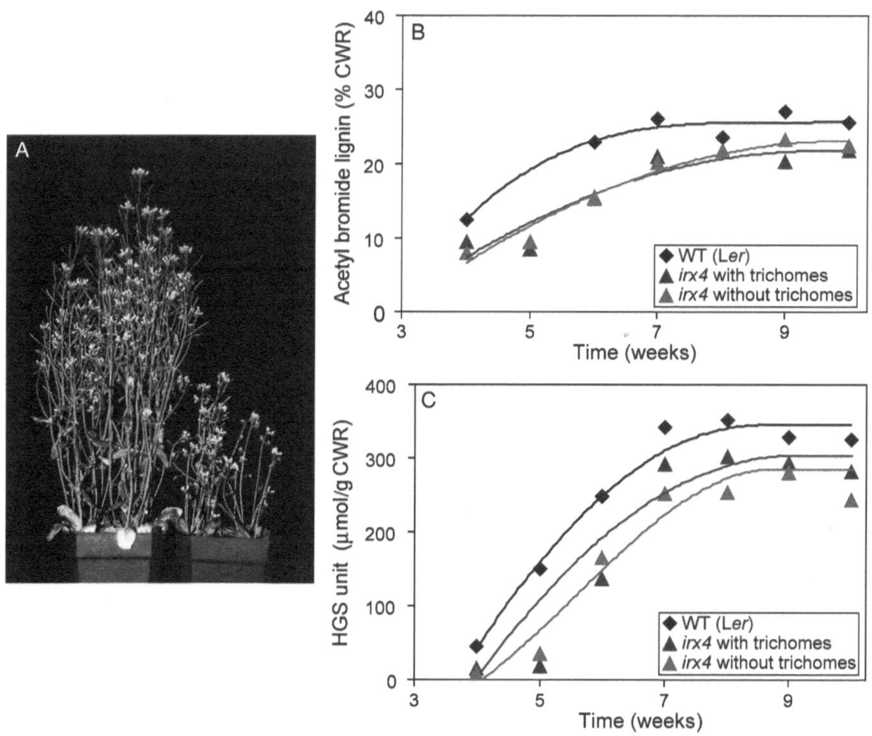

FIGURE 13.6 Effects of mutating CCR in *Arabidopsis thaliana*. (A) Phenotypical differences between wild type (*Ler*, left) and *irx4* mutant (right) plants. (B) Plot of acetyl bromide lignin determinations in stems at various stages of *A. thaliana* growth and development. (C) Estimations of H, G, and S monomer amounts released during thioacidolysis of wild type and *irx4* extractive-free stem tissues. Source: Redrawn from Patten *et al.* (2005). (See Page 25 in Color Section.)

2.3. New opportunities and approaches for renewable sources of bioenergy, biofuels, and bioproducts?

While the extent to which lignin polymer composition and content can ultimately be modified and/or reduced is poorly understood, there are alternative biotechnological opportunities to produce renewable energy biofuels and specialty bioproducts, which have not been examined yet in detail. That is, there are other metabolic outcomes of the phenylpropanoid pathway from its entry point phenylalanine (**1**) (depending upon the species) that can include, for example, differential formation of coumarins, lignans, and flavonoids, as well as allyl/propenyl phenols (see Fig. 13.1). In particular, allylphenols and propenylphenols, which differ in the position of their side-chain double bonds, include the high-value liquids eugenol (**33**), estragole [(**32**), methylchavicol], and anethole (**38**) (Fig. 13.7). These natural products account for much of the aroma present in specialty "essential oils" of various plant species, such as cloves, tarragon, and anise, respectively, and are thought to be produced *in planta* mainly for defense against insects and parasites, as well as for attracting pollinators. In addition, most of these compounds are liquid at room temperature, and their relatively low degree of oxygenation grants them with high heats of combustion; these are potentially desirable characteristics when considering their possible utilization as fuels. Notably, lignins and most lignans, as well as the allyl/propenyl phenols, are all derived from the same monolignol precursors; thus an approach whereby the latter are differentially utilized could impact the production/accumulation of these diverse classes of compounds.

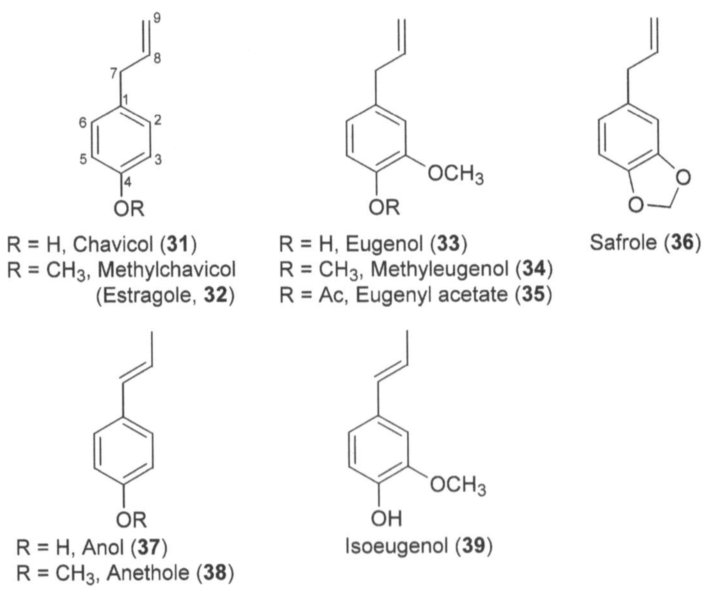

FIGURE 13.7 Selected allylphenols and propenylphenols.

Therefore, biotechnological manipulations of this pathway might be directed not only toward simply reducing lignin levels but also, perhaps, toward retargeting carbon toward related metabolic pathways, for example, through redirection of metabolic (carbon) flux to the production of related phenolic compounds in the main repositories for plant organic carbon storage. The latter could include either oilseed-bearing structures [e.g., flax (*Linum usitatissimum*) seed, Fig. 13.8A] (Ford *et al.*, 2001; Teoh *et al.*, 2003) or heartwood-forming tissues of trees [e.g., western red cedar (*Thuja plicata*), Fig. 13.8B] (Fujita *et al.*, 1999; Kim *et al.*, 2002). Indeed, it is these "repositories" that are largely used as plant renewable resources, whether as sources of (vegetable) oils or for lumber/pulp/paper products. In addition to the structural lignins, heartwood formation is often accompanied by massive deposition of non-structural low molecular weight molecules, such as the monolignol-derived lignan plicatic acid (**30**, see Fig. 13.1 for structure) and its congeners in western red cedar, whose amounts can be approximately 20% of the overall dry weight (Gardner *et al.*, 1959, 1960, 1966). Rational optimization/modification of plant biomass could be done either directly for biofuel/bioenergy/bioproduct generation in specific crops or indirectly as part of (heart)wood processing for pulp/paper, specialty chemicals, etc. Such considerations have *not* yet been explored, although the technologies are now available from previous research studies (Dinkova-Kostova *et al.*, 1996; Fujita *et al.*, 1999; Jiao *et al.*, 1998; Kato *et al.*, 1998; Teoh *et al.*, 2003; Vassão *et al.*, 2006b, 2007). For instance, recent studies have described the formation of some quite well-known phenylpropanoid pathway monomeric metabolites, namely, the liquid allyl/propenyl phenols, chavicol (**31**), eugenol (**33**), and their analogues (**32**, **34–39**) (Fig. 13.7) (Koeduka *et al.*, 2006; Vassão *et al.*, 2006b).

Historically, such allyl/propenyl phenols have commonly been used throughout the world mainly as flavor/fragrance components present in spices, especially cloves, with these being largely imported from Tanzania, Madagascar, and Indonesia. Such material sources are imported simply because it is in these countries that the plant species accumulating these more unusual metabolites are cultivated. The recently described biochemical/biotechnological processes (Koeduka *et al.*, 2006; Vassão *et al.*, 2006a,b, 2007) thus offer yet another possibility

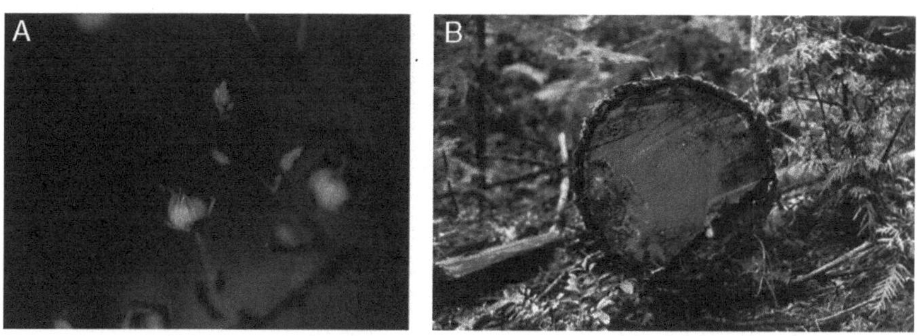

FIGURE 13.8 (A) Flax (*Linum usitatissimum*) oilseed and (B) western red cedar (*Thuja plicata*) heartwood. (See Page 25 in Color Section.)

of diversion of monolignols from either lignin and/or lignan formation in more commonly utilized woody/nonwoody plant species of, for example, North America, to afford instead the liquid allyl/propenyl phenol monomers.

3. CURRENT SOURCES/MARKETS FOR SPECIALTY ALLYL/PROPENYL PHENOLS

The major current source of allyl/propenyl phenols is clove oil, which generally contains eugenol (**33**, 75–85%), eugenyl acetate (**35**, 8–15%), and other minor components. The oil is a product of the clove tree, *Syzygium aromaticum* [*Eugenia caryophyllata*], and is mainly produced in Madagascar, Tanzania (Zanzibar and Pemba Islands), as well as in Indonesia. Worldwide production of this oil approximates 2000 ton year^{-1} (Bauer *et al.*, 2001), with a gross market value of approximately US$ 30–US$ 70 million per annum through its applications in flavors, fragrances, and antibacterials alone (George Uhe Company, 2006). This market size is, however, largely dictated by (1) the labor-intensive gathering of cloves, (2) the costs associated with obtaining the oil, and (3) limited growth/production capabilities. Therefore, a potential exists for the expansion of the source(s) of these molecules through the use of new (bio)technologies.

Eugenol (**33**) and its natural derivative, eugenyl acetate (**35**), are widely used in the perfumery and flavor industries, being considered by the Food and Drug Administration (FDA) as Generally Recognized As Safe (''GRAS'') for use as food additives. For example, eugenol (**33**) is responsible for the majority of clove-like flavors in beverages, ice cream, baked goods, and candy (Maralhas *et al.*, 2006), whereas methyleugenol (**34**) is a flavoring ingredient in ice cream, candy, and cola soft drinks (Smith *et al.*, 2001; U.S. Environmental Protection Agency, 2006a). Eugenol (**33**) is also present in kretek (clove) cigarettes (usually as 40% ground cloves, 60% tobacco) and is used as an industrial source of isoeugenol (**39**) (by alkaline isomerization) and methyleugenol (**34**). Because of its excellent analgesic and antibacterial properties, eugenol (**33**) has long been employed in dentistry in combination with zinc oxide forming a polymerized eugenol cement, which is then used for surgical dressings, temporary fillings, pulp capping agents, and cavity liners (Skinner, 1940; Weinberg *et al.*, 1972). Isoeugenol (**39**) is widely used as a flavor and fragrance additive in baked goods, beverages, chewing gum, and personal products such as perfumes and soaps (Badger *et al.*, 2002). Isoeugenol (**39**) also used to be the main source for the industrial production of vanillin (**40**, Fig. 13.9), but the latter is now more commonly derived either from the petrochemical guaiacol (**41**) or from by-products of the pulp/paper (lignin) industry (Hocking, 1997).

Chemically related products, anethole (**38**) and methylchavicol (**32**), on the other hand, are currently isolated as minor by-products (1–2%) of crude sulfate turpentine (CST) processing (The Flavor and Fragrance High Production Volume, 2002; 2005). This requires a lengthy isolation process of fractional distillation starting from CST, and fractional crystallization from the semipurified terpenoid/phenolic

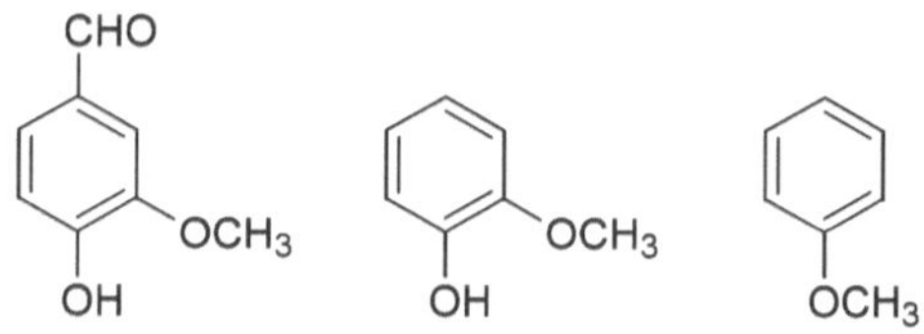

Vanillin (**40**) Guaiacol (**41**) Anisole (**42**)

FIGURE 13.9 Structures of vanillin (**40**), guaiacol (**41**), and anisole (**42**).

mixtures thus produced. Most of the methylchavicol (**32**) obtained in this way is then subjected to alkaline isomerization to generate more anethole (**38**), although chemical synthesis from anisole (**42**, Fig. 13.9) using propionyl chloride or propionaldehyde can also be used. Although CST accounted for 97% of total turpentine production in the United States in 1990, steadily increasing from 27.4% in 1950, the total production of turpentine declined to approximately 20.7 million gallons in 1999 (Haneke, 2002) [containing an estimated 680 and 740 tons methylchavicol (**32**) and anethole (**38**), respectively]. As wages increased, the labor-intensive production of turpentine oil became increasingly less competitive economically, and thus much of the original botanical-derived turpentine was substituted by cheaper petroleum-based solvents in the intervening years. Anethole (**38**) finds application as a perfume in detergents, soaps, and shampoos, and as a flavoring agent in licorice, ice cream, baked goods, and alcoholic beverages (Newberne *et al.*, 1999), while methylchavicol (**32**) is added as a component of root beer and anise-type flavors, as well as condiments and meat seasonings (The Flavor and Fragrance High Production Volume Consortia, 2005).

Some allyl/propenyl phenols have also recently been approved by the Environmental Protection Agency (EPA) for use as insecticides, insect repellents, and/or insect lures, perhaps reflecting the original roles/functions of these natural products *in planta*. Applied as unmodified natural products, they have more specific and less persistent insecticidal activities and therefore pose reduced environmental risk relative to the commonly used organophosphates and pyrethroids. Interestingly, eugenol (**33**) is used as an insecticide not only for various crop plants and ornamentals but also for pets (U.S. Environmental Protection Agency, 2006b); furthermore, it can be used as an attractant in insect traps specific for Japanese beetles (U.S. Environmental Protection Agency, 2006b,c). Anise oil, on the other hand, is approved for use on lawns and ornamentals as a dog and cat repellant (U.S. Environmental Protection Agency, 2006d), whereas methylchavicol (**32**), when applied to trees, repels bark beetles (for example, the southern pine beetle) thereby limiting their aggregation and reproduction abilities (U.S. Environmental Protection Agency, 2006e). Additionally, methyleugenol (**34**) is perhaps the most economically important allyl/propenyl phenol insect attractant, being in widespread use in insect traps for certain fruit flies, including the Oriental fruit fly, Mediterranean fruit fly, and solanaceous fly (Jang *et al.*, 2003; U.S. Environmental Protection Agency, 2006a). (Fruit flies are one of the most destructive pests to Hawaii's agricultural industry, where these low-cost traps are one of the few environment-friendly recommended suppression techniques.)

4. BIOSYNTHESIS OF ALLYL AND PROPENYL PHENOLS AND RELATED PHENYLPROPANOID MOIETIES

Besides lignin, specialized plant metabolism can utilize monolignols in the formation of lignans (phenylpropanoid dimers) and, as recently elucidated, allyl and propenyl phenols. Allylphenols differ from propenylphenols in their side-chain double bond position, with the former having terminal (C-8–C-9) desaturation and the latter having the chemically more stable internal (C-7–C-8) double bond. Several biochemical hypotheses had been created to explain their distinctive lack of a C-9 oxygenated functionality (Canonica *et al.*, 1971; Klischies *et al.*, 1975; Manitto *et al.*, 1974a,b), but experimental support for any was lacking until recently.

Our interest in the allyl/propenyl phenols first began with studies directed at elucidating the biochemical pathway(s) to the lignan nordihydroguaiaretic acid (**43**, NDGA, Fig. 13.10) and its congeners (Cho *et al.*, 2003; Moinuddin *et al.*, 2003), these being abundant metabolites in the creosote bush (*Larrea tridentata*). These substances are of increasing interest due to their potent biological and (potential) medicinal applications. For example, the NDGA derivative **44** is apparently proceeding smoothly through National Institutes of Health trials as an effective chemotherapeutic treatment for the usually hard to treat (refractory) brain and central nervous system tumors (Chang *et al.*, 2004) (see also http://www.clinicaltrials.gov/ct/show/NCT00404248, http://www.clinicaltrials.gov/ct/show/NCT00259818 and http://www.cancer.gov/search/viewclinicaltrials.aspx?cdrid=455645). Interestingly, the creosote bush has long been part of traditional Native American Indian medicine, being used to treat more than 50 diseases, most commonly those of renal and gynecologic origins (Arteaga *et al.*, 2005). These lignans, however, lack oxygenated carbon 9,9′ functionalities that are present in most lignan classes [e.g., podophyllotoxin (**45**), secoisolariciresinol (**46**), Fig. 13.10], as well as in the polymeric lignins and monomeric phenylpropanoids (e.g., monolignols **19–23**) in the vast majority of plant species. Based on previous radiolabeling/stable isotope labeling studies (Moinuddin *et al.*, 2003), it was presumed that the unusual "loss" of the oxygenated functionality occurred at

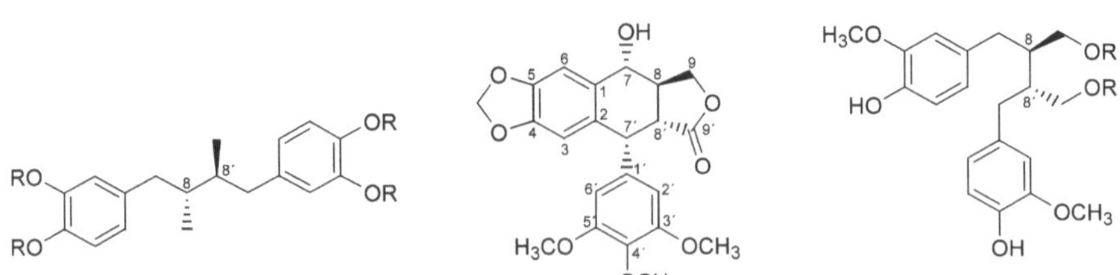

R = H, NDGA (**43**)
R = CH_3, tetra -*O*-methyl NDGA (**44**)

Podophyllotoxin (**45**)

R = H, Secoisolariciresinol (**46**)
R = Glc, Secoisolariciresinol diglucoside (**47**)

FIGURE 13.10 Lignans commonly used in medicinal applications, particularly for cancer treatment/prevention.

the monomer stage; that is, allyl and/or propenyl phenols could be serving as the precursors/substrates for dimerization to form these less common lignans. In this regard, the biosynthetic pathways to the allyl/propenyl phenols had not been elucidated in any organism and, in particular, the precise precursor (substrate) undergoing deoxygenation represented both a long-standing question and a biochemical mystery (Canonica *et al.*, 1971; Klischies *et al.*, 1975; Manitto *et al.*, 1974a,b; Senanayake *et al.*, 1977).

Basil (*Ocimum basilicum*, Thai variety) was used as a suitable study system since it accumulates the simplest allylphenol, methylchavicol (**32**); based on various radiolabeling studies, it was shown that the latter was derived from the corresponding monolignol, *p*-coumaryl alcohol (**19**) (Vassão *et al.*, 2006b). Three potential mechanisms for the conversion of **19** into **32** included reduction of the monolignol side-chain (i.e., saturation) followed by dehydration (Fig. 13.11A); methylation of the phenolic moiety preceding further side-chain modification (Fig. 13.11C); and/or activation of the terminal (C-9) oxygenated functionality prior to side-chain double bond reduction (Fig. 13.11B). Pathways A and C (Fig. 13.11) were eliminated since no experimental evidence in support of either route was obtained.

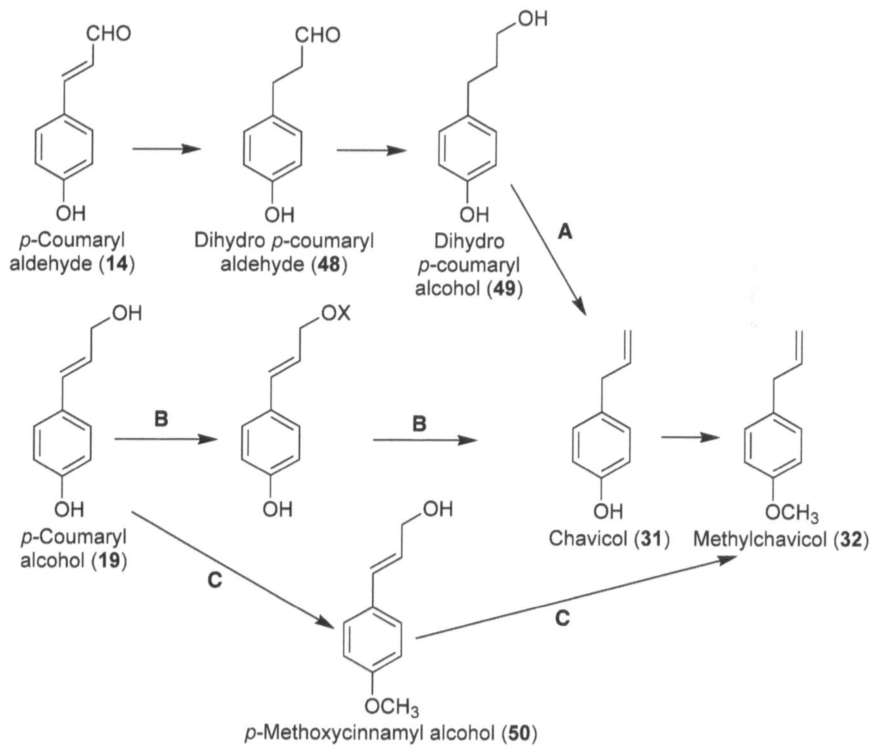

FIGURE 13.11 Hypothetical biosynthetic pathways from *p*-coumaryl alcohol (**19**) to chavicol (**31**) and methylchavicol (**32**), of which pathway B was demonstrated to occur. Source: Vassão *et al.* (2006b).

Interestingly, however, a double-bond reductase was discovered and characterized, which utilized *p*-coumaryl aldehyde (**14**) as the preferred substrate to afford the corresponding side-chain reduced aldehyde (**48**, Fig. 13.11) (Kasahara *et al.*, 2004, 2006; Youn *et al.*, 2006b). This alkenal reductase activity was the first to be reported in the phenylpropanoid pathway, with the corresponding enzymes isolated from *A. thaliana* (AtDBR) and *Pinus taeda* (PtPPDBR) also being homologous to a terpenoid double-bond reductase (pulegone reductase, PulR) from *Mentha piperita* and mammalian alkenal reductases as well (Fig. 13.12). AtDBR and PtPPDBR catalyze the NADPH-dependent reduction of *p*-coumaryl (**14**) and coniferyl (**16**) aldehydes to the corresponding dihydroaldehydes, (Fig. 13.13) and AtDBR has also been shown to catalyze the reduction of 4-hydroxynonenal (4-HNE, **51**), a pro-apoptotic lipid peroxidation product, to 4-hydroxynonanal (Kasahara *et al.*, 2006; Youn *et al.*, 2006b). Based on substrate versatility studies and an X-ray crystal structure for AtDBR, a concerted mechanism involving an enol intermediate was proposed for these zinc-independent alkenal reductases (Youn *et al.*, 2006b). While the corresponding dihydroalcohol product **49** is a well-known plant defense metabolite (Kraus and Spiteller, 1997), it was not, however, converted in basil into either chavicol (**31**) and/or *p*-anol (**37**) (Vassão *et al.*, 2006b). Accordingly, it was not considered as being involved in allyl/propenyl phenol biosynthesis.

Instead, a quite novel metabolic process converting monolignols [such as *p*-coumaryl (**19**) and coniferyl (**21**) alcohols] into allyl/propenyl phenols [chavicol (**31**) and eugenol (**33**), respectively] was discovered (Fig. 13.14), with two enzymes being implicated in their formation *in planta*. The first step is activation of the monolignol side-chain alcohol by conjugation to an activated acid (acyl-CoA), resulting in formation of a monolignol ester. This modification results, in energetic terms, in formation of a more facile leaving group (carboxylate ester), which is more readily displaced by an incoming reducing hydride, for example, in the form of NAD(P)H. Indeed, such coniferyl alcohol acyl transferases have been recently characterized in basil (*O. basilicum*) (Harrison and Gang, 2006) and petunia (*Petunia hybrida*) (Dexter *et al.*, 2007), utilizing acetyl-CoA and coniferyl alcohol (**21**) to afford coniferyl acetate (**53**), and it is anticipated that substrate-versatile acyltransferases may be able to utilize different monolignols and acyl/aroyl-CoA cofactors to generate different esters. One such ester, *p*-coumaryl coumarate (**54**), had been previously shown to serve as substrate for enzyme preparations from *Asparagus officinalis* (Suzuki *et al.*, 2002) and *Cryptomeria japonica* (Suzuki *et al.*, 2004), generating the *nor*-lignans (*Z*)- and (*E*)-hinokiresinol (**56/57;** Fig. 13.15A) respectively. Although the proteins responsible for the latter conversions remain to be fully characterized and/or described, potential mechanisms whereby the departing carboxylate (as CO_2) facilitates formation of the final C8–C7′ bond, without any additional cofactors, can be envisaged (Fig. 13.15A). It is also possible to propose potential mechanisms where *p*-coumaryl coumarate (**54**) [or other *p*-coumaryl alcohol esters, e.g., *p*-coumaryl acetate (**52**)] generates, through the addition of an incoming hydride, chavicol (**31**) and/or its regioisomer *p*-anol (**37**) (Fig. 13.15B). Indeed, the second step in monolignol reduction was shown to be the action of regiospecific reductases that transfer a hydride from NADH or

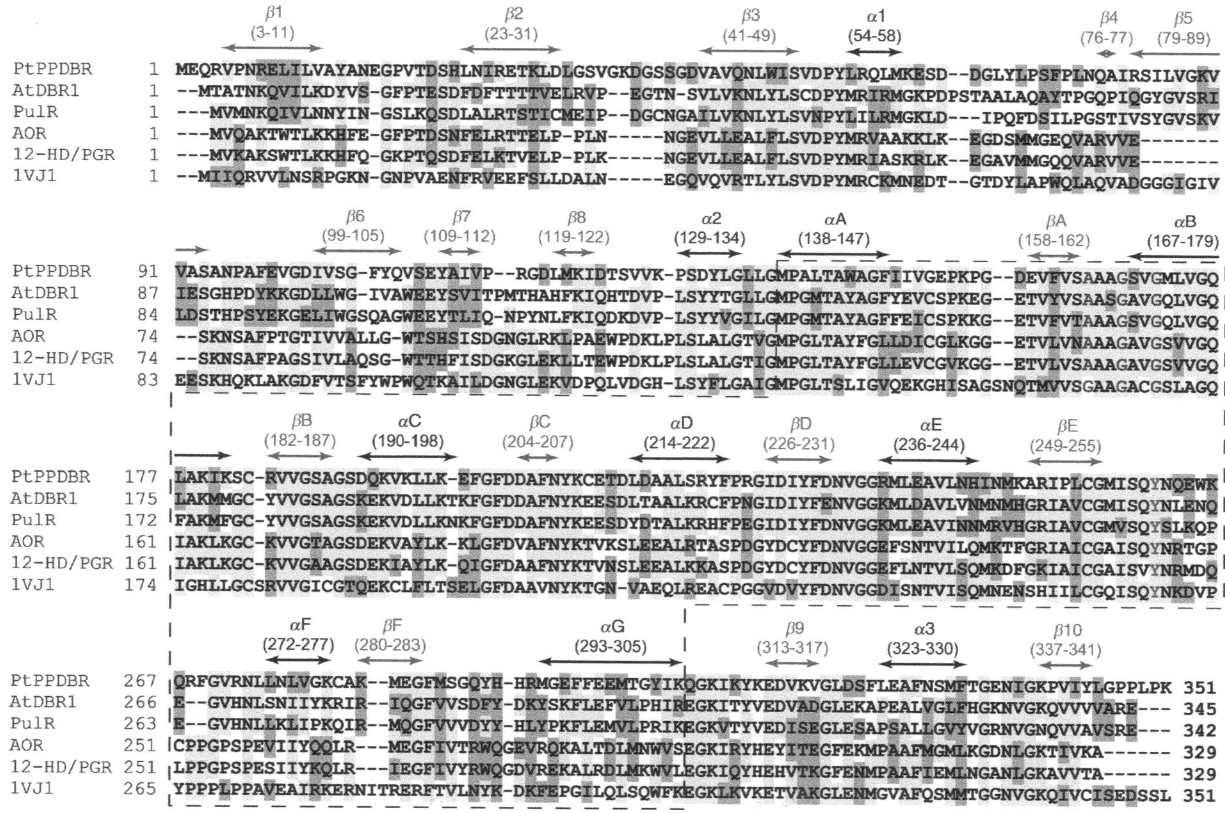

FIGURE 13.12 Alignment of double-bond reductases from *Pinus taeda* (PtPPDBR), *Arabidopsis thaliana* (AtDBR1), and homologues from *Mentha piperita* (pulegone reductase, PulR), rat (*Rattus norvegicus*, AOR), guinea pig (*Cavia porcellus*, 12-HD/PGR), and mouse (*Mus musculus*, 1VJ1). The nucleotide-binding domain is indicated by the dotted line, with conserved AXXGXXG motif in red. Conserved catalytic Tyr residues (Y260 for AtDBR1) are highlighted in light blue, and secondary structural elements are indicated in colored bars (blue for α-helices and orange for β-strands). Source: Redrawn from Youn *et al.* (2006b). (See Page 26 in Color Section.)

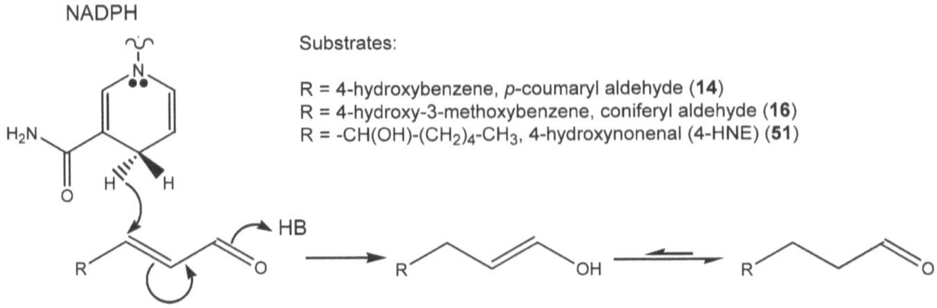

FIGURE 13.13 Possible enzymatic mechanism for AtDBR-mediated conversion of *p*-coumaryl (**14**) and coniferyl (**16**) aldehydes and 4-HNE (**51**) into their corresponding dihydroaldehyde derivatives. Source: Redrawn from Youn *et al.* (2006b).

NADPH into either the C-7 or the C-9 of the corresponding monolignol ester (or a quinone methide derivative thereof), thus forming either an allyl or propenyl phenol, respectively (Figs. 13.14 and 13.15B).

These regiospecific reductases (e.g., chavicol and eugenol synthase, CS/ES, and isoeugenol synthase, IES) have been studied to a larger extent than the monolignol-specific acyltransferases. Computational analyses of CS/ES isolated from basil and IES isolated from petunia indicate greatest homology (~40–45% identity) (Koeduka *et al.*, 2006) to members of the PIP family of reductases (pinoresinol-lariciresinol, isoflavone, and phenylcoumaran-benzylic ether reductases) we have either discovered and/or extensively characterized (Dinkova-Kostova *et al.*, 1996; Fujita *et al.*, 1999; Gang *et al.*, 1999), and for which crystal structures have been determined (Min *et al.*, 2003). Significantly, based on sequence homology, one such PLR/CS/ES homologue from *L. tridentata* (LtCES1) was recently isolated and characterized (Fig. 13.16) (Vassão *et al.*, 2007).

While PLRs are involved in formation of other medicinally important plant metabolites [e.g., podophyllotoxin (**45**) and secoisolariciresinol (**46**)] and various plant defense compounds [e.g., plicatic acid (**30**) in western red cedar heartwood], the biochemical mechanisms of PLR, PCBER, and CS/ES share common properties, including (1) a necessity for a free phenolic functionality in the substrate, indicative of a common quinone methide intermediate (Figs. 13.14 and 13.17) (Kim *et al.*, 2007; Koeduka *et al.*, 2006; Min *et al.*, 2003), and (2) a highly conserved Lys residue (K138 in PLR from *T. plicata*, K133 in its homologue in *L. tridentata*, and K132 in CS/ES from basil) required for catalysis (Min *et al.*, 2003). Figure 13.18 depicts the X-ray crystal structure of one member of this class of reductases, PLR from *T. plicata* (Min *et al.*, 2003). Based on the proposed catalytic mechanisms of CS/ES (Koeduka *et al.*, 2006) and PLR (Min *et al.*, 2003), it is hardly surprising that a high level of similarity was observed between these proteins. All of the PIP reductases, as well as CS/ES, utilize NAD(P)H as the source of a hydride that is regiospecifically (or stereospecifically) added to a carbon that originated from a phenylpropanoid side chain (i.e., either a monolignol derivative or dimer). In fact, a brief phylogenetic analysis indicates that these homologues cluster together with PLR (e.g., from *T. plicata*), PCBER (e.g., from *P. taeda*), IFR (e.g., from

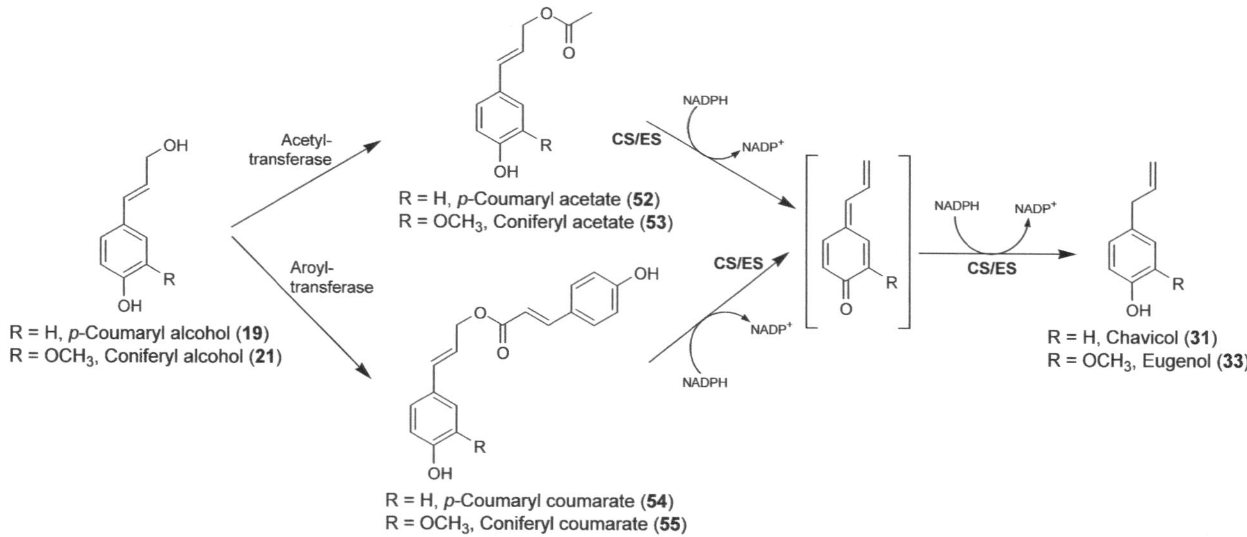

FIGURE 13.14 Biosynthetic pathway to chavicol (**31**) and eugenol (**33**) from the corresponding monolignols, *p*-coumaryl (**19**) and coniferyl (**21**) alcohols. CS, chavicol synthase; ES, eugenol synthase.

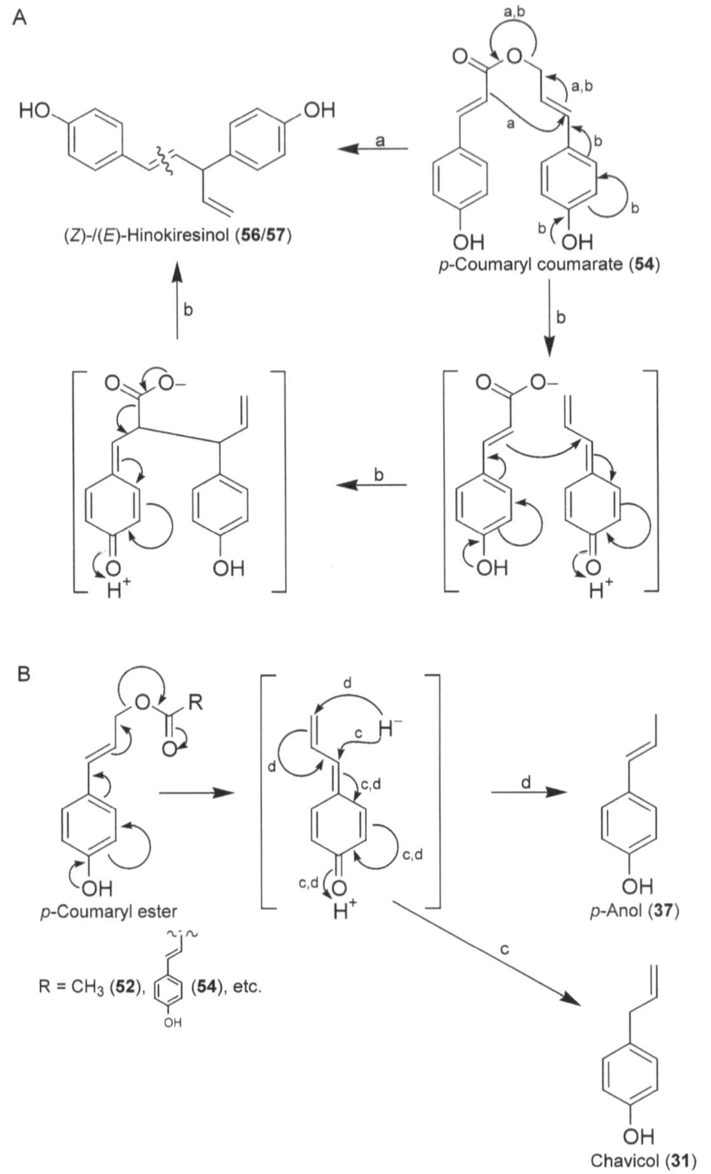

FIGURE 13.15 Possible mechanisms for conversion of *p*-coumaryl alcohol esters into (A) hinokiresinol (**56/57**) and (B) chavicol (**31**) and *p*-anol (**37**). (A) (a) Concerted or (b) through intermediacy of a quinone methide and (B) (c) and (d) formation of a quinone methide intermediate through displacement of the (interchangeable) ester leaving group, with subsequent reduction by hydride [from NAD(P)H] and rearomatization to form either (c) chavicol (**31**) and/or (d) *p*-anol (**37**). The reactions in (B) may also proceed through direct displacement, without intermediacy of the quinone methide, by an incoming hydride at carbons 7 or 9 to form chavicol (**31**) or *p*-anol (**37**), respectively (not shown). Source: Modified from Vassão *et al.* (2006b).

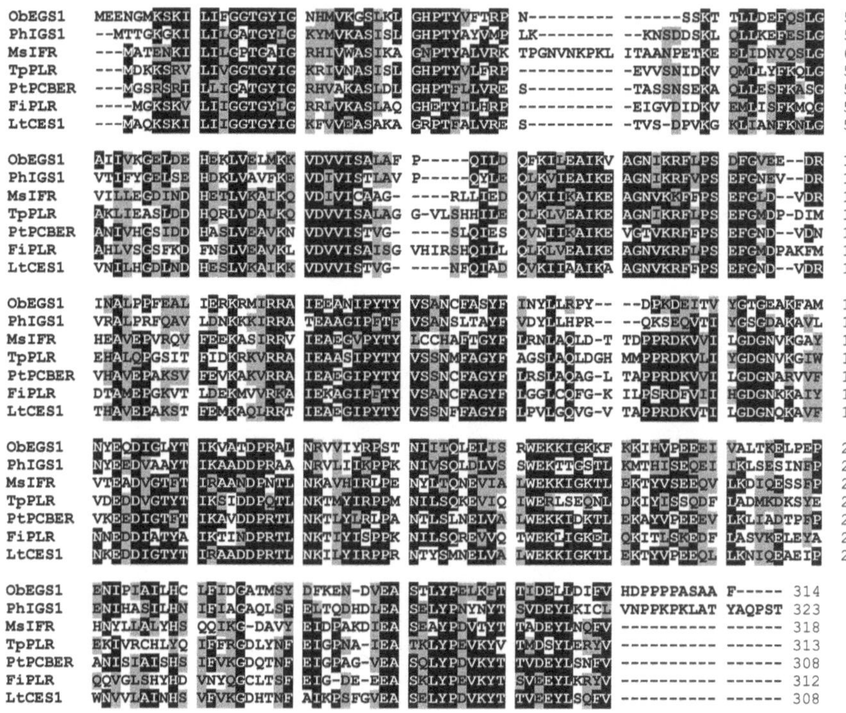

FIGURE 13.16 Amino acid alignment of basil (*Ocimum basilicum*) chavicol/eugenol synthase (ObEGS1), *Petunia hybrida* isoeugenol synthase (PhIGS1), and PIP reductases from *Medicago sativa* (MsIFR), *Thuja plicata* (TpPLR), *Pinus taeda* (PtPCBER), *Forsythia intermedia* (FiPLR), and *Larrea tridentata* (LtCES1).

Medicago), and leucoanthocyanidin reductases (LACR, e.g., from *Vitis vinifera*), with the *L. tridentata* homologue clustering closer to more distant PCBER and IFR homologues (Fig. 13.19).

The biochemical characteristics of these enzymes have been studied, with basil CS/ES and petunia IES reported to have substrate affinities [K_m, coniferyl acetate (**53**)] of 1.6–5.1 mM and V_{max} of 7–20 pkat/μg protein. These are indicative of relatively low substrate affinity, although not far from the range of other enzymes involved in volatile oil biosynthesis (Koeduka *et al.*, 2006). Additionally, the corresponding PLR homologue in the creosote bush (*L. tridentata*) catalyzes similar conversions, but interestingly with apparently higher catalytic efficacy [K_m values of a few hundred micromolar and V_{max} values of a few hundred pkat/μg protein for coniferyl acetate (**53**), *p*-coumaryl acetate (**52**), and *p*-coumaryl coumarate (**54**) (Vassão *et al.*, 2007); see Table 13.5]. We are currently examining the properties of other PIP reductases regarding their abilities to form allyl and propenyl phenols and have thus far seen evidence of some substrate versatility in PLRs acting on monolignol esters (unpublished observations).

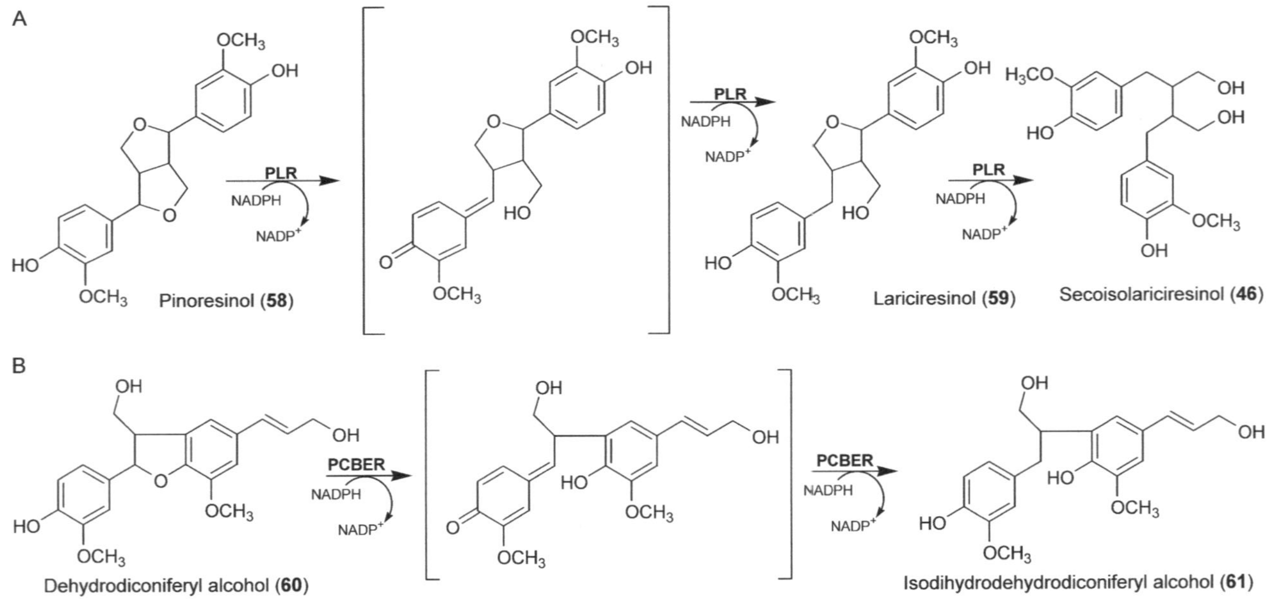

FIGURE 13.17 Reactions catalyzed by (A) PLR and (B) PCBER.

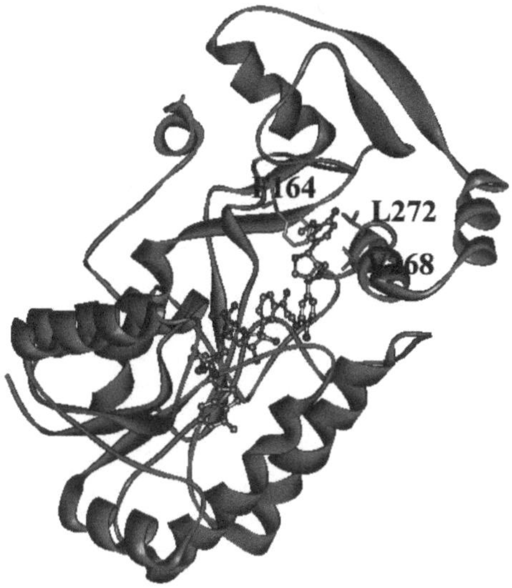

FIGURE 13.18 Schematic representation of the crystal structure of TpPLR1 with NADPH and (−)-pinoresinol (**58**). Source: Reprinted from Min *et al.* (2003). (See Page 27 in Color Section.)

In effect, the long-standing question regarding the biochemical formation of these widely used compounds, allyl and propenyl phenols, has now been elucidated and shown to utilize the same pathway precursors as lignin biosynthesis. Proteins (and their corresponding genes) involved in this process have been isolated and characterized, thus presenting a new approach with which to study and alter the lignification program of woody plants, as well as enabling the production of these compounds in more commonly cultivated plants.

5. POTENTIAL FOR ALLYL/PROPENYL PHENOLS?

We consider that biotechnological approaches utilizing the enzymatic machinery described above should now be explored to produce allyl and propenyl phenols, diverting (to a predetermined extent) some of the carbon flow, for example, from lignin toward these metabolites, while maintaining the resulting plants' capacities to grow and function within acceptable boundaries. The resulting cellulosic biomass could thus become more amenable to pulp/paper manufacture and/or biofuel production due to the lower lignin contents. Alternatively, oilseed metabolism could be manipulated to produce these allyl/propenyl phenol substances in larger amounts. Upon processing and purification, these compounds could potentially find uses as biofuels, biofuel precursors, flavors/fragrances, and/or intermediate chemicals (e.g., as monomers for synthetic polymers).

Moreover, the ability to routinely biotechnologically modify plants and/or cell cultures to produce allyl/propenyl phenols, such as chavicol (**31**) and eugenol (**33**), now offers the opportunity to consider much larger markets for these

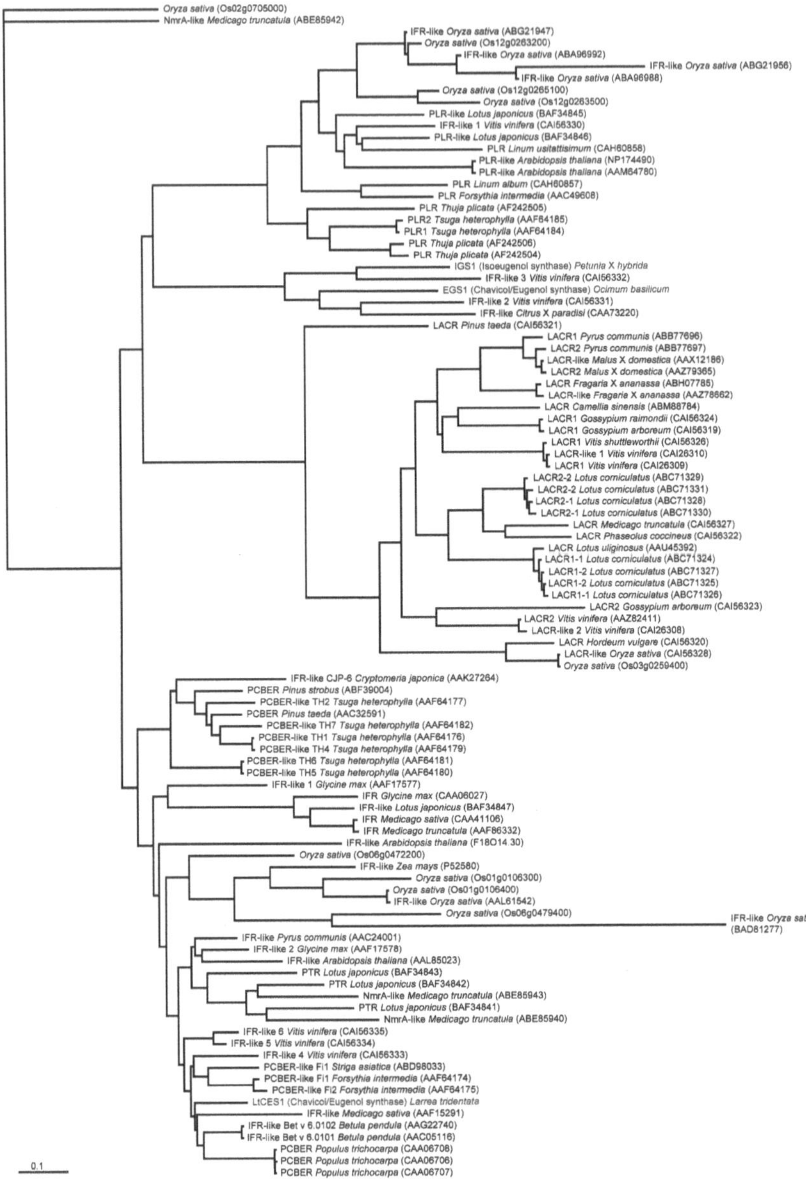

FIGURE 13.19 Currently annotated phylogenetic analysis of several PIP-reductase homologues from different plant species, with relevant homologues in basil (ObEGS1), petunia (PhIGS1), and creosote bush (LtCES1) highlighted. IFR, isoflavone reductase; LACR, leucoanthocyanidin reductase; NmrA, nitrogen metabolite repression regulator; PCBER, phenylcoumaran-benzylic ether reductase; PLR, pinoresinol-lariciresinol reductase; PTR, pterocarpan reductase. Sequences were obtained from the NCBI database and filtered for <0.75 sequence difference, ClustalW-aligned, and subjected to neighbor-joining phylogenetic analysis using PHYLIP (Felsenstein, 1993). (See Page 28 in Color Section.)

TABLE 13.5 Kinetic parameters of LtCES1 from *Larrea tridentata* for different monolignol ester substrates (**52–54**) (Vassão *et al.*, 2007), and chavicol/eugenol synthase from basil and isoeugenol synthase from petunia for coniferyl acetate (**53**) (Koeduka *et al.*, 2006)

	K_m (μM)	V_{max} (pkat/μg prot.)	k_{cat} (s^{-1})	k_{cat}/K_m (M^{-1} s^{-1})
LtCES1				
p-Coumaryl acetate (**52**)	350	200	6.80	19,500
Coniferyl acetate (**53**)	290	190	6.46	22,000
p-Coumaryl coumarate (**54**)	210	75	2.55	12,000
ObEGS1	5100	20	0.7	136
PhIGS1	1600	7	0.3	160

products, that is, in addition to expanding the flavor/fragrance/antiseptic/biocidal markets that currently exist (see above Section 3). For instance, in terms of the flavor/fragrance market, the natural vanillin market can potentially be expanded. Today, only about 0.2% of vanillin (**40**, Fig. 13.9) used originates directly from its botanical source, the vanilla bean, where it commands a sales price of approximately US$ 4000 kg^{-1} as a natural product. The bulk is semisynthetic, being chemically synthesized either from the petrochemical-derived guaiacol (**41**) or from pulp/paper lignin-derivatives (i.e., technical lignins). However, the vanillin (**40**) produced in this way commands a price of only US$ 12 kg^{-1} because it is not "natural."

Driven by consumer preferences toward truly "natural" food products, food additives, and pharmaceuticals, several biocatalytic processes for production of plant-derived metabolites using microbes and plant cell cultures have been developed and patented in recent years (Berger, 1991; Krings and Berger, 1998; Longo and Sanromán, 2006; Priefert *et al.*, 2001; Rabenhorst and Hopp, 1991; Schrader *et al.*, 2004; Shimoni *et al.*, 2000; Yoshimoto *et al.*, 1990). This enthusiasm has been fueled by the promise to generate transgenic cultures with increased efficacy for production of target metabolites. This approach also apparently offers economic advantages over conventional chemical syntheses, including better stereospecificity in product formation and lower amounts of waste products being generated upon processing (Schoemaker *et al.*, 2003). In particular, production of flavors via biotechnological processes offers an additional economic advantage since, unlike their chemically prepared counterparts, the resulting products can be marketed as "natural" under current US and EU legislation (The European Commission, 1991; Food and Drug Administration, 2006; Lesage-Meessen *et al.*, 1996; Shimoni *et al.*, 2000). In this regard, successful microbial production of vanillin (**40**) from various precursors, such as eugenol (**33**) and isoeugenol (**39**), has been reported recently, being achieved at relatively high substrate concentrations (Krings and Berger, 1998; Longo and Sanromán, 2006; Priefert *et al.*, 2001; Rabenhorst and Hopp, 1991; Schrader *et al.*, 2004; Shimoni *et al.*, 2000); for example, transformation of isoeugenol (**39**) (20 g/L) using a strain of *Serratia marcescens* led to vanillin (**40**) accumulation (3.8 g/L). An enzymatic process for conversion of isoeugenol (**39**) into vanillin (**40**) using a ligno stilbene-α,β-dioxygenase from a *Pseudomonas paucimobilis* strain has also been patented (Yoshimoto *et al.*, 1990). Thus, the technologies are now apparently in hand to permit formation of natural vanillin (**40**) through established microbial and genetic manipulations in either plants and/or plant/bacterial cell cultures.

Much larger anticipated potential markets for the allyl/propenyl phenols include the industrial polymers and biofuel/biodiesel. Regarding polymer applications, the expected worldwide production of polystyrenes alone was approximately 25 million metric tons in 2006, representing sales of US$ 31 billion. Allyl/propenyl phenols can be converted into functionalized polystyrene derivatives, and an increased supply creates the potential for their massive usage as intermediate (monomer) chemicals in industrial polymers. Currently, existing applications include eugenol- (**33**) based polymers, which are widely used in dentistry in zinc oxide impression pastes applied as surgical dressings and temporary cements (Skinner, 1940; Weinberg *et al.*, 1972), as well as specialty modifying (i.e., coating) agents in analytical electrodes (Ciszewski and Milczarek, 1998, 1999, 2001, 2003; Rahim *et al.*, 2004).

The functionalized β-methylstyrenes anethole (**38**) and isoeugenol (**39**) can be converted into polymers of several thousand dalton (Bywater, 1963). However, the potential of such conversions has been studied only to a limited extent relative to their vinyl analogues (i.e., styrene and derivatives thereof), in part due to limited supply/availability and because propenylbenzenes do not apparently undergo as efficient free radical polymerization reactions as styrenes—even though their electronic configuration is such that a radical intermediate can also be stabilized by the aromatic ring (Alexander *et al.*, 1981). The most efficient polymerization initiators described thus far for propenylbenzene derivatives are Lewis acids, particularly $AlCl_3$, $SnCl_4$, and BF_3 (Alexander *et al.*, 1981; Cerrai *et al.*, 1969a,b; Secci and Mameli, 1956).

In general, the steric factors on the monomers define, to a large extent, both polymerization rates and molecular weights of the resulting polymers, with anethole (**38**), for example, being more reactive than isoeugenol (**39**) (Alexander *et al.*, 1981). Polymerization reactions can proceed through a conventional 1,2-chain formation, similar to styrene, with the propagating species being a Lewis acid-induced carbocation that is added to the double bond of another monomer. This results in a polymer backbone composed of the carbons 7 and 8 of the original monomers. The molecular weights of the resulting polymers are higher at lower temperatures and, in the case of anethole (**38**) when polymerized by $SnCl_4$, can vary from a few thousand up to about 75,000 Da depending on both temperature and dilution levels (Bywater, 1963; Cerrai *et al.*, 1969a; Secci and Mameli, 1956). For isoeugenol (**39**), the phenolic oxygen moiety also participates in the polymerization reactions, thereby increasing the structural complexity of the resulting polymer(s) so formed (Fig. 13.20A) (Evliya and Olcay, 1974). Additionally, allylphenols, such as methylchavicol (**32**) and eugenol (**33**), can form mixed polymers, resulting from the partial rearrangement of the side-chain double bond upon carbocation formation prior to attachment to the polymer chain (Cihaner *et al.*, 2001; Kennedy, 1964) (Fig. 13.20B).

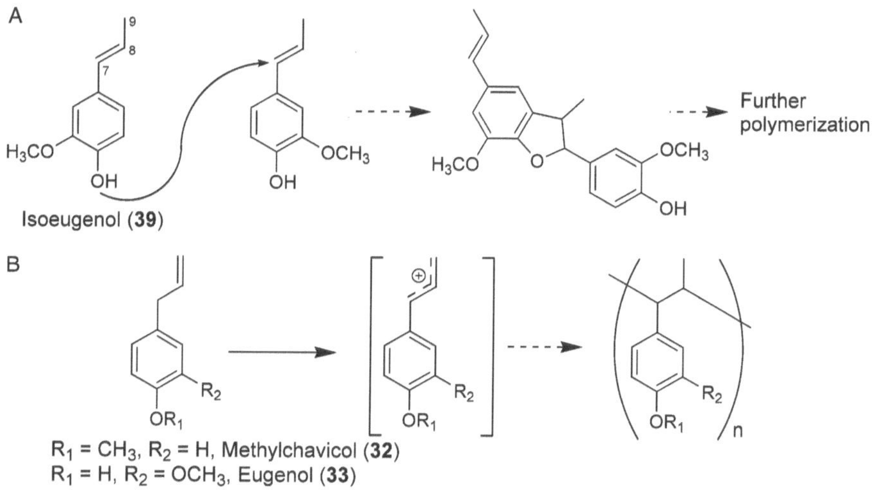

FIGURE 13.20 Polymerization of (A) isoeugenol (**39**) via furanocoumaran intermediacy and (B) methylchavicol (**32**) and eugenol (**33**) through rearrangement prior to polymerization.

In terms of their potential uses as biofuels, it is noteworthy that ~100 billion gallons of gasoline fuel were consumed in the United States in 2005. In addition, the annual consumption of diesel fuel in 2000, including highway diesel, farms, electric power, railroad, fuel oil (residential, commercial, and heating), and kerosene, totaled approximately 57.1 billion gallons. As a measure of the potential scale of production of biodiesel (from vegetable oils, consisting mainly of fatty acid esters), Peterson (1995) recently estimated that if *all* harvested cropland (~363 million acres) in the United States was dedicated exclusively to rapeseed (oilseed) production, then approximately 36.3 billion gallons (assuming 100 gallon/acre) of vegetable oil could be obtained annually. [Note that at present there are approximately 27 billion gallons of vegetable oil produced worldwide annually (Peterson, 1995)]. Yields for other plant species, modified to concurrently synthesize chavicol (**31**), methylchavicol (**32**), or eugenol (**33**), now need to be determined to establish to what extent these productivity numbers can be increased through (for instance) whole plant utilization.

Additionally, if one considers the annual pulp and paper production in the United States (~120 million metric tons/year, 1997 figures), the potential also exists to divert some part of the production of lignins/heartwood lignans and other phenylpropanoid derivatives in commercially important woody plant species away from their natural biosynthetic pathways, that is, to afford allyl/propenyl phenols, etc. In principle, the lignin/lignan substances currently produced annually as by-products of pulp/paper industries (more than 50 million tons) could instead be converted to approximately 15 billion gallons allyl/propenyl phenols per annum, if fully converted. Nevertheless, any reduction in carbon flow to lignin, or reductions/changes in heartwood-forming constituents, could represent a significant increase in biofuel production.

Allyl and propenyl phenols have relatively high heats of combustion at room temperature, with values generally being about 70% (per weight) of medium chain hydrocarbons such as octane and decane. That is, these allyl/propenyl phenols can potentially generate more energy (per weight) than ethanol. In terms of other relevant properties, using two examples only for illustrative purposes, chavicol (**31**) has boiling/flash points of 238/102 °C at normal atmospheric pressure and density approximately 1.01 g/cm^3, whereas eugenol (**33**) values are approximately 253/112 °C and 1.07 g/cm^3 (at 20 °C). Such values are within the ranges needed for biodiesel/biofuel considerations. Their reported freezing points are, however, generally between −10 °C and room temperature, which would reduce their potential as liquid biofuels if used exclusively as such in pure liquid form. This limitation might be circumvented, however, by either blending them into other fuels, similar to the coconut oils added as biofuels to diesel in the Philippines (BBC/PRI/WGBH The World, 2007) or through their chemical derivatization to generate materials of lower freezing point prior to biofuel use (e.g., hydrogenation, which may also help reduce pollutant emission upon combustion). Catalytic hydrogenation of side-chain double bonds of allyl/propenylbenzenes is readily achieved at atmospheric pressures, whereas reduction of the aromatic ring typically requires higher temperatures and pressures using traditional metal catalysts (e.g., supported Pd or Raney Ni). Reduced allyl/propenyl phenols have already been generated by such catalytic hydrogenation reactions; for example, 2-methoxy-4-propyl-cyclohexanol (**62**, Fig. 13.21) was obtained in near-quantitative amounts from eugenol (**33**) (Maillefer, 1990). Newer catalytic systems, however, have the exciting potential to dramatically improve the reduction conditions, for example, as recently reported for the quantitative

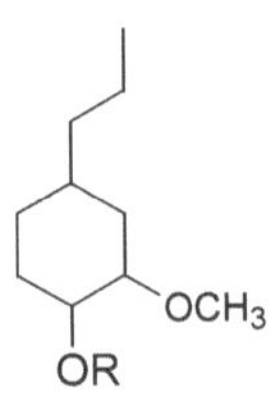

R = H, 2-Methoxy-4-propyl-cyclohexanol (**62**)
R = CH_3, 1,2-Dimethoxy-4-propyl-cyclohexane (**63**)

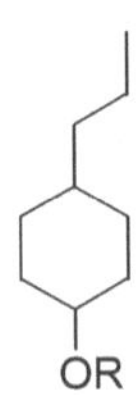

R = H, 4-Propyl-cyclohexanol (**64**)
R = CH_3, 1-Methoxy-4-propyl-cyclohexane (**65**)

FIGURE 13.21 Cyclohexane derivatives formed upon catalytic hydrogenation of the corresponding allyl/propenyl phenols.

hydrogenation of several benzene analogues, at room temperature and atmospheric hydrogen pressure, using ruthenium-containing methylated cyclodextrin catalysts (Nowicki *et al.*, 2006). Thus, this application of biotechnology, if further explored/developed/applied, offers a potentially important new avenue for sources of biofuels/bioenergy.

6. SUMMARY

The biotechnological lignin-reduction endeavors thus far reported have, in general, afforded plants apparently unsuitable for large-scale cultivation, resulting from compromised vasculature and perhaps higher susceptibility to pathogen attacks as well. Similarly, lignin represents a challenge in the processing of plant biomass (e.g., corn, switchgrass, or miscanthus) for biofuel/bioethanol generation using current fermentation techniques. Therefore, novel approaches must be developed if the so-called lignin challenge is to be circumvented for these purposes. Here, we have suggested some new areas of such exploration aiming at tackling these issues through the application of a newly elucidated metabolic process sharing the same lignin biochemical precursors, that is, the biosynthesis of liquid allyl and propenyl phenols. In the past year, the pertinent enzymes, genes, and specific biochemical precursors in this process have been isolated and described, thus lending themselves to exploitation through bioengineering. Further research in these areas will eventually determine the actual (economic) potential of the allyl and propenyl phenols as biofuels, intermediate chemicals for polymer synthesis, and/or specialty chemicals. Success in the research envisaged here could potentially lead to the development of a multibillion dollar per year industry within the United States alone. The challenge will be to identify a viable balance between acceptable structural and defensive properties of cultivated plants and their uses as sources of economically valuable products.

ACKNOWLEDGEMENTS

The authors thank the National Science Foundation (MCB-0417291), the U.S. Department of Energy (DE FG03-97ER20259), the National Institute of General Medical Sciences (5 R01 GM066173–02), McIntire Stennis, and the G. Thomas and Anita Hargrove Center for Plant Genomic Research for support.

REFERENCES

Alexander, R., Jefferson, A., and Lester, P. D. (1981). Cationic oligomerization and polymerization of some propenylbenzene derivatives. *J. Polymer Sci.: Polym. Chem. Ed.* **19,** 695–706.

Anterola, A. M., and Lewis, N. G. (2002). Trends in lignin modification: A comprehensive analysis of the effects of genetic manipulations/mutations on lignification and vascular integrity. *Phytochemistry* **61,** 221–294.

Anterola, A. M., Van Rensburg, H., Van Heerden, P. S., Davin, L. B., and Lewis, N. G. (1999). Multi-site modulation of flux during monolignol formation in loblolly pine (*Pinus taeda*). *Biochem. Biophys. Res. Commun.* **261,** 652–657.

Anterola, A. M., Jeon, J.-H., Davin, L. B., and Lewis, N. G. (2002). Transcriptional control of monolignol biosynthesis in *Pinus taeda*: Factors affecting monolignol ratios and carbon allocation in phenylpropanoid metabolism. *J. Biol. Chem.* **277,** 18272–18280.

Arteaga, S., Andrade-Cetto, A., and Cárdenas, R. (2005). *Larrea tridentata* (creosote bush), an abundant plant of Mexican and US-American deserts and its metabolite nordihydroguaiaretic acid. *J. Ethnopharmacol.* **98,** 231–239.

Atanassova, R., Favet, N., Martz, F., Chabbert, B., Tollier, M.-T., Monties, B., Fritig, B., and Legrand, M. (1995). Altered lignin composition in trangenic tobacco expressing *O*-methyltranferase sequences in sense and antisense orientation. *Plant J.* **8,** 465–477.

Badger, D. A., Smith, R. L., Bao, J., Kuester, R. K., and Sipes, I. G. (2002). Disposition and metabolism of isoeugenol in the male Fischer 344 rat. *Food Chem. Toxicol.* **40,** 1757–1765.

Baedeker, M., and Schulz, G. E. (2002). Autocatalytic peptide cyclization during chain folding of histidine ammonia-lyase. *Structure* **10,** 61–67.

Bauer, K., Garbe, D., and Surburg, H. (2001). Natural raw materials in the flavor and fragrance industry. *In* "Common Fragrance and Flavor Materials: Preparation and Uses" (K. Bauer, D. Garbe, and H. Surburg, eds.), pp. 167–226. Wiley-VCH, Weinheim, Germany.

BBC/PRI/WGBH The World (2007). http://www.theworld.org/?q=node/6595.

Berger, R. G. (1991). Genetic engineering. Part III: Food Flavors. *In* "Encyclopedia of Food Science and Technology" (Y. H. Hui, ed.) Vol. 2 pp. 1313–1320. Wiley-Interscience, New York.

Bywater, S. (1963). Aromatic compounds other than styrene. *In* "The Chemistry of Cationic Polymerization" (P. H. Plesch, ed.) pp. 305–347. Pergamon Press, New York.

Calabrese, J. C., Jordan, D. B., Boodhoo, A., Sariaslani, S., and Vannelli, T. (2004). Crystal structure of phenylalanine ammonia lyase: Multiple helix dipoles implicated in catalysis. *Biochemistry* **43,** 11403–11416.

Camm, E. L., and Towers, G. H. N. (1973). Phenylalanine ammonia lyase. *Phytochemistry* **12,** 961–973.

Canonica, L., Manitto, P., Monti, D., and Sanchez, A. M. (1971). Biosynthesis of allylphenols in *Ocymum basilicum* L. *J. Chem. Soc., Chem. Commun.* 1108–1109.

Cerrai, P., Andruzzi, F., and Giusti, P. (1969a). Polimerizzazione del β-metil-*p*-metossi-stirene (anetolo). I. Cinetica della polimerizzazione in presenza di $BF_3.Et_2O$. *Chim. Ind.* **51,** 681–686.

Cerrai, P., Andruzzi, F., and Giusti, P. (1969b). Polimerizzazione del β-metil-*p*-metossi-stirene (anetolo). II. Effetto di un campo elettrico sulla polimerizzazione in presenza di $BF_3.Et_2O$. *Chim. Ind.* **51,** 687–692.

Chang, C.-C., Heller, J. D., Kuo, J., and Huang, R. C. C. (2004). Tetra-*O*-methyl nordihydroguaiaretic acid induces growth arrest and cellular apoptosis by inhibiting Cdc2 and survivin expression. *Proc. Natl. Acad. Sci. USA* **101,** 13239–13244.

Chen, F., Yasuda, S., and Fukushima, K. (1999a). Evidence for a novel biosynthetic pathway that regulates the ratio of syringyl to guaiacyl residues in lignin in the differentiating xylem of *Magnolia kobus* DC. *Planta* **207,** 597–603.

Chen, F., Yasuda, S., and Fukushima, K. (1999b). Structural conversion of the lignin subunit at the cinnamyl alcohol stage in *Eucalyptus camaldulensis*. *J. Wood Sci.* **45,** 487–491.

Chen, Y.-R., Jacobson, B., Sarkanen, S., and Wang, Y. (2001). The preliminary characterization of lignin depolymerase. *Proc. 11th Internat. Symp. Wood Pulp. Chem.* **1,** 313–316.

Cho, M.-H., Moinuddin, S. G. A., Helms, G. L., Hishiyama, S., Eichinger, D., Davin, L. B., and Lewis, N. G. (2003). (+)-Larreatricin hydroxylase, an enantio-specific polyphenol oxidase from the creosote bush (*Larrea tridentata*). *Proc. Natl. Acad. Sci. USA* **100,** 10641–10646.

Cho, M.-H., Corea, O. R. A., Yang, H., Bedgar, D. L., Laskar, D. D., Anterola, A. M., Moog-Anterola, F. A., Hood, R. L., Kohalmi, S. E., Bernards, M. A., Kang, C., Davin, L. B., and Lewis, N. G. (2007). Phenylalanine biosynthesis in *Arabidopsis thaliana*: Identification and characterization of arogenate dehydratases. *J. Biol. Chem.* doi:10.1074/jbc. M702662200.

Cihaner, A., Testereci, H. N., and Önal, A. M. (2001). Electrochemical polymerization of 4-allylanisole. *Eur. Polym. J.* **37,** 1747–1752.

Ciszewski, A., and Milczarek, G. (1998). A new Nafion-free bipolymeric sensor for selective and sensitive detection of nitric oxide. *Electroanalysis* **10,** 791–793.

Ciszewski, A., and Milczarek, G. (1999). Polyeugenol-modified platinum electrode for selective detection of dopamine in the presence of ascorbic acid. *Anal. Chem.* **71,** 1055–1061.

Ciszewski, A., and Milczarek, G. (2001). Preparation and general properties of chemically modified electrodes based on electrosynthesized thin polymeric films derived from eugenol. *Electroanalysis* **13,** 860–867.

Ciszewski, A., and Milczarek, G. (2003). Electrochemical detection of nitric oxide using polymer modified electrodes. *Talanta* **61,** 11–26.

Cochrane, F. C., Davin, L. B., and Lewis, N. G. (2004). The *Arabidopsis* phenylalanine ammonia-lyase multigene family. Kinetic characterization of the four PAL isoforms. *Phytochemistry* **65,** 1557–1564.

Committee on Biobased Industrial Products; Board on Biology; Commission on Life Sciences; National Research Council (2000). ''Biobased Industrial Products. Priorities for Research and Commercialization.'' Washington, DC: National Academy Press.

Costa, M. A., Bedgar, D. L., Moinuddin, S. G. A., Kim, K.-W., Cardenas, C. L., Cochrane, F. C., Shockey, J. M., Helms, G. L., Amakura, Y., Takahashi, H., Milhollan, J. K., Davin, L. B., Browse, J. A., and Lewis, N. G. (2005). Characterization *in vitro* and *in vivo* of the putative multigene 4-coumarate:CoA ligase network in *Arabidopsis*: Syringyl lignin and sinapate/sinapyl alcohol derivative formation. *Phytochemistry* **66,** 2072–2091.

Dexter, R., Qualley, A., Kish, C. M., Ma, C. J., Koeduka, T., Nagegowda, D. A., Dudareva, N., Pichersky, E., and Clark, D. (2007). Characterization of a petunia acetyltransferase involved in the biosynthesis of the floral volatile isoeugenol. *Plant J.* **49,** 265–275.

Dinkova-Kostova, A. T., Gang, D. R., Davin, L. B., Bedgar, D. L., Chu, A., and Lewis, N. G. (1996). (+)-Pinoresinol-(+)-lariciresinol reductase from *Forsythia intermedia*: Protein purification, cDNA cloning, heterologous expression and comparison to isoflavone reductase. *J. Biol. Chem.* **271,** 29473–29482.

Evliya, H., and Olcay, A. (1974). Oxidative polymerisation of isoeugenol and mild oxidation of synthetic polymers with alkaline cupric hydroxide. *Holzforschung* **28,** 130–135.

Felsenstein, J. (1993). Phylip (Phylogeny Inference Package) version 3.5c. Distributed by the author.

Food and Drug Administration (2006). U.S. Code of Federal Regulations 101.22, 21 http://frwebgate3.access.gpo.gov/cgi-bin/waisgate.cgi?WAISdocID=4697572256+1+1+0&WAISaction=retrieve.

Ford, J. D., Huang, K.-S., Wang, H.-B., Davin, L. B., and Lewis, N. G. (2001). Biosynthetic pathway to the cancer chemopreventive secoisolariciresinol diglucoside-hydroxymethyl glutaryl ester-linked lignan oligomers in flax (*Linum usitatissimum*) seed. *J. Nat. Prod.* **64,** 1388–1397.

Fujita, M., Gang, D. R., Davin, L. B., and Lewis, N. G. (1999). Recombinant pinoresinol-lariciresinol reductases from western red cedar (*Thuja plicata*) catalyze opposite enantiospecific conversions. *J. Biol. Chem.* **274,** 618–627.

Gang, D. R., Kasahara, H., Xia, Z.-Q., Vander Mijnsbrugge, K., Bauw, G., Boerjan, W., Van Montagu, M., Davin, L. B., and Lewis, N. G. (1999). Evolution of plant defense mechanisms: Relationships of phenylcoumaran benzylic ether reductases to pinoresinol-lariciresinol and isoflavone reductases. *J. Biol. Chem.* **274,** 7516–7527.

Gardner, J. A. F., Barton, G. M., and Maclean, H. (1959). The polyoxyphenols of western red cedar (*Thuja plicata* Donn). I. Isolation and preliminary characterization of plicatic acid. *Can. J. Chem.* **37,** 1703–1709.

Gardner, J. A. F., Macdonald, B. F., and Maclean, H. (1960). The polyoxyphenols of western red cedar (*Thuja plicata* Donn). II. Degradation studies on plicatic acid, a possible lignan acid. *Can. J. Chem.* **38,** 2387–2394.

Gardner, J. A. F., Swan, E. P., Sutherland, S. A., and Maclean, H. (1966). Polyoxyphenols of western red cedar (*Thuja plicata* Donn). III. Structure of plicatic acid. *Can. J. Chem.* **44,** 52–58.

George Uhe Company (2006). Flavor and fragrance ingredients market reports newsletter (April), 2006 http://www.uhe.com/mktreport-04_06.htm.

Glenn, J. K., Morgan, M. A., Mayfield, M. B., Kuwahara, M., and Gold, M. H. (1983). An extracellular hydrogen peroxide requiring enzyme preparation involved in lignin biodegradation by the white rot basidiomycete *Phanerochaete chrysosporium*. *Biochem. Biophys. Res. Commun.* **114,** 1077–1083.

Grand, C. (1984). Ferulic acid 5-hydroxylase: A new cytochrome P-450-dependent enzyme from higher plant microsomes involved in lignin synthesis. *FEBS Lett.* **169,** 7–11.

Haneke, K. E. (2002). Turpentine (turpentine oil, wood turpentine, sulfate turpentine, sulfite turpentine) [8006–64–2]. Review of toxicological literature 2002 http://ntp-server.niehs.nih.gov/ntp/htdocs/Chem_Background/ExSumPdf/turpentine.pdf.

Harrison, B., and Gang, D. R. (2006). Characterization of coniferyl acetate acetyl transferase from sweet basil (*Ocimum basilicum*). *The 19th Rocky Mountain Regional Meeting*. Tucson, AZ.

Heller, W., and Kühnl, T. (1985). Elicitor induction of a microsomal 5-*O*-(4-coumaroyl)shikimate 3′-hydroxylase in parsley cell suspension cultures. *Arch. Biochem. Biophys.* **241,** 453–460.

Henniges, O., and Zeddies, J. (2006). Bioenergy in Europe: Experiences and prospects. *In* "Bioenergy and Agriculture. Promises and Challenges" (H. Hazell and R. K. Pachauri, eds.), Vol. 14. International Food Policy Research Institute. http://dx.doi.org/10.2499/2020focus14.

Hocking, M. B. (1997). Vanillin: Synthetic flavoring from spent sulfite liquor. *J. Chem. Educ.* **74,** 1055–1059.

Hoffmann, L., Maury, S., Martz, F., Geoffroy, P., and Legrand, M. (2003). Purification, cloning, and properties of an acyltransferase controlling shikimate and quinate ester intermediates in phenylpropanoid metabolism. *J. Biol. Chem.* **278,** 95–103.

Hoffmann, L., Besseau, S., Geoffroy, P., Ritzenthaler, C., Meyer, D., Lapierre, C., Pollet, B., and Legrand, M. (2004). Silencing of hydroxycinnamoyl-coenzyme A shikimate/quinate hydroxycinnamoyltransferase affects phenylpropanoid biosynthesis. *Plant Cell.* **16,** 1446–1465.

Humphreys, J. M., Hemm, M. R., and Chapple, C. (1999). New routes for lignin biosynthesis defined by biochemical characterization of recombinant ferulate 5-hydroxylase, a multifunctional cytochrome P450-dependent monooxygenase. *Proc. Natl. Acad. Sci. USA* **96,** 10045–10050.

Jang, E., Mcinnis, D., Vargas, R., and Mau, R. (2003). Area-wide integrated pest management (Ipm) of fruit flies in Hawaiian fruits and vegetables 2003 http://www.ars.usda.gov/research/publications/publications.htm?SEQ_NO_115=167070.

Jiao, Y., Davin, L. B., and Lewis, N. G. (1998). Furanofuran lignan metabolism as a function of seed maturation in *Sesamum indicum*: Methylenedioxy bridge formation. *Phytochemistry* **49,** 387–394.

Jourdes, M., Cardenas, C. L., Laskar, D. D., Moinuddin, S. G. A., Davin, L. B., and Lewis, N. G. (2007). Plant cell walls are enfeebled when attempting to preserve native lignin configuration with poly-*p*-hydroxycinnamaldehydes: Evolutionary implications. *Phytochemistry* **68,** 1932–1956.

Jung, E., Zamir, L. O., and Jensen, R. A. (1986). Chloroplasts of higher plants synthesize L-phenylalanine via L-arogenate. *Proc. Natl. Acad. Sci. USA* **83,** 7231–7235.

Kasahara, H., Davin, L. B., and Lewis, N. G. (2004). Aryl propenal double bond reductase. U. S. Patent No. 6,703,229. Filed March 27, 2001. Issued March 9, 2004, 30 pp.

Kasahara, H., Jiao, Y., Bedgar, D. L., Kim, S.-J., Patten, A. M., Xia, Z.-Q., Davin, L. B., and Lewis, N. G. (2006). *Pinus taeda* phenylpropenal double-bond reductase: Purification, cDNA cloning, heterologous expression in *Escherichia coli*, and subcellular localization in *P. taeda*. *Phytochemistry* **67,** 1765–1780.

Kato, M. J., Chu, A., Davin, L. B., and Lewis, N. G. (1998). Biosynthesis of antioxidant lignans in *Sesamum indicum* seeds. *Phytochemistry* **47,** 583–591.

Kennedy, J. P. (1964). Cationic isomerization polymerization of β-methylstyrene and allylbenzene. *J. Polym. Sci.: Part A* **2,** 5171–5176.

Kim, M. K., Jeon, J.-H., Fujita, M., Davin, L. B., and Lewis, N. G. (2002). The western red cedar (*Thuja plicata*) 8–8′ *DIRIGENT* family displays diverse expression patterns and conserved monolignol coupling specificity. *Plant Mol. Biol.* **49,** 199–214.

Kim, S.-J., Kim, M.-R., Bedgar, D. L., Moinuddin, S. G. A., Cardenas, C. L., Davin, L. B., Kang, C., and Lewis, N. G. (2004). Functional reclassification of the putative cinnamyl alcohol dehydrogenase multigene family in Arabidopsis. *Proc. Natl. Acad. Sci. USA* **101,** 1455–1460.

Kim, K. W., Moinuddin, S. G. A., Davin, L. B., Kang, C., and Lewis, N. G. (2007). Defining the molecular basis of differing enantiospecificities of pinoresinol-lariciresinol reductase and homologues thereof in western red cedar using site-directed mutagenesis. (Submitted).

Kirk, T. K., Tien, M., Kersten, P. J., Mozuch, M. D., and Kalyanaraman, B. (1986). Ligninase of *Phanerochaete chrysosporium*. Mechanism of its degradation of the non-phenolic arylglycerol β-aryl ether substructure of lignin. *Biochem. J.* **236,** 279–287.

Klischies, M., Stockigt, J., and Zenk, M. H. (1975). Biosynthesis of the allyphenols eugenol and methyleugenol in *Ocymum basilicum* L. *J. Chem. Soc. Chem. Commun.* 879–880.

Koeduka, T., Fridman, E., Gang, D. R., Vassão, D. G., Jackson, B. L., Kish, C. M., Orlova, I., Spassova, S. M., Lewis, N. G., Noel, J. P., Baiga, T. J., Dudareva, N., *et al.* (2006). Eugenol and isoeugenol, characteristic aromatic constituents of spices, are biosynthesized via reduction of a coniferyl alcohol ester. *Proc. Natl. Acad. Sci. USA* **103,** 10128–10133.

Koukol, J., and Conn, E. E. (1961). The metabolism of aromatic compounds in higher plants. IV. Purification and properties of the phenylalanine deaminase of *Hordeum vulgare*. *J. Biol. Chem.* **236,** 2692–2698.

Kraus, C., and Spiteller, G. (1997). Comparison of phenolic compounds from galls and shoots of *Picea glauca*. *Phytochemistry* **44,** 59–67.

Krings, U., and Berger, R. G. (1998). Biotechnological production of flavours and fragrances. *Appl. Microbiol. Biotechnol.* **49,** 1–8.

Laskar, D. D., Jourdes, M., Patten, A. M., Helms, G. L., Davin, L. B., and Lewis, N. G. (2006). The *Arabidopsis* cinnamoyl CoA reductase *irx4* mutant has a delayed but coherent (normal) program of lignification. *Plant J.* **48,** 674–686.

Lauvergeat, V., Lacomme, C., Lacombe, E., Lasserre, E., Roby, D., and Grima-Pettenati, J. (2001). Two cinnamoyl-CoA reductase (CCR) genes from *Arabidopsis thaliana* are differentially expressed during development and in response to infection with pathogenic bacteria. *Phytochemistry* **57,** 1187–1195.

Lesage-Meessen, L., Delattre, M., Haon, M., Thibault, J.-F., Ceccaldi, B. C., Brunerie, P., and Asther, M. (1996). A two-step bioconversion process for vanillin production from ferulic acid combining *Aspergillus niger* and *Pycnoporus cinnabarinus*. *J. Biotechnol.* **50,** 107–113.

Lewis, N. G., and Yamamoto, E. (1990). Lignin: Occurrence, biogenesis and biodegradation. *Annu. Rev. Plant Phys. Plant Mol. Biol.* **41,** 455–496.

Lewis, N. G., Davin, L. B., and Sarkanen, S. (1999). The nature and function of lignins. *In* "Comprehensive Natural Products Chemistry" (Sir D. H. R. Barton, K. Nakanishi, and O. Meth-Cohn, eds.), pp. 617–745. Elsevier, Oxford.

Li, L., Cheng, X., Leshkevich, J., Umezawa, T., Harding, S., and Chiang, V. (2001). The last step of syringyl monolignol biosynthesis in angiosperms is regulated by a novel gene encoding sinapyl alcohol dehydrogenase. *Plant Cell* **13,** 1567–1585.

Longo, M. A., and Sanromán, M. A. (2006). Production of food aroma compounds: Microbial and enzymatic methodologies. *Food Technol. Biotechnol.* **44,** 335–353.

Louie, G. V., Bowman, M. E., Moffitt, M. C., Baiga, T. J., Moore, B. S., and Noel, J. P. (2006). Structural determinants and modulation of substrate specificity in phenylalanine-tyrosine ammonia-lyases. *Chem. Biol.* **13,** 1327–1338.

Maillefer, E. (1990). Perfuming ingredient, process for its preparation and utilization of said ingredient in perfuming compositions and perfumed products. U.S. Patent No. 4,904,465.

Manitto, P., Monti, D., and Gramatica, P. (1974a). Biosynthesis of phenylpropanoid compounds. Part I. Biosynthesis of eugenol in *Ocimum basilicum* L. *J. Chem. Soc. Perkin Trans.* 1727–1731.

Manitto, P., Monti, D., and Gramatica, P. (1974b). Biosynthesis of anethole in *Pimpinella anisum* L. *Tetrahedron Lett.* **15,** 1567–1568.

Maralhas, A., Monteiro, A., Martins, C., Kranendonk, M., Laires, A., Rueff, J., and Rodrigues, A. S. (2006). Genotoxicity and endoreduplication inducing activity of the food flavouring eugenol. *Mutagenesis* **21,** 199–204.

Meyer, K., Cusumano, J. C., Somerville, C., and Chapple, C. C. S. (1996). Ferulate-5-hydroxylase from *Arabidopsis thaliana* defines a new family of cytochrome P450-dependent monooxygenases. *Proc. Natl. Acad. Sci. USA* **93,** 6869–6874.

Min, T., Kasahara, H., Bedgar, D. L., Youn, B., Lawrence, P. K., Gang, D. R., Halls, S. C., Park, H., Hilsenbeck, J. L., Davin, L. B., Lewis, N. G., and Kang, C. (2003). Crystal structures of pinoresinol-lariciresinol and phenylcoumaran benzylic ether reductases and their relationship to isoflavone reductases. *J. Biol. Chem.* **278,** 50714–50723.

Moinuddin, S. G. A., Hishiyama, S., Cho, M.-H., Davin, L. B., and Lewis, N. G. (2003). Synthesis and chiral HPLC analysis of the dibenzyltetrahydrofuran lignans, larreatricins, 8′-*epi*-larreatricins, 3,3′-didemethoxyverrucosins and *meso*-3,3′-didemethoxynectandrin B in the creosote bush (*Larrea tridentata*): Evidence for regiospecific control of coupling. *Org. Biomol. Chem.* **1,** 2307–2313.

Muzac, I., Want, J., Anzellotti, D., Zhang, H., and Ibrahim, R. K. (2000). Functional expression of an *Arabidopsis* cDNA clone encoding a flavonol 3′-*O*-methyltransferase and characterization of the gene product. *Arch. Biochem. Biophys.* **375,** 385–388.

Newberne, P., Smith, R. L., Doull, J., Goodman, J. I., Munro, I. C., Portoghese, P. S., Wagner, B. M., Weil, C. S., Woods, L. A., Adams, T. B., Lucas, C. D., and Ford, R. A. (1999). The FEMA GRAS assessment of *trans*-anethole used as a flavouring substance. *Food Chem. Toxicol.* **37,** 789–811.

Niggeweg, R., Michael, A. J., and Martin, C. (2004). Engineering plants with increased levels of the antioxidant chlorogenic acid. *Nat. Biotechnol.* **22,** 746–754.

Nowicki, A., Zhang, Y., Léger, B., Rolland, J.-P., Bricout, H., Monflier, E., and Roucoux, A. (2006). Supramolecular shuttle and protective agent: A multiple role of methylated cyclodextrins in the chemoselective hydrogenation of benzene derivatives with ruthenium nanoparticles. *J. Chem. Soc. Chem. Commun.* 296–298.

Pakusch, A.-E., Kneussel, R., and Matern, U. (1989). *S*-Adenosyl-L-methionine:*trans*-caffeoyl-coenzyme A 3-*O*-methyltransferase from elictor-treated parsley cell suspension cultures. *Arch. Biochem. Biophys.* **271,** 488–494.

Patten, A. M., Cardenas, C. L., Cochrane, F. C., Laskar, D. D., Bedgar, D. L., Davin, L. B., and Lewis, N. G. (2005). Reassessment of effects on lignification and vascular development in the *irx4 Arabidopsis* mutant. *Phytochemistry* **66,** 2092–2107.

Patten, A. M., Jourdes, M., Brown, E. E., Laborie, M.-P., Davin, L. B., and Lewis, N. G. (2007). Reaction tissue formation and stem tensile modulus properties in wild-type and *p*-coumarate-3-hydroxylase downregulated lines of alfalfa, *Medicago sativa* (Fabaceae). *Am. J. Bot.* **94,** 912–925.

Peterson, C. L. (1995). Potential production of biodiesel. http://www.uidaho.edu/bioenergy/BiodieselEd/publication/02.pdf.

Piquemal, J., Lapierre, C., Myton, K., O'Connell, A., Schuch, W., Grima-Pettenati, J., and Boudet, A.-M. (1998). Down-regulation of cinnamoyl-CoA reductase induces significant changes of lignin profiles in transgenic tobacco plants. *Plant J.* **13,** 71–83.

Poppe, L., and Rétey, J. (2005). Friedel-Crafts-type mechanism for the enzymatic elimination of ammonia from histidine and phenylalanine. *Angew. Chem. Int. Ed.* **44,** 3668–3688.

Priefert, H., Rabenhorst, J., and Steinbüchel, A. (2001). Biotechnological production of vanillin. *Appl. Microbiol. Biotechnol.* **56,** 296–314.

Rabenhorst, J., and Hopp, R. (1991). Process for the preparation of vanillin, U. S. Patent No. 5,017,388, May 21, 1991.

Rahim, E. A., Sanda, F., and Masuda, T. (2004). Synthesis and properties of novel eugenol-based polymers. *Polym. Bull.* **52,** 93–100.

Rohde, A., Morreel, K., Ralph, J., Goeminne, G., Hostyn, V., De Rycke, R., Kushnir, S., Van Doorsselaere, J., Joseleau, J.-P., Vuylsteke, M., Van Driessche, G., Van Beeumen, J., Messens, E., and Boerjan, W. (2004). Molecular phenotyping of the *pal1* and *pal2* mutants of *Arabidopsis thaliana* reveals far-reaching consequences on phenylpropanoid, amino acid, and carbohydrate metabolism. *Plant Cell* **16,** 2749–2771.

Rubery, P. H., and Fosket, D. E. (1969). Changes in phenylalanine ammonia-lyase activity during xylem differentiation in *Coleus* and soybean. *Planta* **87,** 54–62.

Sarkanen, S., Razal, R. A., Piccariello, T., Yamamoto, E., and Lewis, N. G. (1991). Lignin peroxidase: Toward a clarification of its role *in vivo*. *J. Biol. Chem.* **266,** 3636–3643.

Schoch, G., Goepfert, S., Morant, M., Hehn, A., Meyer, D., Ullmann, P., and Werck-Reichhart, D. (2001). CYP98A3 from *Arabidopsis thaliana* is a 3′-hydroxylase of phenolic esters, a missing link in the phenylpropanoid pathway. *J. Biol. Chem.* **276,** 36566–36574.

Schoemaker, H. E., Mink, D., and Wubbolts, M. G. (2003). Dispelling the myths—Biocatalysis in industrial synthesis. *Science* **299,** 1694–1697.

Schmitt, D., Pakusch, A.-E., and Matern, U. (1991). Molecular cloning, induction, and taxonomic distribution of caffeoyl-CoA 3-*O*-methyltransferase, an enzyme involved in disease resistance. *J. Biol. Chem.* **266,** 17416–17423.

Schrader, J., Etschmann, M. M. W., Sell, D., Hilmer, J.-M., and Rabenhorst, J. (2004). Applied biocatalysis for the synthesis of natural flavour compounds—Current industrial processes and future prospects. *Biotechnol. Lett.* **26,** 463–472.

Schwede, T. F., Retey, J., and Schulz, G. E. (1999). Crystal structure of histidine ammonia-lyase revealing a novel polypeptide modification as the catalytic electrophile. *Biochemistry* **38,** 5355–5361.

Secci, M., and Mameli, L. (1956). Sulla polimerizzazione dell'anetolo con BF_3. *Ann. Chim.* **47,** 580–585.

Senanayake, U. M., Wills, R. B. H., and Lee, T. H. (1977). Biosynthesis of eugenol and cinnamic aldehyde in *Cinnamomum zeylanicum*. *Phytochemistry* **16,** 2032–2033.

Shimoni, E., Ravid, U., and Shoham, Y. (2000). Isolation of a *Bacillus* sp. capable of transforming isoeugenol to vanillin. *J. Biotechnol.* **78,** 1–9.

Sibout, R., Eudes, A., Mouille, G., Pollet, B., Lapierre, C., Jouanin, L., and Séguin, A. (2005). *CINNAMYL ALCOHOL DEHYDROGENASE-C* and *-D* are the primary genes involved in lignin biosynthesis in the floral stem of *Arabidopsis*. *Plant Cell* **17,** 2059–2076.

Skinner, E. W. (1940). "The Science of Dental Materials." Philadelphia: W. B. Saunders Company, Philadelphia.

Smith, R. L., Doull, J., Feron, V. J., Goodman, J. I., Munro, I. C., Newberne, P. M., Portoghese, P. S., Waddell, W. J., Wagner, B. M., Adams, T. B., and McGowen, M. M. (2001). GRAS flavoring substances 20. *Food Technol.* **55,** 34–55.

Suzuki, S., Nakatsubo, T., Umezawa, T., and Shimada, M. (2002). First *in vitro* norlignan formation with *Asparagus officinalis* enzyme preparation. *J. Chem. Soc. Chem. Commun.* 1088–1089.

Suzuki, S., Yamamura, M., Shimada, M., and Umezawa, T. (2004). A heartwood norlignan, (E)-hinokiresinol, is formed from 4-coumaryl 4-coumarate by a *Cryptomeria japonica* enzyme preparation. *J. Chem. Soc. Chem. Commun.* 2838–2839.

Takahashi, H., Costa, M. A., Davin, L. B., and Lewis, N. G. CCOMT (Manuscript in preparation).

Teoh, K. H., Ford, J. D., Kim, M.-R., Davin, L. B., and Lewis, N. G. (2003). Delineating the metabolic pathway(s) to secoisolariciresinol diglucoside hydroxymethyl glutarate oligomers in flaxseed (*Linum usitatissimum*). *In* "Flaxseed in Human Nutrition" (L. U. Thompson and S. C. Cunnane, eds.), 2nd edn., pp. 41–62. AOCS Press, Champaign, IL.

The European Commission (1991). Council Directive 88/388/EEC. http://ec.europa.eu/food/fs/sfp/addit_flavor/flav09_en.pdf.

The Flavor and Fragrance High Production Volume Consortia (2002). Test plan for anethole (isomer unspecificed) and *trans*-anethole, 2002. http://www.epa.gov/chemrtk/pubs/summaries/anethole/c14069tp.pdf.

The Flavor and Fragrance High Production Volume Consortia (2005). Test plan for estragole. http://www.epa.gov/chemrtk/pubs/summaries/estragole/c14022rt.pdf.

Tien, M., and Kirk, T. K. (1983). Lignin-degrading enzyme from the hymenomycete *Phanerochaete chrysosporium* Burds. *Science* **221,** 661–663.

Tien, M., and Tu, C.-P. D. (1987). Cloning and sequencing of a cDNA for a ligninase from *Phanerochaete chrysosporium*. *Nature* **326,** 520–523.

U.S. Department of Energy's Genomics (2006). GTL Bioenergy Research Centers White Paper, August.

U.S. Environmental Protection Agency (2006a). Pesticides: Regulating pesticides. Methyl eugenol (ME) http://www.epa.gov/oppbppd1/biopesticides/ingredients/factsheets/factsheet_203900.htm.

U.S. Environmental Protection Agency (2006b). Pesticides: Regulating pesticides. Floral attractants, repellents, and insecticides products. http://www.epa.gov/pesticides/biopesticides/ingredients/product/prod_florals.htm.

U.S. Environmental Protection Agency (2006c). Pesticides: Biopesticides. Floral attractants, repellents, and insecticides fact sheet. http://www.epa.gov/pesticides/biopesticides/ingredients/factsheets/factsheet_florals.htm.

U.S. Environmental Protection Agency (2006d). Pesticides: Biopesticides. Plant oils fact sheet. http://www.epa.gov/oppbppd1/biopesticides/ingredients/factsheets/factsheet_plant-oils.htm.

U.S. Environmental Protection Agency (2006e). Pesticides: Regulating pesticides. 4-Allyl anisole (estragole) (062150) fact sheet. http://www.epa.gov/pesticides/biopesticides/ingredients/factsheets/factsheet_062150.htm.

Vassão, D. G., Eichinger, D., Kim, S.-J., Davin, L. B., and Lewis, N. G. (2006a). Genes encoding chavicol/eugenol synthase from the creosote bush *Larrea tridentata*. U. S. Patent Application.

Vassão, D. G., Gang, D. R., Koeduka, T., Jackson, B., Pichersky, E., Davin, L. B., and Lewis, N. G. (2006b). Chavicol formation in sweet basil (*Ocimum basilicum*): Cleavage of an esterified C9 hydroxyl group with NAD(P)H-dependent reduction. *Org. Biomol. Chem.* **4,** 2733–2744.

Vassão, D. G., Kim, S.-J., Milhollan, J. K., Eichinger, D., Davin, L. B., and Lewis, N. G. (2007). A pinoresinol–lariciresinol reductase homologue from the creosote bush (*Larrea tridentata*) catalyzes the efficient *in vitro* conversion of *p*-coumaryl/coniferyl alcohol esters into the allylphenols chavicol/eugenol, but not the propenylphenols *p*-anol/isoeugenol. *Arch. Biochem. Biophys.* **465,** 209–218.

Watts, K. T., Mijts, B. N., Lee, P. C., Manning, A. J., and Schmidt-Dannert, C. (2006). Discovery of a substrate selectivity switch in tyrosine ammonia-lyase, a member of the aromatic amino acid lyase family. *Chem. Biol.* **13,** 1317–1326.

Weinberg, J. E., Rabinowitz, J. L., Zanger, M., and Gennaro, A. R. (1972). ^{14}C-Eugenol: I. Synthesis, polymerization, and use. *J. Dent. Res.* **51,** 1055–1061.

Ye, Z., Kneusel, R., Matern, U., and Varner, J. (1994). An alternative methylation pathway in lignin biosynthesis in *Zinnia*. *Plant Cell* **6,** 1427–1439.

Yoshimoto, T., Samejima, M., Hanyu, N., and Komai, T. (1990). Manufacture of benzaldehydes from styrenes with dioxygenase from *Pseudomonas* species. Japan Patent No. JP02200192, 1990.

Youn, B., Camacho, R., Moinuddin, S. G. A., Lee, C., Davin, L. B., Lewis, N. G., and Kang, C. (2006a). Crystal structures and catalytic mechanism of the *Arabidopsis* cinnamyl alcohol dehydrogenases AtCAD5 and AtCAD4. *Org. Biomol. Chem.* **4,** 1687–1697.

Youn, B., Kim, S.-J., Moinuddin, S. G., Lee, C., Bedgar, D. L., Harper, A. R., Davin, L. B., Lewis, N. G., and Kang, C. (2006b). Mechanistic and structural studies of apoform, binary, and ternary complexes of the *Arabidopsis* alkenal double bond reductase At5g16970. *J. Biol. Chem.* **281,** 40076–40088.

Zhang, D., Franceschi, V. R., Davin, L. B., and Lewis, N. G. COMT (Manuscript in preparation).

AUTHOR INDEX

Numbers in bold refer to pages on which full references are listed.

C

L

M

N

S

SUBJECT INDEX

D

E

N

O

P

T

U

V

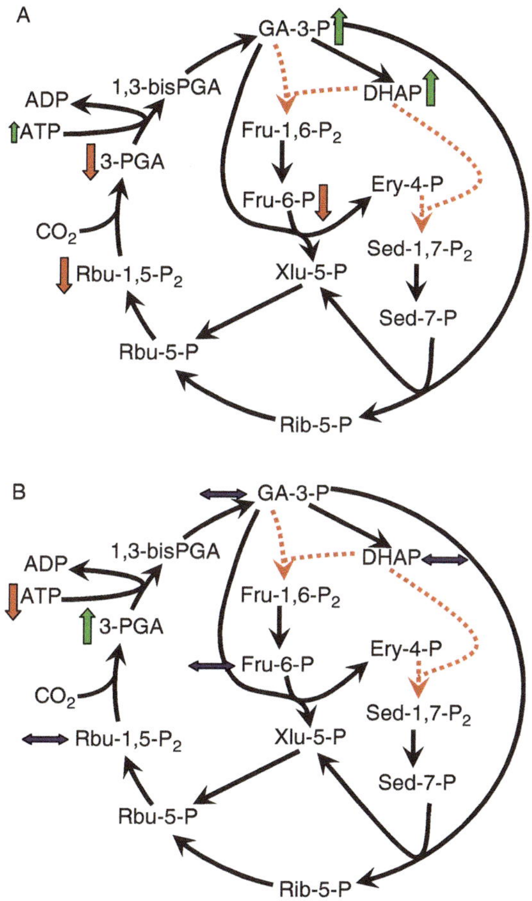

Nicholas J. Kruger and R. George Ratcliffe, FIGURE 1.2 Effect of a decrease in aldolase content on photosynthetic intermediates in potato plants (Haake *et al.*, 1999). Changes in the steady-state levels of Calvin cycle intermediates in aldolase-antisense lines grown under low irradiance (A) or high irradiance in the presence of elevated CO_2 (B) are compared with those in wild-type plants grown under the same conditions. The reactions catalyzed by aldolase are indicated by dotted lines. Symbols refer to the following changes in metabolite content: ↑, increase; ↓, decrease; ↔, roughly similar. (See Page 17 of this Volume.)

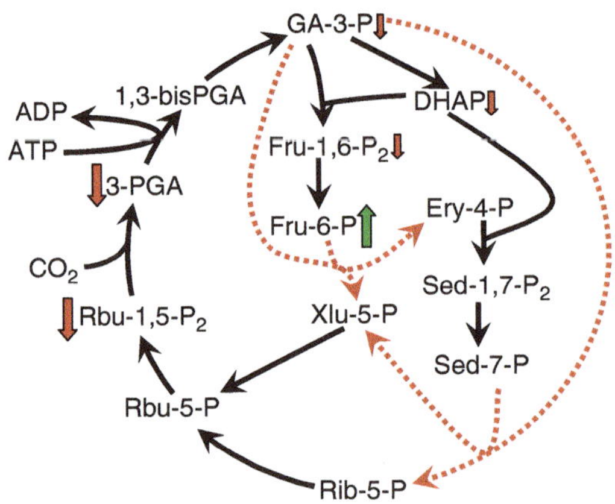

Nicholas J. Kruger and R. George Ratcliffe, FIGURE 1.3 Effect of decreased transketolase content on photosynthetic intermediates in tobacco plants (Henkes *et al.*, 2001). Changes in the steady-state levels of Calvin cycle intermediates in transketolase-antisense lines are compared with those in wild-type plants grown under the same conditions. The reactions catalyzed by transketolase are indicated by dotted lines. Symbols refer to the following changes in metabolite content: ↑, increase; ↓, decrease. (See Page 19 of this Volume.)

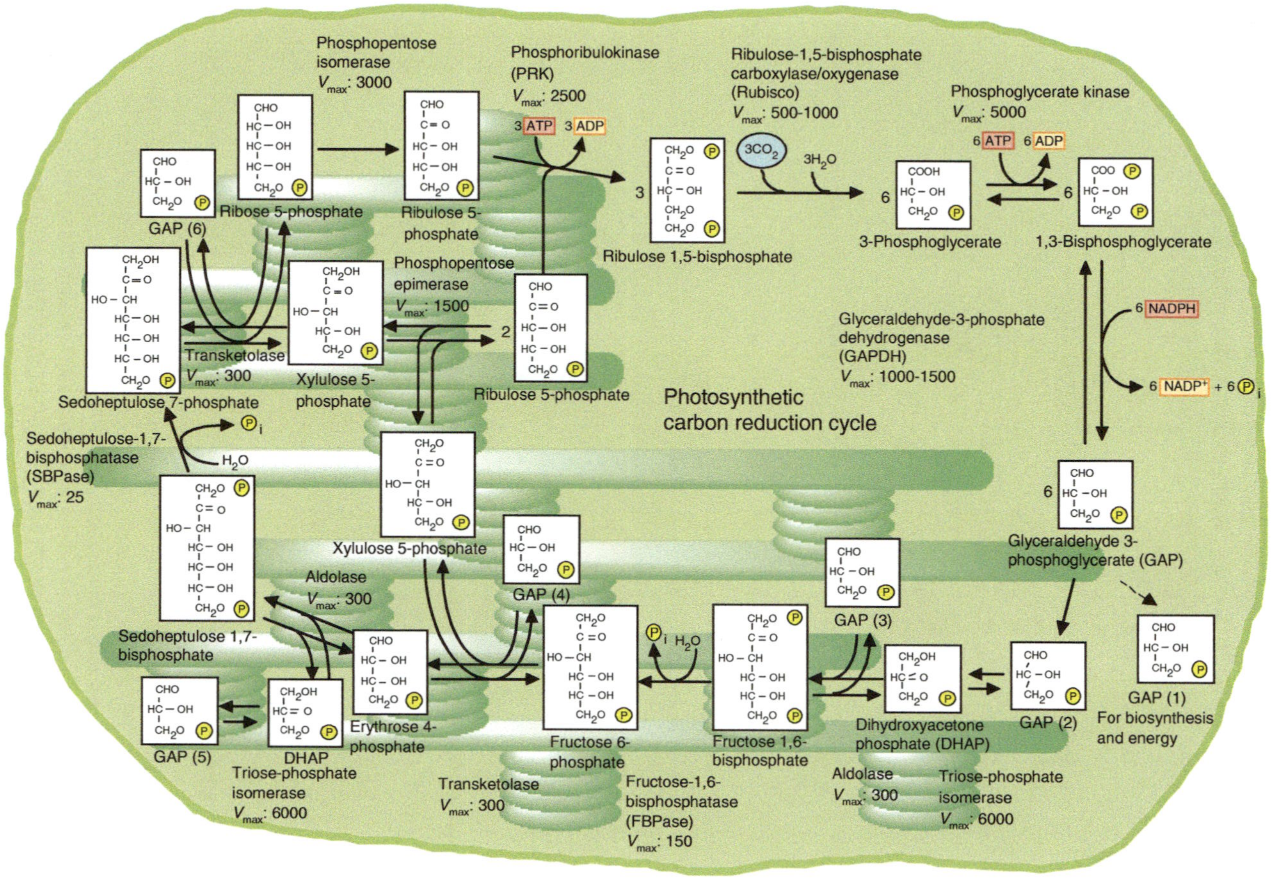

Akiho Yokota and Shigeru Shigeoka, FIGURE 4.1 Photosynthetic carbon reduction cycle. V_{max} of each enzyme is given in micromoles per milligram chlorophyll per hour (Robinson and Walker, 1981). (See Page 84 of this Volume.)

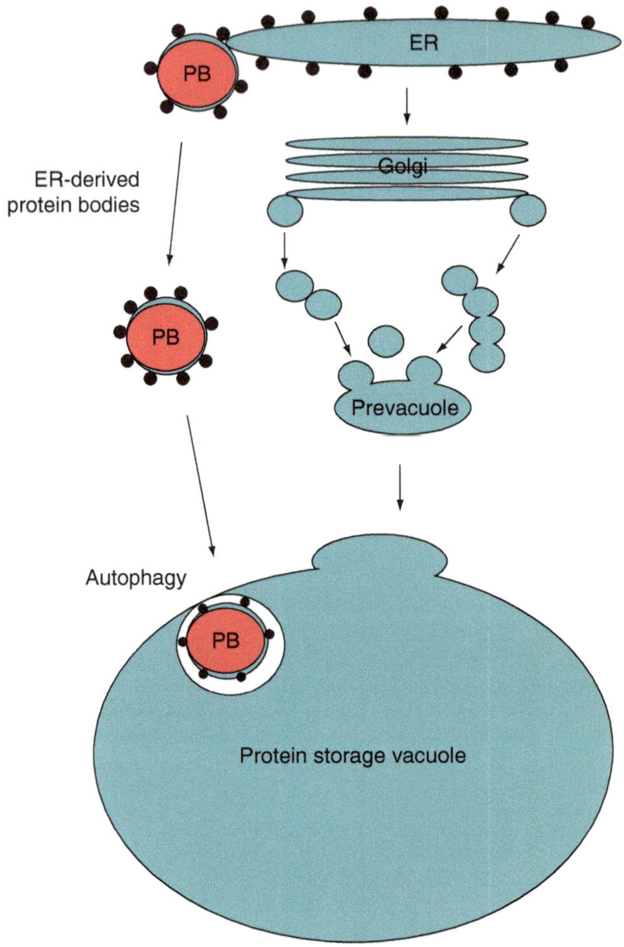

David R. Holding and Brain A. Larkins, FIGURE 5.1 Diagram illustrating the ontogeny of PBs and protein storage vacuoles (PSVs). PBs form through the aggregation of storage proteins within the ER or PSVs. After formation, PBs can either remain attached to the ER or bud off and form separate organelles, that is PSVs. PBs can accumulate in the cytoplasm or become sequestered into PSVs by autophagy. PSVs are formed as the consequence of ER-synthesized storage proteins progressing through the endomembrane secretory system to specialized vacuoles (PSVs) for accumulation. Reprinted from Herman and Larkins (1999) with permission from the ASPB. (See Page 111 of this Volume.)

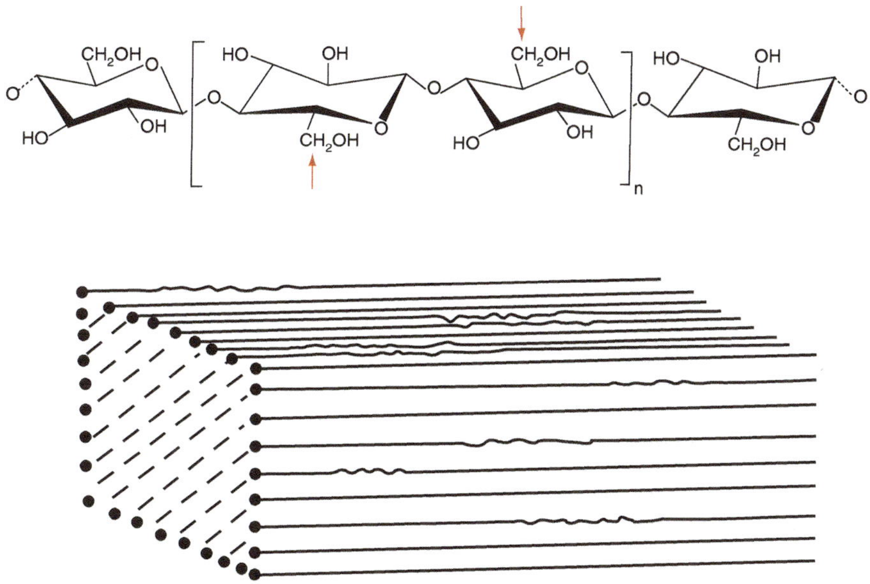

Inder M. Saxena and R. Malcolm Brown, Jr., FIGURE 6.1 Top image is the structural formula for the β-1,4-linked glucan chain of cellulose. The bracketed region indicates the basic repeat unit, cellobiose, in the chain. The glucan chain has a twofold symmetry. The bottom image is a schematic representation of a crystalline cellulose I microfibril. (Reproduced from Brown, Jr. R. M., *J. Poly. Sci. Part A Poly. Chem.* **42**, 489–495.) (See Page 138 of this Volume.)

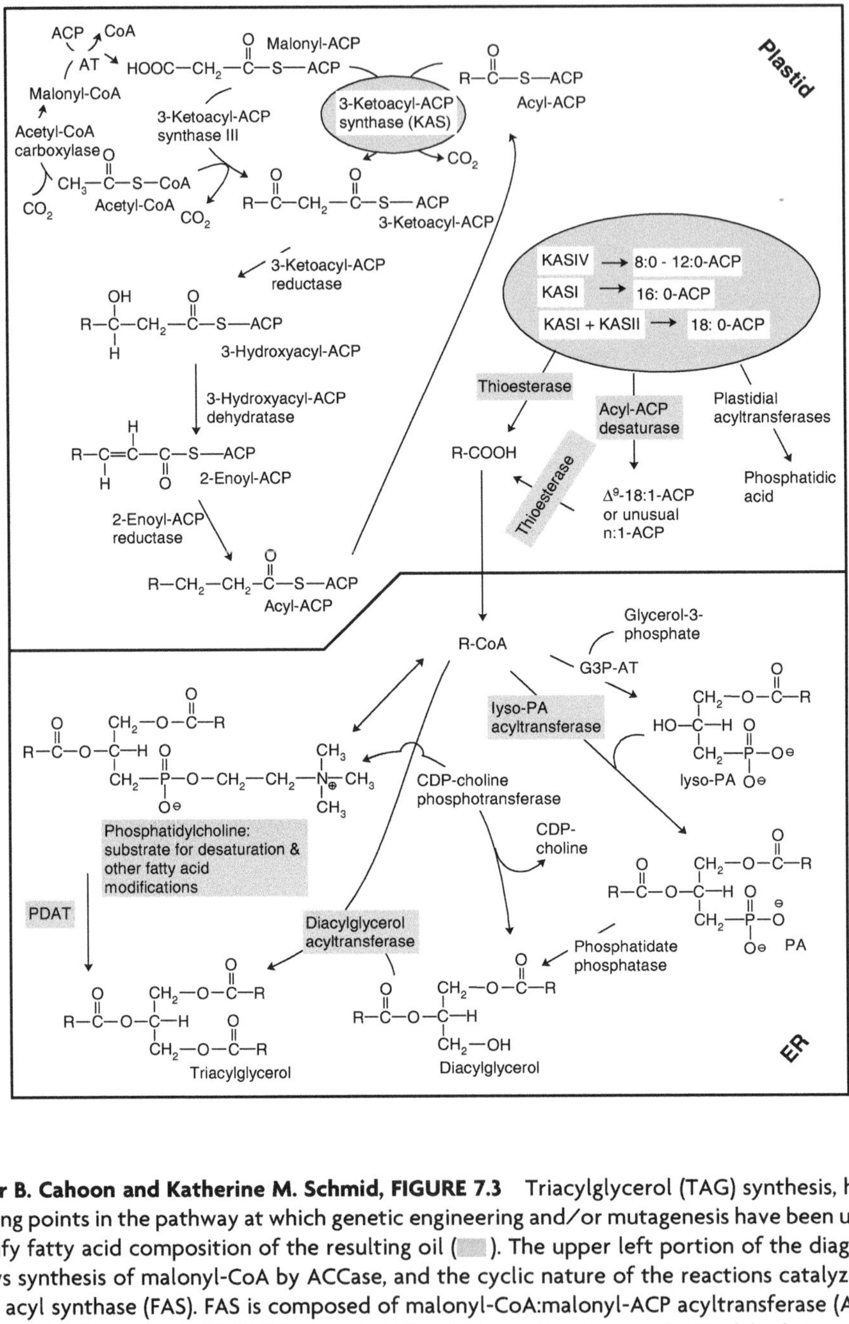

Edgar B. Cahoon and Katherine M. Schmid, FIGURE 7.3 Triacylglycerol (TAG) synthesis, highlighting points in the pathway at which genetic engineering and/or mutagenesis have been used to modify fatty acid composition of the resulting oil ( ). The upper left portion of the diagram shows synthesis of malonyl-CoA by ACCase, and the cyclic nature of the reactions catalyzed by fatty acyl synthase (FAS). FAS is composed of malonyl-CoA:malonyl-ACP acyltransferase (AT), 3-ketoacyl-ACP synthase (KAS), 3-ketoacyl-ACP reductase, 3-hydroxyacyl-ACP dehydratase, and enoyl-ACP reductase. As shown on the right of the diagram, the products of FAS depend on the contributions of various KASes, the substrate and double bond specificities of acyl-ACP desaturases, and the substrate specificities of thioesterases that release fatty acids for export from the

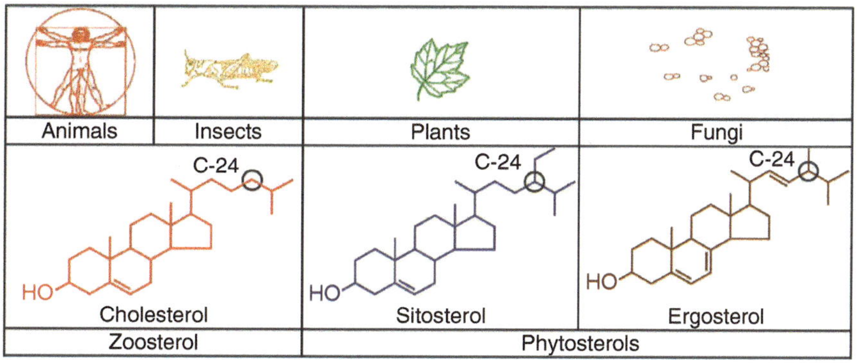

Wenxu Zhou *et al.*, FIGURE 9.1 Major sterols found in living systems. (See Page 242 of this Volume.)

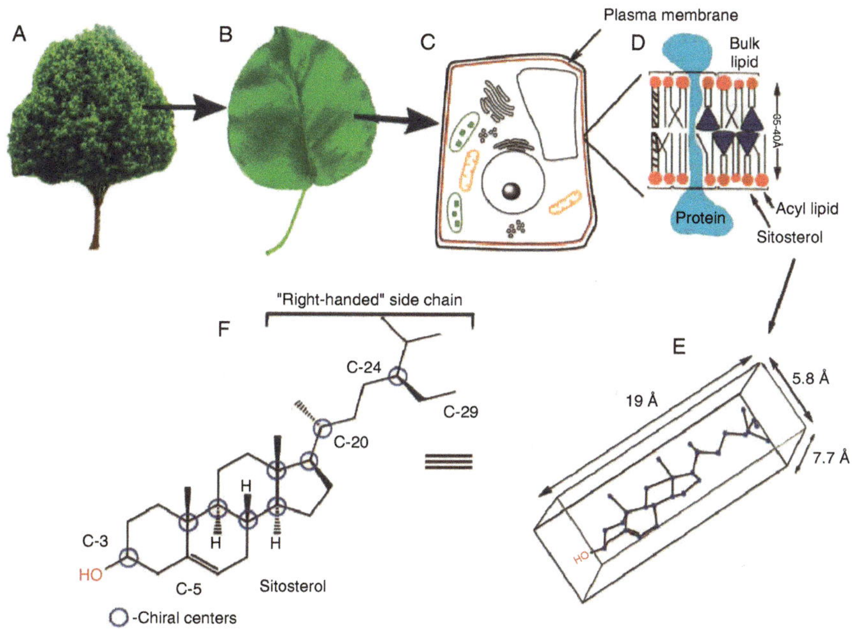

Wenxu Zhou *et al.*, FIGURE 9.2 Sitosterol distribution in the intact plant: (A) whole plant; (B) leaf; (C) cell structure; (D) membrane lipid leaflet; (E) conformational perspective of sterol; (F) structure and stereochemistry in sitosterol. (See Page 243 of this Volume.)

---

plastids. In the ER, phosphatidic acid (PA) is assembled by sequential activities of glycerol-3-phosphate acyltransferase (G3P-AT) and lysophosphatidic acid-acyltransferase (LPAAT). Diacylglycerol (DAG) units released from lyso-PA by phosphatidate phosphatase may be converted directly to triacylglycerol by DGAT. However, a large proportion of TAG fatty acids pass through PC, which serves as a substrate for further fatty acid desaturation and other modifications. Modified fatty acids may then be transferred to TAG: (1) as part of DAG released by the reversible CDP-choline acyltransferase, (2) after return to the acyl-CoA pool, or (3) by direct transfer via PDAT. (See Page 168 of this Volume.)

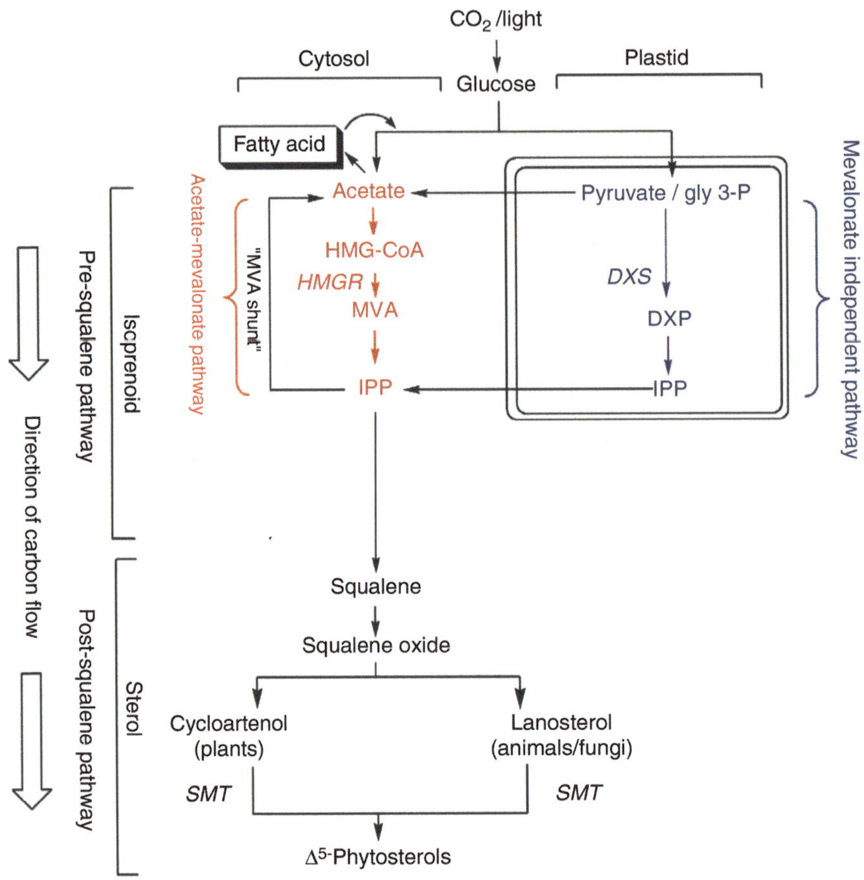

Wenxu Zhou *et al.*, FIGURE 9.3 Stages in the isoprenoid–phytosterol pathway and compartmentation of acetate–mevalonate pathways. (See Page 245 of this Volume.)

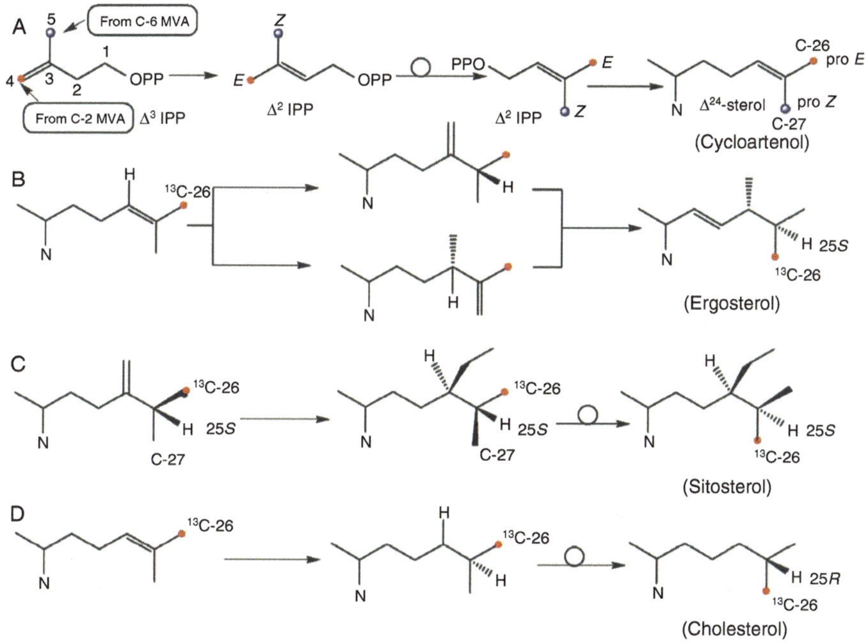

Wenxu Zhou *et al.*, FIGURE 9.4 Determination of labeling patterns in isopentyl diphosphate and sterols synthesized by different pathways. (See Page 247 of this Volume.)

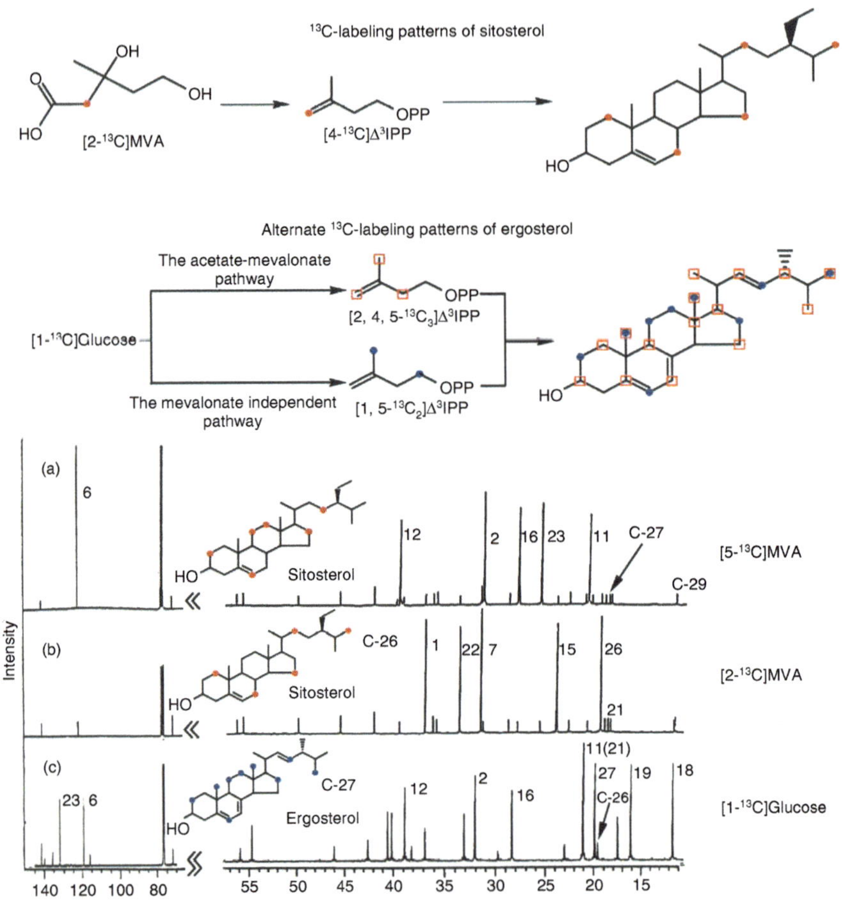

Wenxu Zhou *et al.*, FIGURE 9.5 Stereochemistry of phytosterols at C-25 after ^{13}C labeling of the Pro*E* C-26 of Δ^{24}-sterols. (See Page 248 of this Volume.)

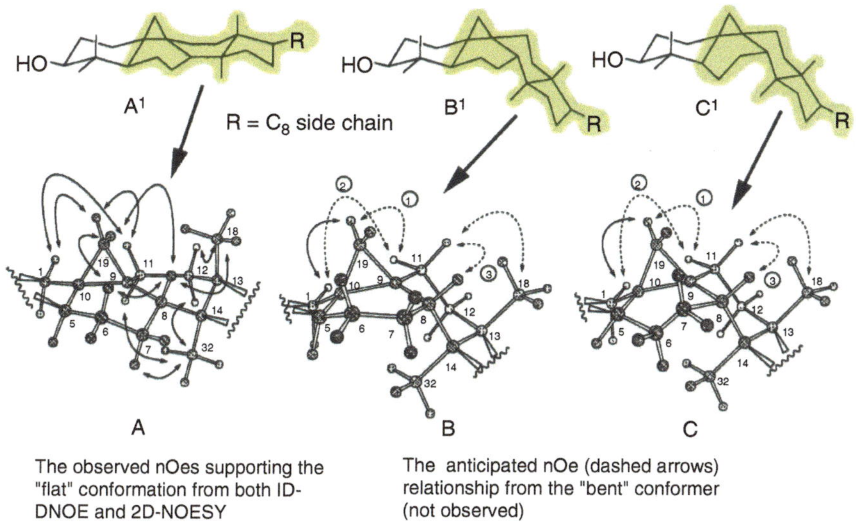

Wenxu Zhou *et al.*, FIGURE 9.8 NOE networks that indicate whether cycloartenol can orient into a flat or bent shape. Adapted from Nes *et al.* (1998b). (See Page 253 of this Volume.)

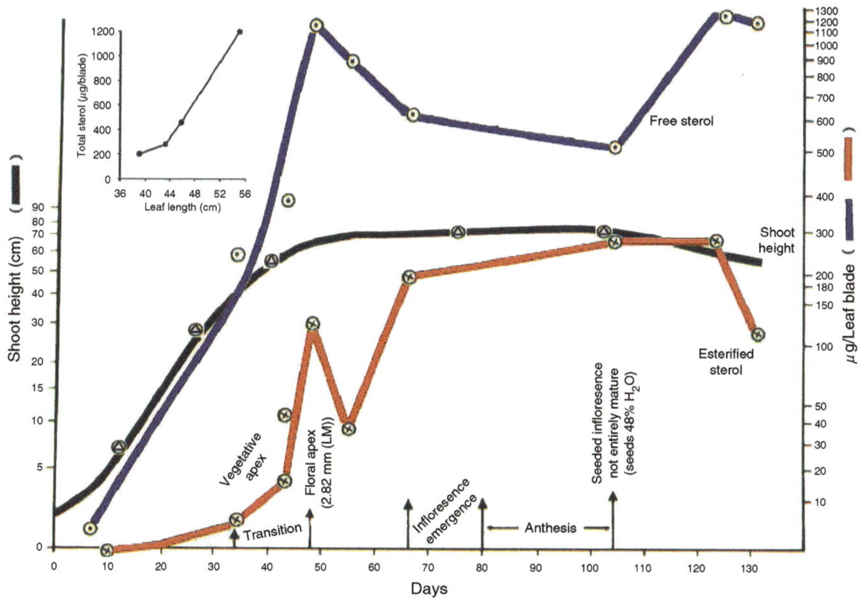

Wenxu Zhou *et al.*, FIGURE 9.11 Correlation of the sterol content with the increase in shoot height of sorghum. Inset in a correlation of the increase in sterol content with maturation of the leaf blade, measured as changes in leaf length. Adapted from Heupel *et al.* (1986). (See Page 256 of this Volume.)

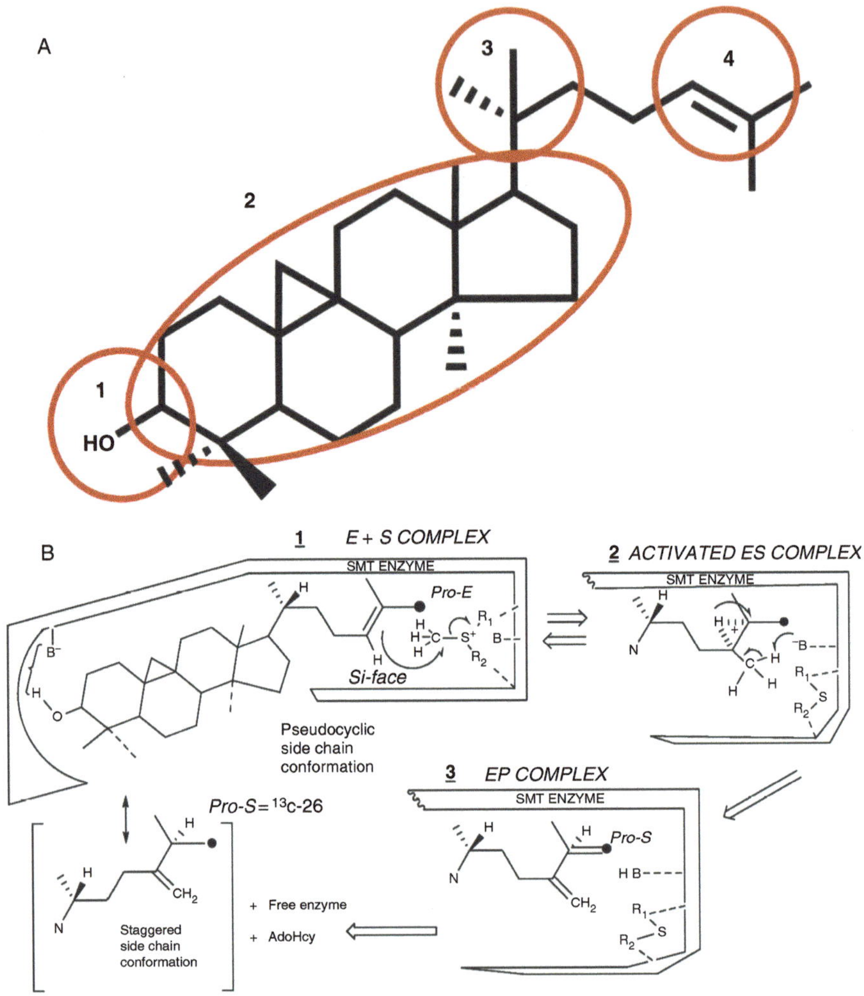

Wenxu Zhou *et al.*, FIGURE 9.15 Postulated domains of sterol molecule (Top, A) recognized by the SMT and the *steric–electric plug* model of SMT catalysis (Bottom, B). (See Page 261 of this Volume.)

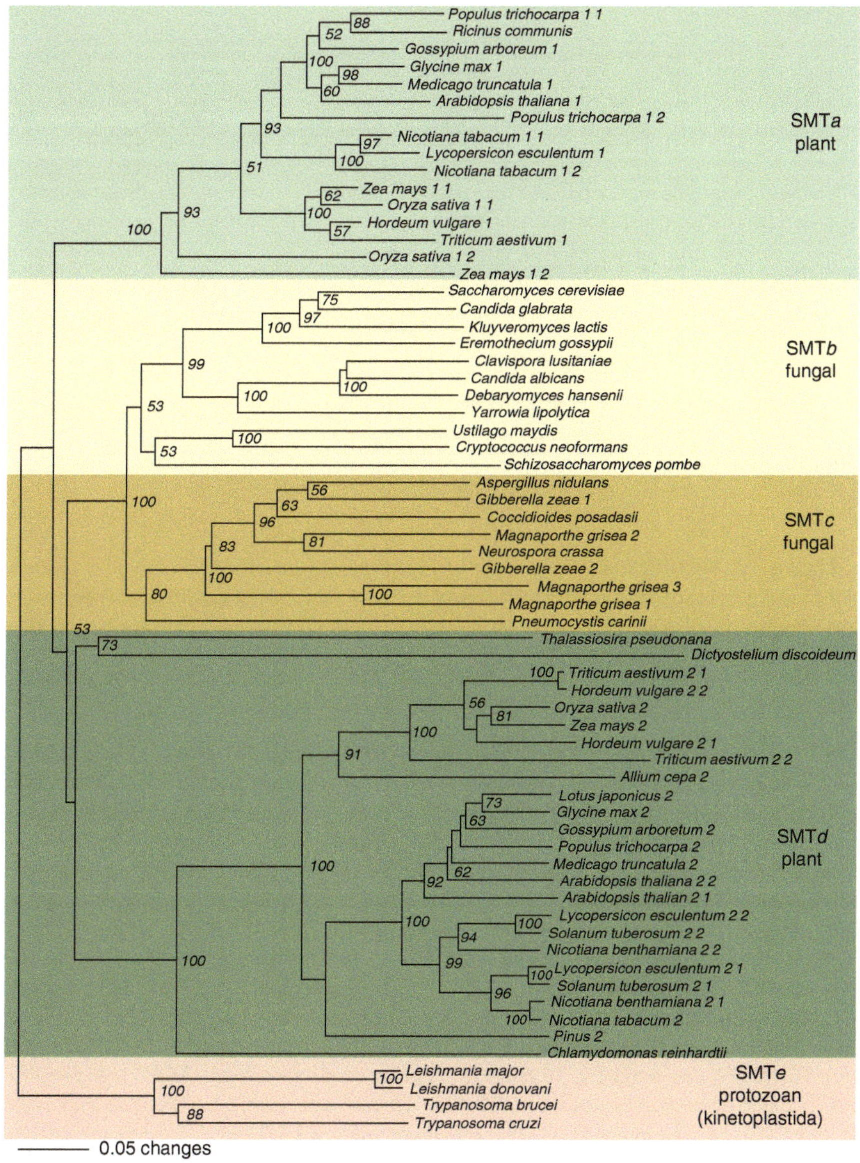

Wenxu Zhou *et al.*, FIGURE 9.17 Rooted phylogenetic tree of eukaryotic SMT was created with PAUP using the neighbor-joining method with kinetoplastida SMTs as the out group. The scale bar represents a distance of 0.05 substitutions per site. Numbers are the percentage bootstrap values for 1,000 replicates. SMTa through SMTe designate SMT subfamilies defined by a minimum of 35% identity between members at the amino acid level. Accession number of SMT sequences obtained from NCBI (http://www.ncbi.nlm.nih.gov): *Ricinus communis*, T10173; *Glycine max* 1, T06780; *Arabidopsis thaliana* 1, AGG28462; *Nicotiana tabacum* 1-1, AAC43951; *Nicotiana tabacum* 1-2,

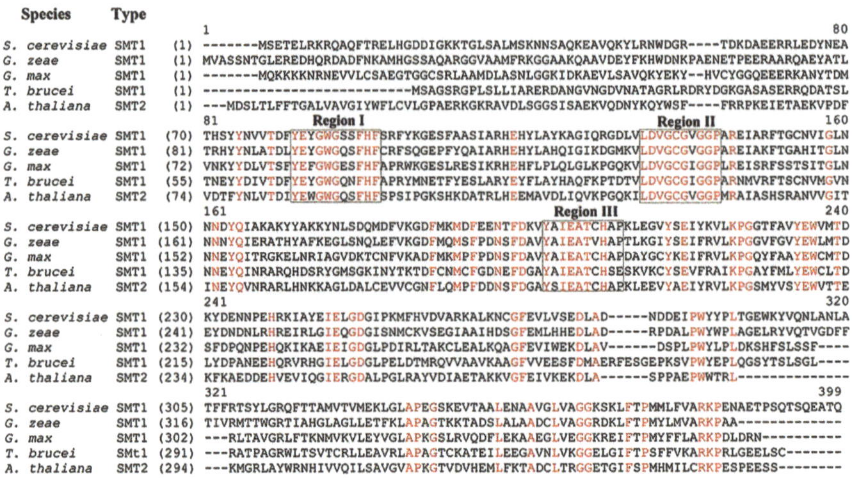

Wenxu Zhou *et al.*, FIGURE 9.18 Alignment of representative deduced amino acid sequences of SMT from fungi, protozoa, and plants that represent SMT1 and SMT2 isoforms. Conserved regions corresponding to sterol (Regions I and III) and AdoMet binding sites (Region II) are boxed. (See Page 267 of this Volume.)

AAC35787; *Zea mays* 1-1, TO4138; *Oryza sativa* 1-1, AAC34988; *Triticum aestivum* 1, ABB49388; *Oryza sativa* 1-2, AAP21419; *Saccharmyces cerevisiae*, NP_013706; *Candida glabrata*, CAG59930; *Kluyveromyces lactis*, AAS52116; *Clavispora lusitaniae*, CAO21936; *Candida albicans*, O74198; *Debaryomyces hansenii*, CAG87427; *Yarrowia lipolytica*, CAG77980; *Ustilago maydis*, EAK84412; *Schizosaccharomyces pombe*, CAB16897; *Gibberella zeae* 1, XP_382959; *Magnaporthe grisa* 2, EAA48309; *Neurospora crassa*, CAB97289; *Gibberella zeae* 2, XP_355916; *Magnaporthe grisa* 3, EAA50587; *Magnaporthe grisa* 1, EAA47049; *Pneumocystis carinii*, AKK54439; *Oryza sativa* 2, ACC34989; *Arabidopsis thaliana* 2-1, ABB62809; *Arabidopsis thaliana* 2-2, CAA61966; *Nicotiana tabacum*, TO3848; *Leishmania donovani*, AAR92098; *Trypanosoma cruzi*, TIGR_5693. Gene indices of SMT obtained from TIGR (http://www.tigr.org): *Populus trichocarpa* 1-1, TC35619; *Medicago trucatula*, TC86500; *Gassypium arboretum* 1, TC20798; *Populus trichocarpa* 1-1, TC36329; *Hordeum vulgare* 1, TC 110279; *Zea mays* 1-2, TC234797; *Cryptococcus neoformans*, TC4573; *Aspergillus nidulans*, TC6295; *Coccidioides posadasii*, TC4513; *Triticum aestivum* 2-1, TC165856; *Hordeum vulgare* 2–2, TC121611; *Zea mays*, TC224796; *Hordeum vulgare* 2-1, TC123636; *Triticum aestivum* 2-2, TC172448; *Allium cape* 2, TC2207; *Lotus japonicus*, TC7995; *Glycine max* 2, TC189052; *Gossypium arboretum* 2, TC 21049; *Populus trichocarpa* 2, TC36329; *Medicago truncatula* 2, TC77751; *Lycopersicon esculentum* 2-2, TC126730; *Solanum tuberosum* 2-2, TC126449; *Nicotiana benthamiana* 2-2, TC7251; *Lycopersicon esculentum* 2-1, TC124648; *Solanum tuberosum* 2-1, TC112127; *Nicotiana benthamiana* 2-1, TC8230; *Pinus* 2, TC52326; *Chlamydomonas reinhardtii*, TC29837. Gene identification number of SMT from The Wellcome Trust Sanger Institute (http://www.sanger.ac.uk): *Leishmanis major*, LM3731Bb05.p1c; *Trypanosoma brucei*, TB10.1520. The SMT gene of *Dictyostelium discoideum* was from IMB (http://genome.imb-jena.de) and gene identification number is pcr25kl1p3887. SMT gene of *Thalassiosira pseudonana* is identified from GRI (http://genome.jgi-psf.org) in scaffold_30, 94659:96009. (See Page 265 of this Volume.)

Wenxu Zhou *et al.*, FIGURE 9.21 Comparison of *smt*1 (*Ac*-mutagenized plant that generates high levels of cholesterol) and *SMT*1 (wild-type that generates high levels of sitosterol) plants. (A) Rosette of 2-week plants; (B) rosette of 2-week-old *smt*1-*smt*1–3 plant; (C) mature (5-week) SMT1 and *smt*1–3 plant. Adapted from Diener *et al.* (2000). (See Page 272 of this Volume.)

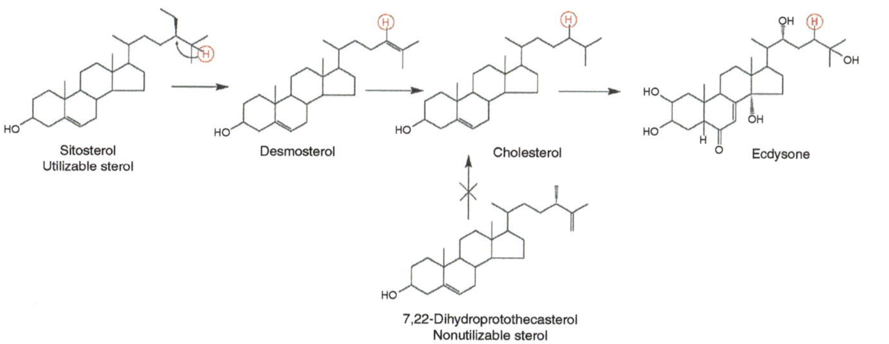

Wenxu Zhou *et al.*, FIGURE 9.22 Conversion of sitosterol to cholesterol and ecdysone by phytophagous insects. The hydrogen circled in red migrates from C-25 in sitosterol to C-24 in cholesterol during the C-24-dealkylation reaction. The methyl group at C-24 in 7,22-dihydroprotothecasterol remains intact for mechanistic reasons. (See Page 275 of this Volume.)

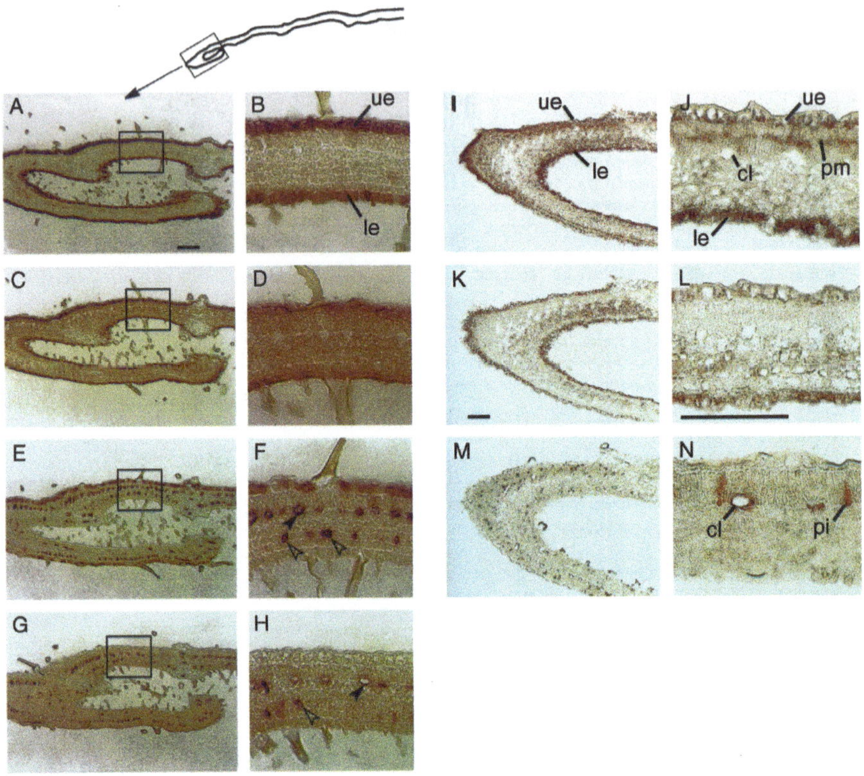

Toni M. Kutchan *et al.*, FIGURE 10.4 Localization of *cyp72a1*, *tdc*, *str1*, *d4h*, and *dat* transcripts in developing *C. roseus* leaves (Irmler *et al.*, 2000; St-Pierre *et al.*, 1999). Panels (A, B, K, L) *tydc*; panels (C, D) *str1*; panels (E, F) *d4h*; panels (G, H, M, N) *dat*; panels (I, J) *cyp72a1*. cl, cross-connecting laticifer cells; le, lower epidermis; pi, idioblast cells associated with palisade mesophyll cells; pm, palisade mesophyll cells; si, idioblast cells associated with spongy mesophyll cells; ue, upper epidermis; *tdc*, tryptophan decarboxylase; *str1*, strictosidine synthase; *cyp72a1*, secologanin synthase; *d4h*, desacetoxyvindoline 4-hydroxlyase; *dat*, deacetylvindoline 4-*O*-acetyltransferase. Bar shown in panels (A, K), 100 μm; in panel (L), 50 μm. Solid arrows in panels (F, H), laticifer cells; open arrows, idioblast cells. (See Page 289 of this Volume.)

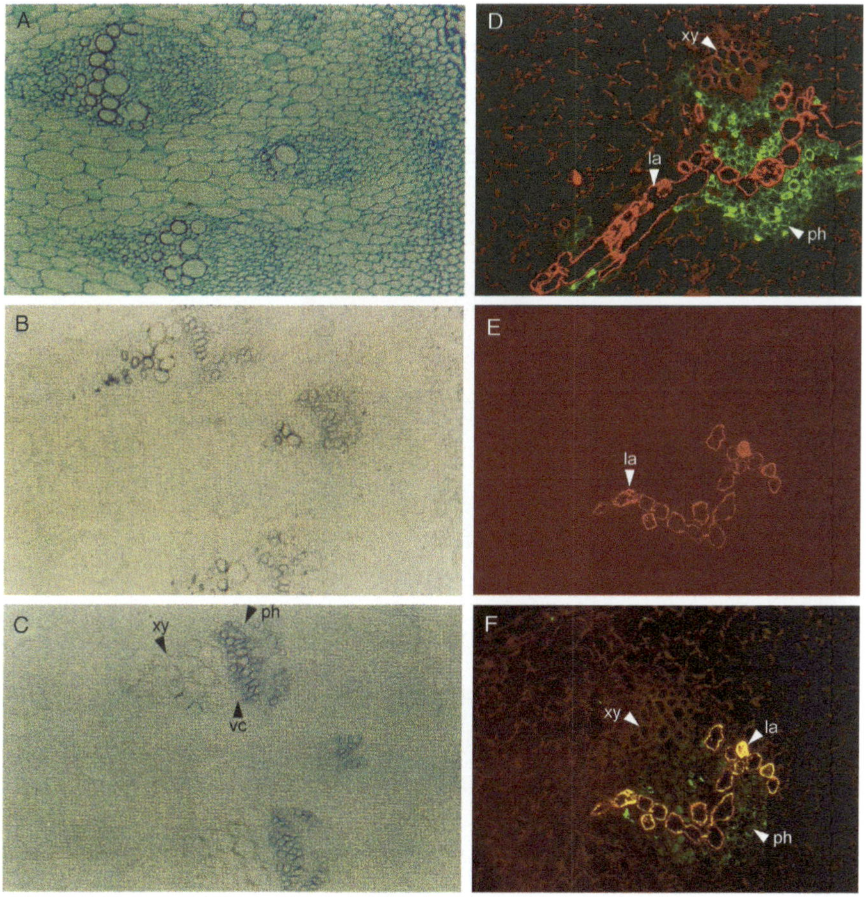

Toni M. Kutchan *et al.*, FIGURE 10.9 Localization of *tydc* (Facchini and De Luca, 1995), SalAT, COR1, and MLP (Weid *et al.*, 2004) in stem of *P. somniferum*. Panel (A) root cross-section stained with aniline safranine and astra blue; (B) *in situ* hybridization of *tydc1*; (C) *in situ* hybridization of *tydc2*; (D) immunolocalization of MLP to laticifers (red fluorescence) and SalAT to phloem parenchyma (green fluorescence); (E) immunolocalization of MLP to laticifers (red fluorescence); and (F) coimmunolocalization of MLP and COR1 to laticifers (yellow fluorescence). Green fluorescence indicates COR1 is present also in phloem parenchyma. xy, xylem; ph, phloem; vc, vascular cambium; la, laticifers; *tydc*, tyrosine/dopa decarboxylase; SalAT, salutaridinol 7-*O*-acetyltransferase; COR1, codeinone reductase; MLP, major latex protein. (See Page 297 of this Volume.)

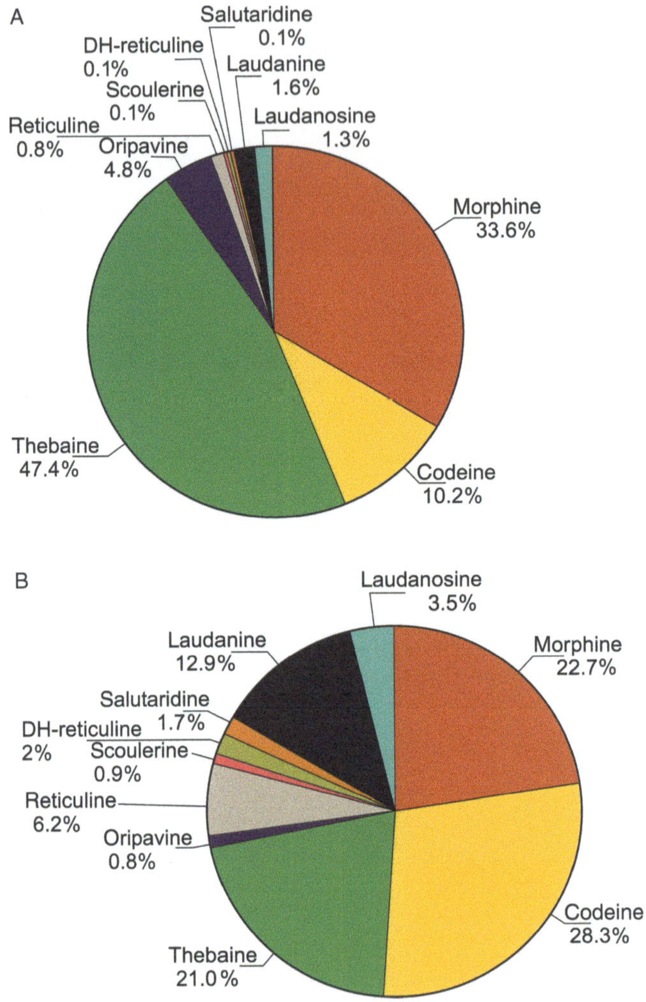

Toni M. Kutchan *et al.*, FIGURE 10.10 Benzylisoquinoline alkaloid quantitation in latex from (A) untransformed *P. somniferum* and (B) transgenic *anti-bbe1* T_2 plants (Frick *et al.*, 2004). The data are presented in percentages (μg alkaloid/100 μg soluble protein in latex). The values given in (A) are the mean of individual analyses of 29 plants. (See Page 300 of this Volume.)

Toni M. Kutchan *et al.*, FIGURE 10.12 Immunohistochemical localization of TR-I, TR-II, and H6H in root cross-sections of *H. niger* (Nakajima and Hashimoto, 1999). Panels (A, B) TR-I; panels (C, D) TR-II; panels (E, F) H6H. Ep, epidermis; OC, outer cortex; IC, inner cortex; En, endodermis; Pe, pericycle; Xy, xylem. (See Page 303 of this Volume.)

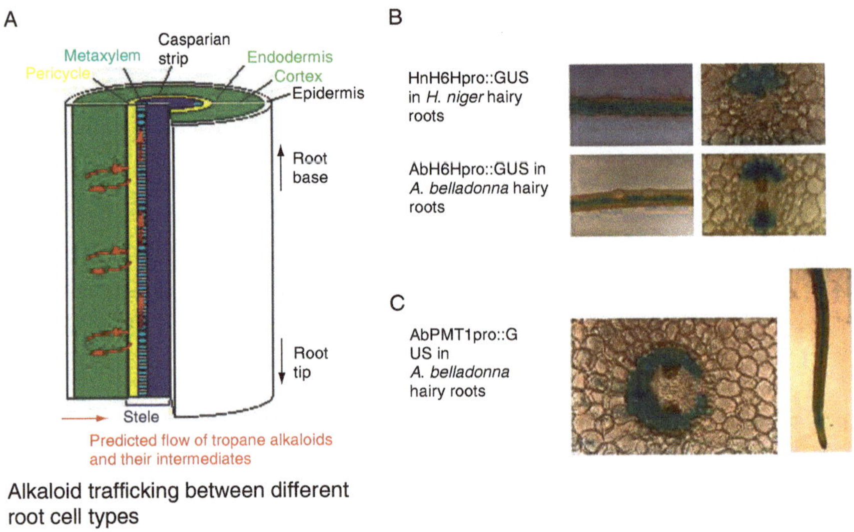

Fumihiko Sato and Yasuyuki Yamada, FIGURE 11.6 Cell-specific gene expression in tropane alkaloids. (A) A model for alkaloid trafficking between different cell types in root. (B) Pericycle cell-specific gene expression of the H6H gene. (C) Cell-specific gene expression of the PMT gene. (See Page 330 of this Volume.)

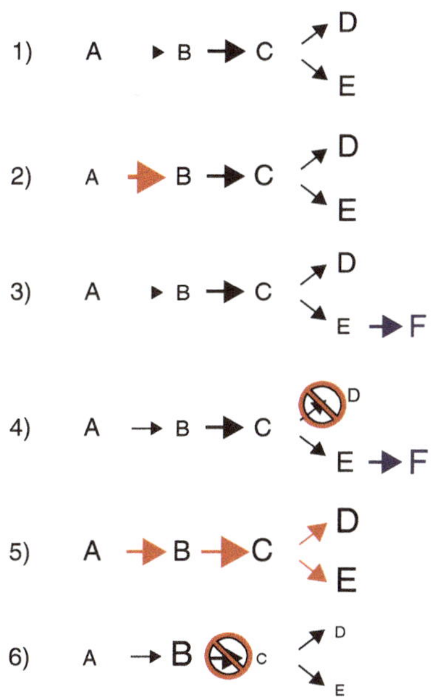

Fumihiko Sato and Yasuyuki Yamada, FIGURE 11.7 Strategies for overcoming the limitations in the production of metabolites in plant cells. (1) Preexisting pathway, (2) overexpression of rate-limiting (early) step to clarify the regulation and to overproduce metabolites, (3) creation of a new branch pathway to produce novel compound F and to increase the sink strength, (4) reducing an undesired pathway to enhance the metabolic flow to a desired path, (5) overexpression of transcriptional factor(s) to activate the entire biosynthetic pathway simultaneously, (6) mutation/ downregulation of gene expression of biosynthetic enzyme to accumulate the intermediate of pathway. (See Page 332 of this Volume.)

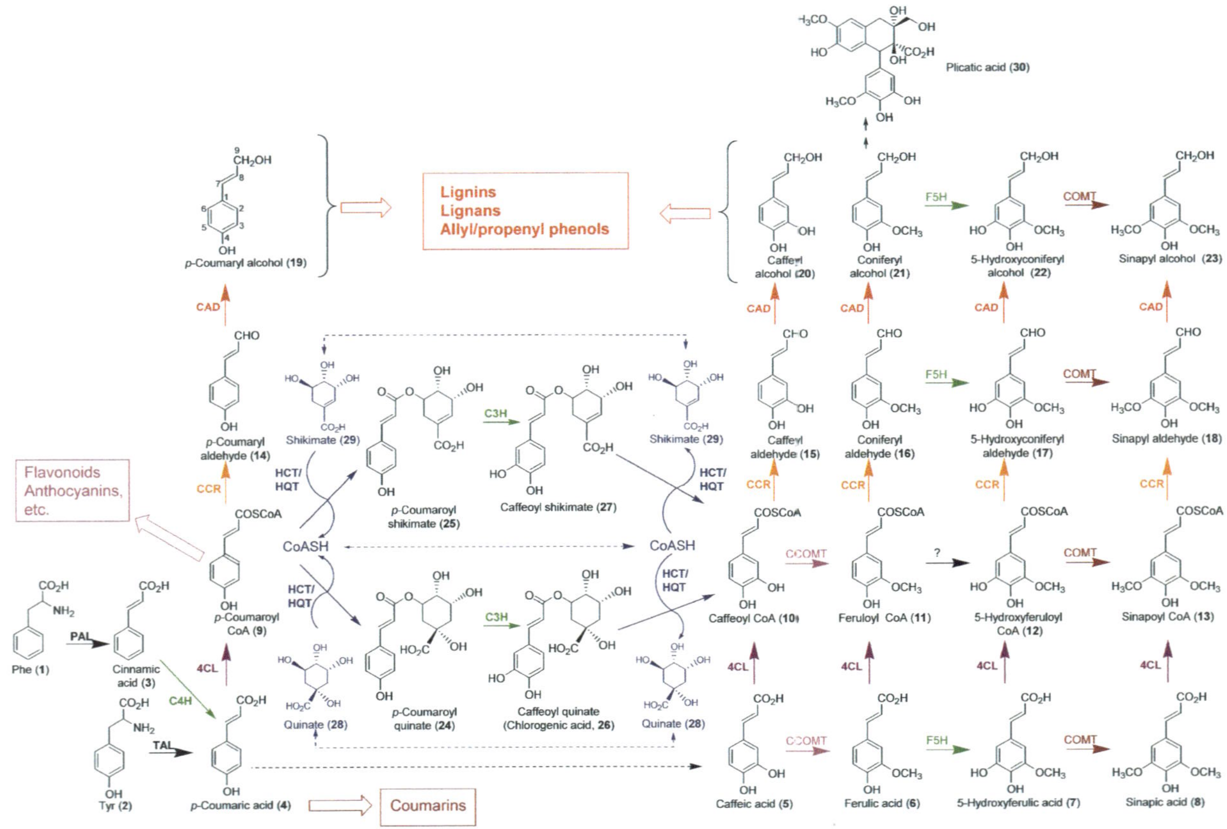

Daniel G. Vassão *et al.*, FIGURE 13.1 Current view of the phenylpropanoid pathway. 4CL, hydroxycinnamoyl CoA ligases; C3H, *p*-coumarate 3-hydroxylase; C4H, cinnamate 4-hydroxylase; CAD, cinnamyl alcohol dehydrogenases; CCOMT, hydroxycinnamoyl CoA *O*-methyltransferases; CCR, cinnamoyl CoA oxidoreductases; COMT, caffeic acid *O*-methyltransferase; F5H, ferulate 5-hydroxylase; HCT/HQT, hydroxycinnamoyl shikimate/quinate transferase; PAL, phenylalanine ammonia lyase; TAL, tyrosine ammonia lyase. (See Page 390 of this Volume.)

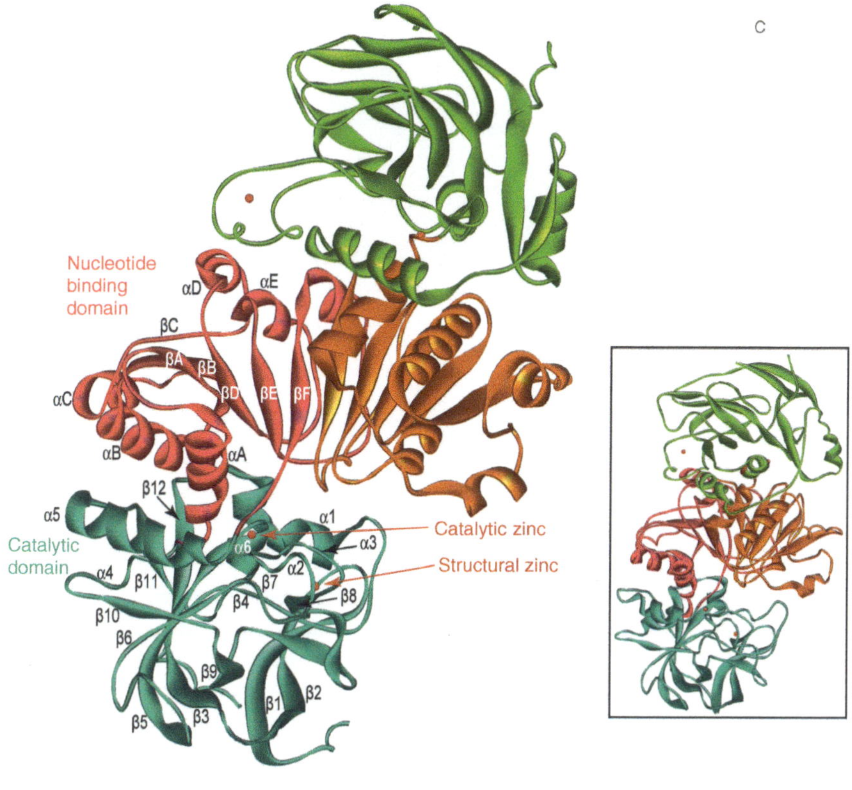

Daniel G. Vassão *et al.*, FIGURE 13.4 Schematic representation of the crystal structure of the AtCAD5 homodimer and energy-minimized model of AtCAD4 (inset). Source: Reprinted from Youn *et al.* (2006a). (See Page 399 of this Volume.)

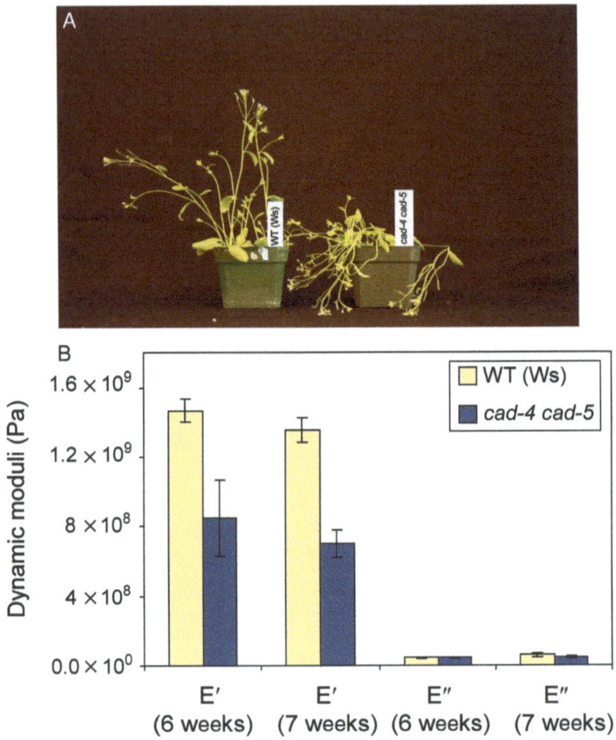

Daniel G. Vassão *et al.*, FIGURE 13.5 Effects of knocking out *Atcad4* and *cad5* (*cad-c cad-d*) in *Arabidopsis thaliana* (ecotype Wassilewskija). (A) Phenotypical differences between 4-week-old wild type (Ws) and lignin-deficient *cad-4 cad-5* double mutant plants. (B) Tensile storage and loss moduli of WT and lignin-deficient (~90%) *cad-4 cad-5* double mutant lines. Source: Redrawn from Jourdes *et al.* (2007). (See Page 400 of this Volume.)

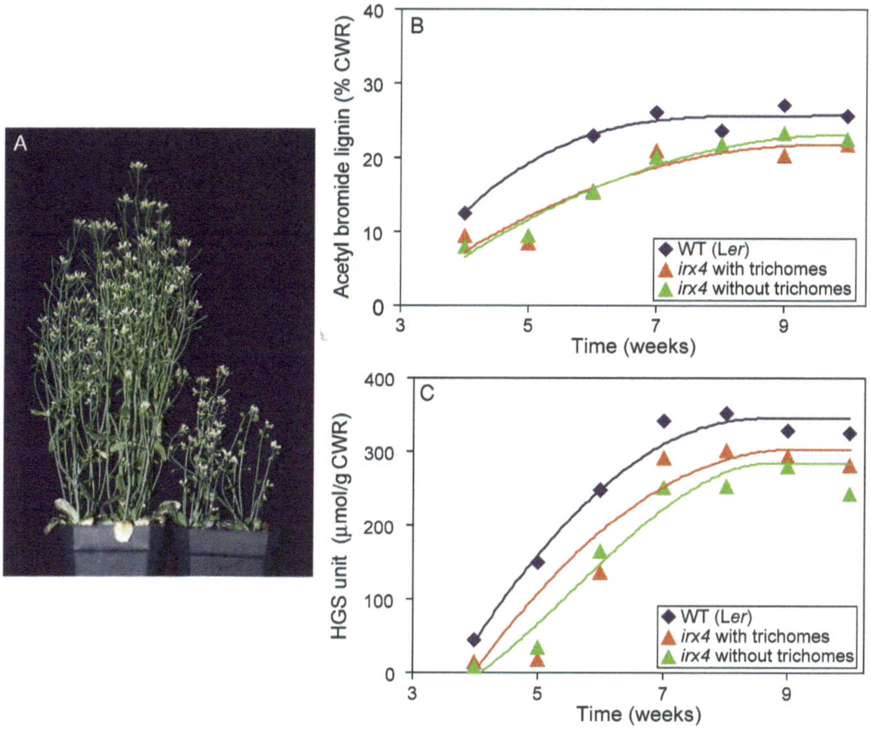

Daniel G. Vassão *et al.*, FIGURE 13.6 Effects of mutating CCR in *Arabidopsis thaliana*. (A) Phenotypical differences between wild type (L*er*, left) and *irx4* mutant (right) plants. (B) Plot of acetyl bromide lignin determinations in stems at various stages of *A. thaliana* growth and development. (C) Estimations of H, G, and S monomer amounts released during thioacidolysis of wild type and *irx4* extractive-free stem tissues. Source: Redrawn from Patten *et al.* (2005). (See Page 401 of this Volume.)

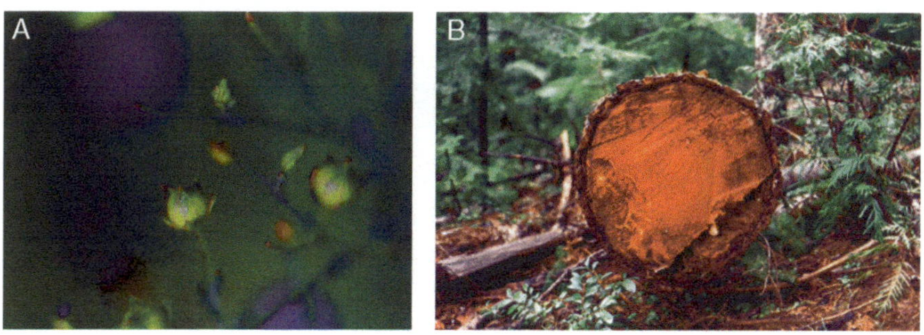

Daniel G. Vassão *et al.*, FIGURE 13.8 (A) Flax (*Linum usitatissimum*) oilseed and (B) western red cedar (*Thuja plicata*) heartwood. (See Page 403 of this Volume.)

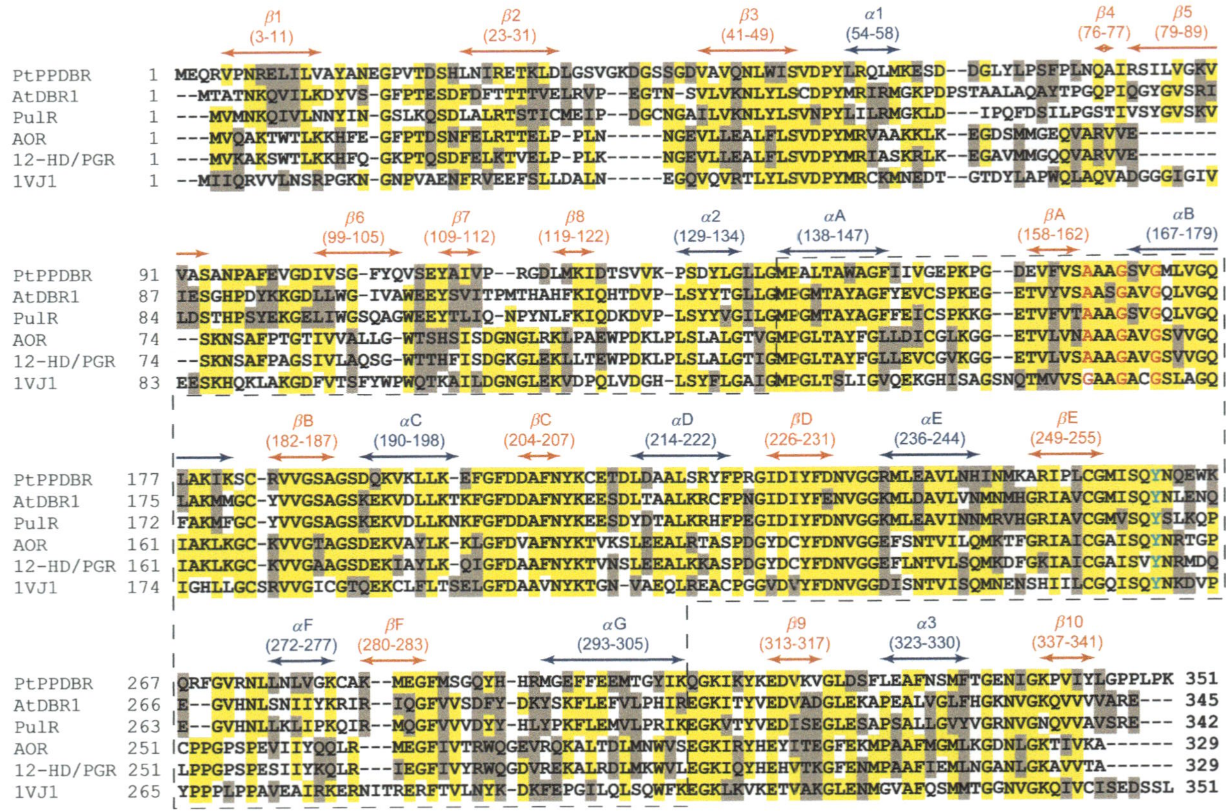

Daniel G. Vassão *et al.*, FIGURE 13.12 Alignment of double-bond reductases from *Pinus taeda* (PtPPDBR), *Arabidopsis thaliana* (AtDBR1), and homologues from *Mentha piperita* (pulegone reductase, PulR), rat (*Rattus norvegicus*, AOR), guinea pig (*Cavia porcellus*, 12-HD/PGR), and mouse (*Mus musculus*, 1VJ1). The nucleotide-binding domain is indicated by the dotted line, with conserved AXXGXXG motif in red. Conserved catalytic Tyr residues (Y260 for AtDBR1) are highlighted in light blue, and secondary structural elements are indicated in colored bars (blue for α-helices and orange for β-strands). Source: Redrawn from Youn *et al.* (2006b). (See Page 409 of this Volume.)

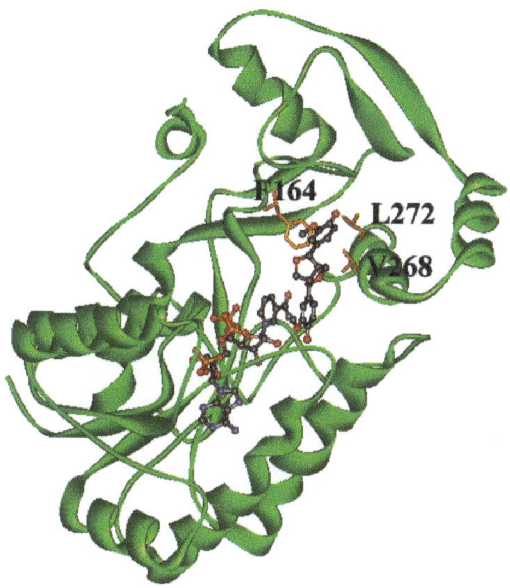

Daniel G. Vassão *et al.*, FIGURE 13.18 Schematic representation of the crystal structure of TpPLR1 with NADPH and (−)-pinoresinol (**58**). Source: Reprinted from Min *et al.* (2003). (See Page 415 of this Volume.)

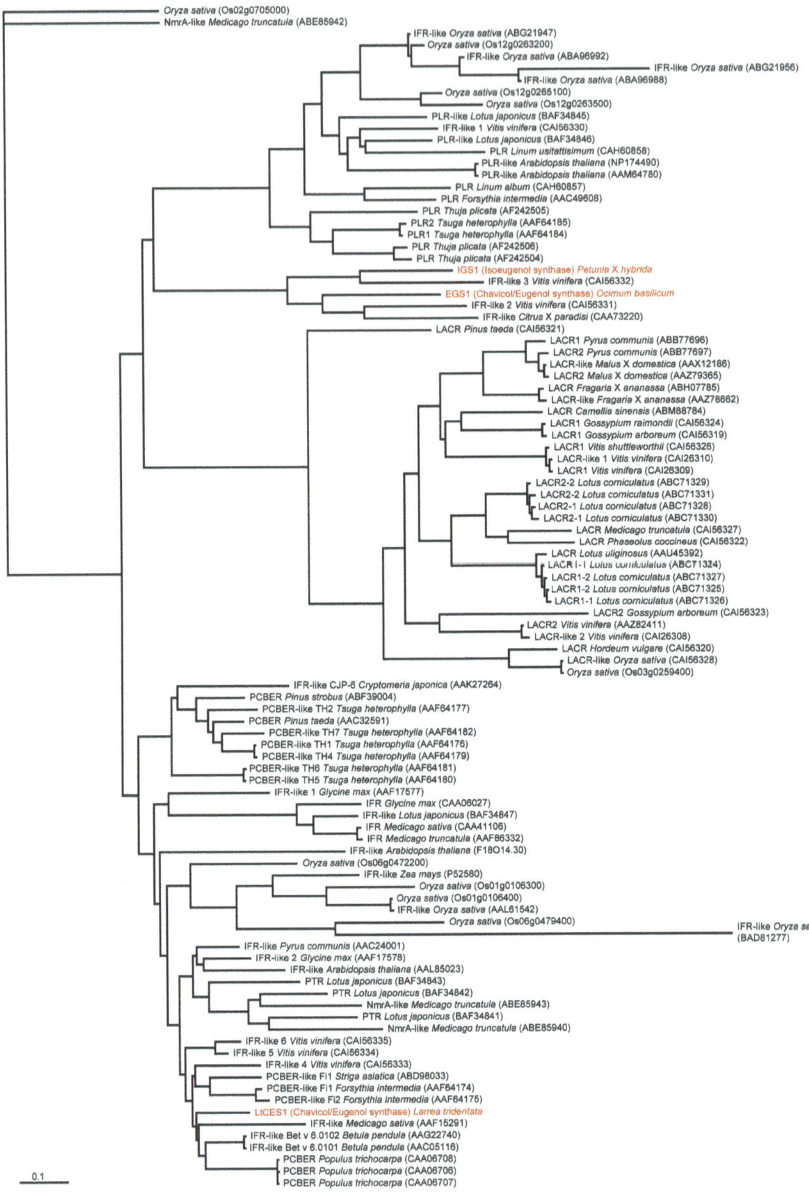

Daniel G. Vassão *et al.*, FIGURE 13.19 Currently annotated phylogenetic analysis of several PIP-reductase homologues from different plant species, with relevant homologues in basil (ObEGS1), petunia (PhIGS1), and creosote bush (LtCES1) highlighted. IFR, isoflavone reductase; LACR, leucoanthocyanidin reductase; NmrA, nitrogen metabolite repression regulator; PCBER, phenylcoumaran-benzylic ether reductase; PLR, pinoresinol-lariciresinol reductase; PTR, pterocarpan reductase. Sequences were obtained from the NCBI database and filtered for <0.75 sequence difference, ClustalW-aligned, and subjected to neighbor-joining phylogenetic analysis using PHYLIP (Felsenstein, 1993). (See Page 416 of this Volume.)

FSC
www.fsc.org
MIX
Papier | Fördert
gute Waldnutzung
FSC® C083411

Kolibri 360 GmbH
Ferdinand-Jühlke-Straße 7
99095 Erfurt, Deutschland
produktsicherheit@kolibri360.de